FUNDAMENTAL NEUROSCIENCE

As seen in this unretouched photograph of a small myelinated axon, mitochondria may assume a variety of sizes, shapes, and orientations.

Edited by

DUANE E. HAINES, PH.D.

Professor and Chairman, Department of Anatomy
and Professor of Neurosurgery at
The University of Mississippi School of Medicine
Jackson, Mississippi

Contributors

M.D. Ard, Ph.D.
J.R. Bloedel, M.D., Ph.D.
P.B. Brown, Ph.D.
N.F. Capra, Ph.D.
R.B. Chronister, Ph.D.
J.J. Corbett, M.D.
J.D. Dickman, Ph.D.
O.B. Evans, M.D.
S.G.P. Hardy, Ph.D.
C.K. Henkel, Ph.D.
J.B. Hutchins, Ph.D.
J.C. Lynch, Ph.D.
T.P. Ma, Ph.D.
P.J. May, Ph.D.
G.A. Mihailoff, Ph.D.
J.P. Naftel, Ph.D.
A.D. Parent, M.D.
F.A. Raila, M.D.
R.W. Rockhold, Ph.D.
M.E. Santiago, M.D.
R.D. Sweazey, Ph.D.
A.C. Terrell, M.S., R.T. (R.) (M.R.)
S. Warren, Ph.D.
R.P. Yezierski, Ph.D.

Illustrators

M.P. Schenk, BS, MSMI, and M.E. Kirkman, BA, MSMI

Photographer

G.W. Armstrong, RBP

Typist

K.M. Squires

FUNDAMENTAL
NEUROSCIENCE

Second Edition

CHURCHILL LIVINGSTONE

New York Edinburgh London Philadelphia

CHURCHILL LIVINGSTONE

The Curtis Center
Independence Square West
Philadelphia, Pennsylvania 19106

Library of Congress Cataloging-in-Publication Data

Fundamental neuroscience/edited by Duane E. Haines; contributors, M.D. Ard . . . [et al.]; illustrators, M.P. Schenk and M.E. Kirkman; photographer, G.W. Armstrong.–2nd ed.

p. ; cm.

Includes bibliographical references and index.

ISBN 0-443-06603-5 (alk. paper)

1. Neurosciences. I. Haines, Duane E. II. Ard, M. D. (March D.)
  [DNLM: 1. Nervous System—anatomy & histology. 2. Nervous System Physiology. 3. Neurons. WL 101 F981 2002]

QP355.2 .F86 2002 612.8–dc21   2001028830

Acquisitions Editor: Jason Malley
Copy Editing Supervisor: Gina Scala
Production Manager: Natalie Ware
Illustration Specialist: Walt Verbitski

FUNDAMENTAL NEUROSCIENCE   ISBN 0-443-06603-5

Last digit is the print number: 9   8   7   6   5   4   3   2   1

Contributors

March D. Ard, Ph.D.
Associate Professor, Department of Anatomy, University of Mississippi School of Medicine, Jackson, Mississippi
The Cell Biology of Neurons and Glia; The Ventricles, Choroid Plexus, and Cerebrospinal Fluid

James R. Bloedel, M.D., Ph.D.
Vice Provost for Research and Advanced Studies and Professor, Departments of Health and Human Performance and Biomedical Sciences, Iowa State University of Science and Technology, Ames, Iowa
The Cerebellum

Paul B. Brown, Ph.D.
Professor, Department of Physiology, West Virginia University School of Medicine, Morgantown, West Virginia
The Electrochemical Basis of Neuronal Integration

Norman F. Capra, Ph.D.
Associate Professor, Department of Oral and Craniofacial Biological Sciences, University of Maryland School of Dentistry, Baltimore, Maryland
The Somatosensory System I: Tactile Discrimination and Position Sense; The Somatosensory System II: Touch, Thermal Sense, and Pain

Robert B. Chronister, Ph.D.
Associate Professor, Department of Cell Biology and Neuroscience, University of South Alabama College of Medicine, Mobile, Alabama
The Hypothalamus; The Limbic System

James J. Corbett, M.D.
McCarty Professor and Chairman of Neurology, University of Mississippi School of Medicine, Jackson, Mississippi; Lecturer in Ophthalmology, Harvard Medical School, Boston, Massachusetts
The Ventricles, Choroid Plexus, and Cerebrospinal Fluid; The Visual System; Visual Motor Systems; The Neurological Examination

J. David Dickman, Ph.D.
Associate Professor, Research Department, Central Institute for the Deaf, St. Louis, Missouri
The Vestibular System

Owen B. Evans, M.D.
Professor and Chairman, Department of Pediatrics, University of Mississippi School of Medicine; Attending Physician, University of Mississippi Hospital and Clinics, Jackson, Mississippi
Development of the Nervous System

Duane E. Haines, Ph.D.
Professor and Chairman, Department of Anatomy, and Professor of Neurosurgery at University of Mississippi School of Medicine, Jackson, Mississippi
Orientation to Structure and Imaging of the Central Nervous System; The Ventricles, Choroid Plexus, and Cerebrospinal Fluid; The Meninges; A Survey of the Cerebrovascular System; The Spinal Cord; An Overview of the Brainstem; The Medulla Oblongata; The Pons and Cerebellum; The Midbrain; A Synopsis of Cranial Nerves of the Brainstem; The Diencephalon; The Telencephalon; Motor System I: Peripheral Sensory, Brainstem, and Spinal Influence on Anterior Horn Neurons; Motor System II: Corticofugal Systems and the Control of Movement; The Cerebellum

S. G. Patrick Hardy, Ph.D.
Professor, Department of Physical Therapy, School of Health Related Professions, University of Mississippi Medical Center, Jackson, Mississippi
Viscerosensory Pathways; Visceral Motor Pathways; The Hypothalamus; The Limbic System

Craig K. Henkel, Ph.D.
Professor, Department of Neurobiology and Anatomy, Bowman Gray School of Medicine of Wake Forest University, Winston-Salem, North Carolina
The Auditory System

James B. Hutchins, Ph.D.
Associate Professor, Department of Anatomy, University of Mississippi School of Medicine, Jackson, Mississippi
The Cell Biology of Neurons and Glia; Development of the Nervous System; The Visual System

James C. Lynch, Ph.D.
Professor, Department of Anatomy, The University of Mississippi School of Medicine, Jackson, Mississippi
The Cerebral Cortex

Terence P. Ma, Ph.D.
Associate Professor, Department of Anatomy, University of Mississippi School of Medicine, Jackson, Mississippi
The Basal Nuclei

Paul J. May, Ph.D.
Associate Professor, Department of Anatomy, University of Mississippi School of Medicine, Jackson, Mississippi
The Midbrain; Visual Motor Systems

Gregory A. Mihailoff, Ph.D.
Professor, Division of Basic Sciences–Anatomy and Neuroscience, Arizona College of Osteopathic Medicine, Midwestern University, Glendale, Arizona
The Spinal Cord; An Overview of the Brainstem; The Medulla Oblongata; The Pons and Cerebellum; The Midbrain; A Synopsis of Cranial Nerves of the Brainstem; The Diencephalon; The Telencephalon; Motor System I: Peripheral Sensory, Brainstem, and Spinal Influence on Anterior Horn Neurons; Motor System II: Corticofugal Systems and the Control of Movement; The Cerebellum

John P. Naftel, Ph.D.
Associate Professor, Department of Anatomy, University of Mississippi School of Medicine, Jackson, Mississippi
The Cell Biology of Neurons and Glia; Viscerosensory Pathways; Visceral Motor Pathways

Andrew D. Parent, M.D.
Professor and Chairman, Department of Neurosurgery, University of Mississippi School of Medicine, Jackson, Mississippi
The Hypothalamus

Frank A. Raila, M.D.
Professor, Department of Radiology, University of Mississippi School of Medicine, Jackson, Mississippi
Orientation to Structure and Imaging of the Central Nervous System

Robin W. Rockhold, Ph.D.
Professor, Department of Pharmacology and Toxicology, University of Mississippi School of Medicine, Jackson, Mississippi
The Chemical Basis for Neuronal Communication

Maria E. Santiago, M.D.
Chief Resident, Department of Neurology, The University of Mississippi Medical Center, Jackson, Mississippi
The Neurological Examination

Robert D. Sweazey, Ph.D.
Associate Professor, Department of Anatomy and Cell Biology, Indiana University School of Medicine, Fort Wayne, Indiana
Olfaction and Taste

Allen C. Terrell, M.S., R.T. (R.) (M.R.)
Chief MRI Technologist, River Regional Health Systems, Vickburg, Mississippi
Orientation to Structure and Imaging of the Central Nervous System

Susan Warren, Ph.D.
Associate Professor, Department of Anatomy, University of Mississippi School of Medicine, Jackson, Mississippi
The Somatosensory System I: Tactile Discrimination and Position Sense; The Somatosensory System II: Touch, Thermal Sense, and Pain

Robert P. Yezierski, Ph.D.
Director, Center for Pain Research, University of Florida College of Dentistry, Gainesville, Florida
The Spinal Cord; The Somatosensory System I: Tactile Discrimination and Position Sense; The Somatosensory System II: Touch, Thermal Sense, and Pain

Preface to the Second Edition

The principles that guided the development of the First Edition of *Fundamental Neuroscience* have remained a central theme in the creation of the Second Edition. These are as follows:

- Providing information that will prepare the user for subsequent steps in the educational process, be they the Board Examination or the clinical years
- Presenting concepts of nervous system development, with emphasis on clinical concepts
- Integrating structural information and functional concepts, with emphasis on their clinical application
- Integrating the discussion of internal blood supply throughout the various chapters and emphasizing its importance to systems neurobiology and clinical issues
- Stressing the value of clinical information and examples in the study of neuron cell biology, regional neurobiology, and systems neurobiology
- Using photographs and color line artwork of exceptional quality
- Integrating anatomy, physiology, and pharmacology throughout, but with particular emphasis on systems neurobiology
- Emphasizing the organization of sensory and motor systems with particular attention to deficits, as seen in the neurologically compromised patient

With these general principles in mind, we have made changes to the Second Edition of *Fundamental Neuroscience* to build on its existing strengths and introduce new information to enhance learning, review, and the application of the basic sciences to clinical medicine. Numerous small—and some large—modifications have been made in all chapters with an eye toward emphasizing the importance of basic neuroscience (regional and systems) in the clinical setting. While it is not possible to describe each change, some of the more significant ones will be mentioned.

First, significantly more information on magnetic resonance imaging (MRI) and computed tomography (CT) techniques has been introduced in Chapter 1, along with additional examples. Also, the important concept of *localizing signs* is introduced early and reinforced by examples in later chapters.

Second, more MRI studies and CT scans are introduced throughout the appropriate chapters. Included are images of normal anatomy and of how brain anatomy is altered in disease states; for example, the alteration of brain structure and its appearance in tumors, after trauma, and with hemorrhage.

Third, numerous new color line drawings have been introduced, or existing ones have been modified, to illustrate a wide variety of clinical conditions. This art includes, but is not limited to, tremors, hemorrhages, rigidity, and the results of cranial nerve lesions. Along this same vein, a number of new photographs of patients with motor or sensory deficits have been added.

Fourth, many changes have been made throughout the chapters to update information, introduce new relevant clinical concepts, clarify points brought to our attention by users of the book, and add important facts or concepts where needed. Every effort has been made to address the suggestions brought to our attention by users of this book.

Fifth, a new chapter, "A Synopsis of Cranial Nerves of the Brainstem" (Chapter 14), has been added. This new offering reviews the central nuclei and functional components of cranial nerves III to XII, but especially emphasizes the peripheral distribution and the peripheral structures innervated, and provides examples of peripheral and central lesions of cranial nerve roots or of their nuclei. This one chapter provides a broad overview of these 10 cranial nerves, with a special emphasis on clinical examples.

Sixth, a second new chapter, entitled "The Neurological Examination" (Chapter 33), has been added. *Fundamental Neuroscience*, 2nd ed., is one of only a very few basic neuroscience books, large or small, to include a chapter on the neurological examination. Although missing in most neuroscience texts, this particular clinical exercise is one of the most important activities that clearly extends the basic science knowledge base into the clinical environment. This interesting and well-illustrated chapter provides practical information on how to perform and interpret the neurological examination. This chapter is a "must-read" for any student who will enter clinical training.

Recognizing that a new official international list of anatomical terms for neuroanatomy has been published (*Terminologia Anatomica*, Thieme Medical Publishers, 1998), we have made every effort to include the most current terminology in this book. This new terminology, having been adopted by the International Federation of Associations of Anatomists, supersedes *all* previous terminology lists. The actual number of changes is modest and is related primarily to directional terms, such as *posterior* for *dorsal*, and *anterior* for *ventral*. It is certain that some changes have eluded detection; these will be made in future printings.

The authors, in an effort to improve the book when and where we can, welcome comments, suggestions, and corrections from students, our colleagues, and from any user of this book.

One feature common to all neuroscience courses is the need to prepare students for the next phase of their education. For example we, as instructors, prepare first-year medical students to become second-year medical students, and we give them information that is *essential* to their successful negotiation of Step 1 of the USMLE. We are not teaching specialists (neurologists, oral surgeons, hand therapists) or, for that matter, even family practitioners. Rather, we provide information to "generalist" students.

The organization of *Fundamental Neuroscience* recognizes two important points. First, many neuroscience courses begin from their own reference point. There is no universally agreed upon topic that should appear first. Some consider the electrical properties of nerve cell membranes first; others start with embryology or with the anatomy of the telencephalon. Second, students in contemporary neuroscience courses must master *regional neuroanatomy* and *systems neurobiology*, as these are the bases for understanding brain and spinal cord function. Students in professional programs will find *regional neuroanatomy* essential to reading MRI and CT scans and will find *systems neurobiology* to be the foundation for the successful diagnosis of the neurologically impaired patient. In an attempt to address these points, the three sections were designed to be used as independent, but interrelated, segments of information, in whatever sequence most closely fits a given educational program.

SECTION I: ESSENTIAL CONCEPTS

This section deals with neuron structure, function, and communication and with development. Each chapter is a free-standing block of information that may be used at the beginning of, or at any time during, a neuroscience course. The instructor who elects to start with principles of neurophysiology may do so, while another may choose to initially emphasize concepts of neurohistology or neuropharmacology. A variety of approaches may be used without compromising continuity.

SECTION II: REGIONAL NEUROANATOMY

Viewing an MRI or CT scan of a *normal* brain is simply looking at brain anatomy *in situ*. The shapes and contours of each structure, and its relationship to adjacent structures, are characteristic and unique. Viewing an MRI or CT scan of an *abnormal* brain is simply evaluating how trauma, or a pathologic process, has altered brain structure. In light of the ever-increasing levels of detail provided by contemporary clinical imaging methods, *understanding brain anatomy in its clinical context* is an essential part of the modern neuroscience educational experience.

Section II covers the ventricles, meninges, and external vasculature of the central nervous system, followed by considerations of the spinal cord and each division of the brain. Some of the unique approaches in chapters of this section are

1. The development of the spinal cord and of each division of the brain is summarized at the beginning of the appropriate chapter.
2. The internal distribution of blood vessels is presented in association with internal structures.
3. Clinical examples are correlated with the morphology of the central nervous system.
4. Photographs and color line drawings are of excellent clarity.

This book introduces some terms that, although not common in many textbooks, are standard fare in the clinical setting. For example, what is termed the *horizontal plane* by morphologists is called the *axial plane* by the clinical neuroscientists. Since many users of this book will find themselves in a clinical curriculum, it seems entirely appropriate to make these correlations early in their training.

SECTION III: SYSTEMS NEUROBIOLOGY

Successful evaluation of the neurologically impaired patient requires a thorough understanding of systems neurobiology. This book describes how systems function under normal conditions, what information they convey and how they do so, and what deficits appear when systems are damaged. The patient with occlusion of a major cerebral vessel may, with time, display increased proteins in the CSF or xanthochromia of the CSF. However, the information that is frequently most valuable for early treatment is gained by identifying neural system dysfunction as soon as possible. This thinking has guided the organization of this section.

Section III provides comprehensive coverage of all sensory, motor, and integrative systems. The innovative approaches used in these chapters include

1. Emphasis on the topographical organization of sensory and motor pathways.
2. A review of the blood supply to most pathways at key levels of the neuraxis.

3. Integration of anatomy, physiology, and pharmacology where appropriate.

4. Original artwork specifically designed to help students understand neural systems.

5. Clinical information that is integrated within the text in close conjunction with the basic science information to which it is related.

Fundamental Neuroscience was written by individuals with significant teaching and research expertise in their respective areas. Consequently, the information is contemporary, integrated across specialties (anatomy, physiology, pharmacology), and attuned to the needs of students taking a modern neuroscience course in professional and graduate programs.

This book is most appropriate for students taking human neurobiology courses in medical, graduate, and dental programs. Its emphasis on basic concepts and systems neurobiology should prove quite useful to students in physical and/or occupational therapy, to other allied health students, or to residents needing a clear, succinct review as they prepare for their specialty board examinations in the clinical neurosciences.

Recognizing that both teaching and learning are multidimensional and on-going processes, the authors welcome comments, suggestions, and corrections from our students, our colleagues, and from other users of this book.

D. E. Haines, Ph.D.

Acknowledgments

A project of this scope is not just the result of the efforts of the various authors—it also reflects suggestions, both great and small, that we have received from many individuals. In this respect, we wish to thank the many faculty who have used the First Edition of this book and have provided valuable insights into how we might improve its usefulness. We especially thank our students, whose probing interest has allowed us to more adequately address their educational needs. We also express our sincere thanks to Drs. V. K. Arand, D. E. Angelaki, R. H. Baisden, A. J. (Tony) Castro, S. C. Crawford, J. L. Culberson, E. Dietrichs, J. T. Ericksen, W. C. Hall, R. Hoffman, J. S. King, W. M. King, G. R. Leichnetz, G. F. Martin, I. J. Miller, R. S. Nowakowski, D. F. Peeler, A. Peters, J. D. Porter, J. A. Rafols, W. A. Roy, L. F. Schweitzer, D. L. Tolbert, and M. L. Woodruff. The photographs of Golgi-stained material are from the Clement A. Fox Collection at Wayne State University and through the courtesy of Dr. Rafols. The photograph on the half-title page was provided by Drs. Ross Kosinski and Greg Mihailoff.

Several individuals who went out of their way to help with the first edition have now joined this project as authors or co-authors of chapters. These include Dr. F. A. Raila (Neuroradiology), Mr. A. C. Terrell (Chief MRI Technologist at the University of Mississippi Medical Center [UMMC], now at River Regional Health Systems, Vicksburg, Mississippi), Dr. A. D. Parent (Neurosurgery), and Dr. Maria Santiago (Neurology). In addition to his co-authorship of Chapter 1, Mr. Terrell identified numerous examples of normal and abnormal scans and was most responsive in finding things on our "want list." Dr. Parent offered us access to some of his pediatric neurosurgery cases and the enthusiastic assistance of his nurse, Ms. T. McMillan. Over the years, Dr. R. B. Harrison has allowed the editor free access to the CT and MRI facilities at UMMC; this kindness is greatly appreciated. We are also indebted to our colleagues in the Department of Neurosurgery (Drs. Parent, Harkey, Lancon, Esposito, Mandybur, and Ross) and the Department of Neurology (Drs. Corbett, Subramony, Wee, Santiago, Sundaram, Lawson, Leis, Fredericks, Rose, and Manning), who have given suggestions or provided information.

All of the artwork and photography (excepting specifically acknowledged photographs) was done in the Department of Medical Illustration and the Department of Biomedical Photography, respectively, at UMMC. The authors are indebted to Mr. Michael P. Schenk (Director of Medical Illustration) and his staff for his modification of some artwork from the First Edition and for generating numerous new pieces of artwork for the Second Edition. The visual impact of the artwork is due largely to the individual skills and collective work of Mr. Schenk and his team. The photography was completed by Mr. G. William Armstrong (Director of Biomedical Photography) and his staff. Many new photographs, especially of clinical examples, were generated for this new edition. The editor is enormously appreciative of their patience and cooperation in getting the best-quality photographs for this book.

Ms. Katherine Squires typed all corrections, changes, and new chapters for the Second Edition. Her good-natured interaction with the editor greatly expedited the timely completion of this project. We would also like to express our appreciation to Ms. Ruth Low for preparing the index.

A special thanks is due Mr. Chris Canavos, proprietor of the Howard Johnson Historic Area Hotel in Williamsburg, Virginia. The editor, while on vacation, did extensive editing and reviewing of proofs for this new edition. Mr. Canavos generously made his office available (photocopy equipment, supplies, staff, and so forth) to assist in this effort, in a very real sense keeping me on schedule.

Production of this finely done and visually appealing book would not have been accomplished without the enthusiastic support of Churchill Livingstone. I am grateful to Mr. William Schmitt (Publishing Director) and his assistant Mr. Tony Galbraith, under whom the project was initiated, and to my editor Mr. Jason Malley, and his assistant Mr. Kevin Kochanski, who smoothed out rough spots in the road. I would also like to express our collective appreciation to Ms. Pat Morrison (Director, Art and Design Department), to Ms. Natalie Ware (Senior Production Manager), to Ms. Gina Scala (Copy Editing Supervisor), to Ms. Ceil Roberts (Art Assistant), to the computer artists Keith Lesko, John Dzedzy, Esteban Cabrera, Karen Giacomucci, and Lisa Weischedel, to Ms. Lynn Hoops (Marketing Manager), and to Mr. Erik Shveima. Last, but certainly not least, the editor expresses a special thanks to his wife, Gretchen; she was an important element in getting everything done.

Figure Acknowledgments

The editor acknowledges the following publishers for permission to use borrowed, and modified, illustrations from some of his publications.

From Haines DE: On the question of a subdural space. Anat Rec 230:3–21, 1991.
Fundamental Neuroscience, 2nd ed., Figure 7–3 (adapted from original)

From Haines DE: Neuroanatomy: An Atlas of Structures, Sections, and Systems. 5th ed. Lippincott Williams & Wilkins, Baltimore, 2000.
Fundamental Neuroscience, 2nd ed., Figures 9–5 (photographs only), 16–10*B*, and 16–12 (MRI only)
Figures 7–2, 8–7, 8–8, 8–14, 31–5, and 31–6 (adapted from originals)
Figures 13–4, 13–15, 15–5, 15–14, 16–14*E* and *F*, 16–17*A* and *B*, 20–14*A*, 26–3*A* and *B*, 30–2*B* (modified from the originals)

Atlas Figure	*Fundamental Neuroscience*, 2nd ed.
2–35	10–4
2–27	12–20, 27–7
4–2	15–7*A*
4–3	15–7*B*
4–5	15–7*C*
4–7	15–7*D*
4–12	15–8
2–9	16–5*A*
2–28	16–5*B*

From Haines DE, Frederickson RG: The meninges. In Al-Mefty O (ed): Meningiomas. Raven Press, New York, 1991.
Fundamental Neuroscience, 2nd ed., Figures 7–4, 7–6, and 7–10 (adapted from original)

The editor also acknowledges Lippincott Williams & Wilkins for permission to modify Figure 11–17 from Parent A: Carpenter's Human Neuroanatomy, 9th ed., 1995, and to use this modified version in Figures 14–1, 14–5, 14–6, and 14–8 to 14–12.

Adapted from Kinnamon SC: Taste transduction: A diversity of mechanisms. Trends Neurosci 11:491–496, 1988.
Fundamental Neuroscience, 2nd ed., Figure 23–11

Adapted from Mistretta CM: Anatomy and neurophysiology of the taste system in aged animals. In Murphy C, Cain WS, Hegsted DM (eds): Nutrition and the Chemical Senses in Aging: Recurrent Advances and Current Research Needs. Ann NY Acad Sci 561:277–290, 1989.
Fundamental Neuroscience, 2nd ed., Figure 23–8

Adapted from Penfield W, Rasmussen T: The Cerebral Cortex of Man: A Clinical Study of Localization of Function. Hafner Publishing, New York, 1968 (facsimile of 1950 ed.).
Fundamental Neuroscience, 2nd ed., Figures 17–10 and 25–3*A*

Adapted from Welker W, Blair C, Shambes GM: Somatosensory projections to cerebellar granule cell layer of giant bushbaby, *Galago crassicaudatus*. Brain Behav Evol 31:150–160, 1988.
Fundamental Neuroscience, 2nd ed., Figure 27–10*B*

Additional borrowed illustrations in *Fundamental Neuroscience*, 2nd ed., are acknowledged in their figure captions.

Contents

Section I: Essential Concepts 1

1 Orientation to Structure and Imaging of the Central Nervous System ▪ 3
   D.E. Haines, F.A. Raila, and A.C. Terrell

2 The Cell Biology of Neurons and Glia ▪ 15
   J.B. Hutchins, J.P. Naftel, and M.D. Ard

3 The Electrochemical Basis of Neuronal Integration ▪ 37
   P.B. Brown

4 The Chemical Basis for Neuronal Communication ▪ 57
   R.W. Rockhold

5 Development of the Nervous System ▪ 71
   O.B. Evans and J.B. Hutchins

Section II: Regional Neurobiology 91

6 The Ventricles, Choroid Plexus, and Cerebrospinal Fluid ▪ 93
   J.J. Corbett, D.E. Haines, and M.D. Ard

7 The Meninges ▪ 107
   D.E. Haines

8 A Survey of the Cerebrovascular System ▪ 121
   D.E. Haines

9 The Spinal Cord ▪ 137
   D.E. Haines, G.A. Mihailoff, and R.P. Yezierski

10 An Overview of the Brainstem ▪ 151
    D.E. Haines and G.A. Mihailoff

11 The Medulla Oblongata ▪ 159
    D.E. Haines and G.A. Mihailoff

12 The Pons and Cerebellum ▪ 173
    G.A. Mihailoff and D.E. Haines

13 The Midbrain ▪ 187
    G.A. Mihailoff, D.E. Haines, and P.J. May

14 A Synopsis of Cranial Nerves of the Brainstem ▪ 199
    D.E. Haines and G.A. Mihailoff

15 The Diencephalon ▪ 219
    G.A. Mihailoff and D.E. Haines

16 The Telencephalon ▪ 235
    D.E. Haines and G.A. Mihailoff

Section III: Systems Neurobiology 253

17 The Somatosensory System I: Tactile Discrimination and Position Sense ▪ 255
    S. Warren, N.F. Capra, and R.P. Yezierski

18 The Somatosensory System II: Touch, Thermal Sense, and Pain ▪ 273
    S. Warren, R.P. Yezierski, and N.F. Capra

19 Viscerosensory Pathways ▪ 293
    S.G.P. Hardy and J.P. Naftel

20 The Visual System ▪ 303
    J.B. Hutchins and J.J. Corbett

21 The Auditory System ▪ 323
    C.K. Henkel

22 The Vestibular System ▪ 341
    J.D. Dickman

23 Olfaction and Taste ▪ 359
    R.D. Sweazey

24 Motor System I: Peripheral Sensory, Brainstem, and Spinal Influence on Anterior Horn Neurons ▪ 373
    G.A. Mihailoff and D.E. Haines

25 Motor System II: Corticofugal Systems and the Control of Movement ▪ 387
    G.A. Mihailoff and D.E. Haines

26 The Basal Nuclei ▪ 405
    T.P. Ma

27 The Cerebellum ▪ 423
    D.E. Haines, G.A. Mihailoff, and J.R. Bloedel

28 Visual Motor Systems ▪ 445
    P.J. May and J.J. Corbett

29 Visceral Motor Pathways ▪ 465
    J.P. Naftel and S.G.P. Hardy

30 The Hypothalamus ▪ 479
    S.G.P. Hardy, R.B. Chronister, and A.D. Parent

31 The Limbic System ▪ 493
    R.B. Chronister and S.G.P. Hardy

32 The Cerebral Cortex ▪ 505
    J.C. Lynch

33 The Neurologic Examination ▪ 521
    M.E. Santiago and J.J. Corbett

Index ▪ 539

Section 1

Essential Concepts

Chapter 1

Orientation to the Structure and Imaging of the Central Nervous System

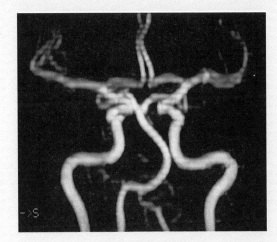

D. E. Haines, F. A. Raila, and A. C. Terrell

Overview 4
Central, Peripheral, and Visceromotor Nervous Systems
Neurons
Reflexes and Pathways

Regions of the Central Nervous System 6
Spinal Cord
Medulla Oblongata
Pons and Cerebellum
Midbrain
Thalamus
Cerebral Hemispheres

Functional Systems and Regions 8
Localizing Signs

Dorsal, Ventral, and Other Directions in the Central Nervous System 9

Clinical Images of the Brain and Skull 10
Computed Tomography
Magnetic Resonance Imaging
Imaging of the Brain and Skull

Our nervous system makes us what we are. Personality, outlook, intellect, coordination (or lack thereof), and the *many* other characteristics that are unique to each of us are the result of complex interactions within our nervous system. Information is received from the environment by sensory receptors and transmitted into the brain or spinal cord. Once inside the brain or spinal cord, this sensory information is processed and integrated, and an appropriate response is initiated.

The nervous system can be viewed as a scale of structural complexity. Microscopically, the individual structural and functional unit of the nervous system is the *neuron*, or nerve cell. Interspersed among the neurons of the central nervous system are supportive elements called *glial cells*. At the macroscopic end of the scale are the large divisions (or parts) of the nervous system that can be handled and studied without magnification. These two extremes are not independent but form a continuum; functionally related neurons aggregate to form small structures, which combine to form larger structures, and so on. Communication takes place at many different levels, the end result being a wide range of productive or life-sustaining nervous activities.

Overview

Central, Peripheral, and Visceromotor Nervous Systems. The human nervous system is divided into the *central nervous system* (CNS) and the *peripheral nervous system* (PNS) (Fig. 1–1*A*). The CNS consists of the brain and spinal cord. Because of their locations in the skull and vertebral column, these structures are the most protected in the body. The PNS is made up of nerves that connect the brain and spinal cord with peripheral structures. These nerves innervate muscle (skeletal, cardiac, smooth) and glandular epithelium and contain a variety of sensory fibers. These sensory fibers enter the spinal cord via the posterior (dorsal) root, and motor fibers exit through the anterior (ventral) root. The *spinal nerve* is formed by the joining of posterior (sensory) and anterior (motor) roots and is, consequently, a *mixed nerve* (see Fig. 1–1*B*). In the case of *mixed cranial nerves*, the sensory and motor fibers are combined into a single root.

The *visceromotor nervous system* is a functional division of the nervous system that has parts in both the CNS and the PNS (see Fig. 1–1). It is made up of neurons that innervate smooth muscle, cardiac muscle, or glandular epithelium or combinations of these tissues. These individual *visceral tissues*, when combined, make up *visceral organs* such as the stomach. The visceromotor nervous system is also called the *autonomic* or *vegetative nervous system* because it regulates motor responses outside the realm of conscious control.

Neurons. At the histologic level, the nervous system is composed of *neurons* and *glial cells*. As the basic structural and functional units of the nervous system, neurons are specialized to receive information, transmit electrical impulses, and influence other neurons or effector tissues. In many areas of the nervous system,

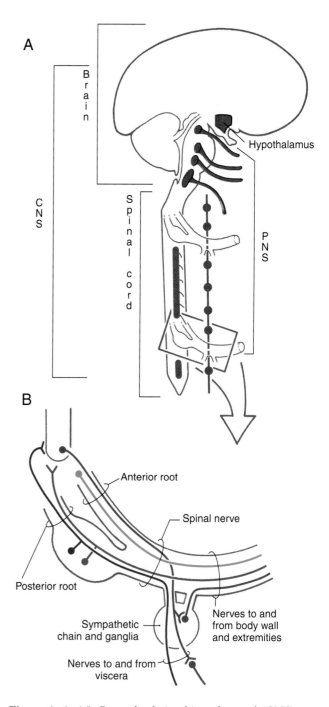

Figure 1–1. (*A*) General relationships of central (CNS), peripheral (PNS), and visceromotor nervous systems. Visceromotor regions of CNS and PNS are shown in red. (*B*) Enlargement of the boxed area in *A* shows the relationships of efferent (outgoing, motor) and afferent (incoming, sensory) fibers to spinal nerves and roots. Motor fibers are general visceral efferent (visceromotor; red) and general somatic efferent (green); sensory fibers are general somatic afferent (blue) and general visceral efferent (black).

neurons are structurally modified to serve particular functions. At this point, however, we shall consider the neuron only as a general concept.

A *neuron* consists of a *cell body* (*perikaryon or soma*) and the processes that emanate from the cell body (Fig. 1–2). Collectively, neuronal cell bodies constitute the *gray matter* of the CNS. Named and usually function-specific clusters of cell bodies in the CNS are called *nuclei* (singular, *nucleus*). Typically, *dendrites* are those processes that ramify in the vicinity of the cell body, whereas a single, longer process called the *axon* carries impulses to a more remote destination. The *white matter* of the CNS consists of bundles of axons that are wrapped in a sheath of insulating lipoprotein called *myelin*. The axon terminates at specialized structures called *synapses* or, in the case of those neurons that innervate muscles, as *motor end plates* (*neuromuscular junctions*), which function much like synapses.

The generalized synapse shown in Figure 1–2 is the most common type seen in the CNS and is sometimes called an *electrochemical synapse*. It consists of a *presynaptic element*, which is usually part of an axon; a gap called the *synaptic cleft*; and the *postsynaptic region* of the innervated neuron or effector structure. Communication across this synapse is accomplished as follows. An electrical impulse (the *action potential*) causes the release of a neuroactive substance (a *neurotransmitter, neuro-*

modulator, or *neuromediator*) from the presynaptic element into the synaptic cleft. This substance is stored in *synaptic vesicles* in the presynaptic element and is released into the synaptic space by the fusion of these vesicles with the cell membrane (see Fig. 1–2).

The neurotransmitter diffuses rapidly across the synaptic space and binds to receptor sites on the postsynaptic membrane. Based on the action of the neurotransmitter at receptor sites, the postsynaptic neuron may be excited (lead to generation of an action potential) or inhibited (prevent generation of an action potential). Neurotransmitter residues in the synaptic cleft are rapidly inactivated by other chemicals found in this space. In this brief example, we see that (1) the neuron is structurally specialized to receive and propagate electrical signals, (2) this propagation is accomplished by a combination of electrical and chemical events, and (3) the transmission of signals across the synapse is in one direction; that is, from the presynaptic neuron to the postsynaptic neuron.

Figure 1–2 shows the convention that will be used for illustrating neurons as elements of reflex arcs and pathways in this book. The dendrites and cell body (the "receiving" parts of the neuron) are represented by a large dot, and the axon (the "sending" part of the neuron) is represented by a line, which terminates in a fork or Y at the synapse.

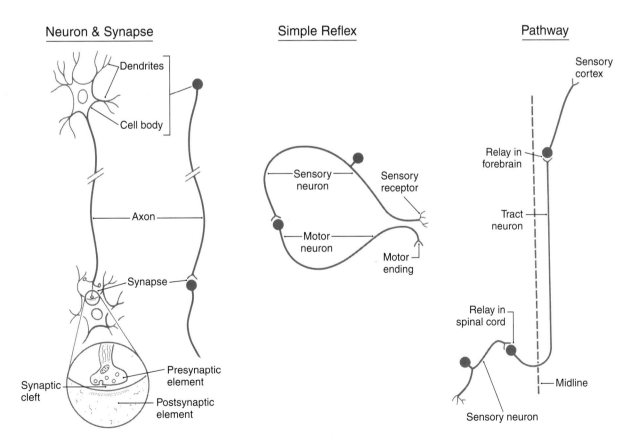

Figure 1–2. A representative neuron and synapse, a simple (monosynaptic) reflex, and a pathway.

Reflexes and Pathways. The function of the nervous system is based on the interactions of neurons with one another. Figure 1–2 illustrates one of the simplest types of neuronal circuits, a reflex arc composed of only two neurons. This is called a *monosynaptic reflex arc*, because only one synapse is involved. In this example, the peripheral end of a sensory fiber responds to a particular type of input. The resulting action potential is conducted by the sensory fiber into the spinal cord, where it influences a motor neuron. The axon of the motor neuron conducts a signal from the spinal cord to the appropriate skeletal muscle, which responds by contracting. *Reflexes are involuntary responses to a particular bit of sensory input.* For example, the physician taps on the patellar tendon and the leg jerks; the patient does not think about it—the motor response just happens. The lack of a reflex (*areflexia*), an obviously weakened reflex (*hyporeflexia*), or an excessively active reflex (*hyperreflexia*) is usually indicative of a neurologic disorder.

Building on these summaries of the *neuron* and of the *basic reflex arc*, we shall briefly consider what neuronal elements constitute a *pathway*. If the patient bumps her knee and not only hits the patellar tendon but also damages the skin over the tendon, two things happen (see Fig. 1–2). First, impulses from receptors in the tendon travel through a reflex arc that causes the leg to jerk (*knee jerk*, or *patellar reflex*). The synapse for this reflex arc is located in the lumbosacral spinal cord. Second, impulses from pain receptors in the damaged skin are transmitted in the lumbosacral cord to a second set of neurons that convey them via ascending axons to the forebrain. As can be seen in Figure 1–2, these axons cross the midline of the spinal cord and form an ascending tract on the contralateral side. In the forebrain, these signals are passed to a third group of neurons that distribute them to a region of the cerebral cortex specialized to interpret them as pain from the knee. This three-neuron chain constitutes a *pathway*, a series of neurons designed to carry a specific type of information from one site to another (see Fig. 1–2). Some pathways carry information to a level of conscious perception (we not only recognize pain but know that it is coming from the knee), and others convey information that does not reach the conscious level.

Regions of the Central Nervous System

Spinal Cord. The spinal cord is located inside the vertebral canal and is rostrally continuous with the medulla oblongata of the brain (Fig. 1–3). An essential link between the peripheral nervous system and the brain, it conveys sensory information originating from the body wall, extremities, and gut and distributes mo-

tor impulses to these areas. Impulses enter and leave the spinal cord through the 31 pairs of spinal nerves (see Fig. 1–1; also see Fig. 9–2). The spinal cord contains sensory fibers and motor neurons involved in reflex activity and ascending and descending pathways

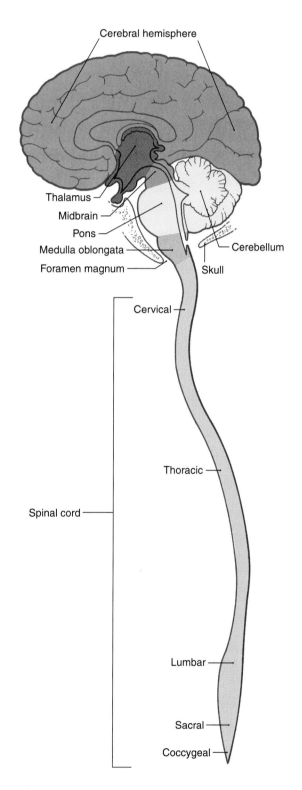

Figure 1–3. The basic divisions of the central nervous system.

(also called *tracts*) that link spinal centers with other parts of the CNS. Ascending pathways convey sensory information to higher centers, whereas descending pathways influence the activity of neurons in the spinal cord gray matter.

Medulla Oblongata. At the level of the foramen magnum, the spinal cord is continuous with the most caudal part of the brain, the *medulla oblongata*, commonly called the *medulla* (see Fig. 1–3). The medulla consists of (1) neurons that perform functions associated with the medulla and (2) ascending and descending tracts that pass through the medulla on their way from or to the spinal cord. In general, fibers that descend through the medulla are involved in motor functions, whereas fibers that ascend through it carry sensory information. Some of the neuronal cell bodies of the medulla are organized into nuclei associated with specific cranial nerves. The medulla contains the nuclei for the glossopharyngeal (cranial nerve IX), vagus (X), and hypoglossal (XII) nerves, as well as portions of the nuclei for the trigeminal (V), vestibulocochlear (VIII), and spinal accessory (XI) nerves. It also contains important relay centers and nuclei that are essential to the regulation of respiration, heart rate, and various visceral functions.

Pons and Cerebellum. Embryologically, the pons and cerebellum originate from the same segment of the developing neural tube. However, in the adult the pons forms part of the *brainstem* (the other parts being the midbrain and medulla), and the cerebellum is a *suprasegmental* structure because it is located posterior (dorsal) to the brainstem (see Fig. 1–3).

Like the medulla, the pons contains many neuronal cell bodies, some of which are organized into cranial nerve nuclei, and it is traversed by ascending and descending tracts. The pons contains the nuclei of the abducens (VI) and facial (VII) nerves and portions of the nuclei for the trigeminal (V) and vestibulocochlear (VIII) nerves. The ventral part of the pons contains large populations of neurons that form a relay station between the cerebral cortex and cerebellum and descending motor fibers that travel to all spinal levels.

The cerebellum is connected with diverse regions of the CNS. Functionally, the cerebellum is considered part of the motor system. It serves to coordinate the activity of individual muscle groups to produce smooth, purposeful, synergistic movements.

Midbrain. Rostrally, the pons is continuous with the midbrain. This latter part of the brain is, quite literally, the link between the brainstem and the forebrain. Ascending or descending pathways to or from the forebrain must traverse the midbrain. The nuclei for the oculomotor (III) and trochlear (IV) cranial nerves, as well as part of the trigeminal (V) complex, are found in the midbrain. Other midbrain centers are concerned with visual and auditory reflex pathways, motor function, the transmission of pain, and visceral functions.

Thalamus. The forebrain consists of the *cerebral hemispheres* and the large groups of neurons that compose the *basal ganglia* and the *thalamus* (see Fig. 1–3). We shall see later that what is commonly called the thalamus actually consists of several regions—for example, the hypothalamus, subthalamus, epithalamus, and dorsal thalamus.

The thalamus is rostral to the midbrain and almost completely surrounded by elements of the cerebral hemisphere. Individual parts of the thalamus can be seen in detail only when the brain is cut in coronal or axial (horizontal) planes.

With the exception of olfaction, all sensory information that eventually reaches the cerebral cortex must pass through the thalamus. One function of the thalamus, therefore, is to receive sensory information of many sorts (temperature, pain, vision, and so on) and to distribute it to the specific regions in the cerebral cortex that are specialized to decode it. Other areas of the thalamus receive input from pathways conveying information on, for example, position sense or the tension in a tendon or muscle. This input is relayed to areas of the cerebral cortex that function to generate smooth, purposeful movements.

Although quite small, the hypothalamus is extremely important. It functions in sexual behavior, feeding, hormonal output of the pituitary gland, body temperature regulation, and a wide range of visceromotor functions. Through descending connections, the hypothalamus influences visceral centers in the brainstem and spinal cord.

Cerebral Hemispheres. The largest and most obvious parts of the human brain are the two cerebral hemispheres. Each hemisphere is composed of three major subdivisions. First, the *cerebral cortex* is a layer of neuronal cell bodies about 0.5 cm thick that covers the entire surface of the hemisphere. This layer of cells is thrown into elevations called *gyri* (singular, *gyrus*) separated by creases called *sulci* (singular, *sulcus*). The second major part of the hemisphere is the *subcortical white matter*, which is made up of myelinated axons that carry information to or from the cerebral cortex. The largest and most organized part of the white matter is the *internal capsule*. This bundle contains fibers passing to and from the cerebral cortex such as *corticospinal* and *thalamocortical* fibers. The third major component of the hemisphere is a prominent group of neuronal cell bodies collectively called the *basal nuclei* (also called the *basal ganglia*). These prominent forebrain centers are involved in motor function. Parkinson disease, a neurologic disorder associated with the basal nuclei, is characterized by a profound impairment of movement.

The gyri and sulci that make up the cerebral cortex are named, and many are associated with particular functions. Some gyri receive sensory input, such as vision or general sensation, whereas others give rise to motor fibers that project to the spinal cord and motor nuclei of cranial nerves. The cerebral cortex also has association areas that are essential for analysis and cognitive thought.

Functional Systems and Regions

A *functional system* is a set of neurons linked together to convey a particular block of information or accomplish a particular task. In this respect, *systems* and *pathways*, in some cases, may be quite similar, and occasionally their meanings may overlap.

Anatomic parts of the CNS, such as the medulla and pons, are commonly called *regions*. The study of their structure and function, called *regional neurobiology*, is the focus of Section II of this book. *Systems* and *pathways*, however, generally traverse more than one region. The system of neurons and axons that allows you to feel the edge of this page, for example, crosses every region of the nervous system between your fingers and the somatosensory cortex. The study of functional systems, called *systems neurobiology*, is the focus of Section III. It is important to remember that the *functional characteristics of regions coexist with those of systems.*

Let us consider an example of how the interrelation of systems and regions can be important clinically. The signals that influence movements of the hand originate in the cerebral cortex. Neurons in the hand area of the *motor cortex* send their axons to cervical levels of the spinal cord, where they influence spinal motor neurons that innervate the muscles of the forearm. These are called *corticospinal fibers*, because their cell bodies are in the cerebral cortex (*cortico-*) and their axons end in the spinal cord (*-spinal*). These fibers pass through the subcortical white matter, the entire brainstem, and upper levels of the cervical spinal cord. En route they pass near nuclei and fiber tracts that are specific to that particular region (Fig. 1–4). In the midbrain, for example, they pass near fibers of the oculomotor nerve, which originate in the midbrain and control certain extraocular muscles. In the medulla, they pass near fibers that originate in the medulla and innervate the musculature of the tongue. An injury to the midbrain, therefore, could cause motor problems in the hand (*systems damage*) combined with partial paralysis of eye movement (*regional damage*). In similar fashion, an injury to the medulla could cause the same hand problem but now in association with partial paralysis of the tongue. As we study the nervous system, we shall see that *successful diagnosis of patients with neurologic disorders will depend on, among other things, a good understanding of both regional and systems neurobiology.*

Localizing Signs. The example (see Fig. 1–4) of corticospinal fibers that innervate spinal motor neurons serving the hand coupled with neuron cell bodies in the midbrain that innervate eye muscles via the oculo-

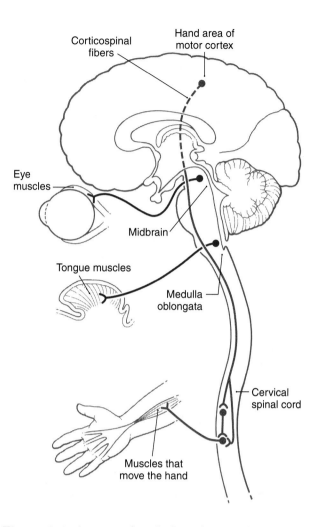

Figure 1–4. An example of the relation of systems to regions. Fibers of the motor system that control hand movement descend from the motor cortex to the cervical spinal cord. In the cord, these fibers influence motor neurons that control hand and forearm muscles. Injury at any point along the way can damage fibers of the system and structures specific to the region. For example, injury to the midbrain could damage both fibers to the hand and fibers to the eye muscles, whereas injury to the medulla could damage both fibers to the hand and fibers to the tongue musculature.

motor nerve also illustrates the concept of *localizing signs.* Brain injury that results in a weakness or paralysis of the upper extremity generally localizes the lesion only to one cerebral hemisphere or perhaps to one side of the brainstem. The clinical examination does not tell us which region of the brain is injured (internal capsule, midbrain, pons, or medulla) or, for that matter, even whether the lesion is in the cervical spinal cord. However, if the paralysis of the upper extremity is coupled with a partial paralysis of eye movement, the lesion can be *specifically localized to the midbrain.* In this example, the lesion in the midbrain damages the fibers of the oculomotor nerve that are specific to this level, while the corticospinal fibers are injured as they traverse the midbrain (see Fig. 1–4). In general, cranial nerve signs are more helpful than long-tract signs in localizing the lesion; that is, they are better *localizing signs*. Many examples of localizing signs are seen in later chapters.

Another general concept of localization states that certain combinations of neurologic deficits may indicate involvement of one of three general locations of the CNS. First, deficits (motor or sensory) located on the same side of the head and body frequently signify lesions in the cerebral hemisphere. Second, deficits on one side of the head and on the opposite side of the body generally indicate a lesion in the brainstem. Such deficits are called *crossed deficits*. Third, deficits of the body only usually suggest a lesion in the spinal cord. Although there are exceptions to these general rules, we shall see that they hold true in many clinical situations.

Dorsal, Ventral, and Other Directions in the Central Nervous System

By convention, directions in the human CNS—such as *posterior* (*dorsal*) and *anterior* (*ventral*), *medial* and *lateral*, and *rostral* and *caudal*—are absolute with respect to the central axis of the brain and spinal cord. In similar manner, the anatomic orientation of the body in space is related to its central axis. For example, if the patient is lying on his stomach, the posterior surface of the trunk is up and its anterior surface is down (Fig. 1–5). If the patient rolls over, his back remains the posterior surface of his body even though it now faces down.

As shown in Figure 1–6, the spinal cord and the brainstem (medulla, pons, and midbrain) form a nearly straight line that is roughly parallel with the superoinferior axis of the body. Therefore, anatomic directions in these regions of the CNS coincide roughly with those of the body as a whole. The pons is rostral to the medulla, for instance, and the cerebellum is posterior (dorsal) to the pons. The situation is different in the forebrain, because during embryonic development, the forebrain rotates (at the cephalic flexure) relative to the midbrain until its rostrocaudal axis corresponds to a line drawn from the forehead to the occiput (from the frontal to the occipital poles of the cerebral hemispheres). This rotation creates a sharp angle in the long axis of the CNS at the midbrain-thalamus junction. Consequently, the long axis of the CNS bends at the midbrain-thalamus junction, and the directions posterior and anterior follow accordingly (see Fig. 1–6). In the cerebral hemisphere (forebrain), *posterior (dorsal)* is toward the top of the brain, *anterior (ventral)* is toward the base of the brain, *rostral* is toward the frontal pole, and *caudal* is toward the occipital pole. Anatomic directions in the forebrain relate to its long

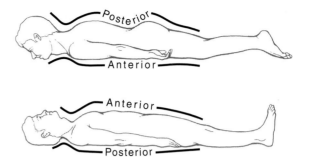

Figure 1–5. An illustration showing that the anatomic directions of the body are absolute with respect to the axes of the body, not with respect to the position of the body in space.

axis; therefore, the posterior side of the forebrain structures faces the vertex of the head, and the anterior aspect of the forebrain faces the base of the skull (see Fig. 1–6). *Posterior and dorsal* and *anterior and ventral* are considered synonymous and *are commonly and frequently used interchangeably*.

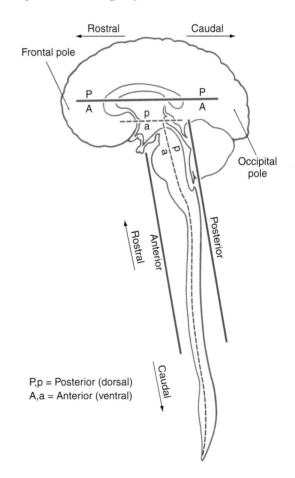

Figure 1–6. The central axis and anatomic directions of the central nervous system (CNS). The dashed line shows the long (rostrocaudal) axis of the CNS. The long axis of the spinal cord and brainstem forms a sharp angle with the long axis of the forebrain. Posterior (dorsal) and anterior (ventral) orientations are also shown.

Clinical Images of the Brain and Skull

Modern technology has given us the tools to view the living brain and skull in some detail. In addition, arteries and veins of the brain and of the meninges can be visualized by mapping the movement of blood through these vascular structures. The resultant images are powerful tools to use in the diagnosis of the neurologically impaired patient.

The most routinely used methods to image the brain and skull are computed tomography (CT) and magnetic resonance imaging (MRI) (Fig. 1–7). As we shall see below, CT is especially useful in visualizing the skull and the brain in the early stages of subarachnoid hemorrhage. On the other hand, MRI, using T1-weighted or T2-weighted techniques, shows brain anatomy in elegant detail, cisternal relationships, cranial nerves, and a wide variety of clinical abnormalities.

Magnetic resonance angiography (MRA) visualizes arteries and veins by measuring the velocity of flow in these structures (Fig. 1–8). The resultant images show detail of vascular structures that, in some situations, may be superior to that seen on angiograms. Arterial structures may be selectively imaged, or combinations of arterial and venous structures or only venous structures can be visualized. Some clinicians refer to these images of venous structures as MRVs (magnetic resonance venograms).

Computed Tomography. CT is an x-ray imaging technique that measures the effects that tissue density and the various types of atoms in the tissue have on x-rays passing through that tissue (Table 1–1; see Fig. 1–7A and B). Changes in the emerging x-ray beam are measured by detectors.

The higher the atomic number, the greater the ability of the atom to attenuate, or stop, x-rays. These attenuation transmission intensities emerging from the

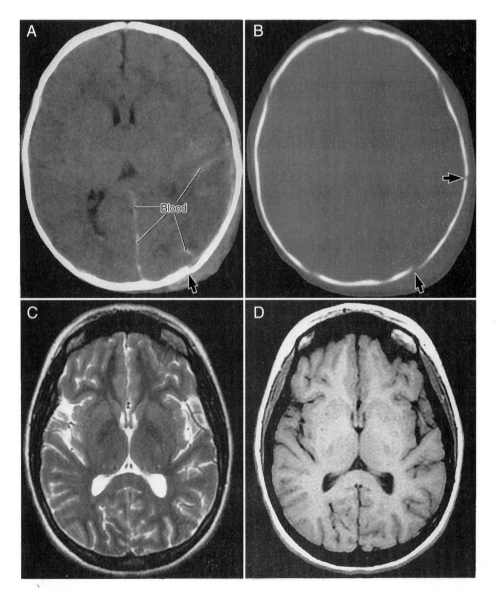

Figure 1–7. Computed tomography (CT) (A, B) of a 2-month-old infant who was a victim of the shaken baby syndrome and magnetic resonance imaging (MRI) of a normal 20-year-old woman (C, D). On the CT study, note that brain detail (A) is less than on the MRI study (C, D) but that the presence of blood (A, in the interhemispheric fissure between the hemisphere and in the brain substance) is obvious. In the same patient, the bone window (B) clearly illustrates the outline of the skull but also clearly shows skull fractures (*arrows* in A and B). In this infant, the ventricles on the left are largely compressed and the gyri have largely disappeared owing to pressure from bleeding into the hemisphere. The pressure results in the effacement of the sulci and gyri on the left side. In a T2-weighted image (C), cerebrospinal fluid is white, internal brain structures are seen in excellent detail, and vessels are obvious. In the T1-weighted image (D), cerebrospinal fluid is dark and internal structures of the brain are somewhat less obvious.

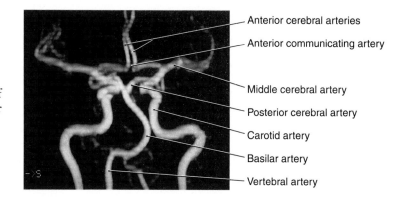

Figure 1–8. Magnetic resonance angiography of the internal carotid artery and the vertebrobasilar systems.

tissue are transformed by a computer into numbers that represent values found in all the points located in the volume of the tissue slice. These values are expressed in Hounsfield Units (HUs). HU values, also known as CT numbers, are used in an arbitrary scale in which bone is specified as +1000 (and is very white; see Fig. 1–7A and B), water as zero, and air as −1000 (and is very black). Approximate numbers are +100 for blood, +30 for brain, and +5 for cerebrospinal fluid (CSF). Using this scale, the HU values, or CT numbers, represent specific shades of gray for each of the various points located in the slice (see Tables 1–1 and 1–2 and Fig. 1–7A and B). The resultant image is seen on a computer monitor, or it can be transferred to x-ray film.

Present-generation CT scanners, known as helical (spiral) scanners, image a continuous spiral slice through a preselected body region very quickly. Computer software converts this information into contiguous slices of a chosen thickness. This technique eliminates movement artifacts and enables reconstruction of soft tissues, bone, or contrast-enhanced vessels into three-dimensional images that can be manipulated in any plane.

CT is a fast and accurate method of detecting recent subarachnoid hemorrhage (Table 1–2; see Fig. 1–7A). An acute subarachnoid hemorrhage in a noncontrast CT scan appears hyperdense (white) in contrast with the subarachnoid spaces and cisterns, which normally are hypodense (dark).

Enhanced CT is a technique using an iodinated contrast material injected intravenously followed by CT examination. Iodine has a large atomic number and attenuates x-rays. As a result, vasculature is visualized as hyperdense (white) structures. This contrast material may also enhance neoplasms or areas of inflammation, because the contrast leaks from the vessels into the cellular spaces owing to a breakdown of the blood-brain barrier. Imaged in this way, the tumor, inflamed meninges, or brain parenchyma will show varying degrees of enhancement or hyperdensity (varying degrees of whiteness).

Magnetic Resonance Imaging (see Tables 1–1 and 1–2 and Fig. 1–7C and D). Protons (hydrogen) constitute a large proportion of body tissue. These atoms have a nucleus and a shell of electrons and a north and a south pole, and they spin around an angulated axis like small planets. As the electrons move with the spinning atom, they induce an electrical current that creates a magnetic field. These atoms function somewhat like little spinning bar magnets. They are aligned randomly because of the changing magnetic effects on

---

Table 1–1.   Appearance of Tissues Imaged by CT and MRI

Modality	Bone	CSF	Gray Matter	White Matter	Fat	Air	Muscle
CT*	↑↑↑	↓↓	↓	↓↓	↓	↓↓↓	↑↑
MRI/T1**	↓↓↓	↓↓↓	↓↓	↓	↑↑	↓↓↓	↓↓
MRI/T2**	↓↓↓	↑↑↑	↓↓	↓↓↓	↑	↓↓↓	↓↓–↓↓↓

* Measures tissue density.
** Measures tissue signal.
↑ ↑ ↑ – ↑ represents very white to light gray:

↓ – ↓ ↓ ↓ represents light gray to very black:

CSF, cerebrospinal fluid; CT, computed tomography; MRI, magnetic resonance imaging.

Table 1–2. Differences in CT Density and MRI Signals in Representative Clinical Examples

Clinical Problem	CT*	MRI T1**	MRI T2**
Acute SAH	↑↑↑	0	0
Subacute SAH	↑↑	0-↑	0
Tumor	0	0	↑-↑↑
Enhanced tumor	↑↑↑	↑↑↑	↑↑↑
Acute infarct	0	0-↓	↑-↑↑
Subacute infarct	0-↓	0-↓↓	↑↑-↑↑↑
Acute ischemia	0	0-↓	↑-↑↑
Subacute ischemia	0-↓	0-↓↓	↑↑-↑↑↑
Edema	0-↓	0-↓	↑-↑↑

* Measures tissue density.
** Measures tissue signal.
↑ ↑ ↑ – ↑ represents very white to light gray:

↓ – ↓ ↓ ↓ represents light gray to very black:

0 represents no change from normal.
CT, computed tomography; MRI, magnetic resonance imaging; SAH, subarachnoid hemorrhage.

each other. When these protons are exposed to a powerful magnet, they stop pointing randomly and align themselves in parallel with the external magnetic field but at different energy levels. The stronger the external magnetic field, the faster the frequency of the spin at that angle. When undergoing an MRI examination, the patient becomes a magnet, with all the protons aligning along the external magnetic field and spinning at an angle with a certain frequency.

A radio wave is an electromagnetic wave. When sent as a short burst into the magnet containing the patient, it is known as a radiofrequency (RF) pulse. This RF pulse can vary in frequency strength. Only when the frequency strength of the RF pulse matches the frequency strength of the angulated spinning proton will the proton absorb energy from the radio wave. This phenomenon is called *resonance* and is the "resonance" in "magnetic resonance imaging." This results in a twofold effect: it cancels out the magnetic effects of certain protons, and it raises the energy levels and magnetic effects of another group of protons. When the radio wave is turned off, the canceled-out protons gradually return to their original state and strength of magnetization, which is called *relaxation* and is described by a time constant known as T1 (see Fig. 1–7D). The protons that aligned themselves at a higher energy level and magnetization also start to lose their energy (relaxation), and this time constant is known as T2 (see Fig. 1–7C). The T1 relaxation time

is longer than the T2 relaxation time. The "deexcited" or relaxed protons release their energy as an "echo" of radio waves. A receiver coil (antenna) absorbs this information, and a computer determines the characteristics of the emitted radio waves from all the specific points in that section of the body. The magnetic resonance image is then constructed and transferred to a computer monitor or x-ray film. T1-weighted or T2-weighted images can be obtained by using varying times to receive the echoes (TE).

Conventional *spin echo* sequences generate images that may be T1-weighted or T2-weighted according to the time interval in milliseconds between each exciting radio wave. This is called *repetition time* (TR). The time interval, in milliseconds, required to collect these radio waves from the relaxing protons is called *echo time* (TE). With spin echo pulse sequences, the shorter the TR and TE, the more the image is considered T1-weighted. The longer the TR and TE, the more the image is considered T2-weighted.

The contrast material used for enhancing tumors and blood vessels is the paramagnetic rare earth gadolinium. It is chelated to a certain molecule and is in solution for intravenous injection. The gadolinium causes an increase in signal by shortening the relaxation time for T1. Owing to a breakdown of the blood-brain barrier, intravascular gadolinium enters the pericellular spaces, where it increases the relaxation state of water protons and generates a bright signal on T1-weighted images.

Acute subarachnoid hemorrhage is poorly imaged by MRI on T1-weighted images but well imaged by CT (see Table 1–2). Some MRI sequences are sensitive for detection of acute bleeding, but other factors may limit this method of examination. Special MRI techniques can also determine if a brain infarct or ischemia is acute (about 1 to 3 hours old) or subacute (about 4 hours old or more). Contraindications for MRI are cardiac pacemakers, cochlear implants, ferromagnetic foreign bodies in the eye, and certain aneurysm clips. Large metallic implants or ferromagnetic foreign bodies in the body may heat up. The general appearance of the brain and adjacent structures in health and disease on MRI and CT are summarized in Tables 1–1 and 1–2.

Imaging of the Brain and Skull. Patients lie on their back (supine) for imaging of the brain or spinal cord and the surrounding bony structures of the skull and vertebral column (Fig. 1–9). In this position, the posterior (dorsal) surface of the brainstem and spinal cord and the caudal aspect (occipital pole) of the cerebral hemispheres face down. The anterior (ventral) surface of the brainstem and spinal cord and the frontal pole are face up (see Fig. 1–9).

Images of the brain are commonly made in *coronal*, *axial* (horizontal), and *sagittal* planes. To illustrate the basic orientation of the CNS in situ, we shall look at

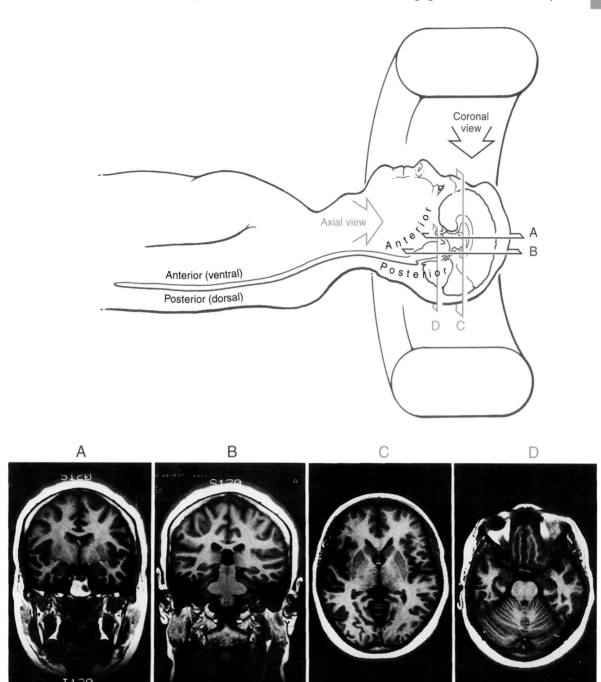

Figure 1–9. The relation of imaging planes to the brain. The diagram shows the usual orientation of a patient in a magnetic resonance imaging machine and the planes of the four scans (T1-weighted images) that are shown. *A* and *B*, Coronal scans; *C* and *D*, axial scans.

examples of coronal and axial images, shown as they would appear in the clinical setting (see Fig. 1–9). Coronal imaging planes are oriented perpendicular to the rostrocaudal axis of the forebrain but are nearly parallel to the rostrocaudal axis of the brainstem and spinal cord. Therefore, a coronal image obtained at a relatively rostral level of the cerebral hemispheres (see Fig. 1–9*A*) will show only forebrain structures,

and these structures will appear in cross section (perpendicular to their long axis). As the plane of imaging is moved caudally, brainstem structures enter the picture (see Fig. 1–9*B*), but the brainstem is cut nearly parallel to its rostrocaudal axis.

Axial images, in contrast, are oriented parallel to the rostrocaudal axis of the cerebral hemispheres but nearly perpendicular to the long axis of the brainstem and spinal cord. Consequently, an axial image obtained midway through the cerebral hemispheres (see Fig. 1–9C) will show only forebrain structures, with the rostral end of the forebrain at the top of the image and the caudal end at the bottom. As the plane of imaging is moved further anteriorly (ventrally) relative to the forebrain, the brainstem appears (see Fig. 1–9D). The brainstem, however, is cut nearly in cross section and is oriented with the anterior (ventral) surface "up" (toward the top of the image) and the posterior (dorsal) surface "down."

A point also needs to be made about how the clinician looks at scans such as those in Figure 1–9. Coronal scans are viewed as though you are looking the patient in the face, whereas axial scans are viewed as though you are standing at the patient's feet looking toward his head while he lies on his back in the machine. Axial scans, in other words, show the cerebral hemispheres from ventral to dorsal, with the patient's orbits at the top of the image and the occiput at the bottom. *In both coronal and axial views, the patient's left side is to the observer's right.*

Sources and Additional Reading

Grossman CB: Magnetic Resonance Imaging and Computed Tomography of the Head and Spine, 2nd ed. Lippincott Williams & Wilkins, Baltimore, 1996.

Jackson GD, Duncan JS: MRI Neuroanatomy: A New Angle on the Brain. Churchill Livingstone, New York, 1996.

Kirkwood JR: Essentials of Neuroimaging. Churchill Livingstone, New York, 1990.

Kretschmann HJ, Weinrich W: Cranial Neuroimaging and Clinical Neuroanatomy: Magnetic Resonance Imaging and Computed Tomography, 2nd ed. Thieme Medical Publishers, New York, 1992.

Chapter 2

The Cell Biology of Neurons and Glia

J. B. Hutchins, J. P. Naftel, and M. D. Ard

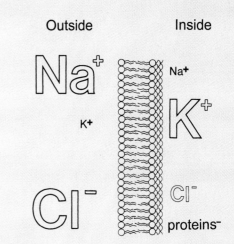

Overview 16

The Structure of Neurons 16
Dendrites
Cell Body
Axons and Axon Terminals
Axonal Transport
Axonal Transport as a Research Tool

Classification of Neurons and Groups of Neurons 22

Electrical Properties of Neurons 23
Ion Channels
Channel-Mediated Changes in Membrane
 Potential
The Sodium/Potassium Pump

Neurons as Information Receivers 25
Sensory Neural Information
Other Neural Information

Neurons as Information Transmitters 25
Synapses
Chemical Synapses

Neurotransmitters 27
Disorders of Neurotransmitter Metabolism

Glia 28

Astrocytes 28
Structural Support and Response to Injury
Growth Factors and Cytokines
Environmental Modulation
Metabolism
Regional Heterogeneity
Astrocytes at the Blood-Brain Barrier
CNS Tumors

Oligodendrocytes 30

Microglia 32

Supporting Cells of the Peripheral Nervous System 33

Degeneration and Regeneration 35

The number of cells in the adult human central nervous system (CNS) has been estimated at 100 billion. All arise from a relatively small population of precursors, yet a diversity of cell types is seen in the adult. Their most basic classification is as *neurons* and *glia* (*glial cells*).

Overview

Nerve cells (neurons) manipulate information. Doing so involves changes in the *bioelectrical* or *biochemical* properties of the cell, and these changes require a vast expenditure of energy for each cell. The nervous system, compared with other organs, is the greatest consumer of oxygen and glucose. These energy requirements arise directly from the metabolic demand placed on cells, which have a large surface area and concentrate biomolecules and ions against an energy gradient. Along with maintaining its metabolism, each neuron (1) *receives information* from the environment or from other nerve cells, (2) *processes information*, and (3) *sends information* to other neurons or effector tissues.

Unlike neurons, *glia* do not receive and transmit information with point-to-point specificity. Rather, their primary function is control of the environment within the CNS. They *shuttle nutritive molecules* from blood vessels to neurons, *remove waste* products, and *maintain the electro-chemical* surroundings of neurons. Glial cells are also essential in the early development of the CNS for *guiding developing neurons* to their correct locations, and, in the adult, glia provide *structural support* for nerve cells.

In order for neurons to carry out the three tasks of receiving, processing, and sending information, they must have specialized structures which are designed to carry out each of these tasks. The basic parts of a neuron are shown in Fig. 2–1. Additionally, specialized mechanisms and structures exist for some special problems faced by neurons. Two such problems are immediately apparent. First, the mix of ions inside neurons is quite different from the mix outside the cell. Maintaining this difference requires huge amounts of energy, since ions must be pumped against ionic and diffusion gradients. The large surface area of neurons compounds this problem. Second, those neurons that send information over long distances must have a way to supply these distant sites with macromolecules and energy. In order to fully understand the cell biology of neurons, it is important to see the biochemical, anatomical, and physiological properties of neurons as part of an integrated whole, the machinery which permits the neuron to do its specialized "jobs." In the following sections, we will examine how neuronal architecture and chemistry are designed to meet these special demands.

The Structure of Neurons

The archetypical neuron is bounded by a continuous plasma membrane and consists of a *cell body*, or *soma*, from which *dendrites* and an *axon* arise (Figs. 2–1 and 2–2). The cell body contains the nucleus surrounded by a mass of cytoplasm that contains the organelles necessary for protein synthesis and metabolic maintenance. Most neurons (*multipolar*) have several dendrites extending from the cell body (see Figs. 2–1 and 2–2). These are usually relatively short processes that taper from a thick base and, in doing so, branch extensively. In contrast, there is a single axon, which is long (extending from a few millimeters to more than a meter) and of a uniform diameter. The axon has few, if any, branches along most of its length, branching extensively only near the distal end (the *terminal arbor*) (see Figs. 2–1 and 2–2). In most neurons, information normally flows from the dendrites to the cell body to the axon (and its terminals), and then to the next neuron or effector tissue. We shall describe these components of the neuron in the order in which information is processed.

Dendrites. Dendrites usually branch extensively in the vicinity of the cell body, giving the appearance of a tree or bush (Fig. 2–3A; also see Figs. 2–1 and 2–2). They receive signals either from other neurons through contacts (*synapses*) made on their surfaces or from the environment via specialized receptors. Information travels from distal to proximal along dendrites to converge at the cell body.

Small budlike extensions (*dendritic spines*) of a variety of shapes are frequently seen on the more distal branches of the dendritic tree (see Figs. 2–1 and 2–3B and C). These are usually the sites of *synaptic contacts* (discussed below). As the dendrite progresses toward the cell body, its thickness increases as it anastomoses with other branches.

Small dendrites contain cytoskeletal elements but appear devoid of ____les. Larger dendrites contain numerous ____filaments (a type of intermediate filament present ____lusively in neurons), *mitochondria*, some saccules of endoplasmic reticulum, and collections of polyribosomes and free ribosomes (see Figs. 2–1 and 2–3D and E). In many nerve cells, the distal dendrites collect into large, trunklike *primary dendrites* that contain the same organelles as those of the cell body. The microtubular and neurofilamentous skeletons of the dendritic tree are continuous throughout its extent and help to maintain its branched structure.

Cell Body. The cell body of a neuron is also called the *soma* (plural *somata*) or *perikaryon* (plural, *perikarya*) (Fig. 2–4; see also Fig. 2–2). The perikaryon is the *metabolic center* of the nerve cell. Abundant mitochondria reflect the high energy consumption of the cell. Active protein synthesis is indicated by other structural

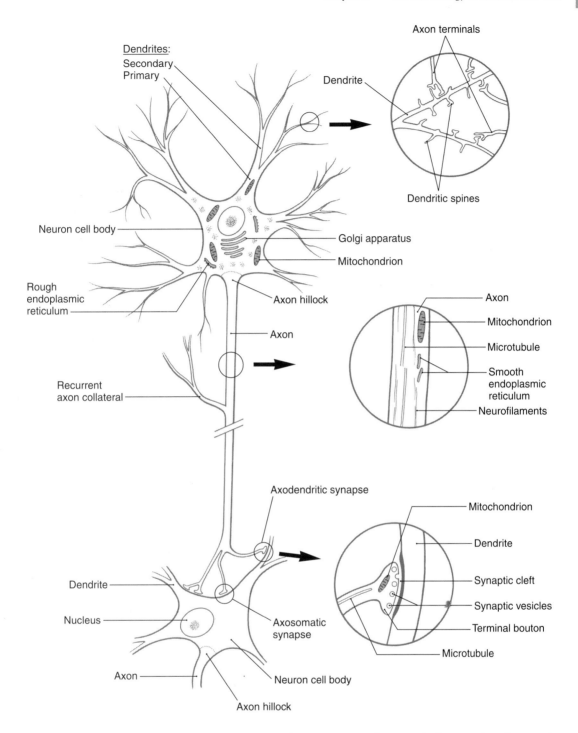

Figure 2–1. Diagrammatic representation of a typical multipolar neuron. Dendrites, with their variety of spines (*upper detail*), branch in the immediate vicinity of the cell body, while the single axon, with its occasional recurrent collaterals, may travel great distances to the next neuron. The cell body contains the organelles essential for neuronal function. Microtubules (*middle and lower details*) are important structures for the transport of substances within the axon. The axon ends as a terminal arbor that forms many terminal boutons (*lower detail*), each containing the necessary machinery for synaptic transmission.

features. A large nucleus contains diffuse chromatin (*euchromatin*) and typically at least one prominent nucleolus. In the cytoplasm, *ribosomes* are abundant, and the *rough endoplasmic reticulum* (rER) and *Golgi complex* are extensive (see Fig. 2–1). The rER is basophilic

(binds basic dyes) as a result of the large amount of ribosomal RNA attached to the endoplasmic membrane. These extensive, stacked layers of rER are seen as patches of basophilic staining (called *Nissl substance*) in histologic preparations of nerve cells.

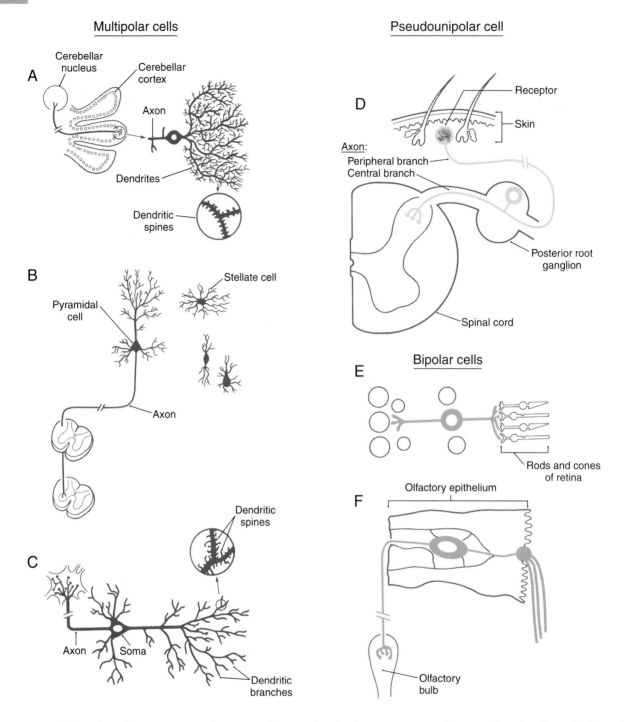

Multipolar cells

A
Cerebellar nucleus
Cerebellar cortex
Axon
Dendrites
Dendritic spines

B
Pyramidal cell
Stellate cell
Axon

C
Axon   Soma
Dendritic spines
Dendritic branches

Pseudounipolar cell

D
Receptor
Skin
Axon:
Peripheral branch
Central branch
Posterior root ganglion
Spinal cord

Bipolar cells

E
Rods and cones of retina

F
Olfactory epithelium
Olfactory bulb

Figure 2–2. Examples of various types of neurons showing the dendrites, somata, and axons of multipolar cells from the cerebellar cortex (*A*) and from the cerebral cortex (*B* and *C*). Compare these with a pseudounipolar cell of the dorsal root ganglion (*D*), and bipolar cells from the retina (*E*) and olfactory epithelium (*F*).

Neurons are classified into three broad types on the basis of the shape of the cell body and the pattern of processes emerging from it. These types are the multipolar, pseudounipolar, and bipolar cells (Table 2–1; see Fig. 2–2). In *multipolar* neurons, multiple dendrites emerge from the cell body, giving it a polygonal shape. A single axon, usually of small diameter, also arises from the cell body. Different kinds of multipolar cells

have characteristic patterns of processes; some are listed in Table 2–1.

The cell body of a *pseudounipolar* (or *unipolar*) neuron gives rise to a single process and is round, with a centrally located nucleus (see Fig. 2–2D). The single process divides close to the cell body into two branches: a *peripheral* branch, which carries sensory information from the periphery, and a *central* branch,

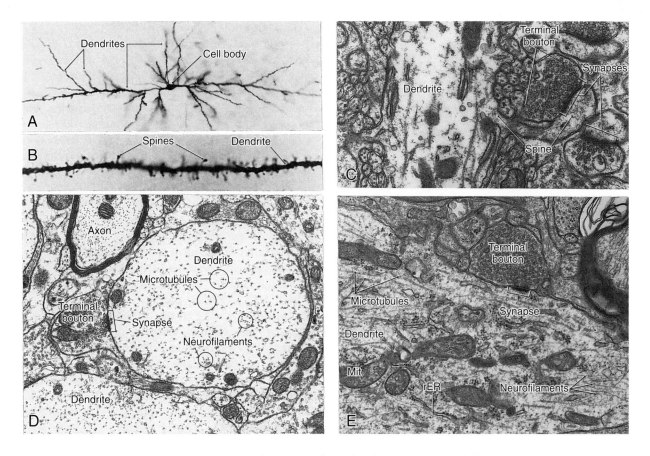

Figure 2-3. Elements of dendrite structure. Dendritic tree of a multipolar neuron (*A*) and dendritic spines (*B*), both in Golgi-stained cortical tissue. Ultrastructural features of dendrites, showing an axonal terminal bouton synapsing on a dendritic spine (*C*), a cross section of a dendrite with characteristic cytoskeletal elements and organelles (*D*), and a longitudinal section of a dendrite in the anterior horn of the spinal cord (*E*). (*A* and *B* courtesy of Dr. Jose Rafols; *D* courtesy of Dr. Alan Peters.)

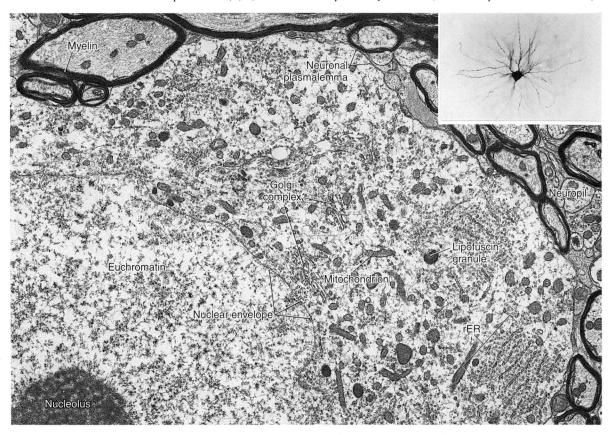

Figure 2-4. The cell body of a multipolar neuron as seen on electron micrograph and in a Golgi-stained preparation (*inset*). (*Inset* courtesy of Dr. Jose Rafols.)

Table 2–1. A Few of the Neuronal Types Found in the Nervous System

Type of Neuron	Location of Cell Bodies
Pseudounipolar	Posterior root or cranial nerve ganglion
Bipolar	Retina
	Olfactory epithelium
	Vestibular ganglion
	Auditory (spiral) ganglion
Multipolar	
Stellate ("star-shaped")	Many areas of CNS
Fusiform ("spindle-shaped")	Many areas of CNS
Pyriform ("pear-shaped")	Many areas of CNS
Pyramidal	Hippocampus; layers II, III, V, and VI of cerebral cortex
Purkinje	Cerebellar cortex
Mitral	Olfactory bulb
Chandelier	Visual areas of cerebral cortex
Granule	Cerebral and cerebellar cortex
Amacrine ("axonless")	Retina

CNS, central nervous system.

which relays the information onward to its target in the CNS. The two processes thus function as a combined axon and dendrite. The distal end of the peripheral process is dendrite-like, and its branches either constitute or contact sensory receptors. The central process terminates in either the spinal cord or the brainstem. The cell bodies of pseudounipolar cells are found primarily in the sensory ganglia of cranial and spinal nerves.

Bipolar neurons have a round or oval-shaped perikaryon, with a single large process emanating from each end of the cell body (see Fig. 2–2E and F). They are commonly found in sensory structures. In the retina, bipolar cells are interposed between receptor cells and output cells. In the olfactory system, they function as both the receptors and the output neurons, with their axons projecting to the olfactory bulb, and in the vestibular and auditory systems, they are the output cells that send information to the brainstem.

Unless special staining methods are used, the cell body of a neuron has the appearance of being the entire cell when viewed in histologic sections. However, as discussed later, the volume and surface area of the cell body of a neuron usually constitute only a

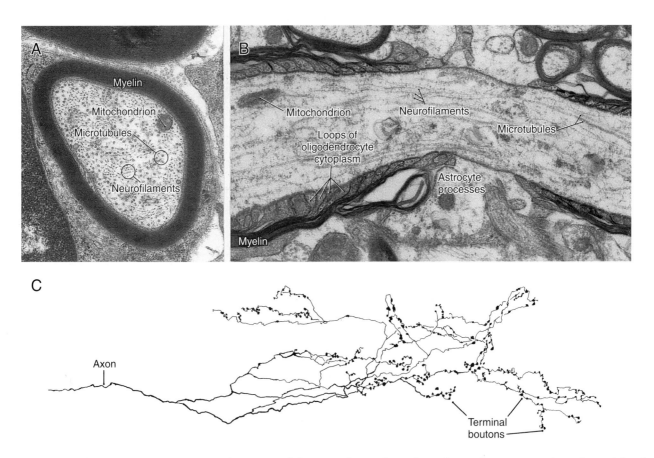

Figure 2–5. Elements of axon structure. Ultrastructural features of a small myelinated axon in a cross section of a peripheral nerve (*A*) and a longitudinal view at a node of Ranvier of a myelinated axon in the central nervous system (*B*). Drawing of the complete terminal arbor of an axon in the thalamus, reconstructed from serial sections (*C*). (*C* courtesy of Dr. Ed Lachica.)

small fraction of the volume and surface area of its processes.

Axons and Axon Terminals. The *axon* arises from the cell body at a small elevation called the *axon hillock*. The proximal part of the axon, adjacent to the axon hillock, is the *initial segment*. The cytoplasm of the axon (axoplasm) contains dense bundles of *microtubules* and *neurofilaments* (Fig. 2–5*A* and *B*; see also Fig. 2–1). These function as structural elements, and they also play key roles in the transport of metabolites and organelles along the axon. Axons are typically devoid of ribosomes, a feature that distinguishes them from dendrites at the ultrastructural level.

In contrast to dendrites, axons may extend for long distances before branching and terminating. For example, the axon of a motor neuron innervating the adductor hallucis muscle in a tall person is well over a meter in length. To place this in scale, if the cell body of this neuron were the size of a softball, the axon (with a diameter of about 6 mm) would extend about 10 km. Although the cell body is the nutritive center of the neuron, its volume represents a small fraction of the collective volume of the dendrites and axon. The surface area of an axon can be several thousand times the surface area of the parent cell body.

Axons in the CNS often end in fine branches known as *terminal arbors* (Fig. 2–5*C*). In most neurons, each axon terminal is capped with small *terminal boutons* (*boutons termineaux*, terminal buttons) (see Figs. 2–1 and 2–3*C* and *E*). These correspond to functional points of contact (synapses) between nerve cells. In some cells, boutons are found along the length of the axon, where they are called *boutons en passant*. Other axons contain swellings, or *varicosities*, which are not button-like but still can represent points of cell-to-cell information transfer.

The site at which an axon terminal communicates with a second neuron, or with an effector tissue, is called a *synapse* (from the Greek word meaning "to clasp"). In general, the synapse can be defined as a contact between part of the neuron (usually an axon) and the dendrites, cell body, or axon of a second neuron. The contact can also be made with an effector cell such as a skeletal muscle fiber. Synapses are considered later in this chapter in the section "Neurons as Information Transmitters."

Axonal Transport. Nerve cells have an elaborate transport system that moves organelles and macromolecules from the cell body out to the axon terminals. Transport in the axon occurs in both directions (Table 2–2; Fig. 2–6). Axonal transport from the cell body toward the terminals is called *anterograde* or *orthograde*; transport from the terminals toward the cell body is called *retrograde*.

Anterograde axonal transport is classified into *fast* and *slow* components. Fast transport, at speeds of up to 400 mm/day, is based on the action of a protein called *kinesin*. Kinesin, an ATPase, moves macromolecule-containing vesicles and mitochondria along microtubules in much the same manner as a small insect crawling along a straw. Slow transport carries important structural and metabolic components from the cell body to axon terminals; its mechanism is less well understood.

Retrograde axonal transport allows the neuron to respond to molecules, for example, growth factors, which are taken up near the axon terminal by either *pinocytosis* or *receptor-mediated endocytosis*. In addition, this form of transport functions in the continual recycling of components of the axon terminal. Retrograde transport along axonal microtubules is driven by the protein dynein rather than by kinesin.

Axonal transport is important in the pathogenesis of some human neurologic diseases. The *rabies virus* replicates in muscle tissue at the site of a bite by a rabid animal and is then transported in a *retrograde* direction to the cell body of neurons innervating the muscle. The infected cells produce and shed copies of the rabies virus, which, in turn, are taken up by the terminals of adjacent cells. In this way the infection becomes distributed throughout the CNS, causing the behavioral changes associated with this disease. From the CNS, the virus travels to the salivary glands by means

Table 2–2. Characteristics of Axonal Transport

Direction of Transport	Speed of Transport	Proposed Mechanism	Substances Carried
Anterograde	Fast (100–400 mm/day)	Kinesin/microtubules	Proteins in vesicles
		Neurotransmitters in vesicles, mitochondria	
	Slow (~1 mm/day)	Unknown	Cytoskeletal protein components (actin, myosin, tubulin) Neurotransmitter-related cytosolic enzymes
Retrograde	Fast (50–250 mm/day)	Dynein/microtubules	Macromolecules in vesicles, "old" mitochondria Pinocytotic vesicles from axon terminal

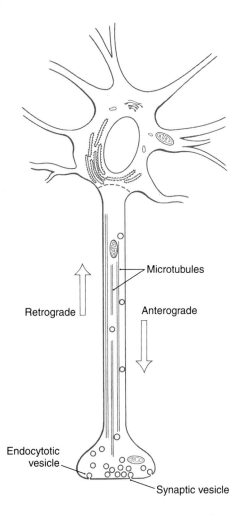

Figure 2–6. Anterograde and retrograde axonal transport.

presence of the label in a cell body suggests that the neuron has axon terminals at the site of injection.

Tracer studies can also exploit the anterograde transport system of neurons. For example, if radioactively labeled amino acids are injected into a group of neuronal cell bodies, they will be incorporated into neuronal proteins and transported in an anterograde direction. The axons containing the labeled proteins can then be detected by autoradiography. Another commonly used anterograde tracer is HRP conjugated to the glycoprotein-binding molecule (lectin) *wheat germ agglutinin* (WGA-HRP). Anterograde tracers are used to identify the distribution patterns of axons arising from a specific population of neuronal cell bodies.

The fact that the cell body is the trophic center of the neuron provides two other methods of studying connections in the nervous system. If the cell body is destroyed, the axon undergoes *anterograde (wallerian)* degeneration. These degenerated axons can be visualized when neural tissue is impregnated with silver nitrate. Variations on this method make it possible to conduct studies on human material obtained at autopsy. Conversely, injury to the axon will result in a set of changes in the cell body that are referred to as *chromatolysis*. The cell body swells, the nucleus assumes an eccentric position, and the Nissl substance disperses. (This breakup of the dye-binding parts of the cell gives chromatolysis its name.) This technique has also been used in animal experimentation and in human autopsy material.

of anterograde axonal transport in neurons innervating these glands. The infected salivary glands shed the virus in the saliva.

The toxin produced by the bacterium *Clostridium tetani* is also transported in a retrograde direction in nerve cells whose axons terminate at the site of infection. *Tetanus toxin* is released from the nerve cell body and taken up by the terminals of neighboring neurons. However, unlike the rabies virus, which is replicated in the cell body, the tetanus toxin is diluted as it passes from cell to cell. In spite of this dilution effect, patients infected with *C. tetani* may suffer from a range of neurologic deficits.

Axonal Transport as a Research Tool. The ability of neurons to transport intracellular materials is exploited in investigations of neuronal connections. For example, when the enzyme *horseradish peroxidase* (HRP) or a *fluorescent substance* is injected into regions containing axon terminals, it is taken up by these processes and transported in a retrograde direction to the cell body. After histologic preparation, the cell bodies containing these retrograde tracers are visualized. The

Classification of Neurons and Groups of Neurons

Functionally related nerve cell bodies and axons are often aggregated to form distinct structures in the nervous system. Table 2–3 lists the main terms used for such structures. In the CNS, a cluster of functionally related nerve cell bodies is most commonly called a *nucleus* (plural, *nuclei*), although a layer of cells may be called a *layer*, *lamina*, or *stratum*, and columnar groups of cells may be called *columns*. The latter term is used for two types of structures. In the cerebral cortex, it refers to a group of cells that are related by function and by the location of the stimulus that drives them. These functional groups form columns oriented perpendicular to the plane of the cortex. The second type of column is found in the spinal cord and refers to a longitudinal group of functionally related cells that extend for part or all of the length of the brainstem or spinal cord.

Bundles of axons in the CNS are called *tracts*, *fasciculi*, or *lemnisci*. These are typically composed of specific populations of functionally related fibers (as in the corticospinal tract and medial lemniscus). A group of several tracts or fasciculi is called a *funiculus* or, in certain

Table 2–3. Terms Used to Describe Groupings of Neuronal Components

Name	Description	Examples
CNS Structures		
Nucleus (plural, nuclei)	A group of functionally related nerve cell bodies in the CNS	Inferior olivary nucleus, nucleus ambiguus, caudate nucleus
Column	In the cerebral cortex, a group of nerve cell bodies that are related in function and in the location of the stimulus that drives them and that form a column oriented perpendicular to the plane of the cortex	The ocular dominance and orientation columns of the visual cortex
	In the spinal cord, a group of functionally related nerve cell bodies that form a longitudinal column extending through part or all of the length of the spinal cord	Clarke's column
Layer, lamina (laminae), stratum (strata)	A group of functionally related cells that form a layer oriented parallel to the plane of the larger neural structure that includes it	Layer IV of cerebral cortex, the stratum opticum of superior colliculus
Tract, fasciculus (fasciculi), lemniscus (lemnisci) (*fasciculus* is Latin for "bundle")	A bundle of parallel axons in the CNS	Optic tract, corticospinal tract, medial longitudinal fasciculus, fasciculus gracilis, medial lemniscus
Funiculus (funiculi) (Latin for "cable")	A group of several parallel tracts or fasciculi	Anterior, posterior, and lateral funiculi of spinal cord
PNS Structures		
Ganglion (ganglia)	A group of nerve cell bodies located in a peripheral nerve or root; it forms a visible knot	Posterior root ganglia, trigeminal ganglion
Nerve, ramus (rami), root	A peripheral structure consisting of parallel axons plus associated cells	Facial nerve, ventral roots of spinal nerves, gray and white rami of spinal nerve roots

CNS, central nervous system; PNS, peripheral nervous system.

cases, a *system*. In the peripheral nervous system (PNS), collections of cell bodies form a *ganglion* (plural, *ganglia*), which may be either *sensory* (dorsal root, cranial nerve) or *motor* (visceromotor or autonomic); and axons make up *nerves*, *rami*, or *roots*.

Neurons may also be classified on the basis of functional characteristics. A neuron that conducts signals from the periphery toward the CNS is called *afferent*; one that conducts signals in the opposite direction is called *efferent*. Neurons with long axons that convey signals to a distant target are called *projection neurons*, whereas neurons that act locally (because their dendrites and axon are limited to the vicinity of the cell body) are called *interneurons* or *local circuit cells*.

Neurotransmitter specificity also can be used to describe neurons and their axons. For example, cells that contain the neurotransmitter *dopamine* are called *dopaminergic* neurons. The neurons whose axons form the corticospinal tracts produce the neurotransmitter *glutamate* and are called *glutamatergic*.

The distinctions among categories based on shape, projection type, or transmitter type are not as clear as those implied in the preceding discussion. For example, most neurons do not resemble the "ideal" multipolar cell. In addition, neurons may overlap several categories of classification. In practice, reference to ganglia, nuclei, and tracts commonly uses a blend of these terms. For example, *dorsal root ganglion* cells are *pseudounipolar* (their

shape), *sensory* (type of input), and *afferent* (information conveyed toward the CNS), and many are *peptidergic* (they contain peptides such as substance P).

Electrical Properties of Neurons

The electrical characteristics of neurons are fully discussed in Chapter 3. Some of these concepts are briefly introduced here in relation to features of the cell membrane.

Nerve cells are electrically charged relative to the fluids bathing them. This is not a unique feature, as virtually all cells in the body have an electrical charge. However, nerve cells are unique in their ability to manipulate the flow of charges across the cell membranes of other neurons or effector cells.

The plasma membrane of a nerve cell controls ion transport so that Na^+ and Cl^- are more concentrated outside the cell than inside it (Fig. 2–7). The interior of the cell ends up with a relative excess of negative charges, so a voltage exists across the cell membrane. This voltage is called the *resting membrane potential*; in a typical neuron it has a value of about -70 mV.

Ion Channels. The voltage across the neural cell membrane is manipulated by the opening and closing of *ion channels* in the membrane. Each kind of channel permits the flow of one species of ion, while largely excluding others (Fig. 2–8). An ion channel is a trans-

Outside   Inside

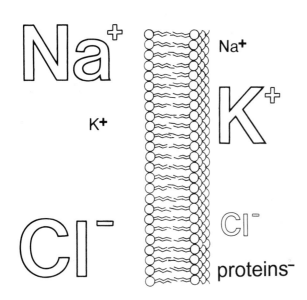

Figure 2–7. Relative ion concentrations in the neuronal cytoplasm (*right*) versus the extracellular fluid (*left*). The concentration of K⁺ is relatively high intracellularly, and the concentrations of Na⁺ and Cl⁻ are relatively high extracellularly.

membrane protein or protein complex that has a central pore that is selective for the ion in question. When the channel is in the "open" conformation, the ion can flow through the pore; when the channel is in the "closed" conformation, the pore is blocked. Channels change from one conformation to the other in response to changes in the electrical or chemical environment (see Chapter 3). In some cases, the transition is rapid in both directions. In other cases, the channel becomes temporarily *refractory* after changing to a particular conformation—that is, it pauses in that state for several milliseconds before being able to change back.

These channels fall into groups, depending on the ions that pass through them (e.g., *sodium* or *potassium channels*). They also may be grouped according to the type of signal that triggers conformational changes (Fig. 2–8). *Voltage-gated ion channels* change their configuration in response to differences in the voltage across the cell membrane. *Ligand-gated ion channels* change their permeability in response to the presence of *neurotransmitters* (chemical messengers) (Table 2–4).

In some instances, a disease can selectively involve one type of ion channel. For example, in *Lambert-Eaton syndrome*, a tumor in the lung (oat cell, or small cell carcinoma) causes the production of antibodies to voltage-gated calcium channels necessary for the function of a synapse (see later). In many cases of *myasthenia gravis*, the patient's immune system produces antibodies to the nicotinic acetylcholine receptor, a

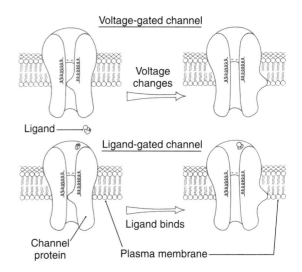

Figure 2–8. Comparison of voltage-gated (*upper*) and ligand-gated (*lower*) ion channels. In a voltage-gated channel, alterations in membrane voltage result in a conformational change opening or closing the channel. In a ligand-gated channel, binding of a specific molecule such as a neurotransmitter induces a similar change.

ligand-gated channel found at the synapse between nerve and muscle cells. These antibodies result in pathologic destruction of *neuromuscular junctions*, which in turn causes the muscle weakness characteristic of the disease.

Channel-Mediated Changes in Membrane Potential. As mentioned, the nerve cell interior has a negative charge with respect to the extracellular fluid, so the resting membrane potential is negative. If the charge difference across the membrane is reduced, so that the membrane potential moves closer to zero, the membrane is said to have undergone *depolarization*. On the other hand, if the charge difference across the

Table 2–4. Voltage Changes That Can Result from Changes in Permeability of Ions

Ion(s)	Direction of Change in Permeability	Effect on Membrane Voltage
K⁺	Increase	Hyperpolarize (IPSP)
	Decrease	Depolarize (EPSP)
Na⁺	Increase	Depolarize (EPSP)
	Decrease	Hyperpolarize (IPSP)
Cl⁻	Increase	Holds membrane near resting voltage
	Decrease	Little change
Na⁺ and K⁺	Increase	Depolarize (EPSP)

All values assume typical values for the resting membrane potential and equilibrium potentials for each ion (see Chapter 3). Each cell's values will vary from these "ideals." At synapses, depolarizations are called excitatory postsynaptic potentials (EPSPs), and hyperpolarizations are called inhibitory postsynaptic potentials (IPSPs).

membrane is increased, making the membrane potential more negative, the membrane is said to be *hyperpolarized.*

Changes in membrane potential are produced by the opening or closing of populations of ion channels. The change is local, in the vicinity of the affected ion channels; a nerve cell membrane may thus have distinct zones of hyperpolarization and depolarization.

There are two functionally different kinds of membrane potential change. A *graded potential* can have any of a range of values, depending on the intensity and duration of the ion channel response and the preexisting membrane potential. It gradually decays back to the resting potential as the membrane restores its resting state. Graded potentials result from the activity in a single kind of ion channel or a group of different kinds of channels. An *action potential*, in contrast, is a large, abrupt spike of depolarization and repolarization. Each action potential is of the same size and shape and is able to propagate itself along the membrane. It results from the opening and closing of voltage-gated sodium and potassium channels that are found only on the axon. An action potential that is triggered anywhere on an axon travels the length of the axon; the resting potential is restored in its wake.

Changes in ionic permeabilities of the cell membrane are the basis for the reception, propagation, and transmission of information within the nerve cell.

The Sodium-Potassium Pump. The resting potential of a neuron is maintained and restored by the activity of a *sodium-potassium pump* located in the plasma membrane. This pump uses ATP energy to actively expel Na^+ while actively importing K^+, thus maintaining the resting concentrations of these ions (Na^+ concentrated on the outside of the cell, K^+ concentrated inside the cell).

Neurons as Information Receivers

Neurons collect, transform, and transmit information. Collection of information by the nerve cell is the first step in this chain of events. Neurons can receive input from other neurons or directly from the environment. *Sensory information* enters the nervous system by the latter of the two routes.

Sensory Neural Information. Neurons that receive information from the environment are called *primary sensory neurons*. These include photoreceptors, chemoceptors, mechanoreceptors, thermoreceptors, and nociceptors. Further information on these receptor types is found in the chapters describing sensory systems. For most receptors, a stimulus results in a *depolarization* of the primary sensory neuron, called a *generator potential.*

The process of converting sensory input into a form interpretable by the nervous system is *transduction.* Each type of sensory receptor transduces a physical stimulus into electrical or chemical changes, which then can be transmitted within the nervous system.

The rod and cone *photoreceptors* of the retina are specialized for transducing light energy in the form of *photons.* As few as three photons (possibly even a single photon!) can be detected by a trained human observer. As a photon strikes the photoreceptor, it sets in motion a complex chain of events culminating in the closing of a large number of normally open sodium channels. As a result, the photoreceptor cell becomes *hyperpolarized.* This makes the photoreceptor unique among sensory cells, in that the membrane potential becomes more negative upon application of the stimulus, rather than more positive.

In humans, the taste and olfactory receptor cells mediate the two primary types of *chemoreception.* Both receptor types respond to the presence of specific chemicals dissolved in a solution. Also included in this category are receptors in the hypothalamus, which sense low blood glucose, low oxygen tension, or changes in blood pH; O_2 and pH receptors are also found in the aortic sinus and the carotid body.

Mechanoreceptors transduce various qualities of physical force into electrical signals that are transmitted by sensory neurons. Such receptors are found in the vestibular, auditory, and somatosensory systems.

Other types of sensory receptors include *nociceptors*, which transduce painful stimuli, and *thermoreceptors*, which sense temperature changes in the skin and viscera.

Other Neural Information. Although sensory neurons transduce external stimuli, most nerve cells rely on other neurons for input. In general, the direction of information flow in a neuron is from dendrites to soma to axon, but most cells also receive information at their cell bodies, and many receive information at the axon terminal. In all these cases, the reception of information is mediated by synapses. Synapses are the points of information transfer from one neuron to another.

Neurons as Information Transmitters

Synapses. The synapse is the location at which a process of one neuron (usually an axon terminal) communicates with a second neuron or an effector (gland or muscle) cell. Although, strictly speaking, the synapse is defined only as a physiologic entity, it has traditionally been defined by its morphologic characteristics. In general, there are two broad morphologic categories of synapses, *chemical* and *electrical* (or *electrotonic*). The vast majority of synapses in the mammalian CNS are of the chemical type.

Chemical Synapses. The most common type of CNS synapse involves an axon that is apposed to a dendrite or dendritic spine of a second neuron. The prototypical chemical synapse consists of a *presynaptic element*, a *postsynaptic element*, and the intervening space (the *synaptic cleft*), which is 20 to 50 nm wide (Fig. 2–9; see also Fig. 2–1). The presynaptic element typically takes the form of an axonal bouton. The bouton contains mitochondria, which supply energy for synaptic function, and also a prominent collection of *vesicles*, which contain the neurotransmitter that will be released into the synaptic cleft. These vesicles are often aggregated near sites on the presynaptic membrane called *active sites* (or *zones*), which are the sites of neurotransmitter release. Directly across the synapse is the postsynaptic membrane, which often appears thick and dark on electron micrographs (see Figs. 2–1 and 2–9). Mitochondria are typically present in its vicinity.

As explained in detail in Chapter 4, communication across the synapse is mediated by the neurotransmitter stored in the presynaptic vesicles (Fig. 2–10; see also Fig. 2–1). Neurotransmitter release is initiated by the arrival of an action potential, which depolarizes the presynaptic terminal. In the terminal, depolarization causes *calcium channels* to open. The resulting influx of Ca^{2+} into the cell causes synaptic vesicles to fuse with the presynaptic plasma membrane and release their neurotransmitter into the cleft (see Chapter 4 for more

detail). The transmitter diffuses across the cleft to bind to specific *receptors* on the postsynaptic membrane, and this event triggers an electrochemical or biochemical change in the postsynaptic cell. This change represents the information as received by the postsynaptic cell.

Two functional properties of chemical synapses should be noted. First, they are *unidirectional*; that is, they transmit information only in the direction from the presynaptic cell to the postsynaptic cell. This directionality results because only the presynaptic cell releases the neurotransmitter, and only the postsynaptic cell expresses the receptor protein that will elicit the normal postsynaptic response to it. Second, the strength of the effect on the postsynaptic membrane is variable and depends partly on the amount of neurotransmitter released into the synapse. Each synaptic vesicle contains a fixed amount of neurotransmitter (called a *quantum*), so the amount of neurotransmitter released depends on the number of vesicles that fuse with the membrane in response to Ca^{2+} influx.

Chemical synapses were once visible only as the terminal boutons of nerve cell axons. In the early years of electron microscopy, two basic morphologic types of synapse became apparent. They were named *Gray's type I* and *type II* synapses. The characteristics of these synapse types are listed in Table 2–5 and illustrated in

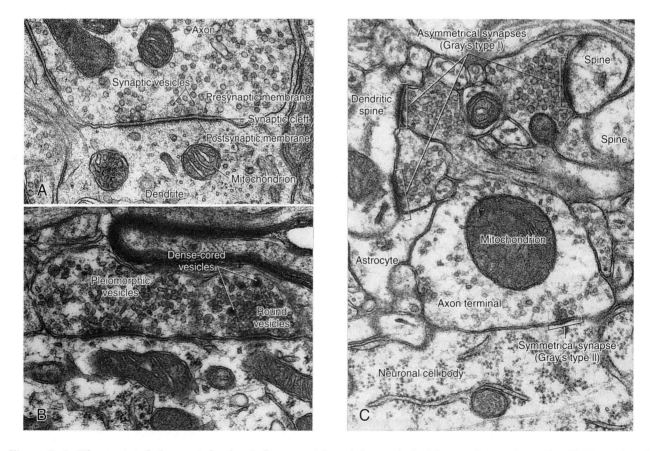

Figure 2–9. Ultrastructural elements of a chemical synapse (*A*), and three principal forms of synaptic vesicles (*B*). Examples of asymmetric and symmetric synapses in visual cortex (*C*). (*C* courtesy of Dr. Alan Peters.)

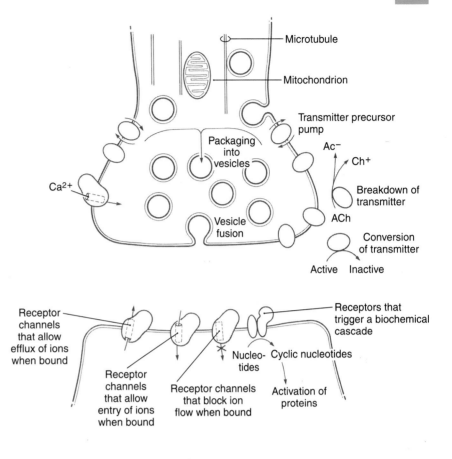

Figure 2-10. Diagrammatic representation of the events occurring at a generalized chemical synapse. Ac⁻, acetate; ACh, acetylcholine; Ch⁺, choline.

Figure 2–9C. At one time it was believed that type I synapses were excitatory in function and type II synapses were inhibitory. We now know that the excitatory or inhibitory function of a synapse depends on the receptors present on the postsynaptic membrane and cannot be reliably predicted from the ultrastructural characteristics of the presynaptic bouton. Nevertheless, the scheme is still a useful way to classify chemical synapses. For example, synapses using acetylcholine often have the Gray type I morphology, whereas those using γ-aminobutyric acid (GABA) usually resemble Gray's type II synapses.

Although the vesicles of Gray's type I and type II synapses differ in size and shape, in both cases the centers of the vesicles appear clear (electron-lucent). Vesicles with an electron-dense core are also seen in some synaptic endings; these dense-cored vesicles are generally thought to contain neuropeptides or serotonin as a neurotransmitter. This type of synapse is not included in the Gray classification.

Neurotransmitters

As we have seen, *neurotransmitters* are a means by which information is exchanged among nerve cells as well as between nerve cells and effector cells. Neurotransmitters are considered fully in Chapter 4 and are mentioned here briefly in relation to the structure of a typical neuron.

Neurotransmitters may be *biogenic amines* (e.g., acetylcholine, dopamine, norepinephrine), *amino acids* (e.g., glutamate, GABA), *nucleotides* (e.g., adenosine), *neuropeptides* (e.g., substance P. cholecystokinin, somatostatin), or even *gases* (e.g., nitric oxide, carbon monoxide). Many of these neurotransmitters are stored in and released from synaptic vesicles in the axon terminal as described previously, but in other cases, such as nitric oxide, nonvesicular release is clearly present.

Biogenic amines (acetylcholine) and amino acid neurotransmitters (GABA) are synthesized in the axon ter-

---

Table 2-5. Morphologic Characteristics of Gray's Type I and Type II Synapses

Gray's Type I Synapses

Dense material present on postsynaptic membrane but not presynaptic membrane (so that the synapse is visibly asymmetric)

Synaptic cleft 30 nm wide

Synaptic vesicles round and large (30–60 nm), with clear centers

Synaptic region up to 1–2 μm long

Gray's Type II Synapses

Dense material present on both the presynaptic and the postsynaptic membranes (so that the synapse appears symmetric)

Synaptic cleft 20 nm wide

Synaptic vesicles oval, flattened, or pleiomorphic (variable) in shape

Synaptic region less than 1 μm long

minal, although the enzymes necessary for their synthesis are produced in the cell body and shipped to the terminal by axonal transport. It should be remembered that the axon lacks the machinery to synthesize proteins (or membrane lipids) and thus must obtain these materials from the cell body. Thus, *axonal transport* is always necessary to support synaptic function.

Disorders of Neurotransmitter Metabolism. Disorders of neurotransmitter metabolism account for a large variety of neurologic and psychiatric illnesses, but in many cases the etiology is not well understood. This category of diseases is under intensive investigation, and three examples are briefly discussed here.

Parkinson disease affects dopamine-synthesizing neurons located in an area of the brainstem known as the *substantia nigra.* For unknown reasons, these dopaminergic cells begin dying at an accelerated rate. The loss of dopamine results in a characteristic tremor and inability to properly control movement. Originally, therapy involved administering supplements of L-dopa, a precursor for dopamine. This treatment increases dopamine synthesis by mass action but loses its effectiveness with time. Currently, therapy involves a combination of L-dopa with carbidopa, which inhibits the enzyme L-aromatic amino acid decarboxylase. Because carbidopa cannot cross the *blood-brain barrier* (defined later), it decreases the metabolism of L-dopa in peripheral tissues, making more L-dopa available to the CNS for dopamine synthesis in the remaining neurons.

Bipolar disorder affects several million Americans and appears to be caused by imbalances in the phosphatidylinositol (PI)-linked neurotransmitter systems. An increase in PI turnover is a biochemical change triggered by some subcategories of acetylcholine, serotonin, norepinephrine, and histamine receptors. It is thought that a pathologic increase in PI turnover beyond normal levels may result in mood changes. The drug lithium carbonate decreases PI turnover, thereby stabilizing the patient's mood.

Alzheimer disease affects more than 1 million Americans. Although the accuracy of diagnosis by psychological testing has improved, a definitive diagnosis can be made only by postmortem microscopic examination of brain tissue. Alzheimer disease is characterized by the degeneration of neurons in basal forebrain nuclei, the loss of synapses in the cerebral cortex and hippocampus, and the presence of pathologic structures called *neurofibrillary tangles* and *senile plaques.* Cortical cells normally receive terminals from *cholinergic* (acetylcholine-releasing) cells in the basal forebrain nuclei. In Alzheimer disease, these terminals are lost, and the activity of choline acetyltransferase (the enzyme responsible for acetylcholine synthesis) in the cortex and hippocampus of diseased patients is extremely low. Other neurotransmitter systems, particularly neuropeptides, are also affected by this disease.

Glia

Unlike neurons, glial cells do not propagate action potentials, and their processes are not specialized to receive and transmit electrical signals. Rather, they provide neurons with structural support and maintain the appropriate microenvironment essential for neuronal function. In addition, astrocytes modulate synaptic activity in their vicinity by releasing small amounts of glutamate.

Glia account for most of the cells in the nervous system, and normal brain function requires them. The major types of glial cells in the CNS (Fig. 2–11; Table 2–6) are *astrocytes* and *oligodendrocytes,* derived from neuroectoderm, and *microglia,* derived from mesoderm. The analogous cell types in the PNS are the satellite cells, Schwann cells, and macrophages.

Astrocytes

Astrocytes occur throughout the CNS. They are highly branched, and many of their processes end in expansions called *end-feet* (Fig. 2–11), which form several kinds of relationships. Most of the free surface of neuronal dendrites and cell bodies, as well as some axonal surfaces, is covered by apposed astrocyte processes or sheaths. Astrocyte end-feet also join together to completely line the interfaces between the CNS and other tissues. The outer surface of the brain and spinal cord, where it meets the inner surface of the *pia mater* (the innermost of the meningeal membranes that enclose the CNS), is covered with a coating of several layers of joined end-feet called the *glia limitans* (or glial limiting membrane). Similarly, every blood vessel in the CNS is jacketed by a layer of end-feet that separate it from the neural tissue.

As shown in Figure 2–11, the astrocytes of gray matter, called *protoplasmic astrocytes,* differ in shape from the astrocytes of white matter, called *fibrous astrocytes* because of their greater content of intermediate filaments. Astrocytes can be distinguished immunohistochemically (for purposes of research and diagnosis) by the presence in intermediate filaments of a distinctive marker protein, *glial fibrillary acidic protein* (GFAP). The GFAP content of protoplasmic astrocytes increases in pathologic conditions.

Structural Support and Response to Injury. During development, astrocytes (in the form of radial glial cells) provide a framework for neuronal migration. In the adult brain, astrocytes frame certain clusters of neurons, for example, the barrels of the somatosensory cortex of rodents. In white matter, they also enclose bundles of unmyelinated axons.

If injury to the CNS results in cell loss, the space created by the breakdown of debris is filled by proliferation or hypertrophy (or both) of astrocytes, resulting

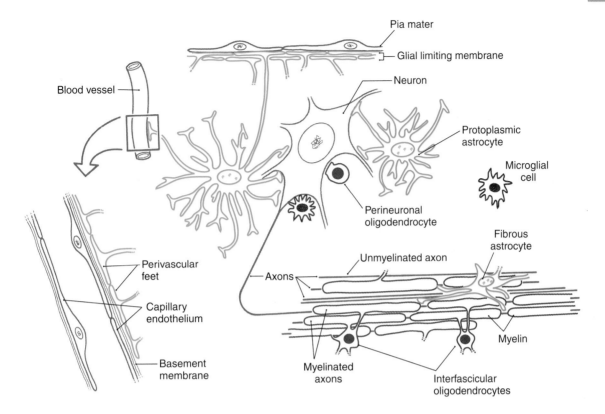

Figure 2–11. Relationships of central nervous system glial cells to neuronal cell bodies, axons, blood vessels, and pia mater.

in the formation of an *astrocytic scar*. That astrocytes retain the ability to proliferate in the mature brain (and thus are susceptible to events that disrupt the control of cell division) explains why the majority of CNS tumors are of astrocytic origin.

Growth Factors and Cytokines. Current research on astrocytes indicates that they secrete growth factors vi-

tal to the support of some neurons. In disease processes, astrocytes may secrete cytokines, which regulate the function of immune cells invading CNS tissue. Microglial cells, neurons, and neighboring astrocytes also have receptors for cytokines such as interleukin-1 (IL-1).

Environmental Modulation. The ionic environment

Table 2–6. Types of Glial Cells and Their Locations and Functions

Cell Type	Location	Function(s)
CNS		
Astrocytes	Throughout the CNS; contact neuronal cells bodies, dendrites, and axons and form a complete lining around the external surfaces of the CNS and around CNS blood vessels; gray matter astrocytes are called *protoplasmic* and white matter astrocytes are called *fibrous*	Maintenance of extracellular ionic environment; secretion of growth factors; structural and metabolic support of neurons
Oligodendrocytes		
Myelinating	Form myelin sheaths around CNS axons	Myelination
Satellite cells	Surround CNS neuronal cell bodies	Unknown
Microglia	Gray and white matter of CNS	Scavenging and phagocytosis of debris following cell injury and death; secretion of cytokines
PNS		
Schwann cells	Form myelin sheaths around myelinated axons and ensheath unmyelinated axons	Myelination; biochemical and structural support of myelinated and unmyelinated axons
Satellite cells	Surround neuronal cell bodies in PNS ganglia	Unknown

CNS, central nervous system; PNS, peripheral nervous system.

and pH of the extracellular space are buffered by astrocytes. These cells have ion channels in their membranes that are different from those in neurons. For example, potassium ions released from neurons during firing of an action potential are removed from the extracellular space by astrocytes via plasma membrane ion channels. Astrocytes are connected to each other by gap junctions and act as syncytia through which excess potassium ions are shunted to perivascular spaces. Astrocytes also propagate calcium waves, which spread through gap junctions between astrocytes to cover broad areas.

Metabolism. Astrocytes also participate in neurotransmitter metabolism. Their membranes have receptors for some neuroactive substances and uptake systems for others. Astrocytic uptake systems terminate the postsynaptic effect of some neurotransmitters by removing them from the synaptic cleft. For example, the amino acid neurotransmitter glutamate is taken up by astrocytes and is then inactivated by the enzymatic addition of ammonia to produce glutamine (catalyzed by the enzyme *glutamine synthetase*). Glutamine released from astrocytes can be reconverted to glutamate in neurons. This astrocytic pathway also detoxifies ammonia in the CNS. In addition, astrocytes release small amounts of glutamate, a transmitter that indirectly modulates synaptic activity by binding to extrasynaptic glutamate receptors on nearby neurons (Fig. 2–12).

Regional Heterogeneity. Astrocytes vary biochemically between gray matter (protoplasmic astrocytes) and white matter (fibrous astrocytes), and from one region of gray matter to another. White matter astrocytes differ from gray matter astrocytes in terms of their ion channels, neurotransmitter receptors and uptake systems, and other special properties. For unknown reasons, astrocyte tumors of particular types occur in characteristic distributions rather than with random frequency throughout the CNS. For example, the malignant tumor glioblastoma multiforme develops most frequently in the frontal or temporal lobe of the cerebral cortex.

Astrocytes at the Blood-Brain Barrier. In many tissues, solutes can pass freely between the capillary plasma and the interstitial space by diffusing through gaps between endothelial cells. In CNS vessels (with a few exceptions), the endothelial cells are sealed together by tight junctions, and solutes can reach the neural tissue only by passing through endothelial cells (Fig. 2–13). The resulting restricted exchange constitutes the *blood-brain barrier*. Water, gases, and lipid-soluble small molecules can diffuse across the endothelial cells; other substances must be carried across by transport systems, and their exchange is highly selective. The blood-brain barrier is of major clinical importance because it largely excludes many drugs from the CNS. CNS vessels are induced to form tight junctions by the surrounding jacket of astrocyte end-feet.

The exchange across the endothelium of CNS vessels is also rendered selective by a reduction in pinocytotic transport. In most parts of the body, endothelial cells transport solutes nonspecifically from the plasma to the perivascular space in pinocytotic vesicles. In CNS endothelial cells, the vesicles are relatively few.

CNS Tumors. Most tumors of CNS tissue arise from astrocytes (Table 2–7); their origin is identifiable by the presence of GFAP as intermediate filaments in tumor cells. The most common types are astrocytomas and glioblastomas. Oligodendrogliomas, arising from oligodendrocytes, and ependymomas, arising from ependymal cells, occur more rarely. Tumors are classified by grade as well as by type. Grades 1 and 2 are slow-growing and have a high potential for cure by means of surgical removal, chemotherapy, or radiation therapy. In contrast, grades 3 and 4 are highly malignant (Fig. 2–14).

Oligodendrocytes

Oligodendrocytes, the other major type of CNS glial cells, also occur in both gray matter and white matter (see Fig. 2–11). The only proven function of oligodendrocytes is *myelination*, that is, the provision of an electrochemically insulating sheath around some axons in the white matter (Fig. 2–15; see also Fig. 2–11). Other oligodendrocytes lie adjacent to and surround neuronal cell bodies in the gray matter, but the significance of this arrangement is not well understood.

A myelin sheath is a membranous wrapping around an axon that greatly increases the speed of conduction of action potentials along the axon. Large-diameter axons have thick myelin sheaths and high conduction velocities; smaller-diameter axons have thinner myelin sheaths and slower conduction velocities; and the smallest axons are unmyelinated and have the slowest conduction velocity.

Myelin is formed by a cell-cell interaction in which an axon destined for myelination is recognized by proteins on the oligodendrocyte surface. The oligodendrocyte responds by producing a flattened, sheet-like process that wraps repeatedly around the axon (see Fig. 2–15). As the layers of membrane accumulate, all cytoplasm is excluded. The mature myelin sheath consists of layers of oligodendrocyte plasma membrane firmly pressed together. Cytoplasm remains only in the innermost and outermost turns of the oligodendrocyte process.

The myelin sheath surrounding an axon is not continuous along its entire length. Rather, the axon is covered by a series of myelin segments, each formed by an interfascicular oligodendroglial cell. The interruptions between segments are called *nodes of Ranvier* (see Fig. 2–5B). Morphologic specializations at the nodes include a dense undercoating of the axonal

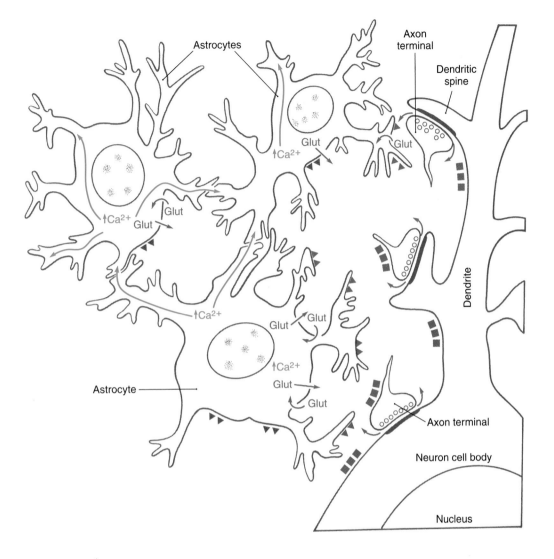

Glutamate released at synapses

Glutamate receptors at extrasynaptic sites on neurons and axon terminals

Glutamate receptors on astrocytes

Norepinephrine receptors on astrocytes

Gutamate transported into astrocytes

Glutamate released from astrocytes

↑Ca²⁺   Intracellular Ca²⁺ is increased by binding of glutamate or norepinephrine to receptors on astrocytes

Ca²⁺ passes from astrocyte to astrocyte through gap junctions

Figure 2–12. Astrocytes participate in neural transmission. Glutamate diffusing from active synapses (*blue arrows*) is taken up by astrocytes (Glut, *curved arrows*) to be metabolized to glutamine. Astrocyte cell surface receptors (*blue and red triangles*) also respond to the neurotransmitters glutamate and norepinephrine, increasing intracellular calcium (*green ascending up arrowheads*). The calcium increase passes via gap junctions to neighboring astrocytes (*green long arrows*). The increase in intracellular calcium causes astrocytes to release small amounts of glutamate (Glut, *straight arrows*), which then affects extrasynaptic glutamate receptors (*blue boxes*) on neighboring neurons. Activation of extrasynaptic glutamate receptors modulates presynaptic transmitter release and postsynaptic neurons' responses (EPSPs and IPSPs) to synaptic transmission.

A

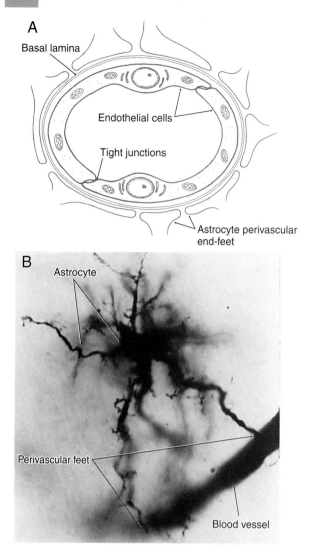

Basal lamina

Endothelial cells

Tight junctions

Astrocyte perivascular
end-feet

B

Astrocyte

Perivascular feet

Blood vessel

Figure 2–13. The relationship of astrocytes to central nervous system (CNS) blood vessels. Perivascular end-feet cover blood vessels of the CNS (*A*). Golgi-stained astrocyte with end-feet apposed to a blood vessel (*B*). (*B* courtesy of Dr. Jose Rafols.)

sponding myelin segments degenerate and are replaced by astrocytic plaques. This loss of myelin results in an interruption of the propagation of the action potential down these axons. Demyelinated axons survive temporarily, and some remyelination is possible by the growth of oligodendrocyte precursor cells that reside in the adult CNS. The array of motor, visual, or general sensory losses in patients with MS reflect the locations of the demyelination.

Microglia

Microglial cells are the immune effector cells of the CNS and are the predominant cells involved in CNS inflammation. The embryonic origin of the microglial cells is controversial, but the prevailing view is that they develop from blood cells of the monocyte-macrophage lineage, which migrate into the CNS during development. They make up about 1% of the CNS cell population (see Fig. 2–11). During health, they are considered to be at rest, or quiescent.

Like macrophages, microglia are able to become phagocytic scavengers. When the CNS suffers injury, activated microglial cells migrate to the site of damage, where they proliferate and phagocytose cell debris.

The cytokines IL-1β and tumor necrosis factor-α (TNF-α), as well as other cytokines and prostaglandins, are secreted by activated microglial cells. Some of the molecules secreted by activated microglial cells are neurotoxic (including glutamate, oxygen radicals, and TNF-α), and neurons may be directly damaged by inflammatory reactions in the CNS. Thus, microglial reaction may be a secondary cause of neuronal damage following stroke or trauma. Excessive secretion of IL-1β and TNF-α by microglial cells also induces endothelial cells to open the blood-brain barrier, allowing leukocyte infiltration into the brain parenchyma. CNS

membrane, as seen at the initial segment of the axon, and contact by an astrocyte process. The rapid ionic exchanges across the axonal membrane essential for generating the action potential and propagating it down the axon occur at these sites. The depolarization is then passively conducted along the axon (as a graded potential) to the next node. This method, *saltatory conduction*, is faster than having ionic exchanges occur continuously along the length of the axon.

The segments of myelin between adjacent nodes of Ranvier are called *internodal segments*, or *internodes*. Although the name *oligodendrocyte* means "cell with few branches," some of these cells give rise to myelinating processes forming internodal segments on as many as 40 axons.

In *demyelinating* diseases, such as *multiple sclerosis* (MS), groups of oligodendrocytes and their corre-

Table 2–7. Classification of CNS Tumors Originating from Glial Cells*	
Astrocytic tumors	Myxopapillary ependymoma
Astrocytoma	Subependymoma
Anaplastic astrocytoma	
Glioblastoma	**Mixed gliomas**
Pilocytic astrocytoma	Mixed oligoastrocytoma
Pleomorphic xanthoastrocytoma	Anaplastic oligoastrocytoma
Subependymal giant astrocytoma	**Choroid plexus tumors**
	Papilloma
Oligodendroglial tumors	Carcinoma
Oligodendroglioma	
Anaplastic oligodendroglioma	**Neuroepithelial, of uncertain origin**
Ependymal tumors	Astroblastoma
Ependymoma	Polar spongioblastoma
Anaplastic ependymoma	Gliomatosis cerebri

* Classification scheme of the World Health Organization.
From Lopes MBS, Vandenberg SR, Scheithauer BW: Molecular Genetics of Nervous System Tumors. Wiley-Liss, New York, 1993, with permission.
CNS, central nervous system.

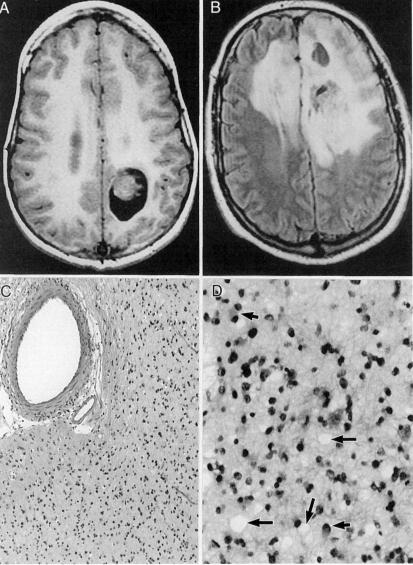

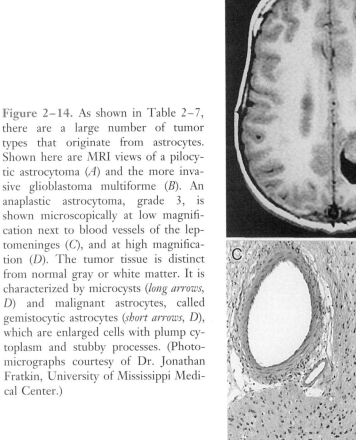

Figure 2–14. As shown in Table 2–7, there are a large number of tumor types that originate from astrocytes. Shown here are MRI views of a pilocytic astrocytoma (*A*) and the more invasive glioblastoma multiforme (*B*). An anaplastic astrocytoma, grade 3, is shown microscopically at low magnification next to blood vessels of the leptomeninges (*C*), and at high magnification (*D*). The tumor tissue is distinct from normal gray or white matter. It is characterized by microcysts (*long arrows, D*) and malignant astrocytes, called gemistocytic astrocytes (*short arrows, D*), which are enlarged cells with plump cytoplasm and stubby processes. (Photomicrographs courtesy of Dr. Jonathan Fratkin, University of Mississippi Medical Center.)

inflammation of such magnitude can be fatal in *bacterial meningitis.* For this reason, immunosuppressive drugs may be given with antibiotics to treat the infection.

Microglial cells are the CNS cells targeted by human immunodeficiency virus (HIV), the virus that causes acquired immunodeficiency syndrome (AIDS). The mechanism by which HIV infection of microglia leads to neuronal damage and dementia is not yet fully understood.

Supporting Cells of the Peripheral Nervous System

The PNS contains supporting cells called *satellite cells* and *Schwann cells*, which are analogous to astrocytes and oligodendrocytes, respectively. Satellite cells surround the cell bodies of neurons in sensory and autonomic ganglia, and Schwann cells ensheath the axons in peripheral nerves (Fig. 2–16).

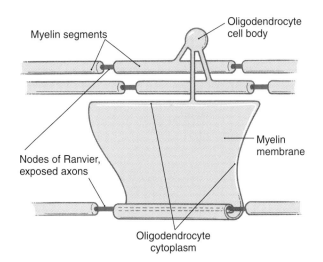

Figure 2–15. Myelin sheath formation by processes of an oligodendrocyte. The cytoplasm of the oligodendrocyte is trapped on the edges of the cell membrane as it wraps around the axon. (Based on data from Butt and Ransom, 1989.)

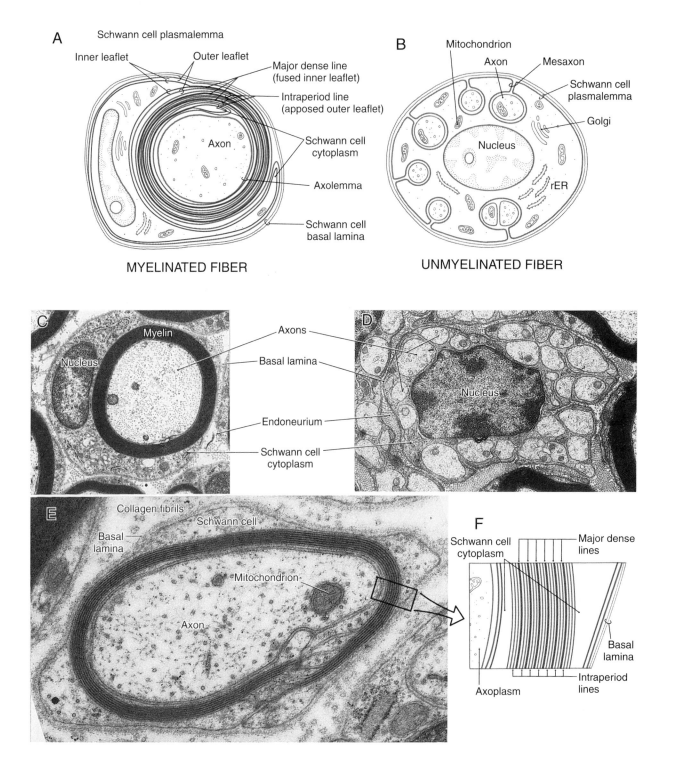

Figure 2–16. Diagrammatic representation of myelinated (*A*) and unmyelinated (*B*) fibers. Schwann cells ensheath all peripheral nerve axons. Multiple wraps of the plasmalemma of a Schwann cell fuse to form compact myelin (see also Fig. 2–15). The inner leaflet of the plasmalemma (red) fuses to form major dense lines, and the outer leaflets (blue) of each adjacent wrap contact each other to form intraperiod lines (*A*). In an unmyelinated fiber, small axons occupy troughs formed by invaginations of the Schwann cell plasmalemma (*B*). Electron micrographs of a myelinated fiber (*C*) and an unmyelinated fiber (*D*) composed of a single Schwann cell supporting more than 20 axons. A small myelinated fiber sectioned through part of a Schmidt-Lantermann cleft reveals the membrane composition of myelin (*E*). The layers of the boxed segment of myelin in *E* are diagrammed in *F*.

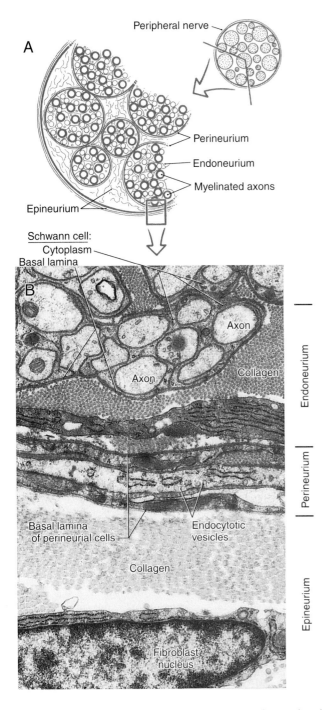

A

Peripheral nerve

Perineurium

Endoneurium

Myelinated axons

Epineurium

Schwann cell:
Cytoplasm
Basal lamina

B

Axon

Axon

Collagen

Endoneurium

Basal lamina
of perineurial cells

Endocytotic
vesicles

Perineurium

Collagen

Epineurium

Fibroblast
nucleus

Figure 2–17. The connective tissue layers of peripheral nerves (*A*). An electron micrograph corresponding to the area enclosed by the box in *A* reveals ultrastructural elements characteristic of each of the three sheaths (*B*).

is attracted to an axon segment and wraps repeatedly around it to produce a compact sheath (Fig. 2–16). As in the CNS, cytoplasm remains only in the innermost and outermost layers of this wrapping (see Fig. 2–16).

There are some differences between CNS and PNS myelin. Small pockets of cytoplasm known as *Schmidt-Lanterman clefts* are found at irregular intervals in PNS myelin. A *basal lamina* covers the external surface of the Schwann cell. This structure is formed by the Schwann cell and may help to stabilize it during the process of myelin formation. In addition, each Schwann cell forms the myelin of only a single internode of a PNS axon, in contrast to the CNS, where oligodendrocytes send out numerous processes that each form a myelin internode. Unmyelinated axons in the PNS are enclosed in canals formed by invaginations in Schwann cells (see Fig. 2–16). These Schwann cells also are covered by a basal lamina.

External to the Schwann cell basal lamina, peripheral nerve fibers are covered by three connective tissue sheaths (Fig. 2–17). The innermost of these, the *endoneurium*, consists of thin type III collagen fibrils and occasional fibroblasts between individual nerve fibers. At the second level, a distinctive sheath, the *perineurium*, surrounds each group (fascicle) of axons. The perineurium is composed of several concentric layers of flattened fibroblasts, which are unusual because they have a basal lamina and an abundance of pinocytotic vesicles (see Fig. 2–17). Perineurial cells also are connected to each other by tight junctions. This forms a protective *blood-nerve* barrier against diffusion of substances into peripheral nerve fascicles. Last, the entire peripheral nerve is covered by *epineurium*, a dense connective tissue sheath of type I collagen and typical fibroblasts.

Tumors of peripheral nerve are usually of Schwann cell origin. The type called a *schwannoma* arises singly and, because it is encapsulated and does not include nerve fibers, is easily excised. The type known as a *neurofibroma* is usually multiple. Neurofibromas are generally difficult to remove because they are unencapsulated and infiltrate nerve bundles.

Degeneration and Regeneration

In the adult mammalian nervous system, neurons lost through disease or trauma are not replaced. Although the adult CNS retains a very small number of neural stem cells, neuronal proliferation is an almost immeasurably rare event outside of the olfactory epithelium. In degenerative diseases, such as Parkinson or Alzheimer disease, death of neurons leads to eventual depopulation of the specific groups of nuclei affected. However, if axons are damaged but the cell bodies remain intact, regeneration and return of function can occur in some circumstances.

In the PNS, large- and intermediate-diameter axons have myelin sheaths, and the smallest-diameter axons are unmyelinated. Schwann cells produce these myelin sheaths and also envelop the unmyelinated axons. The myelin sheaths are similar to the CNS type, consisting of a tight spiral wrapping of fused plasma membrane. They are also formed similarly by a Schwann cell that

The chance of axonal regeneration is best when a peripheral nerve is compressed or crushed but not severed. In milder lesions in which focal demyelination occurs without axonal degeneration (*neurapraxia*), there is loss of conduction in the nerve, but recovery is to be expected. When compression or crushing kills the axons distal to the site of injury, the neuronal cell bodies, which are in the spinal cord or in sensory or autonomic ganglia, usually survive. These cell bodies may undergo chromatolysis in response to the trauma. Days to weeks later, *axonal sprouting* starts at the point of injury, and the axons grow distally. Meanwhile, in the distal part of the nerve, axons die and are removed by macrophages, but the Schwann cells remain. They lose their myelin but keep their basal lamina. Within these tubes of basal lamina, Schwann cells proliferate, forming cordons called *bands of Büngner*. These Schwann cells and basal lamina tubes guide the distally growing axonal sprouts. Regeneration depends on a variety of influences, including neurotrophins (neuronal growth factors) and the basal lamina. The growth rate of sprouting axons is about 1 mm/day.

In a compression injury (*axonotmesis*), the proximal axon sprouts and distal bands of Schwann cells remain in their original orientation, so nerve fibers are lined up just as they were before the injury. Therefore, when the axons regenerate, they will find their original positions within the nerve and are more likely to accurately reconnect with their proper targets.

When a peripheral nerve is severed (*neurotmesis*) rather than crushed, regeneration is less likely to occur.

Sprouting occurs at the proximal end of the axon, and the axon grows, but it may not reach its distal target. As axons grow from the proximal stump toward the distal stump, some may enter appropriate bands of Büngner and may be directed to their correct peripheral targets. These nerve fibers will become functional. Some axons may enter bands of Büngner that lead them to incorrect targets, so normal function does not return. Other axons may fail to enter the Schwann cell tubes, instead ending blindly in connective tissue to form a *neuroma*. Mechanical or chemical stimulation of these blindly ending sensory axons may be the cause of "phantom pain" in persons with amputated limbs.

In axon tracts of the CNS, little or no regeneration can be expected, and in humans there is no regeneration to a functional state. Basal lamina guides, such as those found in the PNS, are not available. When a CNS axon is severed, the neuron mounts a sprouting response. Astrocytes hypertrophy and proliferate at the site of injury and fill any space left by the injury or by degeneration of the damaged nerve tissue. The responding astrocytes grow in a random orientation and form a scar rather than a pathway. Furthermore, astrocytes may not secrete adequate growth factors to sustain regrowing axons. The astrocytic scar appears to be a barrier rather than a guidance mechanism for axonal sprouts. Moreover, specific molecules present in oligodendrocyte myelin may also inhibit axonal regrowth. Eventually the axonal sprouts are retracted, and the loss of function associated with the severed pathway is permanent.

Sources and Additional Reading

Araque A, Carmignoto G, Haydon PG: Dynamic signaling between astrocytes and neurons. Annu Rev Physiol 63:795–813, 2001.

Araque A, Parpura V, Sanzgiri RP, Haydon PG: Tripartite synapses: Glia, the unacknowledged partner. Trends Neurosci 22:208–215, 1999.

Butt AM, Ransom BR: Visualization of oligodendrocytes and astrocytes in the intact rat optic nerve by intracellular injection of lucifer yellow and horseradish peroxidase. Glia 2:470–475, 1989.

Casagrande VA, Hutchins JB: Methods for analyzing neuronal connections in mammals. Methods in Neurosciences, Vol 3: Quantitative and Qualitative Microscopy. Academic Press, San Diego, 1990, pp 188–207.

Dani JW, Chernjavsky A, Smith SJ: Neuronal activity triggers calcium waves in hippocampal astrocyte networks. Neuron 8:429–440, 1992.

Gehrmann J, Matsumoto Y, Kreutzberg GW: Microglia: Intrinsic immuneffector cell of the brain. Brain Res Rev 20:269–287, 1995.

Hertz L: Neuronal-astrocytic interactions in brain development, brain function and brain disease. Adv Exp Med Biol 296:143–159, 1991.

Kettenmann H, Ransom BR (eds): Neuroglia. Oxford University Press, New York, 1995.

Lopes MBS, Vandenberg SR, Scheithauer BW: Molecular Genetics of Nervous System Tumors. Wiley-Liss, New York, 1993.

Peters A, Palay SL, Webster H deF: The Fine Structure of the Nervous System, 3rd ed. Oxford University Press, New York, 1991.

Shepherd GM (ed): The Synaptic Organization of the Brain, 3rd ed. Oxford University Press, New York, 1990.

Stevens CF: The neuron. Sci Am 241:55–65, 1979.

Steward O, Banker GA: Getting the message from the gene to the synapse: Sorting and intracellular transport of RNA in neurons. Trends Neurosci 15:180–186, 1992.

Unwin N: Neurotransmitter action: Opening of ligand-gated ion channels. Cell 72:31–42, 1993.

Chapter 3

The Electrochemical Basis of Neuronal Integration

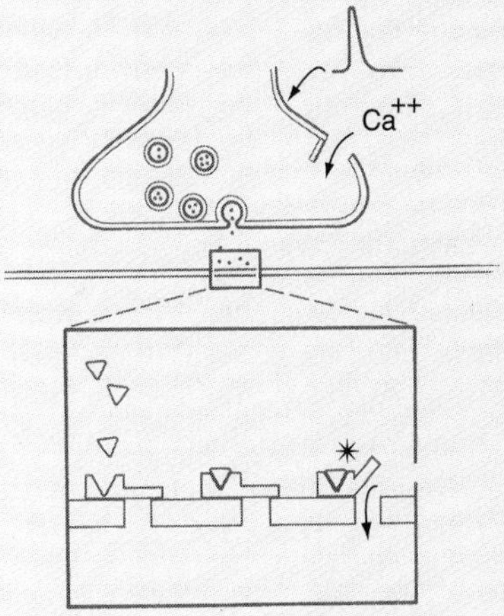

P. B. Brown

Overview 38

Electrical Circuits 38

The Resting Potential 39

Cable Properties of Membranes 40

Action Potentials 42
Ion Channels
The Depolarization Cycle
Repolarization
The Excitability Cycle
Action Potential Propagation
Conduction Block
Accommodation

Receptor Potentials 46
Receptors as Transducers
Ionic Mechanisms
Action Potential Generation
Adaptation

Representation of Information 47
Place Codes
Frequency Coding
Recruitment

Conduction Velocity Groups 48
Axon Diameters and Conduction Velocities
Functional Groupings

Chemical Synapses 50
Presynaptic Mechanisms
Transmitter-Receptor Interactions
Single-Messenger Synapses
Second-Messenger Synapses

Synaptic Information Processing 52
Currents and Potentials
Action Potential Generation
Information Processing

The nervous system is the organ of behavior. Neural circuits process and store information to produce the full range of behavior, including voluntary and involuntary movement, homeostatic processes, and cognitive functions such as perception, emotion, and thinking. Sherrington, the founder of modern neurophysiology, called this information processing the *integrative action of the nervous system*. This diverse repertoire of activities is supported by cellular mechanisms that include molecular and electrochemical processes.

Overview

Neurons are the information-processing elements of the nervous system. Each neuron makes hundreds or thousands of connections with other neurons. Each neuron continuously produces output signals based on its many inputs. Because all of these neural computations take place simultaneously, information processing in the brain is highly *parallel*.

The synaptic connections among neurons are organized so that parallel streams of information are segregated in separate tracts and nuclei. This arrangement leads to *localization of function* within certain combinations of nuclei and tracts. Information is relayed from one nucleus to another in a *serial* fashion, with each nucleus performing a specific set of manipulations on its inputs. This permits the extraction of more and more abstract information from the data provided by sensory receptors, in what is referred to as a *hierarchical* processing scheme.

The storage of information in the central nervous system (CNS) is accomplished in a *distributed* fashion, by modification of the properties of synaptic connections. For example, storage of a particular input/output association involves altering the properties of many synapses in each neuron. Each neuron, in turn, stores many associations through altering its synaptic properties. Typically, many neurons perform similar computations on similar sets of inputs, so the loss of individual neurons has no observable effect on the operation of the nervous system. This *redundancy* is essential because many of the neurons we are born with die in the course of our lifetime and are not replaced.

Although different parts of the nervous system perform diverse operations, they do so by means of a limited repertoire of physiologic mechanisms—namely, the generation and propagation of all-or-none action potentials, synaptic transmission, and the production of graded sensory and synaptic potentials.

Electrical Circuits

The electrical activity of nerve cells is best understood in terms of electrical circuits. In a simple circuit consisting of a *battery* and a *resistor* (Fig. 3–1A), the bat-

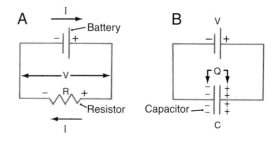

Figure 3–1. Simple circuits showing a battery and resistor (*A*) and a battery and capacitor (*B*).

tery is a *voltage source*, producing a voltage or *potential difference*, V, across the resistor. This voltage causes a flow of charged particles (current, *I*, carried by electrons in metals and by ions in solutions), such that *positive* current moves from the battery's negative pole to its positive pole through the battery, and from the positive pole to the negative pole through the resistor. This closed loop of current is called an *electrical circuit*. Positively charged cations flow in the direction of positive current; negatively charged anions and electrons flow in the opposite direction. Because the charge leaving any point in the circuit is equal to the charge entering it, there is no net transfer, accumulation, or depletion of charge.

The relationship among current, voltage (the term *voltage* is used interchangeably with *potential*), and resistance is expressed by *Ohm's law:*

$$V = IR$$

where *V* is the potential difference across the resistor (*expressed* in *Volts*, V), *I* is the current through the resistor (in *Amperes*, A, and *R* is the resistance of the resistor (in *Ohms*, Ω). Resistance is the tendency to impede the flow of current, similar to the resistance to flow in a blood vessel. *Conductance (G)* is the inverse of resistance:

$$G = 1/R$$

where conductance is expressed in units of *Siemens* (S). A *conductor* is a resistor with a negligible resistance (very high conductance), and an *insulator* is a resistor with a negligible conductance (very high resistance). The lines connecting the battery to the resistor are conductors (e.g., wires), and the empty space surrounding the symbols is an insulator (e.g., the air surrounding a real circuit).

Although there is no net transfer or separation of charge in current through a resistor, charge separation does occur when a voltage is applied across a *capacitor* (see Fig. 3–1B), which consists of two conductors separated by an insulator. When a battery is attached to a capacitor, positive charge leaves the positive pole of the battery and accumulates on the capacitor, repelling positive charge on the opposite side of the capacitor and causing an equal charge to enter the negative pole of the battery. The only net charge transfer is on the two sides of the capacitor. The amount of charge, Q, stored in the capacitor is

$$Q = CV$$

where Q is expressed in *Coulombs*, and capacitance, C, is expressed in *Farads*. When a voltage V is initially attached to a capacitor that has any charge other than $Q = CV$ on it, charge must move; this is a current. In fact, current can be expressed as movement of charge per second: 1 Ampere = 1 Coulomb/second.

The Resting Potential

The plasma membrane separates two conductors that are electrolyte solutions with different ionic concentrations. *Membrane channels* are formed by proteins embedded in the membrane that have hydrophilic cores through which hydrated ions can move (Fig. 3–2A). Because of differences in the size and charge of this central pore, the channels are selective for the passage of certain ions and vary in their permeability to these ions. In the resting membrane, the channels are relatively permeable to K⁺ and Cl⁻ but relatively imper-meable to Na⁺. In addition, the neuron contains proteins to which the membrane is impermeable, and more of these molecules are negatively charged than positively charged.

For each ion, the concentration difference produces a driving force (the *diffusion potential*) on the ion. If an equal and opposite electrical potential is applied across the membrane, it may cancel out the diffusion potential so that there is a net zero force on the ions, and the same number diffuse in each direction, for zero net ion flow (an *equilibrium*). This voltage is the *equilibrium potential*. The ionic gradient acts as a battery, with a voltage equal to the equilibrium potential and an internal resistance whose conductivity is proportional to the permeability of the membrane for that ion (see Fig. 3–2A and B).

In the case of the neuron, three important ions are unequally distributed between the outside and inside of the cell membrane. These concentration differences are supplemented by a membrane *pump*, which uses the energy from ATP to transport three Na⁺ ions out of the cell for every two K⁺ ions it transports into the cell. Because two cations are pumped in for every three pumped out, this is not an electrically neutral process, and the pump current results in a slight hyperpolarization, which is added to the potentials produced by the ion batteries. Although Cl⁻ is not actively pumped, it too is unequally distributed across the membrane owing to the unequal distributions of Na⁺ and K⁺.

Each ion has its own "battery," with a voltage corre-

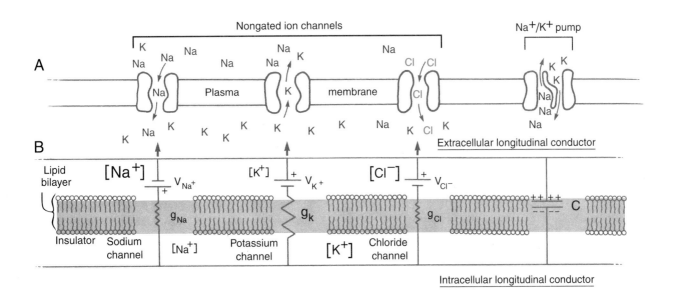

Figure 3–2. Electrical equivalent circuit of nerve cell membrane. The lipid bilayer is an insulator separating the conducting extracellular and intracellular fluids. Na⁺, K⁺, and Cl⁻ channels penetrate the membrane, permitting flow of ions (A). Na⁺ and Cl⁻ have highest concentrations outside, and K⁺ has highest concentration inside the cell (B). These potential gradients determine the magnitudes and polarities of the three-ion "batteries." In the resting membrane, the K⁺ channel has the highest permeability, so the K⁺ potential dominates the membrane potential. The capacitance, C, is composed of the insulating membrane and the conductive extracellular and intracellular electrolytes.

sponding to the ion's equilibrium potential and with its own internal resistance corresponding to the membrane permeability. The equilibrium potential E_{ion} for a monovalent ion is calculated with the *Nernst equation*:

$$E_{ion} = RT/zF \ln [ion]_{out}/[ion]_{in}$$

The $[ion]_{in}$ term is the known concentration of the ion inside the cell, $[ion]_{out}$ is the known concentration outside the cell, R is the gas constant, T is the temperature above absolute zero, F is the Faraday constant, z is the charge of the ion, and ln is the natural log. Substituting and converting to base 10 at body temperature,

$$E_{ion} = 61 (\log_{10} [ion]_{out}/[ion]_{in})$$

For a cell in which Na^+ has external and internal concentrations of 140 mM and 15 mM, respectively, the equilibrium potential comes to 60 mV, inside positive.

Neurons have lower K^+ concentrations outside than inside, with an equilibrium potential of about -70 mV. They also have a higher $[Cl^-]_{out}$ than $[Cl^-]_{in}$ with a Cl^- equilibrium potential of about -60 mV. Their Na^+ equilibrium potential is typically $+55$mV.

There are thus three batteries with different potentials across the membrane. If the resistances of the Na^+ and Cl^- batteries are relatively high (insulators), their voltages do not contribute much to the membrane potential. If the resistance of the K^+ battery is relatively low (a conductor) the K^+ battery voltage dominates the membrane potential. In the resting state, the K^+ permeability is the highest of the three, so the transmembrane potential is close to the K^+ equilibrium potential. The Cl^- is distributed passively, so it is close to its equilibrium potential at the resting potential. The net resting membrane potential RP (also abbreviated V_{rest}) is calculated with the *Goldman-Hodgkin-Katz equation*:

$$RP = 61 \log_{10} \frac{p_K[K^+]_o + p_{Na}[Na^+]_o + p_{Cl}[Cl^-]_i}{p_K[K^+]_i + p_{Na}[Na^+]_i + p_{Cl}[Cl^-]_o}$$

where the p's are membrane permeabilities, the subscript o indicates "out," and the subscript i indicates "in."

It is also convenient to express the Goldman-Hodgkin-Katz equation in terms of *conductance* (g):

$$V_{membrane} = \frac{g_K E_K + g_{Na} E_{Na} + g_{Cl} E_{Cl}}{g_K + g_{Na} + g_{Cl}}$$

It can be seen in this modified equation that if the conductance for Na^+ and Cl^+ drops to zero, the membrane potential equals the potassium equilibrium potential (E_K). This makes intuitive sense: K^+ is then the only ion contributing to the cells membrane potential, the only "battery" wired into the "circuit." However, in a neuron at rest, the K^+ conductance is about 10 times the Na^+ conductance. Therefore, the resting membrane potential is about $^9/_{10}$ of the way between the Na^+ and K^+ equilibrium potentials.

The membrane is not in equilibrium because the ions are not at their equilibrium potentials, and in the absence of the Na^+-K^+ pump there would be a net flow of ions to equalize their equilibrium potentials. However, it is in a steady state with a zero net flow of current.

Intracellular and extracellular electrolytes are good conductors, and the lipid membrane is a good insulator, separating the extracellular and intracellular conductors. This makes the membrane a capacitor, and a charge is stored on the membrane that is proportional to the transmembrane voltage.

Cable Properties of Membranes

If the conductance of the Na^+ channel is abruptly increased (Fig. 3–3A) so that it is greater than the conductance of the K^+ channel (see Fig. 3–3B), then Na^+ begins to flow into the cell (see Fig. 3–3C), discharging the membrane capacitance. The membrane potential moves toward the Na^+ equilibrium potential (see Fig. 3–3D). The membrane potential shift is slowed by the membrane capacitance. The capacitor is initially charged negatively on the inside and positively on the outside, reflecting the polarity of the K^+ battery. As Na^+ flows into the cell, much of the current is used to reverse this capacitive charge. As the capacitor's charge is modified, the voltage changes exponentially with time, approaching the new steady state asymptotically. The rate of change is expressed as the time τ (tau) it takes for an increase or decrease by the factor e, a constant that is also the base of the natural logarithm. For any resistor-capacitor (RC) circuit,

$$\tau = RC$$

This is called the *time constant*.

The tube-shaped axons and dendrites of nerve cells contain more than one ionic battery and associated resistor. Because the ion channels and pumps that generate the resting potential are distributed throughout the membranes of the axons and dendrites, a longitudinal strip of nerve fiber membrane can be thought of as a series of ionic batteries with associated resistors (Fig. 3–4A). The nerve fiber thus is like an undersea cable in which the internal conductor is connected to the conductive seawater by a leaky insulator. For this reason, the passive electrical properties of axons and dendrites are referred to as their *cable properties*.

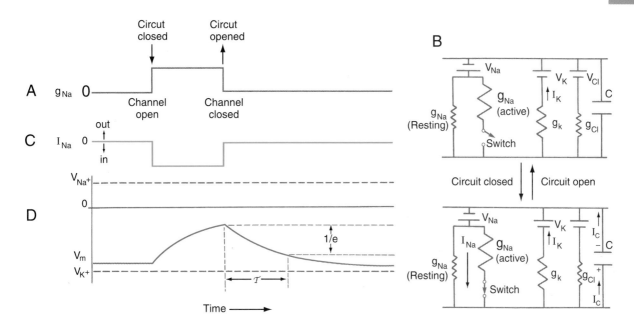

Figure 3–3. The effect of a sudden increase in Na$^+$ permeability. Time course of permeability change *(A)*. Electrical model of permeability change *(B)*; switch on high-conductance path goes from open to closed (a closed switch is the same as an open channel, and an open switch is the same as a closed channel). Na$^+$ current *(C)*. Membrane potential shifts from vicinity of K$^+$ potential toward Na$^+$ potential *(D)*. Membrane capacitance slows the voltage change.

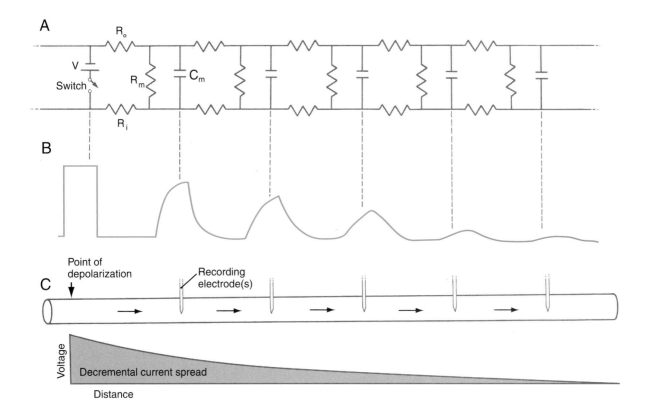

Figure 3–4. Passive electrical properties of the membrane. Electrical model *(A)* where R_0 = external longitudinal resistance, R_i = internal longitudinal resistance, and C_m = membrane capacitance. Voltage pulse is applied at switch, resulting signals are recorded at capacitors *(B)*. Pulse undergoes decremental conduction, decreasing in amplitude with distance from the switch *(B, C)*. Pulse is also slowed progressively with distance.

A current pulse applied across the membrane produces a slower voltage pulse that spreads down the membrane, reaching a peak later and later the further it goes down the membrane (see Fig. 3–4B). The peak amplitude declines exponentially as a function of distance (see Fig. 3–4C). The distance in which it declines to $1/e \times$ the amplitude at the origin is referred to as the *length constant or space constant* λ (lambda). The length constants of dendrites and unmyelinated axons are a few tens or hundreds of microns. Consequently, graded potentials such as receptor and synaptic potentials cannot conduct more than a few hundred microns.

This *decremental* spread of current and voltage does not involve any permeability changes; therefore, it is referred to as *passive*. A synonym for passive conduction is *electrotonic conduction*.

A decrease in the resistance of this cable, the intracellular longitudinal resistance, will cause the space constant to lengthen. An increase in the transmembrane resistance will have the same effect. Similarly, a decrease in membrane capacitance will decrease the time constant. Any of these changes will cause the peak of the voltage pulse to propagate more rapidly and to decline in amplitude more slowly with distance. Mammalian nervous systems use all three effects to speed propagation of action potentials.

Action Potentials

At receptor endings or synapses, a localized process produces a *generator potential*. This potential is referred to as a *graded potential* because it varies continuously in amplitude. If this graded potential passively spreads to an electrically excitable region of membrane called the *trigger zone*, it may produce an *action potential* if it exceeds a *threshold* level. The propagation of all-or-none action potentials permits the transmission of information over much longer distances than would be possible for graded potentials, which are *local* (rapidly diminishing in amplitude with propagation distance).

Ion Channels. The basis for most ionic flow across the lipid plasma membrane is the *ion channel*. Ion channels are intramembranous protein complexes, which are stabilized in the membrane by uncharged lipophilic amino acid residues. These channels have a hollow core or *pore* lined with charged, hydrophilic residues. The permeability to a given ion is determined by the charge distribution and diameter of the pore. Different kinds of channels exhibit differential permeability to specific ions and therefore show considerable ion specificity.

Gated channels open and close, switching rapidly between low ("closed") and high ("open") permeability states. *Voltage-gated* channels change configuration under the influence of depolarization or hyperpolariza-

tion, so that they spend a larger fraction of the time in the open or closed state.

The action potentials of most nerve cells arise from the behavior of voltage-gated Na^+ and K^+ channels. The K^+ channel is relatively simple: It has a single gate that responds to depolarization with a shift in equilibrium toward the open state (Fig. 3–5). This effect, called *activation* of the channel, results in a greater permeability to K^+ ions, which respond by flowing down their concentration gradient (that is, out of the cell). By contrast, the Na^+ channel has *two* voltage-sensitive gates: an activation gate and an inactivation gate (Fig. 3–6). The *activation gate* behaves like the gate of the K^+ channel, responding to depolarization by opening. The *inactivation gate*, by contrast, responds to depolarization by closing. The gates do not simply cancel each other out, however, because the inactivation gate responds more slowly to depolarization than does the activation gate. Depolarization of the membrane thus will produce an initial brief phase in which both gates are open, after which the inactivation gates are closed while the activation gates are still open. The three states of the Na^+ channel are called *resting* (activation gate closed, inactivation gate open), *activated* (both gates open), and *inactivated* (activation gate open, inactivation gate closed). The transition between each pair of states has its own equilibrium. When the channel is in the activated state, Na^+ is free to flow down its concentration gradient into the cell.

The sequence of events underlying an action potential is best described as two regenerative electrochemical cycles: the depolarization and repolarization cycles.

The Depolarization Cycle. An action potential is a self-perpetuating all-or-none wave of depolarization that spreads along the nerve fiber membrane. An action potential is triggered by a depolarization of the resting membrane that reaches or exceeds a certain threshold amplitude; weaker depolarizations remain local and die out passively (Fig. 3–7). The existence of the threshold results from the behavior of the voltage-gated K^+ and Na^+ channels. A weak depolarization (curve 1 in Fig. 3–7A) causes mild activation of the Na^+ and K^+ channels. Because the Na^+ channels respond faster, there is an initial net influx of Na^+, which displaces some of the negative charge on the membrane capacitance, further depolarizing the membrane. However, the resting K^+ permeability is great enough to prevent this process from becoming self-perpetuating. In 1 msec or so, Na^+ inactivation and K^+ activation catch up with the Na^+ activation and overwhelm it, returning the membrane potential to its resting value, and hence returning voltage-dependent Na^+ and K^+ permeabilities to their resting values. This transient process is called a *local response*.

The *threshold* depolarization is defined as the level of depolarization that results in an action potential 50% of the time. A depolarization of this magnitude acti-

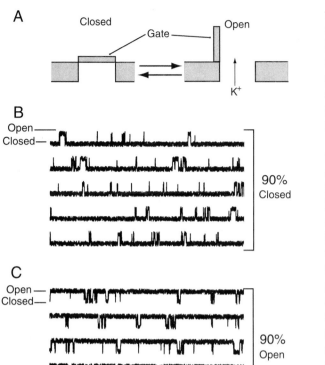

Figure 3–5. A simulated voltage-modulated K$^+$ channel. Closed and open states *(A)*. Recordings from five separate channels, closed 90% of the time *(B)*. The same five channels, open 90% of the time *(C)*.

vates the Na$^+$ channels just strongly enough so that the inward flux of Na$^+$ is not automatically overwhelmed by the growing outward K$^+$ flux, but instead comes into balance with it. A slight perturbation one

way or the other will determine the outcome; either the depolarization will die out (curve 2 in Fig. 3–7), or it will develop into an action potential (curve 3). A suprathreshold depolarization always causes an action potential (curve 4). In this case, the inward Na$^+$ flux equals the outward K$^+$ flux at the moment when the depolarization passes through the threshold value (open arrow in Fig. 3–7B).

Above threshold, the inward Na$^+$ current discharges the membrane capacitance fast enough to result in a depolarization that increases the Na$^+$ activation, keeping it ahead of the developing K$^+$ activation and Na$^+$ inactivation (see Fig. 3–7A). Once this occurs, the process is self-regenerating, and the depolarization cycle (the *Hodgkin cycle*) develops explosively, to the point at which Na$^+$ permeability exceeds K$^+$ permeability and the potential reverses, going positive inside and approaching the Na$^+$ equilibrium potential. Much of the current is absorbed by the membrane capacitance in the active portion of the membrane. Some of the current spreads down the axoplasm in both directions away from the active zone. This is cable current, resulting in a passive spread of the depolarization caused by the action potential.

Repolarization. At the peak of the action potential, inward Na$^+$ current is again equal to outward K$^+$ current (see Fig. 3–7B, closed arrow), but only briefly, because Na$^+$ activation has passed its peak and K$^+$ activation and Na$^+$ inactivation are still increasing. This pulls the membrane back to resting potential. As the membrane approaches the resting potential, Na$^+$ and K$^+$ permeability begin to return to resting levels. If the resting potential is reached before permeabilities return to resting levels, the membrane potential undershoots slightly, approaching the K$^+$ equilibrium potential. This *after-hyperpolarization* persists until permeabili-

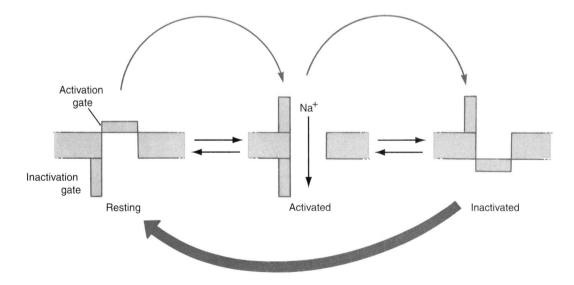

Figure 3–6. A model of the voltage-modulated Na$^+$ channel. Unlike the K$^+$ channel, the Na$^+$ channel has two parts and therefore three possible states.

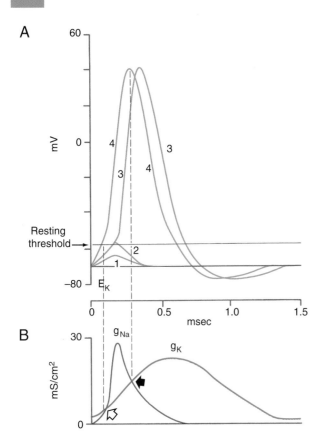

Figure 3-7. Generation of the action potential. Four potentials resulting from different amplitude depolarization pulses *(A)*. Curve 1, local response produced by subthreshold depolarization. Although an active process occurs, K^+ permeability remains greater than Na^+ permeability, and potential returns to resting level without development of an action potential. Threshold depolarization produces a local response (curve 2) half the time and an action potential (curve 3) half the time. At peak of active process, Na^+ permeability is equal to K^+ permeability. Suprathreshold depolarization (curve 4) always produces an action potential the same size as that for curve 3. Na^+ and K^+ permeabilities are equal at threshold *(B, open arrow)* and at the peak of the action potential *(B, closed arrow)*. Panel *B* shows the Na^+ and K^+ permeabilities during the action potential of curve 4 in *A*. Na^+ current is inward at all times, and K^+ current is outward at all times.

ties have returned to their resting states. Most current is absorbed by the membrane capacitance, recharging it.

The Excitability Cycle. The resting threshold is the level of depolarization needed to produce action potentials with a probability of 50% when the membrane permeabilities all start off in their resting states. Once an action potential is generated, the membrane cycles through a stereotyped sequence of changes in threshold. During the rising phase of the action potential and part of the falling phase, additional inward current cannot influence the amplitude of the action potential (it is said to be *all or none*) or produce a second action

potential (Figs. 3-8 and 3-9). The threshold is considered to be infinite, and the membrane is absolutely refractory (unresponsive). This phase of the excitability cycle is therefore called the *absolute refractory period*. At some point in the recovery cycle, it becomes possible to initiate a second action potential if a sufficiently large depolarizing current is used (several times the resting threshold current). During this period, the *relative refractory period*, the continuing Na^+ inactivation and K^+ activation represent an additional barrier that must be overcome, but a new Hodgkin cycle can be initiated with a large enough stimulus. As the recovery from the first action potential proceeds, the threshold continues to come down toward the resting threshold, which is reached once the membrane potential and permeabilities have reached their resting states. Owing to residual Na^+ inactivation and K^+ activation, action potentials generated during the relative refractory period are smaller than those produced from a resting state.

Action Potential Propagation. For the action potential to travel from one end of an axon to the other, the depolarization cycle at the active zone must depolarize adjacent resting membrane to threshold. Then the inactive membrane can develop an action potential, and its depolarization cycle can depolarize the next section of inactive membrane (see Fig. 3-9). This process,

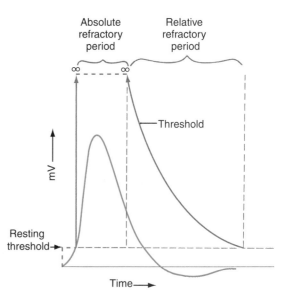

Figure 3-8. Recovery cycle correlated with refractory periods. The *blue curve* shows threshold changes during the recovery cycle and the *red curve* shows an action potential. During the rising and part of the falling phases of the action potential, the threshold is effectively infinite, and the cell is absolutely refractory to depolarization. During the rest of the recovery cycle, the threshold is elevated, but a sufficiently strong depolarization can trigger an action potential. This is the relative refractory period.

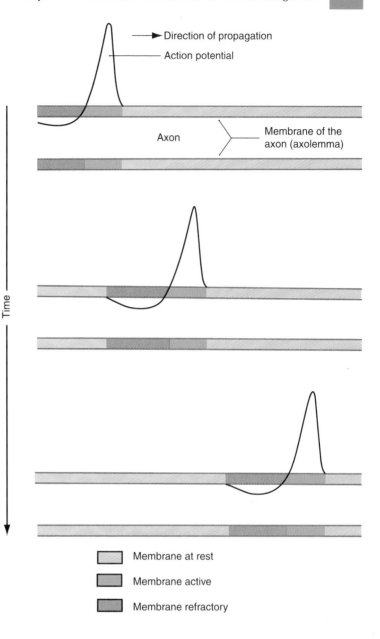

Figure 3–9. A simplified diagram showing that unidirectional propagation is driven by refractory regions of the membrane. See Figures 3–8 and 3–10 for more details.

which proceeds in a wavelike fashion down the axon, is called *propagation* or *conduction* of the action potential. The depolarization of inactive membrane occurs as a result of the net inward current in the active zone (Fig. 3–10; see also Fig. 3–9). Some of the inward current spreads longitudinally by passive spread and flows out of the portion of the membrane that is undergoing repolarization, recharging the membrane capacitance. Some also flows out of the resting membrane in the direction of propagation. This discharges the membrane capacitance, depolarizing the membrane to threshold. The ratio of the amount of current flowing into the inactive membrane to that needed to exceed threshold is referred to as the *safety factor* and must exceed 1 for propagation to succeed. In large axons, it can reach a value of 7, but in smaller ones such as those found at the ends of repeatedly branching termi-

nal axon collaterals, the safety factor may approach 1 or even fall to less than 1 on occasion, resulting in a block of conduction.

The conduction velocities of unmyelinated axons increase with their longitudinal conductances, which are proportional to the squares of their diameter (a larger wire is a better conductor). Above 1 μm diameter, modest conduction velocity increases can only be achieved with large diameter increases. In vertebrates, axons more than about 1 μm in diameter have myelin sheaths. Myelin increases the resistance between the intracellular and extracellular conductances. Capacitance is decreased by enlarging the gap between extracellular and intracellular conductors. These changes result in improved cable properties. Current from an active node bypasses the high-resistance myelinated portions of the membrane, flowing mostly through ad-

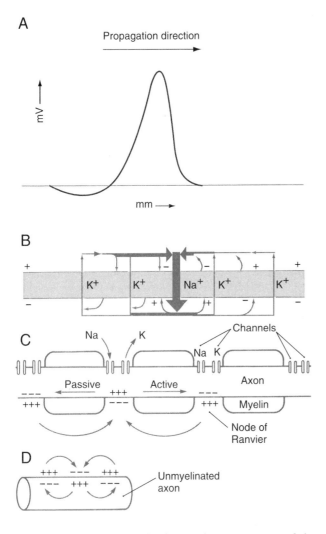

Figure 3–10. Current paths during the propagation of the action potential *(A)* are shown in an electrical circuit of a membrane *(B)* and in myelinated *(C)* and unmyelinated *(D)* axons.

traveling backward from its trailing edge. This failure to backfire is due to the fact that the membrane behind the active zone area is undergoing repolarization and is absolutely refractory or relatively refractory, with a safety factor of less than 1.

Conduction Block. Substances that interfere with the operation of the voltage-gated channels can block conduction. *Tetrodotoxin*, produced by the Japanese puffer fish, blocks the voltage-gated Na$^+$ channels. Local anesthetics also work by blocking action potential propagation. It is also possible to block nerves by applying current. If a hyperpolarizing current is applied, it will oppose the depolarization wave produced by the propagating action potential. This is referred to as a *hyperpolarization block* or *anodal block*. Of interest, a conduction block can also be produced with a cathodal, depolarizing current. Prolonged depolarization holds the K$^+$ channels in an activated state and the Na$^+$ channels in an inactivated state.

Accommodation. In some axons, a prolonged depolarizing generator potential causes K$^+$ activation and Na$^+$ inactivation to develop. The resulting increased threshold (which can also be thought of as decreased *sensitivity*, the reciprocal of threshold) is called *accommodation*. It is similar to depolarization block or the refractoriness found in the wake of an action potential. Accommodation can result in decreased rate of firing in response to a maintained stimulus, or complete inexcitability.

Receptor Potentials

Sensory receptors produce local, graded potentials whose amplitudes parallel the amplitudes of stimuli. Receptors produce their graded potentials by modulating the open times and hence permeabilities of ion channels, shifting the membrane toward or away from the equilibrium potentials for the ions they conduct. In some mechanoreceptors, channel permeability is modulated directly by mechanical stimuli, which deform the channel molecules by stretching the membrane. In other mechanoreceptors, the stimulus initiates a cascade of biochemical reactions that alters channel permeabilities indirectly. In the case of chemical receptors, the specific binding of a chemical substance (*agonist*) to membrane receptor molecules can modulate the permeability of an attached ion channel directly, or through a cascade of biochemical reactions.

Receptors as Transducers. The role of a receptor is to convert physical energy or chemical concentration into receptor current. The conversion of one form of energy into another is called *transduction*. Receptors for different sensory systems are most sensitive to different forms of stimulus energy (or to different chemicals, in the case of taste or olfaction). This specificity ensures that they respond best to the appropriate form of stim-

jacent interruptions between the myelin segments called nodes of Ranvier (see Figs. 3–9 and 3–10*C*). Action potentials therefore jump from node to node, a process called *saltatory conduction.*

Conduction velocity in myelinated axons is proportional to diameter because the internodal distances are proportional to diameter. The ratio of conduction velocity to diameter is 6 m/sec per μm of outer diameter of myelin sheath.

If an action potential is artificially induced in the middle of an axon, for example, by electrical stimulation or mechanical trauma (as happens when the "funny bone" is hit), action potentials course in both directions from the site of initiation. Conduction in the normal direction is referred to as *orthodromic* and in the opposite direction as *antidromic*. The normally evoked action potential starts at one end of the axon. It does not "backfire" and send a stream of action potentials

ulus for the sensory system they serve (the *law of specific nerve energy*). The class of sensation to which stimulation of a receptor gives rise is known as its *modality*. Thus, light receptors in the retina are specific for light and subserve the visual modality. This modality specificity is maintained in the central connections of sensory axons, so stimulus modality is represented by the set of receptors, afferent axons, and central pathways that it activates. Modality representation (or "coding") is therefore a *place code*. The sensory system activated by a stimulus determines the nature of the sensation.

There is also specialization for different types of stimuli within a modality. For example, different cutaneous receptors respond best to different stimuli, evoking the different sensations of touch, itch, tickle, flutter, vibration, fast and slow pain, cold, and heat. These different subclasses of sensations are referred to as *submodalities*.

Ionic Mechanisms. In many receptors, the depolarization that constitutes the receptor potential is produced by a generalized increase in the permeability of the membrane to small ions (e.g., Na^+, K^+, Cl^-). The steady-state potential for such a generalized increase in conductance is close to 0 mV. As the stimulus is increased or decreased, the equilibrium shifts toward more or fewer open channels, producing a graded potential that reflects the stimulus amplitude.

Because the channels producing receptor potentials are not voltage-modulated, they cannot produce an action potential. For this reason, the receptor potential is *local*. Therefore, except for very small cells, whose outputs are less than 1 mm from their inputs, a receptor potential cannot be passed on unless it reaches an area of membrane containing voltage-modulated channels—a trigger zone—and initiates an action potential.

Action Potential Generation. The amplitudes of graded potentials represent information—for example, the magnitude of force applied to a cutaneous mechanoreceptor. The amplitudes of action potentials, however, cannot be used to represent information because they are independent of the amplitudes of generator potentials used to produce them.

The currents produced by receptor channels spread to nearby membrane that contains voltage-modulated channels (the trigger zone), and if the threshold for action potentials is exceeded, an action potential is produced. During a prolonged receptor potential depolarization, action potentials may be generated repetitively at a rate that is related to the depolarization. The action potential frequency "encodes" or represents the magnitude of the receptor potential, and hence the stimulus amplitude.

Adaptation. A constant stimulus may at first result in a rapid discharge, which gradually slows. This is called *adaptation*. Afferent axons that show a substantial drop in discharge rate during a maintained stimulus are called *phasic*, or *rapidly adapting*. Those that show little or no adaptation are *tonic*, or *slowly adapting*. Adapta-

tion may result from accommodation, but it also may have other causes. Some sensory receptors have specializations that cause them to be rapidly adapting. For example, *pacinian corpuscles* have layers of fluid-filled compartments surrounding their transducer membranes, which can couple rapid movements only to the mechanically gated channels. These axons will produce only one or two action potentials at the *beginning* (or in some pacinian corpuscles, at the *end*) of a skin displacement. Some slowly adapting cutaneous receptors will continue to respond indefinitely to a maintained skin contact.

Representation of Information

For the CNS to process information, it must have some means of representing and manipulating it. **Place Codes.** Besides the place codes that represent modality and submodality, a place code is used to represent the location of a stimulus on a sheet of sensory receptors. One such sheet is the skin. Cutaneous sensory axons branch a few times in the skin, with all the branches of a particular axon terminating in the same type of receptor. The area of skin served by the receptor endings of a given axon, the area over which an appropriate stimulus will generate a response in the axon, is called the *receptive field*. In a peripheral nerve, the distribution of responding axons represents the location of a cutaneous stimulus.

The visual system has a sheet of receptors on the retina. The location of a stimulus in the visual field is encoded by the locations of the responding receptors on the retina. The auditory system has a sheet of receptors laid out along a frequency-tuned membrane, such that the location of a receptor along the membrane determines the auditory frequencies to which it will respond best. Thus, the position of responding receptors along this membrane represents auditory frequency (pitch).

In all three sensory systems, the place code representing the locus of stimulation on the sensory sheet is preserved through the ascending pathway to the cortex by means of selective connections. To represent location of a stimulus on the skin, connections from afferent axons are laid out in a spatially ordered fashion in each nucleus, so that their nearest-neighbor relations are the same at each end of the nerve or tract. As a result, the spatial patterns of responding cells in a nucleus resembles the spatial patterns of stimuli applied to receptors on the skin, and the nucleus can be thought of as a *map* of the skin. Electrical stimulation of sites in these maps will elicit sensations that are perceived as originating at the corresponding sites in the receptor sheet. Thus stimulation of a point on the somatosensory cortex would produce a vivid perception of a stimulus located on the skin rather than in the CNS.

Frequency Coding. Although discharge frequency can represent stimulus amplitude, the relationship generally is not linear. In reality, two transformations are involved: from stimulus intensity to receptor potential amplitude and from receptor potential amplitude to frequency of discharge. These two can be combined in a diagram of frequency of discharge as a function of stimulus intensity, as shown in Figure 3–11.

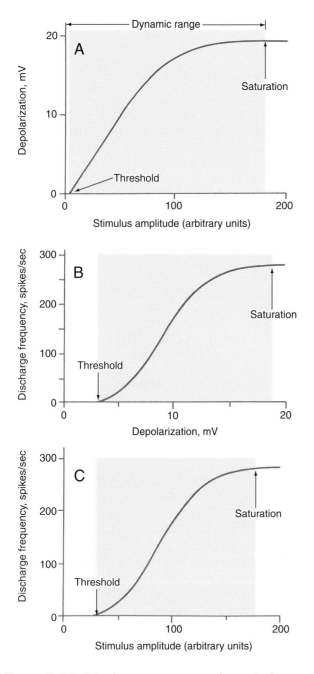

Figure 3–11. Stimulus-response curves for a single axon. Receptor potential as a function of stimulus amplitude *(A)*. Action potential frequency as a function of receptor potential *(B)*. Action potential frequency as a function of stimulus amplitude *(C)*.

Some important generalizations can be reached from these graphs. Receptor potentials have thresholds, but they may be as little as one quantum of energy (e.g., one photon in the visual system). Action potentials also have depolarization thresholds. In some modalities, we can perceive stimuli that elicit as little as one action potential in one sensory axon. Of course, we cannot perceive stimulus intensities below the action potential threshold of the most sensitive axon being stimulated.

Both receptor potential amplitudes and action potential frequencies have upper limits; this is called *saturation*. We cannot perceive changes of stimulus above saturation for the axon with the highest saturation level.

The range of stimuli over which we can perceive variations of stimulus intensity from threshold to saturation is called the *dynamic range* for perception of that stimulus. Every neuron has a dynamic range for its inputs, from its threshold to its point of saturation. Our perceptual dynamic range falls within the limits of the dynamic ranges of all axons capable of responding to the stimulus.

Recruitment. Different afferent axons of the same modality can have different thresholds and different dynamic ranges, so increasing stimulus amplitude causes more afferent axons to discharge, a phenomenon called *recruitment* (Fig. 3–12). Thus recruitment can be added to discharge frequency as a way in which stimulus amplitude can be encoded. Recruitment makes it possible to encode and discriminate stimulus amplitude over a range larger than the dynamic range of a single axon.

Conduction Velocity Groups

Axon Diameters and Conduction Velocities. In a peripheral nerve or in a tract, axon diameters can range from less than 1 μm to more than 20 μm (Fig. 3–13*A*). In general, axon diameters cluster into three or four groups. Because different-diameter nerve fibers have different conduction velocities (cv), synchronous electrical stimulation of all the axons in a nerve will result in their action potentials arriving some distance (x) from the stimulus after different conduction times (t): t = x/cv. In general, axon diameters do not vary continuously; rather, they fall into groups, each forming a *conduction velocity group* that has a specific range of velocities. This is evident in a recording from an entire peripheral nerve. The extracellular waveforms of all of the individual axons' action potentials add together to give a *compound action potential* (see Fig. 3–13*B*). Because the action potentials of different-diameter axons arrive at different times, the compound action potential is longer than the individual action po-

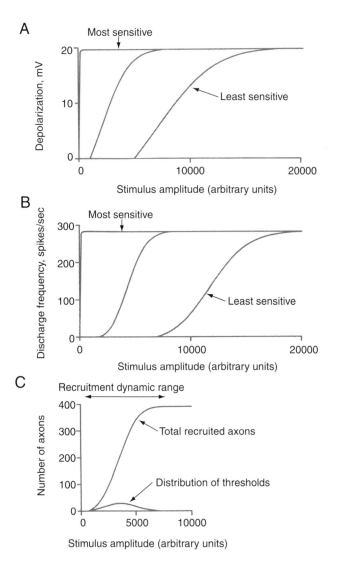

Figure 3–12. Stimulus-response curves for a population of axons. Receptor potential as a function of stimulus amplitude for three axons, including least and most sensitive in population (*A*). Discharge frequency as a function of stimulus intensity for the same three axons (*B*). Distribution of thresholds for population of axons (*bell-shaped curve*), and cumulative number of axons recruited to discharge (*S-shaped curve*), both plotted as a function of stimulus intensity (*C*).

Large axons, therefore, contribute disproportionately to the compound action potential.

Functional Groupings. Different conduction velocity groups have different biophysical properties. The larger the diameter of an axon, the lower its threshold for response to external electrical stimulation. Therefore, if a whole nerve is stimulated with electrical stimuli of gradually increasing amplitude, the largest axons will be recruited first and the smallest last. The smallest axons include many pain fibers and are most sensitive to local anesthesia, which is why local anesthetic agents produce analgesia before complete anesthesia. The largest axons are most sensitive to hypoxia, as demonstrated by the unpleasant sequence of returning sensation after a limb has "fallen asleep."

The axons in different conduction velocity groups also subserve different functions (Table 3–1). Two sys-

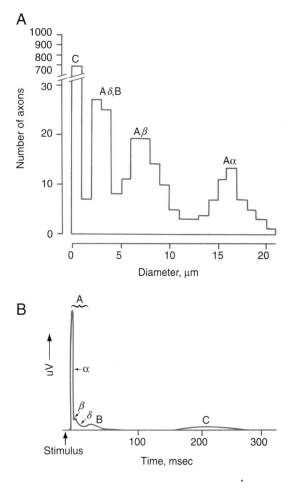

Figure 3–13. The relation of axon diameter to conduction velocity for groups of axons in a nerve. Conduction velocity groups. Distribution of axon diameters for a hypothetical mixed (sensory and motor) nerve (*A*). Compound action potential recorded from the whole nerve, using a stimulus amplitude adequate to stimulate all the axons in the nerve (*B*).

tentials, and it has multiple peaks that correspond to the different conduction velocity groups.

Although small-diameter axons far outnumber large-diameter axons in a nerve, the largest peak in the compound action potential is produced by the largest axon size class. This occurs because the contribution of each action potential to the compound action potential depends on the amount of current the action potential generates. This current is proportional to the surface area of the membrane involved in the action potential.

Table 3–1. Conduction Velocity Groups

Type	Group	Diameter (mm)	Velocity (m/sec)	Function
Sensory Axons				
Ia, Ib	Aα	12–22	72–132	Motor afferent: annulospiral, Golgi tendon organ
II	Aβ	5–12	30–72	Motor afferent: flower spray Sensory afferent: touch, pressure, hair, joint, vibratory receptors
III	Aδ	2–5	12–30	Sensory afferent: hair, free nerve endings: fast pain, temperature
	B	1–3	3–15	Sympathetic
IV	C	0.3–1.3	0.6–2.3	Sympathetic, sensory afferent: slow pain, temperature, mechanoreceptors
Motor Axons				
	α	3.5–8.5	50–100	α-Motor neurons
	γ	2.5–6.5	10–40	γ-Motor neurons

tems are used for classifying peripheral axons. First, these axons may be classified into *groups A, B,* and *C* on the basis of their contribution to a compound action potential in a mixed spinal nerve, where the group A axons produce the first peak and the group C axons produce the last peak (see Fig. 3–13). Group A is divided into subgroups α, β, and δ, where α is fastest and δ slowest. Axons of touch receptors and muscle afferents contribute to the Aα and Aβ peaks. Axons of some hair receptors, thermoreceptors, and nociceptors contribute to the Aδ peak. Second, sensory fibers may also be grouped on the basis of axon diameter and myelin thickness into groups *I, II, III,* and *IV*. Group I axons are large and heavily myelinated; groups II and III are progressively smaller and less myelinated; and group IV axons are small and unmyelinated. Group I is divided into subgroups Ia and Ib, where the fastest Ia fibers supply muscle spindles and the slowest Ib fibers supply Golgi tendon organs. The two schemes can be correlated because axon diameter and myelination determine conduction velocity, which in turn determines the position of a fiber's contribution to the compound action potential. For example, groups C and IV would be equivalent (Table 3–1).

Chemical Synapses

There are two types of synapses, electrical and chemical. At electrical synapses, pairs of ion channels are precisely apposed on opposite sides of the extracellular gap to form structures called *connexons* (Fig. 3–14A). Electrical synapses are rare in mammals.

Chemical synapses share certain common characteristics. The presynaptic cell manufactures a *neurotransmitter*, which is packaged in small membrane-bound *synaptic vesicles* in nerve terminals. When invaded by an action potential, the presynaptic terminal releases a neurotransmitter, which diffuses across the *synaptic cleft* and binds to receptor molecules in the postsynaptic membrane. The transmitter-receptor complex is short-lived, and free transmitter molecules are removed from the synaptic gap by a variety of mechanisms. As a consequence of transmitter-receptor binding, a modification of postsynaptic ion channel permeability occurs. This causes a synaptic current, which produces depolarization or hyperpolarization of the postsynaptic membrane. Because of the specialization of presynaptic and postsynaptic elements, chemical synapses are unidirectional.

The release and diffusion of transmitter take time, so there is a *synaptic delay* between the arrival of the presynaptic action potential and the onset of synaptic current in the postsynaptic neuron. This delay lasts about 0.5 to 1.2 msec. The duration of synaptic currents is at least 1 to 2 msec, because time is needed to remove neurotransmitters from the synapse. The passive membrane properties of postsynaptic neurons slow down the postsynaptic potentials produced by this current, so the fastest potentials last about 5 msec. Thus postsynaptic potentials can be longer than the shortest intervals between action potentials in presynaptic axons (around 1 msec), which has important consequences for information processing at synapses.

Synaptic currents in a postsynaptic neuron can be evoked by activity in thousands of synapses. These currents interact, producing a net hyperpolarization or depolarization. This interaction of graded potentials is the basis for much of the information processing in the nervous system.

Presynaptic Mechanisms. Synaptic vesicles accumulate near the release sites where they bind to the cell membrane (event 1 in Fig. 3–14B). There is a large variety of transmitter molecules (e.g., see Table 4–2).

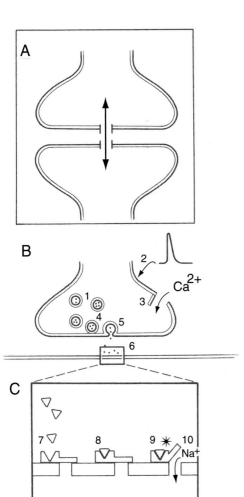

Figure 3–14. General properties of synapses. An electrical synapse *(A)* has a connexon providing a low-resistance bridge between the two nerves. At chemical synapses, presynaptic events *(B)* are as follows: (1) transmitter is synthesized and stored in membrane-bound vesicles, (2) action potential depolarizes the terminal, (3) depolarization causes Ca^{2+} influx through voltage-modulated Ca^{2+} channels, (4) Ca^{2+} influx enables synaptic vesicle docking at release sites, (5) vesicles exocytose their neurotransmitter into the synaptic cleft, and (6) transmitter molecules diffuse across the cleft. Postsynaptic events at chemical synapses *(C)* are as follows: (7) receptor and transmitter dissociated, (8) transmitter-receptor complex formed, (9) activated complex causes channel to open, and (10) ions carrying current pass through channel.

A presynaptic neuron may make one or several transmitters.

The presynaptic action potential produces depolarization of the terminal membrane (event 2 in Fig. 3–14B). Voltage-modulated Ca^{2+} channels open during depolarization, admitting Ca^{2+} ions into the terminal (event 3). The increase of intracellular Ca^{2+} promotes *docking* of synaptic vesicles to release sites (event 4; see

Chapter 4 for more detail) and triggers fusion of the vesicle with the cell membrane to release neurotransmitters into the synaptic cleft (event 5). The smallest unit of neurotransmitter released is the contents of one vesicle. This is called a *quantum*. The total transmitter released is equal to an integral number of quanta.

Once a vesicle has discharged its contents by exocytosis, the vesicle membrane is recycled for reuse. The raised intracellular Ca^{2+} is lowered quickly by an active pump that sequesters Ca^{2+} in the endoplasmic reticulum (ER). The Ca^{2+} is eventually removed from the ER to the extracellular space.

Typically, released transmitter that diffuses into the synaptic cleft (event 6) binds reversibly to postsynaptic membrane receptor molecules. Free transmitter is removed from the synaptic cleft by (1) *inactivation* by membrane-bound enzymes or (2) *reuptake* by endocytosis at the presynaptic membrane. The degradation product produced by enzymatic inactivation is also taken up by the presynaptic membrane, and is used for resynthesis of more transmitter. Recycled and resynthesized transmitter is then reincorporated into vesicles.

Transmitter-Receptor Interactions. A three-state equilibrium is established between the receptor proteins in the postsynaptic membrane and transmitter in the synaptic cleft (see Fig. 3–14C): dissociated (event 7), transmitter-receptor complex (event 8); and activated (event 9). In the simplest postsynaptic mechanisms, the activated receptor causes an associated channel to change configuration, producing altered permeability and synaptic currents (event 10).

When activated, the receptor protein may produce a change in the permeability of a directly coupled channel, or it may initiate a cascade of biochemical reactions. Such biochemical cascades can produce altered ion channel permeabilities, modulation of intracellular Ca^{2+} stores, activation of enzymes, and even regulation of gene transcription.

Therefore, it is convenient to categorize postsynaptic mechanisms in two classes: *single-messenger* synapses, in which the activated transmitter-receptor complex directly modulates channel permeability, and *second-messenger* synapses, in which the activated complex produces a biochemical cascade.

Single-Messenger Synapses. The single messenger in these synapses is the neurotransmitter. Because the activated transmitter-receptor complex directly causes the change in permeability, the synaptic current at a particular channel lasts only as long as the activated complex, usually a fraction of a millisecond (Fig. 3–15).

Second-Messenger Synapses. Second-messenger synapses act through a *guanosine nucleotide–binding protein (G protein)* (Fig. 3–16A) The second messenger is a substance produced as a result of activation of the G protein (see Chapter 4).

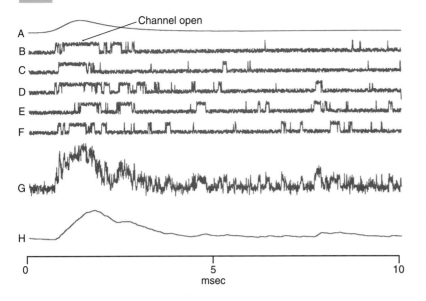

Figure 3–15. Production of a synaptic potential by transmitter-modulated channels. Transmitter concentration in the vicinity of a postsynaptic membrane *(A)*. Currents in five transmitter-modulated channels *(B–F)*. Probability of the open state is proportional to transmitter concentration, so each channel goes from mostly closed in the resting state to mostly open during maximum transmitter concentration. Sum of channel currents for the five simulated channels *(G)*. Membrane capacitance smooths the resulting synaptic potential, which is a slightly delayed reflection of transmitter concentration *(H)*. With larger numbers of channels typically responding during synaptic transmission, the postsynaptic potential is even smoother.

Three types of second messenger G protein systems are most common. In the first (see Fig. 3–16*B*), an adenylyl cyclase coupled to the G protein produces cyclic AMP (cAMP), which acts as the second messenger to produce the permeability change. The second (see Fig. 3–16*C*) couples phospholipase C to the G protein, and the enzyme is used to convert phosphatidylinositol 4,5-biphosphate (PI) to diacylglycerol (DAG) and inositol 1,4,5-triphosphate (IP$_3$), both of which act as second messengers. The third (see Fig. 3–16*D*) couples the G protein to phospholipase A, which initiates a cascade of biochemical reactions starting with the second messenger arachidonic acid and producing a number of subsequent messengers through a variety of enzymatic reactions.

Second-messenger synapses produce many molecules of second messenger per molecule of first messenger, and the second messengers persist after removal of the first messenger, so that the signal is amplified in both strength and duration.

Synaptic Information Processing

Currents and Potentials. Synaptic potentials can be hyperpolarizing (inhibitory, stabilizing) or depolarizing (excitatory). They can also be fast (single messenger) or slow (second messenger). A permeability change causes ions to run down their concentration gradients and bring the local membrane area closer to their equilibrium potentials. To complete a circuit, the active currents produce distributed passive return currents through nearby membrane areas. The area of membrane close to the excitatory synapses that has the lowest threshold for action potential generation (e.g., because it has the most voltage-modulated Na$^+$ channels) is the site in which action potentials are generated, the

trigger zone. In many neurons, it is located at the axon hillock.

Excitatory synapses generally involve increased permeability to Na$^+$ or to a combination of ions whose net combined channel current flow is inward, discharging membrane capacitance and producing depolarization (Fig. 3–17). This brings the membrane closer to the threshold for action potentials at the trigger zone.

At inhibitory synapses, different ionic permeability changes occur that result in a net combined outward current. This can produce a hyperpolarization, which pushes the trigger zone potential further from threshold. If the membrane is already at the steady-state potential for this set of ions, hyperpolarization can result in a lowered membrane resistance with no voltage change. This produces a low-resistance path for the outward flow of passive return current from excitatory synapses. This *shunt* path deflects current away from the trigger zone, decreasing the depolarization produced by excitatory synaptic action (Fig. 3–18).

Potential changes at the trigger zone are thus the effect of interactions of currents generated at all the synapses within passive conduction distance of the trigger zone. For convenience, physiologists subdivide such interactions into *spatial* and *temporal*. In spatial summation (Fig. 3–19*A*), currents from multiple inputs add algebraically. In temporal summation, presynaptic action potentials arrive at intervals shorter than the durations of the postsynaptic potentials they produce (see Fig. 3–19*B*). Thus there is a cumulative effect as postsynaptic potentials overlap in time, and their charges "pile up," or summate, on the membrane capacitance.

Action Potential Generation. The neurons in a given postsynaptic population have varying dynamic ranges relative to the level of activity of neurons in the presynaptic population. Consider the simple case of a single

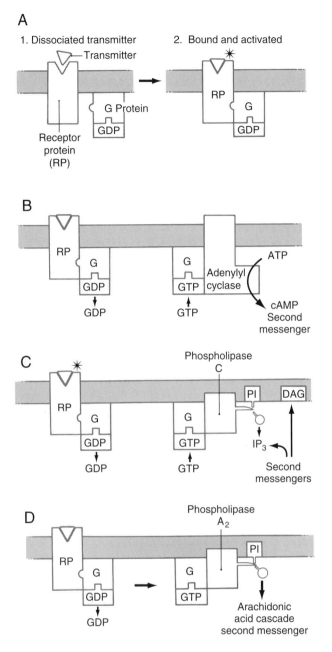

excitatory presynaptic population of sensory afferent axons (Fig. 3–20). An increase in stimulus intensity within the dynamic range for the population results in increased discharge rates and the recruitment of more afferent axons. Together, these cause increased temporal and spatial summation of postsynaptic potentials on their target cells. The net depolarization of neurons in the postsynaptic population will be increased, resulting in increased frequency of discharge for cells that are already firing and a recruitment of additional postsynaptic cells into the discharging population. Therefore, the recruitment and frequency codes resident in the presynaptic population are propagated through the synaptic relay to produce similar recruitment and frequency codes in the output population.

Information Processing. In most CNS nuclei, cells receive input from several different tracts, and local circuits mediate interactions among the cells of the nucleus. Each cell typically receives both excitatory and inhibitory inputs from different sources, and the balance of inhibition and excitation determines the net output. This type of interaction allows control of the sign and magnitude of motor reflexes, detection of features in a visual stimulus such as the locations of edges of visual objects, and comparison of the timing of sounds arriving at the two ears to determine the direction of a sound source.

Figure 3–16. Second-messenger synapses, in which transduction is mediated by a G protein. Interactions between the transmitter (first messenger), the receptor protein, and the G protein (A). Three kinds of second-messenger synapses (B–D). The activated G protein (B) is coupled with adenylate cyclase, which produces cAMP as the second messenger. The activated G protein (C) is coupled with phospholipase C, which cleaves phosphatidylinositol 4,5-bisphosphate (PI_2) to produce the second messengers inositol 1,4,5-trisphosphate (IP_3) and diacylglycerol (DAG). The activated G protein (D) is coupled with phospholipase A_2, which initiates a cascade of biochemical reactions starting with the second messenger arachidonic acid.

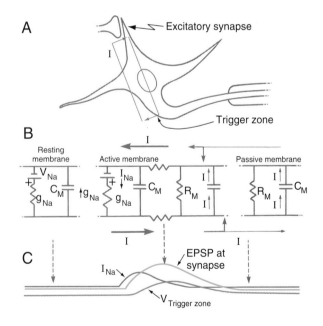

Figure 3–17. The effects of excitatory postsynaptic potentials (EPSPs) on the trigger zone. Current flow between an excitatory synapse and the trigger zone (A). An equivalent circuit (B) showing the change from resting to active membrane during an EPSP (left) and the passive effects at the trigger zone (right). Comparison of the latencies and amplitudes of the current (I_{Na}) synaptic voltage (EPSP) and trigger zone voltage (shown in C).

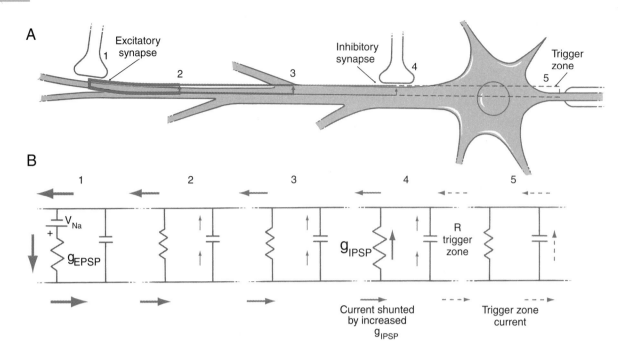

Figure 3–18. Neuron *(A)* and equivalent circuit *(B)* showing a shunt path. The current produced at the excitatory postsynaptic potential (EPSP) (at 1) decrements as it flows passively down the dendrite (1 to 2 to 3 to 4), as represented by the decreasing line and arrow thickness toward the soma. When the inhibitory synapse (at 4) is active, current is shunted across the membrane and does not reach the trigger zone. When this synapse is inactive, the current *(dashed lines, dashed arrows)* spreads to the trigger zone (at 5) and can produce an action potential.

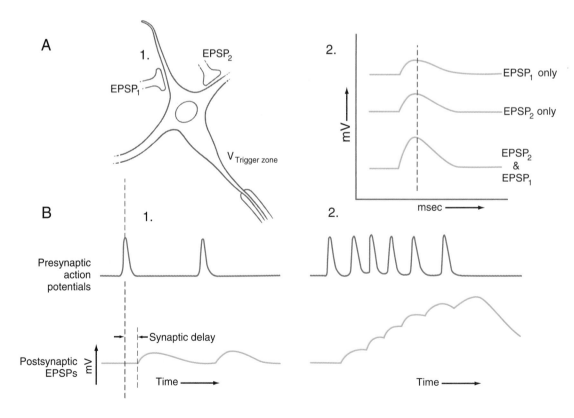

Figure 3–19. Spatial *(A)* and temporal *(B)* summation. Locations of two excitatory synapses on the same postsynaptic neuron *(A, 1)*. EPSP₁ only, EPSP₂ only, and both *(A, 2)*. Currents sum to produce a larger EPSP *(A, 2)*. Low presynaptic action potential frequency produces EPSPs that do not overlap in time and cannot summate *(B, 1)*. High presynaptic action potential frequency produces EPSPs that overlap in time, causing summation *(B, 2)*. EPSP, excitatory postsynaptic potential.

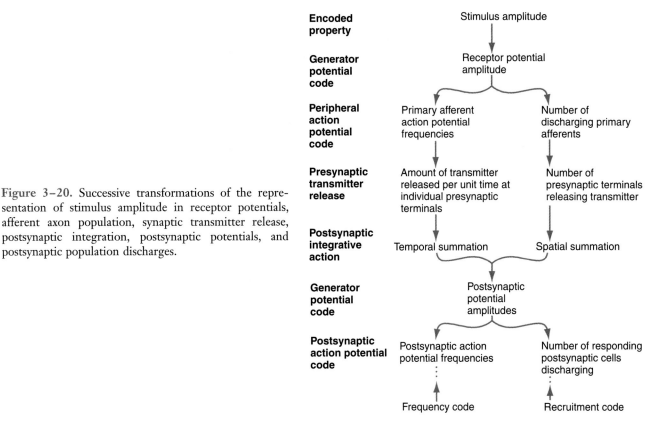

Figure 3–20. Successive transformations of the representation of stimulus amplitude in receptor potentials, afferent axon population, synaptic transmitter release, postsynaptic integration, postsynaptic potentials, and postsynaptic population discharges.

Sources and Additional Reading

Dowling JE: Neurons and Networks. Harvard University Press, Cambridge, Mass, 1992.

Kuffler SW, Nicholls JG, Martin AR: From Neuron to Brain, 2nd ed. Sinauer Associates, Sunderland, Mass, 1984.

Levitan IB, Kaczmarek LK: The Neuron: Cell and Molecular Biology. Oxford University Press, New York, 1991.

Shepherd GM: Neurobiology, 3rd ed. Oxford University Press, New York, 1994.

Chapter 4

The Chemical Basis for Neuronal Communication

R. W. Rockhold

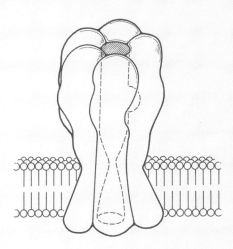

Overview 58

Fundamentals of Chemical Neurotransmission 58
Neurotransmitters
Fast and Slow Synaptic Transmission
Information Flow Across Chemical Synapses

Synthesis, Storage, and Release of Chemical Messengers 60
Composition of Vesicle Membranes
Biosynthesis
Localization
Release

Signal Transduction 63
Receptors and Receptor Subtypes
Structure and Function
Ligand-Gated Ion Channels
G Protein–Coupled Receptors
Effector Proteins
Receptor Regulation

Regulation of Neuronal Excitability 67

Maintenance of the Synaptic Environment 68

Pharmacologic Modification of Synaptic Transmission 68
The Noradrenergic Synapse

Neurons in the human brain communicate primarily by the release of small quantities of *chemical messengers*, most of which are commonly called *neurotransmitters*. These chemicals alter the electrical activity of neurons after they interact with receptors on cell surfaces. Therapeutic alteration of brain function requires an understanding of the processes that regulate the synthesis and release of neurotransmitters, and the means by which receptors alter neuronal electrical activity and biochemical function.

Overview

The brain contains approximately 100 billion (10^{11}) neurons, each of which makes as many as 1000 terminal contacts. Thus, it has been estimated that the human brain contains 10^{14} to 10^{15} connections between neurons. Communication at most of these connections is mediated by *chemical messengers*. The transfer of information between neurons takes place at structurally and functionally specialized locations called *synapses*. Most synapses use chemical messengers that are released in discrete units (*quanta*) from presynaptic axonic or dendritic terminals, in response to depolarization of the terminal.

Rapid diffusion of a chemical messenger across the *synpatic cleft* is followed by binding of this substance to receptors spanning the postsynaptic membrane. There is a resultant alteration in the electrical, biochemical, or genetic properties of that neuron. Less frequently, chemical messengers may also be released at sites without synaptic specializations. These messengers diffuse more widely than do neurotransmitters released at synaptic sites, and they influence receptors located at distant sites and on more than one neuron. Whether synaptic or nonsynaptic, chemical communication in the nervous system depends on (1) the nature of the presynaptically released *chemical messenger*, (2) the type of postsynaptic *receptor* to which it binds, and (3) the mechanism that couples receptors to effector systems in the target cell.

Fundamentals of Chemical Neurotransmission

Neurotransmitters. Specific criteria that define whether a chemical messenger can be identified as a *neurotransmitter* are listed in Table 4–1. Although a wide variety of putative neurotransmitters have been identified, these criteria have been met for only a few chemical substances. Generally, these transmitters can be categorized as *small-molecule messengers* (having fewer than 10 carbon atoms) or larger *neuropeptides* (containing 10 or more carbon atoms).

Small-molecule chemical messengers are classed as

Table 4–1. Criteria Necessary to Define a Substance as a Neurotransmitter

I. Localization	A putative neurotransmitter must be localized to the presynaptic elements of an identified synapse and must be present also within the neuron from which the presynaptic terminal arises.
II. Release	The substance must be shown to be released from the presynaptic element upon activation of that terminal and simultaneously with depolarization of the parent neuron.
III. Identity	Application of the putative neurotransmitter to the target cells must be shown to produce the same effects as those produced by stimulation of the neurons in question.

amino acids, biogenic amines, and nucleotides or nucleosides (Table 4–2). *Amino acid* neurotransmitters include γ-aminobutyric acid (GABA), glycine, aspartate, and glutamate. The vast majority of signaling within the nervous system is carried by amino acid neurotransmitters, specifically GABA and glutamate. For example, it has been estimated that roughly every fifth nerve cell and one of every six synaptic contacts utilizes GABA as a neurotransmitter. The *biogenic amines* include the familiar neurotransmitters acetylcholine, dopamine, norepinephrine, epinephrine, serotonin, and histamine. The nucleotide/nucleoside class includes adenosine and adenosine triphosphate (ATP). *Nitric oxide*, which functions as an endogenous *nitrovasodilator* in the cardiovascular system, has also been identified as a putative neurotransmitter.

More than 40 neuropeptides have been identified in brain tissue. These include methionine enkephalin (met-enkephalin) and leucine enkephalin (leu-enkephalin), as well as larger peptides, such as endorphins, calcitonin gene–related peptide (CGRP), arginine vasopressin, cholecystokinin, and many others (see Table 4–2).

With a few exceptions, one being nitric oxide, *the chemical messengers used by neurons are stored in secretory vesicles and released from them by exocytosis.* In the case of neurotransmitters, these vesicles are found mainly in the presynaptic nerve terminals.

Fast and Slow Synaptic Transmission. The diffusion of a chemical message across the synaptic cleft can be quite rapid. At the neuromuscular junction, for example, it takes only about 50 μsec for acetylcholine to reach the postsynaptic membrane. Total *synaptic delay*, the time from presynaptic release of neurotransmitter to the activation or inhibition of the postsynaptic neuron, is variable. This variability is influenced by the transduction mechanisms in the postsynaptic neuron.

Transduction mechanisms can be divided into fast and slow types. *Fast chemical neurotransmission* operates with a total synaptic delay of only a few milliseconds,

Table 4-2. Substances Believed to Act as Chemical Messengers in the Central Nervous System

Small Molecules	Neuropeptides
Amino acids	**Opioid peptides**
GABA	Methionine enkephalin
Glycine	Leucine enkephalin
Glutamate	β-Endorphin
Aspartate	Dynorphin(s)
Homocysteine	Neoendorphins(s)
Taurine	
	Posterior pituitary peptides
Biogenic amines	Arginine vasopressin
Acetylcholine	Oxytocin
Monoamines	
Catecholamines	**Tachykinins**
Dopamine	Substance P
Norepinephrine	Kassinin
Epinephrine	Neurokinin A
Serotonin	Neurokinin B
Histamine	Eledoisin
Nucleotides and	**Glucagon-related peptides**
nucleosides	Vasoactive intestinal peptide
Adenosine	Glucagon
ATP	Secretin
	Growth hormone–releasing
Other	hormone
Nitric oxide	
	Pancreatic polypeptide–related
	peptides
	Neuropeptide Y
	Other
	Somatostatin
	Corticotropin-releasing factor
	Calcitonin gene–related peptide
	Cholecystokinin
	Angiotensin II

ATP, adenosine triphosphate; GABA, γ-aminobutyric acid.

whereas *slow chemical neurotransmission* usually requires hundreds of milliseconds. In both cases, the receptors on the postsynaptic membranes are glycoproteins that span the lipid bilayer membrane and transduce an extracellular chemical signal into a functional change in the target neuron. The difference relates to the complexity of the transduction mechanism.

In *fast chemical neurotransmission*, the postsynaptic receptor is itself an ion channel. This type of transmission is associated exclusively with *small-molecule neurotransmitters*. The binding of transmitter stimulates the channel to open, permitting a flux of ions across the membrane that alters the membrane potential. The process is fast because it is direct. Ion channels in this type of neurotransmission are called *ligand-gated* or *receptor-gated ion channels*, the ions normally involved are Na^+, K^+, Ca^{2+}, and Cl^-. Movement of these ions causes a change in the *transmembrane electrical potential*, which, if it exceeds threshold, may lead to generation of an action potential.

In *slow chemical neurotransmission*, the signal is transduced by a mechanism involving G protein–coupled receptors. These proteins and their action are discussed later in the chapter. Briefly, the binding of the transmitter (frequently a *neuropeptide*) causes the receptor to activate a G protein, which in turn binds to and influences an effector protein, which elicits the cellular effect. In some cases, the effector protein is an ion channel, which is induced to open or close. Transduction in these cases can be almost as rapid as in fast neurotransmission. More often, the effector is an enzyme that produces an intracellular second messenger, such as cyclic AMP (cAMP), whose cytoplasmic concentration is altered in response to the reception of a signal (binding of the transmitter) at the cell surface and which elicits intracellular responses to the signal. Second messengers can produce a plethora of cellular responses, ranging from the opening or closing of membrane ion channels to alterations in gene expression. These effects are mediated by complex sequences of chemical events, which is why they are relatively slow.

Information Flow Across Chemical Synapses. Transmission of information at a chemical synapse involves the following general sequence of events (Fig. 4–1): (1) secretory vesicle synthesis and transport to the synaptic terminal; (2) for small-molecule neurotransmitters, loading of the transmitter into the vesicle (for neuropeptides, this step accompanies vesicle synthesis); (3) depolarization of the presynaptic terminal; (4) vesicle docking with the presynaptic membrane, exocytosis of its contents, and trans-synaptic diffusion of the transmitter; (5) binding of transmitter to, and activation of, the postsynaptic receptor; (6) transduction of the signal resulting in a postsynaptic response and one or two terminal steps; and (7) active reuptake of the transmitter by the presynaptic cell or (8) enzymatic degradation of the transmitter in the synaptic cleft. These final events eliminate transmitter from the synaptic cleft and thereby terminate its action.

In many synapses, the amount of transmitter that a presynaptic terminal releases in response to an action potential can be regulated from outside the cell. Two regulatory mechanisms are (1) *presynaptic receptor–mediated autoregulation* and (2) *retrograde transmission*. In *presynaptic-mediated autoregulation*, the neuron self-regulates the subsequent quantal release of its own chemical messenger. As a neurotransmitter enters the synaptic cleft, it stimulates not only postsynaptic receptors but also receptors located on the membranes of the terminal from which it was released. This constantly updates the presynaptic neuron concerning neurotransmitter synthesis, release, and the efficiency of information transfer. In most cases, autoregulation is inhibitory. Loss or reduction of this input is interpreted as a reduction in signaling ability and the presynaptic neuron increases the subsequent synthesis and release of stored neurotransmitter.

In *retrograde transmission*, the postsynaptic neuron responds to synaptic activation by releasing a second chemical messenger. This messenger diffuses back

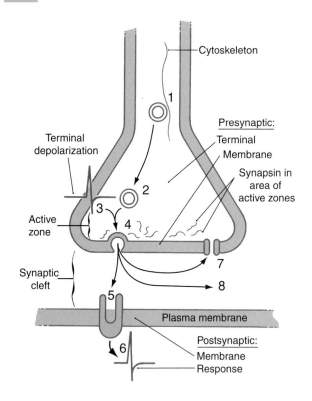

Figure 4–1. A generalized scheme for chemical synaptic transmission. The main steps are as numbered: (1) proximo-distal axonal transport of a secretory vesicle; (2) synthesis and loading of small-molecule messengers in synaptic vesicles (neuropeptides are synthesized and loaded into large dense-cored vesicles in the soma); (3) depolarization of the presynaptic terminal by an arriving action potential, which causes (4) fusion of vesicles with the plasma membrane and exocytosis of the vesicle contents; (5) binding of transmitter with a postsynaptic receptor to produce (6) a postsynaptic response; and finally, elimination of the transmitter from the synapse by either (7) uptake into a cell (here, the presynaptic cell) or (8) enzymatic degradation in the synaptic cleft.

across the synapse and alters the function of the presynaptic terminal. Nitric oxide is currently the best example of a mediator of retrograde transmission.

Synthesis, Storage, and Release of Chemical Messengers

Neuronal chemical messengers are stored in two types of vesicles: *small vesicles* (also called *synaptic vesicles*) and *large dense-cored vesicles*. Synaptic small vesicles (~50 nm in diameter) appear clear and empty in electron micrographs and contain small-molecule chemical messengers such as GABA, glutamate, and acetylcholine. A subset of these small vesicles, with electron-dense cores, are found in both central and peripheral neurons. These vesicles contain the catecholamine family of biogenic amines (dopamine, norepinephrine, and epi-

nephrine). Synaptic vesicles cluster near the exocytotic surface of a presynaptic nerve terminal in regions called *active zones* (see Fig. 4–1).

Large dense-cored vesicles (~75 to 150 nm in diameter) are less numerous and appear in other intraneuronal locations, as well as in the axon terminal. The electron-opaque, dense core is composed of soluble proteins that are mainly one or more neuropeptides. This core may also contain a small chemical messenger—often a biogenic amine, *co-stored* with a neuropeptide.

Neurons in certain hypothalamic nuclei contain a third type of vesicles called the *neurosecretory vesicles*. These vesicles are large (~150 to 200 nm in diameter), contain neurohormones, and are especially concentrated in axon terminals in the neurohypophysis (the posterior pituitary).

Composition of Vesicle Membranes. All vesicles are composed of a lipid bilayer membrane, spanned by a variety of proteins. Some proteins are common to both large dense-cored vesicles and synaptic vesicles, such as those that form *calcium channels*, and the proteins *synaptotagmin* and *SV2*. Other proteins are found in high concentrations only in synaptic vesicles; these include *synaptophysin* and *synaptobrevin*. The differences in protein content reflect the different roles that large dense-cored vesicles and synaptic vesicles play in neurons.

Vesicles also contain proteins that act to accumulate small chemical messengers. These take the form of membrane pumps or transporters, most of which are coupled to the transport of protons. Synaptic vesicles contain at least four classes of *proton-coupled transporters* for chemical messengers, each specific for a different type of messenger. One class drives the accumulation of biogenic amines, including the catecholamines dopamine, norepinephrine, and epinephrine, as well as the monoamine serotonin. Others are specific for acetylcholine, glutamate, and GABA/glycine. Large dense-cored vesicles can also accumulate small chemical messengers in addition to their neuropeptides. However, it is believed that the transporters involved are different from those used by synaptic vesicles.

Biosynthesis. In terms of biosynthesis, an important *difference between synaptic vesicles and large dense-cored vesicles is that the former can be recycled and refilled in the axon terminal, whereas the latter are both made and filled in the neuronal soma and are not recycled.* This reflects the fact that small-molecule neurotransmitters can be synthesized in axon terminals, whereas neuropeptides, because they are synthesized on ribosomes and processed through the endoplasmic reticulum and Golgi complex, can be made only in the soma (Fig. 4–2). The *cis face* of the Golgi complex (also called the *proximal* or *forming face*) is prototypically concave toward the nucleus of the cell, whereas the *trans face* (*distal* or *maturation face*) is convex (see Fig. 4–2). Peptides from the endoplasmic reticulum enter the *cis* face of the

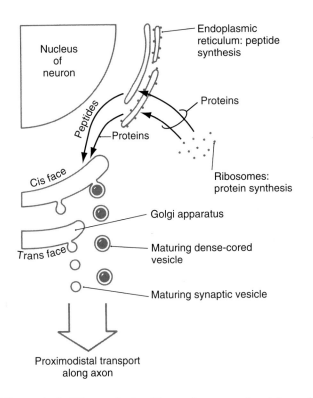

Figure 4–2. The synthesis of large, dense-cored vesicles and of synaptic vesicles in the neuron cell body. Synaptic vesicles are formed without existing stores of neurotransmitter; these are synthesized as the vesicles move into the nerve terminal. Large dense-cored vesicles are formed with existing stores of neuropeptide messengers as electron-dense cores.

Golgi complex and are sorted and packaged into vesicles that bud from its *trans* face.

Large, dense-cored vesicles contain neuropeptide messengers and are filled during the process of vesicle synthesis in the Golgi complex. These vesicles are translocated, by *fast axonal transport* (range 4 to 17 mm/hr), from the cell body to axonal or dendritic release sites (Fig. 4–3). Frequently, neuropeptides are synthesized in the form of large precursor peptides that may be cleaved to yield more than one secreted bioactive neuropeptide. Maturation of neuropeptides can require covalent chemical modification of amino acid side chains, often with the addition of small chemical groups. Examples of the types of chemical modifications include the addition of methyl groups (methylation), sugar moieties (glycosylation), and sulfate groups (sulfation). This process of maturation can occur within the endoplasmic reticulum, during packaging of peptides into large dense-cored vesicles within the Golgi complex, or during axonal transport.

In general, synaptic vesicles are formed initially by budding from the Golgi apparatus within the cell body (Fig. 4–4; see also Fig. 4–2). After transport to and release from the presynaptic terminal, however, the lipoprotein membrane components of the synaptic vesi-

cles are *recycled* in a continuous process that occurs within nerve terminals (see Fig. 4–4). *Synthesis of the chemical messenger in a synaptic vesicle can occur while the vesicle is in the nerve terminal, rather than in the cell body.*

Some small-molecule neurotransmitters are synthesized in the cytosol of the axon and axon terminal and then transported into synaptic vesicles, whereas others are synthesized in the vesicle itself. The synthesis of acetylcholine is an example of the first of these mechanisms. The soluble enzyme *choline acetyltransferase (CAT)* catalyzes the acetylation of choline from acetyl coenzyme A (CoA) to yield the neurotransmitter acetylcholine. A high-affinity vesicular membrane transport protein concentrates this transmitter in cholinergic synaptic vesicles. Synthesis of the catecholamine norepinephrine is an example of the second mechanism. In the case of norepinephrine, synthesis occurs within the synaptic vesicle. The immediate precursor to norepinephrine, dopamine, is concentrated within the *noradrenergic* synaptic vesicle by a transporter specific for biogenic amines. Only then is dopamine converted to norepinephrine by the action of the enzyme, *dopamine β-hydroxylase*, which is attached to the luminal border of the vesicular membrane.

Transporters for small chemical messengers concen-

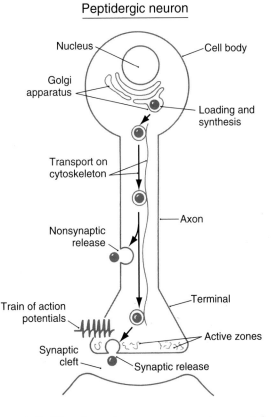

Figure 4–3. The formation, transport, and use of large dense-cored vesicles (containing neuropeptides) in a representative peptidergic neuron.

Small-molecule–secreting neuron

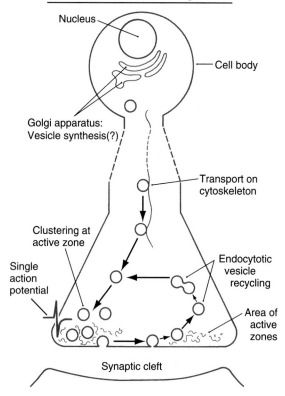

Figure 4–4. The formation, transport, and cycling of synaptic vesicles (containing small-molecule neurotransmitters) in a representative neuron.

trate compounds inside the vesicle to levels 10 to 1000 times higher than those found in the cytosol. The energy required for this transport is derived from an ATP-driven proton pump. The exchange of protons for the chemical messenger allows accumulation of the latter inside the vesicle.

Localization. As mentioned earlier, *synaptic vesicles* are preferentially concentrated in *active zones* of the nerve terminal (Fig. 4–5; see also Fig. 4–4). These zones are biochemically and anatomically specialized for neurotransmitter release. Large numbers of voltage-sensitive calcium channels are clustered in the plasma membrane of active zones. Consequently, depolarization of the axon terminal (or in special cases the dendrites) results in a high local concentration of Ca^{2+}. This calcium causes synaptic vesicles to bind to the plasma membrane and stimulates exocytotic release of vesicle contents into the synaptic cleft. Active zones also contain high concentrations of the filamentous protein *synapsin*, which aids in the clustering of synaptic vesicles.

Although *large dense-cored vesicles* may accumulate in active zones, they also bind to the plasma membrane and release their contents from other sites in the terminal and axon that lack active zones (see Fig. 4–3). As with synaptic vesicles, exocytosis depends on a local

increase in Ca^{2+} concentration. However, the release mechanisms for large dense-cored vesicles appear to be more sensitive to Ca^{2+} than those for synaptic vesicles. Therefore, sites of release do not require the high density of Ca^{2+} channels found in active zones. Release sites outside active zones (such as those associated with large dense-cored vesicles) also do not have anchoring proteins such as synapsins.

Release. The essential structural elements critical for synaptic vesicle release are depicted in Figure 4–5. Proteins in the vesicle wall interact with cytoskeletal proteins to propel vesicles into the active zone. The surface of the synaptic vesicle contains two groups of proteins that are crucial for exocytotic release: *docking proteins* and elements of the *fusion pore*. A rise in intracellular Ca^{2+} levels causes the vesicular docking proteins to interact with docking proteins on the cell membrane, creating a *docking complex* that brings the two membranes into apposition. Complementary proteins on both membranes then interact to form the *fusion pore*, and the lipid bilayers of the two membranes

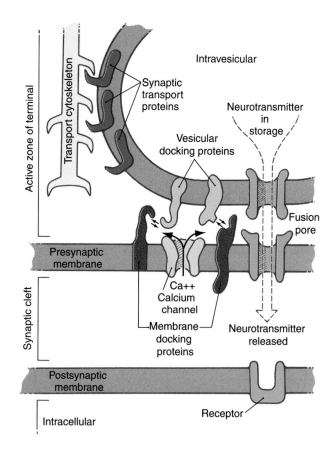

Figure 4–5. The protein components that mediate transport and docking of synaptic vesicles and the probable formation of the fusion pore. Depolarization opens voltage-dependent calcium channels, allowing ingress of Ca^{2+}, which facilitates formation of the docking complex. Once docking is accomplished, additional proteins, under the influence of elevated intracellular Ca^{2+} levels, associate to form a fusion pore.

fuse at this site to form a rapidly expanding hole. Stored neurotransmitter in the vesicle begins to leak out through the fusion pore and exits in bulk (complete exocytosis) as the pore expands.

It is likely that large dense-cored vesicles use somewhat different proteins and mechanisms for docking and exocytosis than those used by synaptic vesicles. Also, while synaptic vesicles undergo exocytosis in response to single nerve impulses (see Fig. 4–4), large dense-cored vesicles respond preferentially to high-frequency trains of impulses (see Fig. 4–3). In experiments using peripheral nerves, a stimulation frequency of 10 Hz is often required to elicit neuropeptide release. Frequencies of that magnitude occur naturally in the autonomic nervous system under conditions of extreme behavioral or physiologic stress. Consequently, neuropeptides may play a role in stress responses.

Signal Transduction

Chemical messengers, once released from a presynaptic site, must interact with a postsynaptic neuron to transmit information. The postsynaptic membrane contains target molecules that exhibit an affinity for individual chemical messengers; these molecules are known as *receptors*. Most receptors are transmembrane glycoprotein chains. The binding of a messenger with its receptor precipitates a change in the architecture (*conformation*) of the glycoprotein chain that begins the process of information transfer. Some exceptions do exist. For example, there are *intracellular receptors* for testosterone. To be activated, drugs such as testosterone must first traverse the plasma membrane to gain access to the receptor.

Receptors and Receptor Subtypes. The *receptor* is capable of altering intracellular function in response to a change in the concentration of a specific chemical messenger in the environment. Thus, a receptor transduces a chemical signal (i.e., the concentration of a chemical messenger) into an intracellular event.

Receptors may be categorized by several means. One simplifying proposal identifies receptors into four general categories: (1) those termed *ligand-gated* channels (also called *transmitter-gated* channels), in which binding of a chemical messenger alters the probability of opening of transmembrane pores or channels; (2) those in which the receptor proteins are coupled to intracellular G proteins as transducing elements; (3) those consisting of single membrane-spanning protein units that have intrinsic enzyme activity (for example, having tyrosine kinase activity); and (4) those termed *ligand-dependent regulators of nuclear transcription* (including receptors for steroids such as testosterone).

On a more specific level, it is common for receptors to be grouped according to the type of native chemical messenger to which they respond. Thus, all the receptors that respond to physiologically relevant concentrations of acetylcholine are called *acetylcholine receptors* (often termed *cholinergic receptors*). Similarly, *adrenoceptors* (often termed *adrenergic receptors*) respond to the catecholamine chemical messengers, epinephrine (previously called adrenaline) and norepinephrine (noradrenaline). Traditionally (and functionally), *receptors are identified by the response of a cell or tissue to a series of chemicals of different but closely allied molecular structures.* Each compound in the series produces identical cell or tissue responses. However, each compound will exhibit a distinct potency (i.e., the concentration required to elicit the desired response) at each different receptor. The rank order of potency for a series of chemicals at each receptor then defines that unique receptor. It is common to rank potencies in terms of the concentration of an agent that produces 50% of the maximal biologic response in the test cell or tissue, or the *effective concentration* (EC_{50}).

Agents that activate a receptor, whether they be native neurotransmitters or exogenous drugs, are termed receptor *agonists*. In contrast, receptor *antagonists* bind to a receptor but do not elicit any response. Rather, by preventing binding of an agonist to its receptor, the antagonist prevents any receptor-mediated signal from being produced. More recently, functional identification of a receptor has been complemented and amplified by molecular cloning techniques, which identify receptor similarities based on the primary amino acid sequence.

The receptors that respond to a given transmitter can often be divided into *subtypes* that elicit different biologic responses. For example, cholinergic receptors are divided into *nicotinic* and *muscarinic* subtypes. A cholinergic synapse with nicotinic receptors is commonly excitatory, whereas one with muscarinic receptors is commonly inhibitory. These receptor subtypes are named after plant compounds that stimulate them selectively and helped lead to their discovery. Nicotinic receptors are named after the nicotine of tobacco, and muscarinic receptors are named after muscarine, a substance found in the toxic mushroom *Amanita muscaria*.

The multiplicity of receptor subtypes can seem overwhelming at first. In the cholinergic system, both nicotinic and muscarinic receptors have subtypes of their own. For example, five different muscarinic receptor subtypes, termed M_1 to M_5, have been recognized. Adrenergic receptors (described in more detail elsewhere) are classified broadly into *α-adrenoceptor* and *β-adrenoceptor* subtypes, each of which is further delineated. Currently, six α-adrenoceptors (α_{1A}, α_{1B}, α_{1D}, α_{2A}, α_{2B}, α_{2C}) and four β-adrenoceptors (β_1, β_2, β_3, and β_4) are recognized. Distinctions between subtypes are conferred by differences in coupling to intracellular second messenger systems, by changes in amino acid sequence of the receptor protein(s), or by insertion of different protein subunits (in receptors in which the integral ion channel is oligomeric, i.e., constructed of multiple, distinct protein subunits). The GABA$_A$ receptor is a good example of this last type of modification. This receptor, like the nicotinic acetylcholine receptor, is a pentamer that forms a transmembrane ion channel. Molecular studies have identified 19 related GABA$_A$ receptor subunits in mammals. Regional differences in the subunit construction of the receptor, which are believed to confer subtle alterations in receptor function, are found in the brain and periphery.

Structure and Function. Transmembrane receptor proteins have a general structure that is based on glycoprotein chains that fold into the neural membrane in multiple loops (Fig. 4–6), although many of the receptors with intrinsic enzyme-associated activity consist of only a single transmembrane subunit. The protein is held in the membrane by several hydrophobic membrane-spanning segments (usually *alpha [α] helices* but in specialized regions having a *beta [β]-pleated sheet* conformation), which are connected by loops that project into the aqueous environment on either side of the

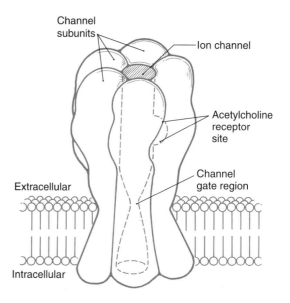

Figure 4–7. A typical ligand-gated ion channel, the nicotinic cholinergic receptor. This receptor is composed of five non-homologous channel subunit proteins, each containing four hydrophobic membrane-spanning regions. Acetylcholine binds to the nicotinic receptor in the central core.

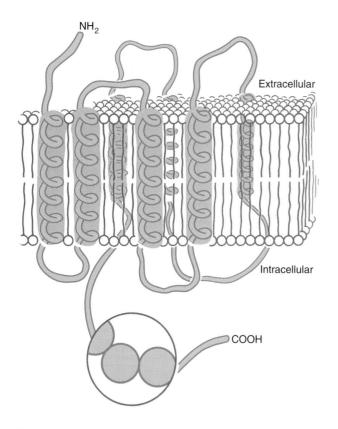

Figure 4–6. A membrane-linked receptor protein (human β-adrenergic receptor), embedded in a neuronal plasma membrane. Hydrophobic transmembrane amino acid sequences are coiled in an α-helical array and form a cluster of seven transmembrane columns. The G protein, although not shown here, would associate with intracellular loops of the receptor protein. The enlarged area denotes the fact that the protein is composed of linked amino acids.

membrane. The N-terminal (NH$_2$-terminal) and C-terminal (COOH-terminal) segments also project into the aqueous environment; they are typically relatively straight. In some transmembrane receptors, such as the β-adrenergic receptor, the N-terminal segment projects extracellularly and the C-terminal segment projects intracellularly (see Fig. 4–6). In contrast, in voltage-gated ion channels, the N-terminal and C-terminal segments usually both project intracellularly. G protein–coupled receptors are formed from a single polypeptide chain, whereas most ligand-gated ion pores are multisubunit structures.

In G protein–coupled receptors, the transmembrane segments of the protein (usually seven in number) form a cluster that contains the binding site or sites for chemical messengers. The site is usually in a relatively hydrophobic pocket in the cluster, although it is sometimes on the extracellular surface of the protein. The receptor binds with its G protein transducer through multiple cationic sites on intracellular hydrophilic regions. The β-adrenergic receptors are the best characterized of the G protein–coupled receptors; their functioning is discussed later in the chapter.

Ligand-Gated Ion Channels. Ligand-gated ion channels are formed by several structurally distinct protein subunits called *channel subunits* (Fig. 4–7). Each channel subunit is a transmembrane glycoprotein (as described previously) with membrane-spanning segments connected by intracellular and extracellular loops. The channel subunits complex to form a roughly cylindrical structure that encloses a water-filled transmembrane channel. As exemplified by the nicotinic cholinergic receptor, the external face of the channel is enlarged and cuplike (see Fig. 4–7). The channel narrows as it crosses the membrane, reducing the inner diameter such that it can selectively pass small cations (Na^+, K^+, and Ca^{2+}) or anions (Cl^-). The internal face of the channel widens again as it emerges from the lipid bilayer. The inner surface of the pore is blocked at rest by amino acid residues that project into the aqueous lumen of the pore and prevent the conductance of charged ions. This part of the channel is termed the *gate*. *Binding sites*, which most commonly occur at relatively hydrophobic regions within the transmembrane region of the channel, are specific for a chemical messenger. When a binding site is filled, conformational changes occur within the channel protein to open the gate and permit selective passage of ions across the membrane.

Two gene superfamilies of ligand-gated ion channels have been identified. One contains nicotinic cholinergic, serotonin (5-hydroxytryptamine), GABA, and glycine receptors; and the other encodes the receptors for the excitatory neurotransmitter glutamate. The segregation of receptors into different superfamilies is based on the degree of homology of amino acid sequences. The subunits of the various receptors in a superfamily have about 20% to 40% sequence homology with each other. The subunits of any given receptor generally have sequence homology of greater than 40%.

G Protein–Coupled Receptors. The G protein receptors introduce a further level of complexity to chemical transmission (Fig. 4–8). *About three fourths of all chemical messengers transmit their information through G protein–coupled receptors.* The basic elements of this system include a *receptor*, which must face the external surface of the membrane, the *guanosine triphosphate*

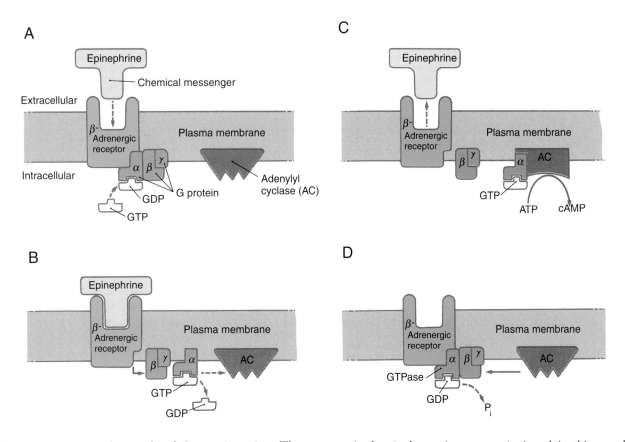

Figure 4–8. A typical example of G protein action. The receptor is the β-adrenergic receptor (activated in this case by epinephrine). It is coupled to the G_s type of G protein, which has a stimulatory action on the effector, adenylyl cyclase. When stimulated, adenylyl cyclase produces the second messenger cAMP from ATP. The cycle of G protein action is as follows: (*A*) In the resting state, the G protein is in the form of an αβγ heterotrimer, and the α subunit carries bound GDP. (*B*) Epinephrine binding to, and conformation change of, the receptor protein and the G protein. GDP, which has been bound to the α subunit, exchanges for GTP as the α subunit dissociates from the βγ complex and from the receptor. (*C*) Binding of the α subunit (with GTP attached) to adenylyl cyclase alters the conformation of the enzyme. This increases enzymatic catalysis of the substrate ATP to cAMP. (*D*) Slow enzymatic dephosphorylation of GTP to GDP by the GTPase allows the α subunit to return to its resting conformation. As this occurs, the α subunit dissociates from adenylyl cyclase and reassociates with the βγ subunit. The heterotrimeric G protein reestablishes a loose association with the receptor protein. GDP, guanosine diphosphate; GTP, guanosine triphosphate.

(GTP)-binding protein, which consists of α, β, and γ subunits, and an *effector protein*, which may be an enzyme that alters the concentrations of intracellular *second messengers* (such as Ca^{2+}, inositol 1,4,5-trisphosphate, diacyl glycerol, or members of the eicosanoid family), or may be an ion channel (see Fig. 4–8A–D). The responses mediated by these receptors are generally slow (hundreds of milliseconds to minutes). The G protein–coupled receptor complex transduces an extremely wide range of chemical messages. About 100 different receptors have been identified that can link to a G protein, and at least 20 distinct G proteins have a similar number of effector proteins. The biogenic amines, bioactive peptides, the eicosanoids, light (one of the first characterized G protein–coupled receptors was rhodopsin in the mammalian photoreceptor), and odorants all interact with G protein–coupled receptors.

The G protein functions to amplify a signal received by a transmembrane receptor, transmitting that message to effector proteins within a neuron. Each G protein exists as a complex (a *heterotrimer*) formed by α, β, and γ subunits (see Fig. 4–8A–D). The $\alpha\beta\gamma$ heterotrimer maintains a loose association with the receptor glycoprotein but is not covalently bound to the receptor. Within the heterotrimer, the α subunit determines the nature of the G protein. It has the ability to bind GTP, detach from the coupled $\beta\gamma$ complex, and alter the activity of an effector protein. The effector proteins whose activity can be modulated by α subunits are diverse and include the enzyme adenylyl cyclase as well as ion channels for calcium and potassium. In contrast, the $\beta\gamma$ complex anchors α subunits to membrane sites and inhibits the GTP–guanosine diphosphate (GDP) exchange that activates the α subunit.

The resting form of a G protein exists as the heterotrimer, with GDP bound to the α subunit and the α subunit bound by the $\beta\gamma$ complex (see Fig. 4–8A). When activated, the α subunit exchanges GDP for GTP and dissociates from the $\beta\gamma$ complex. The α subunit is then free to bind with and alter the activity of the effector protein; in this example it is adenylyl cyclase (see Fig. 4–8B and C). The α subunit has an integral slow GTPase activity, which eventually hydrolyzes the bound GTP to bound GDP (usually in about 3 to 15 seconds). Reassociation of the α subunit–GDP complex with the $\beta\gamma$ complex then completes the cycle (see Fig. 4–8C and D).

The most completely characterized G protein–coupled receptor is the β_2-adrenergic receptor. The endogenous ligand for this receptor is the catecholamine epinephrine. The β_2-adrenergic receptor is coupled to a G protein that stimulates activity of adenylyl cyclase, catalyzing formation of cAMP from intracellular ATP stores (see Fig. 4–8A–D). The stimulatory G protein associated with the β_2-adrenergic receptor is called G_s.

Effector Proteins. As mentioned earlier, G proteins can interact with two main kinds of effector proteins: *ion channels* (called *G protein–coupled ion channels*) and *enzymes that alter the level of intracellular second messenger compounds*. The action of the G protein on its target may be positive or negative. That is, it may cause a channel to open or, more rarely, close, or it may stimulate or inhibit a target enzyme. Most synaptic G protein responses are mediated through second messenger systems. Second messengers can elicit a variety of cellular responses, including the opening or closing of ion channels in the cell membrane (note that this indirect mechanism is in addition to the mechanism by which G proteins can interact directly with ion channels), the release or reuptake of Ca^{2+} from intracellular storage sites, alterations in the activity of key cellular enzymes, and alterations in the expression of specific genes. Many of these effects are mediated by *protein kinases*, enzymes that regulate the activity of other proteins by phosphorylation. A given G protein–coupled receptor may activate more than one mechanism and produce multiple coordinated effects. G protein–coupled systems therefore have the capacity to mediate complex changes in neuronal function.

The best-known second messenger systems involve the enzymes *adenylyl cyclase* (see Fig. 4–8) and *guanylate cyclase*. Adenylyl cyclase produces the second messenger cyclic GMP (cGMP) from cytoplasmic GTP. Another important second messenger system involves the enzyme phospholipase C, which hydrolyzes the membrane phospholipid *phosphatidylinositol 4,5-bisphosphate* to produce the second messengers *inositol 1,4,5-trisphosphate* (IP$_3$) and *diacylglycerol* (DAG). G proteins can also interact with *phospholipase A_2*, which stimulates the formation of members of the eicosanoid family, and with another enzyme called *phospholipase D*.

Receptor Regulation. The postsynaptic receptor is not static, either in terms of response to agonists or in the number of active receptors present on the membrane. The response of postsynaptic receptors to changes in the synaptic environment is a crucial element in neuronal communication. One of the most intensively studied postsynaptic receptor response systems is that of a G protein–coupled receptor, the β-adrenergic receptor—specifically, the β_2-adrenergic receptor subtype. Continuous or repeated exposure of the β_2-adrenergic receptor to an agonist will result in a diminution of the response of that agonist. The mechanisms by which this loss of response occur are characterized by the time frame over which they occur. Exposure to an agonist for seconds to minutes will result in a reduction in the agonist-induced response through processes called *desensitization*. The loss of receptor responsiveness is mediated by agonist-induced changes in the receptor conformation that permit binding of additional intracellular proteins to the receptor. These proteins cause *phosphorylation* (i.e., addition of phosphate moieties) of the intracellular portions of the receptor.

Phosphorylation changes the affinity of the receptors to other intracellular proteins that uncouple activated receptors from their effector proteins.

Homologous desensitization occurs when stimulation of the receptor by an agonist evokes the phosphorylation. Desensitization can also be produced without direct stimulation of the receptor in question, which is termed *heterologous desensitization*. In this latter case, intracellular phosphorylating enzymes are recruited by stimuli other than activation of the receptor. If the stimulus is maintained, the receptor protein may be subsequently sequestered into invaginations of the membrane that undergo internalization, effectively removing the receptor from the membrane surface. Once internalized, a receptor can be degraded or, in some circumstances, recycled back into the membrane as a fully sensitive, active receptor. Receptor *downregulation*, on the other hand, is caused by exposure to agonists for longer periods of time—hours to days—and is characterized by a reduction in the number of active receptors on the cell surface. Downregulation may be achieved by enhanced protein receptor degradation, decreased transcription of the messenger RNA (mRNA) for that receptor, or enhanced degradation of receptor mRNA.

Regulation of Neuronal Excitability

As we have seen, neurotransmitters cause the opening or closing of ion channels in the postsynaptic membrane. If the transmitter signal is transduced by a G protein mechanism, it also may have other effects. The result will be a transient, local change in the polarization of the postsynaptic membrane, called a *synaptic potential*. This potential consists of either a depolarization or a hyperpolarization of the membrane relative to the resting potential. (Membrane potentials and the electrical properties of neuronal cell membranes are discussed in more detail in Chapter 3.) Synaptic potentials are graded in amplitude, reflecting the varying strengths of the incoming synaptic signals that elicit them. They generally do not exceed 20 mV. Local potentials of this type spread passively over the membrane of the postsynaptic cell, gradually losing amplitude and dying out. However, if they reach a *trigger zone*—a locus at which action potentials can be initiated—they may contribute to the production or suppression of action potentials. An action potential is triggered whenever the membrane is depolarized beyond a certain threshold potential. Therefore, *depolarizing* synaptic potentials tend to promote action potentials and are called *excitatory postsynaptic potentials* (*EPSPs*). Conversely, *hyperpolarizing* synaptic potentials inhibit the production of action potentials and are called *inhibitory postsynaptic potentials* (*IPSPs*).

In the central nervous system (CNS), a neuron is constantly bombarded by neurotransmitters, each of which can generate or modify a synaptic potential. Neurotransmitters that move the membrane toward depolarization (by reducing the -70 mV resting potential), with the resultant production of an action potential, are commonly called *excitatory neurotransmitters*. Neurotransmitters that move the membrane away from depolarization (by making the resting membrane more negative, the membrane is hyperpolarized) are frequently referred to as *inhibitory neurotransmitters*. *Because the postsynaptic response is actually elicited by the receptor rather than by the transmitter, the postsynaptic receptor determines whether a given neurotransmitter will be excitatory or inhibitory.* Some neurotransmitters can have either effect, depending on the type of postsynaptic receptor present.

Excitatory neurotransmitters act by promoting the opening of channels selective for cations (either Na^+ or Ca^{2+}) that flux into the cell and depolarize the membrane. In some cases, as in that of certain glutamate receptor subtypes, the neurotransmitter binds directly to a stereospecific site on an associated ion channel. Glutamate receptor subtypes of this ilk are classified as *ionotropic*. Excitatory ionotropic glutamate receptors include the NMDA, AMPA, and kainate subtypes, each of which is named after a selective ligand for that subtype (NMDA, *N*-methyl-D-aspartate; AMPA, α-amino-3-hydroxy-5-methyl-4-isoxazole propionic acid). In other cases, opening of cation channels is accomplished indirectly, following alteration of an intracellular second messenger system or systems. Inhibitory neurotransmitters act by opening channels for K^+ or Cl^-. Important examples of inhibitory neurotransmitters are the amino acids *GABA* and *glycine*.

It is essential to recognize that *a single chemical messenger can evoke either an EPSP or an IPSP, depending on the receptor to which it binds.* A good example is the neurotransmitter norepinephrine. Like glutamate, norepinephrine binds to multiple receptor subtypes. In the CNS, receptors for norepinephrine fall into two categories; α-adrenergic and β-adrenergic receptors. Both types are G protein coupled. The G protein to which β-adrenergic receptors are coupled is of a type called G_s, which *stimulates* the activity of adenylyl cyclase and thus produces a rise in intracellular cAMP. This rise in cAMP leads to an EPSP. In contrast, the G protein to which the α_2-adrenergic receptor subtype is coupled, called G_i, *inhibits* the activity of adenylyl cyclase. The resulting fall in intracellular cAMP leads to an IPSP. In both cases, cAMP acts through enzymes called *cAMP-dependent protein kinases*. In the pathway under discussion, the final targets are membrane ion channels, which open or close in response to the phosphorylation of sites on their cytoplasmic domains. Consequently, norepinephrine can elicit either an excitatory or inhibitory response, depending on the receptor.

Maintenance of the Synaptic Environment

The concentration of a chemical messenger in the synaptic cleft is crucial to information transfer. However, the time frame during which a chemical message is active must be limited if a temporally discrete signal is to be produced. This is particularly true when neurons fire at rates of more than several depolarizations per second. Simple diffusion out of the synaptic cleft is rarely adequate to effectively terminate the postsynaptic signal. Accordingly, active mechanisms exist to reduce or eliminate chemical messengers in the synaptic cleft. The principal mechanisms are *enzymatic degradation of transmitter in the cleft* and *transporter-mediated uptake* across cell membranes.

Acetylcholine and the neuropeptides are examples of transmitters that are neutralized by enzymatic degradation in the cleft. Acetylcholine is cleaved by the enzyme *acetylcholinesterase*, which is synthesized by the neuron and inserted into the postsynaptic membrane near receptor sites. Neuropeptides are degraded through hydrolysis by the action of multiple *peptidases*, which are found in extracellular fluid.

The neurotransmitters whose action is terminated by uptake from the synaptic cleft include the monoamines (such as serotonin, histamine, and the catecholamines) and the amino acid neurotransmitters GABA, glycine, glutamate, and aspartate. This uptake is accomplished by the action of specific membrane-bound *transport proteins*. The monoamine class of biogenic amines (including the catecholamines, serotonin, and histamine) are avidly removed from the synaptic space by such transport proteins.

In the case of norepinephrine, *reuptake* into the cytoplasm of the presynaptic terminal (a process known as *uptake 1*) is primarily responsible for terminating the action of the transmitter (Fig. 4–9). After reuptake, some norepinephrine is enzymatically degraded by the mitochondrial enzyme *monoamine oxidase* (MAO), whereas an additional fraction is retained in a cytoplasmic pool. The norepinephrine in this pool is an important target for drug action. Norepinephrine can also be removed through the action of a transporter on the postsynaptic membrane (*uptake 2*), although this process is usually less effective (see Fig. 4–9). Norepinephrine transported into the postsynaptic neuron is degraded by the enzyme, *catechol-O-methyltransferase (COMT)*.

In the CNS, glial cells, primarily astrocytes, express transporter proteins on their membranes and can remove transmitters from the synaptic cleft. The neurotransmitter dopamine is transported by glial cells in the substantia nigra. These glial cells metabolize the transported dopamine to inactive products by means of a form of MAO unique to the CNS, MAO-B. Metabolism by MAO-B is essential to the activation of a dopaminergic neurotoxin called MPTP (1-methyl-4-phenyl-1,2,3,6-tetrahydropyridine). This substance destroys dopamine neurons in the substantia nigra, resulting in a syndrome markedly similar to that seen in patients with Parkinson disease. The actions of the amino acids, GABA, glycine, glutamate, and aspartate are all terminated by active transport into neurons and glial cells. No active uptake mechanisms have been found that terminate the action of neuropeptides (see Chapter 2).

The mechanisms of termination of some other chemical messengers, such as adenosine, ATP, and nitric oxide, are less well understood. Nitric oxide is very labile; it undergoes redox reactions with membrane and cytoplasmic sulfhydryl moieties, reducing them and becoming oxidized itself. Specific ATPases may terminate the action of ATP functioning as a neurotransmitter.

Pharmacologic Modification of Synaptic Transmission

Drugs can alter virtually every level of neuronal and synaptic function. Therapeutic effects are most commonly achieved by actions of drugs on *neurotransmitter synthesis*, *vesicular uptake and storage*, *depolarization-induced exocytosis*, *neurotransmitter-receptor binding*, and *termination of neurotransmitter action*. Increasingly, drugs are being developed that modify neurotransmitter action through interaction with *postsynaptic effector systems*.

The Noradrenergic Synapse. The noradrenergic synapse is used to illustrate the range of pharmacologic agents that can modify synaptic transmission (see Fig. 4–9). Noradrenergic synapses use the neurotransmitter norepinephrine, and their postsynaptic receptors fall into the two classes introduced earlier, the α-adrenergic and β-adrenergic receptors. In the peripheral nervous system, norepinephrine is a critical neurotransmitter in the regulation of the *sympathetic* division of the *autonomic nervous system*. In the CNS, norepinephrine is synthesized in neurons concentrated in several discrete brainstem regions; these areas send noradrenergic axons throughout the brain and exert widespread effects.

Norepinephrine is synthesized from the amino acid *tyrosine* in a sequence of three enzymatic reactions (see Fig. 4–9). The first two occur in the cytoplasm; the final reaction takes place within the synaptic vesicle. Tyrosine is accumulated in the terminal by a membrane-bound amino acid carrier and is converted to dopa by *tyrosine hydroxylase* (step 1 in Fig. 4–9). The rate of this conversion makes it the rate-limiting step in norepinephrine synthesis. Regulation of tyrosine hydroxylase is accomplished by phosphorylation of the enzyme by intracellular protein kinases, which increases the rate of enzymatic catalysis. The drug *α-methyltyrosine* can limit noradrenergic function by acting as a

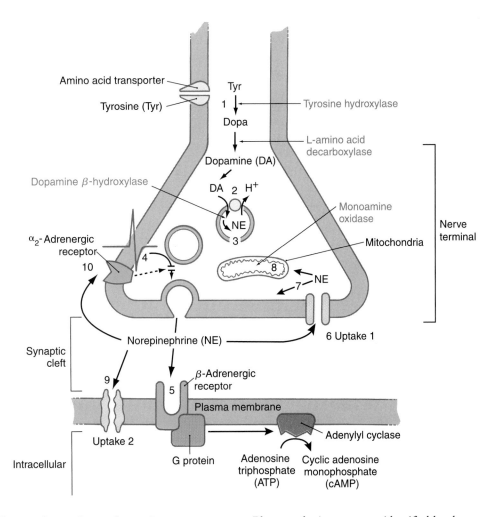

Figure 4–9. The noradrenergic β_2-adrenergic receptor synapse. Pharmacologic agents are identified by the numerals. Synthetic and degradative enzymes are shown in red; membrane receptors, transporters, and ion channels in green; and the postsynaptic effector G protein in blue. (1) *α-Methyltyrosine* competitively inhibits tyrosine hydroxylase; (2) *reserpine* irreversible inhibits the monoamine-H^+ vesicular transport pump; (3) *α-methyldopa* acts as a false transmitter, displacing norepinephrine in the synaptic vesicle; (4) *guanethidine* blocks the ability of membrane depolarization to cause exocytotic release of vesicle contents; (5) *propranolol* is a competitive antagonist at β-adrenergic receptors; (6) *cocaine* blocks synaptic membrane reuptake of norepinephrine (Uptake 1); (7) *tyramine* displaces norepinephrine from a cytoplasmic storage pool back into the synaptic cleft; (8) *pargyline* blocks the degradation of norepinephrine by mitochondrial monoamine oxidase; (9) *corticosterone* prevents uptake of norepinephrine by the postsynaptic membrane (Uptake 2); (10) *yohimbine* is a competitive antagonist at presynaptic, autoinhibitory α_2-adrenergic receptors.

competitive inhibitor of tyrosine hydroxylase, thereby reducing the rate of norepinephrine synthesis. This drug, also called *metyrosine*, is used clinically in management of patients who suffer from symptoms due to the excess production of catecholamines arising from tumors of adrenal chromaffin cell tissue, or *pheochromocytomas*.

Dopa is converted to *dopamine* (a neurotransmitter in its own right) by *L-amino acid decarboxylase*. Dopamine is transported into, and concentrated within, the synaptic vesicle by a monoamine-H^+ transporter (step 2 in Fig. 4–9). The accumulation of dopamine (and ultimately, norepinephrine) can be prevented by *reserpine*, a plant alkaloid that irreversibly inactivates the vesicular transporter (step 2 in Fig. 4–9). The inability to fill vesicles with neurotransmitter results in a pro-

gressive reduction in the level of transmitter in the axon terminal, which inhibits neurotransmission.

Reserpine, one of the earliest therapeutic agents available for the treatment of hypertensive cardiovascular disease, markedly reduces the ability of cardiac noradrenergic neurons to stimulate the heart to increase rate and contractility, thus lowering cardiac output. In addition, noradrenergic neurons that innervate arteriolar smooth muscle produce less norepinephrine, which results in less vasoconstriction and an overall reduction in blood pressure. Unfortunately, reserpine penetrates the blood-brain barrier readily and can cause depletion of neurotransmitter catecholamines, including dopamine, norepinephrine, and epinephrine. This central catecholamine depletion, if severe enough, can cause clinical depression and has been associated with suicide. Today, reserpine is only rarely used to treat hypertension.

Within the vesicle, dopamine is converted to norepinephrine by *dopamine β-hydroxylase*. The accumulation of both dopamine and norepinephrine can also be reduced by administration of *α-methyldopa*. This dopa analog is enzymatically converted in successive steps to α-methyldopamine and α-methylnorepinephrine, which takes the place of the normal synthetic products, resulting in reduction of noradrenergic transmission (see Fig. 4–9). Elevated sympathetic nerve activity contributes to hypertensive cardiovascular disease. Clinically, α-methyldopa is an effective antihypertensive drug and is one of the most commonly used drugs for managing hypertension during pregnancy.

The drug *guanethidine* interferes with the coupling between excitation of the nerve terminal and exocytotic release of norepinephrine, thereby reducing the amount of norepinephrine released. In addition, guanethidine acts like reserpine to inactivate vesicle transport. Unfortunately, guanethidine causes such a profound inhibition of noradrenergic neuron function that its use is associated with undesirable and unpleasant adverse effects, including excessively reduced heart rate, nasal congestion, and *orthostatic hypotension* (a decline in blood pressure that occurs upon standing erect, due to the effects of gravity causing pooling of blood in the lower extremities) (see Chapter 29).

Once released into the synaptic cleft, norepinephrine can bind to two sets of receptors: (1) postsynaptic α-adrenergic or β-adrenergic receptors, which elicit the postsynaptic response, or (2) presynaptic receptors, partially but not exclusively of the α_2-adrenergic subtype, which are involved in autoregulation. Eventually, the norepinephrine is removed from the synapse by reuptake into the presynaptic terminal or uptake into the postsynaptic cell. *Propranolol* is an example of a drug that interferes with the binding of norepinephrine to postsynaptic β-adrenergic receptors. This drug binds competitively to the receptor and prevents its activation, thereby blocking the postsynaptic response (in this case, the rise in intracellular cAMP mediated by G_s activation of adenylyl cyclase). Propanolol and related β-adrenergic receptor antagonists are extremely effective and widely used in cardiovascular medicine. They are indicated for management of hypertension, *angina pectoris* (chest pain due to cardiac ischemia), congestive heart failure, and myocardial infarction.

The autoregulatory presynaptic receptors for norepinephrine exert an inhibitory effect over the amount of norepinephrine released in response to an action potential. They influence both the synthesis of norepinephrine and its exocytotic release. The α_2-adrenergic receptors that are responsible for these effects can be blocked by the drug *yohimbine*. The resulting loss of autoinhibition increases the amount of norepinephrine released and enhances noradrenergic function.

Blockade of α_2-adrenergic receptors and reduction in norepinephrine release reduces postsynaptic effects of norepinephrine. In many peripheral tissues, the actions of the sympathetic (noradrenergic) component of the autonomic nervous system normally are finely balanced by those of the parasympathetic component (which, in most cases, releases acetylcholine) (see Chapter 29). Decline in the function of one component frequently leads to overexpression of the function of its opposing system. This is particularly true in the male urogenital tract, where erection has been linked both to activation of cholinergic nerves and to blockade of α_2-adrenergic receptors. Yohimbine (which has enjoyed a lengthy folk history as an aphrodisiac) has, in recent years, proved modestly effective in promoting erectile function in male patients with impotence of vascular or diabetic origin or of psychogenic origin.

Drugs such as *cocaine* inhibit the presynaptic reuptake of norepinephrine, thus prolonging the synaptic activity of the transmitter and resulting in exaggerated postsynaptic responses. As explained earlier, norepinephrine that has undergone reuptake can either be degraded by the mitochondrial enzyme MAO or be retained in a cytoplasmic pool. Monoamine oxidase inhibitors such as *pargyline* increase the amount of norepinephrine in the cytoplasmic pool, thereby enhancing noradrenergic transmission. Agents such as *tyramine* displace norepinephrine from the cytoplasmic pool back into the synapse, also enhancing noradrenergic activity.

Sources and Additional Reading

Cooper JR, Bloom FE, Roth RH: The Biochemical Basis of Neuropharmacology. Oxford University Press, New York, 1991.

Danner S, Lohse MJ: Regulation of β-adrenergic receptor responsiveness, modulation of receptor gene expression. Rev Physiol Biochem Pharmacol 136:183–223, 1999.

Hall ZW: An Introduction to Molecular Neurobiology. Sinauer Associates, Sunderland, Mass, 1992.

Jessell TM, Kandel ER: Synaptic transmission: A bidirectional and self-modifiable form of cell-cell communication. Cell 72/Neuron 10(Suppl):1–30, 1993.

Kelly RB: Storage and release of neurotransmitters. Cell 72/Neuron 10(Suppl):43–53, 1993.

Kobilka B: Adrenergic receptors as models for G protein–coupled receptors. Annu Rev Neurosci 15:87–114, 1992.

Rahmann H, Rahmann M: The Neurobiological Basis of Memory and Behavior. Springer-Verlag, New York, 1992.

Somogyi P, Tamás G, Lujan R, Buhl EH: Salient features of synaptic organization in the cerebral cortex. Brain Res Rev 26:113–135, 1998.

Trimble WS, Linial M, Scheller RH: Cellular and molecular biology of the presynaptic nerve terminal. Annu Rev Neurosci 14:93–122, 1991.

Zoli M, Agnati LF, Hedlund PB, Li XM, Ferre S, Fuxe K: Receptor-receptor interactions as an integrative mechanism in nerve cells. Mol Neurobiol 7:293–334, 1993.

Chapter 5

Development of the Nervous System

O. B. Evans and J. B. Hutchins

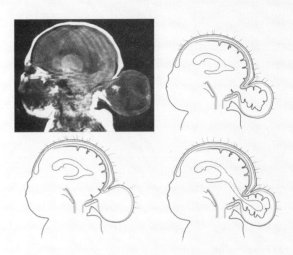

Overview 72

Brain Development 72
Induction
Primary Neurulation
Secondary Neurulation
Primary Brain Vesicles
Secondary Brain Vesicles
Diencephalon and Cerebral Hemispheres
Infectious Diseases Causing Congenital
  Nervous System Defects
Ventricular System

Peripheral Nervous System 79
Neural Crest
Placodes
Cranial Nerve Ganglia
Posterior (Dorsal) Root Ganglia
Visceral Motor System
Schwann Cells

Central Nervous System 81
Basic Features
Spinal Cord
Relationship of Spinal Cord to Vertebral
  Column
Brainstem
Cerebellum
Thalamus
Cerebral Cortex
Gross Abnormalities of Cortical
  Development

**Cellular Events in Brain
Development 87**
Overproduction of Neurons and Apoptosis
Axonal Outgrowth
Synaptogenesis
Plasticity and Competition
Myelination

The central nervous system (CNS) develops from primitive ectoderm, one of the three germ layers of the embryo. From a few dozen cells, which together weigh perhaps a microgram, the brain becomes an organ weighing about 800 g at birth, 1200 g at 6 years of age, and about 1400 g in the adult—about a billion-fold increase. Most but not all neurons undergo their last cell division before birth. The development of a fully functional nervous system requires division and migration of nerve cells and the formation of synaptic connections.

Overview

In view of the complex embryology of the human CNS, it is remarkable that there are so few congenital CNS defects. Although 3% of births are associated with major malformations of the CNS, most fetuses and infants in this category do not survive. About 75% of spontaneously aborted fetuses and 40% of infants who die within the first year of life have major CNS malformations.

The basic form of the human CNS is complete by about the sixth week of gestation. The next phases, which include cellular proliferation and migration, are most prominent in the second trimester of gestation but continue until term. Myelination peaks during the third trimester but continues until adulthood. The development of synaptic connections between neurons and the response of the brain to its experiences result in its functional maturity. This type of development continues throughout life.

From a few primordial cells, about 100 billion neurons develop, each with thousands of contacts with other neurons. Through this network of interconnecting neurons, the human brain is capable not only of directing the movement of the body and sensing the environment but also of thinking, reasoning, experiencing emotions, and dreaming.

Brain Development

The first neural tissue appears at the end of the third week of embryonic development, when the embryonic disc is composed of *ectoderm*, *mesoderm*, and *endoderm*. A specialized part of the ectoderm, the *neuroectoderm*, gives rise to the brain, spinal cord, and peripheral nervous system (Fig. 5–1).

Induction. The *notochord* arises from axial mesoderm at about 16 days and is completely formed by the beginning of the fourth week. It defines the longitudi-

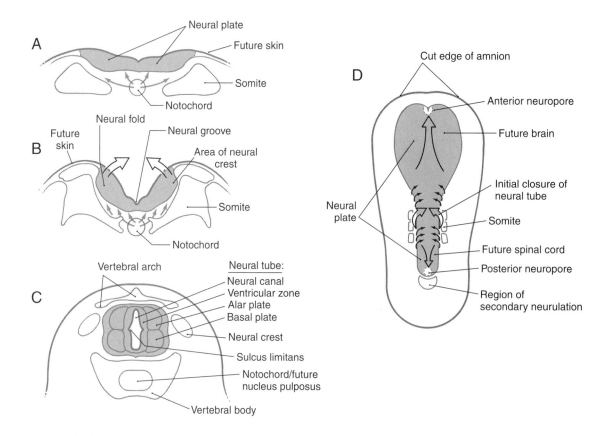

Figure 5–1. Early development of the nervous system. Cross sections (*A–C*) showing the transition from neural plate (*A*) to neural tube (*C*). A dorsal view (*D*) of the neural plate shows the point of initial closure and the direction of closure (*small arrows*) toward anterior and posterior neuropores. Green arrows (*A, B*) represent induction of neural tube formation.

nal axis of the embryo, determines the orientation of the vertebral column, and persists as the *nucleus pulposus* of the intervertebral discs. One important function of the notochord is *induction:* directing the overlying ectoderm to form the neural plate (see Fig. 5–1A and B). Associated with this process is the production of *cell adhesion molecules* in the notochord. These molecules diffuse from the notochord into the neural plate and function to join the primitive neuroepithelial cells into a tight unit.

Within the neuroectoderm, some neuroepithelial cells elongate and become spindle-shaped. This cellular elongation, also induced by the notochord, forms the *neural plate* and is completed by the end of the third week of gestation (see Fig. 5–1A). The neural plate gives rise to most of the nervous system.

Primary Neurulation. The CNS develops from a hollow structure called the *neural tube*, which is produced by *neurulation.* There are two neurulation processes. Most of the neural tube forms from the neural plate by a process of infolding called *primary neurulation.* This part of the neural tube will give rise to the brain and to the spinal cord through lumbar levels. The caudalmost portion of the neural tube, which will give rise to sacral and coccygeal levels of the cord, is formed by a process called *secondary neurulation.* Secondary neurulation is described in the next section. By about day 18 after fertilization, the neural plate begins to thicken at its lateral margins (see Fig. 5–1B). This thickening elevates the edges of the neural plate to form *neural folds.* At about 20 days, the neural folds first contact each other to begin the formation of the *neural tube.* This fusion initially takes place on the dorsal midline at what will become cervical levels of the spinal cord and proceeds, zipper-like, in rostral and caudal directions (see Fig. 5–1C and D). During the process, the lumen of the neural tube, called the neural canal, is open to the amniotic cavity both rostrally and caudally (see Fig. 5–1D). The rostral opening, the *anterior neuropore*, closes at about 24 days, and the caudal opening, the *posterior neuropore*, closes about 2 days later.

Neurulation is brought about by morphologic changes in the *neuroblasts*, the immature and dividing future neurons. As mentioned previously, these cells are elongated and are oriented at right angles to the dorsal surface of the neural plate, which will be the inner wall of the neural canal. Microfilaments in each cell form a circular bundle parallel to the future luminal surface, whereas microtubules extend along the length of the cell. The contraction of the circular bundle of microfilaments causes the microtubules to splay out like the rays of a fan. This forms an elongated conical cell with its apex at the neural groove and its base at the edge of the neural fold. Neurulation does not occur in embryos exposed to colchicine, which depolymerizes microtubules, or to cytochalasin, which inhibits microfilament-based contraction.

Congenital malformations associated with defective neurulation are called *dysraphic defects.* The process of induction also means that the proper development of a structure is dependent on the proper development of its neighbors. There is an intimate relationship of neural tissue to the surrounding bone, meninges, muscles, and skin. Because of this relationship, a failure of neurulation often impairs the formation of these surrounding structures.

Several well-controlled clinical trials have proved that supplementation with the vitamin *folic acid*, found in green, leafy vegetables, can reduce the incidence of neural tube defects. In the MRC Vitamin Study, carried out in Great Britain and published in 1991, women who had previously delivered a child with a dysraphic defect were assigned to either a folic acid supplementation group or a control group during a subsequent pregnancy. Folic acid supplementation reduced the incidence of neural tube defects by about 70% relative to that in untreated controls. The mechanism for this effect is not known at this time. It is thought that women who deliver infants with dysraphic defects have an inborn metabolic problem that is corrected by folate. One research group has suggested that the conversion of homocysteine to methionine, which requires folate as a cofactor, is the critical step. Because it is impossible to identify women at risk, and because the neural plate and tube develop so early in pregnancy, it is important that physicians recommend folic acid supplementation (400 μg/day) to all female patients who intend to have children, whether or not they are pregnant. Additionally, drugs taken for epilepsy, such as valproic acid or carbamazepine, can cause dysraphic defects.

Most dysraphic disorders occur at the location of the anterior or posterior neuropore. Failure of the anterior neuropore to close results in *anencephaly* (Fig. 5–2). In this defect, the brain is not formed, the surrounding meninges and skull may be absent, and there are facial abnormalities. The defect extends from the level of the *lamina terminalis*, the site of anterior neuropore closure, to the region of the *foramen magnum*. Anencephaly occurs in about 5 of every 10,000 live births. Neonatal death is inevitable.

An *encephalocele* is a herniation of intracranial contents through a defect in the cranium (*crania bifidum*) (Fig. 5–3A). The cystic structure may contain only meninges (*meningocele*), meninges plus brain (*meningoencephalocele*), or meninges plus brain and a part of the ventricular system (*meningohydroencephalocele*) (see Fig. 5–3B–D). Encephaloceles are most common in the occipital region, but they may also occur in frontal or parietal locations.

A more subtle defect in the same area is thought to be the cause of the *Arnold-Chiari malformation*, a congenital herniation of the cerebellar vermis through the foramen magnum, which usually causes pressure on the

A   B

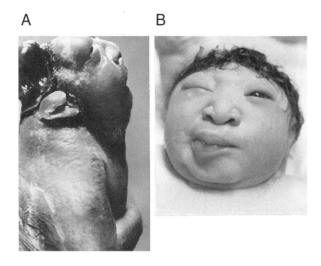

Figure 5–2. Lateral (*A*) and frontal (*B*) views of anencephaly. Note the associated cranial and facial abnormalities. (*A* courtesy of Dr. J. Fratkin.)

medulla oblongata and cervical spinal cord (Fig. 5–4). This defect may go unnoticed until early adulthood and is often associated with a cavitation of the spinal cord (*syringomyelia*) or of the medulla (*syringobulbia*).

Defects in the closure of the posterior neuropore cause a range of malformations known collectively as *myeloschisis*. The defect always involves a failure of the vertebral arches at the affected levels to form completely and fuse to cover the spinal cord (*spina bifida*). If that is the only defect, and the skin is closed over it, the unseen condition is called *spina bifida occulta* (Fig. 5–5*A* and *B*). The site of the defect is usually marked by a patch of dark, coarse hairs. If the skin is not closed over the vertebral defect leaving a patent aperture, the malformation is called *spina bifida aperta*.

As with occipital encephaloceles, a cystic mass (*spina bifida cystica*) may also accompany spina bifida (see Fig. 5–5*C* and *D*). This saccular structure may contain only meninges and cerebrospinal fluid (CSF) (*meningocele*) or meninges and CSF plus spinal neural tissue (*meningomyelocele*). In the latter case, the neural tissue may be the lower part of the spinal cord or, more commonly, a portion of the cauda equina. Infants with meningomyelocele may be unable to move their lower limbs or may

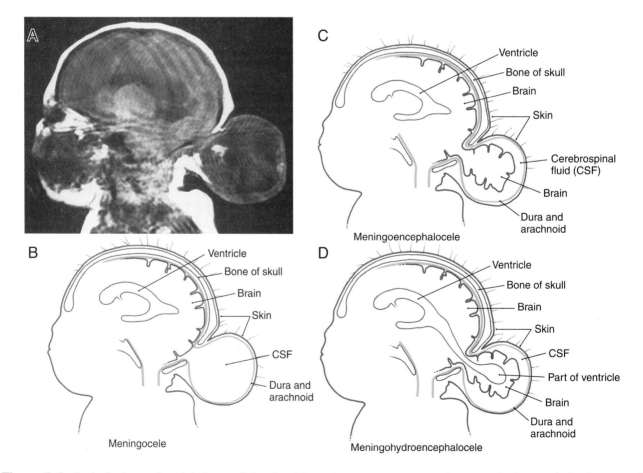

Figure 5–3. Sagittal views of occipital encephaloceles. Magnetic resonance image of meningohydroencephalocele (*A*) and drawings of meningocele (*B*), meningoencephalocele (*C*), and meningohydroencephalocele (*D*).

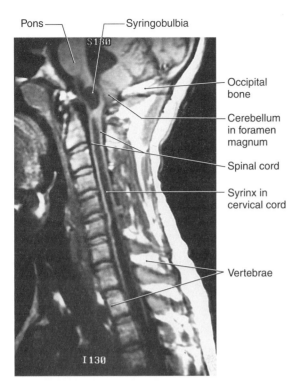

Pons — Syringobulbia

— Occipital bone

— Cerebellum in foramen magnum

— Spinal cord

— Syrinx in cervical cord

— Vertebrae

I 130

Figure 5–4. Sagittal magnetic resonance image of a patient with Arnold-Chiari malformation and with cavitations in the medulla (syringobulbia) and cervical spinal cord (syringomyelia).

not perceive pain sensations from skin innervated by nerves passing through the lesioned area. These infants may also have other CNS malformations, such as *hydrocephalus*. The incidence of meningomyelocele is approximately 5 per 10,000 births.

Secondary Neurulation. The sacral and coccygeal segments of the spinal cord and their corresponding dorsal and ventral roots are formed by *secondary neurulation* (see Fig. 5–1D). This process begins on day 20 and is complete by about day 42. A cell mass, the *caudal eminence*, appears just caudal to the neural tube and then enlarges and cavitates. The caudal eminence joins the neural tube, and its cavity becomes continuous with the neural canal.

Myelodysplasia refers to malformations of the parts of the neural tube formed by secondary neurulation. In most cases the malformation is covered with skin, but the site may be marked by unusual pigmentation, hair growth, *telangiectasias* (large superficial capillaries), or a prominent dimple. A common abnormality is *tethered cord syndrome*, in which the conus medullaris and filum terminale are abnormally fixed to the defective vertebral column. The sustained traction damages the cord, with subsequent loss of sensations from the legs and feet, and problems with bladder control.

Primary Brain Vesicles. During the fourth week after fertilization, in which the anterior neuropore closes,

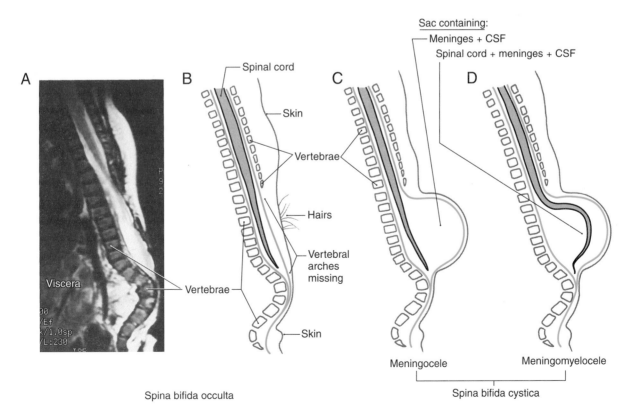

A   Viscera

B   — Spinal cord   — Skin   Vertebrae   — Hairs   — Vertebral arches missing   — Vertebrae   — Skin

Spina bifida occulta

C   Meningocele

D   Sac containing: — Meninges + CSF   Spinal cord + meninges + CSF   Meningomyelocele

Spina bifida cystica

Figure 5–5. Sagittal views of spina bifida malformations. Magnetic resonance image (*A*) and corresponding views showing spina bifida occulta (*A, B*) and spina bifida cystica (*C*, meningocele; *D*, meningomyelocele). CSF, cerebrospinal fluid.

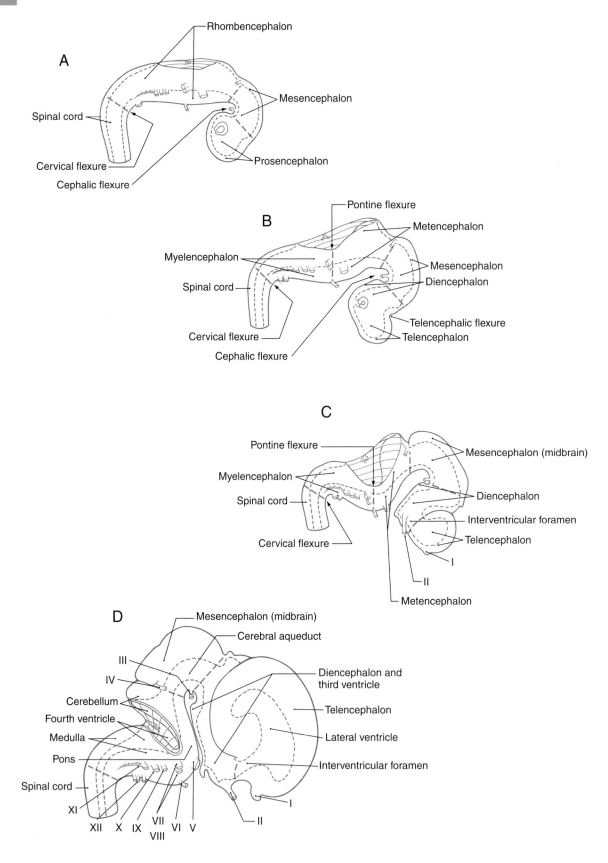

Figure 5–6. Developmental sequence from three primary to five secondary brain vesicles. Three brain vesicles (*A*, at about 4.75 weeks of gestation) divide into five vesicles (*B*, at about 6 weeks of gestation) with the appearance of additional flexures. In subsequent stages (*C*, at about 6.5 weeks of gestation; *D*, at about 8.5 weeks of gestation), there is rapid enlargement of forebrain regions, especially the telencephalon. Note that the ventricular spaces (dashed lines, *A–D*) follow the shape changes in the brain. Cranial nerves are indicated by Roman numerals.

there is rapid growth of neural tissue in the cranial region. The three *primary brain vesicles* formed are *prosencephalon* (forebrain), *mesencephalon* (midbrain), and *rhombencephalon* (hindbrain) (Fig. 5–6*A* and *B*). At the rhombencephalon–spinal cord junction there is a slight bend in the developing neural tube; this is the *cervical flexure*. A second bend in the neural tube at the level of the mesencephalon is the *cephalic* (or *mesencephalic*) *flexure*.

Secondary Brain Vesicles. During the fifth week, the three primary brain vesicles are divided into five *secondary brain vesicles* (see Fig. 5–6*C* and *D*). This requires two additional flexures. The *pontine flexure* divides the hindbrain into the *myelencephalon* caudally and the *metencephalon* rostrally. The mesencephalon does not partition further. The *telencephalic flexure* divides the forebrain into the *diencephalon* caudally and the *telencephalon* rostrally (see Fig. 5–6*C* and *D*). The telencephalon (meaning "end-brain" or "ultimate brain") forms as an outpocketing of the forebrain and expands enormously, with its complex lobes, gyri, and sulci, to become the largest part of the brain.

Diencephalon and Cerebral Hemispheres. The main structures of the forebrain develop during the second month of gestation. Because the mesoderm in this region is simultaneously forming facial structures, abnormalities of forebrain development are often associated with facial defects (see Fig. 5–2). The process of forebrain development is referred to as *central induction*.

At about the end of the fifth week, the telencephalon gives rise to two lateral expansions called the *telencephalic (cerebral) vesicles* (see Fig. 5–6*C* and *D*). These are the primordia of the cerebral hemispheres. Their adult derivatives include the cerebral cortex and the subcortical white matter (including the internal capsule), the olfactory bulb and tract, the basal ganglia, the amygdala, and hippocampus. The *diencephalon* develops into the thalamic nuclei and associated structures and also gives rise to the optic cup, which eventually forms the optic nerve and retina. By 10 weeks of development, the major structures of the CNS are clearly recognizable by their morphologic features, and immature versions of all structures in the brain are present by the end of the first trimester.

The sequence of events by which the primitive prosencephalon differentiates into the diencephalic and telencephalic vesicles is called *prosencephalization*. Failure of the prosencephalon to undergo cleavage results in a malformation called *holoprosencephaly* (Fig. 5–7*A*). In its most severe form (*alobar holoprosencephaly*), no discernible lobes develop. There is a large single forebrain ventricle, the thalamus is poorly developed, and many structures (corpus callosum, longitudinal cerebral fissure and falx cerebri, olfactory structures) are lacking. In *semilobar holoprosencephaly* (see Fig. 5–7*B* and *C*), there is some separation of the forebrain into two discernible lobes (more prominent in occipital areas) and partial development of the falx cerebri. The hemispheres have some visible lobes and gyri, and there are rudimentary but enlarged lateral and third ventricles. Most infants with holoprosencephaly also have facial malformations. These may be as subtle as mild *hypotelorism* (unusually close-set eyes) or as obvious as the presence of only a single, midline eye (*cyclops*) accompanied by a rudimentary nasal structure (*proboscis*). In general, the more severe the brain malformation, the more severe the facial defect.

Infectious Diseases Causing Congenital Nervous System Defects. Exposure to several common disease organisms can produce congenital neural defects through mechanisms that are not well understood. The etiologic agents range from viruses (*rubella* virus and

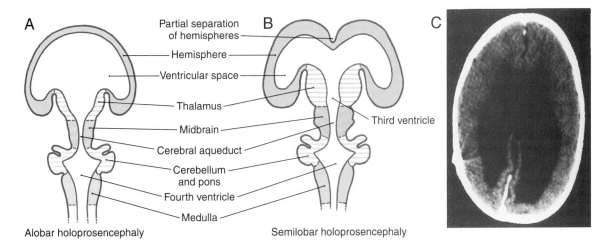

Figure 5–7. Alobar (*A*) and semilobar (*B, C*) holoprosencephaly. In the alobar form (*A*), the single brain vesicle has a horseshoe-shaped ventricle, and many major brain structures are absent. Although ventricles are present in the semilobar form (*B* and *C*, an axial magnetic resonance image), they are enlarged, and many major structures are only partially developed.

cytomegalovirus are the best-known examples), to bacteria (such as the spirochete causing *syphilis*), to protozoans (such as *Toxoplasmas* commonly found in cat feces and garden dirt). Although toxoplasmosis is fairly common, occuring in an estimated 80% of domestic cats, simple steps to reduce *direct* contact with cat feces (such as wearing gloves and handwashing) can greatly reduce the risk, which is highest during the first trimester of pregnancy. Currently, some effects of toxoplasmosis (which can range from mild, such as rashes, to extremely severe, such as hydrocephalus and profound retardation) occur in 0.3% of live births, or 1200 babies out of the approximately 4 million born each year in the United States. It should be emphasized, however, that the benefits of cat ownership outweigh the risks if a few simple precautions are taken to avoid contact with cat feces.

Ventricular System. The ventricular system is an elaboration of the lumen of cephalic portions of the neural tube, and its development parallels that of the brain (Fig. 5–8*A–D*; see also Fig. 5–6*A–D*). This process, also discussed in Chapter 6, is summarized here. The cavities of the telencephalic vesicles become the *lateral ventricles;* the diencephalic cavity becomes the *third ventricle;* and the rhombencephalic cavity becomes the *fourth ventricle.* The cavity of the mesencephalon becomes the narrow *cerebral aqueduct (of Sylvius)* connecting the third and fourth ventricles, and the openings between the lateral ventricles and the third ventricle become the *intraventricular foramina (of Monro).*

The ventricular system is lined with *ependymal cells.* Each ventricle originally has a thin roof composed of an internal layer of ependyma and an outer layer of delicate connective tissue (*pia mater*). In each ventricle, blood vessels invaginate this membrane to form the *choroid plexus.*

Openings that arise in the caudal roof of the fourth ventricle during development form a communication between the ventricular system and the subarachnoid space. These are the midline *medial aperture (foramen of Magendie)* and the paired *lateral foramina of Luschka.* Although these foramina develop slowly, they are patent by the end of the first trimester. Cerebrospinal fluid (CSF) is produced mainly by the choroid plexuses of the lateral and third ventricles. It escapes the ventricular system through foramina of the fourth ventricle and passes into the subarachnoid space. From there, it is absorbed into the venous system.

If the flow of CSF through the ventricles is ob-

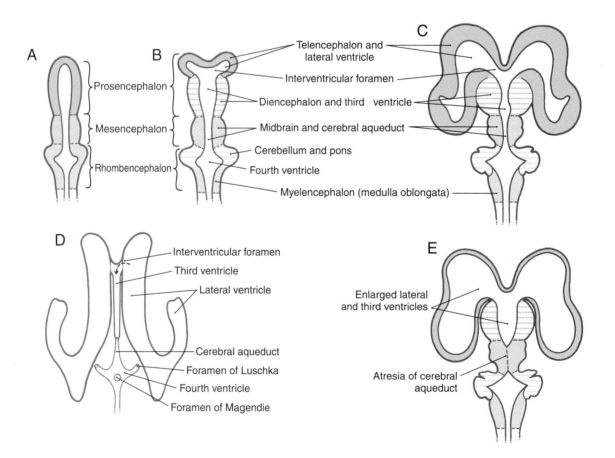

Figure 5–8. Development of the ventricular system and associated brain divisions (*A–C*) and the general adult pattern (*D*) as seen from the dorsal perspective. Failure of the cerebral aqueduct to form causes the third and lateral ventricles to enlarge (*F*).

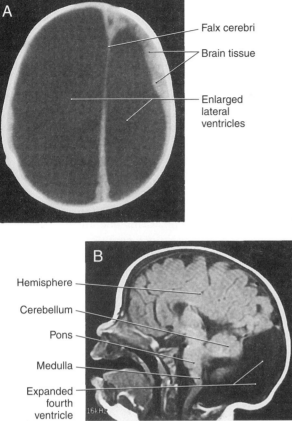

Falx cerebri

Brain tissue

Enlarged lateral ventricles

Hemisphere

Cerebellum

Pons

Medulla

Expanded fourth ventricle

Figure 5–9. Axial magnetic resonance image (*A*) from a 9-month-old boy with congenital hydrocephalus. The brain is compressed and is visible as a thin rim on the inner surface of the skull and on the falx cerebri. Sagittal magnetic resonance image (*B*) from a 5-month-old boy with a Dandy-Walker malformation. In addition to aplasia of the cerebellum and a pronounced enlargement of the fourth ventricle, this patient has almost complete agenesis of the corpus callosum.

Peripheral Nervous System

Neural Crest. The *peripheral nervous system* develops mostly from cells of the *neural crest* (Table 5–1) (see Fig. 5–1). These cells, which arise from the lateral edge of the neural plate, detach and move to locations lateral to the neural tube. The neural crest gives rise to most of the peripheral nervous system, as well as to a number of other structures (see Table 5–1).

Placodes. Specialized epidermal cells called *placodes* are found in the developing head region. These will join neural crest cells, and together placodes and neural crest form the ganglia of cranial nerves V, VII, VIII, IX, and X.

Cranial Nerve Ganglia. Cranial nerves V (trigeminal nerve), VII (facial nerve), IX (glossopharyngeal nerve), and X (vagus nerve) have sensory ganglia that originate from neural crest and placode cells and contain pseudounipolar cell bodies. These are the *trigeminal* or *semilunar* (V) ganglion, the *geniculate* ganglion (VII), the *superior* and *inferior* (IX) ganglia of the glossopharyngeal nerve, and the *jugular* and *nodose* (X) ganglia. The distal processes of these cranial nerves travel as the sensory components of the corresponding cranial nerve, whereas the proximal processes innervate the appropriate cranial nerve nuclei in the brainstem. A notable exception to this pattern is the *mesencephalic nucleus of the trigeminal nerve*. In this cranial nerve nucleus, pseudounipolar cells have failed to migrate with the neural crest and remain inside the central nervous system. They form, in essence, a "ganglion" ectopically trapped inside the mesencephalon.

The cell bodies of the ganglia of cranial nerve VIII (the *vestibulocochlear nerve*) arise primarily from the *otic*

structed during prenatal development, the ventricular system can become markedly dilated, a condition called *congenital hydrocephalus* (Fig. 5–9*A*). The cerebral aqueduct, only 0.5 mm in diameter, is a likely site for such a blockage. Congenital *atresia* (failure to form) of the aqueduct can occur as an isolated event, be inherited, or be associated with CNS deformities (see Fig. 5–8*E*). *Stenosis*, or total obstruction, from cellular debris associated with an infection or from an *intraventricular hemorrhage*, may also occlude this narrow passage.

Enlarged ventricles are also seen in the *Dandy-Walker malformation*. Affected persons have a cystic dilation of the fourth ventricle accompanied by a variable degree of *aplasia* (absence or defective development) of the *cerebellar vermis* (see Fig. 5–9*B*). In some cases, there is also obstruction of the foramina of the fourth ventricle.

Table 5–1. Principal Structures Derived from Neural Crest Cells

Neural Elements

Neurons of:
 Dorsal root ganglia
 Paravertebral (sympathetic chain) ganglia
 Prevertebral (preaortic) ganglia
 Enteric ganglia
 Parasympathetic ganglia of cranial nerves VII, IX, and X
 Sensory ganglia of cranial nerves V, VII, VIII, IX, and X*

Non-neural Elements

Schwann cells
Melanocytes
Odontoblasts
Satellite cells of peripheral ganglia
Cartilage of the pharyngeal arches
Ciliary and pupillary muscles
Chromaffin cells of the adrenal medulla
Pia and arachnoid of the meninges

* Some of the sensory cells in these ganglia arise from placodes.

placode, with a small contribution from neural crest. These ganglion cells retain a bipolar shape in the adult.

Posterior (Dorsal) Root Ganglia. Pseudounipolar cells of *posterior* or *dorsal root ganglia* are derived from neural crest. Each spinal nerve and its corresponding ganglion are associated with a segment (or *somite*) of the developing embryo (Fig. 5–10). As the somites grow out to form portions of the body's connective tissue and musculature, the peripheral processes of the developing pseudounipolar cells of the corresponding dorsal root ganglia grow distally, using the extracellular matrix of the underlying tissue as a guide (see Fig. 5–10).

The matrix molecules *fibronectin* and *laminin* contain the amino acid sequence arginine-glycine-aspartate (called the *RGD sequence* after the one-letter abbreviations for these amino acids). This sequence is recognized by proteins known as *integrins* on the surface of neural crest cells. The selective adhesion of the peripheral process of a neural crest cell to the RGD sequence of the extracellular matrix is probably involved in guiding the distal processes to their correct targets.

The segmental nature of the embryo is reflected in the segmental sensory innervation of the body surface (see Fig. 5–10; see also Fig. 18–4). These segments, known as *dermatomes* (Latin: "skin slices"), are important in the diagnosis of many neurologic disorders.

Visceral Motor System. The postganglionic sympathetic and parasympathetic neurons of the visceral motor system are also derived from the neural crests. Some of these cells remain near their site of origin to form the *sympathetic chain ganglia* adjacent to the vertebral column. Other cells migrate with branches of the aorta to form the sympathetic *prevertebral ganglia*.

Most of the autonomic (visceromotor) neurons of the digestive tract (*Auerbach* and *Meissner* plexuses) are formed by neural crest cells that migrate from the area of the rhombencephalon. Consequently these cells receive vagal innervation in the adult. Visceromotor (autonomic) neurons of the descending colon and pelvic structures are derived from neural crest cells that arise from sacral cord levels during secondary neurulation.

The human syndrome *congenital megacolon* (*Hirschsprung disease*) closely resembles an animal model in which excess extracellular matrix molecules are present in the colon. This results in aberrant migration of neural crest–derived cells. In this animal model, and in

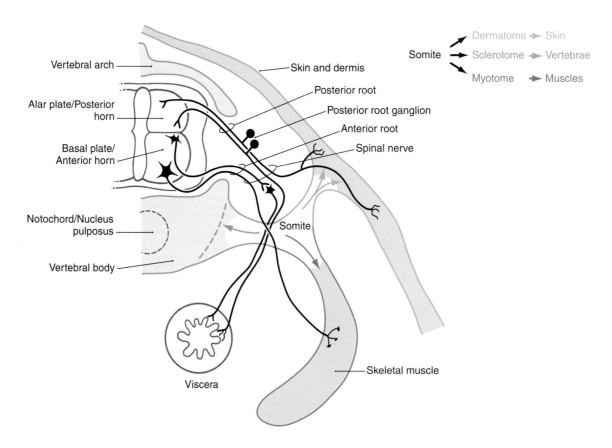

Figure 5–10. Derivatives of a somite and the corresponding innervation of structures that originate from the dermatome and myotome.

Hirschsprung disease, neurons forming the enteric ganglia fail to migrate into the lower bowel. In the absence of these cells, no sensory signal indicating the presence of feces in the colon is sent to the CNS. Therefore, no motor signal is sent to control defecation. Another clinical entity, *familial dysautonomia*, also reflects aberration in the development of neural crest derivatives. Patients with this disorder have both sensory symptoms (impaired pain and temperature perception) and autonomic symptoms (cardiovascular instability, gastrointestinal dysfunction).

Schwann Cells. The Schwann cells, which ensheathe and myelinate axons in the peripheral nervous system, are also derived from neural crest cells (see Chapter 2). Schwann cells migrate in a segmental fashion, accompanying the growing processes of the peripheral nerve fibers they will eventually ensheathe.

Central Nervous System

Basic Features. In general, CNS neuroblasts arise at the ventricular surface of the developing brain (i.e., the luminal surface of the neural tube). Just after the neural tube forms, there is no apparent cell differentiation in a cross section at any level. At this time, the neural tube is a pseudostratified columnar epithelium. As development proceeds and the wall of the neural tube thickens, however, dividing cells cluster at the ventricular surface, leaving a zone without cell bodies at the abluminal surface. This region, with few cell nuclei, is called the *marginal zone.*

As cells undergo their last division, they begin to migrate away from the luminal (ventricular) surface on transient glial cell guides called *radial glia.* As they migrate, they form a moving front of cell bodies between the marginal and ventricular zones called the *intermediate zone.* These features are common to all parts of the developing neuraxis; other elaborations are possible and are outlined in the following sections.

After cells migrate and take up their final positions in the developing brain, they begin to extend processes and form connections with other neurons or muscle cells. Dendritic processes begin to receive information from other developing cells. Meanwhile, an axonal process, tipped by a spadelike extension called the *growth cone*, begins to drive its way through intervening regions to reach distant targets.

Although the idea is controversial, both neurons and glia seem to originate from a single precursor cell population. Two main lineages arise: a *neuroblastic* lineage that generate neurons and a *glioblastic* lineage that includes precursors of radial glial cells, astroglial cells, and oligodendrocytes. The glioblastic lineage is believed to split into three main branches: (1) the *type 1 astrocyte* progenitor, (2) the *oligodendrocyte/type 2 astro-*cyte precursor (called the *O2A progenitor*), and (3) the *radial glia* progenitor. Whereas all other cell types persist into adulthood, radial glial cells in most regions of the brain appear to be converted to astrocytes, *ependymal cells*, or *tanycytes*. There are two notable exceptions. In the cerebellum, radial glial cells retain most of their features as *Bergmann glial cells*, and in the retina, they are seen as *Müller cells*. Differentiation of glial cells is influenced by a variety of *growth factors*, such as platelet-derived growth factor, ciliary neurotrophic factor, and fibroblast growth factor, which are secreted by neighboring glia and neurons.

Spinal Cord. The adult spinal cord gray matter is butterfly-shaped and consists of anterior (ventral) and posterior (dorsal) horns. At some levels of the spinal cord, an *intermediate zone* and a *lateral horn* of the gray matter lie halfway between dorsal and ventral horns.

The spinal cord develops from caudal portions of the neural tube (see Fig. 5–1). The neural canal in this region will become the central canal of the spinal cord (see Fig. 5–6). Neuroblasts that give rise to spinal cord neurons are produced between the fourth and twentieth weeks of development by a burst of proliferation in the ventricular layer lining the neural canal. These cells migrate peripherally to form four longitudinal *plates*, which will become the gray matter of the spinal cord: a pair of anteriorly located cell masses, which constitute the *basal plate*, and a pair of posteriorly located masses, which constitute the *alar plate*. The basal and alar plates on each side are separated by a longitudinal groove called the *sulcus limitans* in the lateral wall of the central canal. The *basal plate* develops into the *anterior (ventral) horn* of the spinal cord and the *alar plate* will become the *posterior (dorsal) horn* of the spinal cord (see Fig. 5–10). Development in the basal plate somewhat precedes that in the alar plate; postmitotic neurons are clearly evident in the basal plate during week 20 of development. That portion of the adult spinal cord commonly called the intermediate zone (and the lateral horn) originate from the interface of the alar and basal plates.

As the basal plate develops, axons of nascent motor neurons form the developing anterior (ventral) roots that will innervate peripheral structures. Anterior horn motor neurons innervate skeletal muscle and are classified as *general somatic efferent (GSE)*. The lateral horn motor neurons project to autonomic (visceromotor) ganglia and are classified as *general visceral efferent (GVE)*. The categories GSE, GVE, and so on are referred to as *functional components*. The cord regions devoted to the GSE and GVE functional components can be thought of as constituting distinct longitudinal *cell columns* in the gray matter (Fig. 5–11). The GSE column runs the full length of the spinal cord. The GVE column extends from T1 through L2, where it is called the *intermediolateral cell column*, and from S2

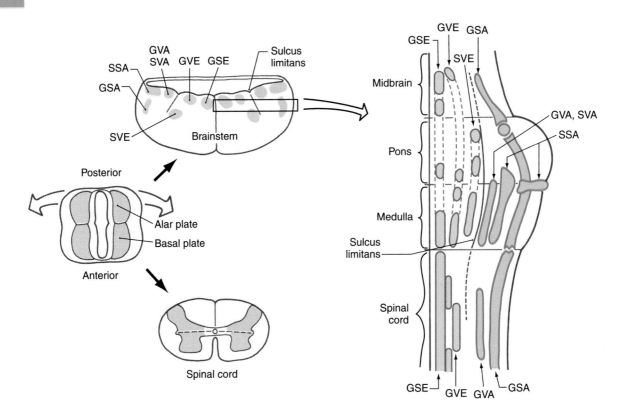

Figure 5–11. The derivatives of the alar (in blue) and basal (in red) plates in the spinal cord and brainstem. Nuclei are grouped into rostrocaudally oriented cell columns that correspond to their functional components. GSA, general sometic afferent; GSE, general somatic efferent; GVA, general visceral afferent; GVE, general visceral efferent; SSA, special somatic afferent; SVA, special visceral afferent; SVE, special visceral efferent.

through S4, where it is called the *sacral visceromotor nucleus.*

Neurons of the alar plate receive the central processes of developing dorsal root ganglion (sensory) cells. Sensory neurons whose peripheral processes innervate the skin and receptors in joint capsules, tendons, and muscles are classified as *general somatic afferent (GSA).* Those that innervate receptors in visceral structures, such as the stomach, are classified as *general visceral afferent (GVA).* Like the GSE and GVE regions, the GSA and GVA functional components constitute separate columns (see Fig. 5–11).

As each somite develops, it subdivides into a *sclerotome,* which forms vertebrae; a *dermatome,* which forms skin and dermis; and a *myotome,* which forms muscles (see Fig. 5–10). The derivatives of the dermatomes and myotomes are innervated, respectively, by the axons of the dorsal (sensory) and ventral (motor) roots of the corresponding spinal cord levels. These roots join at about the level of the future *intervertebral foramina* to form the *spinal nerves* (see Fig. 5–10). The spinal nerves thus show the same segmental pattern as that for the dermatomes and myotomes they innervate. By contrast, the vertebrae develop between the spinal nerves and are thus *intersegmental* in position, even

though they originate from the segmental sclerotomes. This situation comes about because the sclerotomes each split into cranial and caudal halves, and the vertebral rudiments are formed by the union of the caudal half of one sclerotome with the cranial half of the next posterior sclerotome.

Relationship of Spinal Cord to Vertebral Column. Although the spinal cord retains its general shape from the third trimester into adulthood, its physical relationship to the vertebral column alters dramatically (Fig. 5–12). By the end of the first trimester, the spinal cord, its meningeal coverings, and the surrounding vertebral arches are fully formed. The spinal nerves exit at about right angles to the spinal cord and pass through the intervertebral foramina. As development proceeds, the vertebral column grows faster than the spinal cord. The net result is that the cord seems to be drawn rostrally by its attachment to the brain. The intervertebral foramina, containing the spinal nerves, move caudally; and the dorsal and ventral roots from lumbar, sacral, and coccygeal levels are significantly lengthened to form a bundle called the *cauda equina* (see Fig. 5–12).

Brainstem. The brainstem consists of the *myelencephalon* (medulla oblongata), the *pons* (a part of the meten-

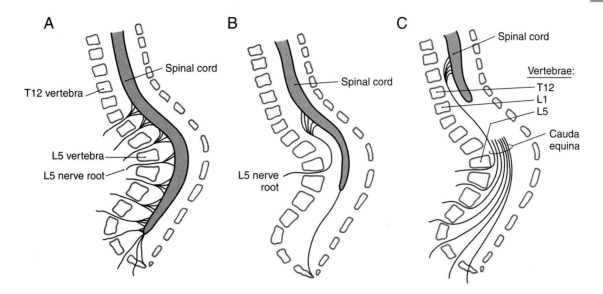

Figure 5–12. The differential growth of the vertebral column and spinal cord forms the cauda equina. The relationship of the cord to the vertebral column is shown diagrammatically at about 12 weeks of gestation (*A*), at 16 to 18 weeks of gestation (*B*), and at about 1 year of age (*C*).

cephalon), and the *mesencephalon* (midbrain). Although developmentally the cerebellum is a part of the meten-cephalon, it is considered a "suprasegmental" structure and not a part of the brainstem.

As one travels from the rostral part of the spinal cord to the candal part of the brainstem (*medulla oblongata*), two features are notable. First, the appearance of the cerebellum and the flaring open of the central canal into the fourth ventricle force the dorsal portion of the neural tube (*alar plate*) to rotate dorsolaterally (see Fig. 5–11). This rotation results in a lateral-to-medial orientation of sensory (*alar plate*) versus motor (*basal plate*) areas of the developing brainstem, in contrast to their dorsoventral relationship in the spinal cord. Second, the *sulcus limitans*, which disappears in the spinal cord during development, is retained as an important landmark in the floor of the fourth ventricle (see Fig. 5–11).

The basal plate in the brainstem give rise to motor cranial nerve nuclei, and the alar plate gives rise to sensory cranial nerve nuclei. As in the spinal cord, these portions of the basal and alar plates differentiate into rostrocaudally oriented cell columns, each of which is associated with a specific functional component. As shown in Figure 5–11, however, there are six cell columns and seven corresponding functional components in the brainstem, as compared with four in the spinal cord. This difference occurs because "special" functional components are unique to the head—special visceral efferent (SVE), special visceral afferent (SVA), and special somatic afferent (SSA). As development proceeds, some of these columns fragment into distinct, separate nuclei. The nuclei derived from a given cell column have the same functional component, and

they generally remain aligned along the same rostrocaudal axis but may lie in different parts of the brainstem (see Fig. 5–11).

Sensory neurons in the brainstem originate from the alar plate. They give rise to four sensory cranial nerve nuclei (see Fig. 5–11; see also Fig. 10–7). The *spinal trigeminal nucleus* forms a continuous cell column from the cord-medulla junction to midpontine levels, whereas the *principal sensory trigeminal nucleus* is found in the rostral pons anterior to the spinal nucleus. Both of these nuclei receive GSA input via cranial nerves V, VII, IX, and X. The *solitary nucleus* extends the entire length of the medulla and receives GVA and taste (SVA) input via cranial nerves VII, IX, and X. Although the *vestibular* and *cochlear nuclei* also arise from the alar plate, the peripheral fibers projecting to these nuclei originate primarily from the *atic placode*. These fibers are classified as SSA. The alar plate also gives rise to other brainstem cell groups, such as the inferior olivary nucleus of the medulla, the basilar pontine nuclei, and the substantia nigra of the midbrain.

Motor neurons in the brainstem originate from the *basal plate*. In contrast to cranial nerve nuclei derived from the alar plate, most of which form continuous cell columns, those cell groups that arise from the basal plate form separate nuclei (see Fig. 5–11) stacked into discontinuous columns. There are three such broken columns: GSE, GVE, and SVE, taken from medial to lateral.

The nuclei of the most medial column have a GSE functional component and innervate muscles that originate from occipital somites (the tongue) or from mesoderm in the vicinity of the optic cup (the eye muscles). These are the *hypoglossal nucleus* (XII, in the medulla),

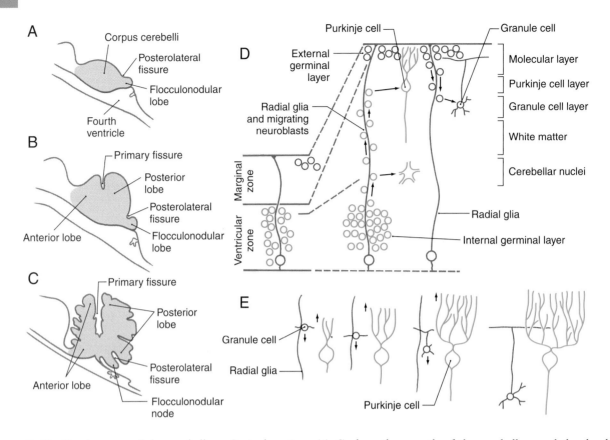

Figure 5–13. Development of the cerebellum. Sagittal sections (*A–C*) show the growth of the cerebellum and the development of the main lobes and fissures. Cytodifferentiation of the cerebellar cortex and nuclei (*D*). Note the migration of neuroblasts on radial glia and the formation of connections between granule cell axons and Purkinje cell dendrites (*E*). As granule cells migrate inward, they trail an axon that forms contacts with the dendrites growing outward from Purkinje cells. These axons become the parallel fibers of the cerebellar cortex.

the *abducens nucleus* (VI, pons), and the *oculomotor* and *trochlear nuclei* (III, IV) of the midbrain.

The next lateral column of nuclei have a GVE functional component and innervate visceral motor ganglia, which, in turn, innervate visceral structures. These nuclei are the *dorsal motor vagal nucleus* (X) and the *inferior salivatory nucleus* (IX) of the medulla, the *superior salivatory nucleus* (VII) of the pons, and the *Edinger-Westphal nucleus* (III) of the midbrain. These nuclei project via the vagal (X), glossopharyngeal (IX), facial (VII), and oculomotor (III) nerves, respectively.

The most lateral and ventral column of nuclei originating from the basal plate innervate muscles that arise from the mesoderm of the *pharyngeal arches*, and, consequently, they are designated as SVE. These cell groups are the *ambiguus nucleus* (IX, X) of the medulla and the *facial nucleus* (VIII) and *trigeminal motor nucleus* (V) of the pons.

Along with the posterolateral-anteromedial division of brainstem into alar and basal plates, there is a rostral-caudal segmentation of the developing rhombencephalon into *rhombomeres*. These are clusters of neuroblasts separated from each other by thin, transversely

oriented bands of neuroepithelial cells. Cells in one rhombomere give rise to a specific motor nucleus (or nuclei) but will not migrate into adjacent rhombomeres. In general, the cell clusters forming the rhombomeres represent the rostral continuation of the alar and basal plates of the developing spinal cord. Indeed, motor nuclei originate from specific rhombomeres, and input from the corresponding sensory ganglia enters the corresponding rhombomere.

Rhombomeres are also sites of *homeobox gene* expression. These genes (abbreviated *Hox*) are "master switches" that control the formation of large blocks of tissue. For example, the gene *Hox 2.1* is expressed only in the rhombomeres that give rise to cranial nerves X and XII; *Hox 2.9* is expressed only in the region of the developing facial nerve. There is considerable sequence similarity between homeobox genes from widely divergent species (such as flies and humans). This similarity implies that expression of homeobox genes is an essential element of neural development that has been preserved in evolution.

Cerebellum. The cerebellum arises from the *rhombic lip*, an alar plate structure that forms part of the wall of the fourth ventricle. The rostral part of the rhombic lip forms the cerebellum, whereas the caudal part gives rise to the *inferior olivary, cochlear,* and *pontine* nuclei.

The rhombic lips join dorsal to the developing fourth ventricle to form the *cerebellar plate.* During the histogenesis of the cerebellar cortex, fissures appear that divide the cerebellum into its main lobes (Fig. 5–13*A–C*). The first, the *posterolateral fissure,* divides the cerebellar plate into the *flocculonodular lobe* and the *corpus cerebelli.* Despite its name, the *primary fissure* is the second to appear, and it divides the corpus cerebelli into *anterior* and *posterior lobes.* The advent of additional fissures divides the anterior and posterior lobes into the lobules characteristic of the adult brain.

The process of cerebellar cortical development involves the migration of neuroblasts to form the cells characteristic of the adult brain (see Fig. 5–13*D*). Initially the cerebellar primordium is composed of the *ventricular zone,* the *intermediate zone,* and the *marginal zone.* By the end of the first trimester, a second layer of neuroblasts has appeared in the outer part of the marginal layer. This is called the *external germinal* (or *granular*) *layer,* and the intermediate zone is now called the *internal germinal* (or *granular*) *layer.*

Radial glial cells extend from the ventricular zone to the surface of the marginal layer and are necessary for the proper migration of developing neurons (see Fig. 5–13*D*). Neuroblasts of the internal germinal layer migrate outward along the radial glia to form the *cerebellar nuclei* and the *Purkinje cells* and *Golgi cells* of the cerebellar cortex. Neuroblasts of the external germinal layer migrate inward along the radial glia to form the *granule cells.* Other cells of the external germinal layer congregate just external to the Purkinje cell layer, where they will differentiate into the *stellate cells* and *basket cells* of the *molecular layer.*

Developing cerebellar neurons participate in other important cell interactions in addition to those with the radial glial cells. For example, as the granule cells migrate inward, they sprout axons that form synaptic contacts with the Purkinje cell dendrites that are growing into the molecular layer (see Fig. 5–13*E*). Continued growth and development of Purkinje cell dendrites depend on these contacts with granule cell axons. Purkinje cell dendrites are stunted in the mutant mouse *weaver,* in which the granule cells die during development. These and other experimental mutant animals with cerebellar defects show the characteristic manifestations of cerebellar disease in humans: *ataxia, hypotonia,* and *tremor.*

Thalamus. The gray matter of the diencephalon develops from a continuation of the brainstem alar plates; there is no homolog of the basal plate in the diencephalon. This alar plate is recognizable at 4 weeks of gestation, and by 6 weeks of gestation it has differentiated into three main areas of the diencephalon: the *epithalamus, thalamus (dorsal thalamus),* and *hypothalamus*

(Fig. 5–14*A* and *B*). These structures are visible as swellings in the wall of the third ventricle, where they are separated from each other by the *epithalamic* and *hypothalamic sulci.* As development progresses, the epithalamic area remains quite small, whereas the hypothalamus and especially the thalamus enlarge. In about 80% of individuals, the two thalami fuse across the third ventricle to form the *interthalamic adhesion (massa intermedia).*

The basic concepts of the development of the thalamus are the same as for other CNS regions. *Radial glia* extend from the third ventricle to the pial surface, and developing neurons migrate along this guide (see Fig. 5–14*B* and *C*). The development of the thalamus occurs in an *"outside-first"* sequence. That is, the first neurons to undergo their final cell division migrate to the outermost portion of the thalamus, where they mature. This means that the most lateral of the thalamic nuclei, such as the geniculate nuclei and the lateral and ventral nuclei, are generated first. The most medial thalamic nuclei, such as the dorsomedial nucleus, are the last to develop.

An important process in the development of the thalamic relay nuclei is the establishment of orderly maps of the sensory world. For example, a retinotopic (vision) map is formed in the lateral geniculate nucleus. As retinal ganglion cells send axons to the lateral geniculate nucleus, the arrangement of axonal contracts on cells in this visual relay center must accurately reflect the positions of ganglion cells in the retina. In this way, the map of visual space on the retina is maintained in the lateral geniculate nucleus and, ultimately, in the visual cortex. Similar maps are formed in the medial geniculate nucleus (tonotopic mapping) and ventral posterolateral nucleus (somatotopic mapping).

Cerebral Cortex. The cerebral cortex is generated using the basic mechanisms described previously for other regions. Except during mitosis and cytokinesis, cortical neuroblasts retain connections to both the ventricular and the pial surfaces of the developing brain and thus have a fusiform shape (Fig. 5–15). The nucleus engages in a peculiar cycle of migration within the cell, however. During the G_1 phase of the cell cycle (before DNA replication), the nucleus travels from near the ventricular pole of the cell to near the pial pole. During the G_2 phase (after DNA replication), it reverses direction and migrates back to a ventricular position. At the start of mitosis, the cell loses contact with the pial surface, but after cytokinesis, the daughter cells grow processes that reconnect with the pial surface (see Fig. 5–15). The cycle is then ready to repeat.

Unlike in other regions of the brain, in the cerebral cortex, the first cells to migrate will disembark from the radial glial cells and take up positions close to the ventricular surface. Successive "waves" of neuroblasts, migrating along radial glia, force their way through the

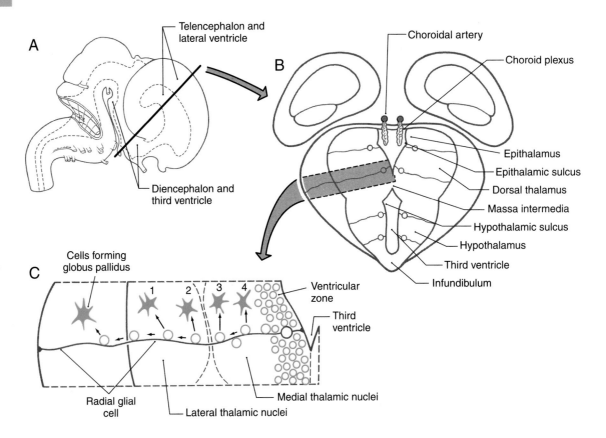

Figure 5-14. Development of the diencephalon. The diencephalon is shown in a lateral view (*A*) and in a coronal view (*B*, plane from *A*) at about 8 weeks of gestation. Details of cell migration (*C*, detail from *B*) show that lateral cell groups (*C*, 1) are formed first, while more medial cell groups (*C*, 4) are formed last.

differentiated cell layers to take up positions progressively closer to the pial surface. This sequence is called an *"inside-out" pattern* of development. These ranks of cells form the *cortical plate* as the axons of previously settled cells grow toward their targets.

The cerebral cortex proper is formed from expansion of the superficial part of the intermediate zone, the *subplate* and *cortical plate* (Fig. 5-16). Developing axons, originating in regions such as the thalamus that innervate the cortex, send out and form transient synaptic contacts in the subplate. The subplate is a transient structure that does not persist into adulthood. Neuronal cell bodies vacate the area between the subplate and the ventricular surface; most of the remaining cell bodies are glial cells. This region forms the *white matter* of the adult nervous system. The ventricular zone is reduced to a single layer of *ependymal cells* that line the lateral ventricles in the adult.

During peak periods of cellular migration, the hemispheric fissures appear and mold the telencephalic surface into the gyri and sulci characteristic of the adult brain. By the end of the first trimester, the *longitudinal cerebral, Sylvian,* and *transverse cerebral fissures* are recognizable. The secondary sulci are completed by 32 weeks of development, with tertiary sulci completed during the last month of gestation.

Gross Abnormalities of Cortical Development. Abnormal patterns of gyri and sulci are caused by disorders of cell migration in the developing cerebral cortex (Fig. 5-17*A*). If gyri fail to form, the cerebral cortex will have a smooth surface, a condition called *lissencephaly*. Unusually large gyri constitute *pachygyria*, and unusually small gyri constitute *microgyria*. Any of these conditions may affect the whole cerebrum or may be localized, and they may coexist in the same patient (see Fig. 5-17*A*).

Abnormal patterns of sulcal and gyral development are seen in *schizencephaly*, a condition in which there are unilateral or bilateral clefts in the cerebral hemispheres (see Fig. 5-17*B*) of almost any size. Small defects may consist of a thin spot in the hemisphere where the pia and ependyma come abnormally close. In severe cases, the defect is large, resulting in a substantial loss of brain tissue and producing an open channel between the ventricular cavity and the subarachnoid space. Schizencephaly may result from a profound failure of cell migration. An alternative explanation, which may apply particularly to severe cases, is that the affected region did not receive an adequate

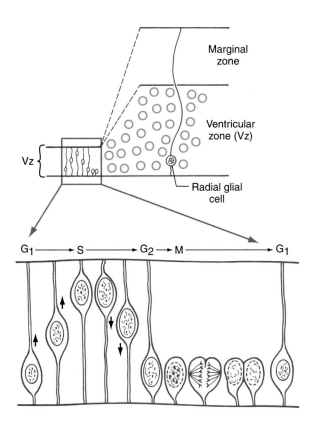

Figure 5–15. Early development of the cerebral cortex. The nuclei of the neuroblasts undergo a cycle of outward and inward migrations as the neuroblasts progress through their cell cycle. Phases of the cell cycle: G_1, first gap phase; S, DNA replication; G_2 second gap phase; M, mitosis.

blood supply during development. The result would be a central area of necrosis, which would become a thin spot or an open channel, surrounded by a zone of abnormal neuroblast migration (see Fig. 5–17B).

Cellular Events in Brain Development

The organization of the brain ultimately determines its function. Three important parameters in brain organization are (1) the density of neurons, (2) the pattern of axon and dendrite branching, and (3) the pattern of synaptic contacts. These characteristics begin to develop toward the end of the peak period of neuronal migration at the sixth month of gestation. Although neuronal density and the basic patterns of axonal and dendritic growth are determined within the first 2 to 3 years after birth, remodeling of synaptic connections continues throughout life.

Overproduction of Neurons and Apoptosis. Embryogenesis produces one and a half to two times more neurons than are present in the mature brain. By 24 weeks of gestation, almost all of these neurons have been produced. Subsequent to this, there is selective death of neurons.

Genetically programmed cell death (*apoptosis*) of neurons is a feature of cellular development in many areas of the brain. In contrast to necrosis (cell death resulting from injury), apoptosis requires protein synthesis and therefore is an active cellular process.

Some growth factors interrupt the normal process of apoptosis. *Nerve growth factor*, *brain-derived neurotrophic factor*, and *fibroblast growth factor* are known to limit cell death. This finding has led to the idea that the administration of growth factors may block the neuronal cell death that occurs in certain degenerative diseases.

Axonal Outgrowth. After neuroblasts complete their final cell division and migrate to their final location, they begin to extend a single axon with one or more distal elaborations known as *growth cones*. This spade-shaped extension of the growing axon is capable of driving through fields of developing nervous or mesen-

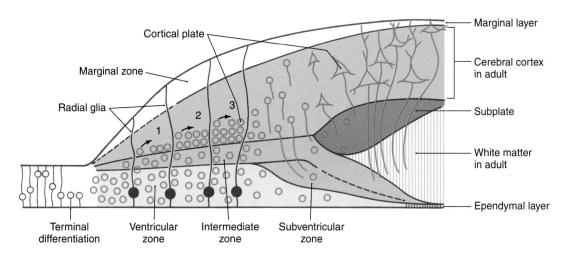

Figure 5–16. Later development of the cerebral cortex. Inner cellular layers (1) are formed first, and progressively more superficial layers (2, 3) are formed later.

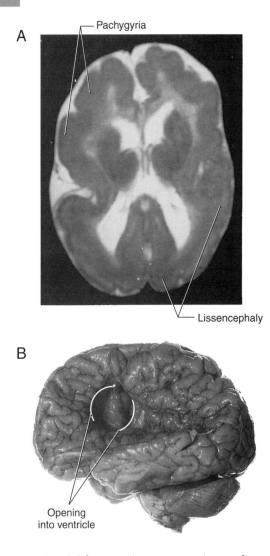

Pachygyria

A

Lissencephaly

B

Opening
into ventricle

Figure 5–17. Axial magnetic resonance image from a 6-month-old female patient with cortical malformations (*A*). Brain of a child with schizencephaly (*B*) showing an open channel from the lateral ventricle out to the brain surface.

chymal tissue to reach distant targets. The guidance of growth cones is influenced by both *tropic factors* (which guide a cell toward a particular target) and *trophic factors* (which maintain the metabolism of a cell or its processes). As an axon grows, it may send out branches, each with its own growth cone. Some branches may terminate in sites that will not ultimately be innervated by the cell. For example, cells of the motor cortex that send axons into the spinal cord to innervate motor neurons also send transient branches to structures of the brainstem; these ectopic connections are not normally maintained.

Synaptogenesis. Once an axonal growth cone arrives at its site of termination, it undergoes biochemical and morphologic changes to become a *presynaptic terminal.* Similarly, the area of the target neuron contacted by the presynaptic process begins expressing the charac-

teristic postsynaptic machinery, such as neurotransmitter receptors and second messenger molecules.

One view of synapse development is that there is competition for the synaptic space available on target neurons. A synapse that forms will persist only if it exchanges the right cues with the target cell. Many synapses that form are subsequently lost and are replaced by other synapses that may be more successful. Thus, only a subset of the large number of synapses that form are ultimately retained. This is the concept of *synaptic stabilization*, and it requires (1) a signal generated by the presynaptic cell, possibly the neurotransmitter to be used at the adult synapse; (2) a means for the postsynaptic cell to respond to the presynaptic signal; and (3) a "retrograde signal" from the postsynaptic cell to the presynaptic cell to indicate which contacts are to remain in maturity.

Plasticity and Competition. An important process related to the development of neuronal organization is *plasticity.* The developing brain is not as vulnerable to injury as is the mature brain. Infants who suffer significant cortical injury in the prenatal or early postnatal period may show surprising functional recovery, ending up with few or no obvious deficits. The mechanism of plasticity relates to alterations in selective neuronal death and axonal simplification and to the retention of transient axonal branches and synapses that would otherwise be lost, as discussed previously.

One example of plasticity and the competition for synaptic space is the development of visual cortical connections. Fibers conveying visual input from each eye arrive in overlapping territories in the visual cortex during the fetal period. Normally the synaptic space within this region is equivalently distributed to terminals carrying input from each eye. However, if the input from one eye is lost or if it is not functionally equivalent to the input from the other eye, the terminals from the "good" eye will experience a competitive advantage and occupy a larger share of the available synaptic space. The period of time during which these types of plastic changes can occur is called the *critical period.* Other areas of cortex have their own critical periods; the duration and time of occurrence of the critical period vary from region to region.

The concept of a critical period has clinical implications. If the input from one eye is dysfunctional during the critical period for visual system development (for example, if one eye is severely myopic), the axon terminals carrying information from that eye are at a disadvantage as they compete for synaptic space. If the causative disorder goes untreated, the "good" eye has exclusive access to visual cortex, and input from the "bad" eye is ignored. This condition is called *amblyopia.* If the myopia is corrected later in life, no signals can pass from the retina to the visual cortex because the appropriate synaptic connections were not formed during the critical period. As a result, the eye remains functionally blind. This blindness can be avoided by implementing clinical interventions that equalize competition for synaptic territory during the critical period.

Synaptic development occurs in parallel with cellular proliferation and migration. Dendritic spines are the site of many synaptic contacts, especially in cortical neurons. The rate of spine formation varies in different parts of the brain but is usually maximal during the sixth month after birth. Many children with mental retardation, including those with *Down syndrome*, have fewer and less complex axonal and dendritic ramifications and fewer dendritic spines than in normal children. In some patients, there is a disturbance of the cytoskeletal structure that supports the architecture of axonal processes. Axonal and synaptic development are especially vulnerable to *perinatal hypoxia*, *malnutrition*, and *environmental toxins*.

Myelination. Oligodendrocytes myelinate neuronal axons in the CNS. Myelination begins at about the sixth month of development and peaks between birth and the first year of life, but it continues into adulthood. A delay in myelination can result in a delay in functional development. The best example is a congenital cortical blindness that resolves during the first year of life.

There is a definite hierarchy in the regional maturation of myelin formation. The motor and sensory tracts throughout the nervous system mature early, whereas the association tracts mature relatively late.

Several neurodegenerative diseases (*leukodystrophies*) affect the formation of myelin. Many other inborn errors of amino and organic acid metabolism impair myelination, notably phenylketonuria. Finally, inadequate nutrition also can impair myelination.

Sources and Additional Reading

Barkovich AJ: Pediatric Neuroimaging, 2nd ed. Raven Press, New York, 1995.

Boulder Committee: Embryonic vertebrate central nervous system: Revised terminology. Anat Rec 166:257–262, 1970.

Evans OB: Manual of Child Neurology. Churchill Livingstone, New York, 1987.

Jacobson M: Developmental Neurobiology, 3rd ed. Plenum Press, New York, 1991.

McConnell SK: The determination of neuronal fate in the cerebral cortex. Trends Neurosci 12:342–349, 1989.

Noden DM: Vertebrate craniofacial development: The relation between ontogenetic process and morphological outcome. Brain Behav Evol 38:190–225, 1991.

Purves D, Lichtman JW: Principles of Neural Development. Sinauer Associates, Sunderland, Mass, 1985.

Rakic P: Principles of neural cell migration. Experientia 46:882–891, 1990.

Scott JM, Weir DG, Molly A, McPartlin J, Daly L, Kirke P: Folic acid metabolism and mechanisms of neural tube defects. Ciba Found Symp 181:180–191, 1994.

Shatz C: The developing brain. Sci Am 267:61–67, 1992.

Walsh C, Cepko CL: Clonally related cortical cells show several migration patterns. Science 241:1342–1345, 1988.

Section II

Regional Neurobiology

The Ventricles, Choroid Plexus, and Cerebrospinal Fluid

J. J. Corbett, D. E. Haines, and M. D. Ard

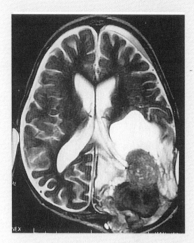

Overview 94

Development 94
Foramina of the Fourth Ventricle
Formation of the Choroid Plexus

Ventricles 97
Lateral Ventricles
Third Ventricle
Cerebral Aqueduct
Fourth Ventricle

Ependyma, Choroid Plexus, and Cerebrospinal Fluid 99
Ependyma
Ependymomas
Choroid Plexus
Cerebrospinal Fluid

Hydrocephalus and Related Conditions 104
Obstructive Hydrocephalus
Aqueductal Stenosis
Communicating Hydrocephalus
Hydrocephalus ex Vacuo
Idiopathic Intracranial Hypertension
Normal Pressure Hydrocephalus

The ventricular spaces of the brain are the adult elaborations of the neural canal of early developmental stages. These spaces, the choroid plexuses in them, and the cerebrospinal fluid (CSF) produced by the choroid plexus are essential elements in the normal function of the brain.

Overview

By about the third week of development, the nervous system consists of a tube closed at both ends and somewhat hook-shaped rostrally (Fig. 6–1). The cavity of this tube, the *neural canal,* eventually gives rise to the *ventricles* of the adult brain and the *central canal* of the spinal cord. The former becomes quite elaborate as the various parts of the brain differentiate, whereas the latter becomes progressively smaller as the spinal cord differentiates into its adult pattern.

The choroid plexus, which secretes the CSF that fills the ventricles and the subarachnoid space, arises from tufts of cells that appear in the wall of each ventricle during the first trimester. These cells are specialized for a secretory function. The production of CSF is an active process that requires an expenditure of energy by the choroidal cells.

Any condition that causes CSF to accumulate, such as overproduction or an obstruction of its movement through the ventricle system, produces serious neurologic deficits. Perhaps the most widely recognized example is *hydrocephalus* as seen in a fetus or in a newborn infant. This condition is usually caused by an obstruction of CSF flow with resultant enlargement of the ventricular spaces. The bones of the developing skull move apart, and the head may enlarge significantly. In most of these cases some type of surgical diversion of CSF flow (a shunting procedure) is necessary.

Development

At about 22 to 26 days of gestational age, the *anterior* and *posterior neuropores* close. At this point, the neural tube is lined by *neuroepithelial cells* undergoing waves of cell division. Some of these precursor cells give rise to the *ependymal cells* that line the developing (and mature) ventricular system and the central canal.

The brain is initially composed of three primary brain vesicles—*rhombencephalon, mesencephalon,* and *prosencephalon*—each containing a portion of the cavity of the neural tube (see Fig. 6–1*A* and *D*). The appearance of the pontine flexure in the rhombencephalon

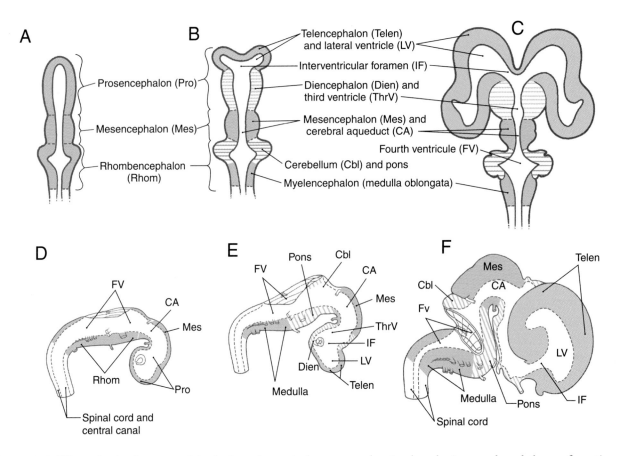

Figure 6–1. The early development of the brain and ventricular system, showing how brain growth and the configuration of the ventricles interrelate. Diagrammatic dorsal views (*A–C*) correlate in general with lateral views (*D–F*) at about 5 weeks (*D*), 6 weeks (*E*), and 8.5 weeks (*F*) of gestation. The outlines of the ventricles are shown in *D–F* as dashed lines.

and the telencephalic flexure in the prosencephalon divides these three vesicles into the five brain vesicles (*myelencephalon, metencephalon, mesencephalon, diencephalon, telencephalon*) characteristic of the adult brain (see Fig. 6–1*B* and *E*). With subsequent development, the telencephalic (cerebral) vesicles enlarge significantly (see Fig. 6–1*C* and *F*). As the brain enlarges from three to five vesicles, each part pulls along a portion of the cavity of the primitive neural tube. These spaces in each brain vesicle form the ventricle of that part of the brain in the adult. Consequently, the shape of the ventricular system conforms, in general, to the changes in configuration of the surrounding parts of the brain.

The lateral ventricles follow the enlarging cerebral hemispheres, and the third ventricle remains a single midline space (see Fig. 6–1*A–C*). The communications between the lateral ventricles and the third ventricle, the *interventricular foramina (of Monro)*, are initially large but become small as development progresses (see Fig. 6–1*C* and *F*).

Proliferation of the neural elements of the mesencephalon results in a reduction in the size of the cavity of this vesicle to form the *cerebral aqueduct* of the adult brain (see Fig. 6–1*C* and *F*). This creates a constricted region in the ventricular system and thus a point at which the flow of CSF may be easily blocked. Occlusion of the cerebral aqueduct during development may be the result of glial scarring (*gliosis*) due to infection or a consequence of developmental defects of the forebrain, a rupture of the amnionic sac in utero, or forking of the aqueduct. The last entity is a genetic sex-linked condition in which the aqueduct is reduced to two or more very small channels that do not properly meet. In addition, the cerebral aqueduct may be reduced to such a small channel that flow is reduced or essentially blocked. Whatever the cause, occlusion of the cerebral aqueduct results in a lack of communication between the third and fourth ventricles and blocks the egress of CSF from the third ventricle. Caudally, the cerebral aqueduct flares open into the fourth ventricle (see Fig. 6–1*B,C,E,* and *F*).

Foramina of the Fourth Ventricle. The ventricles and central canal of the spinal cord form a closed system when they first arise. However, in the second and third months of development, three openings form in the roof of the fourth ventricle, rendering the ventricular system continuous with the subarachnoid space surrounding the brain and spinal cord. The caudal part of the roof of the fourth ventricle consists of a layer of ependymal cells internally and a delicate layer of connective tissue externally (Fig. 6–2). The future apertures first appear in the form of small bulges in the caudal roof and at the lateral extremes of the fourth ventricle. The membrane forming the roof at these points becomes thinned and breaks down. The resultant openings are the medial *foramen of Magendie* (also called the *median aperture*) and the lateral *foramina of Luschka* (see Fig. 6–2).

Formation of the Choroid Plexus. In the adult, the *choroid plexus* is found in both lateral ventricles and in the third and fourth ventricles (see Fig. 6–4). The development of this structure is essentially the same in all of these spaces and is described here for the fourth ventricle.

The caudal roof of the fourth ventricle is composed of ependymal cells on the luminal surface and a delicate external layer of connective tissue, the pia mater. These structures collectively form the *tela choroidea* (Fig. 6–3). Developing arteries in the immediate vicinity invaginate the roof of the ventricle to form a narrow groove, the *choroid fissure*, in the tela choroidea (see Fig. 6–3). The involuted ependymal cells, along with vessels and a small amount of connective tissue, represent the primordial choroid plexus inside the ventricular space. As development progresses, the choroid plexus enlarges, forms many small elevations called *villi*, and begins to secrete CSF (see Fig. 6–3; see also Fig. 6–14). By the end of the first trimester, the choroid plexus is functional, the openings in the fourth ventricle are patent, and there is circulation of CSF through the ventricular system and into the subarachnoid space.

The choroid plexuses of the third and lateral ventri-

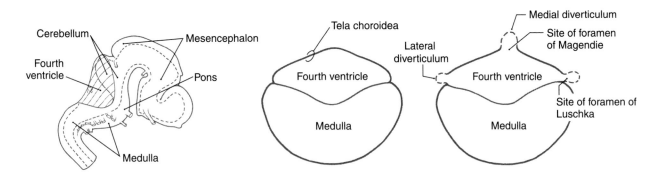

Figure 6–2. Development of the foramina of Luschka and Magendie in the fourth ventricle.

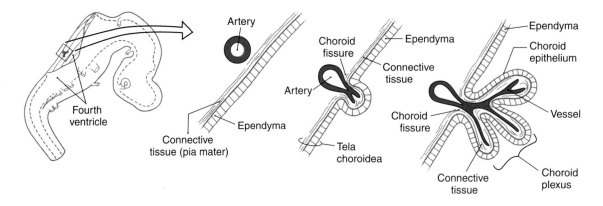

Figure 6–3. Development of the choroid plexus.

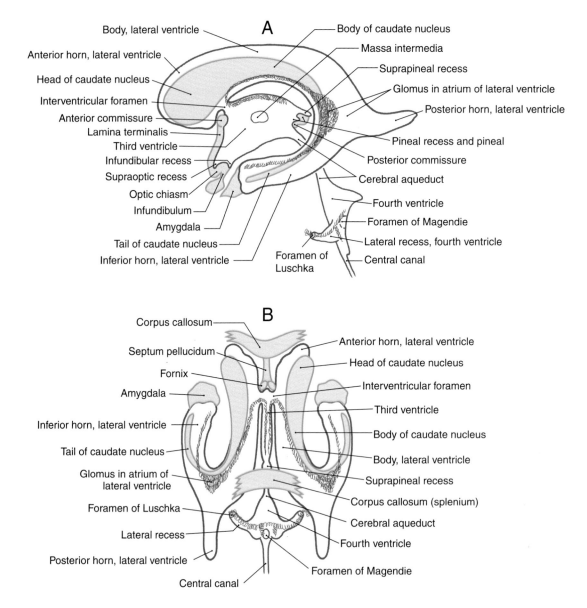

Figure 6–4. Lateral (*A*) and posterior (dorsal) (*B*) views of the lateral, third, and fourth ventricles and the cerebral aqueduct. Structures that border on the various parts of the ventricular system are shown in green and the choroid plexus is shown in red.

cles originate in the same manner. A choroid fissure appears in the roof of the third ventricle and in the medial wall of the lateral ventricle, where the covering of each (the tela choroidea) is thin. The choroid plexus develops along these lines, bulges into the respective space, and is continuous from lateral to third ventricles through the interventricular foramen (see Fig. 6–4).

Ventricles

Lateral Ventricles. The cavities of the telencephalon are the *lateral ventricles*, of which there is one in each hemisphere (see Fig. 6–4*A* and *B*). As the development of the hemispheres creates the frontal, temporal, and occipital lobes, the lateral ventricles are pulled along and thus acquire their definitive adult shape of a flattened **C** with a short tail (see Fig. 6–4). This shape is present by birth. The lateral ventricle consists of an *anterior horn*, a *body*, and *posterior* and *inferior horns* (see Fig. 6–4*A* and *B*). The junction of the body with the posterior and inferior horns constitutes the *atrium of the lateral ventricle.* An especially large clump of choroid plexus, the *glomus* (or *glomus choroideum*), is found in the atrium (see Fig. 6–4*A* and *B*). In adults and especially in elderly persons, the glomus may contain calcifications that are visible (as white spots) on radiographs or computed tomography (CT) scans (Fig. 6–5). Shifts in the position of the glomus, usually accompanied by alterations in the volume or shape of the surrounding ventricle, indicate some type of ongoing pathologic process or space-occupying lesion.

The elaborate shape of the lateral ventricle means that different structures border on different parts of this space. The anterior horn and body of the lateral ventricle are bordered medially by the *septum pellucidum* (at rostral levels) and by a bundle of fibers called the *fornix* (at caudal levels), and posteriorly by the *cor-*

pus callosum (Fig. 6–6; see also Fig. 6–4). The floor of the body of the lateral ventricle is made up of the *thalamus*, and the *caudate nucleus* is characteristically found in the lateral wall of the lateral ventricle throughout its extent (see Figs. 6–4 and 6–6). In the temporal lobe, the inferior horn of the lateral ventricle contains the tail of the *caudate nucleus* in its lateral wall, the hippocampal formation in its medial wall, and a large group of cells (the *amygdaloid complex*) in its rostral end (see Figs. 6–4 and 6–6). The openings between the lateral and third ventricles, the *interventricular foramina* (of Monro), are located between the column of the fornix and the rostral and medial end of the thalamus.

Third Ventricle. The *third ventricle*, the cavity of the diencephalon, is a narrow, vertically oriented midline space that communicates rostrally with the lateral ventricles and caudally with the *cerebral aqueduct* (see Fig. 6–4; see also Fig. 6–8). The third ventricle has an elaborate profile on a sagittal view (see Fig. 6–4*A*), but it is quite narrow in the coronal and axial planes (Fig. 6–7*A* and *B*).

The boundaries of the third ventricle are formed by a variety of structures, the most important being the dorsal thalamus and hypothalamus, and by structures that form small out-pocketings called recesses (Fig. 6–8; see also Fig. 6–4*A*). These are the *supraoptic recess* (above the optic chiasm), the *infundibular recess* (in the infundibulum, the stalk of the pituitary), the *pineal recess* (in the stalk of the pineal), and the *suprapineal recess* (above the pineal). The rostral wall of the third ventricle is formed by a short segment of the *anterior commissure* and a thin membrane, the *lamina terminalis*, that extends from the anterior commissure anteriorly (ventrally) to the rostral edge of the optic chiasm (see Figs. 6–4*A* and 6–8). The floor of the third ventricle is formed by the *optic chiasm* and *infundibulum* and their corresponding recesses, plus a line extending caudally

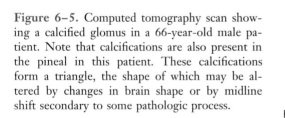

Figure 6–5. Computed tomography scan showing a calcified glomus in a 66-year-old male patient. Note that calcifications are also present in the pineal in this patient. These calcifications form a triangle, the shape of which may be altered by changes in brain shape or by midline shift secondary to some pathologic process.

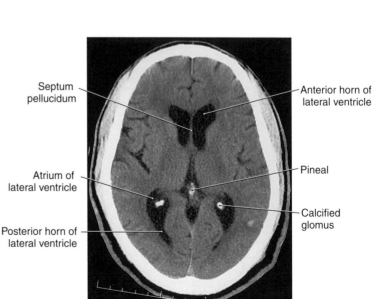

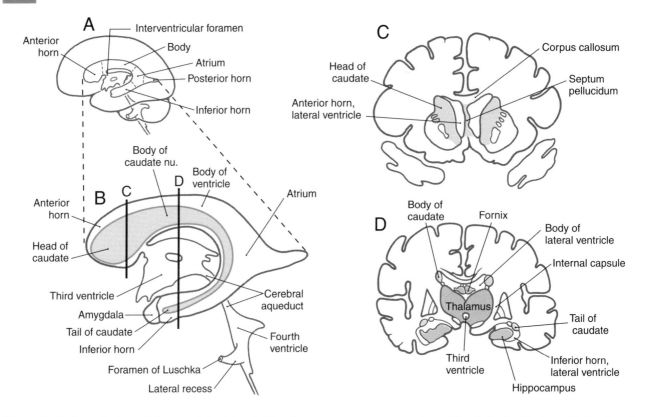

Figure 6–6. Lateral view of the ventricles (*A* and *B*) and representative cross sections (*C* and *D*, details from *B*) showing the lateral and third ventricles and the major structures that border on these spaces.

along the rostral aspect of the midbrain to the cerebral aqueduct. The caudal wall is formed by the *posterior commissure* and the recesses related to the pineal, whereas the roof is the *tela choroidea*, from which the choroid plexus is suspended (see Figs. 6–4*A* and 6–8). **Cerebral Aqueduct.** The *cerebral aqueduct*, the extension of the ventricle through the mesencephalon, com-

municates rostrally with the third ventricle and caudally with the fourth ventricle (see Figs. 6–4*A* and 6–8*B*). This midline channel is about 1.5 mm in diameter in adults and contains no choroid plexus. Its narrow diameter makes it especially susceptible to occlusion. For example, cellular debris in the ventricular system (from infections or *hemorrhage*) may clog the

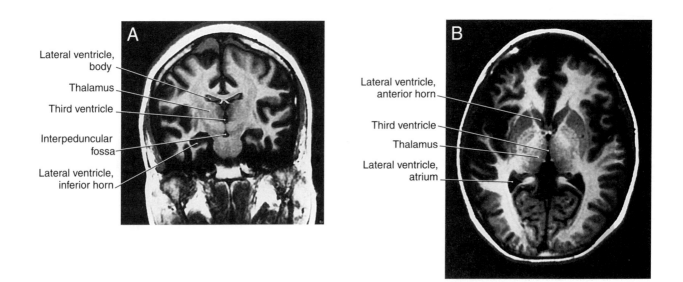

Figure 6–7. Magnetic resonance images of the third ventricle in coronal (*A*) and axial (*B*) views.

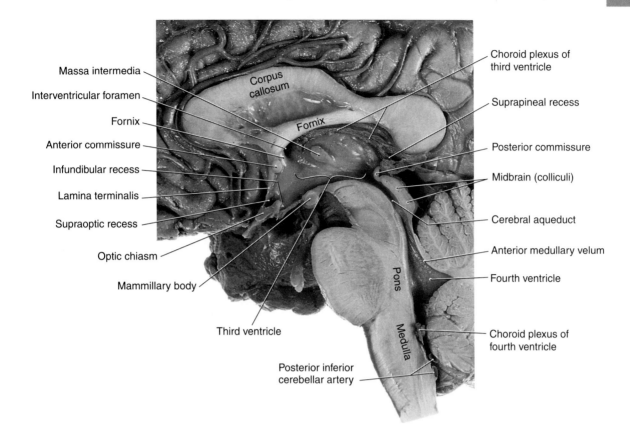

Massa intermedia
Interventricular foramen
Fornix
Anterior commissure
Infundibular recess
Lamina terminalis
Supraoptic recess
Optic chiasm
Mammillary body
Third ventricle
Posterior inferior
cerebellar artery

Corpus callosum
Fornix
Pons
Medulla

Choroid plexus of
third ventricle
Suprapineal recess
Posterior commissure
Midbrain (colliculi)
Cerebral aqueduct
Anterior medullary velum
Fourth ventricle
Choroid plexus of
fourth ventricle

Figure 6–8. Midsagittal view of the brain showing the third ventricle, cerebral aqueduct, and fourth ventricle, and structures closely related to these spaces.

aqueduct. Tumors in the area of the midbrain (e.g., *pinealoma*) may compress the midbrain and occlude the aqueduct. The result is a blockage of CSF flow and enlargement of the third and lateral ventricles at the expense of the surrounding brain tissue. This has been called triventricular hydrocephalus because it simultaneously involves the enlargement of three ventricles. The cerebral aqueduct is surrounded on all sides by a sleeve of gray matter that contains primarily small neurons; this is the *periaqueductal gray* or *central gray*.

Fourth Ventricle. The *fourth ventricle* is a roughly pyramid-shaped space that forms the cavity of the metencephalon and myelencephalon (see Figs. 6–4A and B and 6–8). The apex of this ventricle extends into the base of the cerebellum, and caudally it tapers to a narrow channel that continues into the cervical spinal cord as the central canal. Laterally the fourth ventricle extends over the surface of the medulla as the *lateral recesses*, to eventually open into the area of the pons-medulla-cerebellum junction, the *cerebellopontine angle*, through the *foramina of Luschka* (Fig. 6–9; see also Fig. 6–4). The irregularly shaped *foramen of Magendie* is located in the caudal sloping roof of the ventricle (Fig. 6–10; see also Fig. 6–4). Although the roof of the caudal part of the fourth ventricle and the lateral recesses is composed of *tela choroidea*, the rostral boundaries

of this space are formed by brain structures. These include the cerebellum (covering about the middle third of the ventricle) and the superior cerebellar peduncles and anterior medullary velum (covering the rostral third of the ventricle). The floor of the fourth ventricle, the *rhomboid fossa* (see Fig. 10–3), is formed by the pons and medulla (see Fig. 6–8). *The only openings between the ventricles of the brain and the subarachnoid space surrounding the brain are the foramina of Luschka and Magendie in the fourth ventricle.*

Ependyma, Choroid Plexus, and Cerebrospinal Fluid

Ependyma. The ventricles of the brain and the central canal of the spinal cord are lined by a simple cuboidal epithelium, the *ependyma*. Ependymal cells contain abundant mitochondria and are metabolically active. Their luminal surfaces are ciliated and have microvilli, and the bases contact the subependymal layer of astrocytic processes. There is no continuous basal lamina between ependymal cells and the subjacent glial cell processes (Fig. 6–11). Ependymal cells are attached to each other by *zonulae adherentes* (desmosomes).

In some regions, particularly the third ventricle,

Vestibulocochlear nerve

Facial nerve

Trigeminal nerve

Basilar pons

Abducens nerve

Foramen of Luschka

Glossopharyngeal nerve

Vagus nerve

Accessory nerve

Hypoglossal nerve

Olivary eminence

Pyramid

Figure 6–9. Anterolateral view of the brainstem at the pons-medulla junction showing the foramen of Luschka and the principal structures located in this area. Note the tuft of choroid plexus in the foramen. This area of the subarachnoid space, into which the foramen of Luschka opens, is the lateral cerebellomedullary cistern.

there are patches of specialized ependymal cells called *tanycytes* (see Fig. 6–11). Tanycytes have basal processes that extend through the layer of astrocytic processes to form end-feet on blood vessels and in the neuropil. They may function to transport substances between the ventricles and the blood. In contrast to ependymal cells, tanycytes are attached to each other and to immediately adjacent ependymal cells by tight junctions. Desmosomes are also present between tanycytes.

Ependymomas. This tumor, a type of glial cell neoplasm, originates from the ependymal cells lining the ventricles (Fig. 6–12). Although these tumors may ap-

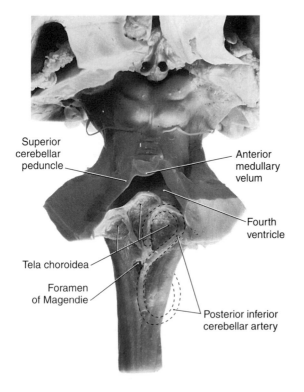

Superior cerebellar peduncle

Anterior medullary velum

Fourth ventricle

Tela choroidea

Foramen of Magendie

Posterior inferior cerebellar artery

Figure 6–10. Posterior (dorsal) view of the brainstem with the cerebellum removed to expose the fourth ventricle, the tela choroidea of the caudal roof of the fourth ventricle, and the route of the posterior inferior cerebellar artery. The choroid plexus on the internal surface of the tela is served by this vessel.

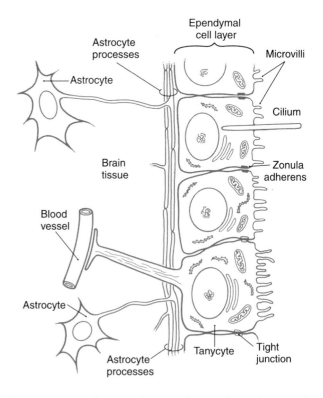

Astrocyte processes

Ependymal cell layer

Microvilli

Astrocyte

Cilium

Brain tissue

Zonula adherens

Blood vessel

Astrocyte

Astrocyte processes

Tanycyte

Tight junction

Figure 6–11. The ependyma and its relationship to the layer of subependymal astrocytic processes, with a representation of a tanycyte.

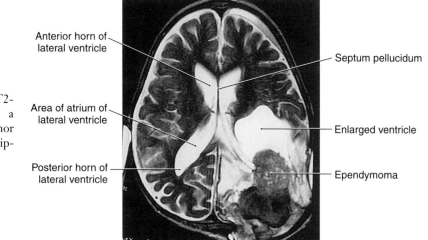

Figure 6-12. Magnetic resonance study (T2-weighted image) of an ependymoma in a 1-year-old male patient. Note that the tumor has extensively invaded the parietal and occipital lobes of the cerebral hemisphere.

Anterior horn of lateral ventricle

Area of atrium of lateral ventricle

Posterior horn of lateral ventricle

Septum pellucidum

Enlarged ventricle

Ependymoma

pear in any ventricle, the majority (60% to 75%) are located in the spaces of the posterior fossa. These neoplasms are seen most frequently in children younger than 5 years of age.

In general, the location of the ependymoma determines the symptoms experienced by the patient. Lesions in supratentorial locations (see Fig. 6-12) may produce signs and symptoms reflecting their location—for example, hydrocephalus in the case of blocked CSF flow, or seizure activity. Lesions in infratentorial locations frequently cause nausea and vomiting, headache, and other signs and symptoms related to hydrocephalus and cranial nerve signs and symptoms indicative of compression of, or tumor infiltration into, the brainstem.

Treatment for patients with ependymoma is primarily with surgical removal followed by focal irradiation. Incomplete removal—for example, in cases with tumor infiltration into the brainstem—reduces survival rates even with radiation therapy and chemotherapy.

Choroid Plexus. The choroid plexus in each ventricle is thrown into a series of folds called *villi* (singular, *villus*). These are covered on their ventricular (luminal) surfaces by a continuum of dome-shaped structures, each with numerous microvilli (Fig. 6-13A and B). Each dome represents the luminal surface of one *choroid epithelial cell*, and the shallow grooves between domes are the points of contact between adjacent *choroid* cells (see Fig. 6-13B and C). Each villus consists of a core of highly vascularized connective tissue derived from the pia mater and a simple cuboidal covering (the *choroid* epithelial cell layer), which is derived from ependymal cells (see Figs. 6-3 and 6-13B and C). The abundant capillaries in the connective tissue core of each villus are surrounded by a *basal lamina*. The endothelial cells of these capillaries have numerous *fenestrations*, which allow a free exchange of molecules between blood plasma and the extracellular fluid

in the connective tissue core (Fig. 6-14). The connective tissue core itself consists of fibroblasts and collagen fibrils. Another basal lamina is formed at the interface between the connective tissue core and the choroid epithelial cells that form the surface of each villus (see Figs. 6-13C and 6-14). These choroid cells have microvilli on their apical (ventricular) surface, interdigitating cell membranes on their sides, and irregular bases. Each is attached to its neighbor by continuous *tight junctions* (*zonulae occludentes*) that seal off the subjacent extracellular space from the ventricular space (Figs. 6-13C and 6-14). This represents the *blood-CSF barrier*. Choroid epithelial cells contain a nucleus, numerous mitochondria, rough endoplasmic reticulum, and a small Golgi apparatus (see Figs. 6-13C and 6-14). Thus, they are specialized to control the flow of ions and metabolites into the CSF.

Although choroid epithelial cells are joined by tight junctions, ependymal cells are not (see Fig. 6-14). Therefore, fluid exchange occurs freely between CSF and the extracellular fluid of the brain parenchyma. The composition of CSF can thus sometimes reflect disease processes occurring in brain tissue. For example, catabolites of catecholamines are reduced in quantity in the CSF of patients with Parkinson disease, a neurodegenerative disorder involving loss of dopaminergic neurons.

In humans the blood supply to the choroid plexuses is via the *choroidal arteries* and the *posterior cerebellar arteries*. Choroid plexus in the inferior horn, atrium of the lateral ventricle, and body of the lateral ventricle is served by the *anterior choroidal artery* (a branch of the internal carotid) and the *lateral posterior choroidal artery* (a branch of P_2). The *medial posterior choroidal artery* (also a branch of P_2) serves the choroid plexus of the third ventricle. The choroid plexus located inside the fourth ventricle is served by branches of the *posterior inferior cerebellar artery* (see Figs. 6-8 and 6-10), and

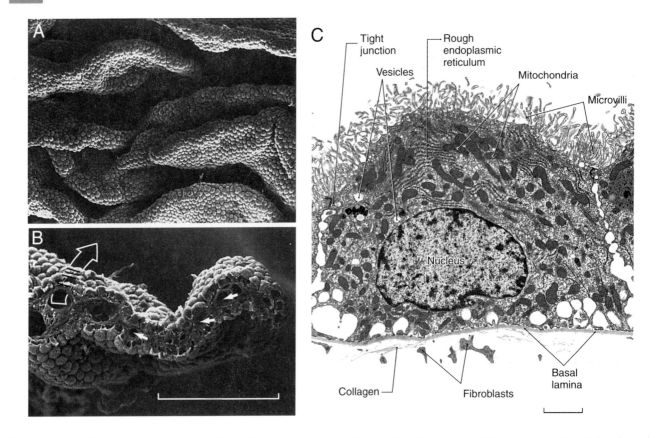

Figure 6–13. Elements of the choroid plexus. Scanning electron micrographs of the surface (*A*) of several villi of the choroid plexus and the cut surface (*B*) of one villus. The characteristic appearance of the luminal surface of the choroidal cells is seen in both micrographs. The *arrows* in *B* mark vessels in the core of the villus (*solid arrow*) and the route of fluid movement from the vessels into the ventricular space (*broken arrow*). Transmission electron micrograph (*C*) showing the internal structure of one choroid epithelial cell. Primate; scale = 100 μm for *A* and *B* and 2 μm for *C*.

the tuft that extends out of the foramen of Luschka into the subarachnoid space (see Fig. 6–9) is served by the *anterior inferior cerebellar artery*.

Cerebrospinal Fluid. Choroid epithelial cells secrete CSF by selective transport of materials from the connective tissue extracellular space (see Fig. 6–14). Sodium chloride is actively transported into the ventricles, and water passively follows the concentration gradient thus established. Other materials, including large molecules, are transported in pinocytotic vesicles from the basal to the apical surface of the epithelium and exocytosed into the CSF. Compared with blood plasma, CSF has higher concentrations of chloride, magnesium, and sodium and lower concentrations of potassium, calcium, glucose, and proteins.

Normal CSF contains very little protein (15 to 45 mg/dL), little immunoglobulin, and only one to five cells (leukocytes) per milliliter. Changes from these normal values are useful in the diagnosis of a variety of disease processes. *Lumbar puncture* is used to collect a sample of CSF for analysis and to measure CSF pressure. A needle is inserted between the third and fourth (or fourth and fifth) lumbar vertebrae into the dural

sac, and a few milliliters of fluid withdrawn. Because the average volume of CSF in the adult is about 120 mL, and the rate of production is about 450 to 500 mL/day, the sample removed is quickly replaced.

The numbers and types of cells found in CSF vary according to the type of disease. In bacterial meningitis or brain abscesses, neutrophils predominate and may reach concentrations of 1000 to 10,000/mL. In syphilitic meningitis, by contrast, 200 to 300 cells/mL would be typical, and most of these would be lymphocytes. Lymphocytes are also the predominant cell type found in active multiple sclerosis, even though there are usually fewer than 50 cells/mL of CSF. The diagnosis of multiple sclerosis also rests on specific changes in the immunoglobulin G content of CSF; immunoglobulin G is both derived from the blood and produced by lymphocytes in the CSF, where it is released during an autoimmune reaction attack.

In marked contrast to the elevated numbers of white blood cells seen in central nervous system (CNS) infections, numerous red blood cells are present in the CSF of patients who have bleeding into the *subarachnoid space (subarachnoid hemorrhage)*. For example, this con-

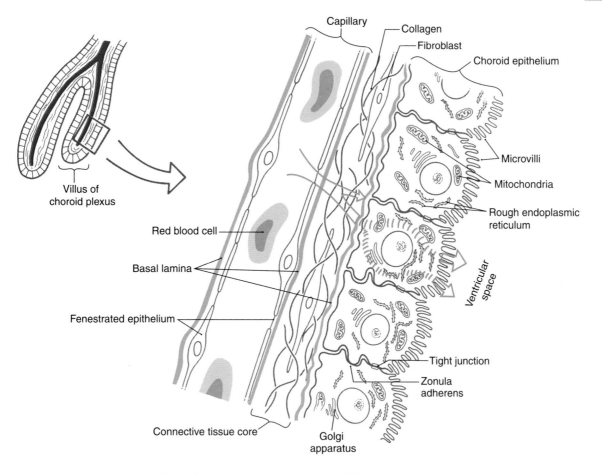

Figure 6–14. The basic structure of the choroid plexus and the route of fluid transport (shown in green) through the choroid epithelium to produce cerebrospinal fluid.

dition may result from rupture of an intracranial aneurysm or arteriovenous malformation. Patients with subarachnoid hemorrhage or those with primary CNS tumors usually have elevated protein levels in their CSF. Elevated CSF protein is also seen in patients with syphilis or meningitis, and in cancer patients in whom the disease has spread (metastasized) into the CSF. The CSF of cancer patients may also contain malignant cells characteristic of their primary lesions.

The CSF produced by the choroid plexuses passes through the ventricular system to exit the fourth ventricle through the foramina of Luschka and Magendie (Fig. 6–15). At this point, the CSF enters the subarachnoid space, which is continuous around the brain and spinal cord. The CSF in the subarachnoid space provides the buoyancy necessary to prevent the weight of the brain from crushing nerve roots and blood vessels against the internal surface of the skull. The weight of the brain, about 1400 g in air, is reduced to about 45 g when it is suspended in CSF. Consequently, the tethers formed by delicate connective tissue strands traversing the subarachnoid space, the *arachnoid trabeculae* (see Fig. 6–15), are adequate to

maintain the brain in a stable position within its CSF envelope.

The movement of CSF through the ventricular system and the subarachnoid space is influenced by two major factors. First, there is a subtle pressure gradient between the points of production of CSF (choroid plexuses in brain ventricles) and the points of transfer into the venous system (arachnoid villi). Because CSF is not compressible, it tends to move along this gradient. Second, CSF is also moved in the subarachnoid space by purely mechanical means. These include gentle movements of the brain on its arachnoid trabecular tethers during normal activities and the pulsations of the numerous arteries found in the subarachnoid space.

After passing through the subarachnoid space, the CSF reaches the arachnoid villi that extend into the superior sagittal sinus (see Fig. 6–15). The subarachnoid space and the CSF it contains extend into the core of each villus. At this point CSF enters the venous circulation through two routes. A limited amount passes between the cells making up the arachnoid villus, whereas most is transported through these cells in membrane-bound vesicles (see also Chapter 7). About

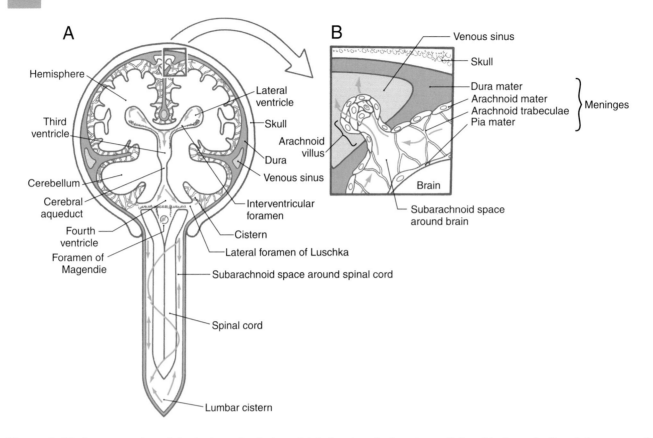

Figure 6–15. Representation of the brain and spinal cord (*A*) showing the locations of choroid plexus (red) and the routes of flow taken by cerebrospinal fluid (CSF) (green) through the ventricles and the subarachnoid space around the central nervous system. The detail (*B*) shows the relationship of an arachnoid villus to the subarachnoid space and the venous sinus. Although not shown here, the venous sinus is lined by an endothelium. CSF enters the venous system primarily by transport through cells of the arachnoid villus (*B, dashed green arrow*), although some fluid moves between these cells (*B, solid green arrow*).

330 to 380 mL of CSF enters the venous circulation per day.

Hydrocephalus and Related Conditions

Blockage of CSF movement or a failure of the absorption mechanism will result in the accumulation of fluid in the ventricular spaces or around the brain (Fig. 6–16). The results, commonly called *hydrocephulus*, are characterized by an increase in CSF volume, enlargement of one or more of the ventricles, and, usually, an increase in CSF pressure. Dilation of the cerebral ventricles may result from a blockage of CSF flow through the system, as in *obstructive hydrocephalus;* from factors not related to impaired flow, as in *communicating hydrocephalus;* or from brain atrophy, as in *hydrocephalus ex vacuo.* Ventricular dilation may also be a sequel to trauma, meningitis, or subarachnoid hemorrhage. Typically there is a moderate to severe increase in intracranial pressure in patients with hydrocephalus.

Obstructive Hydrocephalus. Obstructive hydrocephalus may result from an obstruction somewhere within the ventricular system or within the subarachnoid space. Common intraventricular sites of potential obstruction are the interventricular foramen (or foramina), cerebral aqueduct, caudal portions of the fourth ventricle, and the foramen of the fourth ventricle. Extraventricular obstruction may occur at any place in the subarachnoid space but is more common around the base of the brain, at the tentorium cerebelli and tentorial notch, over the convexity of the hemisphere, and at the superior sagittal sinus.

Aqueductal Stenosis. Aqueductal stenosis may be caused by a tumor in the immediate vicinity of the midbrain (as in *pineoblastoma* or *meningioma*) that compresses the brain and occludes the cerebral aqueduct. This channel may also be occluded by the cellular debris seen following *intraventricular hemorrhage* or bacterial or fungal infections of the CNS. One major sequel of aqueductal blockade is enlargement of the third and both lateral ventricles (see Fig. 6–16). This is sometimes called triventricular hydrocephalus because three ventricles simultaneously enlarge as a result of one lesion or occlusion. Unilateral obstruction of one interventricular foramen, for example by a colloid cyst in one interventricular foramen, results in enlarge-

Normal   Hydrocephalus

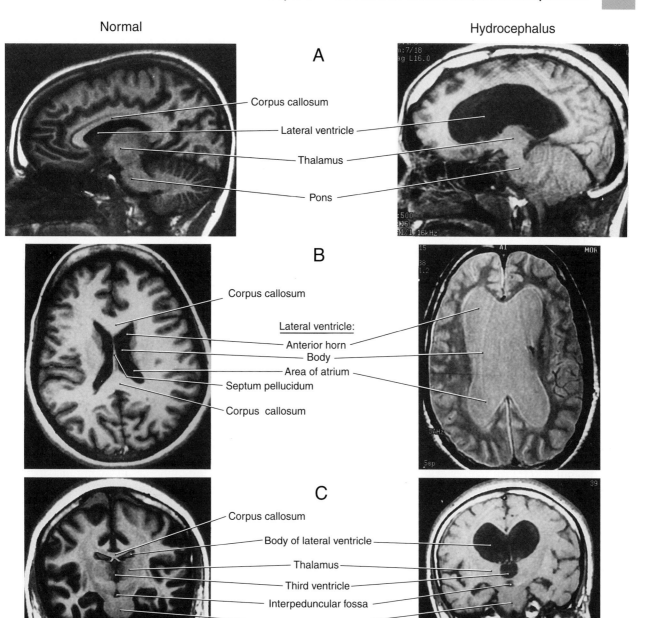

Figure 6–16. Comparison of normal and hydrocephalic brains in sagittal (*A*), axial (*B*), and coronal (*C*) planes as seen on magnetic resonance images.

ment of the lateral ventricle on that side. Blockage of both interventricular foramina will produce enlargement of both lateral ventricles. Obstruction of the exit channels of the fourth ventricle, the foramina of Magendie and Luschka, will result in enlargement of all parts of the ventricular system.

Communicating Hydrocephalus. In *communicating hydrocephalus*, the flow of CSF through the ventricular system and into the subarachnoid space is not impaired. However, movement of CSF through the sub-

arachnoid space and into the venous system is partially or totally blocked. This block may be caused by a congenital absence (agenesis) of the arachnoid villi. Alternatively, these villi may be partially blocked by red blood cells subsequent to a subarachnoid hemorrhage. An exceedingly high level of protein in the CSF (above 500 mg/dL), as seen in patients with CNS tumors or infections, also contributes to communicating hydrocephalus. The high CSF pressure is due partially to the sequestering of protein in the arachnoid villi and subsequent blockage of CSF transport into the venous system.

Additional causes of communicating hydrocephalus include the interruption of CSF movement through the subarachnoid space caused by either subarachnoid hemorrhage or a major CNS infection, such as leptomeningitis, and the subsequent inflammatory response. Overproduction of CSF in patients with *papilloma of the choroid plexus* may also be a factor. In all of these situations there is an enlargement of all parts of the ventricular system. Although rare, hydrocephalus may also be seen in patients with impaired venous flow from the brain.

Hydrocephalus ex Vacuo. This is actually not a true hydrocephalus but rather a general atrophy of the brain resulting in ventricles that are relatively larger owing to the loss of white matter. There is no increase in intracranial pressure, there are no neurologic deficits other than those that may be related to brain atrophy, and treatment is not indicated. Ex vacuo changes may also refer to atrophy with a change in ventricular size that may follow, by several years, an event such as a stroke.

Idiopathic Intracranial Hypertension. Idiopathic intracranial hypertension (pseudotumor cerebri) is an enigmatic condition more commonly seen in obese women of child-bearing age and in persons with chronic renal failure; it is possibly related to vitamin A toxicity. There is an increase in intracranial pressure (25 to 55 cm H_2O), with little evidence of pressure increase on CT or magnetic resonance imaging studies, such as ventricular enlargement or effacement of sulci or cisterns. These patients usually experience headache, a variety of visual deficits (up to blindness), and obvious papilledema (swelling of the optic disc). Treatment includes a program of weight loss, medication, and, if needed, shunting (lumboperitoneal). Surgical fenestration of the optic sheath is usually effective in preserving and improving vision.

Normal Pressure Hydrocephalus. This form of hydrocephalus may develop subsequent to any of several types of chronic meningitis, blood in the subarachnoid space (from trauma or a ruptured aneurysm), or osteitis deformans (Paget disease) of the base of the skull. Affected patients are usually elderly. In most cases the cause is unknown. Although intracranial pressure may initially be elevated and the ventricles enlarged, the pressure may wax and wane over time or even subside to a high-normal level; however, the effects of the increased pressure remain.

Patients with normal pressure hydrocephalus experience urinary problems (frequency, urgency, or incontinence), impaired gait that is most obvious on stepping up as on a curb, and dementia. In some patients the combination of a difficult shuffling gait and dementia may mimic the clinical picture in degenerative disease such as Alzheimer and Parkinson diseases. Treatment is a shunting procedure to reduce CSF pressure and volume. In some cases there is general clinical improvement with lessening of all symptoms including those related to mental status.

Sources and Additional Reading

Davson H, Welch K, Segal MB: Physiology and Pathophysiology of the Cerebrospinal Fluid. Churchill Livingstone, Edinburgh, 1987.

Fishman RA: Cerebrospinal Fluid in Diseases of the Nervous System. WB Saunders, Philadelphia, 1992.

Kida S, Yamashima T, Kubota T, Ito H, Yamamoto S: A light and electron microscopic and immunohisto-chemical study of human arachnoid villi. J Neurosurg 69:429–435, 1988.

North B, Reilly P: Raised Intracranial Pressure: A Clinical Guide. Heinemann, Oxford, 1990.

Pardridge WM (ed): Introduction to the Blood-Brain Barrier. Cambridge University Press, Cambridge, 1998.

Peters A, Palay SL, Webster H deF: The Fine Structure of the Nervous System, Neurons and Their Supporting Cells, 3rd ed. Oxford University Press, New York, 1991.

Russell DS: Observations on the Pathology of Hydrocephalus. SRS265. Her Magesty's Stationery Office, London, 1949.

Segal MB (ed): Barriers and Fluids of the Eye and Brain. CRC Press, Boca Raton, Fla, 1992.

Upton ML, Weller RO: The morphology of cerebrospinal fluid drainage pathways in human arachnoid granulations. J Neurosurg 63:867–875, 1985.

Wood JH (ed): Neurobiology of Cerebrospinal Fluid, vols 1 and 2. Plenum Press, New York, 1980 and 1983.

Yamashima T: Functional ultrastructure of cerebrospinal fluid drainage channels in human arachnoid villi. Neurosurgery 22:633–641, 1988.

Chapter 7

The Meninges

D. E. Haines

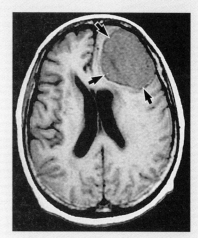

Overview 108

Development of the Meninges 108

Overview of the Meninges 109

Dura Mater 110
  Periosteal and Meningeal Dura
  Dural Border Cell Layer
  Blood Supply
  Nerve Supply
  Dural Infoldings and Sinuses
  Cranial versus Spinal Dura

Arachnoid Mater 113
  Arachnoid Barrier Cell Layer
  Arachnoid Trabeculae and the Subarachnoid Space
  Arachnoid Villi

Meningiomas and Meningeal Hemorrhages 115
  Meningiomas
  Extradural and "Subdural" Hemorrhages
  Hygroma

Pia Mater 116

Cisterns, Subarachnoid Hemorrhages, and Meningitis 117
  Cisterns
  Subarachnoid Hemorrhage
  Meningitis

The human nervous system is extremely delicate and lacks the internal connective tissue framework seen in most organs. For protection the brain and spinal cord are each encased in a bony shell, enveloped by a fibrous coat, and delicately suspended within a fluid compartment. In the living state, the nervous system has a gelatinous consistency, but when treated with fixatives, it becomes firm and easy to handle.

Overview

The brain and spinal cord are surrounded by the skull and vertebral column, respectively. With the exception of the intervertebral foramina, through which the spinal nerves and their associated vessels pass, and the foramina in the skull, which serve as conduits for arteries, veins, and cranial nerve roots, this bony encasement is complete. The membranous coverings of the central nervous system, the *meninges*, are located internal to the skull and vertebral column. The meninges (1) protect the underlying brain and spinal cord; (2) serve as a support framework for important arteries, veins, and sinuses; and (3) enclose a fluid-filled cavity, the *subarachnoid space*, which is vital to the survival and normal function of the brain and spinal cord.

The presence of this bony and meningeal encasement of the central nervous system is a double-edged sword. Although these structures offer maximum protection, in the case of trauma or in a disease process, they can be very unforgiving. For example, growth of a tumor creates a mass that will increase intracranial pressure and compress or displace various portions of the brain. Something has to give inside the skull when a space-occupying lesion develops, and it is the delicate tissue of the brain that gives. The neurologic deficits that result depend on the location of the mass, the rapidity with which it enlarges, and which parts of the brain are damaged.

Development of the Meninges

The meninges develop from cells of the *neural crest* and *mesenchyme* (mesoderm), which migrate to surround the developing central nervous system between 20 and 35 days of gestation (Fig. 7–1A–C). Collectively, these neural crest and mesodermal cells form the *primitive meninges* (*meninx primitiva*). At this stage no obvious spaces (venous sinuses, subarachnoid space) are present in the meninges. Between 34 and 48 days of gestation, the primitive meninges differentiate into an outer, more compact layer called the *ectomeninx*, and an inner, more reticulated layer called the *endomeninx* (see Fig. 7–1D). As development progresses (45 to 60 days of gestation), the ectomeninx becomes more compact, and spaces appear in this layer that correlate with the positions of the future venous sinuses. Concurrently, the endomeninx becomes more reticulated, and the spaces that appear in its inner part correspond to the subarachnoid spaces and cisterns of the adult. In

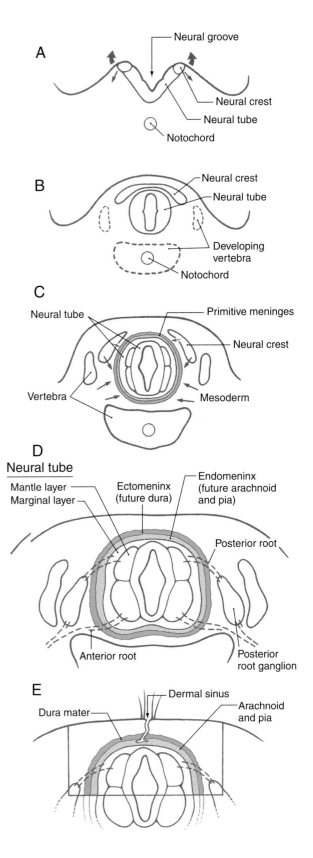

Figure 7–1. Development of the meninges. After the neural tube closes (*A, B*), cells from the neural crest and mesoderm (*C, arrows*) migrate to surround the neural tube and form the primordia of the dura and of the arachnoid and pia (*D*). A dermal sinus (*E*) is a malformation in which there is a channel from the skin into the meninges.

general, the ectomeninx will become the *dura mater* of the adult brain, and the endomeninx will form the *arachnoid mater* and *pia mater* (the *leptomeninges*) of the adult nervous system (see Fig. 7–1D). By the end of the first trimester the meninges have generally reached the overall plan seen in the adult brain and spinal cord.

One development defect associated with closure of the neural tube and formation of the meninges in the lumbosacral area is the *congenital dermal sinus* (also called just dermal sinus) (see Fig. 7–1E). This defect is caused by a failure of the ectoderm (future skin) to completely pinch off from the neuroectoderm and the primitive meninges that envelop it. As a result, the meninges are continuous with a narrow, epithelium-lined channel that extends to the skin surface (see Fig. 7–1E). Dermal sinuses are sometimes discovered in young patients who have recurrent, unexplained bouts of meningitis. These lesions are surgically removed, and recovery is usually complete.

The ectomeninx around the brain is continuous with the skeletogenous layer that forms the skull. This relationship is maintained in the adult, where the dura is intimately adherent to the inner surface of the skull. In the spinal column, the ectomeninx is also initially continuous with the developing vertebrae. However, as development proceeds, the spinal ectomeninx dissociates from the vertebral bodies. A layer of cells remain on the vertebrae to form the periosteum, and the larger part of the ectomeninx condenses to form the spinal dura. The intervening space becomes the spinal *epidural space* (Fig. 7–2). This space is essential for the administration of *epidural anesthetics*.

Overview of the Meninges

In general, the meninges consist of fibroblasts and varying amounts of extracellular connective tissue fibrils. The structural features of each meningeal layer reflect the fact that the fibroblasts of that particular layer are modified to serve a particular function.

The human meninges are composed of the *dura mater*, the *arachnoid mater*, and the *pia mater*. (Fig. 7–3; see also Fig. 7–2). The outermost portion, the *dura mater*, also called the *pachymeninx*, is adherent to the inner surface of the skull, but is separated from the vertebrae by the epidural space (see Fig. 7–2). Around the brain the inner portions of the dura give rise to infoldings or septa, such as the *falx cerebri* or *tentorium cerebelli* (see Fig. 7–2), which separate brain regions from each other. Major venous sinuses are found at the points where these septa originate. Spinal and cranial nerves, as they enter or exit the central nervous system, must pass through a cuff of the dura that is continuous with the connective tissue of the peripheral nerve. Blood vessels traverse the dura in similar fashion.

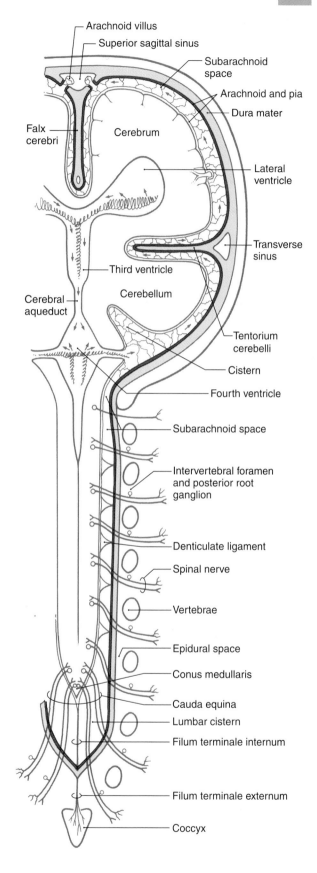

Figure 7–2. The relation of the meninges to the brain and spinal cord and to their surrounding bony structures. The dura is represented in blue, the arachnoid in red.

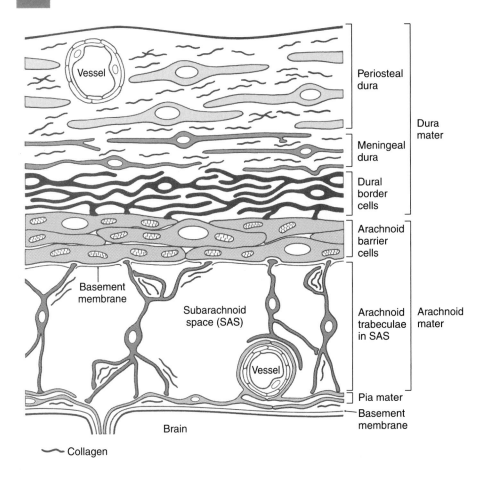

Figure 7–3. The structure of the meninges. Layers of the dura are shown in shades of black, the arachnoid in shades of red, and the pia in green.

The inner two layers of the meninges, the arachnoid mater and the pia mater (see Figs. 7–2 and 7–3), are collectively known as the *leptomeninges*. This term is also commonly used in clinical medicine (as in *leptomeningeal cysts* and *leptomeningitis*). The arachnoid is a thin cellular layer that is attached to the overlying dura but, with the exception of the arachnoid trabeculae, is separated from the pia mater by the *subarachnoid space*. The arachnoid around the brain is directly continuous with the arachnoid lining the inner surface of the spinal dura (see Fig. 7–2). Consequently, the spinal and cerebral subarachnoid spaces are also directly continuous with each other at the foramen magnum. The *subarachnoid space* contains cerebrospinal fluid and vessels and is bridged by fibroblasts of various sizes and shapes that collectively form the *arachnoid trabeculae*. The arachnoid is avascular and does not contain nerve fibers.

The *pia mater* is located on the surface of the brain and spinal cord and closely follows all their various grooves and elevations (see Figs. 7–2 and 7–3). Around the spinal cord the pia mater contributes to the formation of the *denticulate ligaments* and the *filum terminale internum* (or *pial part of the filum terminale*).

Dura Mater

Periosteal and Meningeal Dura. The *dura mater* (*pachymeninx*) is composed of elongated fibroblasts and copious amounts of collagen fibrils (see Fig. 7–3). This membrane contains blood vessels and nerves and is generally divided into outer (*periosteal*), inner (*meningeal*), and *border cell* portions. There is no distinct border between periosteal and *meningeal* portions of the dura (see Fig. 7–3). Fibroblasts of the *periosteal dura* are larger and slightly less elongated than other dural cells. This portion of the dura is adherent to the inner surface of the skull, and its attachment is particularly tenacious along suture lines and in the cranial base. In contrast, the fibroblasts of the *meningeal dura* are more flattened and elongate, their nuclei are smaller, and their cytoplasm may be darker than in periosteal cells. Although cell junctions are rarely seen between dural fibroblasts, the large amounts of interlacing collagen in periosteal and meningeal portions of the dura give these layers of the meninges great strength.

Dural Border Cell Layer. The innermost part of the dura is composed of flattened fibroblasts that have sinuous processes. Collectively, these cells form the *dural border cell layer* (see Fig. 7–3). The extracellular spaces between the flattened cell processes of dural border cells contain an amorphous substance but no collagen or elastic fibers. Cell junctions (desmosomes, gap junctions) are occasionally seen between dural border cells and cells of the underlying arachnoid.

Because of its loose arrangement, enlarged extracellular spaces, and lack of extracellular connective tissue fibrils, *the dural border cell layer constitutes a plane of structural weakness at the dura-arachnoid junction.* This layer is externally continuous with the meningeal dura and internally continuous with the arachnoid. Consequently, *bleeding into this area of the meninges will disrupt and dissect open the dural border cell layer* rather than invading the overlying dura or the underlying arachnoid. We shall consider meningeal hemorrhages after discussing the arachnoid.

Blood Supply. The arterial supply to the dura of the anterior cranial fossa originates from the *cavernous portion of the internal carotid*, the *ethmoidal arteries* (via the ethmoidal foramina), and branches of the ascending pharyngeal artery (via the foramen lacerum). The *middle meningeal artery* serves the dura of the middle cranial fossa and may be compromised when there is trauma to the skull. It is a branch of the maxillary artery and enters the skull through the foramen spinosum. The accessory meningeal artery (via the foramen ovale) and small branches from the lacrimal artery (via the superior orbital fissure) also serve the dura of the middle fossa. The dura of the posterior fossa is served by small meningeal branches of ascending pharyngeal and occipital arteries and by minute branches of the vertebral arteries.

The spinal dura is served by branches of major arteries (e.g., vertebral, intercostal, lumbosacral) that are located close to the vertebral column. These small meningeal arteries enter the vertebral canal via the intervertebral foramina to serve the dura and adjacent structures.

Nerve Supply. The nerve supply to the dura of the anterior and middle fossae is from branches of the *trigeminal nerve*. *Ethmoidal nerves* and branches of the *maxillary* and *mandibular nerves* innervate the dura of the anterior fossa, whereas that of the middle fossa is served mainly by branches from the *maxillary* and *mandibular nerves*. The dura of the posterior fossa receives sensory branches from dorsal roots C2 and C3 (and from C1 when this root is present) and may have some innervation from the vagus nerve. The *tentorial nerve*, a branch of the ophthalmic nerve, courses caudally to serve the tentorium cerebelli. Autonomic fibers to the vessels of the dura originate from the superior cervical ganglia and gain access to the cranial cavity by simply following the progressive branching patterns of the vessels on which they lie.

Nerves to the spinal dura originate as recurrent branches of the spinal nerve located at that level.

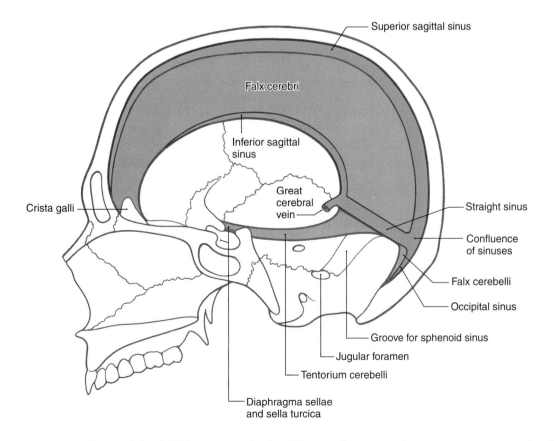

Figure 7–4. Midsagittal view of the skull showing the dural infolding (reflections) and venous sinuses associated with each.

These delicate strands pass through the intervertebral foramina and distribute to the spinal dura and to some adjacent structures.

Dural Infoldings and Sinuses. As noted previously, the dura has *periosteal* and *meningeal* parts. The *periosteal dura* lines the inner surface of the skull and functions as its periosteum. The meningeal dura is continuous with the periosteal dura, but draws away from it at specific locations to form the *dural infoldings* (or *reflections*). The largest of these is the *falx cerebri* (Fig. 7–4). It is attached to the crista galli rostrally, to the midline of the inner surface of the skull, and to the surface of the tentorium cerebelli caudally. The falx cerebri separates the right hemisphere from the left. The *superior sagittal sinus* is found where the falx cerebri attaches to the skull, the *straight sinus* where it fuses with the *tentorium cerebelli*, and the *inferior sagittal sinus* in its free edge (see Fig. 7–4). Many large superficial veins located on the surface of the cerebral hemispheres empty into the superior sagittal sinus.

The *tentorium cerebelli* is the second largest of the dural infoldings (Fig. 7–5; see also Fig. 7–4). Rostrally, it attaches to the clinoid processes, rostrolaterally to the petrous portion of the temporal bone (location of the *superior petrosal sinus*), and caudolaterally to the inner surface of the occipital bone and a small part of the parietal bone (location of the *transverse sinus*) (see Fig. 7–5). The tent shape of the tentorium divides the cranial cavity into *supratentorial* (above the tentorium) and *infratentorial* (below the tentorium) compartments. The supratentorial compartment is divided into right and left halves by the falx cerebri. The sweeping edges of the right and left tentoria, as they arch from the clinoid processes to join at the straight sinus, form the *tentorial notch* (see Fig. 7–5). The occipital lobe is above the tentorium, the cerebellum is below it, and the midbrain passes through the tentorial notch.

The concept of a supratentorial compartment divided into right and left halves and an infratentorial compartment is important to understanding a variety of clinical scenarios involving neurologic problems such as increased intracranial pressure or the presence of a space-occupying lesion. For example, a lesion in one cerebral hemisphere may expand toward the midline, deforming the falx cerebri (*midline shift*), or may cross the midline below the falx. As another example, an increase in intracranial pressure in the supratentorial location may result in herniation of the brain downward through the tentorial notch into the infratentorial compartment. This is the mechanism of *uncal herniation*. Yet a third example is the presence of a space-occupying lesion in the infratentorial compartment, resulting in herniation of the cerebellum through the foramen magnum. It is frequently the tonsil of the cerebellum that enters the foramen magnum; this is *tonsillar herniation*. In all of these examples there is a

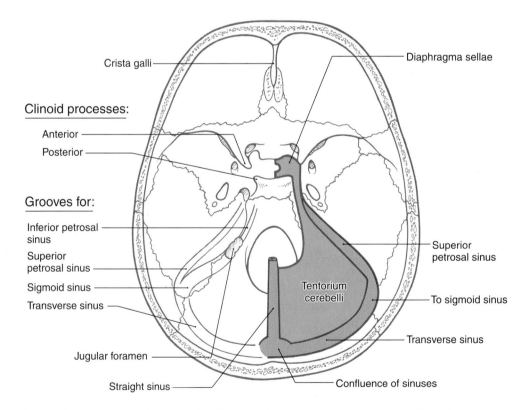

Figure 7–5. View of the cranial base from the dorsal aspect showing the tentorium cerebelli (and its associated sinuses) and the diaphragma sellae. Also indicated are the positions of grooves formed by some of the major sinuses.

characteristic and sometimes predictable change in neurologic status.

Located below the tentorium cerebelli on the midline of the occipital bone is the *falx cerebelli* (see Fig. 7–4). This small dural infolding extends into the space found between the cerebellar hemispheres and usually contains a small *occipital sinus* that communicates with the *confluence of sinuses.*

The smallest of the dural infoldings, the *diaphragma sella* (see Figs. 7–4 and 7–5), forms the roof of the hypophyseal fossa and encircles the stalk of the pituitary. The *cavernous sinuses* are found on either side of the sella turcica, and the *anterior* and *posterior intercavernous sinuses* are found in their respective edges of the diaphragma sella.

The relationships of venous sinuses are discussed in Chapter 8. It should be emphasized, however, that *venous sinuses are endothelium-lined spaces* that communicate with each other. In addition, large veins from the surface of the brain empty into the venous sinuses. As they enter the sinus, these veins are attached to a cuff of dura. Consequently, a blow to the head (or a minor bump to the head in an aged person) may cause the brain to shift just enough in the subarachnoid space to tear a vein at the point where it enters the sinus. This tear may allow venous blood to enter the subarachnoid space or may create a hematoma at the dura-arachnoid interface (see Fig. 7–8).

Cranial Versus Spinal Dura. At the margin of the foramen magnum, the *periosteal dura* essentially stops, but the *meningeal dura* continues caudally in the vertebral canal to eventually attach to the inner aspect of the coccyx as the *filum terminale externum* (*dural part of the filum terminale* or *coccygeal ligament*) (see Fig. 7–2). The *spinal dural sac* is anchored rostrally and caudally and is separated from the adjacent vertebrae by an *epidural space* that contains venous channels, some lymphatics, and fat deposits. There are no dural infoldings around the cord; consequently, there are no venous sinuses in the spinal dura.

Arachnoid Mater

The *arachnoid mater* is located internal to the dural border cell layer and is regarded as having two parts (see Fig. 7–3). The portion of the arachnoid directly apposed to the dural border cells is the *arachnoid barrier cell layer,* and the spindly cells that traverse the subarachnoid space constitute the *arachnoid trabeculae.*

The *subarachnoid space* is located between the arachnoid barrier cell layer and the pial cells located on the surface of the brain or spinal cord. This space contains cerebrospinal fluid, many superficial vessels, and the roots of cranial and spinal nerves as they enter or exit

the nervous system. Enlarged regions of the subarachnoid space are called *subarachnoid cisterns;* these are discussed later in the chapter.

Arachnoid Barrier Cell Layer. Fibroblasts of this layer are more plump than the flattened cells of the dura (see Fig. 7–3). The arachnoid barrier cell layer is tenuously attached to the dural border cell layer by occasional cell junctions. In contrast, arachnoid barrier cells have closely apposed cell membranes and are joined to each other by *numerous tight* (occluding) *junctions*—hence the "barrier" characteristic of this layer. This close apposition of cell membranes excludes any significant extracellular space; consequently, no collagen is found in this layer of the meninges. The tight junctions between these arachnoid cells not only serve as a barrier against the movement of fluids or other substances, but also impart strength to the membrane. In the human, a basement membrane (basal lamina) is found on the surface of the barrier cell layer that faces the subarachnoid space.

Arachnoid Trabeculae and the Subarachnoid Space. The *arachnoid trabeculae* are composed of flattened, irregularly shaped fibroblasts that bridge the subarachnoid space in a random fashion (see Fig. 7–3). Trabecular cells attach to the barrier layer and may attach to each other, to pial cells, or to blood vessels in the subarachnoid space. Although much of the extracellular collagen associated with trabecular cells is confined in the folded processes of these cells, some may be found free in the subarachnoid space. The attachments of the trabecular cells and their framework of collagen fibrils give added strength to the arachnoid mater.

The *subarachnoid space* is located internal to the barrier cell layer and external to the pia mater (see Figs. 7–2 and 7–3). It contains cerebrospinal fluid (CSF), trabecular cells and collagen fibrils, arteries and veins, and the roots of cranial nerves. Although some vessels may lie free in the subarachnoid space, most are covered by a thin layer of the leptomeninges (see Fig. 7–3). These large vessels in the subarachnoid space may rupture, resulting in the spread of blood around the brain; this event is a *subarachnoid hemorrhage. Cerebrospinal fluid* is produced by the *choroid plexuses* of the lateral, third, and fourth ventricles. It exits the ventricular system via the foramina of Magendie and Luschka to enter the subarachnoid space (see arrows in Fig. 7–2). After circulating around the brain and spinal cord, CSF reenters the vascular system primarily through the *arachnoid villi.* The subarachnoid space around the spinal cord is the route used to administer *spinal anesthesia.*

Although it is common to refer to the brain as "floating" in the CSF of the subarachnoid space, it is actually *suspended within this space.* The structural basis for this fact is as follows. The dura is adherent to the skull, the arachnoid to the dura, the arachnoid trabeculae to the pia, and the pia to the surface of the brain. Consequently, the brain is suspended, through this chain, within the fluid milieu of the subarachnoid space by the numerous delicate strands of the arachnoid trabeculae. This is possible because the brain loses about 97% of its weight when it is suspended in CSF. For example, a brain that weighs about 1400 g in air will weigh only about 45 to 50 g in fluid.

Because the arachnoid trabeculae are not rigid, the brain may move within the fluid-filled subarachnoid space. In a closed head injury, the brain may move on its trabecular tethers in response to a sudden blow and be subjected to minor damage (*concussion or contusion*). This injury may result in no, or only momentary, loss of consciousness. Such a minor injury may be found at the point of the blow or at a site opposite the contact (*contrecoup injury*).

Arachnoid Villi. The small specialized portions of the arachnoid that protrude into the superior sagittal sinus through openings in the dura form the *arachnoid villi* or *arachnoid granulations* (Fig. 7–6; also see Fig. 7–2). If they are especially large or calcified (as in older persons), they may be called *pacchionian bodies*.

Arachnoid villi extend into the sinus through tight cuffs in the meningeal dura and are found just off the midline or in cul-de-sacs (the *lateral* or *venous lacunae*) of the sinus. The space in the center of each villus is continuous with the subarachnoid space around the brain. This space is enclosed in a layer of cells that are markedly similar to arachnoid barrier cells, and these arachnoid cells, in turn, are surrounded by a capsule of cells that are essentially the same as dural border cells. These two layers are continuous with their respective meningeal layers through the stalk of the villus (see Fig. 7–6). The *endothelial lining of the sinus* is reflected onto the villus and may cover this structure entirely or may leave a few arachnoid cells exposed; the exposed cells are called *arachnoid cap cells.* The endothelium covering the villus sits on a basement membrane beneath which some extracellular collagen may be found.

Arachnoid villi are structurally adapted for the transport of CSF from the subarachnoid space into the venous circulation (see Fig. 7–6). Cerebrospinal fluid moves only from the villus into the sinus. The two routes of fluid movement are through small intercellular channels located between cells and by way of a vacuole-mediated transport of fluid and other elements (bacteria, blood cells) through villus cells. As CSF traverses the villus, it moves down a pressure gradient from a point of higher pressure (the subarachnoid space) to a point of lower pressure (the venous sinus). If the pressure on the venous side exceeds that on the subarachnoid space side, the flow of CSF will slow or stop. Venous blood, however, never flows from the sinus into the subarachnoid space.

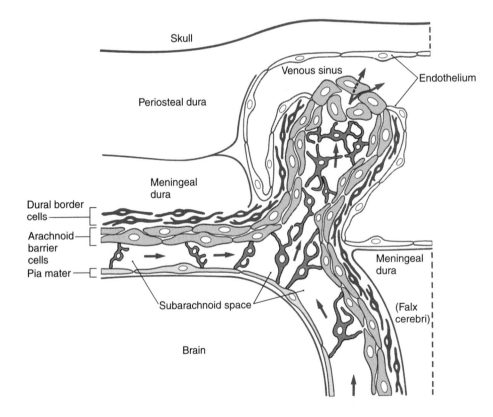

Figure 7–6. Structure of the arachnoid villi. Note the continuity of the cell layers of the villus with those of the meninges. Cerebrospinal fluid (*arrows*) passes from the subarachnoid space into the villus and then into the venous sinus.

Meningiomas and Meningeal Hemorrhages

At this point, it is appropriate to briefly consider meningiomas and meningeal hemorrhages because they are *specifically related to the dura and arachnoid portions of the meninges.*

Meningiomas. Tumors of the meninges, collectively called *meningiomas,* arise primarily from clusters of arachnoid cells found in the villi, at points where cranial nerves or blood vessels traverse the dura, along the base of the skull and at the cribriform plate. An example of a meningioma occurring in the latter location is an olfactory groove meningioma (Fig. 7–7). Although a meningioma may be an incidental finding at magnetic resonance imaging (MRI) performed for another purpose, symptomatic meningiomas are most frequently diagnosed in patients from about 55 to 70 years of age. Some patients, especially those with neurofibromatosis, may present with multiple tumors.

Meningiomas are slowly growing tumors and are classified as benign (typical), atypical, or malignant. The first two types are the most common, are located outside the brain parenchyma, and may invade the adjacent skull but almost never penetrate the brain or

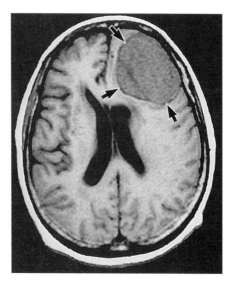

Figure 7–8. Axial magnetic resonance image (T1-weighted) of a meningioma in the frontal lobe of a 62-year-old female patient. Note the sharp interface between the tumor and the brain (*arrows*) and the midline shift. The tumor is clearly external to the brain substance.

spinal cord (Fig. 7–8; see also Fig. 7–7). Malignant meningiomas are rare; they may penetrate the dura and the subjacent brain, thereby complicating their treatment.

Neurologic signs resulting from meningiomas are generally due to compression (see Fig. 7–8) of the brain or spinal cord, or to secondary effects such as edema. Signs and symptoms may also include cranial nerve involvement, long tract signs, or long-standing seizures.

The treatment of choice for meningiomas is surgical removal. In some instances, such as the presence of a tumor in the cavernous sinus, the procedure may be complicated by the concentration of important structures in a small area. Radiation therapy may be used to treat specific types of meningiomas, but chemotherapy has not proved to be effective at the present time.

Extradural and "Subdural" Hemorrhages. If we exclude, for the moment, subarachnoid hemorrhages, *meningeal hemorrhages* can be generally described as *extravasated blood that strips the dura from the skull or dissects open the dural border cell layer* (Fig. 7–9). The most common cause in both situations is an injury to the head, with or without skull fracture. In a head injury, the periosteal dura may be loosened from the skull with consequent damage to a major artery; the middle and accessory meningeal arteries are common victims. Extravascular blood dissects the periosteal dura from the skull and collects to form an *extradural (epidural) hematoma* (see Fig. 7–9). The neurologic deficits seen in patients with epidural hemorrhage are usually those characteristic of increased intracranial pressure. These deficits are, in order of occurrence, headache,

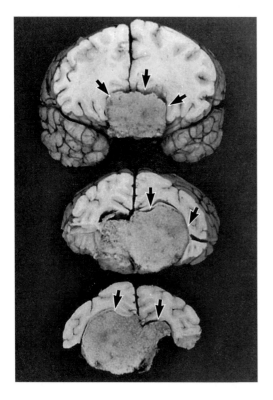

Figure 7–7. A large olfactory groove meningioma. Note that this tumor has significantly compressed (*arrows*) but not invaded the brain in these sequential slices through the frontal lobe. (Courtesy of Dr. Jonathan Fratkin, University of Mississippi Medical Center.)

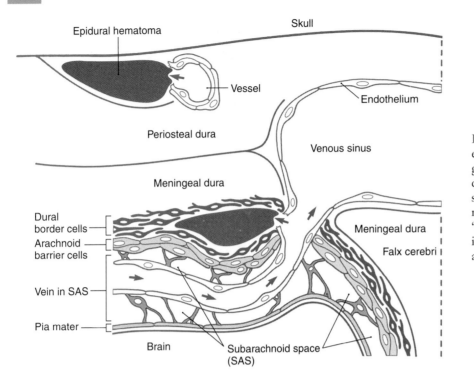

Figure 7-9. The relationship of extravasated blood to the meninges. The epidural hematoma is located between the dura and the skull. Bleeding into the dura-arachnoid interface, classically called a "subdural" hematoma, is actually into a structurally weak cell layer at this juncture.

confusion and disorientation, lethargy, and finally a state of unresponsiveness.

In contrast to extradural hemorrhages, bleeding into the meninges at the junction of the arachnoid with the dura originates mainly from venous structures. A common cause is the tearing of "bridging veins" as they pass through the subarachnoid space and enter a dural venous sinus (see Fig. 7-9). Although these lesions are commonly called "subdural," as noted previously, there is *no naturally occurring space at the arachnoid-dura junction*. Hematomas at this junction are usually caused by extravasated blood that splits open the dural border cell layer (see Fig. 7-9). This extravascular blood does not collect within preexisting space, but rather creates a space at the dura-arachnoid junction. Because these so-called *subdural hematomas* are usually found *within a specific layer of cells*, they actually constitute "*dural border*" hematomas. These lesions generally contain blood in their central area and myofibroblasts, fibroblasts, mast cells, proliferating blood vessels, and dural border cells in the surrounding capsule.

Hygroma. Trauma to the skull may also result in tearing of the arachnoid membrane. In such instances, cerebrospinal fluid, which is under pressure (about 100 to 150 mm H_2O in a recumbent position), also may dissect open and collect with the dural border cell layer. These lesions are called *hygromas*.

Pia Mater

The *pia mater* consists of flattened cells with long, equally flattened processes that closely follow all the surface features of the brain and spinal cord (see Fig. 7-3). The pia and arachnoid together constitute the *leptomeninges*. Vessels in the subarachnoid space (see Fig. 7-3) may be covered by a single layer of pial cells, may be enveloped by several layers of leptomeningeal cells, or may lie free in this space. The pia is separated from the brain surface by a *glial basement membrane* and by occasional places where pial cells pull away from the brain to form a small *subpial space*. Pial cells at the brain surface may be arranged in a single layer or in several layers. Single pial cell processes and their subjacent collagen correspond to the *pia intima*; these closely follow surface features of the brain and spinal cord. When there are several tiers of pial cell processes, the outer layers correspond to the *epipial layer*. In general, the pia is thicker on the spinal cord than on the brain.

Where small vessels penetrate the surface of the brain and spinal cord, they pull along a small envelope of pial cell processes and extracellular space. These *perivascular spaces (Virchow-Robin spaces)* extend for varying distances into the parenchyma of the nervous system and may serve as conduits for the movement of extracellular fluid between the minute spaces around neurons and glial cells and the subarachnoid space.

The spinal cord is anchored in the subarachnoid space by three structures: two pial modifications plus a reticulated septum of arachnoid cell processes that attaches to the posterior midline of the cord. The first of the pial structures, the *denticulate ligaments*, run longitudinally along each side of the spinal cord about midway between the posterior and anterior roots and attach to the inner surface of the arachnoid-lined dural sac (see Fig. 7–2). From each ligament a series of about 20 to 22 structures, shaped much like shark's teeth, extend laterally to attach to the inner surface of the arachnoid-lined dural sac. Second, extending caudally from the conus medullaris is a tough strand composed primarily of pia; this is the *filum terminale internum (pial part of the filum terminale).* The filum terminale internum attaches to the caudal end of the dural sac, which in turn attaches to the coccyx as the *filum terminale externum (coccygeal ligament)* (see Fig. 7–2). Together, these anchoring structures serve a function analogous to that of the arachnoid trabeculae around the brain.

The large space caudal to the conus medullaris, which contains CSF, posterior and anterior roots (constituting the *cauda equina*), and the filum terminale internum, is the *lumbar cistern* (see Fig. 7–2). The retrieval of CSF is an important diagnostic tool for evaluating a variety of central nervous system disorders. A needle introduced into the lumbar cistern (*spinal tap* or *lumbar puncture*) through the third to fourth or fourth to fifth lumbar interspace is the primary method used to collect a sample of CSF from this cistern (see Fig. 9–2).

Cisterns, Subarachnoid Hemorrhages, and Meningitis

The *subarachnoid space* is the thin envelope of space located between the arachnoid and pia (Fig. 7–10). This space has a number of naturally enlarged regions called *subarachnoid cisterns,* which contain CSF, arteries and veins, and in some cases, cranial nerve roots (see Fig. 7–9). Cisterns occur where the brain draws away from the skull as part of its natural variation in shape, thus enlarging the subarachnoid space. In addition to discussing cisterns we shall consider subarachnoid hemorrhage and meningitis at this point, as these clinical problems are most specifically related to the leptomeninges.

Cisterns. Cisterns are usually named according to the structures on which they border. For example, the *interpeduncular cistern* is found in the interpeduncular fossa, the *dorsal cerebellomedullary cistern (cisterna magna)* is found between the cerebellum and the medulla, and so on (see Fig. 7–10). Typically, the shapes of cisterns, as seen on MRI and computed tomography (CT) scans, are determined by the corresponding shapes of surrounding brain structures (Fig. 7–11); this characteristic relationship is useful in diagnosis. The *cisterna magna* is a potential source of CSF if the lumbar cis-

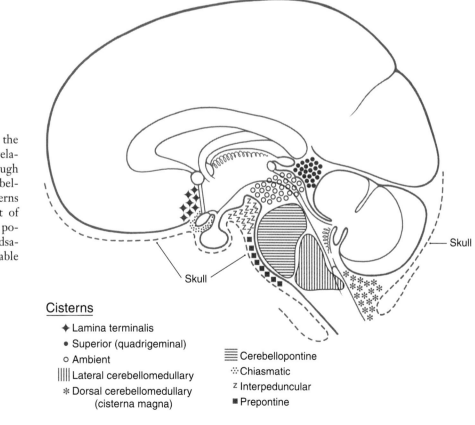

Figure 7–10. The locations of the major subarachnoid cisterns in relation to brain structures. Although the cerebellopontine, lateral cerebellomedullary, and ambient cisterns are located on the lateral aspect of the brainstem, their approximate positions are indicated on this midsagittal view. Compare with Table 7–1.

Cisterns

✦ Lamina terminalis

• Superior (quadrigeminal)

○ Ambient

‖‖‖ Lateral cerebellomedullary

✳ Dorsal cerebellomedullary (cisterna magna)

≡ Cerebellopontine

⋰ Chiasmatic

ᶻ Interpeduncular

■ Prepontine

Skull

Skull

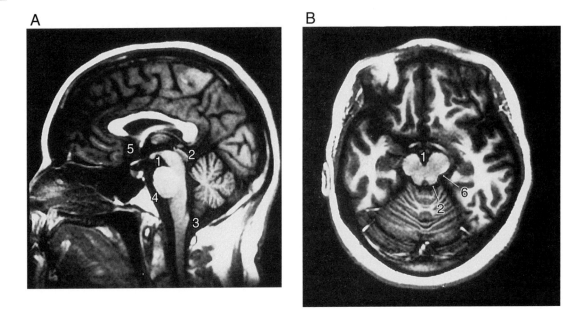

Figure 7–11. Magnetic resonance images in sagittal (*A*) and axial (horizontal, *B*) planes with some of the major cisterns indicated. 1, interpeduncular; 2, superior (quadrigeminal); 3, cisterna magna (dorsal cerebellomedullary); 4, prepontine; 5, of the lamina terminalis; 6, ambient.

tern is not accessible. In a *cisternal puncture*, a needle is carefully introduced into the cisterna magna through the atlanto-occipital membrane and a sample of fluid withdrawn.

Cisterns are bordered by particular brain structures, contain segments of major vessels, and may also contain cranial nerve roots or other structures (Table 7–1). Consequently, a progressively enlarging aneu-

Table 7–1.   Some Principal Cisterns and the Main Arteries, Veins, Cranial Nerves, and Other Structures Associated with Them

Cistern	Artery(ies)	Vein(s)	Cranial Nerve(s)	Structure(s)
Ambient	Portions of posterior cerebral, quadrigeminal, and superior cerebellar arteries	Basal vein (of Rosenthal)	Trochlear	Lateral aspect of crus cerebri
Cerebellopontine (inferior—also called lateral cerebellomedullary)	Vertebral artery and proximal branches of PICA	Retro-olivary and lateral medullary veins	Glossopharyngeal, vagus, spinal accessory, and hypoglossal	Pyramid, inferior olivary eminence, and choroid plexus
Cerebellopontine (superior)	Distal branches of anterior inferior cerebellar, labyrinthine, and basilar arteries	Pontomesencephalic and petrosal veins	Trigeminal, facial, and vestibulocochlear	—
Chiasmatic	Ophthalmic artery and small branches to chiasm and hypophysis	—	Optic nerve and optic chiasm	—
Cisterna magna (also called dorsal cerebellomedullary)	Distal branches of PICA, posterior spinal artery, and branches to choroid plexus of fourth ventricle	Tonsillar and dorsal medullary veins	—	Roots of C1, C2
Interpeduncular	Rostral end of basilar artery and portions of posterior cerebral, choroidal, and thalamogeniculate arteries	Portions of basal vein (of Rosenthal)	Oculomotor root	Mammillary body, medial edge of crus cerebri
Prepontine	Basilar artery and its branches	Pontine veins	Abducens	—
Quadrigeminal	Portions of posterior cerebral, quadrigeminal, and choroidal arteries	Great cerebral vein (of Galen)	Trochlear root	Pineal, superior and inferior colliculi

PICA = posterior inferior cerebellar artery.
Data from Yasargil, 1984.

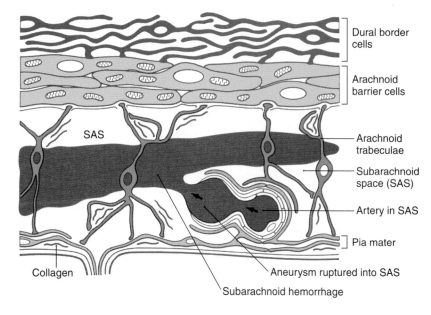

Figure 7–12. Bleeding into the subarachnoid space (subarachnoid hemorrhage) following rupture of an aneurysm into the subarachnoid space.

Labels (clockwise): Dural border cells · Arachnoid barrier cells · Arachnoid trabeculae · Subarachnoid space (SAS) · Artery in SAS · Pia mater · Aneurysm ruptured into SAS · Subarachnoid hemorrhage · Collagen · SAS

rysm or a slow bleed into a particular cistern may result in signs or symptoms related to the structures found in or next to the cistern. For example, an aneurysm protruding into the interpeduncular cistern may affect the oculomotor nerve (see Table 7–1) and, consequently, eye movements or pupil size.

Subarachnoid Hemorrhage. A *subarachnoid hemorrhage* is an extravasation of blood (usually arterial) into the subarachnoid space (Figs. 7–12 and 7–13). *The most common cause of subarachnoid hemorrhage is trauma whereas the most common cause of nontraumatic subarachnoid hemorrhage is rupture of an intracranial aneurysm.* In either case, blood is extruded into the subarachnoid space and may be sequested in cisterns or migrate through the subarachnoid space and subsequently maybe seen on imaging studies to outline structures such as brain divisions or dural reflections (see Fig. 7–13). *Aneurysms* are clearly defined dilations in the walls

of arteries (see Fig. 7–12). Although many aneurysms are thought to be congenital in nature, they may also be caused by an ongoing pathologic process or by trauma, or may be secondary to a general systemic problem such as hypertension. Subarachnoid hemorrhage from a ruptured aneurysm is most common in persons between 40 and 65 years of age. Rupture of an intracranial aneurysm is a catastrophic event with a generally poor prognosis. About one third of affected patients die before or soon after admission to a medical facility, and about one third have permanent disabilities.

The occurrence of a subarachnoid hemorrhage may be signaled by a sudden excruciating headache, neck stiffness, vomiting or nausea, and a depression or loss of consciousness. Bloody CSF obtained by lumbar or cisternal puncture is diagnostic for subarachnoid hemorrhage, and blood can be clearly identified in the

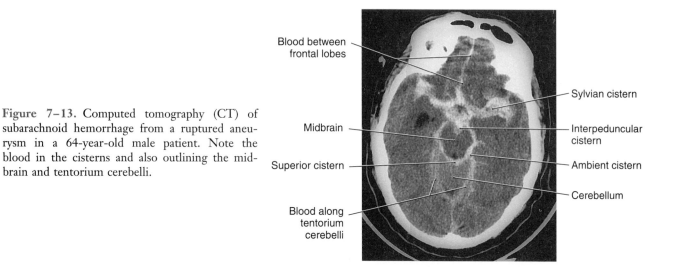

Figure 7–13. Computed tomography (CT) of subarachnoid hemorrhage from a ruptured aneurysm in a 64-year-old male patient. Note the blood in the cisterns and also outlining the midbrain and tentorium cerebelli.

Labels: Blood between frontal lobes · Midbrain · Superior cistern · Blood along tentorium cerebelli · Sylvian cistern · Interpeduncular cistern · Ambient cistern · Cerebellum

subarachnoid space on CT examination (see Fig. 7–13). In some cases a patient may have warning signs and symptoms of an impending subarachnoid hemorrhage (leaking aneurysm). These include intermittent headache, nausea or vomiting, and fainting spells (*syncope*). In persons with neurologic signs that can be traced to an aneurysm, the treatment of choice is to clip the aneurysm or its stalk, thereby separating it from the cerebral circulation.

Meningitis. Meningeal infection may be of either bacterial or viral origin. With *bacterial meningitis*, the meningeal infection is most often located in the subarachnoid space and involves the arachnoid and pia—hence the designation *leptomeningitis*. Such infections may result from a variety of causes including trauma (which may introduce bacteria into the head or spine), septicemia, and metastasis from another site of infection in the body. Bacterial meningitis is generally classified as acute or subacute, depending on how rapidly the disease progresses. The most common causative agents are *Streptococcus pneumoniae* and *Neisseria meningitidis* (accounting for about 70% to 75% of cases).

Signs of acute bacterial meningitis include elevated temperature, alternating chills and fever, and headache; the patient is acutely ill and may have a depressed level of consciousness. These signs and symptoms seen in concert with increased CSF pressure and cloudy CSF containing many white blood cells, increased protein, and bacteria, are diagnostic of the disease. The inflammatory process may result in thickening of the leptomeninges with consequent partial obstruction of CSF flow and signs of hydrocephalus. Although the death rate is low in acute cases with proper treatment, the patient may become ill suddenly and may die within 2 days in rapidly advancing cases.

Subacute bacterial meningitis is usually seen in patients with tuberculosis (tuberculous meningitis) or may be due to mycotic infections. The course of the disease is longer (encompassing weeks rather than days), and the onset is slow and characterized by headache, fever, irritability, and wakefulness at night. In both acute and subacute meningitis, the prognosis is excellent (with about a 90% cure rate) with early diagnosis and proper treatment.

Viral meningitis is caused by a range of viral agents, is most commonly seen in younger patients (under 25 years of age) and is a disease for which no antiviral medications are available. The patient becomes ill over a period of days and experiences fever, headaches of increasing intensity, and confusion and possibly an altered level of consciousness. In a minority of cases, more serious signs and symptoms may be seen, such as seizures, rigidity, or cranial nerve palsies. Treatment in mild cases is supportive and generally focuses on medications for fever, pain, and general discomfort. After an acute period of 1 to 2 weeks, the signs and symptoms moderate, and generally the patient recovers without permanent deficits.

Sources and Additional Reading

Alcolado R, Weller RO, Parrish EP, Garrod D: The cranial arachnoid and pia mater in man: Anatomical and ultrastructural observations. Neuropathol Appl Neurobiol 14–17, 1988.

Al-Mefty O: Meningiomas. Raven Press, New York, 1991.

Frederickson RG: The subdural space interpreted as a cellular layer of meninges. Anat Rec 230:38–51, 1991.

Haines DE: On the question of a subdural space. Anat Rec 230:3–21, 1991.

Haines DE, Harkey LH, Al-Mefty O: The "subdural space": A new look at an outdated concept. Neurosurgery 32:111–120, 1993.

Nabeshima S, Reese TS, Landis DMD, Brightman MW: Junctions in the meninges and marginal glia. J Comp Neurol 164:127–170, 1975.

Nicholas DS, Weller RO: The fine anatomy of the human spinal meninges. J Neurosurg 69:276–282, 1988.

Orlin JR, Osen K, Hovig T: Subdural compartment in pig: A morphologic study with blood and horseradish peroxidase infused subdurally. Anat Rec 230:22–37, 1991.

Peters A, Palay SL, Webster HD: *The Fine Structure of the Nervous System: The Neurons and Supporting Cells*, 3rd ed. WB Saunders, Philadelphia, 1991.

Schachenmayr W, Friede RL: The origin of subdural neomembranes. I. Fine structure of the dura-arachnoid interface in man. Am J Pathol 92:53–68, 1978.

Van Denabeele F, Creemans J, Lambrichts I: Ultrastructure of the human spinal arachnoid mater and dura mater. J Anat 189:417–430, 1996.

Williams PL (ed): Gray's Anatomy, 38th ed. Churchill Livingstone, New York, 1995.

Yasargil MG: Microneurosurgery, I. Microsurgical Anatomy of the Basal Cisterns and Vessels of the Brain, Diagnostic Studies, General Operative Techniques and Pathological Considerations of the Intracranial Aneurysms. Georg Thieme Verlag, Stuttgart, 1984.

Chapter 8

A Survey of the Cerebrovascular System

D. E. Haines

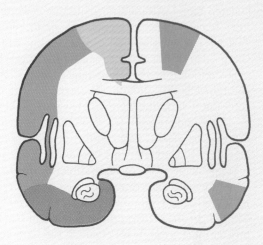

Overview 122

Causes of Vascular Compromise 122

Internal Carotid System 123
Internal Carotid Artery
Anterior Cerebral Artery
Middle Cerebral Artery

Vertebrobasilar System 125
Vertebral Artery
Basilar Artery
Posterior Cerebral Artery

Cerebral Arteries and Watershed Infarcts 128

Circle of Willis 129
Central Branches

Veins and Venous Sinuses of the Brain 130
Cerebral Hemispheres
Basal Aspect of the Brain
Internal Veins of the Hemisphere
Brainstem and Cerebellum

Arteries of the Spinal Cord 134

Veins of the Spinal Cord 135

Blood-Brain Barrier 135

About 50% of the problems that occur inside the cranial cavity and result in neurologic deficits are vascular in origin. Consequently, a good understanding of cerebral vascular patterns is essential to establish an accurate diagnosis. The brain is a voracious consumer of oxygen and therefore requires a great deal of oxygenated blood. Although it makes up only about 2% of total body weight in adults, the brain receives about 15% to 17% of the total cardiac output and consumes about 20% of the oxygen used by the entire body!

An ongoing flow of oxygenated blood is essential for continued brain function. The average person will lose consciousness if the brain is deprived of blood for 10 to 12 seconds; after 3 to 5 minutes, irreparable brain damage or death may result. There are exceptions, however. Individuals who become hypothermic with a subsequent decrease in arterial blood flow to the brain, as in a winter near-drowning, may be revived after 10, 15, or even 20 minutes with little or no permanent damage. In these cases, the reduction in body temperature protects the brain against the consequences of reduced blood flow.

Overview

Blood is supplied to the brain by the *internal carotid* and *vertebral arteries*. The internal carotid arteries enter the skull and then divide into the *anterior* and *middle cerebral arteries*. The vertebral arteries pass through the foramen magnum and join to form the *basilar artery;* hence, the term *vertebrobasilar* is frequently applied to this part of the cerebral circulation. The basilar artery branches into right and left *posterior cerebral arteries.*

Venous outflow from the brain travels through superficial and deep veins, which drain into the dural venous sinuses. Blood in the sinuses, in turn, enters the internal jugular vein. Superficial veins in the scalp and veins in the orbit also may communicate with the dural sinuses, but these are not major conduits for venous drainage from the brain.

This chapter presents an *overview of the cerebrovascular system, with emphasis on the distribution pattern of vessels on the surface of the brain and spinal cord.* Details of the distribution of blood vessels to internal structures are covered when we consider nuclei and tracts within the central nervous system.

Causes of Vascular Compromise

Intracranial hemorrhage originates from arteries or veins and may result from diseases, trauma, developmental defects, or infections. Such bleeding is classified according to its location. *Meningeal hemorrhages* are found in relation to the coverings of the brain (see also Chapter 7), whereas bleeding into the subarachnoid space is called *subarachnoid hemorrhage.* Hemorrhage

may occur into the ventricular spaces (*intraventricular*) or into the substance of the brain (*parenchymatous*). Although many events can lead to cerebral vascular problems with resultant dysfunction, only three examples—aneurysm, cerebral embolism, and *arteriovenous malformation* (Fig. 8–1)—are considered here.

An *aneurysm* is a dilation of a vessel wall, usually an artery. The cavity of the aneurysm is continuous with the lumen of the vessel from which it originates (Fig. 8–1*A*). Cerebral aneurysms may range from small (berry aneurysms) to very large (giant aneurysms >2 cm in diameter); the latter may cause damage by compression. Most intracranial aneurysms (about 85%) are found on branches of the internal carotid artery. Regardless of where they occur, *intracranial aneurysms are frequently located at the branch points of vessels.* The treatment of choice is to clip the stalk of the aneurysm so as to separate its friable sac from the cerebral circulation.

A *cerebral embolism* is the occlusion of a cerebral vessel by some extraneous material (e.g., clot, tumor cells, plaque fragments). This occlusion leads to *ischemia* (a localized anemia) and, if prolonged, ultimately to *infarction* (a localized vascular insufficiency resulting in necrosis) of the area served by the vessel (see Fig. 8–1*B*). An embolus made up exclusively of blood products is called a *thrombus.* The size of the embolus de-

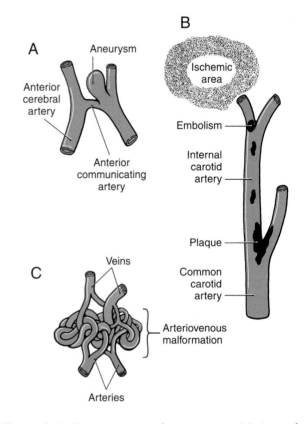

Figure 8–1. Representation of an aneurysm (*A*), an embolism (*B*), and an arteriovenous malformation (*C*).

termines where it lodges. Very small emboli may temporarily occlude small cerebral vessels and give rise to a *transient ischemic attack*, a sudden loss of neurologic function that usually resolves within a few hours to a day. On the other hand, large emboli that suddenly occlude major vessels may result in catastrophic neurologic problems and even death.

An *arteriovenous malformation* (AVM) results when the communications between major arteries and veins do not develop normally (see Fig. 8–1C). These lesions consist of masses of tortuous, interconnecting channels composed of large arteries and veins that communicate directly with each other. An AVM may be located on the surface of the brain or within the substance of the hemisphere or brainstem. Subarachnoid hemorrhage may be the result of bleeding from an aneurysm or AVM. Superficially located AVMs can be surgically removed. However, newer interventional methods now make it possible to pass a small canula into an AVM and inject substances that occlude the larger channels and effectively isolate this vascular malformation from the rest of the cerebral circulation.

AVMs range from small and difficult to detect to quite large; large AVMs may cause damage to adjacent structures (Fig. 8–2). These lesions are commonly identified in the second or third decade of life, although signs and symptoms (focal seizures, hemiparesis) may be noted earlier. Bleeding from AVMs is common and may be "silent" or may result in obvious neurologic deficits.

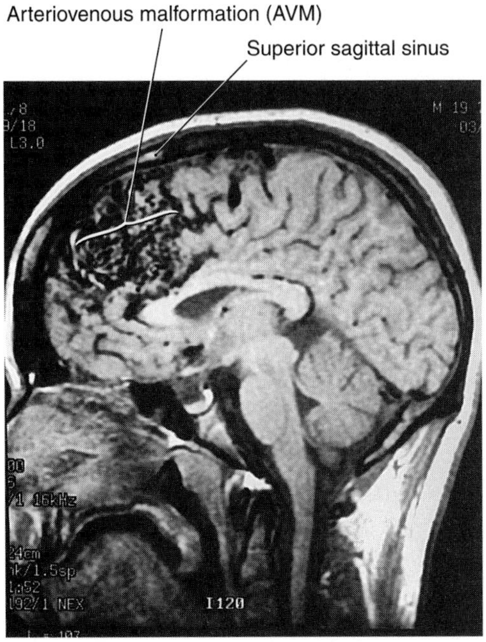

Figure 8–2. Sagittal magnetic resonance image (T1-weighted) near the midline showing an arteriovenous malformation in the frontal lobe. This lesion includes branches of the anterior cerebral artery and drains into the superior sagittal sinus.

Internal Carotid System

Internal Carotid Artery. The *internal carotid artery* consists of a *cervical part* that ascends in the neck, a *petrous part*, a *cavernous part*, and a *cerebral part*. The *petrous part* is located in the carotid canal and has no branches of consequence. The *cavernous part* passes through the cavernous sinus and gives rise to hypophysial and meningeal branches.

The *cerebral part* of the internal carotid begins where this vessel penetrates the dura just anterior (ventral) to the optic nerve. Its branches are the *ophthalmic, posterior communicating*, and *anterior choroidal arteries* (Fig. 8–3). The ophthalmic artery, after entering the orbit via the *optic foramen*, gives rises to the *central artery of the retina* just distal to the foramen. This latter vessel then passes along the ventral aspect of the optic nerve (it may be inside or outside the dura) to eventually enter the nerve about 10 to 15 mm behind the bulb of the eye. The central artery of the retina is an important source of blood supply to the retina. Occlusion of the ophthalmic artery may result in significant visual loss in the ipsilateral eye. Also, aneurysms at the ophthalmic-carotid intersection may cause visual loss because of direct pressure on the optic nerve. The posterior communicating artery joins the *posterior cerebral artery*, and the anterior choroidal artery follows caudolaterally along the optic tract (see Fig. 8–3). The internal carotid artery ends by dividing into the *anterior* and *middle cerebral arteries*.

Anterior Cerebral Artery. The *anterior cerebral artery* passes superiorly over the optic chiasm and is joined to its counterpart by the *anterior communicating artery* (see Fig. 8–3). That part of the anterior cerebral artery between its origin from the internal carotid and the anterior communicating artery is the A_1 segment. The anterior communicating artery and the distal parts of the A_1 segments are located in the *cistern of the lamina terminalis* and give rise to small branches that serve structures in the immediate area.

About 20% to 25% of all intracranial aneurysms are found either on the anterior communicating artery or where this vessel joins the anterior cerebral artery (see Fig. 8–1A). Patients with aneurysms of the anterior communicating artery may have visual deficits because of the close proximity of this vessel to the optic chiasm.

Distal to the anterior communicator, the anterior cerebral artery branches over the medial surface of the hemisphere to about the level of the parieto-occipital sulcus; collectively these branches from the A_2 segment (Fig. 8–4). The main branches of A_2 lie within the *callosal cistern*. Figure 8–4 shows the distribution pat-

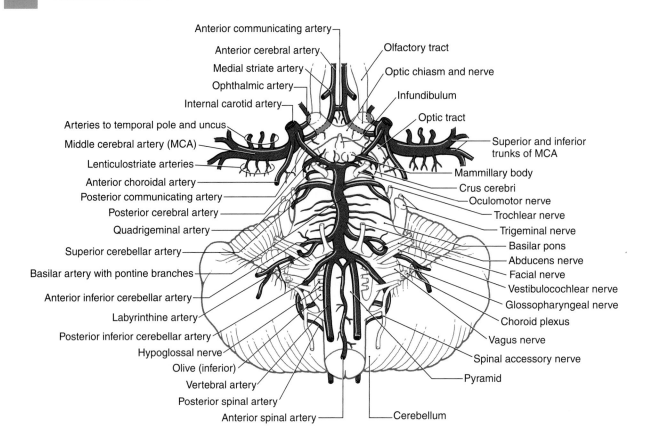

Figure 8–3. Arteries on the base of the brain showing the relationship of vessels to structures and the arrangement of the circle of Willis (see also Fig. 8–10).

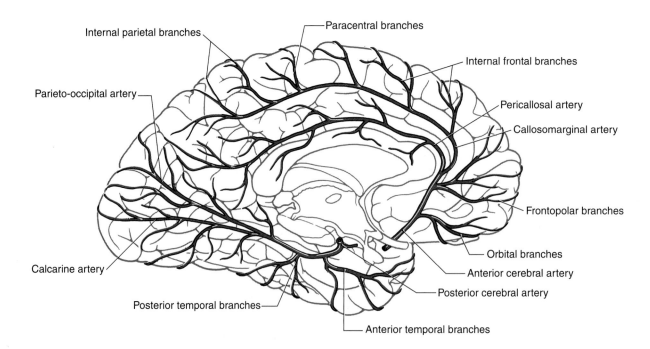

Figure 8–4. Arteries on the medial surface of the cerebral hemisphere and on the inferior surface of the temporal lobe.

tern of the major named branches of the anterior cerebral artery on the medial aspect of the hemisphere. Aneurysms of A_2 are usually located at the branch points of the frontopolar and callosomarginal arteries.

Middle Cerebral Artery. The *middle cerebral artery* (*MCA*) is usually (70% of the time) the larger of the two terminal branches of the internal carotid artery (see Fig. 8–3). The part of this vessel located between its origin from the internal carotid and the point where it branches at the ventromedial aspect of the insula (*the limen insulae*) is the M_1 segment (Fig. 8–5). Branches from M_1 serve adjacent medial and rostral aspects of the temporal lobe and, via *lenticulostriate arteries*, structures located inside the hemisphere. The M_1 segment is located in medial portions of the *sylvian cistern*.

On the ventromedial aspect of the insular cortex, the M_1 segment usually bifurcates into *superior* and *inferior trunks* (see Figs. 8–3 and 8–5). These trunks and their distal branches collectively serve the insular cortex, inner aspects of the opercula, and the lateral surface of the cerebral hemisphere. Those branches on the insular cortex (insular part of the MCA) are the M_2 segment (see Fig. 8–5). As these vessels pass through the *sylvian cistern*, they give rise to small branches that serve the insula. The larger vessels continue onto the inner aspects of the opercula (frontal, parietal, temporal), where they are designated as M_3, the opercular part of the MCA (see Fig. 8–5). Distal branches of the superior and inferior trunks exit the lateral fissure and serve, respectively, *cortical areas located above and below this fissure*. These cortical branches (cortical part of the MCA) collectively form the M_4 segment (see Fig. 8–5). Most of these distal branches are named according to the general area, or structure, they serve; their distribution patterns are shown in Figure 8–6. Aneurysms of the MCA occur most frequently at the bifurcation of the M_1 segment into superior and inferior trunks.

Vertebrobasilar System

Vertebral Artery. The *vertebral artery* leaves the transverse foramen of C1, loops caudally and medially

around the lateral mass of the atlas, and then pierces the atlanto-occipital membrane, to which it is anchored. This circuitous portion of the vertebral artery is vulnerable to injury. For example, hyperextension of the head may compress the vertebral artery between the occipital bone and the posterior arch of the atlas, and extreme rotation of the head may put torsion on this artery and restrict blood flow. The deficits seen in these patients fall under a general classification referred to as *vertebrobasilar insufficiency*. Once inside the subarachnoid space, the vertebral artery is located in the *lateral cerebellomedullary cistern*.

Branches of the vertebral artery supply the medulla, parts of the cerebellum, and the dura of the posterior fossa (Fig. 8–7; see also Fig. 8–3). Its first major branch, the *posterior inferior cerebellar artery* (PICA), arches around the posterolateral medulla and sends branches to this part of the brainstem. Posteriorly, the PICA is located in the *cisterna magna* (dorsal cerebellomedullary cistern). It serves the choroid plexus of the fourth ventricle and then branches over medial parts of the inferior cerebellar surface (see Fig. 8–7). In about 75% of brains, the *posterior spinal artery* is a branch of the PICA; in the other 25%, it arises from the vertebral artery. The posterior spinal artery serves dorsolateral regions of the medulla *caudal to the area served by the PICA* (see Fig. 8–7). The *vertebral artery* supplies the anterolateral medulla and, just before joining its counterpart on the opposite side, gives rise to the *anterior spinal artery*. This vessel usually (85% of cases) originates as two small trunks, which join to form a single artery that courses caudally in the ventral median fissure of the medulla and continues into the spinal cord (see Fig. 8–3; see also Fig. 8–19). The anterior spinal artery is found in the *premedullary cistern*.

Aneurysms of the vertebral artery or its major branches are not common. When present, they are usually found where the PICA branches from the vertebral artery.

Basilar Artery. The *basilar artery* lies in a shallow depression on the ventral surface of the pons in the *prepontine cistern* (see Fig. 8–3). Its first large branch, the *anterior inferior cerebellar artery* (*AICA*), arises from the

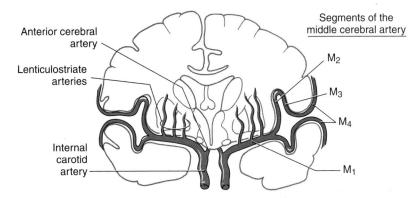

Figure 8–5. A representation of the hemisphere in coronal section showing the segments of the middle cerebral artery (M_1 to M_4).

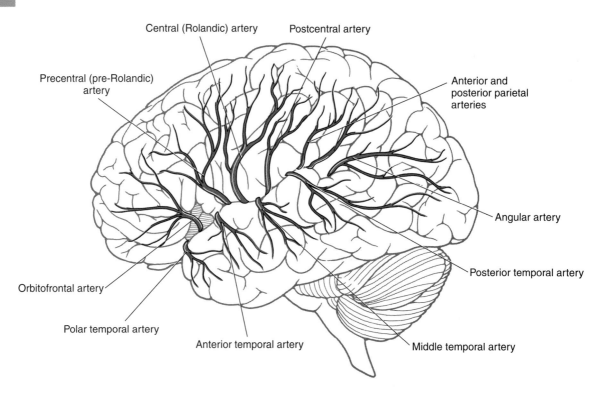

Figure 8–6. Branches of the middle cerebral artery on the lateral surface of the hemisphere. Polar temporal and anterior arteries are branches of M_1; the remaining arteries represent branches of M_4.

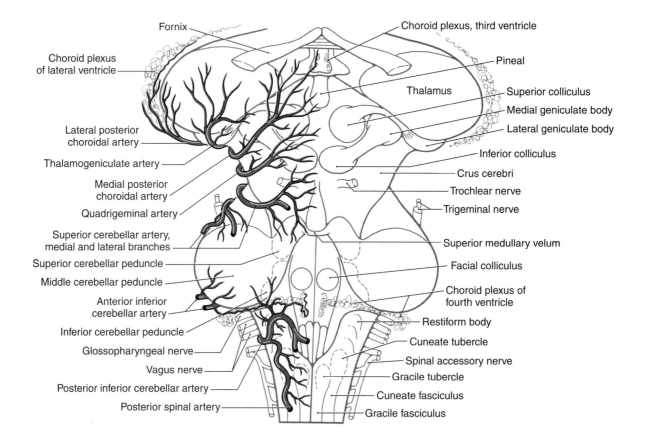

Figure 8–7. Major arteries on the posterior aspect of the brainstem, arteries to the caudal thalamus, and arteries to the choroid plexuses of the lateral, third, and fourth ventricles.

lower one third of the basilar artery and passes through the *cerebellopontine cistern* as it wraps around the caudal aspect of the middle cerebellar peduncle. The AICA serves ventral and lateral surfaces of the cerebellum, parts of the pons, and the portion of choroid plexus that extends out of the foramen of Luschka into the cerebellopontine angle (see Fig. 8–7). The *labyrinthine artery* is usually a branch of the AICA although it may, albeit rarely, originate from the basilar artery. It arises close to the origins of the facial and vestibulocochlear nerves (see Fig. 8–3) and enters the internal acoustic meatus along with these nerves.

The basilar artery gives rise to numerous *pontine arteries* (see Fig. 8–3). These arteries may penetrate the pons immediately as *paramedian branches*, travel for a short distance around the pons as *short circumferential branches*, or pass for longer distances as *long circumferential branches*.

The last major branches of the basilar artery are the *superior cerebellar arteries*. Just distal to their origin each superior cerebellar artery divides into *medial and lateral branches*, which serve their respective regions of the superior surface of the cerebellum (see Figs. 8–3 and 8–7) and most of the cerebellar nuclei. These vessels pass laterally *just caudal to the root of the oculomotor nerve* and wrap around the brainstem in the *ambient cistern* to ultimately serve caudal parts of the midbrain and the

entire superior surface of the cerebellum (see Figs. 8–3 and 8–7).

About 15% of all intracranial aneurysms occur in the vertebrobasilar system. Most are found in relation to the basilar bifurcation and may therefore involve the oculomotor nerve. In similar fashion, an aneurysm of the AICA may produce symptoms of facial or vestibulocochlear nerve involvement, as this vessel travels adjacent to these nerves.

Posterior Cerebral Artery. At the pons-midbrain junction the *basilar artery* bifurcates in the *interpeduncular cistern* and gives rise to the *posterior cerebral arteries*. Each posterior cerebral artery passes laterally *just rostral to the root of the oculomotor nerve* (see Fig. 8–3), wraps around the midbrain in the *ambient cistern*, and then joins the anterior and medial surfaces of the temporal lobes (Fig. 8–8; see also Fig. 8–4). The *posterior cerebral artery* sends branches to the midbrain and thalamus and to the ventral and medial surfaces of the temporal and occipital lobes as far as the level of the parieto-occipital sulcus (see Figs. 8–4 and 8–8).

The posterior cerebral artery is divided into *segments* P_1 to P_4. The P_1 *segment* is located between the basilar bifurcation and the posterior communicating artery. It gives rise to small perforating vessels and to *quadrigeminal* and *thalamoperforating arteries* (see Figs. 8–3 and 8–7). The portion of the posterior cerebral artery be-

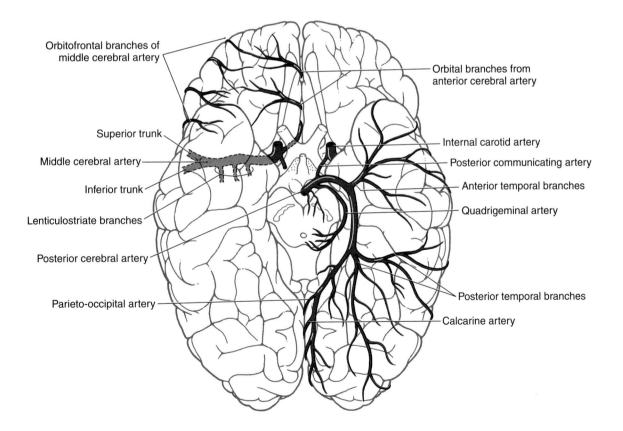

Figure 8–8. Distribution pattern of the major branches of the posterior cerebral artery.

tween the posterior communicator and the inferior temporal branches is the P_2 segment. *Medial* and *lateral posterior choroidal* and *thalamogeniculate arteries*, as well as small perforating branches to the midbrain, originate from P_2. The P_3 segment is the portion of the artery that gives rise to its *temporal branches*, and the *parieto-occipital* and *calcarine arteries* form the P_4 segment (see Figs. 8–4 and 8–8).

Cerebral Arteries and Watershed Infarcts

The territories supplied by branches of the *anterior*, *middle*, and *posterior cerebral* arteries are summarized in Figure 8–9. Each of the functional regions of the cerebral cortex that we shall discuss later lies within the distribution of a particular cerebral artery.

On the lateral surface of the hemisphere, the terminal branches of anterior and middle and middle and posterior cerebral arteries overlap, forming *border zones*

between the areas served by these arteries (see Fig. 8–9). Smaller border zones are also located between the territories of the anterior and posterior cerebral arteries at the parieto-occipital sulcus and between the territories of the cerebellar arteries (see Fig. 8–9). The brain tissue located in these *border zones* is particularly susceptible to damage under conditions of sudden *systemic hypotension* or when there is *hypoperfusion* of the distal vascular bed of a major cerebral artery.

In the case of the cerebral arteries, inadequate perfusion of the *border zones* may result in *watershed infarcts* (see Fig. 8–9). Such lesions represent about 10% of all brain infarcts and may be caused by, for example, hypotension or embolic showers. Damage to the anterior cerebral–middle cerebral border zone in one hemisphere (*anterior watershed infarcts*; see Fig. 8–9) causes a contralateral hemiparesis of the lower extremity and expressive language deficits or behavioral changes. A *posterior watershed infarct* (damage at the middle cerebral–posterior cerebral border zone) commonly produces a partial visual loss accompanied by a variety of language problems.

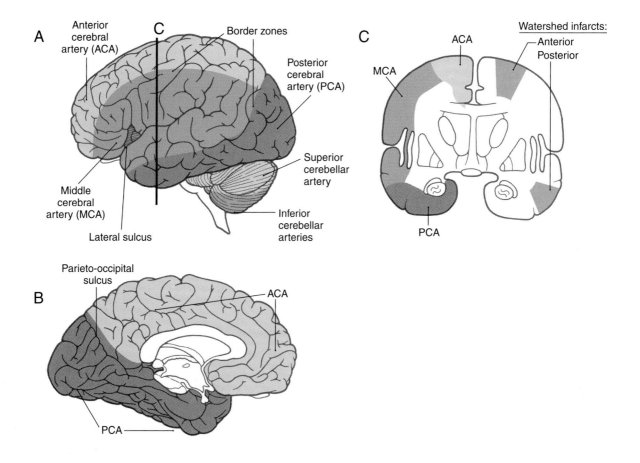

Figure 8–9. Lateral (*A*), medial (*B*), and cross-sectional (*C*) views of the hemisphere showing the regions served by the anterior cerebral (green), middle cerebral (blue), and posterior cerebral (red) arteries. The distal territories of these vessels overlap at their peripheries and create border zones. These zones are susceptible to infarcts (*C*) in cases of hypoperfusion of the vascular bed. Small border zones also exist (*A*) between superior (green) and inferior (blue) cerebellar arteries.

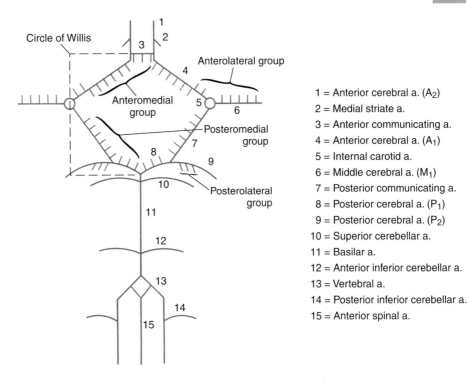

Figure 8–10. Stick drawing of the vertebrobasilar and internal carotid systems showing the configuration of the circle of Willis. The four groups of ganglimic (or central) arteries are shown.

1 = Anterior cerebral a. (A_2)
2 = Medial striate a.
3 = Anterior communicating a.
4 = Anterior cerebral a. (A_1)
5 = Internal carotid a.
6 = Middle cerebral a. (M_1)
7 = Posterior communicating a.
8 = Posterior cerebral a. (P_1)
9 = Posterior cerebral a. (P_2)
10 = Superior cerebellar a.
11 = Basilar a.
12 = Anterior inferior cerebellar a.
13 = Vertebral a.
14 = Posterior inferior cerebellar a.
15 = Anterior spinal a.

Circle of Willis

The *circle of Willis* (*cerebral arterial circle*) is actually a roughly shaped heptagon of arteries located on the anterior (ventral) surface of the brain (Figs. 8–10 and 8–11; see also Fig. 8–3). This loop of vessels passes around the optic chiasm and the optic tract, crosses the crus cerebri of the midbrain, and joins at the pons-midbrain junction. Important structures located inside this circle include the optic chiasm and tracts, infundibulum and tuber cinereum, the mammillary bodies, the hypothalamus, and structures of the interpeduncular fossa (see Figs. 8–3 and 8–11). Arteries forming the circle of Willis give rise to numerous *perforating* (*central or ganglionic*) *branches*, which serve structures located deep to their origin (see Figs. 8–3 and 8–10) and to the large cortical branches (*anterior, middle, and posterior cerebral arteries*) discussed earlier (see Fig. 8–11).

Central Branches. The perforating branches of the circle of Willis are divided into four groups (see Fig. 8–10). The *anteromedial group* originates from A_1 and from the anterior communicating artery. These vessels serve structures in the area of the optic chiasm and anterior parts of the hypothalamus. The *anterolateral group* arises from M_1 with A_1 also sending some branches into this area. Included in this group are the *lenticulostriate arteries*, which serve the interior of the hemisphere. Vessels of the anterolateral group enter the hemisphere via the *anterior perforated substance*. The *posteromedial group* originates from P_1 and from the posterior communicating artery. These vessels supply the crus cerebri and the middle and caudal portions of the hypothalamus, and as they enter the interpeduncular fossa, they form the *posterior perforated substance*. The *thalamoperforating arteries* are part of the posteromedial group, and as their name implies, they serve the thalamus. *The posterolateral group arises from P_2 and is*

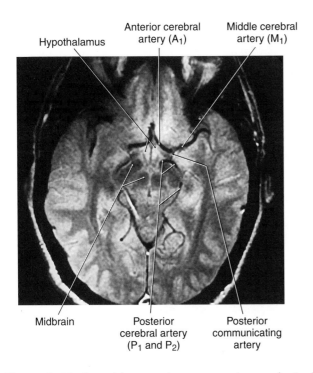

Figure 8–11. An axial magnetic resonance image obtained through the anterior (ventral) aspect of the hemisphere showing most of the vessels forming the circle of Willis (compare with Figs. 8–3 and 8–10).

composed of the *thalamogeniculate* and *posterior choroidal arteries* and some small penetrating branches that enter the midbrain. These vessels serve parts of the thalamus and the choroid plexus.

Veins and Venous Sinuses of the Brain

In contrast to the arterial supply of the brain, which comes from *two major sources*, the venous drainage of the brain exits the skull through *one major vessel*. Venous blood from superficial and deep veins enters the dural sinuses, which, in turn, drain into the *internal jugular vein*. The major venous sinuses are endothelium-lined channels closely associated with the meningeal reflections (see Chapter 7). The *superior* and *inferior sagittal sinuses* are located in the attached and free edges of the falx cerebri, respectively. The *straight sinus* is found where the falx cerebri attaches to the tentorium cerebelli. The other venous sinuses are located adjacent to the inner surface of the skull at specific locations.

Cerebral Hemispheres. The *cerebral veins* on the lateral surface of the hemisphere drain into the superior sagittal and transverse sinuses and into the *superior anastomotic vein (of Trolard)* and the *inferior anastomotic vein (of Labbé)* (Fig. 8–12). These large *anastomotic veins* form channels between the superior sagittal and transverse sinuses and the *superficial middle cerebral vein*. The latter vessel courses medially around the temporal pole to end in the cavernous sinus (Fig. 8–13).

Small vessels on the mid-sagittal surface of the hemisphere drain into the sagittal sinus and, from the medial region of the temporal lobe, into the *basal vein (of Rosenthal)* (Fig. 8–14). The venous blood in these channels, and from the corpus callosum and the interior of the hemisphere (*internal cerebral veins*), drains into the *great cerebral vein (of Galen)* and then into the straight sinus. The *confluence of sinuses (confluens sinuum)* is formed by the junction of the straight sinus, the superior sagittal sinus, and both transverse sinuses (see Figs. 8–13 and 8–14). Rather than a true confluence, the superior sagittal sinus usually drains into the right transverse sinus and the straight sinus into the left.

Basal Aspect of the Brain. Figure 8–13 shows the venous structures on the anterior (ventral) surface of the hemisphere. The *basal vein (of Rosenthal)* begins on the orbital cortex as the *anterior cerebral vein* and in the sylvian fissure as the *deep middle cerebral vein* and proceeds around the medial edge of the temporal lobe to join the straight sinus. It receives venous blood from the midbrain and medial areas of the temporal lobe (see Figs. 8–13 and 8–14). The *transverse* and *sigmoid sinuses* form a shallow groove on the internal surface of the occipital and temporal bones, respectively, and re-

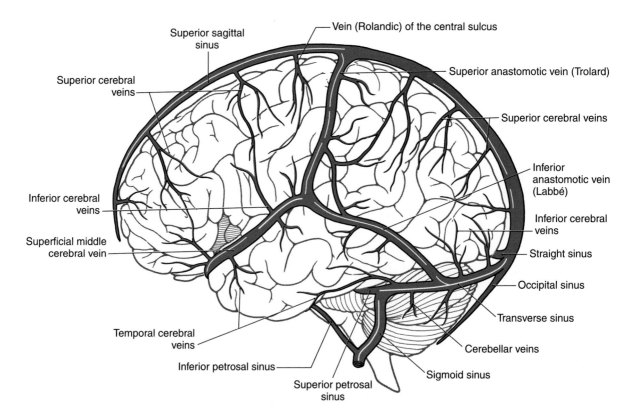

Figure 8–12. Veins and sinuses of the brain from the lateral aspect.

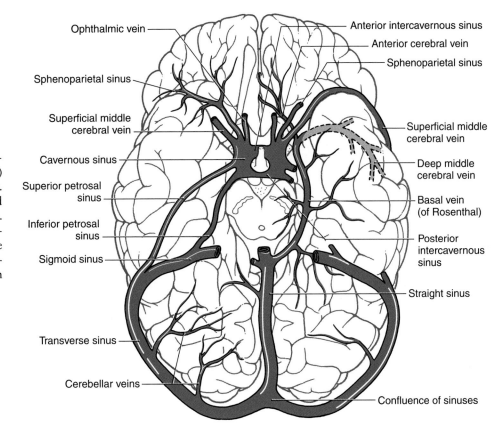

Figure 8-13. Veins and sinuses on the anterior (ventral) surface of the hemisphere. The cerebellum, pons, and caudal midbrain are removed. For clarity, the petrosal sinuses are shown only on the left and the basal vein (of Rosenthal) only on the right. In life, these vessels are bilateral.

Labels: Ophthalmic vein, Sphenoparietal sinus, Superficial middle cerebral vein, Cavernous sinus, Superior petrosal sinus, Inferior petrosal sinus, Sigmoid sinus, Transverse sinus, Cerebellar veins, Anterior intercavernous sinus, Anterior cerebral vein, Sphenoparietal sinus, Superficial middle cerebral vein, Deep middle cerebral vein, Basal vein (of Rosenthal), Posterior intercavernous sinus, Straight sinus, Confluence of sinuses

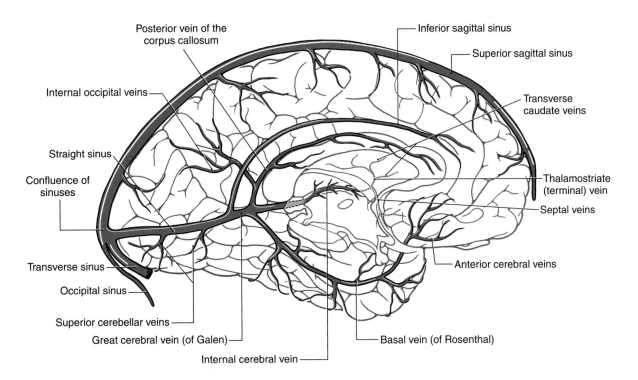

Labels: Posterior vein of the corpus callosum, Internal occipital veins, Straight sinus, Confluence of sinuses, Transverse sinus, Occipital sinus, Superior cerebellar veins, Great cerebral vein (of Galen), Internal cerebral vein, Inferior sagittal sinus, Superior sagittal sinus, Transverse caudate veins, Thalamostriate (terminal) vein, Septal veins, Anterior cerebral veins, Basal vein (of Rosenthal)

Figure 8-14. Veins and sinuses on the medial surface of the hemisphere and anterior (ventral) part of the temporal lobe.

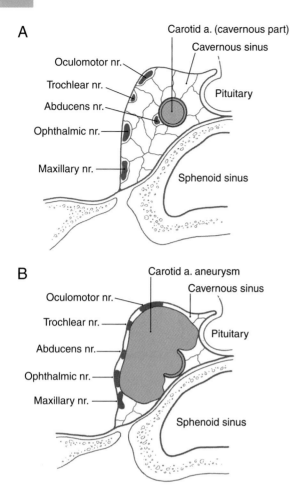

A

Carotid a. (cavernous part)

Cavernous sinus

Oculomotor nr.

Trochlear nr.

Abducens nr.

Pituitary

Ophthalmic nr.

Maxillary nr.

Sphenoid sinus

B

Carotid a. aneurysm

Cavernous sinus

Oculomotor nr.

Trochlear nr.

Pituitary

Abducens nr.

Ophthalmic nr.

Maxillary nr.

Sphenoid sinus

Figure 8–15. The carotid artery and cranial nerves of the cavernous sinus (*A*) in cross section. An aneurysm of the cavernous part of the carotid artery (*B*) will damage some or all of the cranial nerves passing through the sinus.

ceive several tributaries. Venous blood travels from the transverse and petrosal sinuses into the *sigmoid sinus* and then into the *internal jugular vein* at the jugular foramen (see Figs. 8–12 and 8–13).

The *cavernous sinus* (sometimes this is called a venous plexus) is located on either side of the body of the sphenoid bone. It contains the *cavernous part of the internal carotid artery*, the *abducens, oculomotor,* and *trochlear nerves;* and the *ophthalmic* and *maxillary branches* of the trigeminal nerve (Fig. 8–15). These nerves are found internal to the dura surrounding the sinus but are external to its endothelial lining.

The main tributaries of the cavernous sinus are illustrated in Figure 8–13. Each cavernous sinus communicates with its counterpart through the *intercavernous sinuses*. Caudally, the cavernous sinus drains into the *superior* and *inferior petrosal sinuses* and the *basilar plexus* on the ventral aspect of the brainstem.

The neurologic deficits seen in patients with an aneurysm of the cavernous part of the carotid artery are

related to the compact nature of the sinus and the close apposition of cranial nerves III, IV, VI, and V$_1$ and V$_2$ to the carotid artery (see Fig. 8–15*A* and *B*). An expanding aneurysm will affect the adjacent nerves, resulting in a partial or complete paralysis of eye movement, loss of the corneal reflex, and paresthesias or pain within the distribution of the ophthalmic and maxillary nerves. The direct shunting of blood from the internal carotid artery into the cavernous sinus, a *carotid-cavernous fistula*, is rarely the result of a ruptured aneurysm but may occur secondary to trauma (Fig. 8–16).

Internal Veins of the Hemisphere. The main venous channels draining internal structures of the hemisphere are the *internal cerebral veins* (Fig. 8–17; see also Fig. 8–13). They course along the dorsomedial edge of the thalamus and are located in the tela choroidea of the third ventricle. The principal tributaries of the internal cerebral veins and their relationships to adjacent structures are shown in Figure 8–17. Of these, the *thalamastriate vein* (also called the *terminal vein*) merits com-

Anterior cerebral artery

Middle cerebral artery

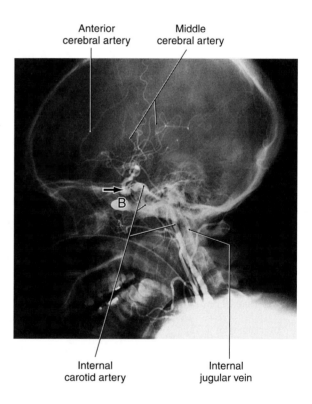

Internal carotid artery

Internal jugular vein

Figure 8–16. Traumatic carotid artery–cavernous sinus fistula (*arrow*). The patient was shot in the face, the bullet (B) entering the orbit and damaging the internal carotid artery in the cavernous sinus. Note that the radiopaque substance injected into the common carotid artery appears in the anterior and middle cerebral arteries *and* internal jugular vein *before* appearing in the veins and sinuses of the head. This means that some blood is passing from the internal carotid into the cavernous sinus and then directly into the internal jugular vein.

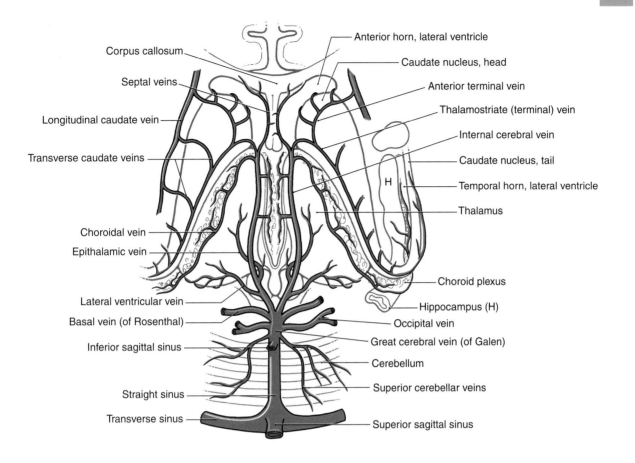

Figure 8–17. Veins draining internal areas of the hemisphere and the tributaries of the great cerebral vein and straight sinus. H, hippocampus.

ment. It is found in association with the stria terminalis and drains the caudate nucleus (via the *transverse caudate veins*) and internal regions of the hemisphere dorsal and lateral to the caudate nucleus.

The two internal cerebral veins join to form the *great cerebral vein* (of Galen). This large venous channel has several notable tributaries (see Figs. 8–13 and 8–17) and is caudally continuous with the *straight sinus.*

Cerebral and spinal veins and dural sinuses lack valves. Consequently, pathologic processes may alter normal venous flow patterns and result in the transport of material *into* the brain. For example, a tumor or infection in the orbit may cause venous blood to flow toward the cavernous sinus rather than away from it. In this way, infectious material or tumor cells may pass from the orbit into the cavernous sinus and, through its connecting channels, to other parts of the brain.

Malformations of the great cerebral vein of Galen are sometimes described as a special type of AVM, or as an aneurysm. This lesion is usually seen in newborns or infants (Fig. 8–18). In these cases, the great cerebral vein is grossly enlarged and fed by large and abnormal branches of the cerebral and cerebellar arteries. Bulging fontanelles, progressive hydrocephalus (resulting from occlusion of the cerebral aqueduct), and

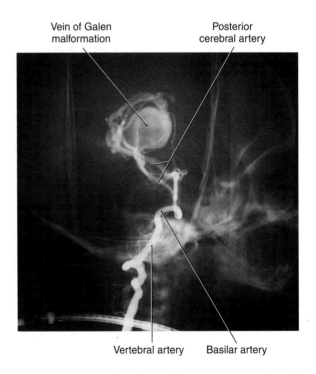

Figure 8–18. Vein of Galen malformation in an infant. This lesion is served by the posterior cerebral arteries.

dilated veins in the face and scalp are characteristic findings.

Brainstem and Cerebellum. The brainstem is drained by a loosely organized network of venous channels located on its surface. In general these vessels enter larger veins or venous sinuses located in the immediate vicinity. For example, veins of the midbrain enter the great cerebral and basal veins, whereas those of the pons and medulla enter the petrosal sinuses, the cerebellar veins, and (from the medulla) the venous channels on the surface of the spinal cord.

Venous drainage from the cerebellum is quite straightforward. The *superior cerebellar veins* enter the straight, transverse, or superior petrosal sinuses. The inferior cerebellar surface is drained by *inferior cerebellar veins*, which enter the inferior petrosal, transverse, or straight sinuses.

Arteries of the Spinal Cord

The blood supply to the spinal cord comes from the *anterior* and *posterior spinal arteries* and from *spinal branches* of segmental arteries (Fig. 8–19). The anterior spinal artery gives off *central* (or *sulcal*) *branches*, which pass alternately to the right and to the left to serve central regions of the spinal cord. The posterior spinal

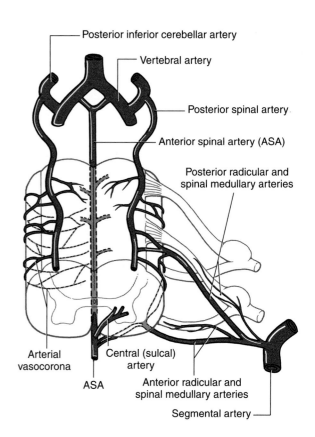

Figure 8–19. Arteries serving the spinal cord.

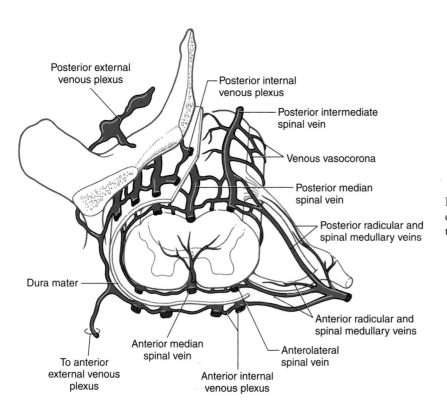

Figure 8–20. Veins draining the spinal cord and the general relationships of internal and external venous plexuses.

arteries course on the surface of the spinal cord medial to the dorsal root entry zone. These arteries serve dorsal parts of the spinal cord and contribute to the *arterial vasocorona* located on the surface of the spinal cord (see Fig. 8–19).

The blood supply to the spinal cord is supplemented at most spinal levels by branches of segmental arteries. These *spinal branches* enter the intervertebral foramina and divide into *posterior (dorsal)* and *anterior (ventral) radicular* and *spinal medullary* arteries (see Fig. 8–19). Because the radicular arteries supply the posterior and anterior roots, each spinal level has these branches. However, the spinal medullary arteries serve to supplement the blood supply to the spinal cord and are therefore not present at every spinal level. The terminal branches of the medullary arteries contribute to the formation of the *arterial vasocorona*. At level T12, L1, or L2, one spinal medullary (or possibly a radicular) artery, usually on the left, is especially large. This is the *artery of Adamkiewicz.* Surgery in this area of the lower back should avoid compromise of this vessel, as it is a major source of blood to lower thoracic and upper lumbar cord levels.

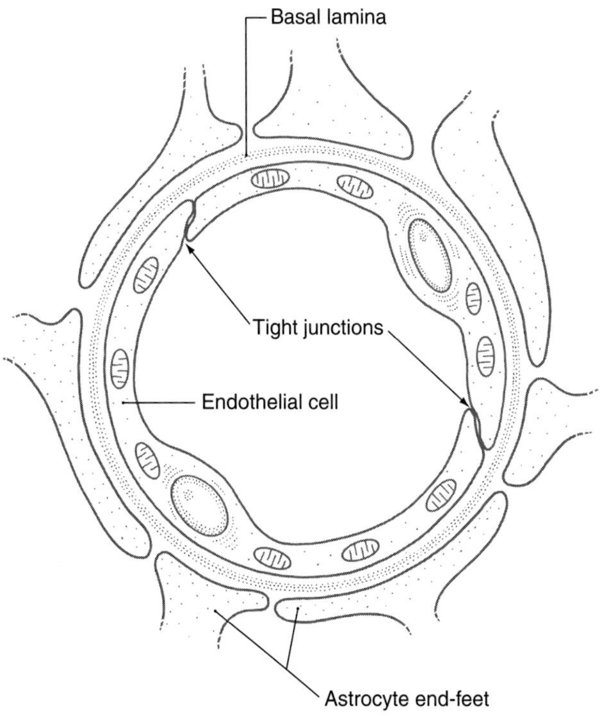

Figure 8–21. Basic structure of the blood-brain barrier.

Veins of the Spinal Cord

In general the venous drainage of the spinal cord mirrors its arterial supply. The location of *anterior* and *posterior spinal veins* and their relationship to other spinal venous structures is shown in Figure 8–20. It is important to note that there is extensive communication between spinal veins and the internal and external venous plexuses found adjacent to the dural sac and vertebral bodies. The veins forming these plexuses apparently lack valves, and the flow in these channels is easily reversed. This represents an important conduit through which metastases from the pelvis, kidney, or lung may spread to the vertebral bodies or into the central nervous system.

Spinal AVMs, although comparable to those of the cranial cavity, do have some unique features. In adults these lesions are usually served by branches of one segmental artery, whereas in children spinal AVMs are usually much larger and have several feeding arteries. Spinal AVMs bleed less frequently than their cranial counterparts and may give rise to more localizing signs and symptoms. For example, in addition to low back pain and occasional sensory or motor problems, patients frequently experience impaired micturition.

Blood-Brain Barrier

Although this chapter is primarily concerned with the distribution of vessels on the surface of the central nervous system, it is appropriate to mention briefly the *blood-brain barrier.* (Fig. 8–21; see Chapter 2 for details). Although this is a physiologic barrier to the movement of many substances into or out of the brain, the blood-brain barrier has anatomic features that correlate with its function.

The endothelial cells of brain capillaries form a continuous lining membrane; they are joined by numerous tight (occluding) junctions and have no intercellular pores or fenestrations (see Fig. 8–21). In contrast, capillaries of the general circulation have fenestrations and intercellular pores. Endothelial cells of brain capillaries rest on a continuous basement membrane (basal lamina), which, in turn, is surrounded by the end-feet of astrocytes (see Fig. 8–19).

Under normal (healthy) conditions, the blood-brain barrier prohibits the movement of high-molecular-weight substances (such as proteins) and vital dyes into the brain. In some disease states, however, the barrier breaks down. In the case of a brain tumor, the new capillaries that proliferate into the lesion do not have a close apposition to astrocytes. As a result, the endothelium of these tumor capillaries develops fenestrations and intercellular pores. These can provide a mechanism for diagnosis. For example, intravascular injection of a radioactive amino acid in such a patient will exit the capillaries in the tumor and will localize therein but will not be found in other (normal) parts of the brain.

Sources and Additional Reading

Crosby EC, Humphrey T, Lauer EW: Correlative Anatomy of the Nervous System. Macmillan Publishing, New York, 1962.

Duvernoy HM: Human Brainstem Vessels. Springer-Verlag, Berlin, 1978.

Gibo H, Carver CC, Rhoton AL Jr, Lenkey C, Mitchell RJ: Microsurgical anatomy of the middle cerebral artery. J Neurosurg 54: 151–169, 1981.

Gillilan L: The correlation of the blood supply to the human brain stem with clinical brain stem lesions. J Neuropath Exp Neurol 23:78–108, 1964.

Gillilan L: The arterial and venous blood supplies to the forebrain (including the internal capsule) of primates. Neurology 18:653–670, 1968.

Gillilan LA: Blood supply of vertebrate brains. In Crosby EC, Schnitzlein HN (eds): Comparative Correlative Neuroanatomy of the Vertebrate Telencephalon. Macmillan Publishing, New York, 1982, pp 266–314.

Hassler O: Deep cerebral venous system in man: A microangiographic study on its areas of drainage and its anastomoses with superficial cerebral veins. Neurology 16:505–511, 1966.

Hassler O: Blood supply to the human spinal cord. Arch Neurol 15: 302–307, 1966.

Hassler O: Venous anatomy of human hindbrain. Arch Neurol 16: 404–409, 1967.

Hassler O: Arterial pattern of human brain stem: Normal appearance and deformation in expanding supratentorial conditions. Neurology 17:368–375, 1967.

Nieuwenhuys R, Voogd J, van Huijzen CHR: The Human Central Nervous System: A Synopsis and Atlas. Springer-Verlag, Berlin, 1988.

Platzer W: Pernkopf Anatomy, Atlas of Topographic and Applied Human Anatomy, vol I, Head and Neck. Urban & Schwarzenberg, Baltimore, 1989.

Smith RR, Zubkov YN, Tarassoli Y: Cerebral Aneurysms, Microvascular and Endovascular Management. Springer-Verlag, New York, 1994, pp. 146–160.

Swash M, Oxbury J (eds): Section 19: Cerebral Vascular Disease. In Clinical Neurology, Churchill Livingstone, Edinburgh, 1991, pp 924–1020.

Yasargil MG: Microneurosurgery, I. Microsurgical Anatomy of the Basal Cisterns and Vessels of the Brain, Diagnostic Studies, General Operative Techniques and Pathological Considerations of the Intracranial Aneurysms. Georg Thieme Verlag, Stuttgart, 1984.

Chapter 9

The Spinal Cord

D. E. Haines, G. A. Mihailoff, and R. P. Yezierski

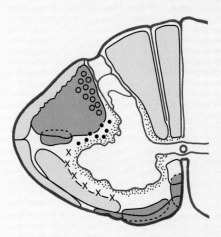

Overview   138

Development   138
Neural Plate
Neural Tube

Spinal Cord Structure   139
Surface Features
Spinal Meninges
White Matter
Gray Matter
Blood Supply

Regional Characteristics   142
Cervical Levels
Thoracic Levels
Lumbar Levels
Sacral Levels

Spinal Nerves   143
Sensory Components of the Spinal Nerve
Neurotransmitters of Primary Sensory Neurons
Motor Components of the Spinal Nerve
Neurotransmitters of Spinal Motor Neurons and Myasthenia Gravis

Spinal Reflexes   146
Tendon Reflex
Flexor Reflex
Crossed Extension Reflex

Pathways and Tracts of the Spinal Cord   148
Ascending Tracts
Descending Tracts

Deficits Characteristic of Spinal Cord Lesions   150

Although small in diameter, the spinal cord is the most important conduit between the body and the brain. It conveys sensory input from the arms, trunk, legs, and most of the viscera and contains fibers and cells that control the motor elements found in these structures. Consequently, injury to the spinal cord, especially at cervical levels, may cause permanent and catastrophic deficits, or even death.

Overview

The spinal cord participates in four essential functions. First, it receives primary sensory input from receptors in skin, skeletal muscles and tendons (*somatosensory fibers*), and from receptors in thoracic, abdominal, and pelvic viscera (*viscerosensory fibers*). Through multisynaptic relays in the spinal cord, much of this sensory input is conveyed to higher levels of the neuraxis.

Second, the spinal cord contains *somatic motor neurons* that innervate skeletal muscles and *visceral motor neurons* that, after synapsing in peripheral ganglia, influence smooth and cardiac muscle and glandular epithelium. Any disease process that damages the somatic motor neuron (as in *poliomyelitis*) or compromises its ability to elicit a response in the skeletal muscle (as in *myasthenia gravis*) will result in weakness or paralysis.

Third, somatosensory fibers enter the spinal cord and influence anterior horn motor neurons either directly, or indirectly through interneurons. These activated motor neurons, in turn, produce rapid involuntary contractions of skeletal muscles. The sensory fiber, the associated motor neuron, and the resultant involuntary muscle contraction constitute the circuit of the *spinal reflex*. Reflexes are essential to normal function and can be used as diagnostic tools to assess the functional integrity of the spinal cord.

Fourth, the spinal cord contains descending fibers that influence the activity of spinal neurons. These fibers originate in the cerebral cortex and brainstem, and damage to them adversely influences the activity of spinal motor and sensory neurons. In many cases the position of a lesion in the brainstem or spinal cord may give rise to a predictable or characteristic series of deficits, such as in *decorticate rigidity* or an *alternating hemianesthesia*.

Although not the specific topic of this chapter, it should also be noted that injury to peripheral nerves will result in motor or sensory deficits distal to the lesion. These are most noticeable in the extremities and may present as motor deficits (*flaccid paralysis*), a loss of sensation (*anesthesia*), or abnormal sensations (*paresthesia*).

Development

Neural Plate. As is explained in more detail in Chapter 5, the spinal cord arises from the caudal portion of the embryonic *neural plate* and from the *caudal eminence*. The neural plate gives rise to the cervical, thoracic, and lumbar levels, whereas the caudal eminence gives rise to the sacral and coccygeal levels. The neural plate appears as a specialized area of ectoderm (*neuroectoderm* or *neuroepithelial cells*) posterior to the notochord at about 18 days (Fig. 9–1*A* and *B*). By 20 days of gestation, the neural plate is an oblong structure that is larger at its rostral area (future brain) and tapered caudally (future spinal cord).

Beginning on day 21, the edges of the neural plate (*neural folds*) enlarge posteromedially to meet on the midline (see Fig. 9–1*B* and *C*). The initial apposition of the neural folds to form the *neural tube* takes place at what will become, in the adult, cervical levels of the spinal cord. This closure simultaneously proceeds in rostral and caudal directions, ultimately creating small openings at either end, between the *cavity of the neural tube* and the surrounding amniotic cavity. These openings are the *anterior* and *posterior neuropores*. The anterior and posterior neuropores close at 24 and 26 days of gestation, respectively.

A variety of defects results from a failure of the neural tube to close (see Fig. 9–1*D–F*). *Rachischisis* occurs when the neural folds do not join and the undifferentiated neuroectoderm remains exposed. In its most extreme form, this deficit results in *anencephaly*, which is a failure of the anterior parts of the neural tube to form and of the anterior neuropore to close. In this situation, the cephalic part of the neural tube (brain) does not form, and there is no skull. Rachischisis and anencephaly are catastrophic developmental defects. In other cases, the neural tube may form normally but the surrounding vertebrae may not, resulting in *spina bifida occulta*, *meningocele*, or *meningomyelocele*.

Neural Tube. The neural tube consists of precursor cells forming the *ventricular zone* (or layer). Proliferating cells of this zone give rise to *ependymal cells* lining the neural tube and to *neuroblasts* and *glioblasts* that form the *mantle zone* (corresponding to the intermediate zone of the forebrain). The *marginal zone* contains glioblasts and the out-growing processes of mantle zone neuroblasts. The *neural crests* detach from the lateral edge of the neural plate and assume a location lateral and ventral to the neural tube.

Structures of the developing neural tube can be correlated with their adult counterparts (see Fig. 9–1*C*). *Neural crest cells* differentiate into cells of the *posterior* (dorsal) *root ganglia*, among other structures. The mantle zone consists of four rostrocaudally oriented columns of neuroblasts, forming the paired *alar plates posteriorly* and the paired *basal plates anteriorly*. The alar plate and basal plate are separated from each other by the *sulcus limitans*. Neuroblasts of the alar plate differentiate into the tract neurons and interneurons of the *posterior horn* of the adult, and those of the basal plate become the motor neurons and interneurons of the

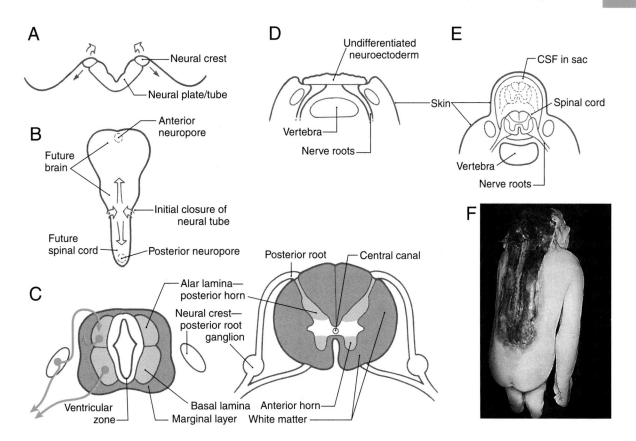

Figure 9–1. Development of the spinal cord. Cross-sectional *(A)* and dorsal *(B)* views of the neural plate and the correlation of neural tube structures with the adult cord *(C)*. The white area between the blue (posterior horn) and red (anterior horn) of the adult spinal cord represents the approximate position of the intermediate gray. Malformations involving defects of the nerve tissue and/or surrounding bone include rachischis *(D)*, meningocele *(E, solid cord)* or meningomyelocele *(E, dashed cord)*, and anencephaly with rachischisis *(F)*. CSF, cerebrospinal fluid. (Photograph courtesy of Dr. Jonathan Fratkin.)

anterior horn. The axons of basal plate neuroblasts, which become somatic motor neurons, extend distally as parts of peripheral nerves. The intermediate gray, an important region of the spinal cord insinuated between the posterior and anterior horns in the adult, originates from portions of both alar and basal plates.

The marginal zone is invaded by processes of neuroblasts located in the mantle zone (see Fig. 9–1C) and by the descending axons of neuroblasts found in the developing brainstem or cerebral cortex. These axons, most of which become myelinated, form the various tracts of the white matter of the adult spinal cord.

Spinal Cord Structure

The adult spinal cord is composed of a butterfly-shaped central area of neuron cell bodies, the *gray matter*, and a surround of myelinated fibers, the *white matter*. Although the cavity of the neural tube was prominent during development, this space is reduced to a small ependymal-lined *central canal* in the adult spinal cord (see Fig. 9–1C).

Surface Features. The human spinal cord extends from the *foramen magnum* to the level of the first or second lumbar vertebra. It consists of 8 cervical, 12 thoracic, 5 lumbar, and 5 sacral levels plus 1 coccygeal level. *Each level (or segment) of the spinal cord is specified by the intervertebral foramina through which the posterior and anterior roots attached to that segment exit the vertebral canal* (Fig. 9–2). Although generally cylindrical in shape, the cord has *cervical* (C4 to T1) and *lumbosacral* (L1 to S2) *enlargements*, which serve, respectively, the upper and lower extremities.

There are eight cervical roots (and spinal cord levels) but only seven cervical vertebrae. So how do the roots relate to their corresponding vertebrae? The C1 root is located between the base of the skull and the C1 vertebra (see Fig. 9–2). Therefore, roots C1 through C7 are located above (rostal to) their respectively numbered vertebrae, and the C8 root is located between the C7 and T1 vertebrae. Beginning with the T1 vertebra and extending caudally, all roots are located caudal to their respectively numbered vertebrae (see Fig. 9–2). It is also important to remember that

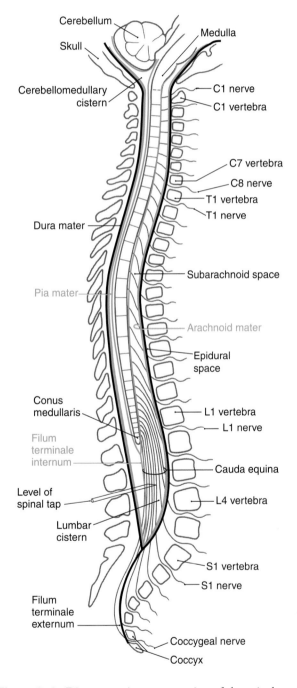

the level of the spinal cord is determined by the interverte-
bral foramen through which the posterior and anterior roots
originating from that cord level exit.

There are few superficial markings on the spinal
cord (Fig. 9–3). The *posterior median sulcus* separates
the posterior portion of the cord into two halves and
contains a delicate layer of pia, the *posterior median*

Cerebellum
Skull
Medulla
Cerebellomedullary
cistern
C1 nerve
C1 vertebra
C7 vertebra
C8 nerve
T1 vertebra
T1 nerve
Dura mater
Subarachnoid space
Pia mater
Arachnoid mater
Epidural
space
Conus
medullaris
L1 vertebra
L1 nerve
Filum
terminale
internum
Cauda equina
Level of
spinal tap
L4 vertebra
Lumbar
cistern
S1 vertebra
S1 nerve
Filum
terminale
externum
Coccygeal nerve
Coccyx

Figure 9–2. Diagrammatic representation of the spinal cord,
the meninges, and other adjacent structures. Note the location for a lumbar puncture (spinal tap).

septum. The *posterolateral sulcus*, which runs the full
length of the cord, represents the entry point of posterior root (sensory) fibers. This area is frequently called
the *posterior (dorsal) root entry zone.* In cervical and upper thoracic regions, a *posterior intermediate sulcus* and
septum are found between the posterolateral and posterior median sulci. This sulcus and septum are insinuated between the medially located *gracile fasciculus* and
the laterally located *cuneate fasciculus* (see Fig. 9–3; see
also Fig. 9–11).

On the anterolateral surface of the spinal cord, the
anterolateral sulcus is the exit point for anterior root
(motor) fibers (see Fig. 9–3). Because the anterior
roots exit in a somewhat irregular pattern, however,
this sulcus is not as distinct as the posterolateral sulcus.

The *anterior median fissure* is a prominent space dividing the anterior part of the cord into halves (see
Fig. 9–3). This fissure contains delicate strands of pia
and, more importantly, the *sulcal branches* of the *anterior spinal artery.*

Spinal Meninges. The tubular dural sac that encloses
the spinal cord is attached cranially to the rim of the
foramen magnum, and its closed caudal end is anchored to the coccyx by the *filum terminale externum*
(see Figs. 9–2 and 9–3). This dural sac is separated
from the vertebrae by the epidural space. The spinal
cord, in turn, is attached to the dural sac by the laterally placed *denticulate ligaments* and the *filum terminale
internum.* This latter structure extends caudally from
the end of the spinal cord, the *conus medullaris,* and
terminates in the attenuated (closed) portion of the
dural sac, which is located adjacent to the S2 vertebrae.
The filum terminale externum extends caudally from
the closed dural sac to its attachment on the inner
aspect of the coccyx (see Fig. 9–2). The *arachnoid mater* adheres to the inner surface of the *dura mater,* and
the *pia mater* is intimately attached to the surface of
the cord. *The subarachnoid space between these layers is
continuous with the subarachnoid space around the brain
and is likewise filled with cerebraspinal fluid.* In adults the
conus medullaris is located at the level of the L1 or L2
vertebral body. Extending caudally from this point to
the end of the dural sac is an enlarged part of the
spinal subarachnoid space, the *lumbar cistern* (see Fig.
9–2). This cistern contains the posterior and anterior
roots from spinal segments L2 to Coc1 as they sweep
caudally. Collectively, these roots form the *cauda
equina.* The method of choice for obtaining a sample
of cerebrospinal fluid for diagnostic purposes is the
lumbar puncture (spinal tap), in which a large-bore needle is introduced between the L3 to L4 or L4 to L5
vertebral arches into the lumbar cistern (see Fig. 9–2).

White Matter. The white matter of the spinal cord is
divided into three large regions, each of which is composed of individual *tracts* or *fasciculi.* The *posterior (dorsal) funiculus* is located between the posterior median

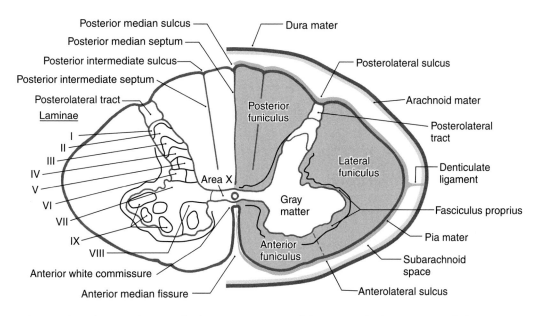

Figure 9–3. The spinal cord at C7 showing the basic organization of the gray and white matter and the meninges (shown only on the *right*). The lamination pattern of the gray matter is shown only on the *left*.

septum and the medial edge of the horn (see Fig. 9–3). At cervical levels this area consists of the *gracile* and *cuneate fasciculi*; collectively, these are commonly referred to as the *posterior (dorsal) columns*.

The *lateral funiculus* is the area of white matter located between the posterolateral and anterolateral sulci (see Fig. 9–3). This region of the cord contains clinically important ascending and descending tracts, the locations of which are shown in Figure 9–11. Those most important in diagnosing the neurologically impaired patient are the *lateral corticospinal tract* and the *anterolateral system* (ALS).

Located between the anterolateral sulcus and the ventral median fissure is a comparatively small region, the *anterior (ventral) funiculus* (see Fig. 9–3). This area contains *reticulospinal* and *vestibulospinal fibers*, portions of the ALS, the *anterior (ventral) corticospinal tract*, and a composite bundle called the *medial longitudinal fasciculus* (MLF).

Two small but important components of the white matter are the *anterior (ventral) white commissure* and the *posterolateral (dorsolateral) tract* (see Fig. 9–3). The former is located on the ventral midline and is separated from the central canal by a narrow band of small cells. The posterolateral (dorsolateral) tract is frequently called the *tract of Lissauer*. It is a small bundle of lightly myelinated and unmyelinated fibers capping the posterior horn.

Gray Matter. The gray matter of the spinal cord is composed of neuron cell bodies, their dendrites and the initial part of the axon, the axon terminals of fibers synapsing in this area, and glial cells. Because this area has few myelinated fibers, it appears distinctly light and has a characteristic shape in myelin-stained sections (see Fig. 9–3; see also Fig. 9–5).

The spinal gray is divisible into a *posterior (dorsal) horn*, an *anterior (ventral) horn*, and the region where these meet, commonly called the *intermediate zone* (or *intermediate gray*). Based on the shape, size, and distribution of neurons located in these areas, the gray matter is divided into *laminae (Rexed laminae) I to IX* and an *area X* around the central canal (see Fig. 9–3). These laminae are also characterized by the input they receive and the trajectory of axons arising therein.

The posterior horn is composed of laminae I to VI (see Fig. 9–3). The most distinct structure in the posterior horn, the *substantia gelatinosa* (lamina II), is capped by cells of the *posteromarginal nucleus* (lamina I). Laminae III to VI are arranged in a series internal to the substantia gelatinosa. Laminae III and IV may also be called the *nucleus proprius* (*posterior* or *dorsal proper sensory nucleus*); their cells have elaborate dendrites that extend into lamina II. Laminae V and VI, which form the base of the dorsal horn, are usually divided into medial and lateral portions.

The intermediate zone, lamina VII, extends from the area of the central canal to the lateral edge of the spinal gray and varies in shape at different levels. Particularly characteristic of lamina VII at thoracic levels are the *posterior thoracic nucleus* (*dorsal nucleus of Clarke*) and the *intermediolateral nucleus*; the latter is frequently called the *intermediolateral cell column* (see Fig. 9–5).

The anterior horn is made up of laminae VIII and IX (see Fig. 9–3). The former contains a population of smaller cells that are interneurons and tract cells. The latter consist of several distinct clusters of large motor neurons whose axons directly innervate skeletal muscle.

Blood Supply. The blood supply to the spinal cord is derived from the *anterior* and *posterior spinal arteries* and from branches of segmental arteries (Fig. 9–4). The segmental branches that serve the posterior and anterior roots and the posterior root ganglia are called *radicular arteries*, and the branches that largely bypass the roots to supplement the blood supply to the cord are called *spinal medullary arteries*. One especially large spinal medullary artery, the *artery of Adamkiewicz*, is most often seen at L2 on the left. This vessel is an important source of blood supply to the cord and must be preserved during surgery in this area; damage to this artery may result in an infarct at lower thoracic and upper lumbar levels of the cord. At each level, terminal branches of the spinal medullary arteries join together to form an arterial network, the *arterial vaso-corona*, on the surface of the spinal cord (see Fig. 9–4).

The posterior columns and peripheral parts of the lateral and anterior funiculi are served by the posterior spinal arteries and arterial vasocorona. Most of the gray matter and the adjacent parts of the white matter are served by the *central branches* of the *anterior spinal artery* (Fig. 9–4). These central branches tend to alternate: one serves the left side of the cord; the next serves the right side.

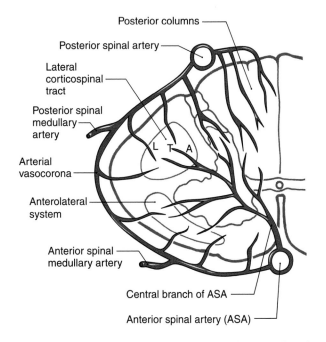

Figure 9–4. Blood supply to the spinal cord. Note that the lateral corticospinal tract and the anterolateral system receive a dual blood supply; also note the topographic arrangement of corticospinal fibers. L, leg-lower extremity; T, trunk; A, arm-upper extremity.

Posterior columns
Posterior spinal artery
Lateral corticospinal tract
Posterior spinal medullary artery
Arterial vasocorona
Anterolateral system
Anterior spinal medullary artery
Central branch of ASA
Anterior spinal artery (ASA)

Trauma, as in hyperextension of the cervical spine, may cause occlusion or spasm of the anterior spinal artery or mechanical injury to the cord. The result is bilateral damage to the cervical cord (*central cervical cord syndrome*). The characteristic features of the syndrome are bilateral weakness of the extremities, primarily evident in the arms, forearm, and hands; a patchy loss of sensation below the lesion; and possible urinary retention.

Regional Characteristics

Although all spinal levels have posterior, lateral, and anterior funiculi and posterior and anterior horns, their shapes and proportion vary between major spinal regions (Fig. 9–5). For example, cervical (C4 to T1) and lumbosacral (L1 to S2) cord levels have prominent posterior and anterior horns because of the extensive sensory input from, and motor outflow to, the upper and lower extremities. In contrast, the posterior and anterior horns at thoracic levels are small; sensory input is less dense, and there is no appendicular musculature at these levels.

Cervical Levels. The cervical cord is oval in shape and proportionately larger than at other spinal levels (see Fig. 9–5). There is a large amount of white matter because a full complement of ascending and descending fiber tracts is present. The gracile and cuneate fasciculi are especially obvious structures at cervical levels. The posterior and anterior horns at levels C1 to C3 are comparatively small, whereas those at C4 to C8 are large. The large size of the latter reflects the sensory and motor innervation of the upper extremity.

Thoracic Levels. In general, the thoracic cord is round, and the posterior and anterior horns are small (see Fig. 9–5). From upper to lower thoracic levels there is a progressive decrease in the amount of white matter. Although both the gracile and cuneate fasciculi are present at upper thoracic levels, only the gracile fasciculus is present at lower thoracic levels. However, the small size of the posterior and anterior horns makes the white matter in thoracic levels appear proportionally large. Two structures especially obvious in the gray matter at thoracic levels are the *posterior thoracic nucleus (dorsal nucleus of Clarke)* and the *lateral horn*. The former, a prominent cell group in medial parts of lamina VII, contains neurons whose axons project to the cerebellum. The latter, also part of lamina VII, is a protrusion into the lateral funiculus formed by the *intermediolateral cell column*. These cells are preganglionic sympathetic neurons whose axons will terminate in either paravertebral or prevertebral ganglia.

Lumbar Levels. At lumbar levels the cord is also round (see Fig. 9–5). The posterior and anterior horns are quite large, and there is less white matter than at

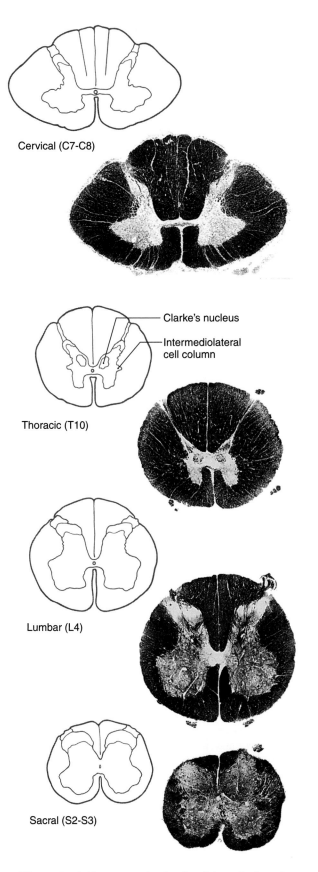

Cervical (C7-C8)

Clarke's nucleus

Intermediolateral cell column

Thoracic (T10)

Lumbar (L4)

Sacral (S2-S3)

Figure 9–5. Representative levels of the spinal cord.

higher levels. Therefore, the posterior and anterior horns appear proportionately large, the reverse of the situation at thoracic levels. The posterior thoracic nucleus (dorsal nucleus of Clarke) is usually obvious at L1 and possibly L2 levels.

Sacral Levels. At sacral levels, the spinal cord is round and is smaller than at lumbar levels (Fig. 9–6). It consists mainly of gray matter, with the white matter forming a relatively thin shell. The intermediate gray matter at levels S2, S3, and S4 contains preganglionic parasympathetic cell bodies (the *sacral visceromotor nucleus*). The substantia gelatinosa (lamina II) is especially obvious at sacral levels.

Spinal Nerves

The spinal nerves are formed by the junction of the posterior and anterior roots of the spinal cord (see Fig. 9–6). As there are 31 spinal cord levels (8 cervical, 12 thoracic, 5 lumbar, 5 sacral, 1 coccygeal), so are there 31 corresponding pairs of spinal nerves. Each spinal nerve contains afferent fibers that convey sensory input from the periphery and efferent fibers arising from spinal motor neurons. These fibers, plus circuits in the spinal gray, are the structural basis for the *spinal reflexes* routinely tested in the neurologic examination (see Chapter 33).

The spinal nerve may contain up to four types of fibers. Two of these are sensory and have their cell bodies in the posterior root ganglion, and two are motor and have their cell bodies in the spinal cord gray matter (see Fig. 9–6).

Sensory Components of the Spinal Nerve. Sensory information is brought to the spinal cord by neuronal processes whose cell bodies reside in the posterior root ganglia. The central processes of these neurons penetrate the spinal cord, and the peripheral processes pass outward in the spinal nerves to innervate body structures. Sensory input originates from (1) the body surface; (2) deep structures such as muscles, tendons, and joints; and (3) internal organs. Fibers conveying input from the first two areas are classified as *general somatic afferent* (GSA), whereas sensory fibers from the gut and other visceral structures are classified as *general visceral afferent* (GVA). The GSA fibers are further classified as either *exteroceptive* or *proprioceptive*.

Exteroceptive (GSA) fibers arise from (1) receptors that are sensitive to mechanical, thermal, or chemical stimuli that may cause tissue damage or (2) receptors sensitive to discriminative touch or vibratory stimuli. The former fibers (A-δ and C) are slowly conducting (0.5–30 meters/sec) and unmyelinated, or lightly myelinated, and they enter the cord via the *lateral division of the posterior root*. These fibers may ascend or descend (or both) in the *posterolateral tract* (*tract of Lissauer*) before entering the posterior horn to terminate primarily in laminae I to V. The latter fibers (A-β) are

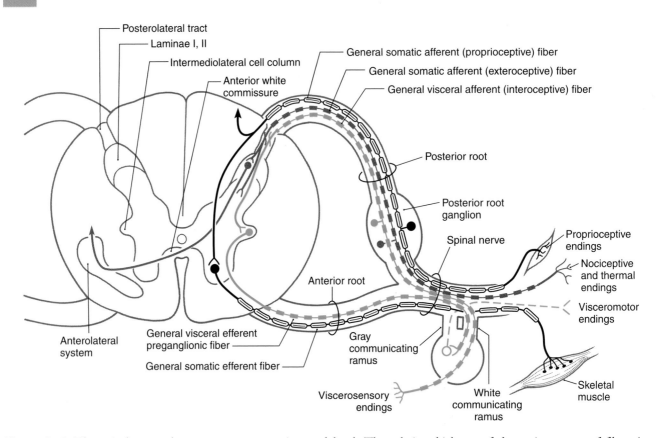

Figure 9–6. The spinal nerve shown on a representative cord level. The relative thickness of the various types of fibers is indicated. The narrow-diameter fiber passing through the gray communicating ramus and terminating in visceromotor endings represents a general visceral efferent (GVE) postganglionic fiber.

rapidly conducting (30–70 meters/sec) and heavily myelinated, and they enter the cord through the *medial division of the posterior root*. After entering the posterior funiculus, these fibers may give rise to ascending or descending collaterals.

Proprioceptive (GSA) fibers originate from receptors located in muscles, tendons, or joints that are sensitive to stretch or pressure; some vibratory sense is also conveyed by these fibers. These are rapidly conducting (70–120 meters/sec; Ia, Ib, A-β), heavily myelinated fibers that also enter the medial division of the posterior root. The central processes of these proprioceptive fibers (and of the heavily myelinated exteroceptive fibers) may directly enter, and ascend in, the posterior columns, or they may branch into the spinal gray to synapse in relay nuclei (such as the posterior nucleus of Clarke) or on cells in the anterior horn that participate in spinal reflexes.

The spinal nerve also conveys sensory information from thoracic, abdominal, and pelvic viscera. This *interoceptive* input originates primarily from receptors that are sensitive to nociceptive stimuli and is conveyed via GVA fibers. These fibers travel through (for example) the *splanchnic nerves* and traverse the sympathetic chain and *white communicating ramus* to enter the spinal

nerve (see Fig. 9–6). Their central processes enter the lateral division of the posterior root and terminate in laminae I and V to VII. These GVA fibers are also lightly myelinated and slowly conducting (1–20 meters/sec).

Neurotransmitters of Primary Sensory Neurons. Although several neuroactive substances have been implicated as transmitters in primary afferent fibers, those having an important role are *substance P* (SP), *calcitonin gene–related peptide* (CGRP), and *glutamate*. Small-diameter (A-δ and C) fibers arising from visceral and somatic structures—that is, GVA and small-diameter GSA fibers—use one or more of these three neurotransmitters, and it is probable that some large-diameter, heavily myelinated GSA fibers use glutamate. Specifically, SP and CGRP can be found in small-diameter GVA and GSA fibers, and these peptides plus glutamate are also found in the smaller cell bodies of the posterior root ganglion, from which these fibers arise. Centrally, fibers and terminals containing these three neurotransmitters can be found in laminae I, II, and V, where the small-diameter axons synapse with cells that relay the information to higher levels of the neuraxis. Conversely, some of the large bodies in posterior root ganglia, which give rise to large-diameter GSA fibers, contain glutamate. This transmitter is also found in the posterior columns in large-diameter, heavily myclinated fibers, indicating that it may function in the relay of proprioceptive information.

Motor Components of the Spinal Nerve. The spinal cord gives rise to two types of motor fibers: (1) those that directly innervate skeletal (striated) muscle and (2) visceromotor (autonomic) fibers that synapse on a second neuron, usually located in a peripheral visceromotor ganglion. The latter (or post-ganglionic) neurons innervate smooth muscle, cardiac muscle, or glandular epithelium (see Fig. 9–6).

The motor cells that innervate skeletal muscle are located in the anterior horn; these cells and their peripheral processes are classified as *general somatic efferent* (GSE). GSE cells from the anterior horn also supply motor innervation to the specialized *intrafusal* muscle fibers of the *muscle spindles* (neuromuscular spindles), sensory structures in muscles that detect muscle length and various aspects of contraction dynamics. Large motor neurons in the anterior horn are organized in two general but overlapping patterns (Fig. 9–7). First, cells innervating proximal muscles are located medially, and cells innervating more distal muscles are located progressively more laterally. This explains why the anterior horn is smaller and narrower at thoracic than at cervical and lumbar levels. At thoracic levels the anterior horn contains motor neurons for only the axial muscles of the trunk, whereas at cervical and lumbar levels it also contains the more lateral groups of motor neurons that innervate the limbs. Second, within the anterior horn at C4 to T1 and L1 to S2, motor neurons innervating extensors tend to be more anteriorly located in the horn, whereas those innervating flexors tend to be found more posteriorly located.

The visceromotor (autonomic) motor neurons of the spinal cord are classified as *general visceral efferent* (GVE) and have their cell bodies in lamina VII. At cord levels T1 to L2, these cells belong to the *intermediolateral cell column* (sympathetic cells), whereas at sacral levels S2 to S4, they belong to the parasympathetic system and form the *sacral visceromotor nucleus* located in the lateral part of lamina VII. Unlike the single-neuron GSE projection, visceromotor pathways consist of two neurons in series (see Fig. 9–6). The spinal cord neuron projects to a visceromotor ganglion and is therefore classified as *GVE-preganglionic*. In the ganglion, it synapses with a *GVE-postganglionic neuron, which innervates the target structure.*

Motor fibers (GSE and GVE) exit in the anterior root and pass into the spinal nerve. GSE fibers continue through the spinal nerve and are conveyed by the progressive branching of peripheral nerves to the skeletal muscles of the body. In contrast, GVE-preganglionic fibers leave the spinal nerve to join the sympathetic trunk via the *white communicating ramus* (see Fig. 9–6). Once they have entered the sympathetic trunk, these preganglionic fibers follow any of several routes, which are considered in detail in Chapter 29. Suffice it to say, GVE-postganglionic fibers from cells of the sympathetic chain ganglia rejoin the spinal nerves via the *gray communicating ramus*, whereas those of the prevertebral ganglia distribute only to the gut. Also, GVE-preganglionic parasympathic neurons are present at S2–S4 levels. Their axons leave the spinal cord in the anterior roots to eventually join branches of the ventral primary rami that form the pelvic nerve.

Neurotransmitters of Spinal Motor Neurons and Myasthenia Gravis. The three populations of spinal motor neurons are (1) large anterior horn cells (α-motor neurons) that innervate extrafusal skeletal muscle cells, (2) smaller cells (*γ-motor neurons*) that innervate only the intrafusal fibers of the muscle spindles, and (3) cells that give rise to preganglionic sympathetic (T1 to L1) or parasympathetic (S2 to S4) fibers, which terminate in peripheral visceromotor (autonomic) ganglia. All three of these cell populations use *acetylcholine* as their neurotransmitter. Consequently, acetylcholine is abundant in axon terminals at the *neuromuscular junction*, and numerous *nicotinic acetylcholine receptors* are present on the postsynaptic junctional folds of the muscle membrane.

Myasthenia gravis, a neurologic disease characterized by moderate to profound muscle weakness, is closely correlated with the presence of circulating antibodies directed against nicotinic receptor sites on the postsynaptic membrane. The result is a blockage of transmission at the neuromuscular junction.

This disease is most frequently seen in patients be-

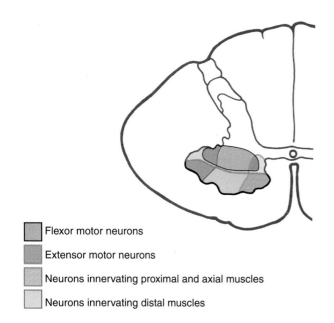

Flexor motor neurons

Extensor motor neurons

Neurons innervating proximal and axial muscles

Neurons innervating distal muscles

Figure 9–7. Representation of the general organization of motor neurons in the anterior horn.

tween 20 and 40 years of age, although younger patients may exhibit symptoms. There are three characteristics of myasthenia gravis. First, muscle weakness may wax and wane over periods of minutes or hours, one day, or several days or weeks. Second, muscles controlling eye movement are frequently involved first (in about 40% of patients) and are ultimately involved in about 85% of all patients. Muscles of the pharynx or larynx, face, and extremities may eventually be involved, but almost always in concert with ocular muscles. Third, the weakness responds to the administration of drugs that enhance cholinergic transmission.

Spinal Reflexes

Afferent fibers in spinal nerves may synapse on tract cells that relay information to higher levels of the neuraxis, or they may terminate on motor neurons or interneurons, both of which may participate in reflex circuits. Reflexes require an afferent fiber, interneurons and/or motor neurons, and a target tissue, usually skeletal muscle. Reflexes may be relatively simple and confined to a single cord level (*intrasegmental*) or complex, involving multiple cord segments (*intersegmental*). Certain disease or central nervous system lesions can affect spinal reflexes, resulting in reflexes that are greatly exaggerated (*hyperreflexia*), diminished (*hyporeflexia*), or absent (*areflexia*). Numerous reflexes are part of the standard neurologic examination (see Chapter 33); only a few examples are given here.

Tendon Reflex. Although a *tendon reflex* may be elicited by tapping any large tendon (e.g., triceps or Achilles), a common example is the *knee-jerk* or *quadriceps stretch reflex* (Fig. 9–8). A brisk tap on the patellar tendon stretches the primary sensory endings in muscle spindles located in the quadriceps femoris muscle, sending an impulse toward the posterior root ganglion via heavily myelinated, rapidly conducting group Ia fibers. The central processes of these afferent axons synapse on and excite motor neurons in the anterior horn that innervate the quadriceps femoris muscle. The result is a sudden contraction of these muscles and an extension (dorsiflexion) of the leg at the knee. Because this reflex requires only one synapse and is a response to muscle stretch, it also may be called a *monosynaptic stretch reflex* or a *myotatic reflex*.

An extension of the simple stretch reflex is seen in *reciprocal inhibition* and *autogenic inhibition* (also called the *inverse myotatic reflex*). In reciprocal inhibition, one group of muscles is excited and the antagonistic group is inhibited (see Fig. 9–8). In this situation, the muscle spindle is stretched by a tap on the patellar tendon, and the impulse enters the spinal cord via a group Ia primary sensory fiber. This fiber branches and has excitatory terminations on quadriceps femoris motor neurons and on group Ia inhibitory (glycinergic) interneu-

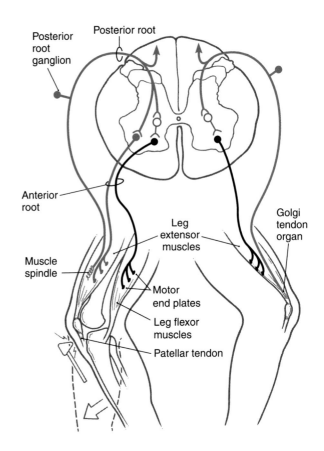

Figure 9–8. Pathway for the patellar tendon reflex and reciprocal inhibition (*left*) and autogenic inhibition (*right*). The inhibitory glycinergic interneurons are represented by the red open cell bodies.

rons. As a result, the quadriceps (extensor) contracts, whereas the interneurons inhibit spinal motor neurons innervating the hamstring (flexor) muscles, which remain passive. This action enhances the effectiveness of the reflex.

The muscle receptor involved in *autogenic inhibition* is the *Golgi tendon organ* (see Fig. 9–8). This receptor responds to relatively high tension (higher than that needed to activate the muscle spindles). Activation causes an increase in the rate of firing of the group Ib sensory fibers that arise from this receptor. In the spinal cord, these fibers terminate on group Ib inhibitory (glycinergic) interneurons, which inhibit motor neurons that innervate the muscle attached to the tendon from which the afferent volley originated.

Flexor Reflex. A further level of complexity in spinal reflexes is seen in the *flexor reflex* (*withdrawal reflex or nociceptive reflex*) (Fig. 9–9). This type of reflex is initiated by cutaneous input, is frequently a response to nociceptive stimuli, and represents an attempt to protect a body part by extricating it from the source of injury. Lightly myelinated or unmyelinated primary sensory fibers (A-δ or C fibers) conveying nociceptive input enter the posterolateral tract (of Lissauer) and branch. Many of these fibers enter the spinal gray, where they form excitatory synaptic contacts with ascending tract cells and with both excitatory and inhibitory interneurons (see Fig. 9–9). While tract neurons relay this nociceptive information to higher levels of the neuraxis, the excitatory glutaminergic interneurons synapse on flexor motor neurons, resulting in activation of the ipsilateral flexor muscles of the thigh (iliopsoas), leg (hamstring muscles), and foot (tibialis anterior) and withdrawal of the extremity. This action is enhanced by the synapse of inhibitory interneurons on extensor (antagonistic) motor neurons and the resultant decreased activity (inhibition) of extensor muscles—for example, the quadriceps femoris muscles. The flexor reflex, considering its afferent and efferent limbs, involves several spinal segments.

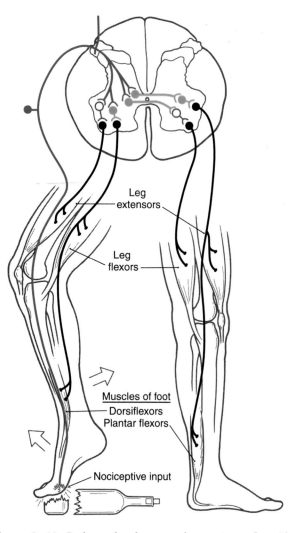

Figure 9–10. Pathway for the crossed extension reflex. Glycinergic interneurons (inhibitory) are represented by the open red cell bodies and glutaminergic interneurons (excitatory) by the closed green cells.

Crossed Extension Reflex. The *crossed extension reflex* builds on the basic circuits of the flexor reflex, but also involves musculature of the contralateral side of the body (Fig. 9–10). By way of interneurons, nociceptive input on A-δ or C fibers excites ipsilateral leg flexor motor neurons and inhibits ipsilateral leg extensor motor neurons. Consequently, the flexors contract, the extensors relax, and the extremity is withdrawn from the painful stimulus. If the reflex occurs during standing or walking, however, the opposite leg must participate in the response to keep the person from falling. The same nociceptive input that resulted in withdrawal on the ipsilateral side is conveyed to interneurons that project to the contralateral anterior horn (see Fig. 9–10). These fibers excite motor neurons polysynaptically, innervating contralateral extensor muscles and inhibiting motor neurons that innervate contralateral flexor muscles. Thus, there is an ipsilat-

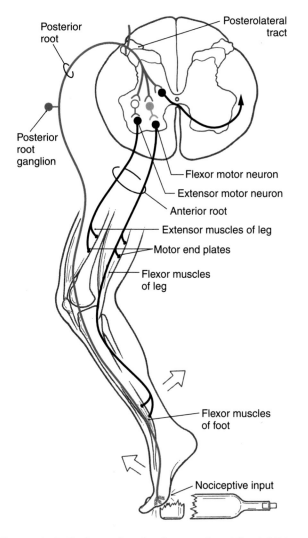

Figure 9–9. Pathway for the flexor reflex. The inhibitory glycinergic interneuron is represented by the red open cell and the excitatory glutaminergic interneuron by the green closed cell. In addition to being involved in reflexes, this nociceptive input is also relayed to higher levels of the neuraxis via the anterolateral system (ascending black fiber).

eral flexion and withdrawal from the stimuli accompanied by an extension of the contralateral leg to support the body.

Pathways and Tracts of the Spinal Cord

The spinal cord white matter consists of (1) long *ascending* and *descending fibers* or *tracts*, which link the spinal cord to higher levels of the neuraxis, and (2) *propriospinal fibers* that project from one spinal level to another (Fig. 9–11). Ascending fibers convey information to higher levels of the neuraxis. Some descending

fibers modulate the transmission of nociceptive information in the posterior horn, whereas others influence the activity of motor neurons. Propriospinal fibers form the basis for a wide variety of intraspinal reflexes.

Many tracts or fibers in the nervous system are *named according to their origin and termination*. For example, the *corticospinal* fibers originate in the cerebral cortex (*cortico-*) and end in the spinal cord (*-spinal*). These are descending fibers because the *cortex* is a more rostral part of the neuraxis than is the *spinal* cord. Likewise, the term *spinothalamic fiber* indicates that the cell of origin is in the spinal cord and the termination in the thalamus; these are ascending fibers. In many situations *the name of the tract or group of fibers indicates three important facts about that fiber population:* (1) whether they are ascending or descending (corticospinal versus spinocerebellar); (2) the location of the cell body of origin (cortex versus spinal cord); and (3) the place where the axons in the tract terminate (spinal cord versus cerebellum). Keeping these basic principles in mind will expedite learning many tracts and pathways.

Ascending Tracts. The *gracile* and *cuneate fasciculi*, collectively called the *posterior columns*, are composed of the central processes of heavily myelinated primary sensory fibers that convey proprioceptive, tactile, and vibratory information from the ipsilateral side of the body (see Fig. 9–11). Fibers in the gracile fasciculus originate from sacral, lumbar, and lower thoracic (below T6) levels; those in the cuneate fasciculus originate from upper thoracic (above T6) and cervical levels. Injury to the posterior columns on one side results in an *ipsilateral loss of proprioception, discriminative touch, and vibratory sense below the level of the lesion*.

The *posterior (dorsal) spinocerebellar* and *anterior (ventral) spinocerebellar tracts* are located on the lateral surface of the cord, meeting at approximately the level of the denticulate ligament (see Figs. 9–3 and 9–11). The fibers of the former tract arise from the posterior thoracic nucleus (of Clarke) in lamina VII at T1 to L2, and the fibers of the latter tract arise primarily from cells of laminae V to VIII and from large ventral horn neurons called *spinal border cells*, both at lumbosacral levels.

In the anterolateral area of the spinal cord there is a large composite bundle called the *anterolateral system* (ALS) (see Fig. 9–11). This system encompasses those regions of the white matter that were classically divided into anterior and lateral spinothalamic tracts. The ALS contains *spinothalamic*, *spinomesencephalic* (spinotectal, spinoperiaqueductal), *spinohypothalamic*, and *spinoreticular fibers*. The fibers of the ALS originate primarily from posterior horn cells, but it is now known that some arise from neurons in the anterior horn. Those fibers that coalesce to form the ALS cross in the anterior white commissure, ascending about two levels as they do so.

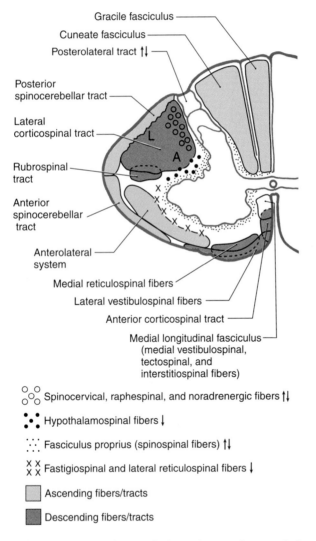

Gracile fasciculus
Cuneate fasciculus
Posterolateral tract ↑↓
Posterior spinocerebellar tract
Lateral corticospinal tract
Rubrospinal tract
Anterior spinocerebellar tract
Anterolateral system
Medial reticulospinal fibers
Lateral vestibulospinal fibers
Anterior corticospinal tract
Medial longitudinal fasciculus (medial vestibulospinal, tectospinal, and interstitiospinal fibers)

○○ / ○○ Spinocervical, raphespinal, and noradrenergic fibers ↑↓
•• / •• Hypothalamospinal fibers ↓
∴∴ Fasciculus proprius (spinospinal fibers) ↑↓
X X / X X Fastigiospinal and lateral reticulospinal fibers ↓
▢ Ascending fibers/tracts
▢ Descending fibers/tracts

Figure 9–11. Ascending and descending pathways of the spinal cord shown in cross section as they are organized at cervical levels. Corticospinal fibers from the lower extremity region of the motor cortex (L) are located in lateral portions of the tract, whereas fibers from the arm–upper extremity regions of the cortex (A) are medially located in the tract. ↓, descending fibers; ↑, ascending fibers.

In general, fibers of the ALS convey nociceptive, thermal, and touch information to higher levels of the neuraxis. Consequently, *injury to the spinal cord that involves ALS fibers will result in a loss of pain, temperature, and crude touch sensations on the contralateral side of the body beginning about one or two segments below the lesion.* The ALS is somatotopically organized; this means that in the spinal cord, lower portions of the body are represented more posterolaterally and upper levels are represented anteromedially.

Nociceptive input and some discriminative touch are also carried by *postsynaptic posterior column fibers* and by the *spinocervicothalamic tract.* The former originate from laminae III to VIII (mainly IV) and ascend ipsilaterally in the dorsal columns. The latter fibers arise from the same laminae but ascend as a diffuse population in the posterior part of the lateral funiculus to end in the lateral cervical nucleus at levels C1 to C3. The existence of these minor fiber populations in humans may explain the recurrence of pain perception in some patients who have had an *anterolateral cordotomy* for intractable pain.

Other, more diffusely arranged, ascending fibers include *spino-olivary, spinovestibular,* and *spinoreticular fibers.* These are discussed in later chapters in relation to the functional systems they serve.

Descending Tracts. The lateral funiculus (see Figs. 9–3 and 9–11) contains the *lateral corticospinal* and *rubrospinal* tracts, as well as other fiber populations that are more diffuse in their distribution (*reticulospinal, fastigiospinal, raphespinal, hypothalamospinal*). Corticospinal fibers arise from the cerebral cortex and descend through the brainstem. At the medulla–spinal cord junction, most cross to form the *lateral corticospinal tract,* but some remain uncrossed as the *anterior corticospinal tract.* Lateral corticospinal fibers are somatotopically arranged; fibers that originate from lower extremity areas of the cerebral cortex and project to lumbosacral levels are lateral, whereas those traveling to cervical levels from upper extremity areas of the cortex are medial (see Fig. 9–11). One important function of this tract is to influence spinal motor neurons, *especially those controlling fine movements of the distal musculature.* Consequently, lesions of lateral corticospinal fibers on one side of the cervical cord result in *ipsilateral paralysis of the upper and lower extremities on that side* (*hemiplegia*). In contrast, a lesion of corticospinal fibers above (rostral to) the spinal cord–medulla junction, and therefore above the decussation of these fibers, will result in a *hemiplegia on the opposite (contralateral) side of the body.*

Rubrospinal fibers arise from the red nucleus of the midbrain, cross at that level, and descend in the spinal cord with lateral corticospinal fibers (see Fig. 9–11). In general, rubrospinal fibers, as well as lateral corticospinal fibers, excite flexor motor neurons and inhibit extensor motor neurons.

Although diffusely arranged, other descending fibers in the lateral funiculus serve important functions (see Fig. 9–11). *Reticulospinal fibers* in this area originate from the medullary reticular formation, and *fastigiospinal fibers* from the fastigial nucleus of the cerebellum. At spinal levels, the former are uncrossed and the latter are crossed. Because their function is to help maintain posture, these fibers tend to excite extensor motor neurons and inhibit flexor motor neurons. *Raphespinal fibers* originate mainly from the nucleus raphe magnus of the brainstem, descend bilaterally in posterior areas of the lateral funiculus, and function to modulate the transmission of nociceptive information at spinal levels. The activity of GVE motor neurons of the intermediolateral cell column is influenced by *hypothalamospinal fibers,* which descend through lateral areas of the brainstem and spinal cord. Lesions in the brainstem or cervical spinal cord that interrupt these fibers result in ipsilateral *ptosis, miosis, anhidrosis,* and *enophthalmos* (the *Horner syndrome*).

The *anterior funiculus* (see Fig. 9–11) contains *reticulospinal* and *vestibulospinal* fibers, the *anterior corticospinal tract,* and the *medial longitudinal fasciculus* (MLF). Reticulospinal fibers in this area arise in the pontine reticular formation of the brainstem, whereas vestibulospinal fibers originate from the vestibular nuclei. *Lateral vestibulospinal fibers* arise from the lateral vestibular nucleus, and *medial vestibulospinal fibers* originate primarily from the medial vestibular nucleus. Reticulospinal and vestibulospinal fibers of the anterior funiculus function in postural mechanisms through their general excitation of extensor motor neurons and inhibition of flexor motor neurons. Fibers of the *anterior corticospinal tract* are uncrossed, but most of these fibers cross in the ventral white commissure before terminating on medial motor neurons that innervate axial muscles.

The MLF, although quite small, is generally regarded as a composite bundle containing *medial vestibulospinal fibers* (from the medial vestibular nucleus), *tectospinal fibers* (from the superior colliculus of the midbrain), *interstitiospinal fibers* from the interstitial nucleus of the rostral midbrain, and some *reticulospinal fibers.* Tectospinal and vestibulospinal fibers are found only at cervical levels; the other fibers extend to lower cord levels. These fibers terminate primarily in laminae VII and VIII but ultimately influence motor neurons innervating primarily axial and neck musculature.

The comparatively simple structure of the spinal cord belies its functional importance. Although the cord is smaller around than the little finger, descending motor control of the body below the neck and all sensory input from the same areas must traverse it. Consequently, lesions in the spinal cord that would be considered of little consequence in larger parts of the brain may cause global deficits or death. As the cord merges into the brainstem, the organization and function of the central nervous system become progressively more complex.

Deficits Characteristic of Spinal Cord Lesions

The functional characteristics of ascending and descending tracts of the spinal cord are described in later chapters. However, it is appropriate at this point to touch on some general features that correlate with the structure of the spinal cord.

Cavitation of the central regions of the spinal cord, as in *syringomyelia*, will frequently damage fibers crossing in the anterior white commissure (see Fig. 9–3). This bundle conveys fibers from the posterior horn across the midline to enter the ALS on the opposite side (see Fig. 9–6). Consequently, a lesion of this structure will damage fibers going in both directions, resulting in a bilateral loss of pain and thermal sensations according to the levels of the spinal cord involved.

A functional *hemisection of the spinal cord (the Brown-Séquard syndrome)* results in a clinical picture that reflects damage to the lateral corticospinal tract, the ALS, and the posterior columns. A lesion on the right at C4 will result in a weakness or paralysis (*hemiparesis, hemiplegia*) on the right side (corticospinal damage), loss of pain and thermal sensations on the left side (ALS damage—these fibers cross in the anterior white commissure), and a loss of proprioception, vibratory sense, and discriminative touch on the right (gracile and cuneate fasciculi injury).

Variations on these main themes may occur. For example, a spinal cord hemisection at T8 would affect the body below that level but would spare the upper trunk and upper extremity. A lesion involving the posterior columns bilaterally would result in proprioceptive and discriminative touch losses below the level of the lesion but would spare pain and thermal sensations. In our study of systems neurobiology we shall explore these and other examples of dysfunction resulting from spinal cord lesions.

Sources and Additional Reading

Brown AG: Organization in the Spinal Cord: The Anatomy and Physiology of Identified Neurons. Springer-Verlag, Berlin, 1981, pp 1–238.

Dado RJ, Katter JT, Giesler GJ: Spinothalamic and spinohypothalamic tract neurons in the cervical enlargement of rats. I. Locations of antidromically identified axons in the thalamus and hypothalamus. J Neurophysiol 71:959–980, 1994.

Quencer RM, Bunge RP, Egnor M, Green BA, Puckett W, Naidich TP, Post MJD, Norenberg M: Acute traumatic central cord syndrome: MRI-pathological correlations. Neuroradiology 34:85–94, 1992.

Rexed B: The cytoarchitectonic organization of the spinal cord in the cat. J Comp Neurol 96:415–495, 1952.

Rexed B: A cytoarchitectonic atlas of the spinal cord in the cat. J Comp Neurol 100:297–379, 1954.

Schoenen J, Faull RLM: Spinal cord: Cytoarchitectural, dendroarchitectural, and myeloarchitectural organization. In Paxinos G (ed): The Human Nervous System. Academic Press, San Diego, 1990, pp 19–53.

Willis WD: The pain system, the neural basis of nociceptive transmission in the mammalian nervous system, vol 8. In Gildenberg PL (ed): Pain and Headache. S Karger, Basel, 1985, pp 1–346.

Willis W, Coggeshall RE: Sensory Mechanisms of the Spinal Cord, 2nd ed. Plenum Press, New York, 1991, pp 1–575.

Yezierski RP: Spinomesencephalic tract: Projections from the lumbosacral spinal cord of the rat, cat, and monkey. J Comp Neurol 267:131–146, 1988.

Chapter 10

An Overview of the Brainstem

D. E. Haines and G. A. Mihailoff

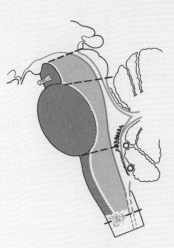

Basic Divisions of the Brainstem 152
Medulla Oblongata
Pons
Midbrain
Tegmental and Basilar Areas

Ventricular Spaces of the Brainstem 153
The Rhomboid Fossa

Cranial Nerve Nuclei and Their Functional Components 154

The term *brainstem* (sometimes written *brain stem*) is used in two ways: it can mean either the portion of the brain that consists of the medulla oblongata, pons, and midbrain, or the portion that consists of these structures plus the diencephalon. This book follows the former convention. For our purposes, therefore, *the brainstem consists of the rhombencephalon (excluding the cerebellum) and the mesencephalon.* These regions of the brainstem all share a basic organization, which is the topic of this chapter. The medulla, pons, and midbrain are discussed in detail in Chapters 11 to 13.

Basic Divisions of the Brainstem

Medulla Oblongata. At about the level of the foramen magnum, the spinal cord merges into the caudal-most portion of the brain, the *medulla oblongata* or *myelencephalon*, commonly called the medulla. The foramen magnum marks the approximate location of the *pyramidal (motor) decussation of the medulla* (Fig. 10–1*A*). The medulla is slightly cone-shaped and enlarges in diameter as it extends rostrally toward the pons-medulla junction. Posteriorly (dorsally), this junction is represented by the caudal edge of the middle and inferior cerebellar peduncles, whereas anteriorly (ventrally), this border is formed by the caudal edge of the basilar pons (see Fig. 10–1).

The cranial nerves associated with the medulla include the *hypoglossal* (XII, motor), parts of the *accessory* (XI, motor), *vagus* (X, mixed), and *glossopharyngeal* (IX, mixed) *nerves* (see Fig. 10–1*A*). The *abducens* (VI, motor), *facial* (VII, mixed), and *vestibulocochlear* (VIII, sensory) *nerves* are frequently called the *cranial nerves of the pons-medulla junction* because they exit the brainstem at this particular location (see Fig. 10–1*A*).

Pons. The pons (the anterior part of the *metencephalon*) extends from the pons-medulla junction to an imaginary line drawn from the exit of the trochlear nerve posteriorly to the rostral edge of the basilar pons anteriorly (see Fig. 10–1*A–C*). What we commonly call the pons is actually composed of two portions, the *pontine tegmentum* (located internally; see Fig. 10–1) and the *basilar pons*. This latter structure is bulbous and quite characteristic of the anterior aspect of the pons. The cerebellum, although part of the metencephalon, is *not* part of the brainstem. It is joined to the brainstem by three large, paired bundles of fibers called the *cerebellar peduncles*. These are the *inferior cerebellar peduncle*, the *middle cerebellar peduncle*, and the *superior cerebellar peduncle*, connecting the cerebellum to the medulla oblongata, basilar pons, and midbrain, respectively. The *trigeminal nerve* (V, mixed) emerges from the lateral aspect of the pons (see Fig. 10–1*A*).

Midbrain. The *midbrain* (*mesencephalon*) extends ros-

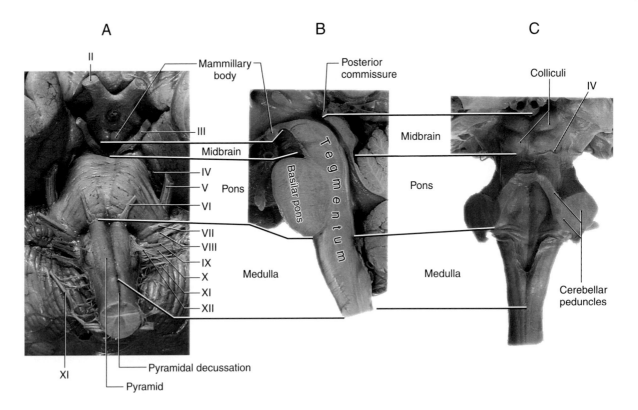

Figure 10–1. Anterior (ventral) (*A*), midsagittal (*B*), and posterior (dorsal) (*C*) views of the brainstem. Cranial nerves are labeled by their corresponding Roman numerals. In *C*, the cerebellum is removed to expose the posterior surface of the brainstem and the fourth ventricle.

trally from the pons-midbrain junction to join the diencephalon (thalamus). This latter interface is usually described as a line drawn from the posterior commissure posteriorly to the caudal edge of the mammillary bodies anteriorly (see Fig. 10–1*B*). The *oculomotor nerve* (III, motor) exits the anterior aspect of the midbrain, whereas the *trochlear nerve* (IV, motor) exits its posterior aspect (see Fig. 10–1*A* and *C*). The posterior (dorsal) aspect of the midbrain is characterized by the *superior and inferior colliculi* and the anterior (ventral) aspect by the *crus cerebri* and *interpeduncular fossa*.

Tegmental and Basilar Areas. The central core of the midbrain and the pons is called the *tegmentum*, and their anterior (ventral) parts are the *basilar* areas. These regions are continuous with each other and with comparable areas of the medulla (Fig. 10–2; see also Fig. 10–1*B*). Although usually not considered part of the tegmentum *per se*, the central portion of the medulla shares structural and functional similarities with the former region. The *tegmentum of the pons and midbrain* and the *contiguous central portion of the medulla* contain ascending and descending tracts, many relay nuclei, and the nuclei of cranial nerves III to XII.

The basilar part of each brainstem division is ante-

rior to the tegmentum (of the midbrain and pons) and to the central portion of the medulla (see Fig. 10–2). Consequently, these basilar structures also form a rostrocaudal continuum. Basilar structures of the brainstem include the descending fibers of the *crus cerebri* (midbrain), *basilar pons*, and *pyramid* (medulla), and specific populations of neurons in the midbrain and pons that originate from the alar plate of the embryonic brain.

Ventricular Spaces of the Brainstem

The ventricular spaces of the brainstem are the cerebral aqueduct in the mesencephalon and the fourth ventricle in the rhombencephalon (see Fig. 10–2). The *cerebral aqueduct* is a narrow channel, 1 to 3 mm in diameter, that connects the third ventricle (the cavity of the diencephalon) with the fourth ventricle. The cerebral aqueduct contains no choroid plexus; its walls are formed by a continuous mantle of cells collectively called the *periaqueductal gray*. The roof of the midbrain is the *tectum*.

The *fourth ventricle* is the cavity of the *rhombencephalon*. Its rostral portion lies between the pons and cerebellum, and its caudal part is located in the medulla (see Fig. 10–2). The fourth ventricle is continuous rostrally with the cerebral aqueduct and caudally with the central canal of the cervical spinal cord. It also communicates with the subarachnoid space via three openings: the midline *foramen of Magendie* and the two lateral *foramina of Luschka*. The foramen of Magendie is located in the caudal roof of the ventricle and opens into the *dorsal cerebromedullary cistern* (*cisterna magna*) (see Fig. 10–2). The foramina of Luschka are located at the ends of the lateral recesses of the fourth ventricle and open into the subarachnoid space at the cerebellopontine angles (see Fig. 6–9). The *lateral recesses* are horn-shaped widenings of the fourth ventricle that extend around the brainstem at the pons-medulla junction (see Fig. 10–4; see also Fig. 6–9).

The roof of the fourth ventricle is formed mainly by the *anterior* (or *superior*) *medullary velum* rostrally, by the thin membranous *tela choroidea* caudally, and by a small part of the cerebellum in the middle (see Figs. 10–1*B* and 10–2). From rostral to caudal, the walls of the fourth ventricle are formed by the superior cerebellar peduncles, the middle and inferior cerebellar peduncles, and the attachment of the tela choroidea to the medulla (Fig. 10–3). The tela arises from the inferior surface of the cerebellum and sweeps caudally to attach to the V-shaped edges of the medullary portion of the ventricular space. The choroid plexus of the fourth ventricle is suspended from the inner surface of the tela, and parts of it protrude outward through the foramina of Luschka (see Figs. 6–4 and 6–9).

The Rhomboid Fossa. The floor of the fourth ventri-

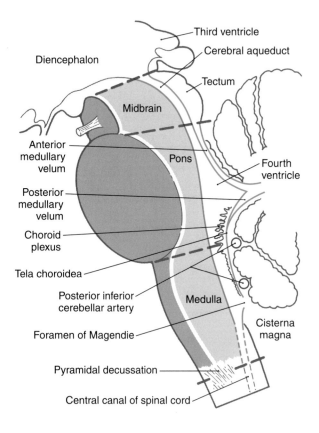

Figure 10–2. Midsagittal drawing of the brainstem. Ventricular spaces of the brainstem are outlined in green. The *tegmental* and *basilar* areas and *contiguous areas of the medulla* are shown in light and dark gray, respectively. Compare with Figure 10–1*B*.

Third ventricle
Cerebral aqueduct
Diencephalon
Tectum
Midbrain
Anterior medullary velum
Pons
Posterior medullary velum
Fourth ventricle
Choroid plexus
Tela choroidea
Posterior inferior cerebellar artery
Medulla
Foramen of Magendie
Cisterna magna
Pyramidal decussation
Central canal of spinal cord

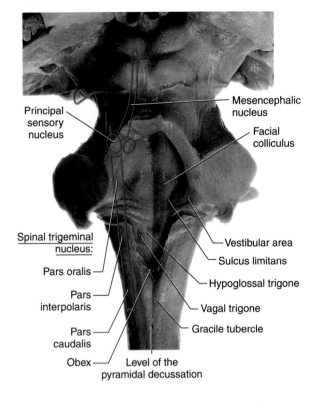

Principal sensory nucleus

Mesencephalic nucleus

Facial colliculus

Spinal trigeminal nucleus:

Pars oralis

Pars interpolaris

Pars caudalis

Obex

Vestibular area

Sulcus limitans

Hypoglossal trigone

Vagal trigone

Gracile tubercle

Level of the pyramidal decussation

Figure 10–3. Posterior view of the brainstem. The approximate locations of the trigeminal nuclei are shown. The cerebellum is removed to expose the posterior aspect of the medulla and midbrain and the rhomboid fossa. The trigeminal motor nucleus, located medial to the principal sensory nucleus, is not labeled.

cle is called the *rhomboid fossa*. It is divided into two halves by a deep *median sulcus*, and each half is traversed rostrocaudally by a groove called the *sulcus limitans* (see Figs. 10–3 to 10–6). There are two slight

depressions along the course of the sulcus limitans, somewhat like deep spots within this sulcus. The rostral depression, the *superior fovea*, is laterally adjacent to the *facial colliculus*, and the caudal depression, the *inferior fovea*, is laterally adjacent to the *vagal and hypoglossal trigones* (see Figs. 10–3 and 10–4). In some surgical procedures involving the fourth ventricle or medulla, the sulcus limitans and foveae represent important landmarks. The *striae medullares* of the fourth ventricle are a series of fiber bundles running from the midline laterally into the lateral recess (see Fig. 10–4). The rostral edge of these fibers is generally regarded as the pons-medulla junction in the floor of the fourth ventricle.

Elevations in the floor of the fourth ventricle indicate the locations of underlying cranial nerve nuclei and associated fiber bundles (see Figs. 10–4 to 10–6). In general, the cranial nerve nuclei that are located between the medial sulcus and the sulcus limitans are motor in function, whereas those located lateral to the sulcus are sensory in function (see Figs. 10–5 and 10–6). Medial to the sulcus limitans, the *hypoglossal* and *vagal trigones* represent the underlying *hypoglossal* and *dorsal motor vagal nuclei*. In the caudal pontine region, the *facial colliculus*, located medial to the sulcus limitans, marks the location of the underlying *abducens motor nucleus* and the *internal genu* of the facial nerve. Lateral to the sulcus limitans in the medulla and caudal pons is a flattened region called the *vestibular area*, which marks the location of the *vestibular nuclei*.

Cranial Nerve Nuclei and Their Functional Components

Cranial nerves, like spinal nerves, contain sensory or motor fibers or a combination of these fiber types.

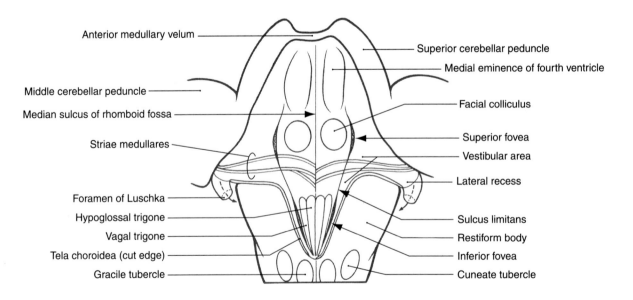

Anterior medullary velum

Middle cerebellar peduncle

Median sulcus of rhomboid fossa

Striae medullares

Foramen of Luschka

Hypoglossal trigone

Vagal trigone

Tela choroidea (cut edge)

Gracile tubercle

Superior cerebellar peduncle

Medial eminence of fourth ventricle

Facial colliculus

Superior fovea

Vestibular area

Lateral recess

Sulcus limitans

Restiform body

Inferior fovea

Cuneate tubercle

Figure 10–4. The rhomboid fossa (floor of the fourth ventricle), elevations and depressions in the floor, and structures bordering on the fossa. Compare with Figures 10–1C and 10–3.

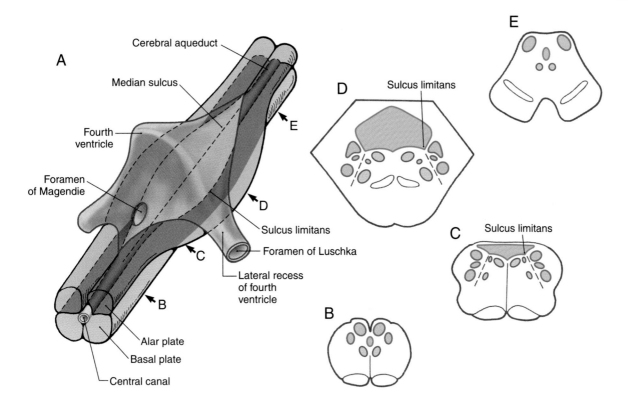

Figure 10–5. Diagram showing the alar and basal plates in relation to the ventricular spaces. The alar plates shift laterally (*A*), where the fourth ventricle flares open at the obex, and then shift back to a posterior position, where the ventricle funnels into the cerebral aqueduct. The position of structures derived from the alar and basal plates in relation to the sulcus limitans and the ventricular space is shown for the medulla (*B*, *C*), the pons (*D*), and the midbrain (*E*).

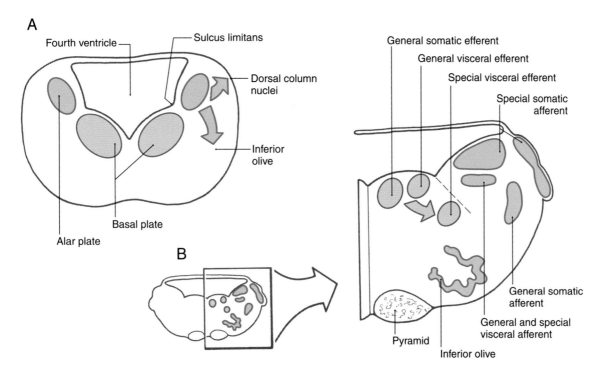

Figure 10–6. Development of the alar and basal plates at early (*A*) and later (*B*) stages, showing their relation to functional components of the cranial nerve nuclei in the brainstem.

These various fibers are classified on the basis of their embryologic origin or common structural and functional characteristics. Primary sensory fibers, somatic motor neurons, and preganglionic and postganglionic visceromotor neurons that exhibit ". . . like anatomical and physiological characters so that they . . . act in a common mode . . ." (Herrick) are classified as having a specific *functional component*. For example, fibers conveying sharp pain, a specific type of input, from two widely separated body parts (the hand and the leg) have the same functional component. This principle, already introduced in relation to spinal nerves (see Chapter 9), is also directly applicable to cranial nerves.

Early in development, the rostrocaudally oriented cell columns forming the alar and basal plates essentially extend throughout the brainstem. As development progresses, neuroblasts in alar and basal plates begin to migrate to form their adult structures, and the caudocephalic continuity of the cell columns may be disrupted. In this respect, *the primitive cell column retains its relative position as it differentiates, but it may become discontinuous as the individual nuclei derived from the same column are formed* (see Figs. 10–5 and 10–7). Motor nuclei of cranial nerves arise from basal plate neuroblasts, whereas the nuclei that receive primary sensory input via cranial nerves originate from the alar plate.

In the caudal medulla, the rostral continuation of the central canal is small; therefore, basal and alar plates are located anterior and posterior, respectively, to this space (see Fig. 10–5). As the fourth ventricle flares open *at the level of the obex*, the alar plate shifts laterally and the basal plate retains an anterior (and now medial) position (see Figs. 10–5 and 10–6). Rostrally, as the fourth ventricle funnels into the cerebral aqueduct of the midbrain, the alar plate rotates back to a posterior position, and the basal plate again assumes an anterior position (see Fig. 10–5). The *sulcus limitans*, a persistent embryologic landmark in the medulla and pons, separates structures derived from the basal plate from those derived from the alar plate.

These points are clearly illustrated by first considering the basal plate. Some of these neuroblasts retain their position adjacent to the midline but become segmented into the *hypoglossal, abducens, trochlear,* or *oculomotor nuclei* (Figs. 10–6 and 10–7). The skeletal muscles innervated by these motor neurons originate from occipital myotomes (tongue musculature) and from mesenchyme in the orbit (extraocular muscles); therefore, their functional component is *general somatic efferent* (GSE). Lateral to the GSE cell groups, a second population of neuroblasts forms the *dorsal motor vagal nucleus, inferior salivatory nucleus, superior salivatory nucleus,* or the *Edinger-Westphal (visceromotor) nucleus* (see Figs. 10–6 and 10–7). Because the axons of these cells synapse in peripheral ganglia that, in turn, innervate smooth muscle, cardiac muscle, or glandular epithe-

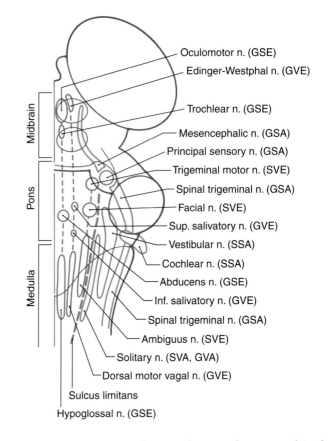

Figure 10–7. Diagram showing the cranial nerve nuclei of the brainstem and their functional components in the brainstem. The various nuclei are unrolled onto a single plane (see also Fig. 10–5*B*) with basal plate derivatives (red) and alar plate derivatives (blue) located medially and laterally, respectively, to the sulcus limitans. GSE, general somatic efferent; GVE, general visceral efferent; SVE, special visceral efferent; GSA, general somatic afferent; SSA, special somatic afferent; GVA, general visceral afferent; SVA, special visceral afferent.

lium, the functional component of these motor neurons is *general visceral efferent* (GVE). The third set of cranial nerve motor nuclei to originate from the basal plate is represented by neuroblasts that migrate anterolaterally in the brainstem to form the *nucleus ambiguus,* the *facial nucleus,* or the *trigeminal motor nucleus* (see Figs. 10–6 and 10–7). These motor neurons innervate skeletal muscles that originate from the pharyngeal arches rather than from occipital myotomes. Consequently, their functional component is *special visceral efferent.*

Simultaneous with these developments in the basal plate, the alar plate gives rise to cell groups that will receive sensory input via cranial nerves. Taste (*special visceral afferent* [SVA]) and *general visceral afferent* (GVA) sensations, such as pain from the gut, enter the brainstem with cranial nerves VII, IX, and X. The central processes of these sensory fibers form the *solitary tract* and end in the surrounding *solitary nucleus* (see Fig. 10–7). Consequently, the functional components SVA and GVA are associated with these sensory fibers, which enter the solitary nucleus and tract. The eighth cranial nerve, the vestibulocochlear, transmits signals concerned with balance, equilibrium, and hearing. These sensory fibers end in the *vestibular* and *cochlear nuclei* of the medulla and pons, respectively (see Figs. 10–6 and 10–7). Because of the unique embryologic origin of the peripheral receptors of these nerves, the functional component associated with these fibers and nuclei is *special somatic afferent* (SSA) (see Fig. 10–7). Sensory input from the face, oral cavity, and scalp to the apex of the head enters the brainstem via the trigeminal nerve. Centrally some of these fibers form the *spinal trigeminal tract* and synapse in the adjacent *spinal trigeminal nucleus*. Others end in the *principal* sensory nucleus or form the *mesencephalic tract*. In the latter case the cell bodies form the adjacent *mesencephalic nucleus*. Because these cell groups receive general sensory input, the functional component associated with these fibers and nuclei is *general somatic afferent* (GSA) (see Figs. 10–6 and 10–7). In addition, cranial nerves VII, IX, and X contribute GSA fibers to the spinal trigeminal tract and nucleus.

The spinal trigeminal nucleus extends caudally from about mid-pontine levels to the spinal cord–medulla junction. On the basis of its cytoarchitecture and connections, the *spinal trigeminal nucleus* is divided into a *pars caudalis* (between the level of the cervical spinal cord and obex), a *pars interpolaris* (between the level of the obex and the rostral end of the hypoglossal nucleus), and a *pars oralis* (rostral to the level of the hypoglossal nucleus) (see Fig. 10–3).

The blood supply to the brainstem is via branches of the *vertebral* and *basilar arteries*. As we shall see in the next three chapters, branches of the vertebrobasilar system serve not only the medulla, pons, and most of the midbrain but also the entire cerebellum.

Sources and Additional Reading

Readings for the brainstem chapters are listed at the end of Chapter 13.

Chapter 11

The Medulla Oblongata

D. E. Haines and G. A. Mihailoff

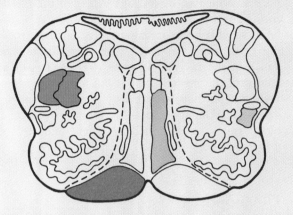

Development 160
  Basal and Alar Plates

External Features 161
  Anterior Medulla
  Lateral Medulla
  Posterior Medulla
  Vasculature

Internal Anatomy of the Medulla 162
  Summary of Ascending Pathways
  Summary of Descending Pathways
  Spinal Cord–Medulla Transition
  Caudal Medulla: Level of the Motor Decussation
  Caudal Medulla: Level of the Sensory Decussation
  Mid-Medullary Level
  Rostral Medulla and Pons-Medulla Junction
  Reticular and Raphe Nuclei
  Vasculature

The medulla oblongata, or *myelencephalon*, is the most caudal segment of the brainstem. It extends rostrally from the level of the foramen magnum to the pons. The cavity of the medulla consists of a narrow, caudal part, which is the continuation of the central canal of the cervical spinal cord, and a flared, rostral portion, which is the medullary part of the *fourth ventricle*. The modest size of the medulla (0.5% of total brain weight) belies its importance. All the tracts passing to or from the spinal cord traverse the medulla, and 7 of the 12 cranial nerves (VI to XII) are associated with the medulla or the pons-medullary junction. Also, the medullary reticular formation contains cell groups that influence heart rate and respiration. The blood supply to the medulla is via branches of the *vertebral arteries*.

Development

The basic structural plan of the medulla is an elaboration of that seen in the spinal cord (Figs. 11–1 and 11–2). The basal and alar plates give rise to specific nuclei, and the surrounding mantle layer is invaded by axons originating from other levels. Beginning in the medulla, however, the basic derivatives of the primitive neural tube are augmented by the appearance of other structures that characterize each brainstem level.

Basal and Alar Plates. *Basal plate* neuroblasts of the medulla give rise to the *hypoglossal nucleus* (general so-

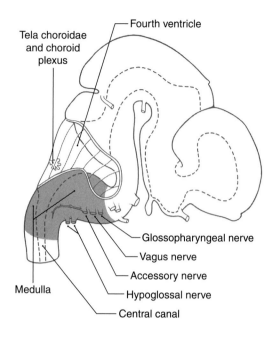

Figure 11–1. Lateral view of the brain at about 7 weeks of gestational age. The medulla is highlighted.

matic efferent [GSE] cells), the *dorsal motor vagal nucleus* and the *inferior salivatory nucleus* (both contain general visceral efferent [GVE] cells), and the nucleus ambiguus (special visceral efferent [SVE] cells) (Fig. 11–2A and B). Caudal to the obex, the hypoglossal

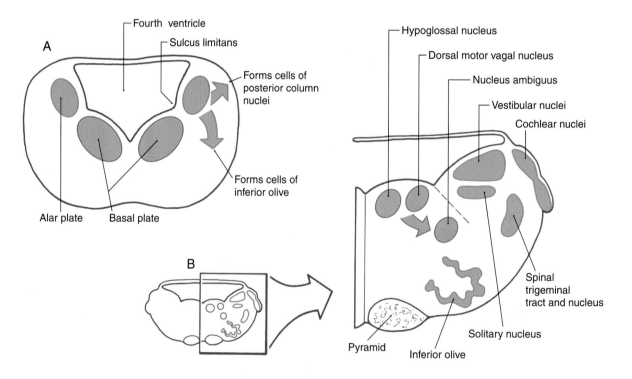

Figure 11–2. Development of the medulla at early *(A)* and later *(B)* stages showing the relationships of alar and basal plates and their adult derivatives in the medulla.

and dorsal motor vagal nuclei are quite small and are found in the central gray surrounding the central canal. Rostral to the obex, all of these nuclei are located medial to the sulcus limitans (see Fig. 11–2*B*).

The cranial nerve nuclei derived from the *alar plate* in the medulla, and their corresponding functional components, include the *vestibular* and *cochlear nuclei* (special somatic afferent [SSA]), the *solitary nucleus* (general visceral afferent [GVA] and special visceral afferent [SVA], and the *spinal trigeminal nucleus* (general somatic afferent, GSA) (see Fig. 11–2). Alar plate neuroblasts caudal to the obex give rise to the *gracile* and *cuneate nuclei*. Rostral to the obex, some alar plate cells migrate ventromedially to form the nuclei of the *inferior olivary complex*.

Concurrent with these events, developing fibers traverse the medulla. An especially prominent bundle of axons collects on the anterior (ventral) surface of the medulla to form the *pyramids* (see Fig. 11–2*B*).

External Features

Anterior Medulla. The anterior (ventral) aspect of the medulla is characterized by an *anterior (ventral) median fissure*, two laterally adjacent longitudinal ridges, the *pyramids*, and the *olive (inferior olivary eminence)* (Fig. 11–3). The pyramids issue from the basilar pons and extend caudally to the *pyramidal decussation*, where about 90% of their fibers cross. Because most of the fibers that form the pyramid arise in the motor cortex as corticospinal fibers, their crossing is frequently called the *motor decussation*. Rootlets of the *hypoglossal nerve* (cranial nerve XII) exit the medulla via the *preolivary sulcus*, a shallow groove located between the pyramid and the olive. The *abducens nerve* (cranial nerve VI) emerges at the pons-medullary junction, generally in line with the rootlets of cranial nerve XII.

Lateral Medulla. On the lateral aspect of the medulla, a shallow trough, the *postolivary sulcus*, is located between the *restiform body* and the large eminence formed by the underlying *inferior olivary nucleus* (Fig. 11–4*A* and *B*). Cranial nerves IX (*glossopharyngeal*), X (*vagus*), and the so-called medullary part of XI (*accessory*) emerge from the postolivary sulcus. The *facial nerve* (VII), along with the *intermediate root* of the facial nerve (VIIi; see Chapter 12), and the *vestibulocochlear nerve* (VIII) emerge from the posterolateral medulla at the pons-medulla interface. The general region of the exit of the facial and vestibulocochlear nerves is clinically regarded as the *cerebellopontine angle*. Indeed, an *acoustic neuroma* is a tumor of the eighth cranial nerve

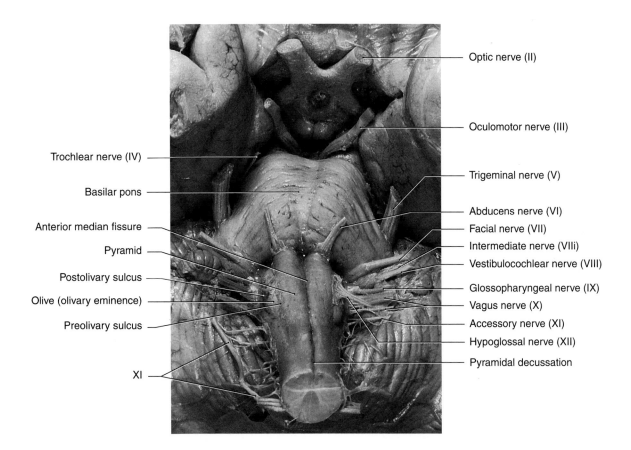

Trochlear nerve (IV)

Basilar pons

Anterior median fissure

Pyramid

Postolivary sulcus

Olive (olivary eminence)

Preolivary sulcus

XI

Optic nerve (II)

Oculomotor nerve (III)

Trigeminal nerve (V)

Abducens nerve (VI)
Facial nerve (VII)
Intermediate nerve (VIIi)
Vestibulocochlear nerve (VIII)

Glossopharyngeal nerve (IX)
Vagus nerve (X)
Accessory nerve (XI)
Hypoglossal nerve (XII)
Pyramidal decussation

Figure 11–3. Anterior (ventral) view of the brainstem with emphasis on structures of the medulla.

A

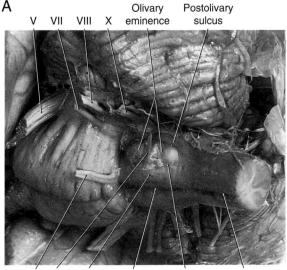

V   VII   VIII   X   Olivary eminence   Postolivary sulcus

VI   XII   Pyramid   Anterior median fissure   Preolivary sulcus   Pyramidal (motor) decussation

B

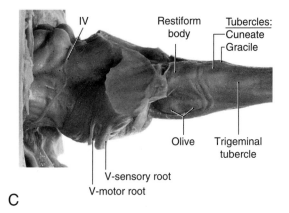

IV   Restiform body   Tubercles:
—Cuneate
—Gracile

Olive   Trigeminal tubercle

V-sensory root
V-motor root

C

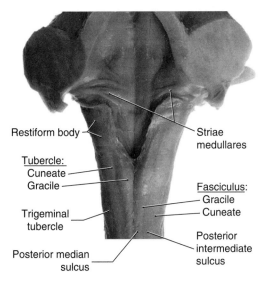

Restiform body   Striae medullares

Tubercle:
Cuneate
Gracile   Fasciculus:
—Gracile
—Cuneate

Trigeminal tubercle   Posterior intermediate sulcus

Posterior median sulcus

Figure 11–4. Anterolateral *(A)*, lateral *(B)*, and posterior *(C)* views of the medulla. Cranial nerves are indicated by Roman numerals, and the cerebellum has been removed from *B* and *C*. The same specimen is used in *B* and *C*.

and is a lesion located at the cerebellopontine angle. On the lateral medullary surface caudal to the level of the obex, fibers of the spinal trigeminal nucleus and tract assume a superficial location and form the *trigeminal tubercle* (*tuberculum cinereum*) (see Fig. 11–4*B* and *C*). Rostral to the obex, these trigeminal fibers are located internal to a progressively enlarging *restiform body*.

Posterior Medulla. At and caudal to the level of the obex, the posterior (dorsal) surface of the medulla is characterized by the *gracile* and *cuneate fasciculi* and their respective *tubercles* (see Fig. 11–4*C*). These tubercles are formed by the underlying *gracile* and *cuneate nuclei*. Rostrolateral to the *gracile* and *cuneate tubercles* and forming a prominent elevation on the posterolateral aspect of the medulla is the *restiform body*. This structure contains a variety of afferent cerebellar fibers and becomes progressively larger as it extends toward the pons-medulla junction. In the caudal pons, fibers of the restiform body join with a much smaller bundle, the *juxtarestiform body*, to form the *inferior cerebellar peduncle*.

Vasculature. The blood supply to the entire medulla and to the choroid plexus of the fourth ventricle arises from branches of the *vertebral arteries* (see Fig. 11–16). In general, the medial medulla is served by the *anterior spinal artery*, the anterolateral medulla by small branches from the *vertebral*, and the posterolateral medulla rostral to the obex by the *posterior inferior cerebellar artery*. Caudal to the obex, the posterior medulla is served by the *posterior spinal artery*. The internal distribution of these vessels is discussed below.

Internal Anatomy of the Medulla

Summary of Ascending Pathways. The ascending tracts that originate from the spinal cord gray matter (*anterolateral system*, *posterior* and *anterior spinocerebellar tracts*, and so on) and from posterior root ganglion cells (*gracile* and *cuneate fasciculi*) continue into the medulla (Fig. 11–5). Some anterolateral fibers terminate in the medulla, as *spinoreticular fibers*, and others convey pain and temperature input to more rostral levels, including the thalamus as *spinomesencephalic* and *spinothalamic fibers*. Posterior column fibers synapse in the medulla, but the tactile and vibratory information carried by these fibers continues rostrally via the *medial lemniscus* (see Fig. 11–5). Spinocerebellar axons enter the cerebellum through the restiform body (posterior tract) or the superior cerebellar peduncle (anterior tract). Other ascending bundles, such as *spino-olivary* and *spinovestibular* fibers, terminate in the medulla.

Summary of Descending Pathways. The descending tracts that originate from the cerebral cortex (*corticospinal*; see Fig. 11–5) and from the midbrain (*rubrospinal*, *tectobulbospinal*), and pons (*reticulospinal*, *vestibulospinal*) traverse the medulla en route to the spinal cord. The medulla contributes additional fibers to the latter two fiber systems. At this level, the *medial longitudinal fasciculus* contains only descending fibers. The majority of these descending axons influence, either directly or indirectly, the discharge patterns of motor neurons in the spinal cord gray matter.

Spinal Cord–Medulla Transition. The spinal cord–medulla transition is characterized by changes that begin at the caudal level of the *pyramidal (motor) decussation* (Figs. 11–6 and 11–7). The spinal cord gray matter is replaced by the motor (crossing of corticospinal fibers) decussation; the central gray matter enlarges; the posterolateral tract (dorsolateral fasciculus) and

substantia gelatinosa of the spinal cord merge, respectively, into the spinal trigeminal tract and nucleus; and nuclei characteristic of the medulla appear. The caudal medulla is described in the following sections beginning at the levels of the motor and sensory decussations.

Caudal Medulla: Level of the Motor Decussation. At the level of the *motor decussation (pyramidal decussation)*, about 90% of corticospinal fibers cross the anterior midline to form the contralateral *lateral corticospinal tract* of the cord (see Figs. 11–5 to 11–7). Posteriorly, at this level the *gracile* and *cuneate nuclei* first appear in their respective fasciculi (see Figs. 11–6 and 11–7). Because the *gracile* and *cuneate fasciculi* are collectively called the *posterior* (or *dorsal*) *columns*, their respective nuclei are frequently referred to as the *posterior (dorsal) column nuclei*. Laterally, the *spinal trigeminal tract* (visible on the surface of the medulla as the *trigeminal tubercle* or *tuberculum cinereum*) is located on the medullary surface. Internal to the spinal tract is the *spinal trigeminal nucleus, pars caudalis* (see Fig. 11–6).

The spinal trigeminal tract is composed of central processes of primary sensory fibers that enter the brain mainly in the trigeminal nerve. These fibers terminate on cells of the spinal trigeminal nucleus, which, in turn, projects to the contralateral thalamus as the *anterior (ventral) trigeminothalamic tract*.

In the lateral medulla, the anterolateral system and *rubrospinal tract* are found medial to the superficially located *posterior* and *anterior spinocerebellar tracts* (see Figs. 11–6 and 11–7). It is appropriate to emphasize that anterolateral system fibers (conveying pain and temperature input from the contralateral side of the body) and spinal trigeminal tract fibers (conveying pain and temperature from the ipsilateral face) are located adjacent to each other throughout the anterolateral medulla.

The anterior medulla contains the most rostral part of the *accessory nucleus* (cranial nerve XI), remnants of the medial motor cell column of C1, and the *medial longitudinal fasciculus* and *tectobulbospinal system* (see Figs. 11–6 and 11–7). At this level, the tectospinal fibers in the TBS are incorporated into the medial longitudinal fasciculus. These small bundles are displaced laterally by the motor (pyramidal) decussation compared to their position at more rostral levels.

The *central gray* surrounds the central canal of the medulla and contains the caudal extremes of the hypoglossal (XII) and dorsal motor vagal nuclei (X) (see Fig. 11–6). When the ventricle flares open at the level of the obex, these nuclei occupy the medial floor of the ventricular space.

Caudal Medulla: Level of the Sensory Decussation. Cells of the posterior column nuclei (gracile and cuneate nuclei) give rise to axons that swing anteromedially, as *internal arcuate fibers*, to cross the midline immediately rostral to the motor decussation (see Fig. 11–5). This constitutes the *sensory decussation*, so

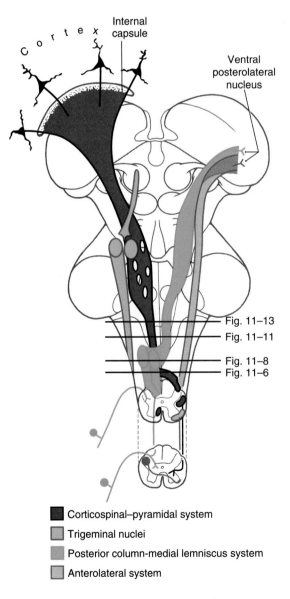

Internal
capsule

Cortex

Ventral
posterolateral
nucleus

Fig. 11–13
Fig. 11–11

Fig. 11–8
Fig. 11–6

■ Corticospinal–pyramidal system

▨ Trigeminal nuclei

▨ Posterior column-medial lemniscus system

▨ Anterolateral system

Figure 11–5. Diagram of the brain showing the location and trajectory of three important pathways and the trigeminal nuclei. The color coding for each is continued in Figures 11–6, 11–8, 11–11, and 11–13.

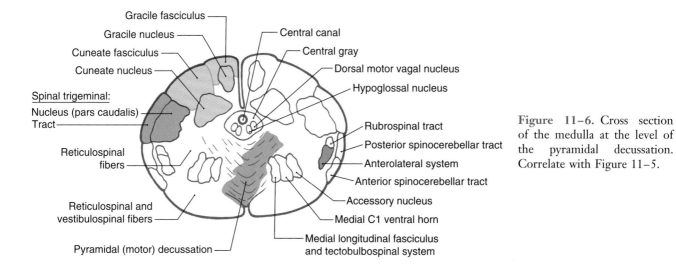

Figure 11-6. Cross section of the medulla at the level of the pyramidal decussation. Correlate with Figure 11-5.

named because it is the point at which a major ascending sensory pathway (posterior column–medial lemniscus) crosses the midline.

At this level the posterior columns (gracile and cuneate fasciculi) are largely replaced by the *gracile* and *cuneate nuclei* (Figs. 11-8 and 11-9). Fibers conveying tactile and vibratory sensations from lower and upper levels of the body terminate, respectively, in the gracile and cuneate nuclei. The axons of these cells, in turn, form the *internal arcuate fibers*, which cross the midline

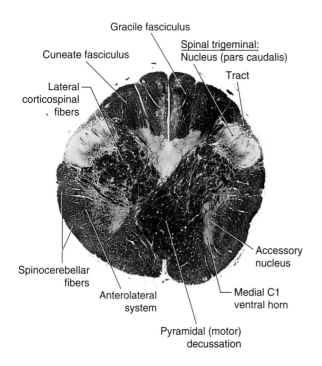

Figure 11-7. A fiber (myelin)-stained cross section of the medulla at the level of the pyramidal decussation. Compare with Figure 11-6.

as the sensory decussation and collect as the *medial lemniscus* on the contralateral side (see Figs. 11-5, 11-8, and 11-9). Information from lower extremities (gracile cell axons) is conveyed in the anterior (ventral) part of the medial lemniscus, and information from the upper extremities (cuneate cell axons) is conveyed in the posterior (dorsal) part of the medial lemniscus. The *accessory cuneate nucleus* is located lateral to the cuneate nucleus (see Fig. 11-8). Its cells receive primary sensory input via cervical spinal nerves and project to the cerebellum as *cuneocerebellar fibers*. In doing so, they represent the upper extremity equivalent of the posterior spinocerebellar tract.

The *spinal trigeminal tract* and *nucleus (pars caudalis)* maintain their position in the lateral medulla. At this level, however, posterior spinocerebellar fibers have migrated posteriorly to cover the spinal tract, heralding the beginnings of the *restiform body* (see Figs. 11-8 and 11-9). Just medial to the spinal trigeminal nucleus, a small column of motor neurons, the *nucleus ambiguus*, appears (see Fig. 11-8). The axons of these SVE cells travel in the glossopharyngeal (IX) and vagus (X) nerves. Fibers of the *anterolateral system* and *rubrospinal tract* are located in the anterolateral medulla (see Fig. 11-8). The *lateral reticular nucleus*, a distinct cell group adjacent to the anterolateral system, receives spinal input and projects to the cerebellum.

Structures characteristic of the anterior surface of the medulla at this level include the *pyramid*, fibers of the *hypoglossal nerve*, and the caudal end of the inferior olivary complex (see Figs. 11-8 and 11-9). The inferior olivary nuclei (internal to the olivary eminence), which become larger at more rostral levels, receive input from a variety of areas and project primarily to the cerebellum. Internal to the pyramid, and along the midline from anterior to posterior, are the *medial lemniscus, tectobulbospinal fibers,* and *medial longitudinal fascic-*

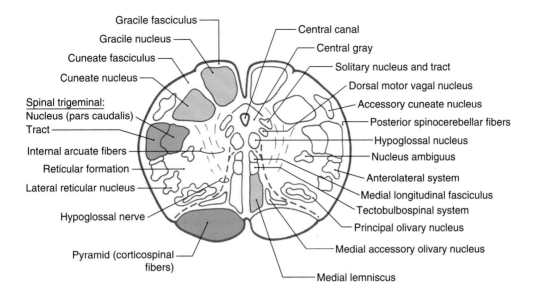

Figure 11–8. Cross section of the medulla at the level of the sensory decussation. Correlate with Figure 11–5.

ulus (see Fig. 11–8). At this level, medial longitudinal fasciculus fibers are characteristically found adjacent to the midline and anterior to structures of the central gray.

The *central gray* is larger than at the level of the motor decussation, and caudal parts of the *hypoglossal* and *dorsal motor vagal nuclei* and the *solitary nucleus* and *tract* can be clearly identified along its perimeter (see Figs. 11–8 and 11–9). Hypoglossal (GSE) motor neurons innervate the ipsilateral half of the tongue. These

fibers course anterolaterally along the lateral edge of the medial lemniscus and pyramid and share a common blood supply with these structures. The GVE cells of the dorsal motor vagal nucleus send their axons to autonomic ganglia associated with viscera in the thorax and abdomen. The *solitary tract* and *nucleus* receive GVA and SVA (taste) input from cranial nerves VII, IX, and X. At this level of the medulla, GVA input to the solitary nucleus comes mainly from thoracic and

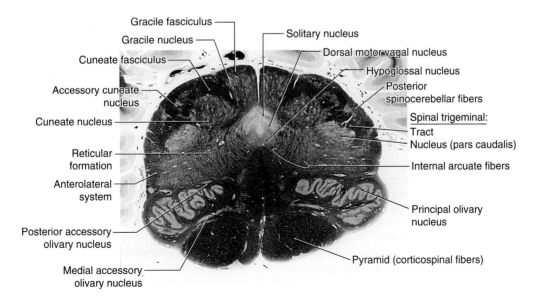

Figure 11–9. A fiber (myelin)-stained cross section of the medulla at the level of the sensory decussation. Compare with Figure 11–8.

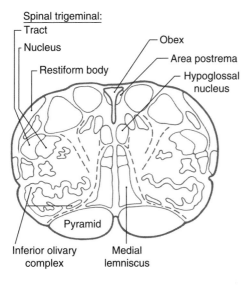

Spinal trigeminal:
- Tract
- Nucleus
- Restiform body
- Obex
- Area postrema
- Hypoglossal nucleus

Pyramid

Inferior olivary complex   Medial lemniscus

Figure 11–10. Cross section of the medulla at the level of the obex.

abdominal viscera (via cranial nerve X) and the carotid sinus (via cranial nerve IX).

The fourth ventricle flares open at the level of the *obex* (Fig. 11–10). The *area postrema* is an emetic (vomiting) center located in the wall of the ventricle at this level. Especially noticeable changes at this point compared with more caudal levels include enlargement of the inferior olivary complex and restiform body.

Mid-Medullary Level. Rostral to the obex, the struc-

tures in the medial floor of the fourth ventricle are the *hypoglossal* and *dorsal motor vagal nuclei* and, lateral to the *sulcus limitans*, the *vestibular nuclei* (Figs. 11–11 and 11–12). The latter cell groups consist, at this level, of *medial* and *inferior* (or spinal) *vestibular nuclei*. They receive input from cranial nerve VIII and interconnect with areas of the brain concerned with balance and eye movement. The *solitary tract* and *nucleus* occupy their characteristic position immediately anterior to the vestibular nuclei.

Laterally, the *restiform body* forms a prominent elevation on the posterolateral aspect of the medulla (see Figs. 11–11 and 11–12). This structure contains *posterior spinocerebellar, cuneocerebellar, olivocerebellar, reticulocerebellar,* and other cerebellar afferents. In the base of the cerebellum, these fibers join with the *juxtarestiform body* to form the *inferior cerebellar peduncle*.

The *spinal trigeminal tract* and *nucleus* (*pars interpolaris*) are internal to the restiform body (see Figs. 11–11 and 11–12). Other structures in the lateral medulla are comparable with those seen more caudally. These include the *nucleus ambiguus* and the *lateral reticular nucleus* as well as the *anterolateral system, anterior spinocerebellar tract,* and *rubrospinal tract* (see Fig. 11–11). At all medullary levels, neurons of the nucleus ambiguus contribute axons to cranial nerves IX and X, which innervate pharyngeal and laryngeal muscles, including those of the vocal folds.

Anterolaterally, the inferior olivary complex is prominent at mid-medullary levels and is composed of a large, saccular *principal olivary nucleus* and diminutive *medial* and *posterior* (*dorsal*) *accessory olivary nuclei* (see

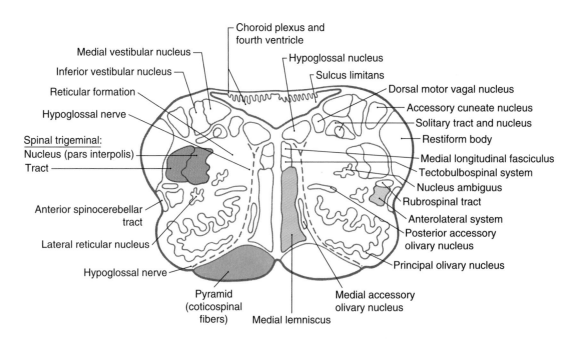
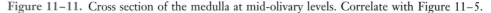

Figure 11–11. Cross section of the medulla at mid-olivary levels. Correlate with Figure 11–5.

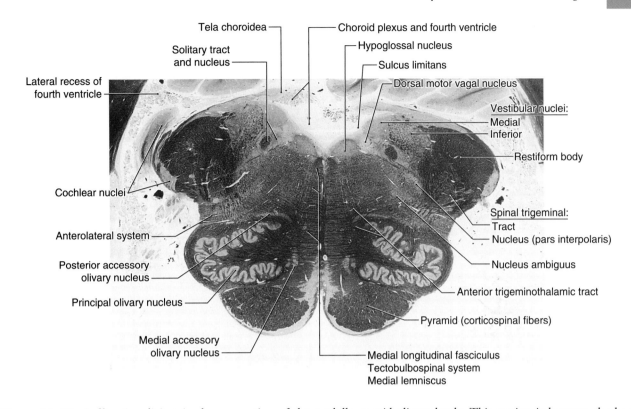

Figure 11–12. A fiber (myelin)-stained cross section of the medulla at mid-olivary levels. This section is between the levels represented in Figures 11–11 and 11–13.

Figs. 11–11 and 11–12). These cell groups receive input from a variety of central nervous system nuclei and project primarily to the contralateral cerebellum (as *olivocerebellar fibers*) through the restiform body. Anteriorly and medially, the orientation of the *pyramid, medial lemniscus, medial longitudinal fasciculus,* and *tectobulbospinal system* remains essentially the same as at more caudal levels (see Figs. 11–11 and 11–12).

Rostral Medulla and Pons-Medulla Junction. Comparison of Figures 11–11 and 11–13 shows that many of the structures seen in the mid-medulla are present in essentially the same locations in the rostral medulla. Therefore, we shall emphasize the features that are different in the rostral medulla.

In the floor of the fourth ventricle, the positions occupied by the hypoglossal and dorsal motor vagal nuclei at more caudal levels are taken by the *prepositus (hypoglossal) nucleus* and the *inferior salivatory nucleus* (see Fig. 11–13). The prepositus nucleus is a small, somewhat flattened cell group that is easily distinguished from the hypoglossal nucleus. The GVE cells of the inferior salivatory nucleus are located immediately anterior to the medial vestibular nucleus and medial to the solitary tract and nucleus. Axons of these cells dis-

tribute to the otic ganglion via peripheral branches of the glossopharyngeal nerve.

The *medial* and *inferior* (or *spinal*) *vestibular nuclei* are prominent at this level and are joined, in this plane of section, by the *posterior* and *anterior cochlear nuclei* (see Fig. 11–13; see also Fig. 11–12). The latter nuclei are located on the posterior and lateral aspects of the restiform body at the pons-medullary junction. The medial vestibular nucleus appears homogeneous in fiber-stained sections, and the inferior vestibular nucleus has a salt-and-pepper appearance (see Fig. 11–12). This appearance results because small descending bundles of myelinated fibers (pepper) are intermingled with cells (salt) in the inferior nucleus. Medial to the restiform body is the spinal trigeminal tract and the *pars oralis* of the *spinal trigeminal nucleus* (see Fig. 11–13; see also Fig. 10–3).

Although structures in anterior and medial areas of the medulla are unchanged from those at mid-medullary levels, some changes take place at the pons-medullary junction that merit comment (Fig. 11–14). Fibers of the *restiform body* arch posteriorly to enter the cerebellum where they are joined by fibers of the *juxtarestiform body* to form (collectively) the *inferior cerebellar peduncle*. The *facial motor nucleus* (SVE cells) appears

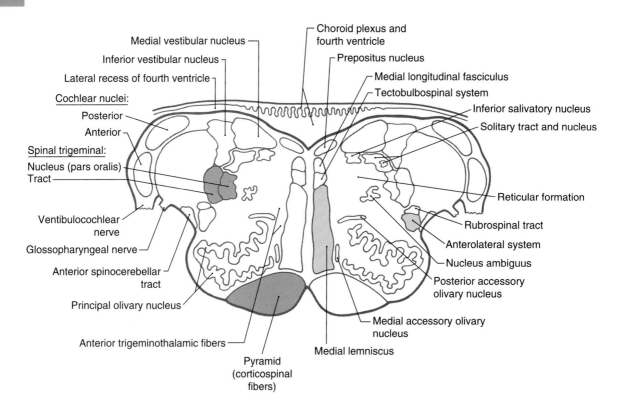

Figure 11–13. Cross section of the medulla at rostral olivary (and medullary) levels. Correlate with Figure 11–5.

anterolaterally, and the *trapezoid body* and *superior olivary nucleus* (both conveying auditory information) appear adjacent to the facial nucleus and the spinal trigeminal tract and nucleus (see Fig. 11–14). The inferior olivary complex disappears, and the *central tegmental tract*, one source of input to the inferior olive, appears about where the latter cell group was located (see Fig. 11–14). Finally, the *medial lemniscus* begins to shift anterolaterally and to rotate from a posteroanterior orientation, which it exhibits in the medulla, toward a horizontal orientation more characteristic of the pons (see Fig. 11–14). At the pons-medulla junction, the cross section of the medial lemniscus is oriented obliquely (posteromedial to anterolateral); by the level of the mid-pons, it is horizontal.

Reticular and Raphe Nuclei. The word *reticulum* is Latin for "little net" (diminutive of *rete*, net) and denotes meshlike structures. The *reticular nuclei* of the brainstem are diffuse and ill-defined and have little apparent internal organization. Collectively, they make up the *reticular formation*, which may be thought of simply as including all of the cells that are interspersed among the more compact and named structures of the brainstem.

Raphe is a Greek word for "suture" or "seam." Thus, the *raphe nuclei* are bilaterally symmetrical cell groups in the brainstem that are located directly adjacent to the midline.

The *medial medullary reticular area* consists of the *central nucleus of the medulla* at caudal medullary levels and the *gigantocellular reticular nucleus* rostrally; the latter cell group extends into the pons (Fig. 11–15). The *lateral medullary reticular area* contains a compact col-

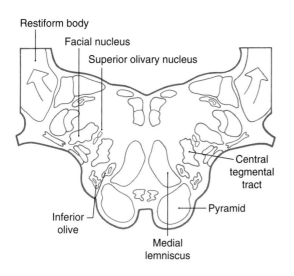

Figure 11–14. Cross section of the medulla at the pontomedullary junction. Fibers of the restiform body sweep up (*arrows*) into the cerebellum at this level.

umn of cells, the *lateral reticular nucleus*, and a diffuse population of cells that forms the *parvocellular nucleus* and the *ventrolateral reticular area* (area reticularis superficialis ventrolateralis) (see Fig. 11–15). The latter cells function in the control of heart rate and respiration. Consequently, a sudden onset of *central apnea*, indicating damage to these respiratory areas, is often a prime early sign of medullary compression. For example, apnea may occur when an increase in intracranial pressure forces the tonsils of the cerebellum downward into the foramen magnum (*tonsillar herniation*) so that they press against the medulla.

The raphe nuclei of the medulla are the *nucleus raphes pallidus* and *nucleus raphes obscurus* and, at rostral levels, the *nucleus raphes magnus* (see Fig. 11–15). The nucleus pallidus and nucleus obscurus are located at mid- to rostral medullary levels along the posterior and anterior midline, respectively. The nucleus raphes magnus begins in the rostral medulla and extends into the caudal pons (see Fig. 11–15). Cells of these raphe nuclei receive input from several areas, including the central gray of the mesencephalon, and project to the spinal cord. *Raphespinal* fibers from raphe magnus are especially important for the inhibition of pain transmission in the posterior horn of the spinal cord.

Vasculature. The blood supply to the medulla is via branches of the *vertebral arteries* (Fig. 11–16). These branches are the *anterior spinal artery* and the *posterior inferior cerebellar artery* (PICA). The *posterior spinal artery* is usually a branch of the PICA.

Medial structures of the medulla at all levels, including the pyramid, medial lemniscus, and hypoglossal nucleus and roots, are served by penetrating branches of the *anterior spinal artery* (see Fig. 11–16). Vascular insufficiency of these branches may result in a contralateral hemiparesis (pyramidal and corticospinal damage), contralateral loss of proprioception and vibratory sense (medial lemniscus), and a deviation of the tongue to the ipsilateral side when protruded (hypoglossal root or nucleus injury).

The posterior medulla caudal to the obex is served by branches of the *posterior spinal artery* (see Fig. 11–16A). Major structures in this area include the posterior column (gracile and cuneate) nuclei and the spinal trigeminal tract and nucleus. Although vascular lesions of the posterior spinal artery are rare, they produce an ipsilateral loss of proprioception and vibratory sense on the body (damage to posterior columns and nuclei) coupled with an ipsilateral loss of pain and temperature sensation from the face (spinal trigeminal tract).

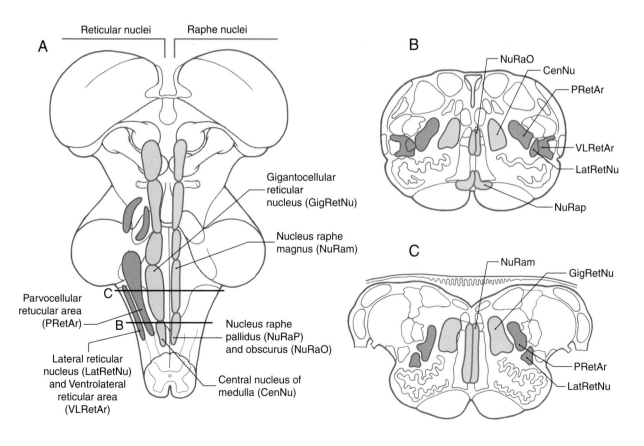

Figure 11–15. Posterior (dorsal) (*A*) view of the brainstem and caudal (*B*) and rostral (*C*) cross sections showing the raphe and reticular nuclei of the medulla.

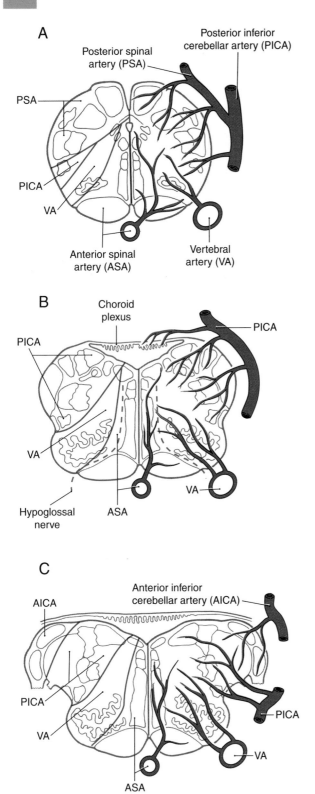

Rostral to the obex, the entire posterolateral medulla is served by branches of the *posterior inferior cerebellar artery* (PICA) (Fig. 11–17; see also Fig. 11–16*B* and *C*). Included in the territory served by this vessel are the anterolateral system, spinal trigeminal tract and nucleus, vestibular nuclei, solitary tract and nucleus, and the nucleus ambiguus. Vascular insufficiency of the PICA (or blockage of one vertebral artery) gives rise to a characteristic set of sensory and motor deficits commonly called the *lateral medullary syndrome, PICA syndrome,* or *Wallenberg syndrome* (see Fig. 11–17). The deficits seen and the corresponding structures involved are (1) contralateral loss of pain and temperature sensation from the body (anterolateral system), (2) ipsilateral loss of pain and temperature sensation from the face (spinal trigeminal tract and nucleus), (3) some vertigo and nystagmus (vestibular nuclei), (4) loss of taste from the ipsilateral half of the tongue (solitary tract and nucleus), and (5) hoarseness and dysphagia (nucleus ambiguus or roots of cranial nerves IX and X) (see Fig. 11–17*C*). Patients with the lateral medullary syndrome may also have the *Horner syndrome*, due to injury to hypothalamospinal fibers descending through the lateral areas of the medulla. We shall explore the details of these clinical syndromes in later chapters.

In addition to this broad expanse of the medulla, branches of the PICA also serve the choroid plexus of the fourth ventricle. At the pons-medulla junction, the cochlear nuclei and a small adjacent part of the restiform body are served by branches of the anterior inferior cerebellar artery (see Fig. 11–16*C*).

Figure 11–16. Blood supply of the medulla caudal to the obex (*A*), at mid-medullary (*B*) and rostral medullary (*C*) levels. Arteries are shown on the right, and the territories served by each are indicated on the left.

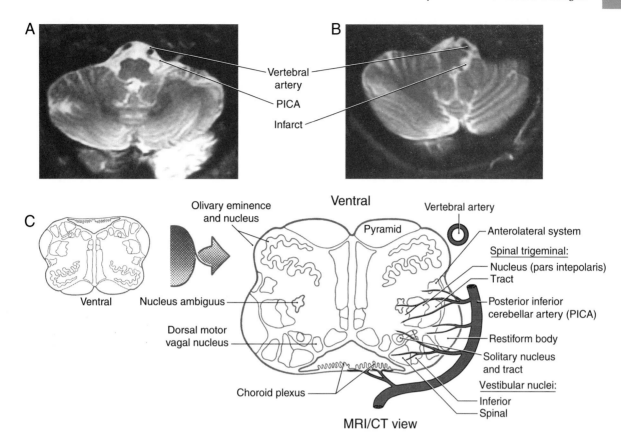

Figure 11–17. Lateral medullary (Wallenberg) syndrome. A normal magnetic resonance image (MRI) study (*A*) shows the vertebral and posterior inferior cerebellar (PICA) arteries in relation to the medulla. The patient whose MRI study is shown in *B* had an occlusion of the PICA, which resulted in an infarct of the territory of the medulla served by this vessel. The structures damaged in this lesion are shown in *C*. Compare with Figure 11–16. CT, computed tomography.

Sources and Additional Reading

Readings for the brainstem chapters are listed at the end of Chapter 13.

Chapter 12

The Pons and Cerebellum

G. A. Mihailoff and D. E. Haines

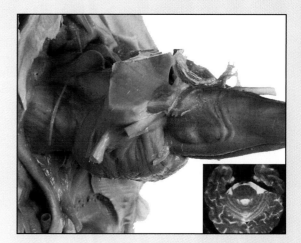

Development 174
Basal and Alar Plates
Cerebellum

External Features 175
Basilar Pons
Rhomboid Fossa of the Pons
Cerebellum
Vasculature

Internal Anatomy of the Pons 178
Summary of Ascending Pathways
Summary of Descending Pathways
Caudal Pontine Level
Mid-Pontine Level
Rostral Pontine Level
Reticular and Raphe Nuclei
Vasculature

Internal Anatomy of the Cerebellum 184
Cerebellar Cortex
Cerebellar Nuclei
Vasculature

The *metencephalon* consists of the pons and cerebellum. The pons is the middle segment of the brainstem, the caudal part being the medulla and the rostral portion being the midbrain. Although comprising only about 1.3% of the brain by weight, the pons has many important functions. The motor and sensory nuclei and the exit points of cranial nerves V to VIII are associated with the pons. The *cerebellum is not part of the brainstem* but rather is considered a suprasegmental structure because it is located posterior (dorsal) to the brainstem. The cerebellum is comparatively large, comprising about 10.5% of the total brain weight. Functionally, the cerebellum is part of the motor system. The blood supply to the pons and cerebellum arises from branches of the basilar and cerebellar arteries.

Development

The pons and cerebellum are considered together in this chapter because they arise from the same region of the developing neural tube. The metencephalon is rostral to the myelencephalon and extends from the pontine flexure to the mesencephalic isthmus (Fig. 12–1). At this level, the cavity of the neural tube is enlarged, forming the parts of the fourth ventricle associated with the pons and cerebellum.

Basal and Alar Plates. The *basal* and *alar plates* of the brainstem extend from the medulla rostrally into the developing pons. The cranial nerve *motor* nuclei found in the pons (trigeminal, abducens, facial, and superior salivatory) originate from the basal plate and are located medial to the sulcus limitans (Fig. 12–2A and B).

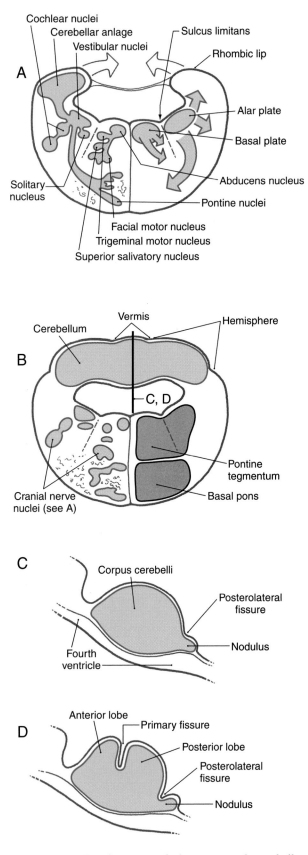

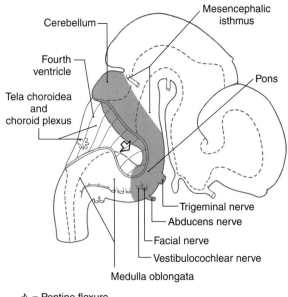

Figure 12–1. Lateral view of the brain at about 7 weeks of gestational age. The pons and cerebellum are highlighted.

Figure 12–2. Development of the pons and cerebellum. The pontine alar and basal plates give rise to cranial nerve nuclei and the pontine nuclei, and the cerebellum originates from the rhombic lips (*A, B*). Sagittal views of the cerebellum (*C, D*; plane of section from *B*) show the relationships of posterolateral and primary fissures.

The functional components of these motor neurons include *special visceral efferent* (SVE) (trigeminal and facial), *general somatic efferent* (GSE) (abducens), and *general visceral efferent* (GVE) (superior salivatory).

The cranial nerve *sensory* nuclei located in the pons include portions of the trigeminal and vestibulocochlear nuclei and the rostral tip of the solitary nucleus. These nuclei originate from the alar plate and are found lateral to the sulcus limitans (see Fig. 12–2*A* and *B*). Their functional components are *general somatic afferent* (GSA) for the trigeminal, *special somatic afferent* (SSA) for vestibular and cochlear nuclei, and *special* and *general visceral afferent* (SVA, GVA) for the solitary nucleus. The portion of the posterior (dorsal) pons that contains these motor and sensory nuclei, as well as the reticular formation and several ascending and descending tracts, is the *pontine tegmentum* (see Fig. 12–2*B*).

The anterior (ventral) area of the developing pons is invaded by large numbers of fibers. Although some will terminate here, others pass through to more caudal targets. *Alar plate* neuroblasts also migrate into this anterior pontine region to form the *basilar pontine nuclei*. These nuclei, their axons, and the descending fibers passing to and through this area collectively form the *basilar pons* (see Fig. 12–2*B*).

Cerebellum. The cerebellum develops from the *rhombic lips* of the pontine alar plates. These lips expand posteromedially toward each other until they meet at the midline and fuse to form the *cerebellar plate* (see Fig. 12–2*A* and *B*), which is the rudiment of the cerebellum. As development progresses, the cerebellum is divided by transverse fissures into lobes and lobules. The first of these fissures to appear is the *posterolateral fissure*, which separates the *flocculonodular lobe* caudally

from the *corpus cerebelli* rostrally. The *primary fissure*, the second groove to appear, divides the corpus cerebelli into the *anterior* and *posterior lobes* (see Fig. 12–2*B*–*D*). Internal changes, such as development of the cerebellar cortex and nuclei, take place concurrently with these external events.

External Features

Basilar Pons. The portion of the brainstem lying between the midbrain rostrally and the medulla caudally is the pons (*pons* is the Latin word for "bridge"). Anteriorly and laterally (Fig. 12–3), the pons consists of a massive bundle of transversely oriented fibers that enter the cerebellum as the *middle cerebellar peduncle* (*brachium pontis*). The exit of the trigeminal nerve marks the transition from the basilar pons, which is anterior (ventral) to the trigeminal root, to the middle cerebellar peduncle, which lies posterior (dorsal) to the exit of the trigeminal nerve (Fig. 12–4; see also Fig. 12–3). Rostrally, the large bundles forming the *crus cerebri* of the midbrain extend into the basilar pons. Caudally, some of these descending axons emerge to form the *pyramids* of the medulla (see Fig. 12–4*A*).

The cranial nerves that emerge from the pons are the *trigeminal* (V), *abducens* (VI), *facial* (VII), and *vestibulocochlear* (VIII) nerves. The trigeminal nerve exits laterally and is composed of a large sensory root and a small motor root. The portion of the trigeminal nerve that traverses the subarachnoid space between the pons and the trigeminal ganglion forms a landmark that is visible on magnetic resonance images at this level (see Fig. 12–4*B*). The abducens, facial, and vestibulococh-

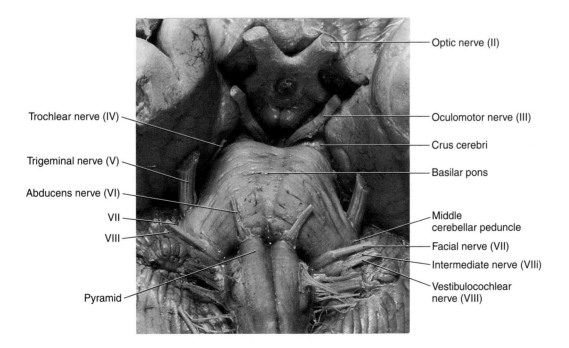

Figure 12–3. Anterior (ventral) view of the brainstem with emphasis on the pons.

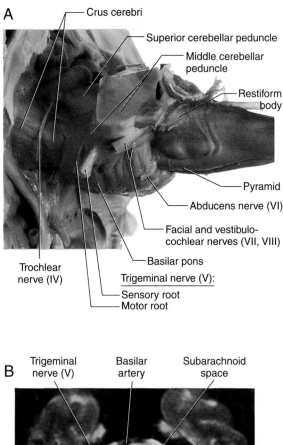

A

- Crus cerebri
- Superior cerebellar peduncle
- Middle cerebellar peduncle
- Restiform body
- Pyramid
- Abducens nerve (VI)
- Facial and vestibulo-cochlear nerves (VII, VIII)
- Basilar pons
- Trigeminal nerve (V):
- Trochlear nerve (IV)
- Sensory root
- Motor root

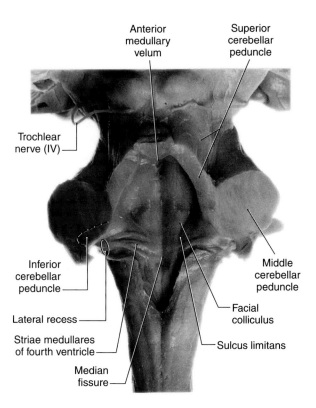

- Anterior medullary velum
- Superior cerebellar peduncle
- Trochlear nerve (IV)
- Inferior cerebellar peduncle
- Lateral recess
- Striae medullares of fourth ventricle
- Median fissure
- Middle cerebellar peduncle
- Facial colliculus
- Sulcus limitans

Figure 12–5. Pontine part of the fourth ventricle and rhomboid fossa. Also see Figure 10–4 for further details of the rhomboid fossa.

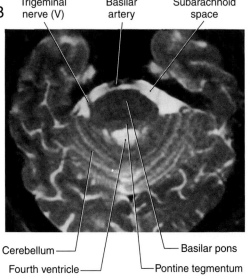

B

- Trigeminal nerve (V)
- Basilar artery
- Subarachnoid space
- Cerebellum
- Fourth ventricle
- Basilar pons
- Pontine tegmentum

Figure 12–4. A dissection with the cerebellum removed showing the cranial nerves and relationships of the pons *(A)* and a magnetic resonance image of the pons and root of the trigeminal nerve *(B)*.

lear nerves emerge in medial to lateral sequence along the pons-medulla junction (see Fig. 12–3). Although cranial nerve VII is commonly called the facial nerve, it is actually composed of two roots, the *facial nerve* (SVE fibers) and the *intermediate nerve* (SVA, GVE, and GSA fibers). The *vestibulocochlear nerve* (SSA fibers) emerges posterolaterally and, with the facial and intermediate nerves and *labyrinthine artery*, occupies the internal acoustic meatus.

Rhomboid Fossa of the Pons. The *rhomboid fossa* forms the floor of the fourth ventricle. Its caudal por-

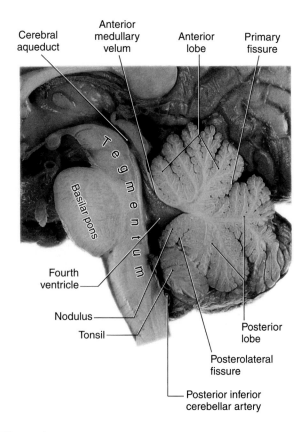

- Cerebral aqueduct
- Anterior medullary velum
- Anterior lobe
- Primary fissure
- Tegmentum
- Basilar pons
- Fourth ventricle
- Nodulus
- Tonsil
- Posterior lobe
- Posterolateral fissure
- Posterior inferior cerebellar artery

Figure 12–6. Sagittal view of the brainstem (with emphasis on the pons) and cerebellum.

tion is located in the medulla, and its larger, more rostral area is in the pons. The posterior (dorsal) surface of the *pontine tegmentum*, which forms the floor of the fourth ventricle, is visible only when the cerebellum is detached from the brainstem (Fig. 12–5; see also Fig. 10–4). This part of the ventricular floor is characterized by an elevation called the *facial colliculus* located between the median fissure and the *superior fovea* of the *sulcus limitans* and by an area called the *vestibular area* located lateral to the sulcus limitans. The facial colliculus is formed by the underlying abducens nucleus and internal genu of the facial nerve (see below), and the vestibular area marks the location of the vestibular nuclei. The *brachium pontis* and the *brachium conjunctivum* form the lateral walls of the fourth ventricle in the pons; the roof is formed by the anterior medullary velum, by a small part of the cerebellum, and by a portion of the tela choroidea (Fig. 12–6).

Cerebellum. The cerebellum is located posterior (dorsal) to the brainstem and fills much of the posterior fossa. It is attached to the brainstem by three pairs of cerebellar peduncles (*superior, middle,* and *inferior*). In sagittal section, the human cerebellum appears wedge-shaped (see Fig. 12–6), with its superior surface apposed to the tentorium cerebelli and its inferior surface curving toward the foramen magnum.

The cerebellum consists of *anterior, posterior,* and *flocculonodular lobes;* each lobe, in turn, is composed of lobules (Fig. 12–7; see also Fig. 12–6). Lobes and lobules are separated from each other by *fissures.* Lobules are made up of yet smaller folds of cerebellar cortex called *folia* (singular, *folium*). Cerebellar folia, lobules, and lobes can often be followed across the midline from one side of the cerebellum to the other.

Each cerebellar lobe (and lobule) is also divided into rostrocaudally oriented *vermis* (medial), *intermediate* (paravermis), and *hemisphere* (lateral) areas of cortex (see Fig. 12–7). The vermal cortex is approximately 1.0 cm across at its widest point. The hemisphere is expansive in the human cerebellum and is separated from the vermis by a somewhat ill-defined intermediate cortex.

Four *cerebellar nuclei* are located in the white matter core of each hemisphere. From medial to lateral, they are the *fastigial, globose, emboliform,* and *dentate* nuclei (see Fig. 12–7). These cells receive input from branches of cerebellar afferent fibers and from Purkinje cells located in the cerebellar cortex. In turn, axons of cerebellar nuclear cells provide the main output signals of the cerebellum. The structure, function, and connections of the cerebellar cortex and nuclei are considered in greater detail in Chapter 27.

Vasculature. The *basilar artery* and its branches serve basilar and tegmental areas of the pons. The internal distribution of the basilar artery and its branches is discussed later in this chapter. The *superior cerebellar*

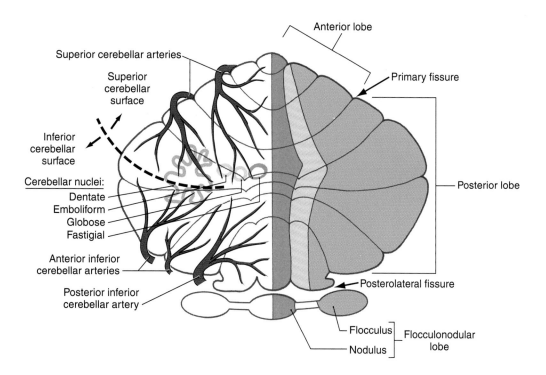

Figure 12–7. Unfolded view of the cerebellar surface with lobes and fissures labeled on the right side and the blood supply to the various lobes and their underlying nuclei indicated on the left. The colors of the rostrocaudal regions (zones) of the lobules correlate with the cerebellar nuclei to which they are related; gray, medial (vermis) area; green, intermediate (paravermis) area; blue, lateral (hemisphere) area.

artery distributes to the superior surface of the cerebellum and most of the cerebellar nuclei; the inferior surface is served by *anterior* and *posterior inferior cerebellar arteries* (see Fig. 12–7).

Internal Anatomy of the Pons

Summary of Ascending Pathways. The major ascending pathways seen in the medulla continue into the pons (Fig. 12–8). These include the *medial lemniscus, anterolateral system, anterior (ventral) trigeminothalamic fibers,* and the *anterior spinocerebellar tract.* Although most

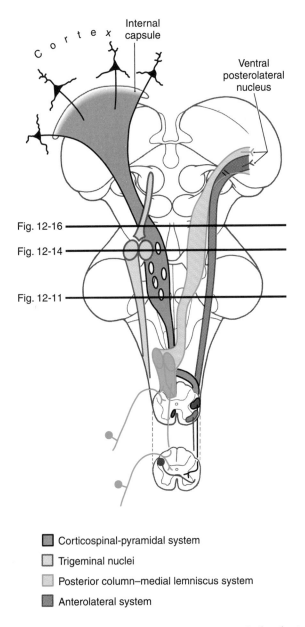

Figure 12–8. Diagrammatic representation of the brain showing the location and trajectory of three important pathways and the trigeminal nuclei. The color coding for each is continued in Figures 12–11, 12–14, and 12–16.

of these fibers continue through the pons, some anterolateral system fibers terminate in the pontine reticular formation (as *spinoreticular* fibers), and anterior spinocerebellar axons enter the cerebellum. The *restiform body,* a prominent structure in the rostral medulla, sweeps posteriorly into the cerebellum in the caudal pons.

Summary of Descending Pathways. The most prominent groups of descending fibers arise from cells located in the midbrain or forebrain and therefore traverse the pons (see Fig. 12–8). These include the *corticospinal fibers,* the *central tegmental* and *rubrospinal tracts,* and the *tectobulbospinal system.* The *medial longitudinal fasciculus* occupies a characteristic position near the midline in the floor of the fourth ventricle. At the pons-medullary junction, this bundle contains mainly descending fibers; in the rostral pons it is made up of ascending fibers.

Caudal Pontine Level. As noted earlier, the pons is divided into a posterior part, the *tegmentum,* and an anterior region, the *basilar pons.* This section and the following two sections describe the anatomy of the pons at three levels: caudal pontine, mid-pontine, and rostral pontine. Each level is described from posterior (dorsal) to anterior (ventral), beginning with the tegmentum and proceeding to the basilar pons.

At caudal pontine levels, the *facial colliculus* is formed by the underlying *abducens nucleus* and fibers comprising the *internal genu* of the facial nerve (see Figs. 12–10 to 12–12). Axons from the GSE cells of the abducens nucleus course anteriorly through the tegmentum, pass adjacent to the corticospinal fibers in the basilar pons, and exit the brainstem at the pons-medullary junction as the abducens nerve (see Figs. 12–10 and 12–12). The internal genu of cranial nerve VII is composed of the axons of SVE cells from the facial nucleus. These axons loop around the abducens nucleus from caudal to rostral, as the internal genu, then course anterolaterally to exit the brainstem (see Fig. 12–12). Anterolateral to the abducens nucleus, these SVE fibers are surrounded by cells of the *superior salivatory nucleus,* the axons of which exit the brainstem as the GVE component of the intermediate nerve (see Fig. 12–12).

Medial to the abducens nucleus is the *medial longitudinal fasciculus* and the *tectobulbospinal system* (see Fig. 12–11). As in the medulla, these bundles are internal to the ventricular space and adjacent to the midline.

The posterolateral tegmentum contains the vestibular nuclei and the solitary tract and nucleus (Fig. 12–9; see also Fig. 12–11). The *lateral, medial,* and *inferior vestibular nuclei* are present at this level, whereas the *superior vestibular nucleus* becomes prominent more rostrally. The small bundles of fibers coursing between the vestibular nuclei and the cerebellum in the wall of the fourth ventricle form the *juxtarestiform body* (Figs. 12–10 and 12–11). This structure is composed of vestibulocerebellar and cerebellovestibular fibers and,

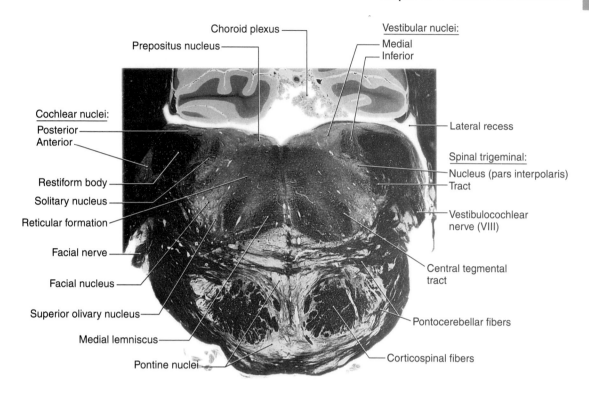

Figure 12–9. A fiber (myelin)-stained cross section at the level of the facial motor nucleus (caudal pons).

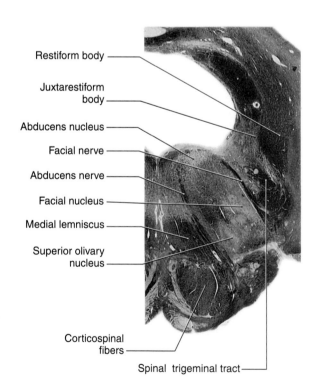

Figure 12–10. A fiber (myelin)-stained cross section of the pons at the level of the facial colliculus. Compare with Figure 12–11.

along with *the laterally adjacent restiform body*, constitutes the *inferior cerebellar peduncle*. The rostral portions of the *solitary tract* and *nucleus* are located anterior (ventral) to the vestibular nuclei and consist of a core of primary sensory fibers (tract) surrounded by cell bodies (nucleus). This part of the solitary complex receives mainly taste input (SVA fibers) and is sometimes called the *gustatory nucleus.*

The central portion of the pontine tegmentum at caudal levels contains, from medial to lateral, the *central tegmental tract*, the *superior olivary nucleus*, the *facial motor nucleus*, and the *spinal trigeminal tract* and *nucleus* (Fig. 12–12; see also Figs. 12–9 to 12–11). A major part of the central tegmental tract includes fibers coursing from the red nucleus of the midbrain to the inferior olive of the medulla (*rubro-olivary fibers*). Cells of the superior olive receive input from the anterior cochlear nucleus and send their axons into the lateral lemniscus on both sides. The route followed by motor facial fibers is shown in Figure 12–12; these axons innervate the ipsilateral muscles of facial expression. The spinal trigeminal tract is composed of general sensory (GSA) fibers from the ipsilateral half of the face, oral cavity, and much of the scalp. Although most of this input is via the trigeminal nerve (hence the name of the tract), cranial nerves VII, IX, and X also make modest contributions to the spinal trigeminal tract and nucleus. Axons of the spinal trigeminal tract synapse in the spinal trigeminal nucleus, the cells of which project to the contralateral thalamus as *anterior (ventral) trigeminothalamic fibers.*

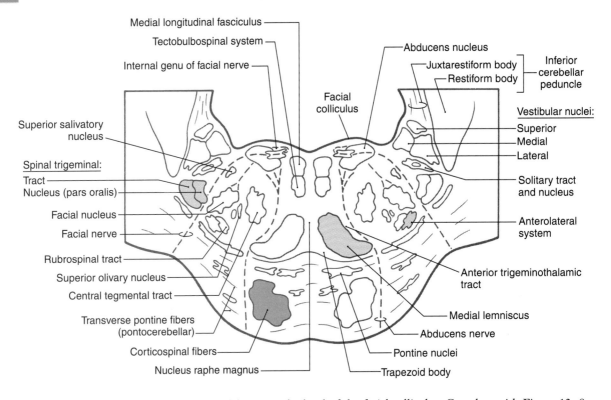

Medial longitudinal fasciculus
Tectobulbospinal system
Internal genu of facial nerve
Abducens nucleus
Juxtarestiform body
Restiform body
Inferior cerebellar peduncle
Facial colliculus
Superior salivary nucleus
Spinal trigeminal:
Tract
Nucleus (pars oralis)
Facial nucleus
Facial nerve
Rubrospinal tract
Superior olivary nucleus
Central tegmental tract
Transverse pontine fibers (pontocerebellar)
Corticospinal fibers
Nucleus raphe magnus
Vestibular nuclei:
Superior
Medial
Lateral
Solitary tract and nucleus
Anterolateral system
Anterior trigeminothalamic tract
Medial lemniscus
Abducens nerve
Pontine nuclei
Trapezoid body

Figure 12–11. Cross section of the caudal pons at the level of the facial colliculus. Correlate with Figure 12–8.

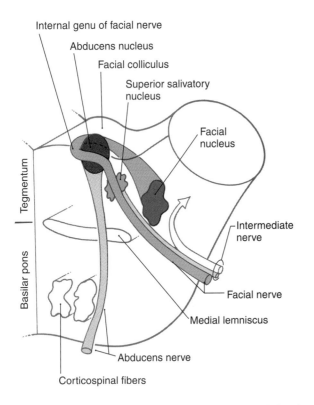

Internal genu of facial nerve
Abducens nucleus
Facial colliculus
Superior salivary nucleus
Tegmentum
Basilar pons
Facial nucleus
Intermediate nerve
Facial nerve
Medial lemniscus
Abducens nerve
Corticospinal fibers

Figure 12–12. Diagrammatic representation of the left side of the pons, as viewed from rostral to caudal, showing the relationships of the abducens and facial nerves. The open arrow indicates the general somatic afferent and special visceral afferent parts of the intermediate nerve; these fibers course caudally to enter the spinal trigeminal and solitary tract and nuclei, respectively.

The *anterolateral system*, *rubrospinal tract*, and *trapezoid body* are located in the anterolateral tegmentum (see Fig. 12–11). Although pain and temperature signals from the contralateral side of the body are conveyed by anterolateral system fibers, some of these axons end in the pontine reticular formation as *spinoreticular fibers*. The trapezoid body is composed of decussating axons from the cochlear nuclei. After crossing, these fibers ascend to form the *lateral lemniscus* and convey auditory signals to the midbrain.

The *medial lemniscus* was oriented vertically in the medulla, but in the caudal pons it begins to shift to a horizontal position (Fig. 12–13; see also Fig. 12–9). At this level, the anterior (ventral) part of the medial lemniscus (lumbosacral representation) shifts somewhat laterally, and its posterior (dorsal) portion (cervicothoracic representation) assumes a more medial location. The anterior surface of the medial lemniscus lies along the border between the tegmentum and basilar pons.

The anterior (ventral) pons contains the *basilar pontine nuclei*, longitudinally running *corticospinal* and *corticopontine fibers*, and transversely oriented *pontocerebellar fibers* (see Figs. 12–9 and 12–11). On each side of the midline, the corticospinal fibers located in the pyramid of the medulla are, in the basilar pons, completely surrounded by the *basilar pontine nuclei*. These pontine cells receive input from diverse regions of the neuraxis. In turn, most of their axons cross the midline and enter the cerebellum via the *middle cerebellar peduncle* (*brachium pontis*) as pontocerebellar fibers.

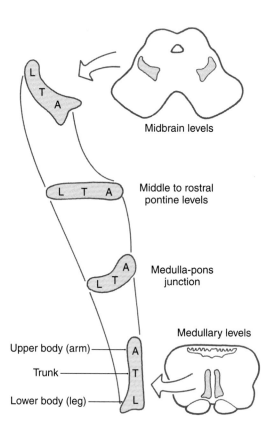

Figure 12–13. The orientation of the medial lemniscus at all brainstem levels. In this illustration, "arm" denotes the upper extremity, and "leg" denotes the lower extremity.

Mid-Pontine Level. Prominent features of the tegmentum at this level are the *principal* (or *chief*) *sensory trigeminal nucleus*, the *trigeminal motor nucleus*, and the *mesencephalic tract* and *nucleus* (Figs. 12–14 and 12–15). The principal sensory and motor trigeminal nuclei are located in the lateral tegmentum, and the mesencephalic tract and nucleus extend rostrally in the lateral wall of the central gray. Cells of the principal sensory nucleus receive GSA input from the ipsilateral trigeminal nerve and project to the thalamus via *posterior (dorsal) trigeminothalamic* (uncrossed) and *anterior (ventral) trigeminothalamic* (crossed) *fibers*. The SVE cells of the trigeminal motor nucleus innervate the masticatory muscles on the ipsilateral side. Last, the unipolar cell bodies of the mesencephalic nucleus and their laterally adjacent processes, the mesencephalic tract convey proprioceptive input to a variety of nuclei, including the trigeminal motor nucleus.

The *locus (nucleus) ceruleus* is located in the lateral floor of the fourth ventricle at this level (see Fig. 12–14). Ceruleus neurons contain pigment (hence the name *nucleus pigmentosus pontis*) and constitute the largest single source of noradrenergic axons in the central nervous system.

A comparison of Figures 12–11 and 12–14 reveals that most major tracts in the pontine tegmentum (*medial longitudinal fasciculus, tectobulbospinal system, medial lemniscus, anterolateral system,* and *anterior trigeminothalamic fibers*) occupy positions comparable to those seen at more caudal levels. Consequently, this section emphasizes only the features that are new at mid-pontine levels. The medial lemniscus is oriented horizontally at this point (see Figs. 12–13 and 12–15), and the *rubro-*

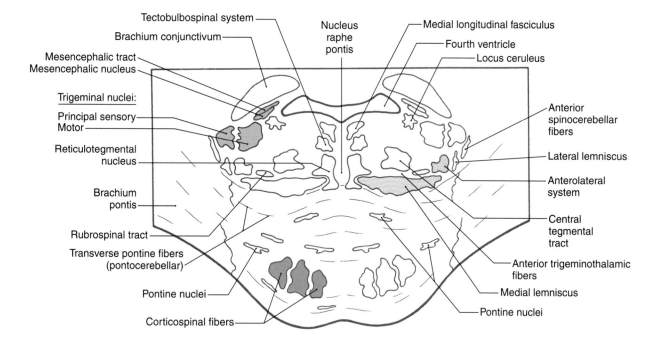

Figure 12–14. Cross section of the mid-pons at the level of the principal sensory and motor trigeminal nuclei. Correlate with Figure 12–8.

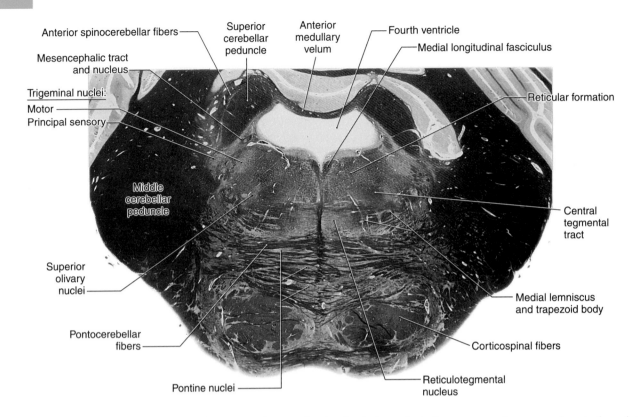

Anterior spinocerebellar fibers
Superior cerebellar peduncle
Anterior medullary velum
Fourth ventricle
Medial longitudinal fasciculus
Mesencephalic tract and nucleus
Trigeminal nuclei:
Motor
Principal sensory
Reticular formation
Middle cerebellar peduncle
Central tegmental tract
Superior olivary nuclei
Medial lemniscus and trapezoid body
Pontocerebellar fibers
Corticospinal fibers
Pontine nuclei
Reticulotegmental nucleus

Figure 12–15. A fiber (myelin)-stained cross section at the level of the principal sensory and motor trigeminal nuclei. Compare with Figure 12–14.

spinal fibers have shifted medial to the anterolateral system. Most auditory fibers are now concentrated in the *lateral lemniscus*, and rostral parts of the *superior olivary nucleus* appear just lateral to the central tegmental tract. *Anterior spinocerebellar fibers* migrate posteriorly and enter the cerebellum by coursing over the surface of the superior cerebellar peduncle. The *brachium conjunctivum (superior cerebellar peduncle)* arises from the cerebellar nuclei, sweeps rostrally, forming the lateral wall of the fourth ventricle (see Figs. 12–14 and 12–15), and enters the caudal midbrain tegmentum, where it decussates.

Neurons in the ventral tegmentum close to the midline extend into posterior portions of the basilar pons and constitute the *reticulotegmental nucleus* (see Figs. 12–14 and 12–15). This cell group is continuous with the basilar pontine nuclei, and its axons enter the cerebellum through the contralateral *brachium pontis*. These cells also share similar cytologic features and afferent projections with neurons of the basilar pons.

Rostral Pontine Level. The only cranial nerve structures present in the pontine tegmentum at rostral levels are the *mesencephalic nucleus* and *tract*. These structures are located in the lateral aspect of the periaqueductal gray and remain in this position into the midbrain (Fig. 12–16; see also Fig. 12–14). Anterior (ventral) to the mesencephalic tract and nucleus is the *locus (nucleus) ceruleus*; this noradrenergic cell group also extends into the caudal midbrain.

The *brachium conjunctivum (superior cerebellar pedun-*

cle) converges toward its decussation in the caudal midbrain, and most other major tracts in the tegmentum occupy positions comparable to those seen at mid-pontine levels (compare Fig. 12–14 with Fig. 12–16). The rubrospinal tract is shifted even more medially, and the *lateral lemniscus* is close to the posterolateral surface of the brainstem at this point. Also, the composition of the *basilar pons* is essentially the same as that seen at mid-pontine levels (see Figs. 12–14 and 12–16).

Reticular and Raphe Nuclei. Much of the pontine tegmentum is occupied by the reticular formation. This central core is generally divided into a medial area of primarily large neurons (magnocellular region) and a lateral area of mainly small neurons (parvocellular region) (Fig. 12–17). The magnocellular reticular nuclei of the pons are, from caudal to rostral, the *gigantocellular reticular nucleus* and the *caudal* and *oral pontine reticular nuclei*. The parvocellular area nuclei of the pons contain a diffuse *lateral reticular formation* at caudal and mid-pontine levels and include the *medial* and *lateral parabrachial nuclei* at rostral levels. The latter cell groups are located adjacent to the brachium conjunctivum.

The *raphe nuclei* are symmetrically distributed on either side of the midline (see Fig. 12–17). Next to the medial lemniscus at caudal pontine levels is the *nucleus raphes magnus* (the *raphe magnus*). This cell cluster extends caudally into the rostral medulla and is an important synaptic station for signals involved in the inhi-

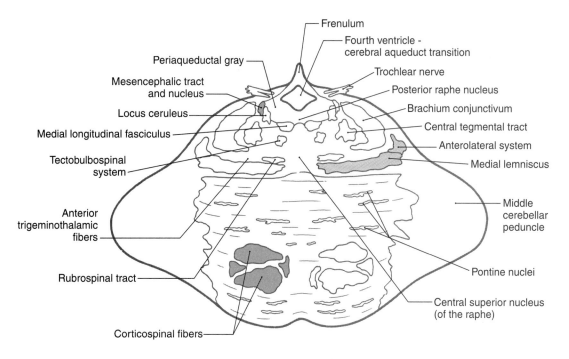

Figure 12–16. Cross section of the rostral pons.

bition of pain at medullary and spinal levels. In the caudal one third of the tegmental pons, this cell group is replaced by the *nucleus raphes pontis*, which extends a little beyond mid-pontine levels. The *superior central nucleus* and *posterior (dorsal) raphe nucleus* are found in the rostral pons; the latter extends into the caudal midbrain (see Fig. 13–16).

Vasculature. Internal areas of the tegmental and basi-

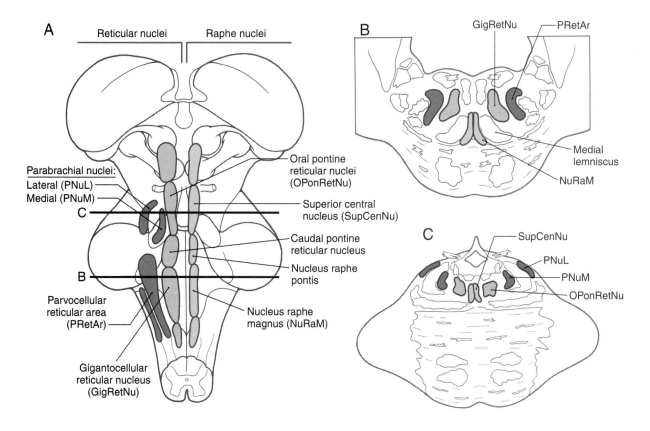

Figure 12–17. Posterior (dorsal) view of the brainstem *(A)* and caudal *(B)* and rostral *(C)* cross sections showing the raphe and reticular nuclei of the pons.

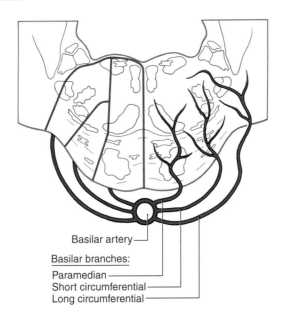

Figure 12–18. Blood supply of the pons. Arteries are shown on the right, and the general territories served by each on the left.

lar pons are served by branches of the *basilar artery* (Fig. 12–18). *Paramedian branches* distribute to medial areas of the basilar pons, including corticospinal fibers and the exiting fibers of the abducens nerve. The lateral part of the basilar pons is served by *short circumferential branches*, and the entire tegmental area plus a wedge of the middle cerebellar peduncle receives blood via the *long circumferential branches*. At caudal levels (levels of the facial colliculus), the long circumferential supply is supplemented by branches of the *anterior inferior cerebellar artery*. Rostrally, beginning at about the level of the principal sensory and motor trigeminal nuclei, the blood supply to the pontine tegmentum is supplemented by branches of the *superior cerebellar artery*.

Internal Anatomy of the Cerebellum

Cerebellar Cortex. Each folium of the cerebellum is organized into three layers oriented parallel to the cortical surface. From external to internal, these are the *molecular, Purkinje cell*, and *granular layers* (Figs. 12–19 and 12–20). Internal to the granular layer, and forming the core of each folium, is a layer of subcortical white matter composed of all fibers arriving (afferents to the cortex) or leaving (efferents of the cortex) the cerebellar cortex.

The *molecular layer* contains *stellate cells, basket cells*, and the expansive dendrites of *Purkinje cells* (see Fig. 12–19). In addition, the molecular layer contains two other major fiber populations: *climbing fibers* and *paral-*

lel fibers. The former originate from the contralateral inferior olive and the latter from the granule cells (see Fig. 12–19).

Purkinje cells are the efferent neurons of the cerebellar cortex. Their large somata form a single layer at the molecular layer–granular layer interface (see Figs. 12–19 and 12–20). Each Purkinje cell has a complex dendritic tree that extends into the molecular layer and exhibits a fanlike orientation that is perpendicular to the long axis of the folium. Axons of Purkinje cells pass through the subcortical white matter to end in the cerebellar and vestibular nuclei.

The granular layer contains *granule cells, Golgi cells*, and *mossy fibers* (see Figs. 12–19 and 12–20). The last originate from cells located in many nuclei throughout the brainstem and spinal cord. Traversing the granular layer are climbing fibers, en route to the molecular layer, and Purkinje cell axons leaving the cortex.

Cerebellar Nuclei. The cerebellar nuclei are, from medial to lateral, the *fastigial, globose, emboliform*, and *dentate* (see Fig. 12–7). They receive input from Purkinje cell axons and from collaterals of cerebellar afferent fibers. In general, Purkinje cells of the vermis relate to the fastigial nucleus, those of the intermediate cortex to the globose and emboliform nuclei, and those of the lateral cortex to the dentate nucleus (see Fig. 12–7). Some Purkinje cells of the vermis and the floc-

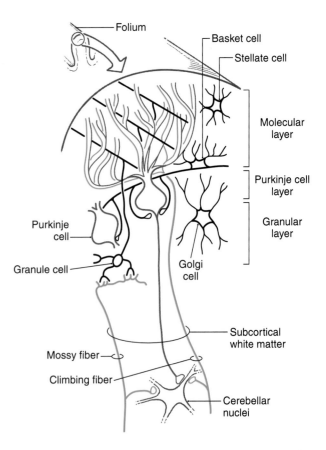

Figure 12–19. Diagrammatic representation of the cells and fibers of the cerebellar cortex.

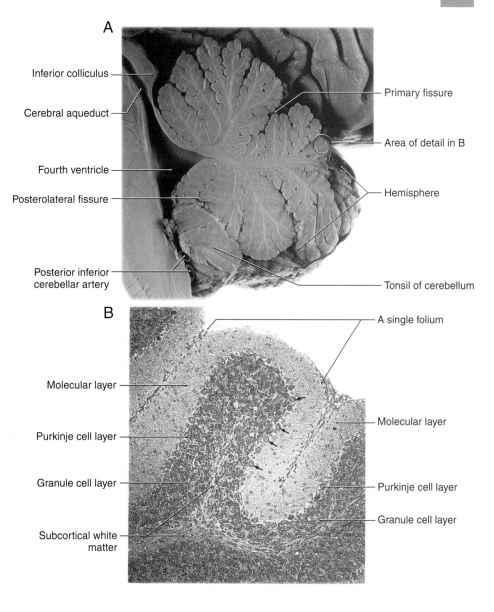

A

Inferior colliculus

Cerebral aqueduct

Fourth ventricle

Posterolateral fissure

Posterior inferior cerebellar artery

Primary fissure

Area of detail in B

Hemisphere

Tonsil of cerebellum

B

Molecular layer

Purkinje cell layer

Granule cell layer

Subcortical white matter

A single folium

Molecular layer

Purkinje cell layer

Granule cell layer

Figure 12–20. Midsagittal view *(A)* of the cerebellum and a detail *(B)* showing the histologic layers of the cerebellar cortex. Arrowheads point to the cell bodies of Purkinje cells. Compare with the diagrammatic representation in Figure 12–19.

culonodular lobe send axons directly to the vestibular nuclei. Cerebellar nuclear cells project to a variety of cell groups throughout the neuraxis, primarily via the brachium conjunctivum.

Vasculature. The blood supply to the cerebellar cortex and nuclei arises via branches of the *superior cerebellar artery* and the *anterior* and *posterior inferior cerebellar arteries* (see Fig. 12–7). The superior cerebellar artery serves the superior surface of the cerebellum and, by way of penetrating branches, the fastigial, globose, and emboliform nuclei and most areas of the dentate nucleus. This vessel also distributes to the superior and middle cerebellar peduncles.

Branches of the anterior inferior cerebellar artery distribute to the more lateral areas of the inferior cerebellar surface, including the flocculus (see Fig. 12–7). They also serve parts of the middle cerebellar peduncle, caudal regions of the dentate nucleus, and the choroid plexus in the subarachnoid space at the cerebellopontine angle.

The posterior inferior cerebellar artery (PICA) branches to more medial regions of the inferior cerebellar surface and to the nodulus (see Fig. 12–7). The choroid plexus of the fourth ventricle receives branches from the PICA. This vessel is also an important source of blood to posterolateral regions of the medulla.

Sources and Additional Reading

Readings for the brainstem chapters are listed at the end of Chapter 13.

The Midbrain

G. A. Mihailoff, D. E. Haines, and P. J. May

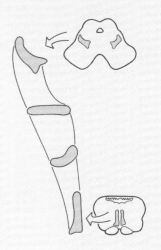

Development   188
Basal and Alar Plates

External Features   189
Anterior (Ventral) Midbrain
Posterior (Dorsal) Midbrain
Vasculature

Internal Anatomy of the Midbrain   190
General Regions: Tectum, Tegmentum, and Basis Pedunculi
Summary of Ascending Pathways
Summary of Descending Pathways
Caudal Midbrain Levels
Rostral Midbrain Levels
Midbrain-Diencephalon Junction
Reticular and Raphe Nuclei
Vasculature

The *mesencephalon*, or midbrain, is the most rostral portion of the brainstem. It gives rise to cranial nerves III and IV, conducts ascending and descending tracts, and contains nuclei that are essential to motor function. Caudally, the midbrain is continuous with the pons, and rostrally it joins the diencephalon. The *cerebral aqueduct*, the cavity of the midbrain, is continuous rostrally with the third ventricle and caudally with the fourth ventricle. The blood supply to the mesencephalon is primarily from proximal branches of the posterior cerebral arteries (P$_1$ or P$_2$) and from small penetrating branches of the posterior communicating artery.

Development

The mesencephalon (Fig. 13–1) arises early in development as one of the three primary brain vesicles. Neuroblasts of the *mantle layer* give rise to *alar* and *basal plates*, which are continuous with the alar and basal plates of the rhombencephalon (Fig. 13–2). These cell groups are the rostral continuations of the same primitive cell columns described for the metencephalon. The surrounding *marginal layer* contains the developing axons of cells located in other levels of the neuraxis. The *cerebral aqueduct* is narrow relative to the fourth ventricle; therefore, the basal and alar plates lie anterior (ventral) and posterior (dorsal) to it as in the spinal cord and caudal medulla.

Basal and Alar Plates. The primitive neurons of the *alar plate* give rise to the quadrigeminal plate, from which the *superior* and *inferior colliculi* arise (see Fig. 13–2A, upper arrow). In humans, the superior colliculus consists of alternating layers of cells and fibers, whereas the inferior colliculus appears more homoge-

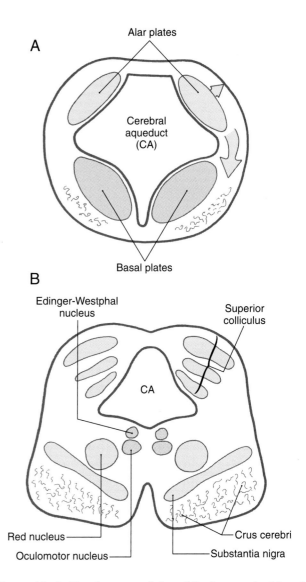

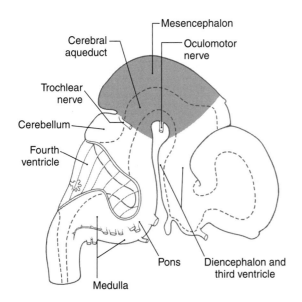

Figure 13–1. Lateral view of the human brain at about 7 weeks of gestation. The midbrain is highlighted.

Figure 13–2. Development of the midbrain at early (A) and later (B) stages showing the alar and basal plates and the structures derived from each. The level shown at B is diagrammatic of the rostral (superior colliculus) midbrain.

neous, but consists of a large central nucleus and several small, peripherally located nuclei. Alar plate neuroblasts also migrate into anterior areas of the developing midbrain to form the *red nucleus* and the *substantia nigra* (see Fig. 13–2A, lower arrow).

Basal plate neuroblasts give rise to the *general somatic efferent* (GSE) neurons of the *oculomotor* and *trochlear nuclei*. In addition, the *general visceral efferent* (GVE) cells of the *Edinger-Westphal nucleus*, a visceral motor cell group associated with the oculomotor complex, also arise from the basal plate (see Fig. 13–2).

As the basal and alar plates differentiate, the marginal layer is invaded by axons originating from cells located outside the midbrain. These fibers collect in the anterolateral area of the developing mesencephalon to form an especially prominent bundle, the *crus cerebri* (plural, *crura cerebri*; see Fig. 13–2B).

External Features

Anterior (Ventral) Midbrain. The presence of a pair of large axon bundles, the crura cerebri, is a characteristic feature of the anterior (ventral) aspect of the midbrain. These bundles emerge from the cerebral hemispheres caudal to the optic tracts, converge slightly toward the midline as they course through the midbrain, and disappear into the basilar pons (Fig. 13–3). The *oculomotor nerves* exit the medial edge of each crus and pass through the space between the crura, which is called the *interpeduncular fossa* (see Fig. 13–3). Anteriorly (ventrally), the rostral limit of the midbrain is marked by the exit of the crura cerebri from the cerebral hemispheres and by the caudal edge of the mammillary bodies. The caudal border of the midbrain is formed where each crus enters the basilar pons.

The subarachnoid space of the interpeduncular fossa is called the *interpeduncular cistern*. This cistern contains the oculomotor nerves and the upper part of the basilar artery, including its bifurcation and proximal branches. Numerous vessels penetrate the roof of this fossa and create many small perforations (see Fig. 13–3B). Consequently, this area is frequently called the *posterior perforated substance.*

Posterior (Dorsal) Midbrain. The posterior surface of the adult midbrain is characterized by four elevations collectively called the *corpora quadrigemina* (Fig. 13–4). The rostral two elevations are the *superior colliculi*, and the caudal two are the *inferior colliculi*. Just caudal to the inferior colliculus, the exit of the trochlear nerve marks the pons-midbrain junction on the posterior surface of the brainstem, whereas the midbrain-diencephalic boundary is formed by the posterior commissure (Fig. 13–5).

Rostrolaterally, the inferior colliculus is joined to the *medial geniculate body* of the diencephalon by a fiber bundle called the *brachium of the inferior colliculus* (see Fig. 13–4). The inferior colliculus and the medial geniculate body are part of the auditory system. The *brachium of the superior colliculus* extends from the optic tract to the superior colliculus in a groove located between the *medial geniculate body* and *pulvinar* of the diencephalon (see Fig. 13–4; see also Fig. 13–14). The superior colliculus, pulvinar, and lateral geniculate body are parts of the visual and visual-motor systems.

On the midline, the *pineal gland*, a diencephalic structure, extends posteriorly above and between the superior colliculi (see Fig. 13–5). Tumors of the pineal produce non-communicating (obstructive) hydrocephalus because of compression of the colliculi of the midbrain and resulting occlusion of the cerebral aqueduct.

The subarachnoid space immediately posterior (dorsal) to the colliculi is the *quadrigeminal cistern*. This cistern contains the exiting trochlear nerves, the great vein of Galen, and distal branches of the posterior cerebral arteries.

Vasculature. The primary blood supply to the mesencephalon arises via branches of the *basilar artery*, with smaller branches from the *superior cerebellar, anterior choroidal, medial posterior choroidal,* and *posterior communicating arteries.* An important source of blood to the posterior (dorsal) portion of the midbrain is the *quadri-*

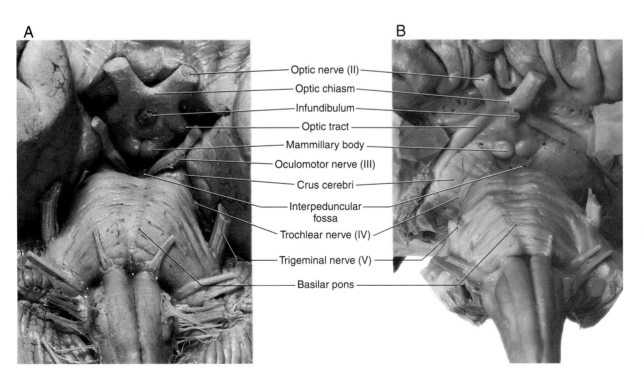

A   B

- Optic nerve (II)
- Optic chiasm
- Infundibulum
- Optic tract
- Mammillary body
- Oculomotor nerve (III)
- Crus cerebri
- Interpeduncular fossa
- Trochlear nerve (IV)
- Trigeminal nerve (V)
- Basilar pons

Figure 13–3. Anterior (ventral) views (*A*, undissected; *B*, dissected) of the brainstem with emphasis on the midbrain.

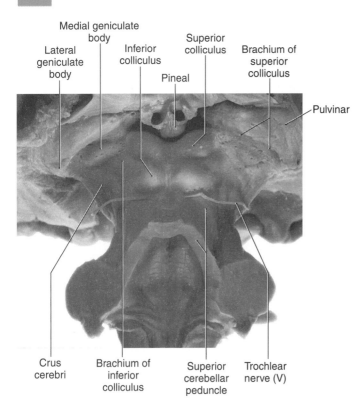

Medial geniculate body

Lateral geniculate body

Inferior colliculus

Superior colliculus

Brachium of superior colliculus

Pineal

Pulvinar

Crus cerebri

Brachium of inferior colliculus

Superior cerebellar peduncle

Trochlear nerve (V)

Figure 13–4. Posterior (dorsal) view (dissected) of the brainstem with emphasis on the midbrain and its junction with the diencephalon.

geminal artery, a branch of the posterior cerebral artery (P₁ segment). Also, the *superior cerebellar artery* gives rise to branches that serve caudal parts of the posterior midbrain and adjacent regions of the pons.

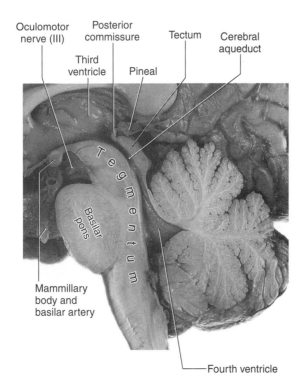

Oculomotor nerve (III)

Posterior commissure

Third ventricle

Pineal

Tectum

Cerebral aqueduct

Tegmentum

Basilar pons

Mammillary body and basilar artery

Fourth ventricle

Figure 13–5. Sagittal view of the brainstem with emphasis on midbrain structures.

Internal Anatomy of the Midbrain

General Regions: Tectum, Tegmentum, and Basis Pedunculi. The midbrain is divisible into three regions, which can be appreciated best in cross section. Posterior to the cerebral aqueduct is the *tectum* (roof) of the midbrain (Fig. 13–6; see also Fig. 13–5). The characteristic structures of this area are the superior and inferior colliculi. The *periaqueductal gray* (*central gray*) is a sleeve of neuron cell bodies that completely surrounds the cerebral aqueduct. The *tegmentum of the midbrain* extends from the base of the tectum to, but

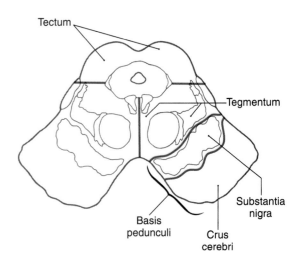

Tectum

Tegmentum

Substantia nigra

Basis pedunculi

Crus cerebri

Figure 13–6. Major subdivisions of the midbrain.

does not include, the substantia nigra. The anterolateral portion of the midbrain on either side is formed by the *basis pedunculi,* which consists of the *substantia nigra* and the *crus cerebri* (see Fig. 13–6). In turn, the *crus cerebri* is composed primarily of descending fibers. The term *cerebral peduncle* is sometimes used for the crus cerebri but actually represents the entire midbrain below the tectum (tegmentum + basis pedunculi).

Summary of Ascending Pathways. The long ascending pathways that traverse the pons and medulla continue through the midbrain (Fig. 13–7). These include the *medial lemniscus, anterolateral system,* and the *posterior* and *anterior trigeminothalamic tracts.* The *lateral*

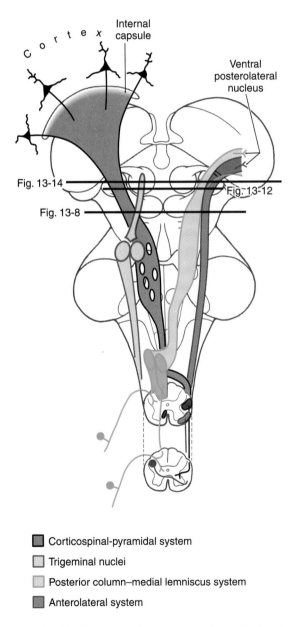

Figure 13–7. Diagrammatic representation of the brain showing the location and trajectory of three important pathways and the trigeminal nuclei. The color coding for each is continued in Figures 13–8, 13–12, and 13–14.

■ Corticospinal-pyramidal system

□ Trigeminal nuclei

▨ Posterior column–medial lemniscus system

■ Anterolateral system

lemniscus is also prominent in the caudal midbrain. Other, smaller bundles, such as the *medial longitudinal fasciculus,* are also present and occupy positions comparable to those seen in lower levels of brainstem.

Summary of Descending Pathways. Descending fibers from the cerebral cortex pass through the midbrain and, as we have seen, are also prominent features of the pons, medulla, and spinal cord (see Fig. 13–7). At midbrain levels these *corticospinal, corticonuclear (corticobulbar),* and *corticopontine* fibers are parts of the crus cerebri. Some of the descending fiber bundles that were discussed in the pons and medulla, such as the *tectobulbospinal system* and the *rubrospinal* and *central tegmental tracts,* originate from nuclei of the midbrain.

The following sections describe the anatomy of the midbrain at caudal and rostral brain levels and at the level of the midbrain-diencephalon junction. The anatomy is presented from posterior (dorsal) to anterior (ventral) at each level, starting with the tectum and proceeding through the tegmentum to the basis pedunculi.

Caudal Midbrain Levels. Transverse sections through the caudal midbrain are characterized by the presence of the *inferior colliculus, trochlear nucleus,* and *decussation of the superior cerebellar peduncle* (Figs. 13–8 to 13–10). The inferior colliculus is composed of a large *central nucleus* bordered posteriorly by a smaller *pericentral (dorsal) nucleus* and laterally by an *external (lateral) nucleus. Lateral lemniscus* fibers enter the inferior colliculus from an anterior (ventral) direction and give this area a goblet-like appearance (see Figs. 13–8 and 13–9). The lateral lemniscus provides auditory input to the nuclei of the inferior colliculus, which then transmit this information to the medial geniculate body via fibers of the *brachium of the inferior colliculus.* A complete tonotopic representation (frequency map) of the cochlea is found in several of the central auditory nuclei (see Chapter 21).

The *periaqueductal gray,* or *central gray,* is a prominent collection of small neurons surrounding the cerebral aqueduct at all midbrain levels (see Figs. 13–8 to 13–10; see also Figs. 13–12 and 13–13). It receives somatosensory input, is interconnected with the hypothalamus and thalamus, and projects caudally to brainstem nuclei. Its connections and the presence of a high level of opiate receptor binding activity indicate that the periaqueductal gray plays an important role in the brain mechanisms responsible for the suppression and modulation of pain (analgesia). The cell bodies of the *mesencephalic nucleus* and their laterally adjacent fibers, the *mesencephalic tract,* are located in the lateral edge of the periaqueductal gray, and the *posterior (dorsal) raphe nucleus* is located anteriorly on the midline, adjacent to the trochlear nucleus and medial longitudinal fasciculus (see Fig. 13–8).

The central area of the tegmentum at the level of the inferior colliculus is occupied by the decussating

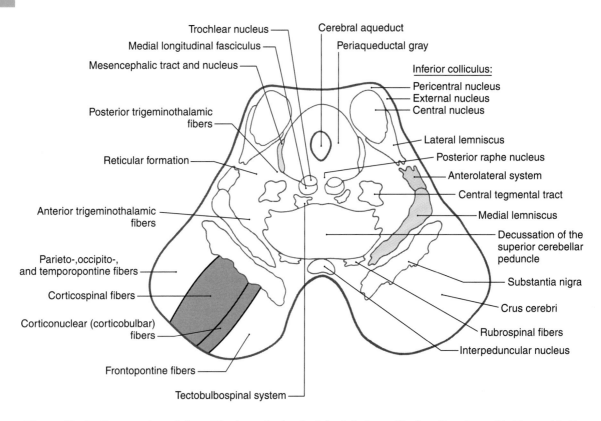

Figure 13-8. Cross section of the midbrain at the level of the inferior colliculus. Correlate with Figure 13-7.

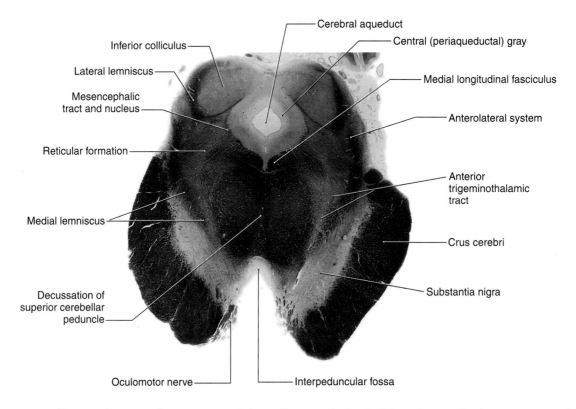

Figure 13-9. A fiber (myelin)-stained cross section of the midbrain at the level of the inferior colliculus. Compare with Figure 13-8.

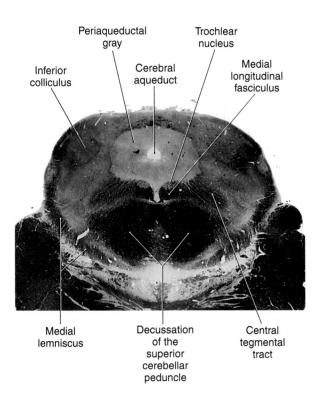

Periaqueductal gray

Trochlear nucleus

Inferior colliculus

Cerebral aqueduct

Medial longitudinal fasciculus

Medial lemniscus

Decussation of the superior cerebellar peduncle

Central tegmental tract

Figure 13–10. A fiber (myelin)-stained cross section of the midbrain at an intercollicular level showing the trochlear nucleus. Compare with Figure 13–8.

pass through the midbrain, but it also contains *spinomesencephalic fibers* that end in the midbrain. Some of these fibers synapse in the reticular nuclei of the midbrain and pons (*spinoreticular fibers*), some synapse in the tectum (*spinotectal fibers*), and some synapse in the periaqueductal gray. The latter fibers represent an important link in the pathways involved in the inhibition of pain at medullary and spinal levels. Medial lemniscal fibers have shifted from a horizontal position, characteristic of the rostral pons, to the anteromedial to posterolateral orientation of the midbrain. The somatotopy of fibers in the medial lemniscus follows accordingly (Fig. 13–11).

Posterior (*dorsal*) and *anterior* (*ventral*) *trigeminothalamic tracts* are located adjacent to the central tegmental tract and the medial lemniscus, respectively (see Figs. 13–8 and 13–9). These bundles are somewhat diffusely arranged and convey crossed (anterior) and uncrossed (posterior) somatosensory information from the face.

The base of the midbrain on either side is formed by the *basis pedunculi*, which consists of the *substantia nigra* and the *crus cerebri* (see Fig. 13–8). The substantia nigra is related to motor function and is considered in the next section of this chapter. The crus cerebri con-

fibers of the *superior cerebellar peduncle* (see Figs. 13–8 and 13–9). From this point, the majority of these cerebellar efferent axons pass rostrally to targets in the midbrain and thalamus, although some turn caudally and enter the pons and medulla. At the level of this decussation, the *general somatic efferent* (GSE) cell bodies of the *trochlear nucleus* form an oval-shaped cell group nestled in the fibers of the *medial longitudinal fasciculus* (see Figs. 13–8 and 13–10). The axons of trochlear motor neurons pass laterally and posteriorly around the periaqueductal gray to cross the midline before exiting the brainstem just caudal to the inferior colliculus. They innervate the superior oblique muscle. *Tectobulbospinal fibers* are ventral to the *medial longitudinal fasciculus*, and, as their name states, the fibers of the *central tegmental tract* occupy the center of the tegmentum (see Figs. 13–8 to 13–10).

Immediately anterior to the decussation of the superior cerebellar peduncle are the *interpeduncular nucleus* and *rubrospinal fibers* (see Fig. 13–8). The interpeduncular nucleus is related to the limbic system, a part of the brain that functions in the control of emotional behavior.

Located in the anterolateral and lateral portions of the midbrain tegmentum are fibers of the *anterolateral system* and *medial lemniscus* (see Fig. 13–8). The anterolateral system conveys pain and temperature signals from the contralateral side of the body. Most of its fibers terminate in the dorsal thalamus and therefore

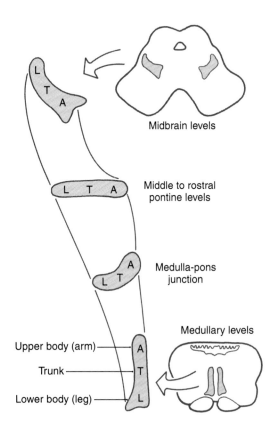

Midbrain levels

Middle to rostral pontine levels

Medulla-pons junction

Medullary levels

Upper body (arm)

Trunk

Lower body (leg)

Figure 13–11. The orientation of the medial lemniscus in the midbrain as compared with that in the pons and medulla. In this illustration, "arm" denotes the upper extremity, and "leg" denotes the lower extremity.

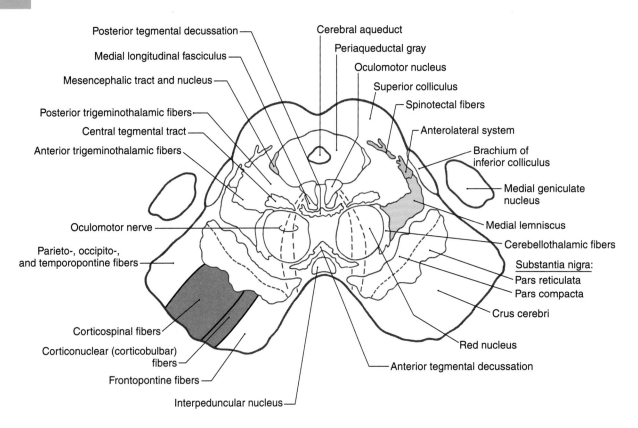

Figure 13–12. Cross section of the midbrain at the level of the superior colliculus. Correlate with Figure 13–7.

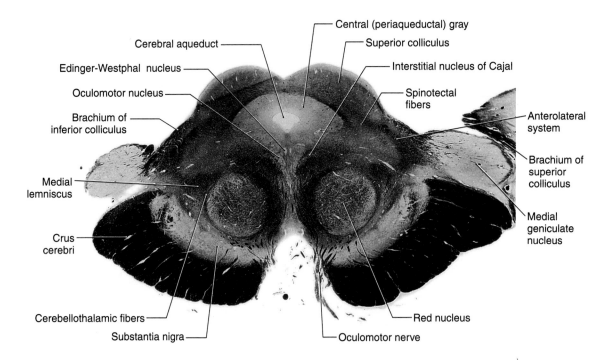

Figure 13–13. A fiber (myelin)-stained cross section of the midbrain at the level of the superior colliculus. This plane of section is between those illustrated in Figures 13–12 and 13–14.

tains *corticospinal, corticonuclear (corticobulbar)*, and *corticopontine* fibers. The former two fiber populations are found in the middle third of the crus cerebri. Corticospinal fibers projecting to lumbosacral cord levels (lower extremity representation) are found laterally, and corticonuclear (corticobulbar) fibers that project to cranial nerve nuclei (face representation) are located more medially. Corticospinal fibers projecting to cervical cord levels (upper extremity representation) occupy an intermediate position. Corticopontine fibers in the medial third of the crus arise from the frontal lobe (*frontopontine*), whereas those in the lateral third originate from parietal, occipital, and temporal lobes (*parietopontine, occipitopontine*, and *temporopontine*).

Rostral Midbrain Levels. Transverse sections through the rostral midbrain are characterized by the presence of the *superior colliculus, red nucleus*, and *oculomotor nuclei* (Figs. 13–12 and 13–13). The paired *superior colliculi* are composed of alternating layers of gray matter (cells) and white matter (fibers). The "superficial" three layers (I to III) receive input from the retina and visual cortices and project to the thalamus. The "deep" four layers (IV to VII) subserve gaze changes including eye movements and project to the thalamus, brainstem, and spinal cord. Descending crossed projections from the tectum pass to a variety of brainstem areas (e.g., as *tectoreticular* and *tecto-olivary fibers*) and—although this element is minor in humans—to the cervical spinal cord as *tectospinal fibers*. Together, these fibers constitute the *tectobulbospinal system* of the brain-

stem; the name "tectobulbospinal" reflects the diversity of the fibers in this bundle. *Rubrospinal fibers* originate from the contralateral red nucleus; these fibers descend to medullary and spinal levels, where they influence the activity of motor neurons innervating skeletal muscles.

Anterior to the periaqueductal gray, the oculomotor complex forms a V-shaped region between the medial longitudinal fasciculi (Fig. 13–14; see also Fig. 13–12). This complex consists of the *general somatic efferent* (GSE) cells of the oculomotor nucleus and the *general visceral efferent* (GVE) cells of the Edinger-Westphal nucleus. Oculomotor fibers arch through and medial to the red nucleus, emerge from the brainstem at the medial edge of the basis pedunculi, and innervate four of the six extraocular muscles. The *Edinger-Westphal nucleus* (see Figs. 13–13 and 13–14) lies posterior to the oculomotor nucleus and provides the preganglionic parasympathetic fibers that travel to the ciliary ganglion via the oculomotor nerve. These parasympathetic fibers are located on the perimeter of the oculomotor nerve and, consequently, are the first to be affected in a compression injury to this nerve. The postganglionic fibers from the ciliary ganglion innervate the sphincter pupillae and ciliary muscles.

Several other cell groups are located close to the main oculomotor nucleus (Fig. 13–15; see also Fig. 13–13). These include (1) the *interstitial nucleus of Cajal* located adjacent to the fibers of the medial longitudinal fasciculus (MLF), (2) the *nucleus of Darkschewitsch* situated within the ventrolateral border of the periaque-

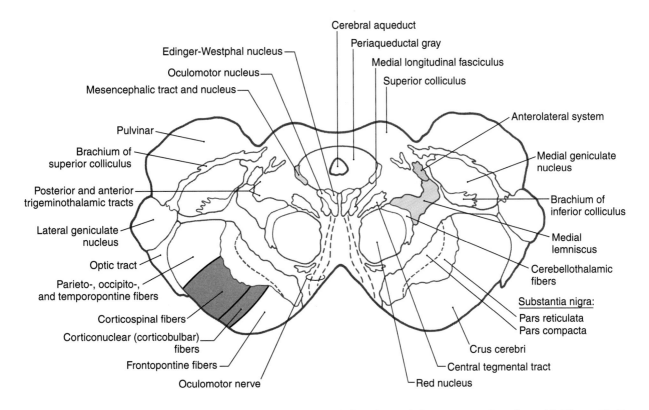

Figure 13–14. Cross section of the midbrain at the mesencephalon-diencephalon junction. Correlate with Figure 13–7.

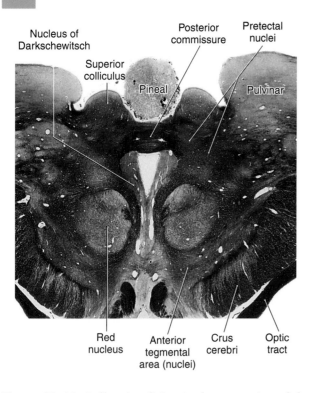

Figure 13–15. A fiber (myelin)-stained cross section of the midbrain at the mesencephalon-diencephalon junction, showing the pretectal area.

ductal gray, and (3) the *nucleus of the posterior commissure*, which is one of the *pretectal nuclei*. Most of these nuclei are involved in the control of eye movements.

The *red nucleus*, a prominent structure in the midbrain tegmentum at this level (see Figs. 13–12 to 13–15), is so named because in the unfixed brain the dense vascularity of the region gives it a pink color. It is composed of a caudal magnocellular and a rostral parvocellular region, but these subdivisions are less distinct in primates and humans than in most other animals. The red nucleus is involved in motor function and has extensive connections throughout the neuraxis. Its efferents include the *rubrospinal tract*, which travels to the contralateral spinal cord, and *rubro-olivary fibers*, which descend in the *central tegmental tract* to the ipsilateral inferior olivary complex. Afferents to the red nucleus arise from the contralateral cerebellar nuclei and the ipsilateral cerebral cortex.

The major tracts of the caudal midbrain tegmentum are present in similar locations at rostral midbrain levels (see Fig. 13–12). *Posterior (dorsal)* and *anterior (ventral) tegmental decussations* cross the midline at the levels of the posterior and anterior limits of the red nuclei, respectively. The posterior decussation carries tectobulbospinal fibers, and the anterior decussation carries rubrospinal fibers. Immediately lateral to the red nucleus are *cerebellorubral* and *cerebellothalamic* fibers. These are crossed ascending fibers from the decussa-

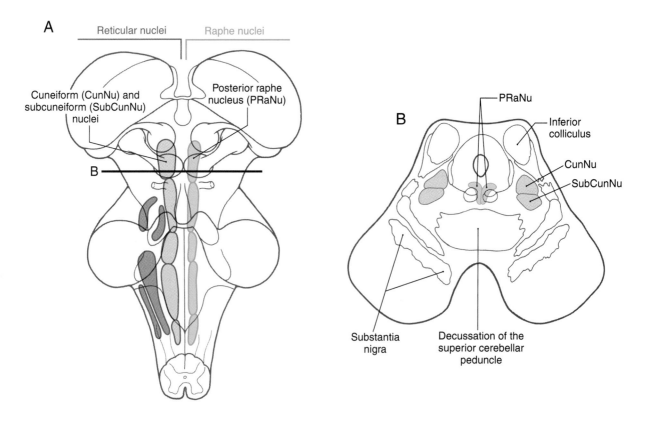

Figure 13–16. Posterior *(A)* view of the brainstem and a cross section *(B)* at the level of the inferior colliculus showing the raphe and reticular nuclei of the brainstem.

tion of the superior cerebellar peduncle. *Posterior (dorsal)* and *anterior (ventral) trigeminothalamic fibers* and the *medial lemniscus* occupy central and more lateral areas of the tegmentum (see Fig. 13–12). Fibers of the *anterolateral system* are located at the lateral extreme of the medial lemniscus and, at this level, *spinotectal fibers* enter the deep layers of the superior colliculus (see Fig. 13–13).

The structure of the *basis pedunculi* and the organization of fibers in the *crus cerebri* are the same in the rostral midbrain as more caudally (see Fig. 13–12). The *substantia nigra* is functionally associated with the basal nuclei and is commonly divided into a *compact part* (pars compacta) and a *reticular part* (pars reticulata) (see Figs. 13–12 and 13–14). Cells of the reticular part project to the superior colliculus, thalamus, and pontine reticular formation, whereas those of the compact part project diffusely to the *caudate nucleus* and *putamen* where their synaptic terminals release dopamine. Parkinson's disease, a deficit characterized by tremor and difficulty in initiating or terminating movement, is associated with the loss of dopamine-containing cells in the compact part.

Midbrain-Diencephalon Junction. The major tracts and nuclei seen at the level of the superior colliculus are still present in their same locations, at the mesencephalic-diencephalic interface. A comparison of Figures 13–12 and 13–14 reveals these similarities.

Groups of cells related primarily to the visual system and collectively referred to as the *pretectal nuclei* are located at the rostral extent of the superior colliculus laterally adjacent to the posterior commissure (see Fig. 13–15). Indeed, the nucleus of the posterior commissure is one of the pretectal nuclei. The pretectal nuclei include a controlling center for the *pupillary light reflex*. Neurons involved in this reflex receive bilateral retinal inputs and project bilaterally to visceral motor (parasympathetic) neurons in the Edinger-Westphal nucleus.

Several structures appear at this junction that signal the transition from midbrain to thalamus (see Fig. 13–14). Laterally, a small part of the *pulvinar* (a thalamic nucleus) is present, and portions of the *medial* and *lateral geniculate nuclei* also appear in the same plane. Fibers of the *brachium of the inferior colliculus* convey auditory signals to the *medial geniculate nucleus*. The optic tract contains visual fibers, some of which terminate in the lateral geniculate nucleus, while others continue into the superior colliculus and pretectal area via the *brachium of the superior colliculus* (see Figs. 13–14 and 13–15; see also Fig. 13–4). The *anterior (ventral) tegmental nucleus (of Tsai)* is a diffuse cell group located anteromedially to the red nucleus and rostrally continuous with the lateral hypothalamic area. Many of the cells of this nucleus utilize dopamine as a neurotransmitter and receive input from and project to hypothalamic and limbic structures that function in emotional behavior.

Reticular and Raphe Nuclei. The reticular formation of the midbrain tegmentum is composed of the *cuneiform* and *subcuneiform nuclei* (Fig. 13–16). The midbrain reticular formation participates in the ascending systems that regulate states of consciousness. Many of these neurons project to the thalamus, especially the thalamic reticular nucleus, and the hypothalamus. This ascending fiber system is largely responsible for maintaining an alert, wakeful state and thus forms part of the *ascending reticular activating system*. Lesions involving the midbrain reticular formation can result in *hypersomnia*, which is characterized by slow respiration and an electroencephalographic pattern (large-amplitude slow waves) indicative of a sleep state.

Located in anterior parts of the periaqueductal gray, the *posterior (dorsal) raphe nucleus* extends from the rostral pons into the caudal midbrain (see Fig. 13–16). The serotonergic cells of this nucleus project to wide areas of the cerebral cortex, where they modulate neuronal activity involved in sleep/dream cycles.

Vasculature. The blood supply to the midbrain originates from the *basilar artery* and its major branches (the *quadrigeminal* and *superior cerebellar arteries*) and from the *anterior choroidal artery*, which is a branch of the internal carotid (Fig. 13–17), and the *medial posterior*

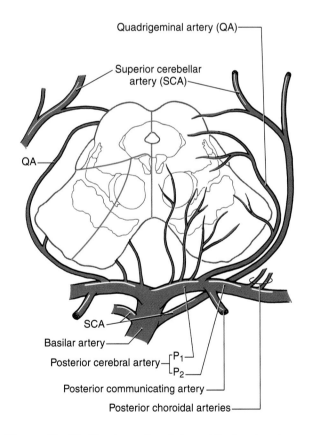

Figure 13–17. Blood supply of the midbrain. Arteries are shown mainly on the right and the territories served by each on the left. The anterior choroidal artery, which is a branch of the internal carotid and follows the general route of the optic tract (see Fig. 13–3B), also sends branches to the lateral portions of the midbrain.

choroidal artery, which is usually a branch of P_2. Medial regions of the midbrain receive numerous small branches from the P_1 segment of the posterior cerebral artery and from the *posterior communicating artery*. These paramedian branches constitute the *posteromedial group* of branches from the circle of Willis (see Fig. 8–10). Included in their territory are the oculomotor, trochlear, and Edinger-Westphal nuclei; the exiting oculomotor fibers; the red nucleus; and medial aspects of the substantia nigra and crus cerebri (see Fig. 13–17). Occlusion of the paramedian branches that serve only the crus cerebri and oculomotor roots results in a contralateral hemiplegia (corticospinal involvement) and an ipsilateral oculomotor palsy with pupillary dilation (oculomotor involvement). This combination of deficits is referred to as the *Weber syndrome*.

Ventrolateral regions of the midbrain are served by penetrating branches of the *quadrigeminal artery*, the *anterior choroidal artery* (see Fig. 13–17), and the *medial posterior choroidal artery*. The region served by these branches includes the lateral parts of the crus and substantia nigra and the medial lemniscus.

The posterior midbrain is served primarily by the *quadrigeminal artery* (*collicular artery*), which typically arises from P_1 (see Fig. 13–17). Much of the periaqueductal gray, the nuclei of the superior and inferior colliculi, the anterolateral system, and the brachium of the inferior colliculus are served by quadrigeminal branches. Additional blood supply to the area surrounding the exit of the trochlear nerve and the inferior colliculus arises from medial branches of the *superior cerebellar artery*.

Sources and Additional Reading

Bobillier P, Seguin S, Petitjean F, Salvert D, Tovret M, Jouvet M: The raphe nuclei of the cat brain stem: A topographical atlas of their efferent projections as revealed by autoradiography. Brain Res 113:449–486, 1976.

Bogerts B: A brainstem atlas of catecholaminergic neurons in man, using melanin as a natural marker. J Comp Neurol 197:63–80, 1981.

Brodal A: The Reticular Formation of the Brainstem, Anatomical Aspects and Functional Correlations. Charles C Thomas, Springfield, Ill, 1958.

Brodal A: Neurological Anatomy, 3rd ed. Oxford University Press, New York, 1981.

Crosby EC, Humphrey T, Lauer EW: Correlative Anatomy of the Nervous System. Macmillan Publishing, New York, 1962.

Duvernoy HM: Human Brainstem Vessels. Springer-Verlag, Berlin, 1978.

Duvernoy H: The Human Brain Stem and Cerebellum, Surface, Structure, Vascularization, and Three-Dimensional Sectional Anatomy with MRI. Springer-Verlag, Vienna, 1995.

Haines DE: Neuroanatomy, An Atlas of Structures, Sections, and Systems, 5th ed. Lippincott Williams & Wilkins, Philadelphia, 2000.

Hobson JA, Brazier MAB (eds): The Reticular Formation Revisited: Specifying Function for a Nonspecific System. Int Brain Res Organization Monogram Series, vol 6. Raven Press, New York, 1980.

Hubbard JE, DiCarlo V: Fluorescence histochemistry of monoamine-containing cell bodies in the brain stem of the squirrel monkey (*Saimiri sciureus*) III. Serotonin-containing groups. J Comp Neurol 153:385–398, 1974.

Jenkins TW, Truex RC: Dissection of the human brain as a method for its fractionation by weight. Anat Rec 147:359–366, 1963.

Kretschmann H-J, Weinrich W: Cranial Neuroimaging and Clinical Neuroanatomy, Magnetic Resonance Imaging and Computed Tomography, 2nd ed. Thieme Medical Publishers, New York, 1992.

Larsell O, Jansen J: The Comparative Anatomy and Histology of the Cerebellum, The Human Cerebellum, Cerebellar Connections, and Cerebellar Cortex. The University of Minnesota Press, Minneapolis, 1972.

Leblanc A: The Cranial Nerves: Anatomy, Imaging, Vascularisation. Springer, New York, 1995.

Nieuwenhuys R: Chemoarchitecture of the Brain. Springer-Verlag, Berlin, 1985.

Nieuwenhuys R, Voogd J, van Huijzen CHR: The Human Central Nervous System, A Synopsis and Atlas, 3rd ed. Springer-Verlag, Berlin, 1988.

Olszewski J, Baxter D: Cytoarchitecture of the Human Brain Stem, 2nd ed. S. Karger, Basel, 1982.

Palay SL, Chan-Palay V: Cerebellar Cortex, Cytology and Organization. Springer-Verlag, New York, 1974.

Parent A: Carpenter's Human Neuroanatomy, 9th ed. Williams & Wilkins, Baltimore, 1996.

Paxinos G (ed): The Human Nervous System. Academic Press, San Diego, Calif, 1990, Chapters 7–14.

Taber E, Brodal A, Walberg F: The raphe nuclei of the brain stem in the cat. I. Normal topography and cytoarchitecture and general discussion. J Comp Neurol 114:161–187, 1960.

Tatu L, Moulin T, Bogousslavsky J, Duvernoy H: Arterial territories of human brain: Brainstem and cerebellum. Neurology 47:1125–1135, 1996.

Weber JT, Martin GF, Behan M, Huerta MF, Harting JK: The precise origin of the tectospinal pathway in three common laboratory animals: A study using the horseradish peroxidase method. Neurosci Lett 11:121–127, 1979.

Chapter 14

A Synopsis of Cranial Nerves of the Brainstem

D. E. Haines and G. A. Mihailoff

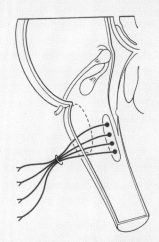

Overview 200

Motor Cell Columns and Nuclei 200

Sensory Cell Columns and Nuclei 201

Cranial Nerves of the Medulla Oblongata 202
Hypoglossal Nerve
Accessory Nerve
Vagus Nerve
Glossopharyngeal Nerve
The Jugular Foramen

Cranial Nerves of the Pons-Medulla Junction 208
Vestibulocochlear Nerve
Facial Nerve
Abducens Nerve

The Cranial Nerve of the Pons 213
Trigeminal Nerve

Cranial Nerves of the Midbrain 215
Trochlear Nerve
Oculomotor Nerve

Although the brainstem is quite small, comprising only about 2.6% of total brain weight, the size of this structure belies its importance. First, all ascending and descending tracts linking the spinal cord and the forebrain traverse the brainstem. Second, there are important ascending fibers (e.g., spinoreticular, spinoperiaqueductal gray) and descending fibers (e.g., rubrospinal, vestibulospinal, reticulospinal) that interconnect the brainstem with the spinal cord. These tracts, and the more expansive connections they make, are essential to the successful function of the nervous system. Third, the nuclei and the exit and entrance points of 10 of the 12 cranial nerves are associated with the brainstem.

Lesions of the brainstem, regardless of their origin (vascular, tumor, trauma), frequently involve cranial nerves. Indeed, in patients with brainstem lesions who have long tract signs, the accompanying cranial nerve deficits represent good localizing signs.

Overview

In general, the exit of a cranial nerve from the brainstem ("exit" is used here with reference to both efferent and afferent fibers of the nerve) is associated with the same brainstem area in which the nuclei of that nerve are found. The obvious exception is the trigeminal nerve, the sensory nuclei of which form a continuous cell column from rostral regions of the midbrain to the spinal cord–medulla interface.

This chapter reviews cranial nerves of the brainstem from caudal (hypoglossal) to rostral (oculomotor) and presents a number of clinical examples. The goal here is not simply to review the information covered in the last three chapters but to consider the cranial nerves in a somewhat broader perspective. Structure, function, and dysfunction are described in an integrated manner because this is how cranial nerves are evaluated in the clinical setting.

Motor Cell Columns and Nuclei

Early in development the derivatives of the *basal plate* that form motor nuclei of the cranial nerves in the brainstem tend to form rostrocaudally oriented cell columns. As the brainstem enlarges, these cell columns become discontinuous. That is, they are in line with, but are separated from, each other in the adult brain (see Figs. 10–4 to 10–6). As we have seen, and shall review here, those nuclei that are in line with each other, and that have arisen from the same original cell column, have developmental, structural, and functional characteristics in common.

The most medial cranial nerve motor nuclei in the brainstem are the *hypoglossal* (XII), *abducens* (VI), *trochlear* (IV), and *oculomotor* (III) nuclei (Fig. 14–1; see also Fig. 10–7). These nuclei share three characteristics.

First, they are located adjacent to the midline and anterior to the ventricular space of their particular brain division. Second, the motor neurons in these nuclei innervate skeletal muscle that originates from head mesoderm in the occipital region (tongue muscles) and in the area of the orbit (extraocular muscles). Third, the functional component of the lower motor neurons in these nuclei is general somatic efferent (GSE)—reflecting the fact that these motor neurons innervate skeletal muscles that originate from head mesoderm *not* located in a pharyngeal arch.

Laterally adjacent to the GSE cell column are the nuclei that collectively constitute the cranial part of the craniosacral division (parasympathetic) of the visceromotor nervous system. These nuclei—with the cranial nerves on which the preganglionic fibers travel—are (1) the *dorsal motor vagal nucleus*—vagus nerve; (2) the *inferior salivatory nucleus*—glossopharyngeal nerve; (3) *the superior salivatory nucleus*—the facial nerve–intermediate part; and (4) the *Edinger-Westphal nucleus*—oculomotor nerve (see Figs. 14–1 and 10–7). These nuclei share the following characteristics. First, they form a discontinuous column located slightly lateral to the GSE nuclei. Second, the neurons in these nuclei give rise to preganglionic axons that terminate in a peripheral ganglion, the cells of which give rise to postganglionic fibers that innervate a visceral structure. Third, because these motor neurons are part of a pathway that innervates a visceral structure (tissue composed of smooth muscle, glandular epithelium, or cardiac muscle or a combination of these), they are classified as general visceral efferent (GVE). They can also be called GVE-preganglionic parasympathetic, as this term completely identifies their relationships.

The most lateral motor cell column in the medulla and in the pontine tegmentum is formed by the *nucleus ambiguus*, the efferents of which travel on the vagus and glossopharyngeal nerves, and by the *facial motor nucleus* and the *trigeminal motor nucleus*, related to facial and trigeminal nerves, respectively (see Figs. 14–1, 10–7). These motor nuclei also share common characteristics. First, they form a discontinuous column in the more lateral part of the medulla and pontine tegmentum. Their position, as is the case for the GSE and GVE cell columns, reflects the differentiation of the basal plate in the brainstem. Second, the muscles innervated by these lower motor neurons originate from mesenchyme specifically located within the pharyngeal arches. The *muscles of mastication* (trigeminal nerve innervation) originate in arch I; the *muscles of facial expression* (facial nerve innervation), in arch II; the *stylopharyngeus muscle* (glossopharyngeal nerve innervation), in arch III; and the *constrictors of the pharynx, intrinsic laryngeal muscles, palatine muscles (except the tensor veli palatine) and the vocalis* (vagal nerve innervation), in arch IV. Third, owing to the fact that these lower motor neurons innervate *skeletal muscles arising within*

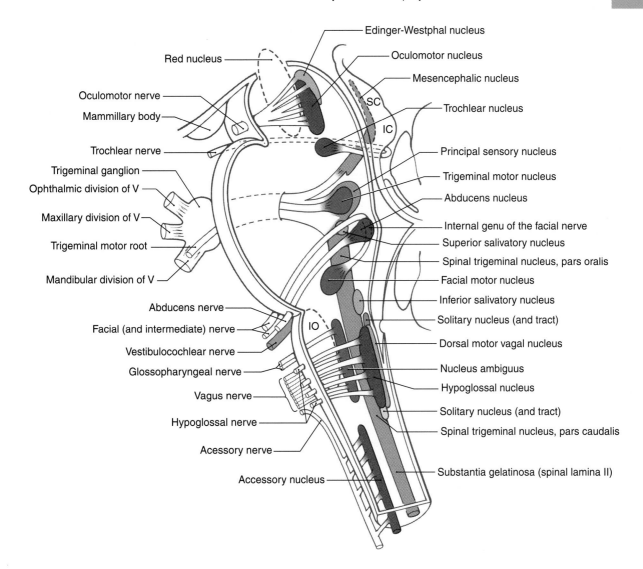

Figure 14–1. The location of cranial nerve nuclei and the course of their fibers within the brainstem. The functional component associated with each nucleus and with the fibers to or from that nucleus are color-coded—in this figure and in other figures showing connections in this chapter—as follows: general somatic efferent (GSE), red; general somatic afferent (GSA), light red; special visceral efferent (SVE), blue; general visceral efferent (GVE), light blue; special visceral afferent (SVA), green; general visceral afferent (GVA), light green; and special somatic afferent (SSA), gray. (Modified, with permission, from M.B. Carpenter and J. Sutin: Human Neuroanatomy, 8th ed., Williams & Wilkins, 1983.)

pharyngeal arches, they are classified as special visceral efferent (SVE).

Sensory Cell Columns and Nuclei

The derivatives of the *alar plate* that give rise to cranial nerve sensory nuclei of the brainstem are located lateral to the sulcus limitans (see Figs. 10–5 and 10–6). In contrast to the motor nuclei, which all form rostrocaudally oriented but discontinuous cell columns, all three of the sensory nuclei in the brainstem form what can arguably be described as continuous cell columns in the adult. These sensory nuclei–cell columns are located in the lateral aspects of the brainstem.

The most medial of these cell columns is the *solitary tract and nucleus*, which is the visceral afferent center of the brainstem (see Figs. 14–1 and 10–7). *No matter what cranial nerve returns visceral afferent information to the brainstem, the central processes of these primary afferent fibers always form the solitary tract, the fibers of which terminate in the solitary nucleus* (Fig. 14–2). This point is emphasized throughout the chapter in discussions of individual cranial nerves. Visceral afferent information is conveyed centrally on the facial, glossopharyngeal, and vagus nerves and consists of taste fibers (special visceral afferent [SVA]) and fibers conveying general visceral sensation (general visceral afferent [GVA]). The majority of taste input reaches rostral portions of

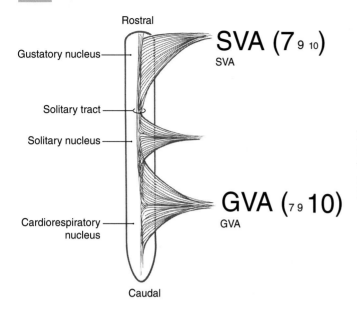

Rostral

Gustatory nucleus

SVA (7 9 10)
SVA

Solitary tract

Solitary nucleus

GVA (7 9 10)
GVA

Cardiorespiratory
nucleus

Caudal

Figure 14–2. Diagrammatic representation of the solitary tract and nucleus. The functional components associated with rostral versus caudal portions of the tract and nucleus, and the cranial nerves conveying this input, are shown in letters and numbers of proportionate size for each area.

the solitary nucleus (sometimes called the gustatory nucleus), whereas most general visceral sensation enters the caudal portion of the solitary nucleus (sometimes referred to as the cardiorespiratory nucleus) (see Fig. 14–2). Because the most rostral cranial nerve that contributes to the solitary tract and nucleus is the facial nerve (a nerve of the pons-medulla junction), the solitary tract and its nucleus are found throughout the medulla but do not extend rostrally beyond the pons-medulla junction.

Immediately and posteriorly (dorsally) adjacent to the solitary tract and nucleus are the *medial and spinal vestibular nuclei* (see Fig. 22–12). These continue rostrally, are joined by the *anterior and posterior cochlear nuclei* at the pons-medulla junction, and interface in the caudal pons with the *superior and lateral vestibular nuclei* (see Figs. 14–9 and 22–12). This cell column receives sensory input from the vestibulocochlear nerve (cranial nerve VIII) only and subserves the sense of hearing (SSA, exteroceptive functional component) and balance and equilibrium (SSA, proprioceptive functional component).

The nuclei of the trigeminal sensory system form a continuous cell column extending from the spinal cord–medulla junction to the rostral midbrain (see Figs. 14–1 and 10–7). The trigeminal sensory nuclei are divided into (1) the *spinal trigeminal nucleus* (consisting of a *pars caudalis, pars interpolaris,* and *pars oralis*), located in the lateral medulla and extending into the caudal pons; (2) the *principal sensory nucleus,* located at the mid-pontine level; and (3) the *mesencephalic nucleus,* extending rostrally into the midbrain at the lateral aspect of the periaqueductal gray (see Figs. 14–1 and 13–8). As is the case for the solitary tract and nucleus (the visceral receiving center of the brainstem), the principal sensory nucleus and especially the spinal trigeminal nucleus constitute the general sensory receiv-

ing center of the brainstem. *Although general somatic afferent (GSA) pain and thermal sensations enter the brainstem on four different cranial nerves (trigeminal, facial, glossopharyngeal, and vagus), the central processes of these primary afferent fibers enter the spinal trigeminal tract and terminate in the medially adjacent spinal trigeminal nucleus.* This theme is revisited in the review of individual nerves later in the chapter. In the medulla, the spinal trigeminal tract is laterally adjacent to the spinal nucleus.

Cranial Nerves of the Medulla Oblongata

The cranial nerves that are commonly identified as exiting the medulla are the hypoglossal nerve (cranial nerve XII) through the abducens nerve (cranial nerve VI) (Figs. 14–3 and 14–4). However, in the subsequent discussion, the abducens (VI), facial (VII), and vestibulocochlear (VIII) nerves are considered as the nerves of the pons-medulla junction. Consequently, the cranial nerves that are generally associated with *only* the medulla are the *hypoglossal* (XII), *accessory* (XI), *vagus* (X), and *glossopharyngeal* (IX) nerves (see Figs. 14–3 and 14–4). The unique situation of the accessory nerve is addressed further on.

Hypoglossal Nerve. The *hypoglossal nucleus* is located internal to the hypoglossal trigone. Axons of hypoglossal motor neurons pass anteriorly in the medulla along the lateral aspect of the medial lemniscus and the pyramid (see Fig. 11–11) to exit via the preolivary fissure (see Fig. 14–3). They continue through the *hypoglossal canal* and distribute to the intrinsic muscles of the tongue plus the hypoglossus, palatoglossus, and genioglossus muscles (Fig. 14–5). In addition to the hypoglossal nerve, the hypoglossal canal may also contain an emissary vein and a

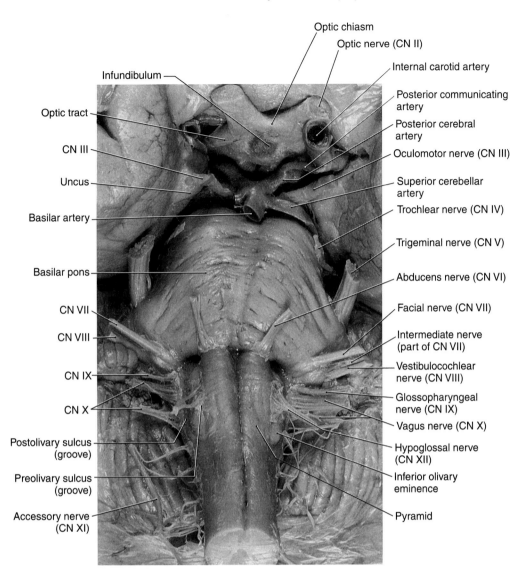

Figure 14–3. An anterior (ventral) view of the brainstem with particular emphasis on cranial nerves (CNs).

small meningeal branch to the dura of the posterior fossa from the ascending pharyngeal artery.

The blood supply to the hypoglossal nucleus and its exiting fibers is via penetrating branches of the anterior spinal artery. Occlusion of these branches (as in the medial medullary syndrome) may result in paralysis of the genioglossus muscle with *deviation of the tongue toward the side of the lesion (the weak side) on protrusion.* In addition, the afflicted person experiences a *contralateral hemiparesis* (corticospinal tract involvement) and a *contralateral loss of position sense, vibratory sense, and two-point discrimination* (medial lemniscus involvement), because the anterior spinal artery also serves these structures.

Other lesions that may affect hypoglossal function include a lesion of the root of the nerve only (causing tongue deviation to the side of the lesion with no other deficits) or injury to the internal capsule. In the latter case, corticonuclear (corticobulbar) fibers to hypoglossal motor neurons innervating the genioglossus muscle are predominantly crossed. Consequently, internal capsule lesions may result in a deviation of the tongue to the contralateral side (side opposite the lesion) on protrusion, in concert with other deficits such as a contralateral hemiplegia and a drooping of the facial muscles in the lower quadrant of the contralateral side of the face. See Figures 25–12 and 25–13 for examples of lesions that result in hypoglossal nerve dysfunction.

Accessory Nerve. This so-called cranial nerve was historically described as having a cranial part and a bulbar part. However, experimental studies have shown that the neurons that innervate the sternocleidomastoid and trapezius muscles are located in the *cervical cord* only; these muscles are not innervated by motor neurons located in the medulla. However, for consistency,

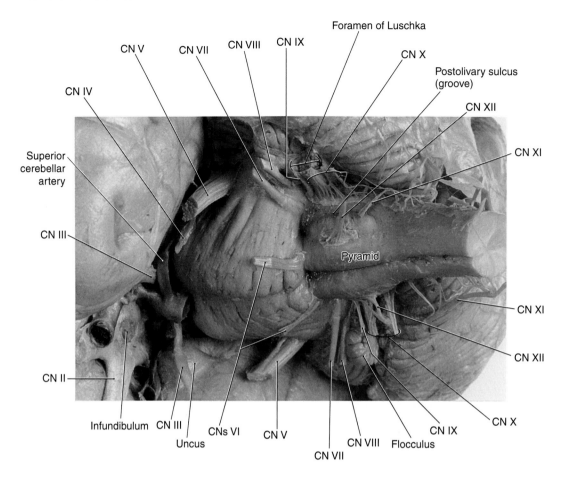

Figure 14–4. An anterolateral (ventrolateral) view of the brainstem with special emphasis on cranial nerves (CNs). Note the position and relationships of the foramen of Luschka.

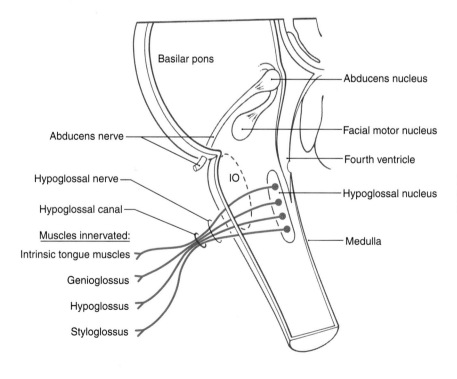

Figure 14–5. The central origin and peripheral distribution of the hypoglossal nerve (cranial nerve XII).

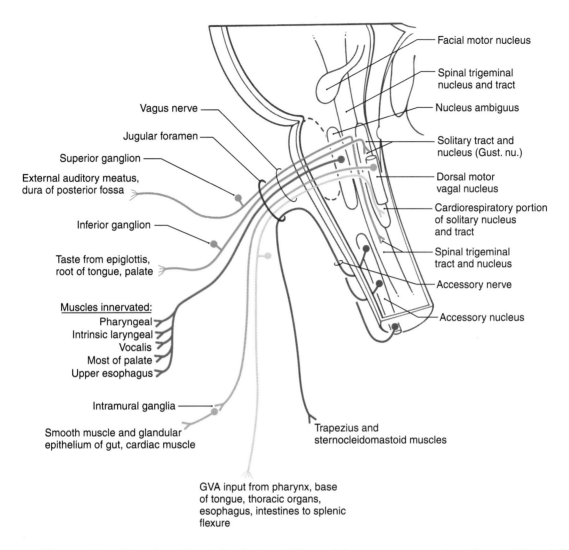

Vagus nerve

Jugular foramen

Superior ganglion

External auditory meatus, dura of posterior fossa

Inferior ganglion

Taste from epiglottis, root of tongue, palate

Muscles innervated:
Pharyngeal
Intrinsic laryngeal
Vocalis
Most of palate
Upper esophagus

Intramural ganglia

Smooth muscle and glandular epithelium of gut, cardiac muscle

Facial motor nucleus

Spinal trigeminal nucleus and tract

Nucleus ambiguus

Solitary tract and nucleus (Gust. nu.)

Dorsal motor vagal nucleus

Cardiorespiratory portion of solitary nucleus and tract

Spinal trigeminal tract and nucleus

Accessory nerve

Accessory nucleus

Trapezius and sternocleidomastoid muscles

GVA input from pharynx, base of tongue, thoracic organs, esophagus, intestines to splenic flexure

Figure 14–6. The central nuclei and peripheral distribution of fibers of the accessory nerve (cranial nerve XI) and the vagus nerve (cranial nerve X). Visceral afferent cell bodies (SVA, GVA) collectively form the inferior ganglion, and GSA cell bodies collectively form the superior ganglion of cranial nerve X. Gust. nu., rostral portions of solitary nucleus–gustatory nucleus.

and in recognition of wide usage, the accessory nerve is considered here as a cranial nerve of the medulla.

The accessory nerve originates from motor neurons in the cervical spinal cord (Fig. 14–6). The axons of these neurons exit the lateral aspect of the cord, co-alesce to form the nerve (see Fig. 14–3), ascend to enter the cranial cavity via the foramen magnum, and exit the posterior fossa through the jugular foramen (Fig. 14–7). En route through the posterior fossa these fibers briefly join the caudal portions of the vagus

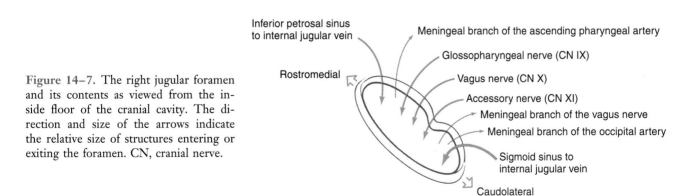

Figure 14–7. The right jugular foramen and its contents as viewed from the inside floor of the cranial cavity. The direction and size of the arrows indicate the relative size of structures entering or exiting the foramen. CN, cranial nerve.

Inferior petrosal sinus to internal jugular vein

Rostromedial

Meningeal branch of the ascending pharyngeal artery

Glossopharyngeal nerve (CN IX)

Vagus nerve (CN X)

Accessory nerve (CN XI)

Meningeal branch of the vagus nerve

Meningeal branch of the occipital artery

Sigmoid sinus to internal jugular vein

Caudolateral

nerve (see Fig. 14–6) and then leave the vagus to exit the skull as the accessory nerve.

The relationship of the accessory nerve to the vagus nerve is very similar to the relationship of the seventh nerve to the trigeminal nerve via the chorda tympani. In the latter case the taste fibers from the anterior two thirds of the tongue travel on the trigeminal nerve and then join the seventh nerve via the chorda tympani. However, throughout their extent, both central and peripheral, these taste fibers are considered as part of the seventh nerve, not the fifth. In like manner, fibers of the accessory nerve temporarily join the vagus and then leave it to exit the skull (see Fig. 14–6). These accessory nerve fibers do not originate from the medulla, do not distribute peripherally with the vagus, and have their cells of origin in the cervical spinal cord. What has classically been called the bulbar part of the accessory nerve is actually a misnomer; these fibers are the caudal portions of the vagus nerve to which the accessory nerve temporarily relates. Reflecting the fact that the sternocleidomastoid and trapezius muscles in the human originate from mesoderm caudal to the fourth arch, the functional component associated with these motor neurons is GSE.

Lesions of the root of the accessory nerve result in *drooping of the shoulder* (trapezius paralysis) on the ipsilateral side and *difficulty in turning the head* to the contralateral side (sternocleidomastoid paralysis) against resistance. Weakness of these muscles is not especially obvious in cervical cord lesions, because a hemiparesis or hemiplegia (indicating damage to corticospinal fibers) is the overwhelmingly obvious deficit. However, a lesion of the internal capsule may also result in deficits similar to those described above, owing to interruption of the corticonuclear (corticobulbar) fibers to the accessory nucleus; these fibers are primarily uncrossed.

Vagus Nerve. The *vagus nerve* is located at an intermediate position between the midline and lateral aspect of the medulla, exits the postolivary sulcus (see Figs. 14–3 and 14–4), and contains both motor and sensory components. This cranial nerve exits the cranial cavity via the jugular foramen (see Fig. 14–7) and exhibits two ganglia immediately external to the foramen. The *superior ganglion* contains the cell bodies of GSA fibers, whereas the *inferior ganglion* contains the cell bodies of GVA and SVA fibers.

The motor cells in the medulla that distribute their axons on the vagus nerve are located in the *dorsal motor nucleus of the vagus* (GVE–parasympathetic preganglionic) and in the *nucleus ambiguus* (SVE) (see Fig. 14–6). Preganglionic parasympathetic GVE cells send their axon, via branches of the vagus nerve, to end in *terminal (intramural) ganglia* located adjacent to, or within, visceral structures of the trachea and bronchi of the lungs, the heart, and the digestive system to the level of the splenic flexure of the colon (see Fig. 14–6). In

general, vagal influence causes constriction of the bronchioles, decreases heart rate, and increases blood flow, peristalsis, and secretions in the gut (see Chapter 29 for details). Axons of SVE motor neurons in the nucleus ambiguus distribute on branches of the vagus nerve to the constrictor muscles of the pharynx, intrinsic laryngeal muscles (including the vocalis muscle), the palatine muscles (except the tensor veli palatini, which is innervated by the fifth nerve), and the skeletal muscle in about the upper half of the esophagus (see Fig. 14–6). These muscles originate from the fourth pharyngeal arch—hence their innervation by neurons with an SVE functional component.

The sensory fibers conveyed on the vagus nerve relay general somatic, general visceral, and special visceral sensations. The general somatic sensations are represented by GSA input (recognized as pain and thermal sensations) from a small area on the ear and part of the external auditory meatus and from the dura of the posterior cranial fossa (see Fig. 14–6). These fibers have their cell bodies in the superior ganglion of the vagus nerve and enter the medulla as part of the vagus, but the central processes of these GSA primary afferent fibers enter the spinal trigeminal tract and synapse in the medially adjacent spinal trigeminal nucleus. The general and special visceral sensations conveyed by the vagus are represented by GVA and SVA fibers (see Fig. 14–6). General visceral sensations from the heart, pharynx and larynx, lungs, and gut to the level of the splenic flexure are conveyed by these GVA fibers. Their cell bodies are in the inferior ganglion of the vagus nerve, whereas the central processes of these GVA fibers enter the solitary tract and terminate in the surrounding solitary nucleus. The same trajectory is also followed by taste fibers (SVA fibers) on the vagus. These fibers originate from scattered taste buds on the epiglottis and base of the tongue, have their cell bodies in the inferior ganglion, enter the brainstem on the vagus nerve, and centrally distribute to the solitary tract and nucleus. Both general sensory (GSA, GVA) and taste (SVA) information conveyed on the vagus nerve is eventually relayed to the sensory cortex, where it is interpreted as, for example, pain from the external auditory meatus (GSA), a sense of fullness from the gut (GVA), or taste (SVA). Details of these central pathways are described in later chapters.

A lesion of the root of the vagus nerve will result in *dysphagia*, due to a unilateral paralysis of pharyngeal and laryngeal musculature, and *dysarthria*, due to a weakness of laryngeal muscles and the vocalis muscle. There are, however, no lasting demonstrable symptoms specifically related to visceromotor (autonomic) dysfunction. Taste loss is not detectable and cannot be tested, and the small somatosensory (GSA) loss involving the external auditory meatus and canal is usually of no consequence.

Unilateral injury inside the medulla, as with tumors,

vascular lesions, or syringobulbia, may give rise to similar deficits (as described above) owing to damage to the nucleus ambiguus. Bilateral lesions of the medulla, although rare, result in *aphonia*, *aphagia*, *dyspnea*, or *inspiratory stridor*. Such lesions may be life-threatening, especially if they involve the dorsal motor nucleus. Dysarthria may also be seen in patients after thyroid surgery if the recurrent laryngeal nerve has been damaged.

Glossopharyngeal Nerve. The *glossopharyngeal nerve* exits the medulla at the postolivary sulcus immediately rostral to the vagus nerve (see Figs. 14–3 and 14–4) and exits the skull via the jugular foramen (see Fig. 14–7), along with the vagus and accessory nerves. Like the vagus, the glossopharyngeal nerve has two ganglia: an *inferior ganglion* containing visceral afferent cell bodies (for GVA and SVA fibers) and a *superior ganglion* containing GSA cell bodies.

Motor fibers that distribute on the glossopharyngeal nerve originate from the *inferior salivatory nucleus* (GVE–preganglionic parasympathetic fibers) and from the *nucleus ambiguus* (SVE fibers) (Fig. 14–8). Axons of cells of the inferior salivatory nucleus exit on the glossopharyngeal nerve and join the tympanic nerve and then the lesser petrosal nerve to synapse with GVE-postganglionic neurons in the otic ganglion. These postganglionic parasympathetic cells supply secretomotor input to the parotid gland. The contribution of the nucleus ambiguus to the glossopharyngeal nerve serves to innervate the *stylopharyngeus muscle* (see Fig. 14–8). This muscle assists in swallowing and participates in the efferent part of the gag reflex.

Again, as with the vagus nerve, the sensory fibers of the glossopharyngeal nerve are general sensory (GSA) and general and special visceral (GVA, SVA) (see Fig. 14–8). The GSA fibers originate from cutaneous receptors on part of the pinna and the external auditory canal and have their cell bodies in the superior ganglion, and their central processes join the spinal trigeminal tract before terminating in the medially adjacent spinal trigeminal nucleus. The general visceral (GVA) fibers convey information from the parotid gland, the epithelial lining of the middle ear and the pharynx, with an especially important input from the carotid body. The carotid body contains chemoreceptors that are sensitive to changing levels of oxygen and

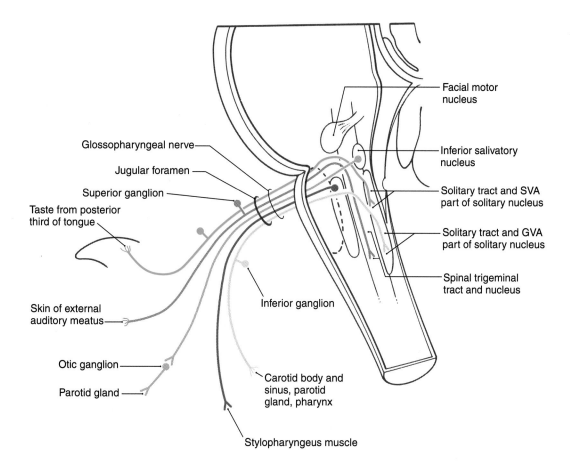

Facial motor nucleus

Glossopharyngeal nerve

Jugular foramen

Superior ganglion

Taste from posterior third of tongue

Inferior salivatory nucleus

Solitary tract and SVA part of solitary nucleus

Solitary tract and GVA part of solitary nucleus

Spinal trigeminal tract and nucleus

Skin of external auditory meatus

Inferior ganglion

Otic ganglion

Parotid gland

Carotid body and sinus, parotid gland, pharynx

Stylopharyngeus muscle

Figure 14–8. The central nuclei and peripheral distribution of fibers of the glossopharyngeal nerve (cranial nerve IX). Visceral afferent cell bodies (SVA, GVA) collectively form the inferior ganglion and GSA cell bodies collectively form the superior ganglion of cranial nerve IX.

carbon dioxide in the blood. The primary afferent GVA fibers have their cell bodies in the inferior ganglion of the glossopharyngeal nerve, and centrally they enter the solitary tract to terminate in the solitary nucleus. Taste from the posterior one third of the tongue (see Fig. 14–8) is conveyed centrally by SVA fibers that also have their cell bodies of origin in the inferior ganglion of the glossopharyngeal nerve. As is the case for all taste fibers, the central processes of these afferent fibers enter the solitary tract and terminate on cells of the surrounding solitary nucleus. All afferent inputs on the glossopharyngeal nerve (GSA, GVA, SVA) are transmitted to relay nuclei of the thalamus and on to the sensory cortex, where the information is fully appreciated and interpreted.

Lesions of the ninth cranial nerve are relatively rare but may occur in combination with the vagal and accessory roots at the jugular foramen (see Fig. 14–7). Deficits related to only the glossopharyngeal nerve are largely restricted to a loss of taste from the posterior one third of the tongue and a loss of the gag reflex on the side of the lesion. In the latter case, both the sensory and motor limbs of the reflex are affected. In addition, the ninth nerve is subject to *glossopharyngeal neuralgia*. This disorder is characterized by attacks of intense idiopathic pain arising from the GVA sensory distribution of the nerve (pharynx, caudal parts of the tongue, tonsil, and possibly areas of the middle ear). The attacks may be spontaneous or may result from artificial stimulation of the back of the oral cavity or from swallowing or even talking. Glossopharyngeal neuralgia may be quite severe and can be seriously disabling.

The Jugular Foramen. In addition to containing cranial nerves IX, X, and XI, which pass through approximately its middle third, the jugular foramen also serves as a conduit for other important structures (see Fig. 14–7). In general, the foramen can be divided into a rostral and slightly medial area, a middle portion (containing cranial nerves IX, X, and XI), and a caudal and somewhat lateral portion.

The rostromedial portion (see Fig. 14–7) contains the continuation between the *inferior petrosal sinus* and the *internal jugular vein*. The inferior petrosal sinus, although small, represents a communication between the cavernous sinus and the internal jugular vein. In addition the rostromedial portion of the jugular foramen also contains a *meningeal branch of the ascending pharyngeal artery*. This small vessel is one source of arterial blood to the meninges of the posterior fossa.

The caudolateral portion of the jugular foramen (see Fig. 14–7) is the point at which the *sigmoid sinus* is continuous with the *internal jugular vein*. The sigmoid sinus is large and represents an important route for venous drainage from the brain. This part of the jugular foramen also contains the *meningeal branch of the occipital artery*, another source of arterial blood to the meninges of the posterior fossa. The *meningeal branch of the vagus nerve* enters the cranial cavity through this part of the foramen and serves as one of the nerves for the sensory innervation of the dura of the posterior fossa.

Cranial Nerves of the Pons-Medulla Junction

The cranial nerves of the pons-medulla junction are the *vestibulocochlear nerve* (the most lateral and, for the purposes of this discussion, exclusively sensory), the *facial nerve* (intermediate in location and a mixed nerve), and the *abducens nerve* (the most medial and exclusively motor) (see Figs. 14–3 and 14–4).

There are three reasons why cranial nerves VI, VII, and VIII are considered as nerves of the pons-medulla junction. First, the abducens motor nucleus and the facial motor nucleus are located at the pons-medulla junction (see Figs. 14–1, 12–9, and 12–10). Second, in humans the exit of all three of these nerves is located at the caudal edge of the pons (see Fig. 14–3). Indeed, two of the three (facial and vestibulocochlear) nerves exit the skull via the same foramen. Third, in a broader comparative sense, cranial nerves VI, VII, and VIII do not exit from the medulla. For example, in animals with a more modest cerebral cortex and cerebellum (and consequently a smaller pons), these cranial nerves exit the brain in association with the trapezoid body, which is exposed on the surface of the brainstem. In humans, because of the larger size of the pons, the trapezoid body is located internally *but still* at the pons-medulla junction. For these reasons it is appropriate to consider these three cranial nerves in the manner presented here.

Vestibulocochlear Nerve. The eighth cranial nerve is the most lateral (see Fig. 14–3) of the cranial nerves of the brainstem and is centrally related to the *posterior* and *anterior cochlear nuclei* (SSA, exteroception—hearing) and to the *medial, spinal (inferior), superior,* and *lateral vestibular nuclei* (SSA proprioception—balance and equilibrium). The two portions of cranial nerve VIII originate from highly specialized receptors that are well protected in the petrous portion of the temporal bone. These receptors are described in Chapters 21 and 22. Although the two nerve roots form separate fascicles at their origin, they move to form essentially one nerve root by the time they enter the brainstem.

The cochlear portion of cranial nerve VIII originates from cells in the spiral ganglion located in the cochlea (Fig. 14–9; see also Fig. 21–2). This ganglion is made up of bipolar cells, the central processes of which travel on the cochlear division through the internal acoustic meatus to terminate in the posterior and anterior cochlear nuclei. The cochlear nuclei, in turn, project to several brainstem relay nuclei, which

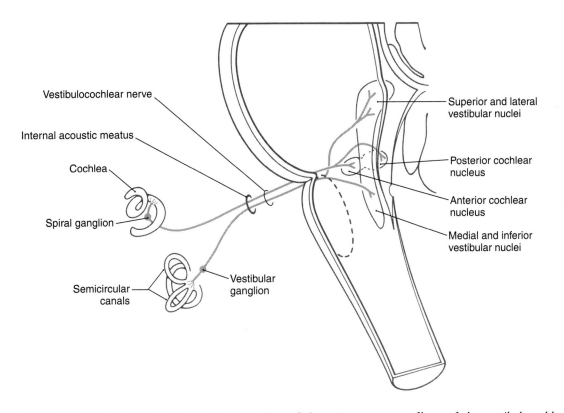

Figure 14–9. The peripheral origin and central termination of the primary sensory fibers of the vestibulocochlear nerve (cranial nerve VIII).

ultimately convey auditory information to the medial geniculate nucleus and from here to the auditory cortex and participate in auditory reflexes. These central pathways are described in Chapter 21.

Although the cochlear part of cranial nerve VIII is classically described as exclusively afferent (sensory), it is appropriate to note that cholinergic cells adjacent to the principal and accessory olivary nuclei give rise to axons that exit the brainstem on the cochlear nerve. These small bundles are called the olivocochlear tract (sometimes also called the efferent cochlear bundle). These efferent fibers of the cochlear nerve synapse in relation to the inner and outer hair cells and function to inhibit (dampen) the ability of the hair cell to respond to stimuli.

The vestibular division of cranial nerve VIII originates from the bipolar cells of the *vestibular ganglion* located medial to the *semicircular canals* (see Figs. 14–9, 22–2 and 22–13). The central processes of these cells travel on the root of the nerve, traverse the internal acoustic meatus, enter the brainstem at the pons-medulla junction, and distribute centrally to the vestibular nuclei located in the medulla and caudal pons (see Fig. 14–9). It is important to remember that the internal acoustic meatus houses not only the vestibulocochlear nerve but also the facial nerve (including its intermediate part) and the labyrinthine artery. After receiving input from the ampullae of the semicircular canals and

the utricle and saccule, the vestibular nuclei have important central connections to the cerebellum, to oculomotor nuclei (nuclei of cranial nerves VI, IV, and III), and to other brainstem centers related to balance, position sense, equilibrium, and the coordination of eye movements with head movements. Specific information related to these central pathways is described in Chapter 22.

Lesions of, or injury to, the vestibulocochlear nerve may result from a wide range of causes. The clinical manifestations include hearing loss, tinnitus, vertigo, dizziness, and related problems such as ataxia. Injury to the cochlea, spinal ganglion, or cochlear fibers in cranial nerve VIII results in loss of hearing on the side of the lesion; this type of hearing loss is *sensorineural hearing loss*. Lesions within the brainstem (or at higher levels) may affect the patient's ability to precisely interpret or to localize a sound in space, but such lesions do not result in deafness in one ear. *Conductive hearing loss* results from a failure of conduction through the middle ear, usually involving the ossicles. Tinnitus is a ringing, hissing, or roaring sensation perceived by the patient. It is related to the auditory part of cranial nerve VIII and may result from injury to the peripheral part of the nerve or may follow central lesions that damage auditory fibers or structures. Injury to the vestibular fibers of cranial nerve VIII result in *vertigo* (a perception of movement) and nystagmus that may be accompanied by *nausea* and *vomiting*. The vertigo may be subjective (the patient perceives that his or her body is moving) or objective (the patient perceives that objects in the environment are moving). Nystagmus (rhythmic movements of the eyes) results from the interruption of vestibular influence over the brainstem motor neurons controlling eye movement, and the nausea and vomiting are a natural consequence of the disconnection between body or eye movements and the environment.

Lesions of the vestibular nuclei and their main central connections, especially the cerebellum, result in a sensation of being dizzy (in which objects or the environment is perceived as moving or spinning), ataxia and unsteady gait (the patient feels that his or her "balance is off"), and nystagmus. These signs and symptoms may range from mild to severe and may be accompanied by nausea and vomiting. Signs and symptoms of vestibular dysfunction may result from a wide range of causes such as toxicity from certain medications, trauma, diabetes, cerebellar lesions, and acoustic neuroma. The *Ménière syndrome* is characterized by hearing loss and sound distortion as well as vertigo and a sensation of dizziness or unsteadiness on walking or standing. The cause is unknown, but there seems to be an increase in endolymphatic pressure with an increase

in size of the utricle, saccule, and cochlear duct.

Facial Nerve. The *facial motor nucleus* is located in the pons just rostral to the pons-medulla transition (Fig. 14–10; see also Fig. 14–1). Axons from the SVE motor neurons of the facial nucleus course posteromedially to arch around the abducens nucleus from caudal to rostral before turning anterolaterally to exit the brainstem (see Fig. 14–10). In their passage through the anterolateral pons these fibers are joined by GVE preganglionic parasympathetic axons from neurons of the superior salivatory nucleus. These GVE fibers, along with SVA taste fibers from the anterior two thirds of the tongue, GSA fibers from the pinna and a few GVA fibers, form the intermediate nerve (see Fig. 14–10). The facial nerve fibers and intermediate nerve fibers are intermingled within the pons but emerge from the brainstem as two separate nerve bundles.

After exiting the brainstem into the cerebellopontine angle, the facial and intermediate nerves merge and are joined by the vestibulocochlear nerve. These nerve trunks enter the petrous portion of the temporal bone through the *internal acoustic meatus* (see Fig. 14–10) accompanied by the *labyrinthine artery*. Coursing posterolaterally through the facial canal in the petrous temporal bone, the facial nerve approaches the middle ear cavity, where it turns (*internal genu*) sharply poste-

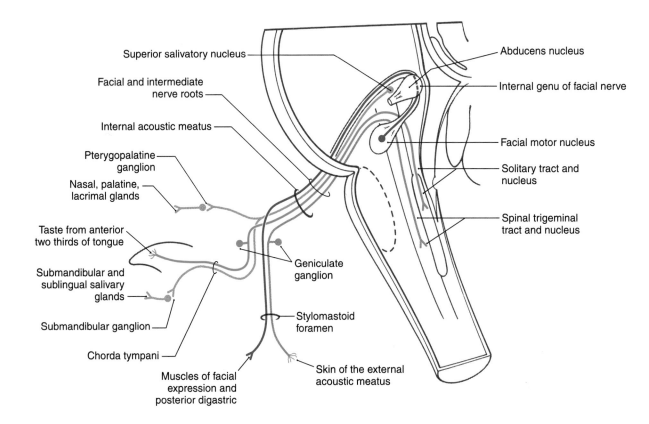

Figure 14–10. The central nuclei and peripheral distribution of fibers of the facial nerve (cranial nerve VII). The few GVA fibers from the nasopharynx, palate, and submandibular and sublingual salivary glands are not shown here; they have cell bodies of origin in the geniculate ganglion and project to more caudal regions of the solitary nucleus.

riorly and continues superior and posterior to the middle ear cavity to eventually exit the temporal bone via the *stylomastoid foramen*. The *geniculate ganglion* (see Fig. 14–10) of the facial nerve is located at the internal genu, and it is here that the *greater petrosal nerve* is formed by GVE–preganglionic parasympathetic fibers that leave the facial nerve. The greater petrosal nerve courses anteromedially through and on the petrous temporal bone to reach the area just superior to the foramen lacerum, where it joins the *deep petrosal nerve* to form the *nerve of the pterygoid canal*. From here, preganglionic parasympathetic axons enter the *pterygopalatine ganglion*. Postganglionic parasympathetic fibers from this ganglion supply the mucous membrane of the hard and soft palates, nasal cavity, and paranasal sinuses (see Fig. 14–10). Other postganglionic parasympathetic fibers join the maxillary division (V_2) of the trigeminal nerve and continue on branches that enter the orbit to provide parasympathetic innervation of the lacrimal gland.

En route through the temporal bone, the facial nerve gives off a small branch containing SVE motor fibers that supply the stapedius muscle and a larger branch, the *chorda tympani* (see Fig. 14–10). The latter nerve enters the middle ear cavity, passes across the inner surface of the tympanic membrane, and then exits the middle ear through the petrotympanic fissure to reach the infratemporal fossa. Here it joins the lingual branch of V_3 to distribute preganglionic parasympathetic fibers to the *submandibular ganglion* and to collect SVA taste afferent fibers from the anterior two thirds of the tongue (see Fig. 14–10). Postganglionic fibers from the submandibular ganglion innervate the submandibular and sublingual salivary glands and portions of the mucosal surfaces of the tongue and oral cavity.

Upon exiting the stylomastoid foramen, the fibers of the facial nerve pass through, and around, the parotid gland as the nerve divides into its five terminal branches. Most of the fibers that remain in the facial nerve at this point provide motor innervation (SVE) to the muscles of facial expression (see Fig. 14–10), the posterior belly of the digastric muscle, and the stylohyoid muscle. In addition, the facial nerve contains a relatively small number of cutaneous fibers (GSA) that supply portions of the external ear and external auditory canal.

The facial nerve includes three varieties of sensory fibers. The first, taste fibers (SVA), pass centrally from the anterior two thirds of the tongue first on the lingual branch of V_3. These SVA fibers then leave the lingual nerve via the chorda tympani to join the facial nerve to reach their cell bodies in the geniculate ganglion (see Fig. 14–10). The central processes of these primary afferent fibers continue with the facial nerve, pass into the cranial cavity through the internal acoustic meatus, and enter the brainstem in the intermediate nerve. These SVA fibers enter the solitary tract and terminate in rostral portions of the solitary nucleus, the central receiving area for all taste sensory signals (see Fig. 14–2).

The second variety of sensory fibers in the facial nerve is relatively small in number. Cutaneous sensory fibers (GSA) from regions of the external ear and external auditory canal course centrally on the facial nerve (see Fig. 14–10). Moreover, it appears that the muscles of facial expression contain relatively few muscle spindles; therefore, the contingent of muscle afferent fibers normally found within the GSA population is small or nonexistent in the facial nerve. The cutaneous fibers reach their cell bodies in the geniculate ganglion (with SVA and GVA sensory neurons), and their central processes course into the brainstem with the intermediate nerve. In the brainstem, these GSA fibers enter the spinal trigeminal tract and terminate in the spinal trigeminal nucleus (see Fig. 14–10); these signals are transmitted rostrally to the thalamus or used in local reflex circuits.

The third variety of sensory fibers is equally small in number. Receptors in the mucous membranes of the palate and nasopharynx give rise to primary sensory fibers (GVA) that have their cell bodies in the geniculate ganglion and enter the brainstem on the intermediate nerve. These fibers convey general sensory information from caudal regions of the oral cavity and will enter the more caudal portions of the solitary tract and terminate in the adjacent nucleus.

Corticonuclear (corticobulbar) fibers used in the performance of voluntary movements involving the muscles of facial expression distribute to the facial nuclei in a bilateral manner. The ipsilateral face motor cortex distributes *bilaterally* to those facial motor neurons that control muscles in the upper face (i.e., frontalis, orbicularis oculi) (see Fig. 25–11). In contrast, the face motor cortex projects only *contralaterally* to those facial motor neurons that control the lower facial expression muscles, such as those located near the angle of the mouth that are used to smile voluntarily (see Figs. 25–11 and 33–15). Lesions of the efferent fibers from the face motor cortex or of the internal capsule (supranuclear) result in drooping or sagging of the corner of the mouth contralateral to the lesion when the patient is asked to smile voluntarily. Such lesions are referred to as *central seven* lesions. Of interest, although some patients cannot smile when asked to do so by the neurologist, they can sometimes smile "involuntarily" or spontaneously in response to an amusing comment or situation.

Signs and symptoms due to peripheral lesions of the facial nerve (infranuclear; lower motor neuron) depend on the location of the damage. If the injury occurs proximal to the geniculate ganglion (see Fig. 14–10) and the origin of the greater petrosal nerve, the patient exhibits a *loss of voluntary control of ipsilateral facial expression muscles in the upper and lower portions of the face*.

This motor (the Bell palsy) deficit is accompanied by decreased mucosal secretion in the nasal and oral cavities and decreased tear fluid production and salivary gland output, all on the ipsilateral side. Cutaneous sensation of the external ear and external auditory canal is also diminished, but innervation of this territory is difficult to assess because cranial nerves IX and X contribute as well. In addition, there is *decreased taste sensation (SVA) on the anterior two thirds of the tongue* (but general sensation on the face, GSA, is preserved—why?) and *hyperacusis* on the side ipsilateral to the lesion.

If the lesion occurs distal to the ganglion (see Fig. 14–10) but proximal to the origin of the chorda tympani and stapedial nerve, decreased salivation and taste and hyperacusis may be present in conjunction with decreased facial expression throughout the ipsilateral side of the face. However, tear fluid production and the mucosal surfaces of the nasal and oral cavities are unaffected because the greater petrosal nerve is intact. It follows, then, that decreased function of all facial expression muscles on one side of the face in combination with the *absence of any deficits involving parasympathetic function or the sense of taste* serves to localize the lesion at, or distal to, the stylomastoid foramen (see Fig. 14–10).

The SVE fibers originating in the facial nucleus also form the efferent limb of the *corneal reflex*. The afferent limb of this reflex travels via the ophthalmic division of cranial nerve V. The central processes of these fibers have their cell bodies in the trigeminal ganglion, enter the spinal trigeminal tract, and terminate in the spinal nucleus. Trigeminothalamic fibers originating in the spinal nucleus send collaterals into the facial motor nucleus and then continue rostrally to the thalamus; this collateral connection completes the reflex circuit.

Abducens Nerve. The *abducens nucleus* is located internal to the facial colliculus (see Fig. 14–1). This structure, in turn, is found in the floor of the rhomboid fossa just lateral to the median sulcus and rostral to the striae medullares of the fourth ventricle. The abducens nucleus contain motor neurons and interneurons.

The axons of the GSE motor neurons of the *abducens nucleus* course through the tegmental and basilar pons, exit at the pons-medulla junction generally in line with the preolivary sulcus (Fig. 14–10; see also Figs. 14–3 and 14–4), and enter the dura just caudal to the cavernous sinus. These axons then enter and pass through the lumen of the sinus in close association with the cavernous portion of the internal carotid artery (see Fig. 8–15) and enter the orbit via the superior orbital fissure. The abducens nerve innervates the ipsilateral lateral rectus muscle (see Fig. 14–11).

The interneurons in the abducens send their axons into the contralateral *medial longitudinal fasciculus*, where they ascend to the oculomotor nucleus on that side (see Fig. 14–11). These crossed ascending fibers form excitatory synapses on oculomotor neurons that

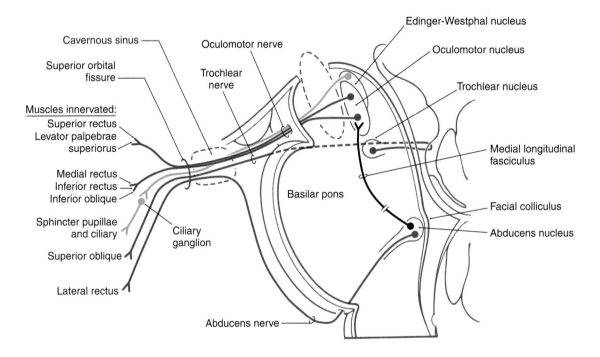

Figure 14–11. The central nuclei and peripheral distribution of fibers of the abducens nerve (cranial nerve VI), trochlear nerve (cranial nerve IV), and oculomotor nerve (cranial nerve III) nuclei. Interneurons in the abducens nucleus (black) enter the contralateral medial longitudinal fasciculus and distribute to medial rectus motor neurons in the nucleus of cranial nerve III. Disruption to this pathway results in internuclear ophthalmoplegia.

innervate the medial rectus muscle on that side. The connection thus formed is such that when the patient looks to the right (in the horizontal plane), the right lateral rectus contracts (this eye abducts) and the left medial rectus contracts (this eye adducts). Lesions that involve this pathway lead to the clinical condition known as *internuclear ophthalmoplegia*.

Lesions of the abducens nerve, the abducens nucleus, or the internuclear axons in the medial longitudinal fasciculus have similar, yet individually unique features. Injury to abducens fibers in the pons (as in a medial pontine syndrome) or in its course outside the brain results in a flaccid paralysis of the ipsilateral lateral rectus muscle; the affected eye is slightly introverted and does not abduct on attempted lateral gaze to that side. However, the opposite eye adducts because the internuclear neurons are intact.

A lesion of the abducens nucleus, such as a fourth ventricle tumor invading the facial colliculus, damages both the motor neurons and the internuclear neurons. The result is paralysis of the lateral rectus muscle on the side of the lesion accompanied by failure of the opposite medial rectus muscle to contract on attempted gaze to the side of the lesion. This particular example combines a lower motor neuron lesion of the lateral rectus on the side of the lesion with an intranuclear ophthalmoplegia—a paralysis of the medial rectus on the opposite side seen in attempted gaze toward the side of the lesion.

Damage only to internuclear axons in the left medial longitudinal fasciculus (as in multiple sclerosis) results in an inability to adduct the left eye on attempted gaze to the right. In contrast with lesions of the nucleus or nerve, the right eye can abduct, because the motor neurons in the right nucleus (and their axons) are intact.

One important source of cerebral cortical influence over abducens motor neurons originates from cells in the frontal eye field (see Fig. 28–14). This area of cortex projects both to the paramedian pontine reticular formation (PPRF), also called the *horizontal gaze center*, and to the superior colliculus, which also projects to the PPRF. Each PPRF, in turn, projects to the abducens nuclei on both sides of the brainstem. Sudden cortical damage (such as stroke or trauma) involving the frontal eye field results in an involuntary conjugate deviation of the eyes to the side of the lesion.

The Cranial Nerve of the Pons

The largest of the cranial nerves of the brainstem, the trigeminal nerve, exits the lateral aspect of the pons (see Figs. 14–3 and 14–4). It consists of a large sensory root (*portio major*), which is attached to the *trigeminal ganglion*, and a small motor root (*portio minor*), which bypasses the ganglion (Fig. 14–12). The exit

point of this nerve is also the border between the *basilar pons* and the *middle cerebellar peduncle*, which are located anterior (ventral) and posterior (dorsal), respectively, to the nerve roots.

At the most general level of organization the trigeminal nerve consists of *ophthalmic* (V_1), *maxillary* (V_2), and *mandibular* (V_3) divisions; these are the large sensory branches originating from the trigeminal ganglion (see Fig. 14–12). These branches serve the upper eyelid, forehead, and scalp to a point just beyond the vertex of the skull (V_1); the upper eyelid, upper lip and teeth, maxillary aspect of the face, and a lateral strip in the temporal area (V_2); and the lower lip and teeth, much of the oral cavity, and the skin over the mandible and continuing as a band up the lateral side of the head rostral to the ear (V_3) (see Figs. 18–4 and 18–13). The ophthalmic division exits the cranial cavity via the *superior orbital fissure* (see Fig. 14–12) and distributes to the orbit, with some branches exiting the orbit through the *supraorbital notch* (or foramen). The maxillary division exits the cranial cavity through the *foramen rotundum* (see Fig. 14–12) and then immediately passes through the *pterygopalatine fossa* and the *inferior orbital fissure* into the orbit; some of these branches exit the orbit via the *infraorbital foramen* onto the maxillary region of the face. The mandibular division exits the cranial cavity through the *foramen ovale* (see Fig. 14–12), as does the motor root of the fifth nerve.

Trigeminal Nerve. The *trigeminal nerve* is a mixed nerve, having sensory and motor components. The sensory nuclei form a continuous cell column that extends from the spinal cord–medulla junction to rostral levels of the mesencephalon. The motor nucleus is located medially adjacent to the principal sensory nucleus at about mid-pontine levels.

The sensory nuclei of the trigeminal nerve are the *spinal trigeminal nucleus*, extending from caudal to rostral throughout the lateral medulla and into the caudal pons; the *principal sensory nucleus*, located in the lateral parts of the pontine tegmentum at about mid-pontine levels at the rostral end of the spinal nucleus; and the *mesencephalic nucleus* and *mesencephalic tract*, which extend rostrally from the principal sensory nucleus into the midbrain along the lateral aspect of the periaqueductal gray (see Fig. 14–12). The spinal trigeminal nucleus consists of a *pars caudalis* (from the spinal cord–medulla junction to the obex), a *pars interpolaris* (from the obex to the rostral end of the hypoglossal nucleus), and the *pars oralis* (to the caudal aspect of the principal sensory nucleus).

The trigeminal nerve, along with small contributions from cranial nerves VII, IX, and X from the external ear, conveys sensory input from the entire face from the edge of the mandible to just caudal to the vertex of the head, the cornea and conjunctiva, mucosa of the nasal cavity and sinuses, the upper and lower

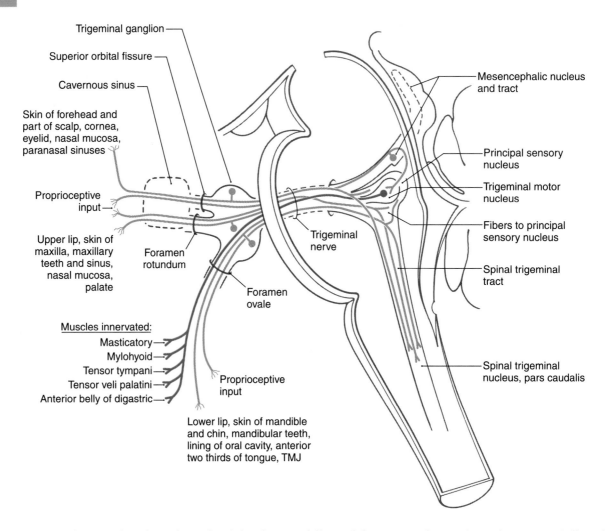

Figure 14–12. The central nuclei and peripheral distribution of fibers of the trigeminal nerve (cranial nerve V). Collaterals to the trigeminal motor nucleus from mesencephalic fibers connect the afferent and efferent limbs of the pathway underlying the jaw jerk reflex. TMJ, temporomandibular joint.

teeth, lining of the oral cavity including the tongue, and the lining of the pharynx and larynx (cranial nerves IX and X), and proprioceptive inputs from the muscles of mastication and extraocular muscles. The sensory information relayed on the fifth nerve is concerned with pain and thermal sensations, discriminative sensations, and proprioception.

Primary afferent fibers conveying pain, thermal sense, and nondiscriminative touch (GSA exteroception) have their cell bodies in the trigeminal ganglion and in the geniculate ganglion (cranial nerve VII) and superior ganglia of cranial nerves IX and X (the lateral three are related to the external ear only) (see Figs. 14–6, 14–8, 14–10, and 14–12). Their central processes form the spinal trigeminal tract and terminate in the medially adjacent nucleus. Fibers conveying discriminative touch from the face and oral cavity (GSA exteroception) follow a similar trajectory, but centrally these fibers terminate in the principal sensory nucleus. Those fibers conveying GSA proprioceptive informa-

tion from the masticatory muscles, extraocular muscles, and periodontal ligament receptors enter the brainstem via the trigeminal nerve. However, in clear contrast to other primary sensory fibers, these fibers have their cell bodies in the *mesencephalic nucleus, not in the trigeminal ganglion*. Sensory inputs to the spinal trigeminal nucleus and to the principal sensory nucleus are relayed to the thalamus via the *anterior (ventral)* and *posterior (dorsal) trigeminothalamic tracts* and from the thalamus to the somatosensory cortex, where the sensations are perceived and interpreted.

The motor nucleus of the trigeminal nerve innervates muscles that originate from the first pharyngeal arch—hence their SVE functional component. The motor root of the trigeminal nerve exits the skull via the foramen ovale, along with V_3 (see Fig. 14–12). These motor neurons innervate the muscles of mastication (including the medial and lateral pterygoid muscles), the tensor tympani, the tensor veli palatini, the mylohyoid, and the anterior belly of the digastric, all on the ipsilateral side (see Fig. 14–12). The motor nucleus receives afferent fibers involved in the *jaw jerk reflex.*

The reflexes associated with the trigeminal nerve are the *jaw jerk reflex* and the *corneal reflex.* The afferent limb of the *jaw jerk reflex* originates at receptors in the muscles of mastication, courses centrally on the mandibular division, and has its cell bodies in the mesencephalic nucleus. Collaterals from these afferent fibers project bilaterally to the trigeminal motor nucleus, the neurons of which give rise to the efferent limb of the reflex. The afferent limb of the *corneal reflex* originates from pain and touch receptors in the cornea. These fibers travel on the ophthalmic division of cranial nerve V, have their cell bodies in the trigeminal ganglion, and terminate centrally in the spinal trigeminal nucleus. Trigeminothalamic fibers en route to the thalamus send collaterals into the facial motor nucleus, which is the origin of the efferent limb of this reflex; in response to a stimulus that touches the cornea, the eyes blink.

The primary deficits in patients with lesions of the trigeminal nerve, or its central nuclei, are sensory symptoms in the peripheral distribution of the nerve (or its individual division) and paralysis of the masticatory muscles. The sensory deficits are a complete loss of pain, thermal, and tactile sensations on the ipsilateral side of the face and much of the scalp; a loss of the same sensations in the oral cavity (and from the teeth) on the same side; and a loss of the corneal reflex occurring in response to corneal touch, also on the same side. However, because the efferent limb of the corneal reflex travels via the facial nerve, touching the cornea on the side opposite a trigeminal root lesion results in a blink on that side as well as a blink on the side of the root lesion. An injury to one trigeminal nerve does not cause a loss of the jaw jerk reflex.

Another sensory disorder associated with the trigeminal nerve is *tic douloureux (trigeminal neuralgia).* This condition is characterized by severe, unexpected, lancinating pain restricted to one or more of the divisions of the nerve. Paroxysms of intense pain may result from stimulation of a trigger zone frequently located around the lip or nose or on the cheek. A wide variety of stimuli such as shaving, putting on makeup, talking, chewing, eating, a breeze on the face, or a sudden facial expression may precipitate an attack. Indeed, some patients may become malnourished because they avoid eating, chewing, and swallowing for fear of an attack. Trigeminal neuralgia is seen more frequently in patients older than 35 years of age, usually involves the maxillary division more often than the mandibular (second most commonly involved) or ophthalmic (rarely involved) division, and can severely compromise the quality of life. The causes are unknown. However, trigeminal neuralgia may be seen in patients with multiple sclerosis, degenerative changes in the trigeminal ganglion, vascular malformations, or aberrant loops of small arteries that may compress the nerve root.

Motor deficits resulting from trigeminal root lesions result in paralysis of the masticatory muscles on that side. The patient has difficulty chewing food, and the jaw deviates toward the side of the lesion (the weak side) on closing. This deviation occurs because of the unopposed action (pull of the mandible toward the midline) of the pterygoid muscles on the opposite (unlesioned) side.

Sensory or motor deficits related to the trigeminal nerve can also occur from central lesions. Such lesions include but are not limited to tumors or vascular lesions in the medulla or pons, metastatic lesions, and the obstruction of vessels. A well-known example of a lesion due to vascular obstruction is the *lateral medullary syndrome* (also called the *posterior inferior cerebellar artery syndrome* or *Wallenberg syndrome*), in which there is an alternating sensory loss. This lesion (see Fig. 18–13) results in an ipsilateral loss of pain and thermal sense on the face (spinal trigeminal tract involvement) and a contralateral loss of the same sensations on the body (anterolateral system involvement), in concert with motor deficits most frequently related to damage to the nucleus ambiguus or vestibular nuclei.

Cranial Nerves of the Midbrain

The cranial nerves of the midbrain are the trochlear nerve (cranial nerve IV) and the oculomotor nerve (cranial nerve III) (see Figs. 14–3 and 14–4). The nuclei of both nerves are located adjacent to the midline just ventral to the periaqueductal gray, both nerves are exclusively motor, and, like other exclusively motor nerves, they exit the midbrain just lateral to the midline. These two nerves, along with cranial nerve VI, innervate extraocular muscles.

Trochlear Nerve. The trochlear nerve is the only motor cranial nerve formed entirely by axons that cross the midline before their exit. For example, axons arising from the left trochlear nucleus arch caudally along the lateral edge of the periaqueductal gray to decussate in the most anterior aspect of the anterior medullary velum and then exit the brainstem immediately caudal to the right inferior colliculus (see Fig. 28–6).

The trochlear nucleus is situated within and slightly posterior to the medial longitudinal fasciculus at a cross-sectional level through the inferior colliculus (see Fig. 14–11). After arching around the periaqueductal gray, decussating, and exiting from the posterior surface of the midbrain, the axons of these GSE motor neurons have a long intracranial course. They exit into the superior cistern and then course laterally and anteriorly through the ambient cistern on the lateral aspect of the midbrain. The nerve then enters the dura, courses rostrally within the lateral wall of the cavernous sinus (see Fig. 8–14), and enters the orbit via the superior orbital fissure (see Fig. 14–11). Cranial nerve IV innervates the superior oblique muscle, which normally functions to direct the eye inferolaterally—that is, downward and outward.

It is important to remember that trochlear motor neurons innervate the contralateral superior oblique muscle, as the clinical findings may reflect this innervation. For example, a lesion at the root of the nerve in the ambient cistern or cavernous sinus, or at the superior orbital fissure, results in paralysis of the superior oblique muscle on that side. If the lesion is on the left side, the left eye cannot rotate slightly downward and outward. On the other hand, in a patient with multiple sclerosis involving the medial longitudinal fasciculus (MLF), the damage to which has extended to the trochlear nucleus, a variation on this theme may be seen. In this situation, a lesion in the right MLF with trochlear nucleus involvement results in paralysis of the left superior oblique muscle, and the left eye cannot rotate downward and outward to the left. In addition, and in clear contrast to the trochlear nerve root lesion, the patient also has internuclear ophthalmoplegia on the right; on attempted lateral gaze to the left, the right medial rectus muscle does not adduct the eye on that side.

An important source of cerebral cortical input to the trochlear nucleus is from neurons located in the frontal eye field. These projections pass to the rostral interstitial nucleus of the medial longitudinal fasciculus (riMLF), also called the *vertical gaze center*, and to the superior colliculus, which also projects to the riMLF. The riMLF sends a large projection to the ipsilateral trochlear nucleus and a smaller projection to the nucleus on the contralateral side. Sudden cortical damage (such as from stroke or trauma) involving the frontal eye field results in an involuntary conjugate deviation of the eyes to the side of the lesion.

Oculomotor Nerve. The oculomotor nucleus is located within the ventral portion of the periaqueductal gray just posterior to the medial longitudinal fasciculus and is present in about the rostral half of the midbrain (see Fig. 14–11). It is divided into several smaller subnuclei that contain GSE motor neurons that innervate all the extraocular muscles except the superior oblique and lateral rectus. This innervation involves ipsilateral muscles except for the superior rectus motor neurons, whose axons decussate within the nucleus to enter the contralateral oculomotor nerve. Although seemingly significant, this crossed pathway is typically ignored in the clinical setting, because the effect of losing the innervation to the (contralateral) superior rectus muscle is usually masked by the actions of the functionally intact muscles in that orbit.

Also located in the midbrain periaqueductal gray matter immediately posterior to the oculomotor complex is the Edinger-Westphal nucleus (see Fig. 14–11).

This group of visceral motor cells is composed of preganglionic parasympathetic GVE neurons whose axons project to the ciliary ganglion via the oculomotor nerve.

Oculomotor nerve fibers pass anteriorly (ventrally) through and around the red nucleus to eventually exit the midbrain in the interpeduncular fossa. Emerging from the midbrain, the nerve passes between the posterior cerebral and superior cerebellar arteries, enters the interpeduncular cistern, and then penetrates the dura lateral to the sella turcica to course within the lateral wall of the cavernous sinus (see Fig. 8–15). The nerve then exits the dura and passes through the superior orbital fissure, along with cranial nerves IV, VI, and V_1, to enter the orbit (see Figs. 14–11 and 14–12). In the orbit the oculomotor nerve divides into superior and inferior divisions, each division forming a few small communicating branches to the ciliary ganglion in addition to their muscular branches. The ciliary ganglion then gives off several small, short ciliary nerves, which reach the posterior aspect of the globe of the eye, where they penetrate the sclera to eventually reach the sphincter pupillae and ciliary muscles (see Fig. 14–11).

There are four smooth muscles related to each orbit that require visceromotor innervation. The GVE fibers in the oculomotor nerve enter the *ciliary ganglion*, the postganglionic parasympathetic fibers of which innervate the *sphincter pupillae* and *ciliary muscles.* In contrast, the dilator pupillae muscle and the *superior tarsal muscle* are activated by sympathetic innervation. Postganglionic sympathetic fibers exit the superior cervical ganglion and course, via the internal carotid plexus, to join the ophthalmic artery. Coursing into the orbit with the latter artery via the optic canal, sympathetic postganglionic fibers may join the ciliary ganglion directly, may join the nasociliary nerve (a branch of V_2 from which the *long ciliary branches* originate), or may join the oculomotor nerve and then enter the ciliary ganglion. Once in the ciliary ganglion, the sympathetic fibers continue, *without synapsing*, into the *short ciliary nerves* to reach the *dilator pupillae muscle.* Some of the sympathetic postganglionic fibers that travel via the oculomotor nerve continue beyond the ciliary ganglion and levator palpebrae superioris to reach the superior tarsal muscle.

Although it is known that the extraocular muscles contain muscle spindles, the oculomotor, trochlear, and abducens nerves are regarded as purely motor nerves. The spindle afferent fibers appear to join sensory nerves in the orbit, such as the frontal and nasociliary nerves, and eventually pass through V_1 to reach their cell bodies in the trigeminal ganglion. From here, sensory information enters the brainstem via the large sensory root of the trigeminal nerve.

The oculomotor nucleus does not receive direct cortical projections via the corticonuclear (corticobulbar) system. The cortex of the frontal eye field exerts its control over oculomotor neurons via projections to the rostral interstitial nucleus of the medial longitudinal fasciculus (riMLF—i.e., the vertical gaze center) and the superior colliculus. This part of the tectum also projects to the riMLF, which, in turn, sends numerous fibers to the ipsilateral oculomotor nucleus and a smaller contingent to the contralateral side. Consequently, cortical and capsular lesions have an effect on the actions of muscles innervated by the oculomotor nerve, but this effect is indirect and results from the loss of cortical input to the brainstem gaze control centers. Cortical damage (such as from stroke or trauma) involving the frontal eye fields produces an involuntary conjugate deviation of the eyes toward the side of the injury. One easy way to remember this correlate is that the patient "involuntarily looks to the lesioned side." If the cortical lesion is large enough to involve corticospinal fibers, the deviation of the eyes is toward the side of the cortical damage but away from the side of the resulting hemiparesis.

Lesions involving the oculomotor nucleus, the oculomotor nerve in the interpeduncular cistern, or the nerve in the lateral wall of the cavernous sinus all generally have the same result. Loss of the GSE motor fibers paralyzes all of the extraocular muscles in the ipsilateral orbit except the superior oblique and lateral rectus muscles. As a result, the ipsilateral eye assumes an *abducted and depressed position* (down and out), owing to the unopposed action of the lateral rectus and superior oblique muscles. The patient also experiences *diplopia (double vision)* because the image seen by each eye cannot be directed to corresponding portions of each retina, as most of the extraocular muscles in one eye are not functional. Furthermore, interruption of the preganglionic parasympathetic fibers in the oculomotor nerve results in characteristic signs and symptoms in the ipsilateral eye. First, the pupil is dilated (*mydriasis*) and unreactive to light because the sphincter pupillae muscle is denervated (the dilator pupillae muscle, innervated by sympathetic fibers, is intact). Second, the lens in the ipsilateral eye cannot accommodate because the ciliary muscle is also denervated. Third, although the innervation to the superior tarsal muscle is intact because the course of the sympathetic fibers does not involve the oculomotor nerve outside the orbit, the upper eyelid exhibits *ptosis* (droop) because the levator palpebrae has been denervated by the oculomotor nerve lesion.

Because the parasympathetic fibers are located near the periphery (outer surface) of the oculomotor nerve, visceromotor signs and symptoms such as a subtle ptosis or mildly diminished pupil reactivity can appear before the onset of, or in the absence of, any extraocular muscle dysfunction with external compressive injury to the oculomotor nerve. The external compression affects the superficially located, smaller-diameter visceromotor fibers first. In contrast, in diabetic patients, the onset of an eye movement disorder may not be accompanied by visceromotor signs or symptoms. This is due to the fact that diabetes is a vascular problem and the larger vessels inside the nerve are compromised first, thereby affecting the internally located, larger-diameter GSE motor axons but sparing the smaller and more superficially located visceromotor fibers. Isolated lesions of the oculomotor nerve distal to its passage through the superior orbital fissure are relatively rare and produce variable symptoms depending on the location of the lesion.

Sources and Additional Reading

Carpenter MB, Sutin J: Human Neuroanatomy, 8th ed. Williams & Wilkins, Baltimore, 1983.

Crosby EC, Humphrey T, Lauer EW: Correlative Anatomy of the Nervous System. Macmillan Publishing, New York, 1962.

Duvernoy H: The Human Brain Stem and Cerebellum: Surface, Structure, Vascularization, and Three-Dimensional Sectional Anatomy with MRI. Springer-Verlag, New York, 1995.

Haerer AF: DeJong's The Neurologic Examination. Lippincott, Philadelphia, 1993.

Haines DE: Neuroanatomy: An Atlas of Structures, Sections, and Systems, 5th ed. Lippincott Williams & Wilkins, Philadelphia, 2000.

Kretschmann H-J, Weinrich W: Cranial Neuroimaging and Clinical Neuroanatomy: Magnetic Resonance Imaging and Computed Tomography, 2nd ed. Thieme Medical Publishers, New York, 1992.

LeBlanc A: The Cranial Nerves: Anatomy, Imaging, Vascularisation. Springer-Verlag, New York, 1995.

Moore KL, Dalley AF: Clinically Oriented Anatomy, 4th ed. Lippincott Williams & Wilkins, Philadelphia, 1999.

Nieuwenhuys R, Voogd J, van Huijzen CHR: The Human Central Nervous System: A Synopsis and Atlas, 3rd ed. Springer-Verlag, Berlin, 1988.

Parent A: Carpenter's Human Neuroanatomy, 9th ed. Williams & Wilkins, Baltimore, 1996.

Rowland LP: Merritt's Neurology, 10th ed. Lippincott Williams & Wilkins, Philadelphia, 2000.

Victor M, Ropper AH: Adams and Victor's Principles of Neurology, 7th ed. McGraw-Hill, New York, 2001.

Chapter 15

The Diencephalon

G. A. Mihailoff and D. E. Haines

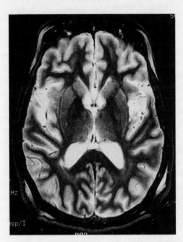

Overview 220

Development of the Diencephalon 221

Basic Organization 221

Dorsal Thalamus (Thalamus) 223
Anterior Thalamic Nuclei
Medial Thalamic Nuclei
Lateral Thalamic Nuclei
Intralaminar Nuclei
Midline Nuclei
Thalamic Reticular Nucleus
Summary of Thalamic Organization

The Internal Capsule 229

Hypothalamus 229
Lateral Hypothalamic Zone
Medial Hypothalamic Zone
Afferent Fiber Systems
Efferent Fibers

Ventral Thalamus (Subthalamus) 230

Epithalamus 232

Vasculature of the Diencephalon 233

Although considered by some investigators to be part of the brainstem, the diencephalon is treated here as a portion of the forebrain. The diencephalon includes the *dorsal thalamus, hypothalamus, ventral thalamus,* and *epithalamus,* and it is situated between the telencephalon and the brainstem. In general, the diencephalon is the main processing center for information destined to reach the cerebral cortex from all ascending sensory pathways (except those related to olfaction) and numerous other subcortical cell groups. The right and left halves of the diencephalon, for the most part, contain symmetrically distributed cell groups separated by the space of the third ventricle.

Overview

The *dorsal thalamus,* or *thalamus* as it is commonly called, is the largest of the four principal subdivisions of the diencephalon and consists of pools of neurons that collectively project to all areas of the cerebral cortex. Some of the thalamic nuclei receive somatosensory, visual, or auditory input and transmit this information to the appropriate area of the cerebral cortex.

Other thalamic nuclei receive input from subcortical motor areas and project to those parts of the overlying cortex that influence the successful execution of a motor act. A few thalamic nuclei receive a more diffuse input and accordingly relate in a more diffuse way to widespread areas of cortex.

The *hypothalamus* is also composed of multiple nuclear subdivisions and is connected primarily to the forebrain, brainstem, and spinal cord. This part of the diencephalon is involved in the control of visceromotor (autonomic) functions. In this respect, the hypothalamus regulates functions that are "automatically" adjusted (such as blood pressure and body temperature) without our being aware of the change. In contrast, conscious sensation and voluntary motor control are mediated by the dorsal thalamus.

The *ventral thalamus* and *epithalamus* are the smallest subdivisions of the diencephalon. The former includes the subthalamic nucleus, which is linked to the basal nuclei of the forebrain and functions in the motor sphere; lesions in the subthalamus give rise to very characteristic involuntary movement disorders. The epithalamus is functionally related to the limbic system.

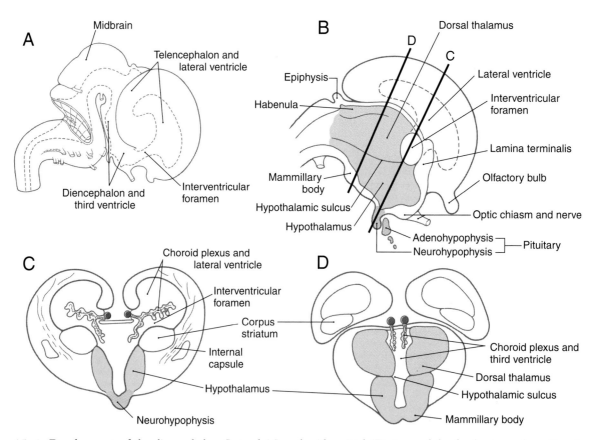

Figure 15–1. Development of the diencephalon. Lateral (*A*) and mid-sagittal (*B*) views of the forebrain at about 8 to 9 weeks of gestational age. The cross-sectional views (*C, D*) are taken from the planes shown in *B* and emphasize diencephalic structures.

Development of the Diencephalon

The cell groups that give rise to the diencephalon form in the caudomedial portion of the prosencephalon, bordering on the space that will become the third ventricle. The developing brain at this level consists initially of a roof plate and the two alar plates; it lacks a well-defined floor plate and basal plates.

A shallow groove appears in the wall of the third ventricle and extends rostrally from the developing cerebral aqueduct to the ventral edge of the interventricular foramen (Fig. 15–1*A* and *B*). This groove, the *hypothalamic sulcus*, divides the alar plate into a posterior (dorsal) area, the future *dorsal thalamus*, and an anterior (ventral) portion, the future *hypothalamus*. The third ventricle, which is the vertically oriented, midline space of the diencephalon, is continuous with the paired lateral ventricles via the *interventricular foramina* (see Fig. 15–1*B*). The dorsal thalamus on each side of the third ventricle increases rapidly in size and, in many brains, will partially fuse across the space of the third ventricle to form the *massa intermedia*, or *interthalamic adhesion* (see Fig. 15–3). This structure is present in about 80% of the general population.

The epithalamus develops from the caudal portion of the roof plate (see Fig. 15–1*B*). By the seventh week, a small thickening of the roof plate forms. It gradually increases in size and evaginates to form the *epiphysis*, which develops into the *pineal gland* of the adult. The portion of the roof plate immediately rostral to the epiphysis gives rise to the *habenula*, a small thickening in which the habenular nuclei will develop.

Just anterior to the habenular region, the roof plate epithelium and adjacent pia mater give rise to the choroid plexus of the third ventricle which, in the adult, remains suspended from the roof of this space (Fig. 15–1*B–D*). This choroid plexus is continuous through the interventricular foramina with that of the lateral ventricles. Elsewhere, in locations around the perimeter of the third ventricle, specialized patches of ependyma lie on the midline and form unpaired structures called the *circumventricular organs*. These structures include the subfornical organ, the organum vasculosum of the lamina terminalis, the subcommissural organ, and the pineal gland (Fig. 15–2). These cellular regions are characterized by the presence of fenestrated capillaries, which implies an absence of the blood-brain barrier. These structures are thought to release metabolites and neuropeptides into the cerebrospinal fluid or into the cerebrovascular system.

The development of the pituitary gland during the third week is linked to that of the diencephalon (see Fig. 15–1*B* and *C*). A downward extension of the floor of the third ventricle, the *infundibulum*, meets *Rathke's pouch*, an upward outpocketing of the stomodeum, the

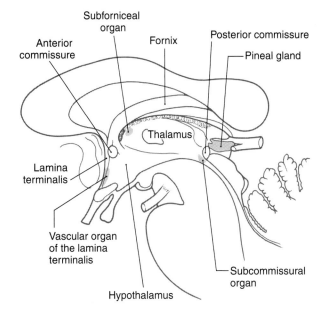

Figure 15–2. Mid-sagittal view showing the locations of circumventricular organs.

primitive oral cavity. By the end of the second month, Rathke's pouch loses its connection with the developing oral cavity but maintains its attachment to the infundibulum. As development continues, Rathke's pouch gives rise to the *anterior lobe* (*adenohypophysis*) and *pars intermedia* of the pituitary gland, whereas the infundibulum differentiates into the *posterior lobe* of the pituitary gland, or *neurohypophysis* (see Fig. 15–1*B*). A *craniopharyngioma* (Rathke's pouch tumor) can arise from a portion of Rathke's pouch that fails to undergo proper migration and apposition to the infundibulum. These tumors mimic lesions of the pituitary and may cause visual problems, diabetes insipidus, and increased intracranial pressure.

Basic Organization

The junction between the diencephalon and midbrain lies along a line extending from the posterior commissure to the caudal edge of the mammillary body on the medial aspect of the hemisphere (Fig. 15–3). On the surface of the hemisphere, this interface is represented by a line following the caudal edge of the optic tract (Fig. 15–4). The boundary between the diencephalon and surrounding telencephalon is less distinct and is represented laterally by the internal capsule (discussed later) and rostrally by the interventricular foramen, lamina terminalis, and optic chiasm (see Fig. 15–3).

The cavity of the diencephalon, the *third ventricle*, is a narrow, vertically oriented midline space located between the paired dorsal thalami and hypothalami of the

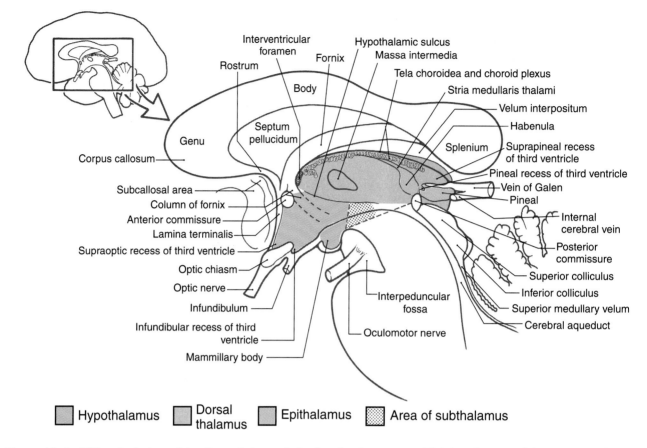

Figure 15–3. Mid-sagittal view of the diencephalon and closely related structures. This is a drawing of the specimen shown in Figure 15–5.

two sides (Figs. 15–5 and 15–6). In addition to its connections with the lateral ventricles and the cerebral aqueduct, the third ventricle has small evaginations or *recesses* associated with the optic chiasm, the infundibulum, and the pineal gland (see Figs. 15–1 and 15–3).

All four diencephalic subdivisions can be approximated in a mid-sagittal section of the forebrain (see Figs. 15–3 and 15–5). The *dorsal thalamus* is located posterior (dorsal) to the hypothalamic sulcus and extends from the interventricular foramen caudally to the

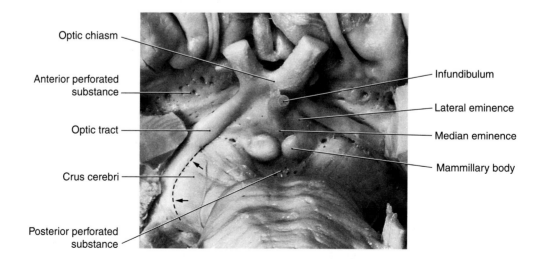

Figure 15–4. Anterior (ventral) view of the hemisphere emphasizing diencephalic structures visible on the surface and showing the diencephalic-mesencephalic interface as represented by the caudal edge of the optic tract (*arrows*).

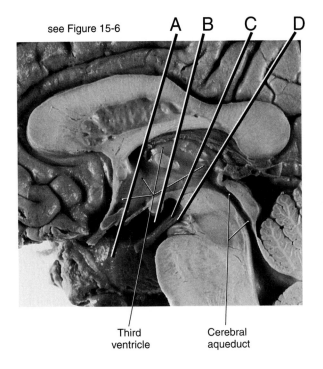

see Figure 15-6

A B C D

Third ventricle

Cerebral aqueduct

Figure 15-5. Mid-sagittal view of the diencephalon. This view correlates with the drawing in Figure 15-3. Lines A through D indicate the planes of the stained sections in Figure 15-6.

level of the splenium of the corpus callosum. The *hypothalamus* lies anterior (ventral) to the hypothalamic sulcus and is bordered rostrally by the lamina terminalis and caudally by a line that extends from the posterior aspect of the mammillary body posteriorly (dorsally) to intersect with the hypothalamic sulcus. The only diencephalic structures visible on the anterior surface of the hemisphere are those related to the hypothalamus, including the optic chiasm, infundibulum, medial and lateral eminences, and mammillary bodies (see Fig. 15-4). The *ventral thalamus* (*subthalamus*) does not border on the ventricle; rather, it occupies a position caudal to the hypothalamus, rostral to the diencephalon-midbrain junction, and *lateral to the midline* (see Figs. 15-3 and 15-6B). Epithalamic structures are located posteriorly and caudally, in close apposition to the posterior commissure, and include the pineal gland, the habenular nuclei, and the main afferent bundle of these nuclei, the stria medullaris thalami.

Dorsal Thalamus (Thalamus)

The *dorsal thalamus* (or *thalamus*) (Figs. 15-7 and 15-8; see also Figs. 15-3 and 15-6) is a massive collection of neuronal cell groups that participate in a widely diverse array of functions involving motor, sensory, and limbic systems. Typically, thalamic output neurons project to the cerebral cortex; in fact, very little information reaches the cerebral cortex without

first being processed by thalamic neurons. As a result, the thalamus is often regarded as the functional "gateway" to the cerebral cortex. In turn, nearly all regions of the cerebral cortex give rise to reciprocal projections that return to the thalamic region from which they originally received input.

The thalamus is covered on its lateral aspect by a layer of myelinated axons, the *external medullary lamina*, which includes fibers that enter or leave the subcortical white matter (see Fig. 15-6B and C). Within the external medullary lamina are clusters of neurons that form the *thalamic reticular nucleus*. The medial surface of the thalamus borders the third ventricle, and the external medullary lamina and thalamic reticular nucleus blend with the thalamic fasciculus and zona incerta, respectively, to form an interface between dorsal and ventral thalami (see Fig. 15-6B).

An *internal medullary lamina*, also consisting of myelinated fibers, extends into the substance of the thalamus, where it forms partitions or boundaries that divide the thalamus into its principal cell groups (Fig. 15-9; see also Fig. 15-8): the *anterior, medial, lateral,* and *intraluminar nuclear groups*. The latter are located in the portion of the internal medullary lamina that separates the lateral and medial nuclear groups. In addition, there are *midline thalamic nuclei* located just superior to the hypothalamic sulcus and a diffuse population of neurons located in the external medullary lamina, the *thalamic reticular nucleus*.

Finally, attached to the caudolateral portion of the thalamus are the *medial* and *lateral geniculate bodies* (and their *nuclei*) (see Figs. 15-6D, 15-7D, and 15-9). Although considered here as components of the lateral nuclear group, the geniculate nuclei are sometimes considered as a separate part of the thalamus, the *metathalamus*.

Anterior Thalamic Nuclei. This group of cells consists of a large principal nucleus and two smaller nuclei; collectively, these cell groups form the *anterior nucleus of the thalamus* (see Figs. 15-6A, 15-7B, 15-8, and 15-9). The anterior nucleus forms a prominent wedge on the rostral aspect of the dorsal thalamus just caudolateral to the interventricular foramen; this wedge is the *anterior thalamic tubercle*. Rostrally, the internal medullary lamina divides to partially encapsulate the anterior nucleus. The cells of this nucleus receive dense limbic-related projections from (1) the mammillary nuclei via the mammillothalamic tract and (2) the medial temporal lobe (hippocampus) via the fornix. The output of this nucleus is primarily directed to the cingulate gyrus through the anterior limb of the internal capsule (Fig. 15-10).

Medial Thalamic Nuclei. This region of the dorsal thalamus comprises the *dorsomedial nucleus* (Figs. 15-7C and 15-8). This expansive group of neuronal cell bodies is composed of large parvicellular (located caudally) and magnocellular (located rostrally) parts and a

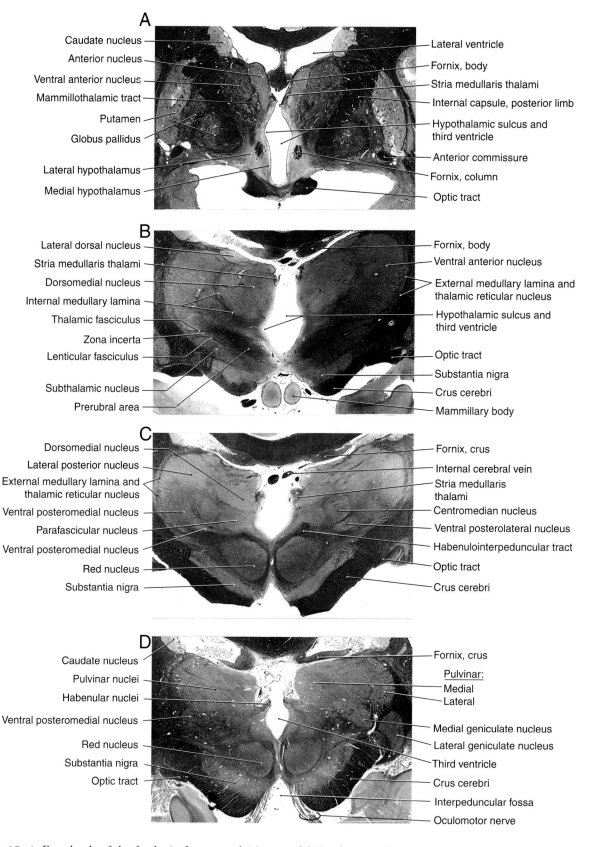

Figure 15–6. Four levels of the forebrain from rostral (*A*) to caudal (*D*) showing the internal structure of the hemisphere with emphasis on the diencephalon. These levels correlate with those shown in Figure 15–5 and with the planes represented in the exploded view in Figure 15–9. Weil stain.

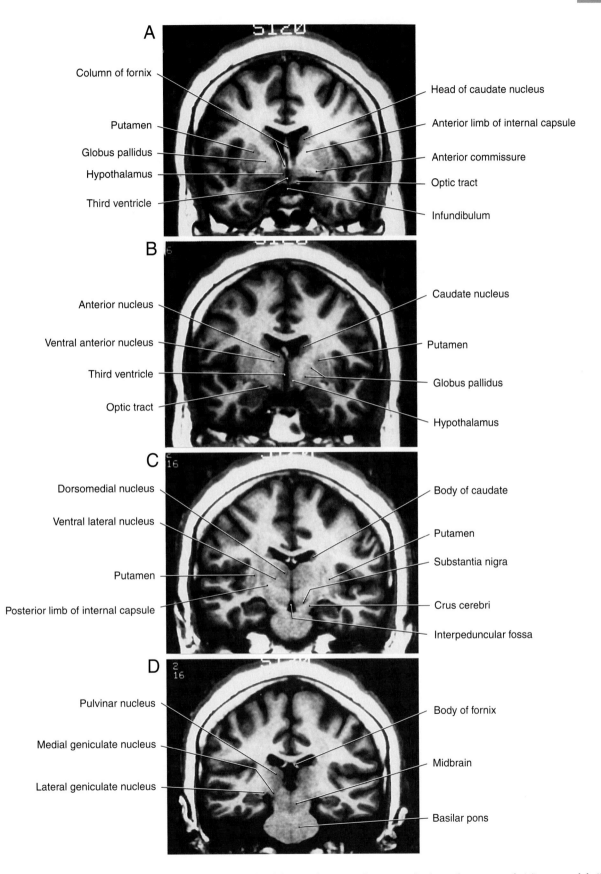

A
- Column of fornix
- Putamen
- Globus pallidus
- Hypothalamus
- Third ventricle
- Head of caudate nucleus
- Anterior limb of internal capsule
- Anterior commissure
- Optic tract
- Infundibulum

B
- Anterior nucleus
- Ventral anterior nucleus
- Third ventricle
- Optic tract
- Caudate nucleus
- Putamen
- Globus pallidus
- Hypothalamus

C
- Dorsomedial nucleus
- Ventral lateral nucleus
- Putamen
- Posterior limb of internal capsule
- Body of caudate
- Putamen
- Substantia nigra
- Crus cerebri
- Interpeduncular fossa

D
- Pulvinar nucleus
- Medial geniculate nucleus
- Lateral geniculate nucleus
- Body of fornix
- Midbrain
- Basilar pons

Figure 15–7. Magnetic resonance images of the cerebral hemisphere in the coronal plane from rostral (*A*) to caudal (*D*). Although many structures are clearly seen, emphasis in the labeling is placed on diencephalic structures. All images are T$_1$-weighted.

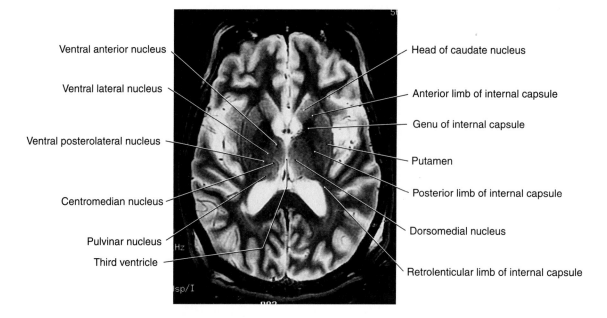

Figure 15–8. T2-weighted magnetic resonance image of the cerebral hemisphere in the axial plane. Emphasis in labeling is on diencephalic structures. Compare with Figure 15–11.

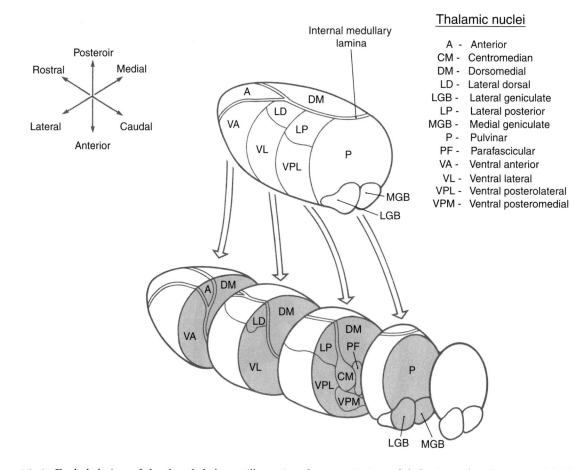

Figure 15–9. Exploded view of the dorsal thalamus illustrating the organization of thalamic nuclei. Compare with Figure 15–6A–D and with Figure 15–7.

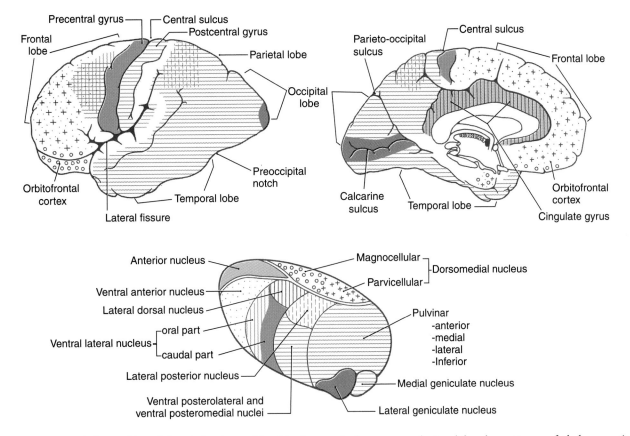

Figure 15–10. Relationship of the thalamic nuclei with the cerebral cortex as depicted by the patterns of thalamocortical connections. Each thalamic nucleus is pattern-coded or color-coded to match its target area in the cerebral cortex.

small paralaminar part adjacent to the internal medullary lamina (see Figs. 15–6B–D, 15–9). The two larger portions are linked to parts of the frontal and temporal lobes and to the amygdaloid complex (see Fig. 15–10). Cells of the paralaminar subdivision receive input from the frontal lobe and substantia nigra and may play a role in the control of eye movement.

Lateral Thalamic Nuclei. This large collection of thalamic neurons is grouped into *dorsal* and *ventral tiers.* The relatively small group of *dorsal tier* nuclei includes the *lateral dorsal* and *lateral posterior nuclei* along with the much larger *pulvinar nucleus* (pulvinar) (see Figs. 15–6B–D, 15–7D, 15–8, and 15–9). The connections of the lateral dorsal and lateral posterior nuclei are formed with the cingulate gyrus and parietal lobe respectively (see Fig. 15–10). The large *pulvinar nucleus* consists of anterior, medial, lateral, and inferior subdivisions. The inferior division receives input from the superior colliculus and projects to visual association cortex. Other portions of the pulvinar project to areas of the temporal, parietal, and frontal lobes that are especially concerned with visual function and eye movements (see Fig. 15–10).

The large *ventral tier* of the lateral group consists of three separate nuclei (see Figs. 15–6A–D, 15–7B,C, and 15–8). The *ventral anterior nucleus* (VA) and the

slightly more caudal *ventral lateral nucleus* (VL) are important motor-related nuclei whereas the *ventral posterior nucleus,* consisting of *ventral posterolateral* (VPL) and *ventral posteromedial* (VPM) *nuclei,* conveys somatosensory information to the cerebral cortex. The VA is composed of a large parvocellular portion and a small magnocellular part. The former receives input from the medial segment of the globus pallidus, and the latter receive afferents from the reticular portion of substantia nigra. The efferent projections from VA are diffuse and appear to include selected parts of the frontal lobe (see Fig. 15–10).

The VL (see Figs. 15–6B and 15–9) is also composed of three subdivisions: a pars oralis, a pars medialis, and a pars caudalis. The largest of these, the pars oralis, receives a dense projection from the internal segment of the globus pallidus; some of these afferents enter the caudal subdivision. In contrast, the medial subdivision receives its main input from the cerebellar nuclei. Consequently, pallidal and cerebellar projections are largely *segregated* within this nucleus. The output of VL reflects its segregated input in that the oral and medial parts project to separate areas of the frontal lobe (see Fig. 15–10).

The larger and more laterally located VPL nucleus and the comparatively smaller and more medially located VPM nucleus both receive somatosensory input from the contralateral side of the body (see Figs. 15–6C and 15–9). The medial lemniscus and spinothalamic fibers terminate in a somatotopic manner (cervical fibers medial, sacral fibers lateral) within VPL, whereas trigeminothalamic fibers from the spinal trigeminal nucleus and the principal trigeminal sensory nucleus terminate in VPM. Both VPL and VPM project to the postcentral gyrus of the parietal lobe (see Fig. 15–10).

A small group of cells called the *ventral posterior inferior nucleus* is situated ventrally between VPL and VPM. These cells process vestibular input and project to lateral areas of the postcentral gyrus that are located in the depths of the central sulcus. Similarly, a small group of cells forming the rostral (oral) portion of VPL receives cerebellar input and projects to the precentral gyrus of the frontal lobe; this nucleus probably represents a few cells that have been displaced from the slightly more rostrally located VL. This cell group is also called the *ventral intermediate nucleus* because of its location between VL and VPL.

The *lateral* and *medial geniculate nuclei* are considered parts of the lateral thalamic nuclear group (see Figs. 15–6D, 15–7D, and 15–9). The medial geniculate nucleus receives ascending *auditory* input via the brachium of the inferior colliculus and projects to the primary auditory cortex in the temporal lobe. The lateral geniculate nucleus receives *visual* input from the retina via the optic tract and in turn projects to the primary visual cortex on the medial surface of the occipital lobe (see Fig. 15–10).

Located in the posterior thalamus at about the level of the pulvinar and geniculate nuclei is a cluster of cell groups collectively called the *posterior nuclear complex.* This complex consists of the suprageniculate nucleus, nucleus limitans, and the posterior nucleus. These nuclei are positioned dorsal to the medial geniculate and medial to the rostral pulvinar. The posterior nuclear complex receives and sends to the cortex nociceptive cutaneous input that is transmitted over somatosensory pathways.

Intralaminar Nuclei. Embedded within the internal medullary lamina are the discontinuous groups of neurons that form the *intralaminar nuclei.* These cells are characterized by their projections to the neostriatum to other thalamic nuclei, along with diffuse projections to the cerebral cortex. Two of the most prominent cell groups are the *centromedian* and *parafascicular nuclei* (see Figs. 15–8 and 15–9). The centromedian nucleus projects to the neostriatum and to motor areas of the cerebral cortex, whereas the parafascicular nucleus projects to rostral and lateral areas of the frontal lobe. Other intralaminar nuclei receive input from ascending pain pathways and project to somatosensory and parietal cortex.

Midline Nuclei. The midline nuclei are the least understood components of the thalamus. The largest is the *paratenial nucleus,* which is located just ventral to the rostral portion of the stria medullaris thalami; other cells are associated with the interthalamic adhesion (massa intermedia). Although inputs are poorly defined, efferent fibers reach the amygdaloid complex and the anterior cingulate cortex, suggesting a role in the regulation of visceral function.

Thalamic Reticular Nucleus. The cells of this nucleus are situated within the external medullary lamina and between this lamina and the internal capsule (see Fig. 15–6B and C). Axons of these cells project medially into the nuclei of the dorsal thalamus or to other parts of the reticular nucleus, but not into the cerebral cortex. Afferents are received from the cortex and from nuclei of the dorsal thalamus via collaterals of thalamocortical and corticothalamic axons. It appears that thalamic reticular neurons modulate, or gate, the responses of thalamic neurons to incoming cerebral cortical input.

Summary of Thalamic Organization. Each thalamic nucleus (with a few exceptions) gives rise to efferent projections (thalamocortical axons) that target some portion of the cerebral cortex. That region of cortex then provides a reciprocal projection (corticothalamic axons) that returns to the original thalamic nucleus. Figure 15–10 reviews the major thalamocortical relationships.

The nuclei of the thalamus have been classified according to their connections as either *relay nuclei* or *association nuclei.* A *relay nucleus* is one that receives input predominantly from a single source such as a sensory pathway or a cerebellar nucleus, or from the basal nuclei. The incoming neural information is processed and then sent to a localized region of either sensory, motor, or limbic cortex. *Relay nuclei* include the medial and lateral geniculate nuclei, VPL, VPM, VL, VA, and the anterior thalamic nuclei. It is important to note that these nuclei do not *merely* relay neural signals; in fact, considerable neural processing also takes place in these nuclei. However, their position in a modality-specific pathway linking one particular source to one particular destination makes the term "relay" a useful designation. In contrast, an *association nucleus* receives input from a number of different structures or cortical regions and usually sends its output to more than one of the *association areas* of the cerebral cortex (i.e., areas that are neither sensory nor motor cortex; see Chapter 32). *Association nuclei* include DM, LD, LP, and the nuclei of the pulvinar complex.

A thalamic nucleus can also be designated *specific* or *nonspecific* on the basis of thalamocortical signals generated in response to electrical stimulation delivered to a localized site in that thalamic nucleus. Focal electrical stimulation of a *specific* nucleus produces a rapidly conducted, sharply localized evoked response in the ipsilateral cerebral cortex. All relay nuclei and association nuclei are *specific nuclei*. Focal electrical stimulation of a *nonspecific* nucleus produces widespread activity in the cortex of *both* hemispheres, at a significantly longer time delay than with stimulation of a specific nucleus. It is thought that *nonspecific* nuclei play a role in modulating the excitability of large regions of cortex. *Nonspecific* nuclei include the midline nuclear group, the intralaminar nuclear group (such as the CM nucleus), and a portion of the VA. Chapter 32 contains additional information on the connections of thalamic nuclei.

The Internal Capsule

Axons pass between the diencephalon, particularly the dorsal thalamus, and the cerebral cortex in a fan-shaped mass of fibers, the *internal capsule*, that courses from the central core of the hemisphere into the brainstem (Fig. 15–11; see also Fig. 15–6A–D). Even though this structure consists mostly of axons that reciprocally link the thalamus and cerebral cortex, it also contains cortical efferent fibers that project to the brainstem (corticorubral, corticoreticular, corticonuclear/corticobulbar) or spinal cord (corticospinal).

Although the internal capsule is described in detail in Chapter 16, it is summarized here because of its important relationship to the thalamus. As seen in axial section (see Fig. 15–11) the internal capsule consists of an *anterior limb*, *genu*, and *posterior limb*. The genu is located immediately lateral to the anterior thalamic nucleus, at about the same level as the interventricular foramen. The anterior limb extends rostrolateral from the genu and is insinuated between the caudate and lenticular nuclei. The posterior limb extends caudolateral from the genu and consists of a large part separating the thalamus from the globus pallidus and smaller portions that link temporal and occipital lobes with thalamic nuclei.

Hypothalamus

Unlike the thalamus, which is primarily related to somatic functions, the hypothalamus (see Figs. 15–7A and B and 15–8) is mainly involved in *visceromotor*, *viscerosensory*, and *endocrine* activities. The hypothalamus and related limbic structures receive sensory input regarding the internal environment and, in turn, regulate through four mechanisms the motor systems that modify the internal environment. *First*, the hypothalamus is a principal modulator of autonomic nervous system function. *Second*, it is a viscerosensory transducer, containing neurons with specialized receptors capable of responding to changes in the temperature or osmolality of blood, as well as to specific hormonal levels in the general circulation. *Third*, the hypothalamus regulates the activity of the anterior pituitary through the production of releasing factors (hormone-releasing hormones), and, *fourth*, it performs an endocrine function by producing and releasing oxytocin and vasopressin into the general circulation within the posterior pituitary.

The hypothalamus can be divided into lateral, medial, and periventricular zones (Fig. 15–12; see also Fig. 15–6A). The *lateral zone*, often called the *lateral hypothalamic area*, extends the full rostrocaudal length of the hypothalamus and is separated from the medial zone by a line drawn through the fornix in the sagittal plane (Fig. 15–13C). The *medial zone* is divided from rostral to caudal into three regions: the chiasmatic, the tuberal, and the mammillary regions (see Figs. 15–12 and 15–13A–D). The *periventricular zone* includes the neurons that border the ependymal surfaces of the third ventricle.

Lateral Hypothalamic Zone. The *lateral hypothalamic area* (see Fig. 15–13A–D) is composed of diffuse clusters of neurons intermingled with longitudinally oriented axon bundles. The latter, which form the *medial*

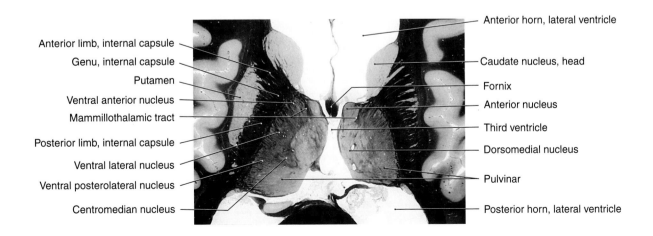

Figure 15–11. Axial view of the forebrain showing the relationship of the thalamus to the limbs of the internal capsule. Weil stain.

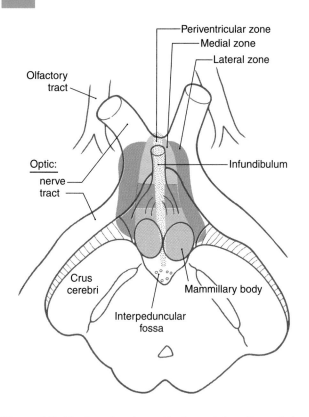

Figure 15–12. Anterior (ventral) view of the diencephalon illustrating the three zones of the hypothalamus as superimposed on external structures. The colors used for medial and lateral zones correlate with those in Figure 15–13.

forebrain bundle, are diffusely organized in the human brain. No discrete named nuclei are present in this lateral area, although the supraoptic nucleus (see later on) is considered by some authorities to be part of it. Cells of the lateral hypothalamic area are involved in cardiovascular function and in the regulation of food and water intake.

Medial Hypothalamic Zone. In contrast to the lateral zone, the medial hypothalamus contains well-defined groups of neurons whose function and connections are established. Within the *chiasmatic* (anterior) *region* are five nuclei: the *preoptic, supraoptic, paraventricular, anterior*, and *suprachiasmatic nuclei* (see Fig. 15–13*A* and *B*). Nuclei in the chiasmatic region are generally involved in regulating hormone release (preoptic, supraoptic, periventricular), cardiovascular function (anterior), circadian rhythms (suprachiasmatic), and body temperature and heat loss mechanisms (preoptic). In the *tuberal region* are the *dorsomedial, ventromedial*, and *arcuate nuclei* (see Fig. 15–13*A* and *C*). The *ventromedial nucleus* is regarded as the food intake (satiety) center. Bilateral lesions of this hypothalamic region produce *hyperphagia*, a greatly increased food intake with resultant obesity. Cells of the *arcuate nucleus* deliver peptides to the portal vessels and, through these channels, to the anterior pituitary. Some of these peptides are *releasing*

factors, which cause an increase in the secretion of specific hormones by the anterior pituitary, and some are *inhibiting factors*, which inhibit the secretion of specific hormones by the anterior pituitary.

At caudal levels, the *mammillary region* is composed of the *posterior nucleus* and the *mammillary nuclei* (see Fig. 15–13*A* and *D*). In humans, the mammillary nuclei consist of a large medial and a small lateral nucleus. Although both of these nuclei receive input via the fornix, only the medial nucleus projects to the anterior nucleus through the mammillothalamic tract. This latter bundle traverses the internal medullary lamina as it enters the anterior nucleus (see Figs. 15–6*A* and 15–11). The neurons of the posterior nucleus are involved in activities that include elevation of blood pressure, pupillary dilation, and shivering or body heat conservation. The mammillary nuclei are involved in the control of various reflexes associated with feeding, as well as mechanisms relating to memory formation.

Afferent Fiber Systems. Although many axonal systems extend into the hypothalamus, only four inputs are mentioned here; the entire group is discussed in Chapter 30. The fornix and the stria terminalis are two major afferent fiber bundles that reach the hypothalamus (see Fig. 15–6*A–D*). The *fornix* consists of axons that largely originate in the hippocampus, and the *stria terminalis* arises from neurons in the amygdaloid complex (see Fig. 16–16). Fibers composing the *ventral amygdalofugal bundle* exit the amygdala and course through the substantia innominata to enter the hypothalamus and thalamus (see Fig. 16–16). As mentioned earlier, the *medial forebrain bundle* passes bidirectionally through the lateral hypothalamic region. This composite fiber bundle consists of ascending axons that originate in areas throughout the neuraxis and terminate in the hypothalamus and other axons that exit the hypothalamus to reach forebrain and brainstem targets.

Efferent Fibers. The hypothalamus is the source of a diverse array of efferent fibers (see Chapter 30). Several nuclei give rise to descending fibers that contribute to the dorsal longitudinal fasciculus and the medial forebrain bundle, and to diffuse projections that pass into the tegmentum. These fiber systems project directly to numerous brainstem nuclei, as well as to preganglionic sympathetic and parasympathetic neurons in the spinal cord. Other projections reach the thalamus and frontal cortex, and still others extend to the posterior pituitary or to the tuberohypophysial portal system for delivery of substances to the anterior pituitary.

Ventral Thalamus (Subthalamus)

The *ventral thalamus* (also called *subthalamus*) includes the large *subthalamic nucleus*, the medially adjacent *prerubral area* (*field H of Forel*), and, posteriorly, the *zona incerta* (see Figs. 15–3 and 15–6*B*). As the term *ventral*

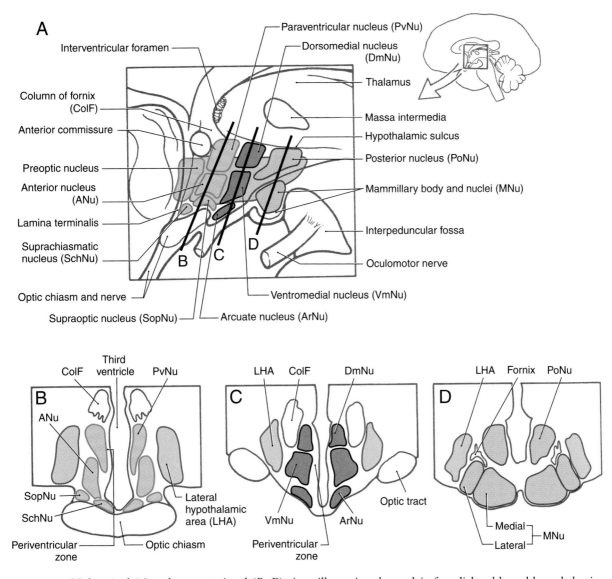

Figure 15–13. Mid-sagittal (*A*) and cross-sectional (*B–D*) views illustrating the nuclei of medial and lateral hypothalamic zones and the nuclei associated with chiasmatic (*B*), tuberal (*C*), and mammillary (*D*) regions. The colors used here correlate with those in Figure 15–12. (*A* adapted from Haymaker W, Anderson E, Nauta WJH: The Hypothalamus. Charles C Thomas, Springfield, Ill., 1969, with permission.)

thalamus (or subthalamus) implies, these cell groups are located ventral (anterior) to the large expanse of the dorsal thalamus. The *subthalamic nucleus* is a lens-shaped cell group situated rostral and posterior to the substantia nigra and immediately anterior to a distinct myelinated fiber bundle, the *lenticular fasciculus* (see Fig. 15–6B). The cells of the subthalamic nucleus receive input from motor areas of the cerebral cortex, project to the substantia nigra, and are reciprocally connected with the globus pallidus. The subthalamic nucleus can be affected by vascular lesions involving posteromedial branches of the posterior cerebral or posterior communicating arteries, which results in a characteristic clinical condition known as *hemiballismus*. Patients with this involuntary movement disorder ex-

hibit rapid and forceful flailing movements, which usually involve the contralateral upper extremity. These movements can be very debilitating because the patient has no control over their initiation or duration.

The *zona incerta* is located posterior to the subthalamic nucleus and is separated from it by the *lenticular fasciculus* (see Fig. 15–6B). Posterior to the zona incerta are the myelinated axons of the thalamic fasciculus. The zona incerta contains output neurons that project to a variety of locations, including the cerebral cortex, the superior colliculus, the pretectal region, and the basilar pons. Afferent projections arise from the motor cortex and as collaterals from the medial lemniscus.

The *prerubral area* (*Forel's field H*) is located just rostral to the red nucleus and medial to the subthal-

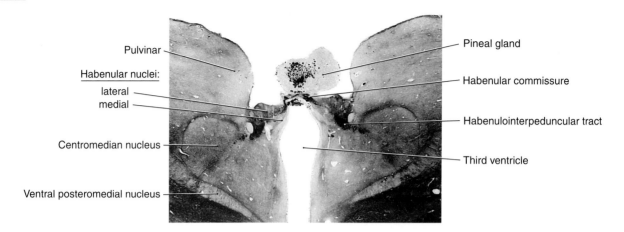

Figure 15–14. Caudal level of the diencephalon showing the habenular nuclei, pineal gland, and related structures. The stria medullaris thalami has disappeared at this level because its fibers have dispersed to end in the habenular nuclei.

amic nucleus (see Fig. 15–6B). There are scattered neurons in this region, and traversing the prerubral area are fibers from the *lenticular fasciculus* (Forel's field H_2) that enter the *thalamic fasciculus* (Forel's field H_1).

Epithalamus

The *pineal gland, habenular nuclei,* and *stria medullaris thalami* are the principal components of the epithalamus (Fig. 15–14; see also Fig. 15–3). The pineal gland consists of richly vascularized connective tissue containing glial cells and pinealocytes but no true neurons. Mammalian pinealocytes are related to the photoreceptor elements found in this gland in lower forms, such as amphibians. In humans, however, they remain only indirectly light sensitive and receive information concerning photic stimuli through a multisynaptic neural circuit.

Pinealocytes have clublike processes that are apposed to blood vessels but do not have direct synaptic contacts with central nervous system neurons. These cells synthesize melatonin from serotonin via enzymes that are sensitive to diurnal fluctuations in light. Levels of serotonin *N*-acetyltransferase increase during the night (in the absence of photic stimulation), and the synthesis of melatonin is enhanced. Exposure to light turns off the enzymatic activity, and melatonin production is diminished. Thus, the production of melatonin by pinealocytes is rhythmic and calibrated to the 24-hour cycle of photic input to the retina. This is called a *circadian rhythm*.

Photic stimulation of pinealocytes occurs through an indirect route. Retinal ganglion cells project to the suprachiasmatic nucleus of the hypothalamus, which in turn influences neurons of the intermediolateral cell column in the spinal cord through descending connec-

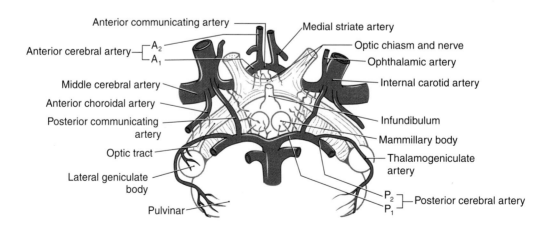

Figure 15–15. Anterior (ventral) view of the diencephalic region showing the arterial circle of Willis and the distribution of central branches to hypothalamic structures.

tions. These preganglionic sympathetic neurons project to the superior cervical ganglion, which in turn innervates the pineal gland via postganglionic fibers that travel on branches of the internal carotid artery.

Pinealocytes also produce serotonin, norepinephrine, and neuroactive peptides, such as thyrotropin-releasing hormone (TRH), which are normally associated with the hypothalamus. These secretory products are released into the general circulation or the cerebrospinal fluid.

Pinealomas (tumors with large numbers of pinealocytes) are accompanied by depression of gonadal function and delayed puberty, whereas lesions that lead to the *loss* of pineal cells are associated with precocious puberty. This indicates that pineal secretory products exert an inhibitory influence on gonadal formation.

The habenular nuclei are located just anterior to the pineal gland and consist of a large lateral nucleus and a small medial nucleus (see Fig. 15–14). Both nuclei contribute axons to the *habenulo-interpeduncular tract* (fasciculus retroflexus), which terminates in the midbrain interpeduncular nucleus. The *stria medullaris thalami*, which arches over the medial aspect of the dorsal thalamus near the midline, conveys input to both habenular nuclei. The *habenular commissure*, a small bundle of fibers riding on the upper edge of the posterior commissure, connects the habenular regions of the two sides.

Vasculature of the Diencephalon

The diencephalon is supplied by smaller vessels that branch from the various arteries making up the cerebral arterial circle (circle of Willis) and by larger arteries that originate from the proximal parts of the posterior cerebral artery (Figs. 15–15 and 15–16A and B). The hypothalamus and subthalamus are supplied by *central (performing or ganglionic) branches* of the circle. Anterior parts of the hypothalamus are served by central branches (*anteromedial group*) arising from the anterior communicating artery and the A_1 segment of the anterior cerebral artery and from branches of the proximal part of the posterior communicating artery. Caudal hypothalamic regions and the ventral thalamus are supplied by branches of the *posteromedial group;* these branches arise from the posterior communicating artery and the P_1 segment of the posterior cerebral artery.

Arising from the P_1 segment near the basilar bifurcation are the *thalamoperforating arteries*. These large vessels (of which there may be more than one on each side) penetrate deeply to supply rostral areas of the thalamus (see Figs. 15–15 and 15–16A and B). If these vessels are occluded during surgery in this region, as can occur, for example, when an aneurysm of the basilar bifurcation is clipped, the patient can be rendered permanently comatose. Slightly more distal branches, which usually arise from the P_2 segment, are the *poste-*

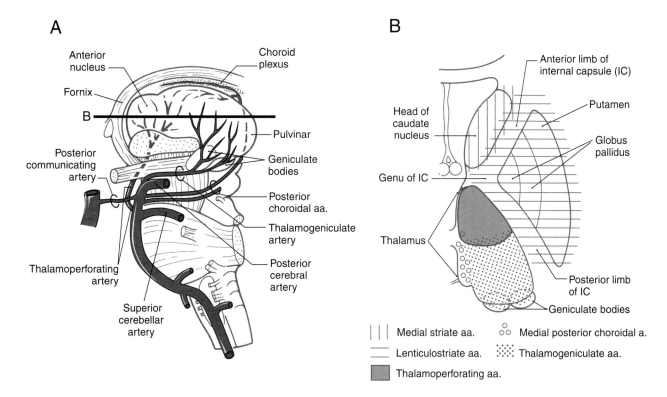

Figure 15–16. Blood supply to the dorsal thalamus. The lateral aspect (*A*) shows the general distribution of the main arteries. A view in the axial plane (*B*) through the hemisphere shows the internal territories served. The distribution of the main striate arteries, especially to the internal capsule, is also shown.

rior choroidal and *thalamogeniculate* arteries. These arteries also supply portions of the diencephalon (see Figs. 15–15 and 15–16*A* and *B*). A narrow portion of the caudal and medial thalamus bordering on the third ventricle is supplied by the *medial posterior choroidal artery*, whereas the thalamogeniculate branches irrigate the caudal thalamus, including the pulvinar and the geniculate nuclei (see Figs. 15–15 and 15–16*B*). In addition, branches of the medial posterior choroidal artery also serve the choroid plexus of the third ventricle.

Although the thalamus receives a blood supply largely separate from that of the internal capsule (see Fig. 15–16*B*), vascular lesions in the thalamus may extend into the internal capsule or vice versa. Ischemic or hemorrhagic strokes in the hemisphere may result in *contralateral hemiparesis* in combination with *hemianesthesia*. These losses correlate with damage to corticospinal and thalamocortical fibers in the internal capsule. On the other hand, strokes involving the larger thalamic arteries may result in total or dissociated sensory losses. These patients may subsequently experience persistent, intense pain (*thalamic pain, Dejerine-Roussy syndrome*) as the lesion in the thalamus resolves with time.

Sources and Additional Reading

Alexander GE, Crutcher MD, DeLong MR: Basal ganglia-thalamocortical circuits: Parallel substrates for motor, oculomotor, "prefrontal" and "limbic" functions. In Uylings HBM, Van Eden CG, DeBruin JPC, Corner MA, Feenstra MGP (eds): Progress in Brain Research, vol 85: The Prefrontal Cortex: Its Structure, Function, and Pathology. Elsevier Biomedical Press, Amsterdam, 1990.

Burt AM: Textbook of Neuroanatomy. WB Saunders, Philadelphia, 1993.

Hirai T, Jones EG: A new parcellation of the human thalamus on the basis of histochemical staining. Brain Res Rev 14:1–34, 1989.

Jones EG: The Thalamus. Plenum Press, New York, 1986.

Nieuwenhuys R, Voogd J, van Huijzen C: The Central Nervous System, A Synopsis and Atlas, 3rd ed. Springer-Verlag, Berlin, 1988.

Parent A: Carpenter's Human Neuroanatomy, 9th ed. Williams & Wilkins, Baltimore, 1996.

Schell GR, Strick PL: The origin of thalamic inputs to the arcuate premotor and supplementary motor areas. J Neurosci 4:539–560, 1984.

Ungerleider LG, Galkin TW, Mishkin M: Visuotopic organization of projections from striate cortex to inferior and lateral pulvinar in rhesus monkey. J Comp Neurol 217:137–157, 1983.

Walker AE: The Primate Thalamus. University of Chicago Press, Chicago, 1938.

Chapter 16

The Telencephalon

D. E. Haines and G. A. Mihailoff

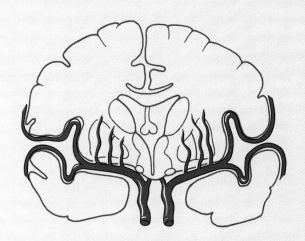

Overview 236

Development 237

Lobes of the Cerebral Cortex 238
Frontal Lobe
Parietal Lobe
Temporal Lobe
Insular Lobe
Occipital Lobe
Limbic Lobe
Vasculature of the Cerebral Cortex

White Matter of the Cerebral Hemisphere 242
Association Fibers
Commissural Fibers: The Corpus Callosum
Projection Fibers: The Internal Capsule
Vasculature of the Internal Capsule

Basal Nuclei (Ganglia) 245
Caudate and Lenticular Nuclei
Nucleus Accumbens and Substantia Innominata
Subthalamic Nucleus and Substantia Nigra
Major Connections of the Basal Nuclei
Vasculature of the Basal Nuclei and Related Structures

Hippocampus and Amygdala 250
Temporal Lobe Lesions
Vasculature of the Hippocampus and Amygdala

The telencephalon is the largest part of the human brain, constituting about 85% of total brain weight, and is that portion in which all modalities are represented. Various sensory inputs (e.g., vision, hearing) are localized in some areas, whereas motor functions are represented in other regions and are modulated by subcortical nuclei. The telencephalon contains circuits that interrelate regions that have specific functions, such as motor or visual, with other regions called *association areas*. Seeing a familiar image may precipitate a cascade of neural events having olfactory, emotional, sensory, and motor components. Damage to association areas results in complex neurologic deficits. The patient may not be blind or paralyzed but may be unable to recognize sensory input (*agnosia*), express ideas or thoughts (*aphasia*), or perform complex goal-directed movements (*apraxia*).

Overview

The telencephalon consists of two large *hemispheres* separated from each other by a deep *longitudinal cerebral fissure*. Each hemisphere has an outer surface, the *cerebral cortex*, which is composed of layers of cells.

The cortex is thrown into elevations called *gyri* (singular, *gyrus*) that are separated by grooves called *sulci* (singular, *sulcus*). Internal to the cortex are large amounts of *subcortical white matter*, along with aggregates of gray matter that form the *basal nuclei* (*ganglia*) and the *amygdala*. Although not parts of either the telencephalon or the basal nuclei, the *subthalamic nucleus* (of the diencephalon) and the *substantia nigra* (of the mesencephalon) have important connections that functionally link them with the basal nuclei.

Information passing into or out of the cerebral cortex must traverse the subcortical white matter. The myelinated fibers forming the white matter are organized into (1) association bundles that connect adjacent or distant gyri in one hemisphere; (2) commissural bundles that connect the two hemispheres, the largest of these being the *corpus callosum*; and (3) the *internal capsule*. The internal capsule contains axons projecting to numerous downstream nuclei (*corticofugal fibers*) and axons conveying information to the cerebral cortex (*corticopetal fibers*). The terms corticofugal and corticopetal are umbrella terms that include all efferent and all afferent fibers, respectively, of the cerebral cortex. In later chapters, specific populations of cortical

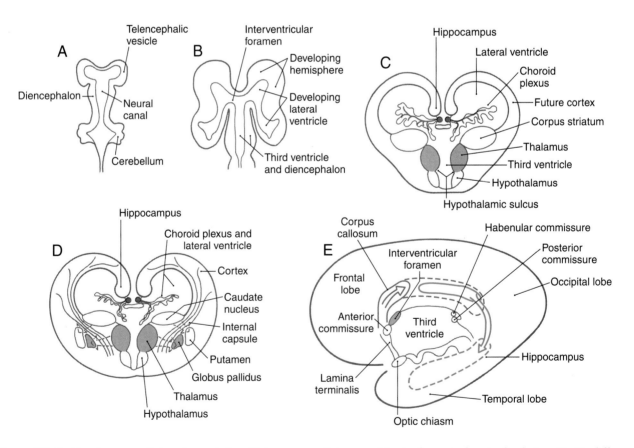

Figure 16–1. Development of the telencephalon. Enlargement of the ventricles is shown in longitudinal view (*A, B*), differentiation of the basal nuclei and internal capsule in cross section (*C, D*), and growth of the corpus callosum (in red) and hippocampus (in green) in the sagittal plane (*E*).

efferent fibers (such as *corticospinal* or *corticostriatal fibers*) and of cortical afferent fibers (such as *thalamocortical fibers*) are discussed.

The *hippocampal complex* and the *amygdala* are located in the walls of the temporal horn of the lateral ventricle. The axons of cells in these structures coalesce to form the *fornix* and *stria terminalis*, respectively.

Development

Enlargements of the prosencephalon, the *telencephalic (cerebral) vesicles*, appear at about 5 weeks of gestation. As the cerebral vesicles enlarge in all directions, they pull along portions of the neural canal that will form the cavities of the telencephalon, the *lateral ventricles* (Fig. 16–1*A* and *B*). The primitive lateral ventricles extend into frontal, parietal, temporal, and occipital areas as they develop and form that portion of the ventricle found in each of these lobes in the adult. The *interventricular foramina*, which connect each lateral ventricle to the midline *third ventricle* (cavity of the diencephalon), are initially large but become smaller as development progresses. In the adult brain, each interventricular foramen is bordered rostromedially by the column of the fornix and caudolaterally by the anterior tubercle of the dorsal thalamus (see Fig. 15–11).

Cells forming the *corpus striatum* appear in the floor of the developing lateral ventricle at the time when primordial cell groups in the wall of the third ventricle are giving rise to diencephalic structures (see Fig. 16–1*C* and *D*). As development progresses, the corpus striatum is bisected by axons growing to and from the cerebral cortex. These axons form the *internal capsule* of the adult and divide the corpus striatum into a medially located *caudate nucleus* and a laterally located *putamen*. As the diencephalon enlarges, it gives rise to the thalamus and hypothalamus *and to cells that migrate across the developing internal capsule to assume a position medial to the putamen* (see Fig. 16–1*D*). These cells become the *globus pallidus* of the adult and, in combination with the putamen, form the *lenticular nucleus*.

The initial development of the major commissural bundles and of the hippocampus takes place along the medial aspect of the hemisphere (see Fig. 16–1*C–E*). In the adult brain there are three major interhemispheric commissures: the *anterior commissure*, the *hippocampal commissure*, and the *corpus callosum*. The first of these to appear, the *anterior commissure*, arises within the lamina terminalis. The latter is a membrane-like structure that extends from the anterior commissure anteriorly (ventrally) to the rostral edge of the optic chiasm (see Fig. 15–3). The second to form, the *hippocampal commissure*, develops along with the hippocampal primordium. As growth occurs, the hippocampus, which originates in the posteromedial part of the hemi-

sphere, is displaced into the temporal lobe where it assumes a position characteristic of the adult (see Fig. 16–1*E*). In the process, fibers from one side cross to the other side, as the *hippocampal commissure*, just anterior (ventral) to the area that will be occupied by the corpus callosum. The third commissure to develop, the *corpus callosum*, originates from the area of the lamina terminalis as a structure initially composed of astrocytic processes. Axons from developing neurons in each hemisphere traverse this glial structure to access the contralateral side. As this takes place, the corpus callosum enlarges in a caudal direction to form the prominent structure found in the adult (Fig. 16–2*A*; see also Fig. 16–1*E*).

Failure of the corpus callosum to develop (*agenesis of the corpus callosum*) may be accompanied by an absence of the anterior and hippocampal commissures (see Fig. 16–2*A* and *B*). Although some patients with this condition may experience focal seizures and have mental retardation; others live for many years with few or no obvious neurologic deficits. These individuals frequently have developmental abnormalities in other parts of the nervous system.

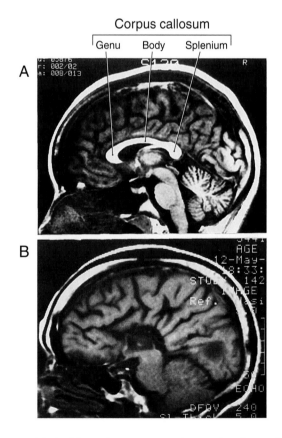

Figure 16–2. Magnetic resonance image in the sagittal plane of a normal adult *(A)* and of another adult with agenesis of the corpus callosum *(B)*. A comparison of *B* with *A* reveals an absence of the corpus callosum, aberrant gyri on the medial surface of the hemisphere, and other structural defects.

Lobes of the Cerebral Cortex

On the basis of the arrangement of major sulci, the cerebral cortex is divided into six lobes, five of which are exposed on the surface of the cerebral hemisphere and one (the insular lobe) located internal to the lateral sulcus. Four of these lobes are named according to the overlying bones of the skull.

On the lateral surface of the hemisphere the major sulci are the *central sulcus*, the *lateral (sylvian) sulcus*, and a small lateral end of the *parieto-occipital sulcus* (Fig. 16–3; see also Fig. 16–5). The *preoccipital notch* is the high point along the ventrolateral margin of the hemisphere. An imaginary line connecting the terminus of the parieto-occipital sulcus with the preoccipital notch intersects with another line drawn caudally from the lateral sulcus. These lines, along with the central and lateral sulci, divide the lateral surface into *frontal, parietal, temporal,* and *occipital lobes* (Fig. 16–3).

On the medial surface of the hemisphere, the major sulci separating lobes are the *cingulate, parieto-occipital,* and *collateral* (Fig. 16–4; see also Fig. 16–5). Two imaginary lines also separate lobes on the medial surface. One connects the medial end of the central sulcus with the cingulate sulcus; the other joins the parieto-occipital sulcus with the preoccipital notch. This combination of sulci and lines separate the four lobes noted

previously, plus the *limbic lobe*, on the medial surface of the hemisphere (see Fig. 16–4).

Deep to the lateral (sylvian) sulcus is an infolded region of cortex called the *insula*. This region is separated from the opercula of the frontal, parietal, and temporal lobes by the *circular sulcus of the insula*. Consequently, this region satisfies the definition of a subdivision of the cerebral cortex in that it is separated from the adjoining cortical structures by a named sulcus.

Frontal Lobe. The lateral surface of the frontal lobe is divided by *inferior* and *superior frontal sulci* into *inferior, middle,* and *superior frontal gyri* (see Fig. 16–3), the latter folding onto the medial aspect of the hemisphere (Fig. 16–5B; see also Fig. 16–4). The inferior frontal gyrus is divided into a *pars opercularis, pars triangularis,* and *pars orbitalis* (see Fig. 16–5A). The anterior (ventral) surface of the frontal lobe is composed of the *gyrus rectus,* the *olfactory sulcus,* and a series of *orbital gyri* (Fig. 16–6). The rostral-most point of this lobe is the *frontal pole* of the brain.

The *olfactory bulb* and *tract,* which relay sensory information, lie on the anterior surface of the frontal lobe in the olfactory sulcus (see Fig. 16–6). At the point where the olfactory tract attaches to the hemisphere, it bifurcates into *medial* and *lateral striae* (Fig. 16–7). The triangle formed by this bifurcation is called the *olfactory trigone.* Immediately caudal to this

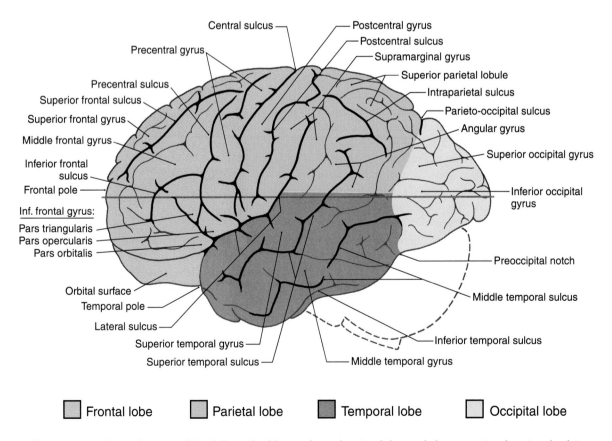

Figure 16–3. Lateral aspect of the left cerebral hemisphere showing lobes and their associated gyri and sulci.

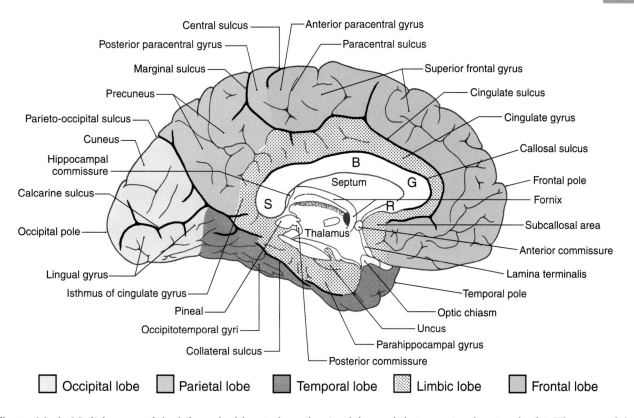

Figure 16–4. Medial aspect of the left cerebral hemisphere showing lobes and their associated gyri and sulci. The parts of the corpus callosum are R, rostrum; G, genu; B, body (or trunk); and S, splenium.

trigone, the surface of the hemisphere is characterized by numerous small holes formed by vessels (lenticulostriate arteries) as they enter the brain; this is the *anterior perforated substance* (see Fig. 16–7). The olfactory tract and striae and the cell groups associated with the anterior perforated substance are functionally related to the limbic system.

The *precentral gyrus* is continuous on the medial surface of the hemisphere with the *anterior paracentral gyrus;* the latter is separated from the superior frontal gyrus by the *paracentral sulcus* (see Fig. 16–4). These two gyri collectively form the *primary motor cortex.*

Parietal Lobe. The parietal lobe consists of the *postcentral gyrus,* located between the *central* and *postcentral sulci,* and the *superior* and *inferior parietal lobules,* which are separated by the *intraparietal sulcus* (see Fig. 16–3). Gyri forming the superior parietal lobule extend onto the medial surface of the hemisphere as the *precuneus,* whereas the inferior parietal lobule is made up of the *angular* and *supramarginal gyri.* The latter is a crescent-shaped ridge of cortex around the caudal terminus of the lateral sulcus.

As the *postcentral gyrus* extends onto the medial surface of the hemisphere, it is continuous with the *posterior paracentral gyrus* (see Figs. 16–4 and 16–5). This cortical area is bordered rostrally by an imaginary line that connects the central sulcus to the cingulate sulcus and caudally by the *marginal ramus of the cingulate sulcus.* The latter is frequently called the *marginal sulcus.* Taken together, the postcentral gyrus and posterior paracentral lobule constitute the *primary somatosensory cortex.*

Temporal Lobe. The gyri that form the temporal lobe are found on lateral and anterior (ventral) aspects of the hemisphere between the *lateral sulcus* and the *collateral sulcus* (see Figs. 16–3 to 16–6). These gyri are, beginning at the lateral sulcus, the *superior, middle,* and *inferior temporal gyri* and a broad area of cortex, the *occipitotemporal gyri,* extending from the *temporal pole* to the occipital lobe. The *superior temporal sulcus* ends in the loop of cortex forming the angular gyrus of the inferior parietal lobule. An *inferior temporal sulcus* may be found between the inferior temporal and occipitotemporal gyri, or it may be absent, in which case these gyri blend around the inferior margin of the hemisphere.

On the upper edge of the temporal lobe, and extending into the depths of the lateral fissure, are the *transverse temporal gyri* (of Heschl). These gyri (Fig. 16–8) form the *primary auditory cortex.*

Insular Lobe. The oval region of cortex located deep inside the lateral fissure is the *insular lobe* (Figs. 16–5C and 16–8). This area is characterized by a set of long gyri in its caudal part (the *gyri longi*) and a set of short gyri in its rostral part (the *gyri breves*). These are separated from each other by the *central sulcus of the insula.* The insular cortex is continuous, at the circular sulcus

A

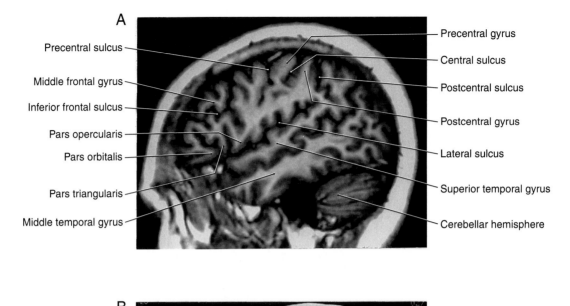

Precentral sulcus —
Middle frontal gyrus —
Inferior frontal sulcus —
Pars opercularis —
Pars orbitalis —
Pars triangularis —
Middle temporal gyrus —

— Precentral gyrus
— Central sulcus
— Postcentral sulcus
— Postcentral gyrus
— Lateral sulcus
— Superior temporal gyrus
— Cerebellar hemisphere

B

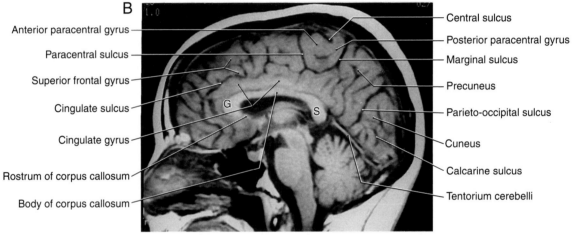

Anterior paracentral gyrus —
Paracentral sulcus —
Superior frontal gyrus —
Cingulate sulcus —
Cingulate gyrus —
Rostrum of corpus callosum —
Body of corpus callosum —

— Central sulcus
— Posterior paracentral gyrus
— Marginal sulcus
— Precuneus
— Parieto-occipital sulcus
— Cuneus
— Calcarine sulcus
— Tentorium cerebelli

C

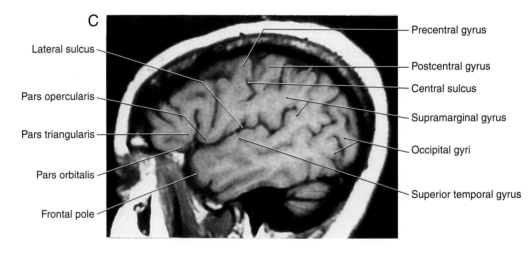

Lateral sulcus —
Pars opercularis —
Pars triangularis —
Pars orbitalis —
Frontal pole —

— Precentral gyrus
— Postcentral gyrus
— Central sulcus
— Supramarginal gyrus
— Occipital gyri
— Superior temporal gyrus

Figure 16–5. T_1-weighted magnetic resonance images in the sagittal plane at about the midline *(B)*, at the lateral-most aspect of the cerebral hemisphere *(C)*, and through approximately the lateral two thirds of the hemisphere *(A)*. Gyri and sulci of the cerebral cortex are labeled. G, genu of corpus callosum; S, splenium of corpus callosum.

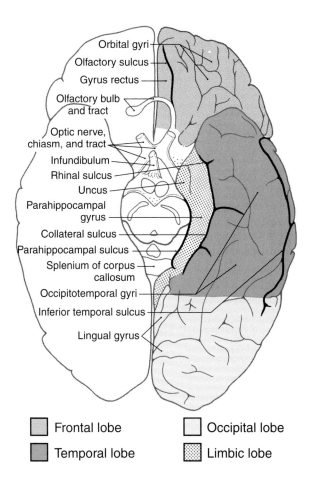

Figure 16–6. Anterior (ventral) aspect of the left cerebral hemisphere showing lobes and their associated gyri and sulci.

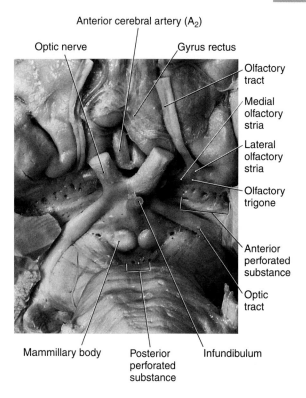

Figure 16–7. Anterior (ventral) aspect of the hemisphere in the area of the anterior perforated substance.

of the insula, with that of the adjacent frontal, parietal, and temporal lobes. This continuity forms lips on each lobe that overlie the insula to form the *frontal, parietal,* and *temporal opercula* (see Fig. 16–8; see also Fig. 16–10). The *limen insulae* (threshold to the insula) is the area in which the anterior (ventral) surface of the hemisphere is continuous with the insular cortex (see Fig. 16–8). Although the function of the insula is still somewhat unclear, it is known that the insular cortex receives *nociceptive* and *viscerosensory input.*

Occipital Lobe. The occipital lobe forms the caudal end of the hemisphere; its caudal extreme is the *occipital pole* of the brain (see Figs. 16–3 and 16–4). The irregular collection of gyri on the lateral surface of the occipital lobe constitutes the *occipital gyri.* An important landmark on the medial aspect of the occipital lobe is the *calcarine sulcus.* This sulcus separates the cuneus, which is posterior (dorsal) to it, from the *lingual gyrus,* which is anteriorly (ventrally) located. The *primary visual cortex* is located in the portions of these gyri that border directly on the calcarine sulcus.

Limbic Lobe. The *limbic lobe* is part of a more complex entity commonly called the *limbic system.* As discussed in Chapter 31, the limbic system includes this

lobe plus its afferent and efferent connections with other telencephalic, diencephalic, and brainstem nuclei.

The limbic lobe is a ring of cortex that makes up the medial-most rim of the hemisphere. Although not specified as a separate lobe in some texts, it is designated here as such because of its unique functional

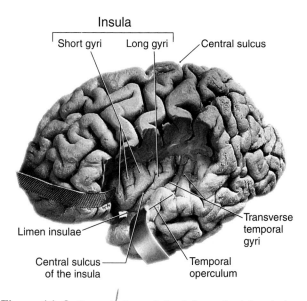

Figure 16–8. Lateral view of the left cerebral hemisphere. The frontal and parietal opercula are removed and the temporal operculum is retracted to expose the insula and transverse temporal gyri.

characteristics. Beginning anteriorly (and just ventral to the rostrum of the corpus callosum) this ring of cortex consists of the *subcallosal area, cingulate gyrus, isthmus of the cingulate gyrus, parahippocampal gyrus*, and *uncus* (see Figs. 16–4 to 16–6). The limbic cortex is separated from adjacent cortical areas by the *cingulate sulcus* and the *collateral sulcus* and from the *corpus callosum* by the *callosal sulcus*. On the anterior (ventral) surface of the temporal lobe, the *rhinal sulcus* is located between the rostral extreme of the parahippocampal gyrus and the laterally adjacent occipitotemporal gyri (see Fig. 16–6).

The function of the limbic lobe is complex and cannot be designated as, for example, primarily sensory or motor. Rather, it is linked to circuits that influence complex functions such as memory, learning, and behavior.

Vasculature of the Cerebral Cortex. The lobes composing the cerebral cortex receive their blood supply via terminal branches of the *anterior, middle*, and *posterior cerebral arteries*. Details of individual branches are given in Chapter 8; only general points are reviewed here.

The *anterior cerebral artery*, the smaller of the terminal branches of the internal carotid artery, passes rostromedially and is joined to its counterpart by the *anterior communicating artery*. The anterior cerebral artery arches posteriorly (dorsally) and caudally to serve the medial surface of the hemisphere to about the level of the parieto-occipital sulcus (Fig. 16–9). The part of the anterior cerebral artery between the internal carotid and the anterior communicator is called the A_1 segment. Branches of A_1 serve structures in the immediate vicinity, including the optic chiasm and anterior parts of the hypothalamus. The branches of the anterior cerebral distal to the anterior communicator collectively form the A_2 segment. These serve the medial surface of the frontal and parietal lobes, including the *anterior* (lower extremity part of motor cortex) and *posterior* (lower extremity part of somatosensory cortex) *paracentral gyri*.

The *middle cerebral artery*, the larger of the terminal branches of the internal carotid artery, passes laterally to the *limen insula*, where it generally branches into *superior* and *inferior trunks* (Fig. 16–10; see also Fig. 16–9). This initial part of the middle cerebral, the M_1 segment, gives rise to the *lenticulostriate (lateral striate) arteries*. The superior and inferior trunks and their distal branches collectively form the M_2, M_3, and M_4 segments. The insular lobe itself is served by branches of M_2. As the distal branches of the superior and inferior trunks pass onto the inner surface of the opercula they become the M_3 segment, and upon exiting the lateral sulcus they become the M_4 segment (see Fig. 16–10*A*; see also Chapter 8). Important cortical regions served by the M_4 segment include the trunk, upper extremity, and head areas of the motor cortex (*precentral gyrus*), somatosensory cortex (*postcentral gyrus*), auditory cortex (*transverse temporal gyri*), parietal

association cortex (*parietal lobules*), and large parts of the posterolateral (dorsolateral) surface of the frontal lobe.

The *posterior cerebral artery* begins at the basilar bifurcation (see Fig. 16–9). The first and second parts of this vessel (P_1 and P_2 segments) are located, respectively, between the *basilar bifurcation* and the *posterior communicating artery* and just distal to the latter vessel. Midbrain and diencephalic structures are the main targets of branches of P_1 and P_2 segments. The P_3 segment of the posterior cerebral artery is the part from which the *temporal branches* originate, and P_4 is the segment that gives rise to *parieto-occipital* and *calcarine arteries*. The calcarine artery is the source of blood to the *primary visual cortex*, which borders on the calcarine sulcus. Other important telencephalic structures in the domains of P_3 and P_4 include the *parahippocampal gyrus* and the *precuneus*.

White Matter of the Cerebral Hemisphere

All information entering or leaving the cerebral cortex or connecting one part of the cortex with another must

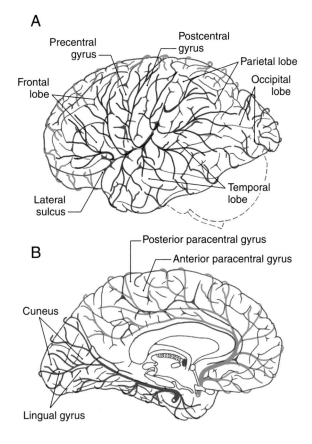

Figure 16–9. Lateral (*A*) and medial (*B*) views of the left cerebral hemisphere showing the distribution of the anterior (in green), middle (in red), and posterior (in blue) cerebral arteries.

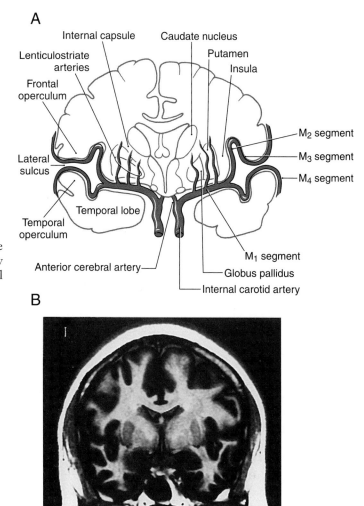

Figure 16–10. Cross section of the cerebral hemisphere showing the main branches of the middle cerebral artery *(A)* and a magnetic resonance image *(B)* in the coronal plane at a comparable level.

pass through the subcortical white matter. In general, the white matter core of the hemisphere contains *association fibers*, *commissural fibers*, and *projection fibers*.

Association Fibers. The *association fibers* interconnect different areas of cortex within the same hemisphere. These may be *short association fibers* that connect the cortices of adjacent gyri or *long association fibers* that interconnect more distant areas of cortex (Fig. 16–11). Important examples of the latter are the *cingulum* located internal to the cingulate gyrus and continuing into the parahippocampal gyrus, the *inferior longitudinal fasciculus* (temporal-occipital interconnections), and the *uncinate fasciculus* (frontal-temporal interconnections). The *superior longitudinal fasciculus*, located in the core of the hemisphere, interconnects frontal, parietal, and occipital cortices, whereas the *arcuate fasciculus* interconnects frontal and temporal lobes (see Fig. 16–11). In the white matter of the temporal lobe, fibers passing between the frontal and occipital areas make up the *inferior fronto-occipital fasciculus*.

The *claustrum*, a thin layer of neuron cell bodies located internal to the insular cortex, is sandwiched between two small association bundles (Fig. 16–12).

The *external capsule* is insinuated between the claustrum and putamen, and the *extreme capsule* is located between the claustrum and the insular cortex.

Commissural Fibers: The Corpus Callosum. In general, commissural fibers interconnect corresponding structures on either side of the neuraxis. The largest bundle of commissural fibers is the *corpus callosum* (see Figs. 16–2 and 16–4). This huge bundle is located posterior (dorsal) to the diencephalon and forms the roof of much of the lateral ventricles. The corpus callosum consists of, from rostral to caudal, a *rostrum*, *genu*, *body* (also called *trunk*), and *splenium* (see Figs. 16–4 and 16–5B). Many of the fibers passing through the genu arch rostrally to interconnect the frontal lobes; these form the *minor* (or *frontal*) *forceps*. The fibers interconnecting the occipital lobes loop through the splenium of the corpus callosum, forming the *major* (or *occipital*) *forceps*. The *tapetum*, which is located in the lateral wall of the atrium and posterior horn of the lateral ventricle, is also composed of fiber bundles that cross in the splenium.

Smaller commissural bundles are the *anterior commissure* and the *hippocampal commissure* (see Fig. 16–4).

In sagittal views, the anterior commissure is located caudal to the rostrum of the corpus callosum but rostral to the main part of the fornix. This bundle interconnects various parts of the frontal and temporal lobes. The *hippocampal commissure* is formed by fibers that originate in the hippocampal formations and cross the midline as a thin layer anterior (ventral) to the splenium of the corpus callosum.

The *posterior commissure* and the *habenular commissure* are caudal parts of the diencephalon (see Figs. 16–1 and 16–4). The former crosses the midline at the base of the pineal gland and just posterior (dorsal) to the cerebral aqueduct. The latter is a small fascicle running along the upper aspect of the posterior commissure and interconnecting the habenular nuclei.

Projection Fibers: The Internal Capsule. The projection fibers of the hemispheres include both the axons that originate outside the telencephalon and project to the cerebral cortex (*corticopetal*) and the axons that arise from cerebral cortical cells and project to downstream targets (*corticofugal*). A prime example of the former are projections from the thalamus to the cerebral cortex (*thalamocortical fibers*); examples of the latter are *corticospinal*, *corticopontine*, and *corticothalamic fibers*. Projection fibers are organized into a large, com-

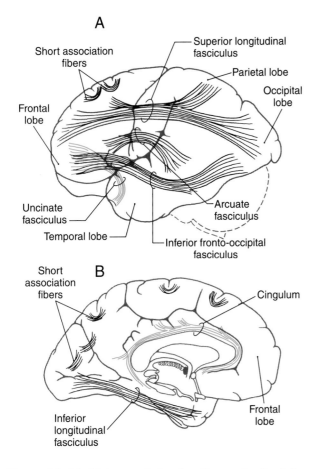

Figure 16–11. Main association bundles as visualized from lateral (*A*) and medial (*B*) aspects of the left hemisphere.

pact bundle called the *internal capsule* (see Fig. 16–12), which has intimate structural associations with the diencephalon and basal nuclei. Consequently, to divide the internal capsule into its constituent parts, reference must be made to these adjacent cell groups.

In an axial plane through the hemisphere, the internal capsule appears as a prominent V-shaped structure with the V pointing medially (Fig. 16–13; see also Fig. 16–12). It is divided into three parts: (1) an *anterior limb* insinuated between the head of the caudate nucleus and the lenticular nucleus, (2) a *posterior limb* located between the dorsal thalamus and the lenticular nucleus, and (3) a *genu* located at the intersection of the anterior and posterior limbs, which is approximately at the level of the interventricular foramen (see Fig. 16–12).

The *anterior limb of the internal capsule* contains *thalamocortical/corticothalamic fibers* (collectively called the *anterior thalamic radiations*), which interconnect the dorsomedial and anterior thalamic nuclei with areas of the frontal lobe and the cingulate gyrus. *Frontopontine fibers*, especially those from the prefrontal areas, also pass through this structure. The *genu of the internal capsule* contains *corticonuclear (corticobulbar) fibers* that arise in the frontal cortex just rostral to the precentral sulcus and from the precentral gyrus (primary motor cortex).

The *posterior limb of the internal capsule* is larger and more complex (see Fig. 16–12). It is sometimes divided into a *thalamolenticular part* (located between the thalamus and the lenticular nucleus), a *sublenticular part* (fibers passing ventral to the lenticular nucleus), and a *retrolenticular part* (fibers located caudal to the lenticular nucleus). However, contemporary terminology and common usage refer to the "thalamolenticular part" as the *posterior limb*, the "sublenticular part" as the *sublenticular limb*, and the "retrolenticular part" as the *retrolenticular limb*. This latter terminology is considerably less cumbersome and much easier to remember and is the convention followed here (see Figs. 16–12 and 16–13). By this scheme the internal capsule consists of five parts: an *anterior limb*, *genu*, *posterior limb*, *sublenticular limb*, and *retrolenticular limb*.

The major fiber populations passing through the posterior, sublenticular, and retrolenticular limbs of the internal capsule are summarized in Figure 16–12. Included in the posterior limb are *corticospinal fibers* arising from the motor cortex and projecting to the contralateral spinal cord and *thalamocortical/corticothalamic fibers (central thalamic radiations)* that interconnect nuclei of the dorsal thalamus with the overlying cortex. Studies in humans have revealed that corticospinal fibers are somatotopically arranged in about the caudal half of the posterior limb. *Geniculotemporal radiations (auditory radiations)* convey auditory information from the medial geniculate nucleus to the transverse temporal gyri through the sublenticular limb. Visual input from the lateral geniculate body to the occipital cortex is conveyed via *geniculocalcarine radiations* through the

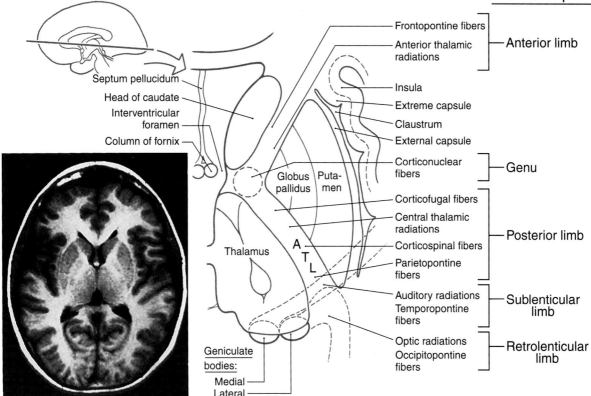

Internal capsule

- Frontopontine fibers — Anterior limb
- Anterior thalamic radiations
- Insula
- Extreme capsule
- Claustrum
- External capsule
- Corticonuclear fibers — Genu
- Corticofugal fibers
- Central thalamic radiations — Posterior limb
- Corticospinal fibers
- Parietopontine fibers
- Auditory radiations — Sublenticular limb
- Temporopontine fibers
- Optic radiations — Retrolenticular limb
- Occipitopontine fibers

Septum pellucidum
Head of caudate
Interventricular foramen
Column of fornix
Globus pallidus
Putamen
Thalamus
Geniculate bodies:
Medial
Lateral

Figure 16–12. Diagram of the internal capsule in the axial plane with a magnetic resonance image at a comparable level. The various parts of the internal capsule labeled in the drawing can be identified. Corticofugal is an umbrella term under which specific populations of descending fibers, such as corticoreticular, corticorubral, and corticotectal, are included. Although not labeled here, specific types of corticofugal fibers are also present in the sublenticular and retrolenticular limbs of the internal capsule. A, upper extremity; L, lower extremity.

retrolenticular limb (see Fig. 16–12). The geniculocalcarine radiations, *optic radiations*, form a distinct lamina of fibers immediately lateral to the tapetum as they course caudally into the occipital lobe.

Fibers of the internal capsule flare out into the hemisphere as they pass distal to the caudate and putamen. This abrupt divergence of internal capsule fibers forms the *corona radiata* ("radiating crown"), which contains *converging corticofugal fibers*, as well as *diverging corticopetal fibers* (see Fig. 16–18).

Vasculature of the Internal Capsule. The blood supply to the genu and most of the posterior limb of the internal capsule is via the *lenticulostriate arteries*; these are branches of the M_1 segment of the middle cerebral artery (see Figs. 16–10 and 16–13). Branches of the *anterior choroidal artery* supply the anterior (ventral) region of the posterior limb and the immediately adjacent retrolenticular limb. The anterior limb receives somewhat of a dual blood supply, in that lenticulostriate arteries and branches of the *medial striate artery* (usually a branch of A_2) serve this area. Occlusion (ischemic stroke) or rupture (hemorrhagic stroke) of the vessels serving the posterior limb may result in

motor and sensory deficits. Most commonly seen is a contralateral hemiparesis, indicating damage to corticospinal fibers. Hemiparesis may be accompanied by contralateral somatosensory losses, which result from the interruption of thalamocortical fibers relaying sensory information from the thalamus to the cortex.

Basal Nuclei (Ganglia)

The term "basal ganglia" is somewhat of a misnomer. The cells forming these structures are not "ganglia"— a term usually reserved to describe aggregations of neuronal somata in the peripheral nervous system— but are "nuclei" in the central nervous system. In addition, the definition of what cell groups make up the basal nuclei has been revised over the years, with contemporary views focusing on the functional characteristics of these nuclei. In this respect, *the basal nuclei consist of (1) the caudate and lenticular nuclei (together forming the dorsal [or posterior] basal nuclei), (2) the nucleus accumbens plus parts of the adjacent olfactory tubercle (the ventral striatum), and (3) the substantia innominata (ventral pallidum)* (Fig. 16–14). As described previously, the *subthalamic nucleus* and *substantia nigra* are not components of the basal nuclei, but they are included in this discussion because of their important structural and functional relationship with the caudate and lenticular nuclei.

The basal nuclei function primarily in the motor sphere. Damage to these nuclei, as in vascular lesions, degenerative genetic disorders, or problems of unknown etiology, results in a variety of motor deficits, some of which are recognized as characteristic involuntary movements.

Caudate and Lenticular Nuclei. Collectively, the caudate and lenticular nuclei form the *corpus striatum*. The corpus striatum, in turn, is divided into the *neostriatum*, consisting of the *caudate nucleus* and the *putamen*, and the *paleostriatum*, or *globus pallidus* (see Fig. 16–14). Collectively, the globus pallidus and putamen form the *lenticular nucleus*.

The caudate nucleus is characteristically located in the lateral wall of the lateral ventricle and consists of three parts; a *head*, *body*, and *tail* (Figs. 16–15 and 16–16). The *head of the caudate nucleus* forms a prominent bulge in the anterior horn of the lateral ventricle. At about the level of the interventricular foramen, the caudate diminishes in size but continues caudally as the *body of the caudate nucleus* in the lateral wall of the body of the lateral ventricle. In the lateral wall of the atrium

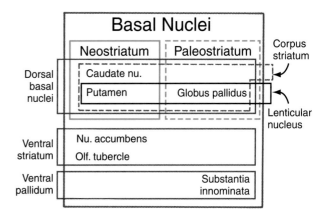

Figure 16–14. A series of stacked boxes showing which nuclei form the various parts of the basal nuclei and the terms used to describe them.

of the lateral ventricle, the body of the caudate nucleus turns anteriorly (ventrally) and then rostrally to continue as the *tail of the caudate nucleus* in the posterolateral (dorsolateral) wall of the temporal horn of the lateral ventricle. Thus, the **C** shape of the caudate nucleus faithfully follows the **C** shape of the lateral ventricle (excluding the posterior horn) (see Figs. 16–15A and B and 16–16A, B, and E–H).

The *lenticular nucleus* is located within the base of the hemisphere and is surrounded by white matter (see Figs. 16–15 and 16–16). The internal capsule borders the lenticular nucleus medially, and the external capsule separates it from the claustrum laterally. The lenticular nucleus consists of a larger lateral part, the *putamen*, and a smaller medial portion, the *globus pallidus* (or *pallidum*). The putamen extends more posterior (dorsal), more rostral, and more caudal when compared with the globus pallidus and is clearly the larger part of the lenticular nucleus when viewed in axial or coronal planes (see Fig. 16–15B–E). The *globus pallidus* is located internal to the putamen and is smaller in all dimensions (see Figs. 16–15 and 16–16). It is divided into *medial (internal)* and *lateral (external)* parts by thin sheets of vertically oriented white matter. The globus pallidus is also separated from the putamen by a thin lamina of white matter.

Nucleus Accumbens and Substantia Innominata. The *nucleus accumbens* is located rostrally and anteriorly in the hemisphere where the putamen is continuous with the head of the caudate nucleus (Fig. 16–17). This cell group is also closely apposed to the septal nuclei and the nucleus of the diagonal band, both of which extend into the base of the septum pellucidum. At a slightly more caudal level, the *substantia innominata* (basal nucleus of Meynert) is located internal to the anterior perforated substance in the area anterior (ventral) to the anterior commissure (see Fig. 16–17). This area contains clusters of small and large cells and

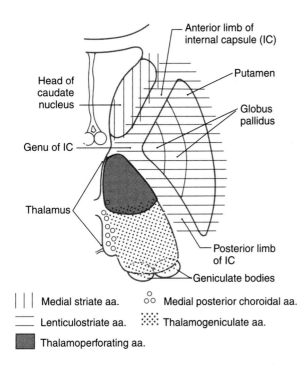

Figure 16–13. Blood supply to the basal nuclei, internal capsule, and thalamus shown in the axial plane.

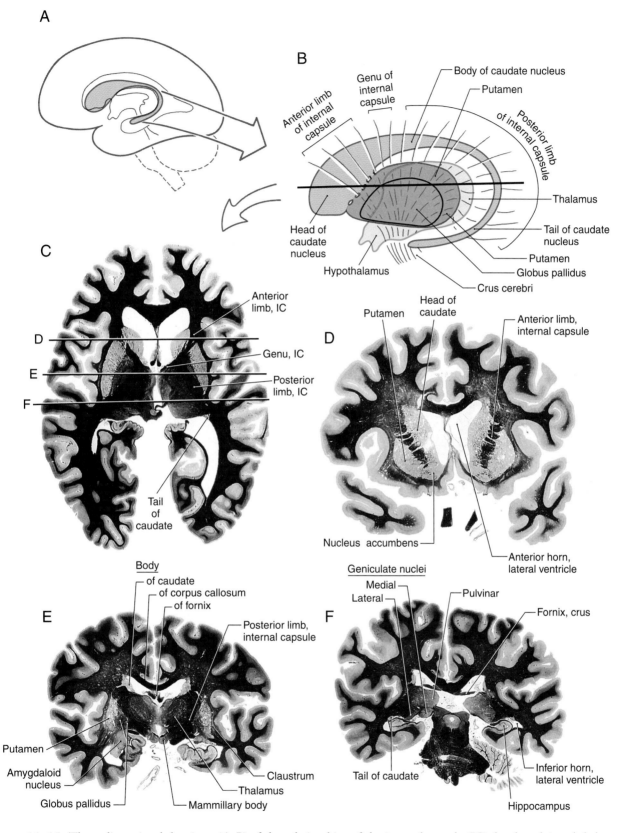

Figure 16–15. Three-dimensional drawings *(A, B)* of the relationships of the internal capsule (IC), basal nuclei, and thalamus. The axial section *(C)* represents the approximate plane shown in *B;* the coronal sections *(D–F)* are taken from the three levels indicated in *C. C–F,* Weil stains.

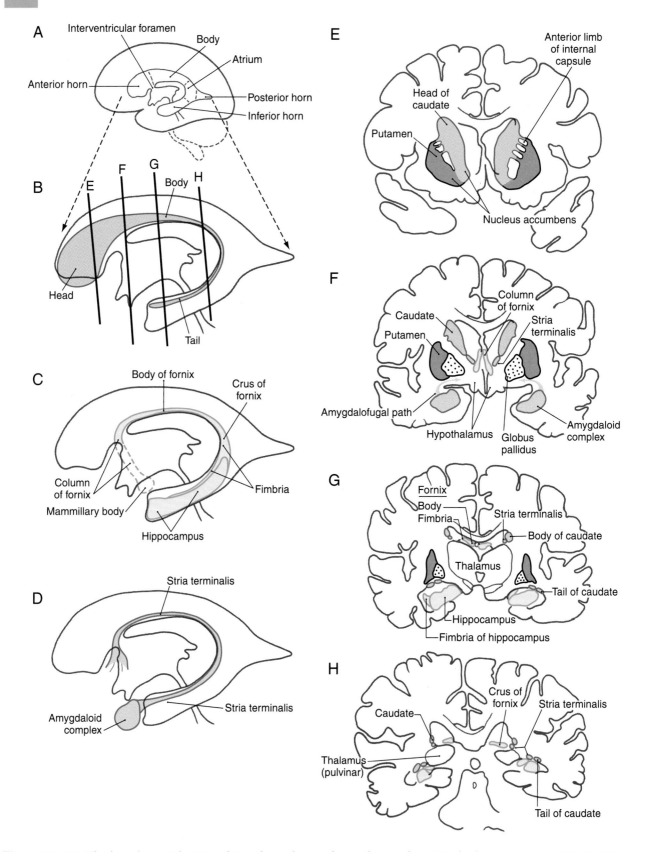

Figure 16–16. The lateral ventricle *(A)* and its relationship to the caudate nucleus *(B)*, the hippocampus and fornix *(C)*, and the amygdala and stria terminalis *(D)*. The coronal levels in *E* through *H* correlate with the planes indicated in *B* and are color-coded to match the corresponding structure in *B, C,* or *D*.

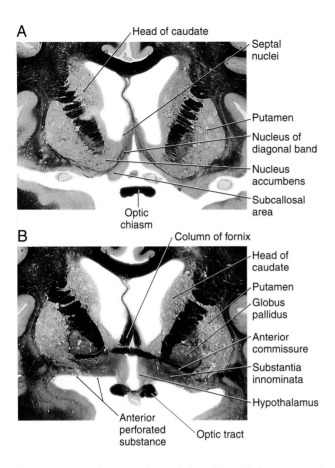

A

Head of caudate
Septal nuclei
Putamen
Nucleus of diagonal band
Nucleus accumbens
Subcallosal area
Optic chiasm

B

Column of fornix
Head of caudate
Putamen
Globus pallidus
Anterior commissure
Substantia innominata
Hypothalamus
Optic tract
Anterior perforated substance

Figure 16–17. Cross sections of the telencephalon at rostral levels showing the nucleus accumbens (*A*) and the substantia innominata (*B*) and adjacent structures. Weil stain.

emphasis on the major fiber bundles of the basal nuclei (see Fig. 16–18).

The two largest bundles of efferent fibers exiting the basal nuclei are the *lenticular fasciculus* and the *ansa lenticularis*. The former leaves the globus pallidus, traverses the posterior limb of the internal capsule, and forms a thin sheet of fibers insinuated between the subthalamic nucleus and the zona incerta. These fibers loop around the medial aspect of the zona incerta and pass laterally (and posteriorly) as the *thalamic fasciculus* (see Fig. 16–18). The fibers of the *ansa lenticularis* originate at a slightly more rostral level, arch around the anteroventral aspect of the internal capsule, and pass caudally to join with the fibers of the lenticular fasciculus as they enter the thalamic fasciculus (see Fig. 16–18).

Two smaller but equally important bundles are the *subthalamic fasciculus* and *connections between the substantia nigra and neostriatum* (see Fig. 16–18). The subthalamic fasciculus is composed of bidirectional connections between the globus pallidus and the subthalamic nucleus; these fibers also traverse the posterior limb of the internal capsule. The connections between the substantia nigra and the neostriatum are named according to the origin and termination of the fiber. Axons that project from nigral cells to the neostriatum are

diffuse bundles of fibers. Although many areas of the nervous system are affected in Alzheimer disease, there is an especially noticeable loss of larger neurons in the substantia innominata.

Subthalamic Nucleus and Substantia Nigra. Although not a part of the telencephalon either developmentally or geographically, the subthalamic nucleus and the substantia nigra are intimately allied with the basal nuclei based on their connections (Fig. 16–18). The *subthalamic nucleus*, a component of the diencephalon, is a flattened, lens-shaped cell group located rostral to the substantia nigra. It is medial to the internal capsule and is capped by a thin sheet of fibers called the *lenticular fasciculus*. The *substantia nigra*, a part of the midbrain, is found internal to the crus cerebri and immediately caudal to the subthalamic nucleus. It is divided into a reticulated part (*pars reticulata*) and a compact part (*pars compacta*); the latter is characterized by numerous melanin-containing neuron cell bodies.

Major Connections of the Basal Nuclei. The connections of the basal nuclei are discussed in their entirety in Chapter 26. It is appropriate, however, to briefly review these relationships here, with particular

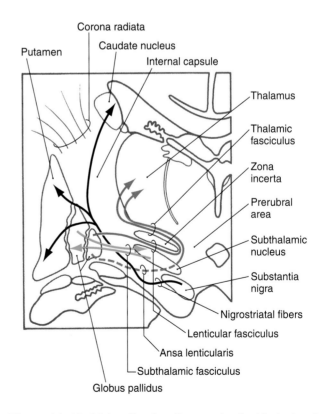

Putamen
Corona radiata
Caudate nucleus
Internal capsule
Thalamus
Thalamic fasciculus
Zona incerta
Prerubral area
Subthalamic nucleus
Substantia nigra
Nigrostriatal fibers
Lenticular fasciculus
Ansa lenticularis
Subthalamic fasciculus
Globus pallidus

Figure 16–18. Major fiber bundles associated with the basal nuclei and with the subthalamic nucleus and substantia nigra.

nigrostriatal projections, and fibers that project from striatal cells to the substantia nigra are *striatonigral* fibers.

Vasculature of the Basal Nuclei and Related Structures. The blood supply to the caudate and putamen is provided by branches of the *medial striate artery*, *lenticulostriate branches* of M_1, and the *anterior choroidal artery* (see Figs. 16–10 and 16–13). The medial striate artery, usually a branch of A_2, serves much of the head of the caudate nucleus. Most of the lenticular nucleus and the surrounding internal and external capsules are supplied by the lenticulostriate branches of M_1. Anterior (ventral) and medial portions of the head of the caudate, as well as the body of the caudate, are also served by these arteries. The tail of the caudate, adjacent portions of the lenticular nucleus, and adjacent temporal lobe structures (hippocampus, choroid plexus) receive their blood supply via the anterior choroidal artery, a branch of the internal carotid artery. The blood supply to the subthalamic nucleus and the substantia nigra arises from the *posteromedial branches* of the P_1 segment and branches of the posterior communicating artery. These vessels penetrate the brain at the midbrain-diencephalon junction through the posterior perforated substance.

Hippocampus and Amygdala

The *hippocampal formation* and the *amygdaloid complex* are located in the temporal lobe. The former lies in the anteromedial floor of the temporal horn of the lateral ventricle and the latter in the rostral end of this space. Through a variety of pathways (see Chapter 31), these structures interconnect with numerous telencephalic and diencephalic centers.

Developmentally, the hippocampus is formed by a rolling-in of primitive cortex to form the curved, multilayered structure characteristic of the adult brain (see Fig. 16–15F). The hippocampal formation is found internal to the parahippocampal gyrus and is composed of the *subiculum*, the *hippocampus* (also called *Ammon's horn*), and the *dentate gyrus*. The cortex of the parahippocampal gyrus is continuous with the subiculum, which, in turn, is continuous with the hippocampus. The dentate gyrus forms a reverse loop around the hippocampus and, in doing so, presents a serrated surface that is medially exposed to the subarachnoid space. Details of the internal structure of the hippocampal formation are discussed in Chapter 31 and summarized in Figure 31–4.

Axons of hippocampal neurons converge to form a prominent bundle that arches around caudal, posterior (dorsal), and rostral aspects of the thalamus. This bundle, the *fornix*, is a major efferent path of the hippocampal formation (see Fig. 16–16A–C and F–H). It is composed of a flattened caudal part, the *crus*; a compact posterior (dorsal) portion, the *body*; and a part that arches around the rostral part of the thalamus and passes through the hypothalamus to terminate in the mammillary body—this is the *column of the fornix*. Located along the edge of the dentate gyrus and continuing on the lateral edge of the crus and body of the fornix is a thin fringe of fibers called the *fimbria*.

The *amygdaloid nuclear complex* (commonly called the *amygdala*) is located internal to the cortex of the uncus (see Figs. 16–4 and 16–16D–F). Two major efferent bundles are related to the amygdala. First, the *stria terminalis* follows a looping trajectory that shadows, in a reverse direction, the orientation of the caudate nucleus (see Fig. 16–16D). In the temporal horn, the stria terminalis is located just medial to the tail of the caudate nucleus. As the stria terminalis arches rostrally, it assumes a position in the shallow groove between the caudate nucleus and the dorsal thalamus (see Fig. 16–16G and H), where it is accompanied by the terminal vein (superior thalamostriate vein). At about the level of the interventricular foramen, the fibers of the stria terminalis fan out to enter and terminate in the hypothalamus, the septal area, and the neostriatum.

The second major efferent bundle of the amygdala is the diffusely arranged *ventral amygdalofugal pathway*. These fibers leave the amygdaloid complex, pass medially through the *substantia innominata*, and continue medially to enter hypothalamic and septal nuclei, or turn caudally and distribute to the brainstem (see Fig. 16–16F).

Cell groups located internal to the subcallosal area collectively form the *septal nuclei* (see Fig. 16–17). Consequently, the subcallosal area together with a small strip of cortex located adjacent to the lamina terminalis, the *paraterminal gyrus*, is commonly called the *septal area*. These septal nuclei are medially adjacent to the nucleus accumbens and continuous with sheets of neuronal cell bodies that extend into the *septum pellucidum*. The latter structure extends, in general, from the fornix to the inner surface of the corpus callosum. It forms the medial wall of the anterior horns and a small part of the bodies of the lateral ventricles (see Figs. 16–12 and 16–15). In general, the septal nuclei have complex interconnections with hippocampal, amygdaloid, and limbic structures.

Temporal Lobe Lesions. Injury to the temporal lobe, especially bilateral damage, almost always involves the hippocampus and amygdala. Deficits most directly linked to trauma to these structures include profound changes in eating and sexual behavior, in aggression levels, and in memory function. With deficits in memory function, the patient may demonstrate a loss of recent memory or show an inability to acquire new memory (learn new tasks), while memory of events that took place in the distant past remains intact.

Vasculature of the Hippocampus and Amygdala. The blood supply to the hippocampal formation and amygdaloid complex is primarily via the *anterior choroidal artery*. This vessel arises from the internal carotid, passes along the medial edge of the temporal horn, and sends branches into the hippocampus and amygdala. It also serves the tail of the caudate, the choroid plexus of the temporal horn, and anterior (ventral) regions of the lenticular nucleus. The cortex of the uncus and that of the parahippocampal gyrus are served by superficial branches of the *middle cerebral* and *posterior cerebral arteries*, respectively.

Sources and Additional Reading

Bailey P, von Bonin G: The Isocortex of Man. University of Illinois Press, Urbana, 1951.

Crosby EC, Humphrey T, Lauer EW: Correlative Anatomy of the Nervous System. Macmillan Publishing, New York, 1962.

Kretschmann H-J, Weinrich W: Cranial Neuroimaging and Clinical Neuroanatomy, Magnetic Resonance Imaging and Computed Tomography, 2nd ed. Thieme Medical Publishers, New York, 1992.

Kuhlenbeck H: The Central Nervous System of Vertebrates: A General Survey of Its Comparative Anatomy with an Introduction to the Pertinent Fundamental Biologic and Logical Concepts, vol 5, part I: Derivatives of the Prosencephalon: Diencephalon and Telencephalon. S. Karger, Basel, 1977.

Kuhlenbeck H: The Central Nervous System of Vertebrates: A General Survey of Its Comparative Anatomy with an Introduction to the Pertinent Fundamental Biologic and Logical Concepts, vol 5, part II: Mammalian Telencephalon: Surface Morphology and Cerebral Cortex. S. Karger, Basel, 1978.

Nieuwenhuys R, Voogd J, van Huijzen C: The Human Central Nervous System, A Synopsis and Atlas, 3rd ed. Springer-Verlag, Berlin, 1988.

Parent A: Carpenter's Human Neuroanatomy, 9th ed. Williams & Wilkins, Baltimore, 1996.

Paxinos G (ed): The Human Nervous System. Academic Press, San Diego, 1990, pp 439–755.

Varnavas GG, Grand W: The insular cortex: Morphological and vascular anatomic characteristics. Neurosurgery 44:127–138, 1999.

Zola-Morgan S, Squire LR: Neuroanatomy of memory. Annu Rev Neurosci 16:547–563, 1993.

Section III

Systems Neurobiology

The Somatosensory System I: Tactile Discrimination and Position Sense

S. Warren, N. F. Capra, and R. P. Yezierski

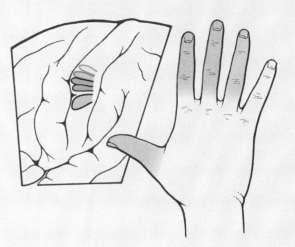

Overview 256

Posterior Column–Medial Lemniscal System (PCMLS) 256
Peripheral Mechanoreceptors
Primary Afferent Fibers
Spinal Cord and Brainstem
Ventral Posterolateral Nucleus
Primary Somatosensory (SI) Cortex
Additional Cortical Somatosensory Regions

Trigeminal System 264
Overview
Trigeminal Nerve

Anterior and Posterior Trigeminothalamic Tracts 265
Peripheral Receptors

Receptive Field Properties of Cortical Neurons 267

Neuroimaging and Functional Localization 268

Plasticity and Reorganization in the Primary Somatosensory Cortex 268

Nonconscious Proprioception: Spinocerebellar Pathways 269
Posterior Spinocerebellar Tract
Cuneocerebellar Tract
Anterior Spinocerebellar Tract
Rostral Spinocerebellar Tract
Trigeminocerebellar Connections

If you reach into your pocket to determine the types of coins present, you are gathering information through the activation of specialized receptors of the *somatosensory system*. Specifically, the size of a coin is determined by noting the joint angles when the coin is held between the forefinger and thumb. "Heads and tails" may be identified using *slowly adapting* receptors sensitive to stimuli that indent the skin. Dimes can be distinguished from pennies by stroking their edges with the fingertips and activating *rapidly adapting* receptors. This information is transmitted to the cerebral cortex by a multisynaptic pathway called the *posterior column–medial lemniscal system*. At the same time, much of this information, along with information concerning muscle tension and length, is also transmitted to the cerebellar cortex, where it is used to regulate muscle activity that allows manipulation of the coins. The *spinocerebellar pathways* are among those that subserve these nonconscious somatosensory functions.

Overview

In general, the somatosensory system transmits and analyzes *touch* or *tactile* information from external and internal locations on the body and head. The result of these processes leads to the appreciation of somatic sensations, which can be subdivided into the submodalities *discriminative touch*, *flutter-vibration*, *proprioception* (*position sense*), *crude* (*nondiscriminative*) *touch*, *thermal* (*hot and cold*) *sensation*, and *nociception* (*pain*). The following anatomically and functionally discrete pathways transmit these signals: (1) the *posterior column–medial lemniscal* pathway, (2) the *trigeminothalamic* pathways, (3) the *spinocerebellar* pathways, and (4) the *anterolateral system*.

This chapter describes pathways that transmit discriminative touch, flutter-vibration, and proprioceptive information. These pathways are the *posterior column–medial lemniscal pathway*, portions of the trigeminothalamic pathways originating in the *principal trigeminal sensory nucleus*, and the *spinocerebellar* pathways. The pathways subserving the submodalities of pain, thermal sense, and crude touch, itch, and tickle compose the anterolateral system. These and portions of the trigeminothalamic pathways are described in Chapter 18.

Posterior Column–Medial Lemniscal System (PCMLS)

The posterior column–medial lemniscal pathway (see Figs. 17–6 and 17–8) is involved with the perception and appreciation of mechanical stimuli. It underlies the capacity for fine form and texture discrimination, form recognition of three-dimensional shape (*stereognosis*),

and motion detection. This pathway is also involved in transmitting information related to conscious awareness of body position (*proprioception*) and limb movement (*kinesthesia*) in space.

Characteristic features of the PCMLS include transmission in general somatic afferent (GSA) fibers that have fast conduction velocities, a limited number of synaptic relays in which processing of the signal occurs, and a precise somatotopic organization. These features provide the basis for the accurate localization of the body region touched. There is only limited convergence along the pathway; consequently, the signal is transmitted with high fidelity and a high degree of spatial and temporal resolution. This pathway signals somatic sensations using *frequency* and *population codes*. In frequency coding, a cell's firing rate signals stimulus intensity or temporal aspects of the tactile stimulus. In population coding, the distribution in time and space of activated cells in the central nervous system signals location of the stimulus, as well as its motion or direction if any.

The high degree of resolution in the PCMLS is the result of inhibitory mechanisms such as *lateral* (*surround*) *inhibition*. This mechanism, which sharpens and enhances the discrimination between separate points on the skin, is critical for *two-point discrimination*. The ability to discriminate between two points simultaneously applied varies widely over different parts of the body.

Peripheral Mechanoreceptors. The first step in evoking somatic sensations is the activation of peripheral mechanoreceptors. Mechanical pressure, such as skin deformation, is *transduced* into an electrical signal in the peripheral process of a primary afferent neuron. This leads to a depolarizing *graded membrane potential* across the membrane of the neuron. If this potential depolarizes the *trigger zone*, located at the first myelin segment of the axon, to *threshold*, an *action potential* is produced (see Chapter 3). In most receptors, transduction occurs between the mechanoreceptor and subjacent primary afferent membrane. In contrast, Merkel cells may influence their associated primary afferent axon by vesicular release of a transmitter substance.

Each morphologic type of receptor responds to different tactile stimuli. *Cutaneous tactile receptors* (Table 17–1; Fig. 17–1) are located in the basal epidermis and dermis of *glabrous* (palms, soles, lips) and *hairy* skin. These low-threshold mechanoreceptors may be encapsulated, such as *Meissner* and *Pacinian corpuscles*, or unencapsulated, such as *Merkel cell neurite complexes* and *hair follicle receptors*. Meissner corpuscles, hair follicle receptors, and Pacinian corpuscles respond to transient, phasic, or vibratory stimuli. These receptors respond to each initial application or removal of a stimulus but fail to respond during maintained stimulation. Conse-

Table 17-1.   Cutaneous Mechanoreceptors and Their Associated Fiber Types and Sensations

Receptor Type (Adaptation Rate)	Sensation Produced by Microstimulation	Fiber Type (Group*)	Receptive Field Size (Average)	Number per cm²	
				Fingertip	*Palm*
Meissner corpuscle (RA)	Tap, flutter 5–40 Hz	II	Small (54.9 ± 8.6 mm²)	>100	40
Hair follicle receptors (RA, SA)	Motion, direction	II	N/A	N/A	N/A
Pacinian corpuscle (RA)	Vibration 60–300 Hz	II	Large	20	10
Merkel cell (SA)	Touch—pressure	II	Small (44.7 mm²)	70	30
Ruffini complex (SA)	Unknown	II	Large	50	15

RA, rapidly adapting; SA, slowly adapting.

quently, they are *rapidly adapting* (RA) *receptors* (Fig. 17–2*A*). Hair follicle receptors are also capable of signaling motion, its direction or orientation, and its velocity.

Merkel cell neurite complexes and some hair follicle receptors signal tonic events such as discrete small indentations in the skin. They provide input related to both the displacement and velocity of a stimulus. They are also capable of encoding stimulus intensity or duration because they are *slowly adapting* (SA) and are active so long as the stimulus is present (see Fig. 17–2*A*).

For example, Merkel cell complexes are crucial to reading Braille.

Deep tactile mechanoreceptors are found within the dermis of the skin and in the fascia surrounding muscles and bone and in the peridontium. These receptors include *Pacinian corpuscles, Ruffini endings,* and other encapsulated nerve endings located in the periosteum, the deep fascia, and the mesenteries. The receptors of this group respond to pressure, vibration (see Fig. 17–2*B* and Table 17–1), or skin stretch and distention or tooth displacement.

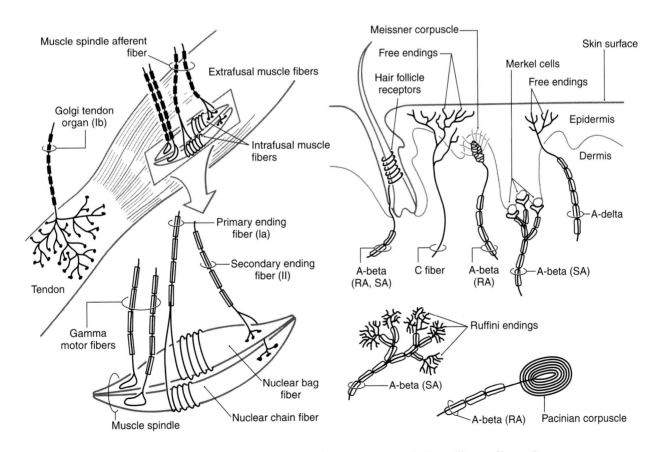

Figure 17-1. Proprioceptive receptors and cutaneous mechanoreceptors and their afferent fibers. Cutaneous receptors are either rapidly adapting (RA) or slowly adapting (SA).

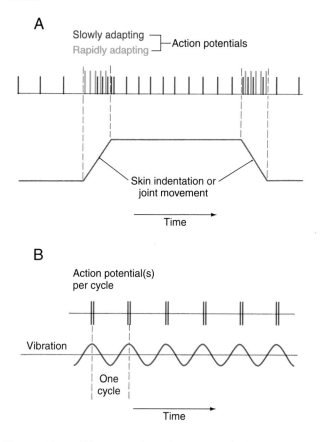

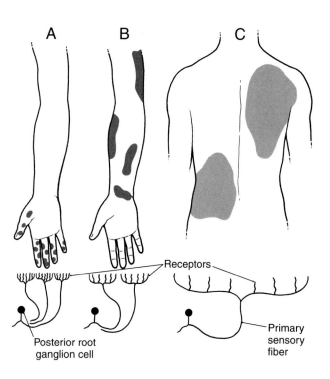

Figure 17–3. Variation in the size of receptive fields as a function of peripheral innervation density. The greater the density of receptors, the smaller the receptive fields of individual afferent fibers.

Figure 17–2. Diagrammatic action potentials (top trace, *A*) evoked by skin indentation and removal of a cutaneous stimulus or of joint movement (bottom trace, *A*) in primary afferent fibers innervating slowly adapting (SA:red) and rapidly adapting (RA:green) cutaneous mechanoreceptors. Diagrammatic action potentials (blue, *B*) evoked in a Pacinian corpuscle afferent fiber by sinusoidal stimulation of the skin surface (bottom trace, *B*).

Proprioceptive receptors (Table 17–2; see Fig. 17–1) are located in muscles, tendons, and joint capsules. These receptors include the nuclear bag and nuclear chain intrafusal muscle fibers of *muscle spindles* and their associated nerve fibers. The *Golgi tendon organs* and their group Ib fibers and the encapsulated Ruffini-type joint receptors also function in this capacity. They

respond to static limb and joint position or to the dynamic movement of the limb (kinesthesia) and are important sources of information for balance, posture, and limb movement.

The accuracy with which a tactile stimulus is detected depends on the density of receptors and the size of receptive fields (Fig. 17–3). The greatest density of cutaneous tactile receptors is found on the tips of the glabrous digits and in the perioral region. Other regions, like the back, have much lower density, thus creating a receptor density gradient between various body parts. The *receptive field* is the area of skin innervated by branches of a GSA fiber, the stimulation of which activates its receptors (see Fig. 17–3). *Small receptive fields* are found in areas, such as the finger tips,

Table 17–2.	Muscle and Joint Proprioceptors and Their Associated Fiber Types and Sensations		
Receptor Type (Adaptation Rate)	**Sensation**	**Functional/Signal**	**Fiber Type (Group)**
Nuclear bag fiber (SA—primary annulospiral endings)	High dynamic sensitivity	Length and rate of change; length and velocity	Ia
Nuclear chain fiber (SA—secondary flower spray ending)	Low dynamic sensitivity	Length; tension	II
Golgi tendon organ (SA)	Tension	Muscle force; tension	Ib
Ruffini endings (SA)	Limb position	Joint movement and pressure	I

SA, slowly adapting.

where receptor density is high and each receptor serves an extremely small area of skin. In such regions the individual is able to discriminate small variations in a variety of sensory inputs. In other regions, receptor density is low, and each receptor serves an expansive area of skin, creating *large receptive fields* with resultant reduction in discriminative ability.

At all levels of the tactile pathway, densely innervated body parts are represented by greater numbers of neurons and take up a disproportionately large part of the somatosensory system's body representation. As a result, the finger tips and lips provide the central nervous system with the most specific and detailed information about a tactile stimulus.

Primary Afferent Fibers. As initially described in Chapter 9, primary afferent GSA fibers consist of (1) *a peripheral process* extending from the posterior root ganglion to either contact peripheral mechanoreceptors or end as free nerve endings, (2) *a central process* extending from the posterior root ganglion into the central nervous system, and (3) *a pseudounipolar cell body* in the posterior root ganglion. The peripheral distribution of the afferent nerves associated with each spinal level delineates the segmental pattern of *dermatomes*. In clinical testing, these ribbon-like strips of skin are associated primarily with fibers and pathways that convey pain and thermal information; they are considered in Chapter 18.

Peripheral nerves are classified by two schemes. One is based on their contribution to a compound action potential (A, B, and C waves) recorded from an entire mixed peripheral nerve (e.g., sciatic nerve) after electrical stimulation of that nerve. The other scheme specific to cutaneous fibers (e.g., lateral antebrachial cutaneous nerve, sural nerve) is based on fiber diameter, myelin thickness, and conduction velocity (classes I, II, III, and IV) (Table 17–3; Fig. 17–4). The two schemes are related because conduction velocity determines a fiber's contribution to the compound action potential. Discriminative touch, vibratory sense, and position sense are transmitted by group Ia, Ib, and II fibers (see Tables 17–1 and 17–2).

Spinal Cord and Brainstem. On the basis of cell and fiber diameter, primary sensory fibers are categorized as *large* and *small*. Large-diameter fibers subserve discriminative touch, flutter-vibration, and proprioception (groups Ia, Ib, II, and Aβ; see Tables 17–1 and 17–2). They enter the spinal cord via the *medial division of the posterior root* (see Chapter 9) and then branch (Fig. 17–5). One set of branches terminates on second-order neurons in the spinal cord gray matter at, above, and below the level of entry. These branches contribute to a variety of spinal reflexes and to ascending projections such as *postsynaptic posterior column fibers*. The largest set of branches ascends cranially and contributes to the formation of the *gracile* and *cuneate fasciculi*. These fiber bundles are collectively termed *posterior columns* owing to their position in the spinal cord (Fig. 17–6; see also Fig. 17–5).

Within the posterior columns, fibers from different dermatomes are organized topographically. Sacral level fibers assume a medial position, and fibers from progressively more rostral levels (up to thoracic level T6) are added laterally to form the *gracile fasciculus*. Thoracic fibers from above T6 and cervical fibers form the laterally placed *cuneate fasciculus* in the same manner. Thus, the leg is represented medially and the arm lat-

Table 17–3. Peripheral Sensory and Motor Fibers: Groups, Diameters, and Conduction Velocities

Electrophysiologic Classification of Peripheral Nerves	Classification of Afferent Fibers ONLY (Class/Group)	Fiber Diameter (μm)	Conduction Velocity (m/sec)	Receptor Supplied
Sensory Fiber Type				
Aα	Ia and Ib	13–20	80–120	Primary muscle spindles, Golgi tendon organ
Aβ	II	6–12	35–75	Secondary muscle spindles, skin mechanoreceptors
Aδ	III	1–5	5–30	Skin mechanoreceptors, thermal receptors, and nociceptors
C	IV	0.2–1.5	0.5–2	Skin mechanoreceptors, thermal receptors, and nociceptors
Motor Fiber Type				
Aα	N/A	12–20	72–120	Extrafusal skeletal muscle fibers
Aγ	N/A	2–8	12–48	Intrafusal muscle fibers
B	N/A	1–3	6–18	Preganglionic autonomic fibers
C	N/A	0.2–2	0.5–2	Postganglionic autonomic fibers

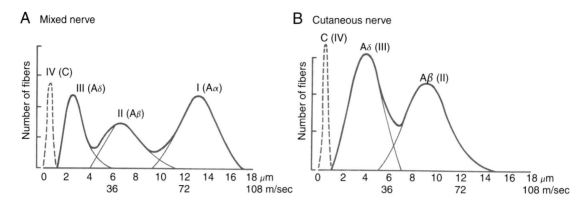

Figure 17–4. Compound action potential evoked in a mixed nerve (*A*) and a cutaneous nerve (*B*) in response to electrical stimulation. Note the increase in the number of small-diameter fibers and the absence of the *A*α fibers in the cutaneous nerve (*B*).

erally within the posterior columns (see Fig. 17–6). Compromise of blood flow in the posterior spinal artery, which supplies the posterior funiculus, or mechanical injury to the posterior columns (as in *Brown-Séquard syndrome*) results in an *ipsilateral reduction or loss of discriminative, positional, and vibratory tactile sensations at and below the segmental level of the injury.*

The *posterior column nuclei*, the *gracile* and *cuneate nuclei*, are found in the posterior medulla at the rostral end of their respective fasciculi. They are supplied by the posterior spinal artery (see Fig. 17–6). The cell bodies of the gracile and cuneate nuclei are the *second-order neurons* in the PCMLS. They receive input from *first-order neurons,* having cell bodies in the ipsilateral

posterior root ganglia. The gracile nucleus receives input from sacral, lumbar, and lower thoracic levels via the *gracile fasciculus;* the cuneate nucleus receives input from upper thoracic and cervical levels through the *cuneate fasciculus.*

In addition to the somatotopic organization of projections to the posterior column nuclei, there is a submodality segregation of tactile inputs within these nuclei. The relay neurons that receive excitatory input from primary afferent fibers form submodality-specific rostrocaudal bands (Fig. 17–7). Rapidly adapting inputs terminate centrally and caudally within the nuclei. Slowly adapting cutaneous input and muscle spindle and joint inputs project to the rostral pole of the cune-

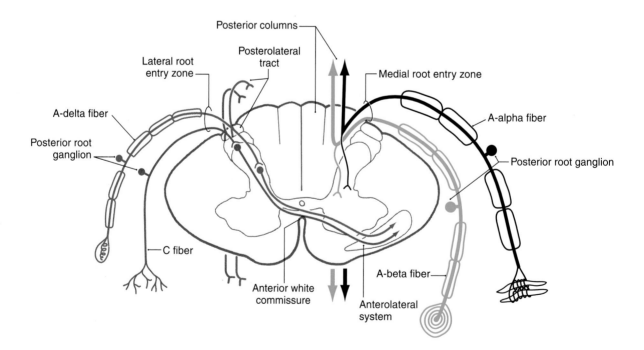

Figure 17–5. A representative section of the cervical spinal cord showing large-diameter Aα and Aβ fibers on the right and small-diameter Aδ and C fibers on the left.

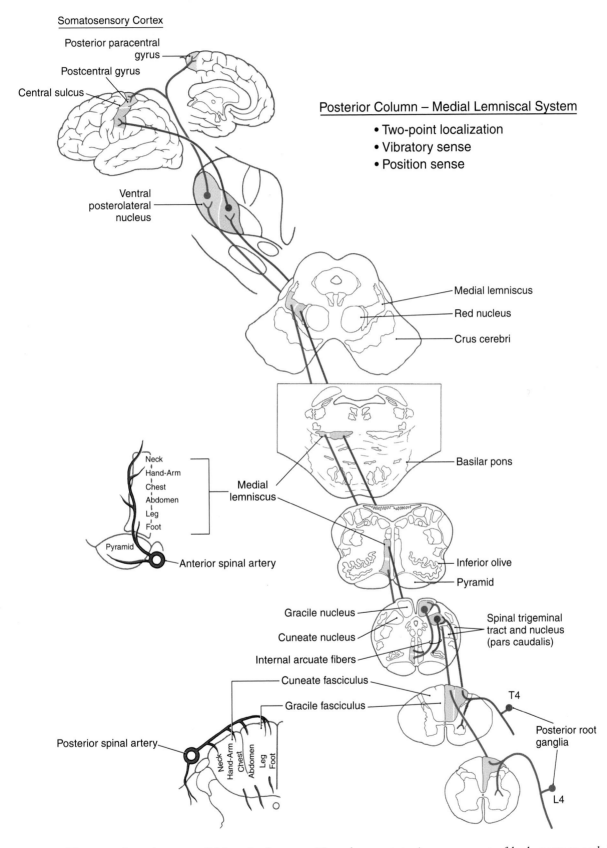

Somatosensory Cortex

Posterior paracentral gyrus

Postcentral gyrus

Central sulcus

Ventral posterolateral nucleus

Posterior Column – Medial Lemniscal System

- Two-point localization
- Vibratory sense
- Position sense

Medial lemniscus

Red nucleus

Crus cerebri

Basilar pons

Medial lemniscus

Neck
Hand-Arm
Chest
Abdomen
Leg
Foot

Pyramid

Anterior spinal artery

Inferior olive

Pyramid

Gracile nucleus

Cuneate nucleus

Internal arcuate fibers

Spinal trigeminal tract and nucleus (pars caudalis)

Cuneate fasciculus

Gracile fasciculus

T4

Posterior spinal artery

Neck
Hand-Arm
Chest
Abdomen
Leg
Foot

Posterior root ganglia

L4

Figure 17–6. The posterior column–medial lemniscal system. Note the somatotopic arrangement of body parts at each level of this pathway.

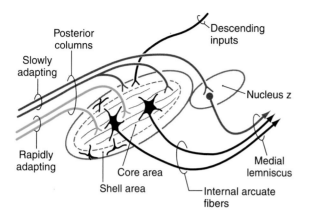

Figure 17–7. Schematic representation of the posterior column nuclei and nucleus z. Slowly adapting inputs, including those from joint and muscle receptors, terminate preferentially in the shell region. Rapidly adapting inputs, including Meissner corpuscles and some hair follicle receptors, project to the core region. The output of the posterior column nuclei is influenced by descending inputs arising in other brainstem and cortical areas.

ate and gracile nuclei and to the rostrally adjacent *nucleus z*. The posterior column nuclei also receive descending axons from the contralateral primary somatosensory cortex and from the medullary reticular formation (nucleus reticularis gigantocellularis) (see Fig. 17–7).

The posterior column nuclei have an "inner core" region of large projection neurons surrounded by a diffuse "shell" of small fusiform and radiating cells (see Fig. 17–7). The latter are interneurons responsible for feedback inhibition in the posterior column nuclei. This feedback alters activity of projection neurons of the inner core. In addition, the presence of non–posterior column inputs to these projection cells suggests that information received by the posterior column nuclei is not simply relayed but undergoes signal processing.

The *second-order cells* in the core regions of the posterior column nuclei send their axons to the contralateral thalamus (see Fig. 17–6). In the medulla, the *internal arcuate fibers*, axons of cells in the posterior column nuclei, arc anteromedially toward the midline, decussate, and ascend as the *medial lemniscus* on the opposite side. Fibers in the medial lemniscus that arise in the cuneate nucleus are located posterior to those that originate from the gracile nucleus (see Fig. 17–6). The anterior spinal artery supplies the medial lemniscus in the medulla, and penetrating branches of the basilar artery supply it in the pons. Vascular damage at these brainstem levels leads to deficits in discriminative touch, vibratory, and positional sensibilities over the contralateral body. As the medial lemniscus moves rostrally through the brainstem, it rotates laterally so that the arm representation comes to lie medially and the

leg laterally (see Fig. 17–6). This somatotopic organization is maintained as the lemniscus ascends through the brainstem and terminates on cells in the *ventral posterolateral nucleus* (VPL) of the thalamus.

The *postsynaptic posterior column pathway*, a small supplemental pathway in humans that relays nondiscriminative tactile signals to supraspinal levels, consists of non-primary afferent axons carrying tactile signals in the posterior columns (Fig. 17–8). The cells of origin of this pathway are located in laminae III and IV of the posterior horn. Axons of these second-order *postsynaptic posterior column fibers* travel in the posterior columns and, together with other tactile primary afferent fibers, terminate in the posterior column nuclei. Cells of these nuclei relay this postsynaptic posterior column input to the contralateral thalamus via the medial lemniscus. Although this pathway is small, it may provide the morphologic basis for the return of some tactile sensation after vascular lesions involving the PCML system (see Fig. 17–8).

Ventral Posterolateral Nucleus. The *ventral posterior nucleus*, sometimes called the *ventrobasal complex*, is a wedge-shaped cell group located caudally in the thalamus. Its lateral border abuts the internal capsule, and ventrally it borders on the external medullary lamina. The ventral posterior nucleus is composed of the later-

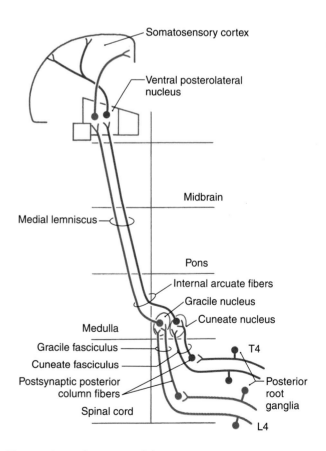

Figure 17–8. Summary of the postsynaptic posterior column pathway.

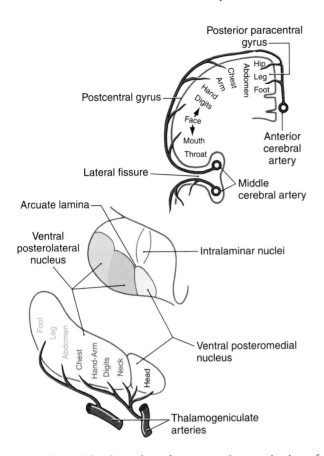

Figure 17–9. Blood supply and somatotopic organization of the body in the ventral posterolateral and posteromedial nuclei and in the primary somatosensory (SI) cortex.

functional properties. Rapidly and slowly adapting inputs terminate on different cell groups within the "core" region of VPL. Pacinian inputs and inputs arising from joints and muscles are confined to a "shell" region on the posterior, rostral, and anterior edges of the nucleus. Individual lemniscal axons arborize in the sagittal plane to terminate on longitudinal cell clusters, called *rods*, in the VPL. This arrangement of inputs and target cells creates isorepresentations consisting of neurons with similar receptive fields and submodalities arranged along a rostrocaudal axis.

The ventral posterior nucleus contains two populations of identified neurons. The first consists of large-diameter multipolar cells that give rise to axons that traverse the posterior limb of the internal capsule and terminate mainly in the primary (SI) and secondary (SII) somatosensory cortices. These *thalamocortical cells* and *fibers* are the *third-order neurons* in the PCMLS that provide excitatory input (glutaminergic) to the cortex. The second population consists of inhibitory (GABAergic) *local circuit interneurons*, which receive excitatory corticothalamic inputs and influence the firing rates of thalamocortical cells. In addition, these thalamocortical cells are also influenced by GABAergic input from the thalamic reticular nucleus and by excitatory (glutaminergic) corticothalamic fibers that arise in layer VI of the primary and secondary somatosensory cortices.

Primary Somatosensory Cortex (SI). Axons from third-order thalamic neurons terminate in the *primary somatosensory cortex* (SI) (Fig. 17–10; see also Figs. 17–6

ally located *ventral posterolateral nucleus* (VPL) and the medially located *ventral posteromedial nucleus* (VPM). In humans, these nuclei have also been termed the ventralis caudalis externus and ventralis caudalis internus, respectively. The VPL is separated from the VPM by fibers of the *arcuate lamina*. The ventral posterior nucleus is supplied by thalamogeniculate branches of the posterior cerebral artery, and compromise of these vessels can result in loss of all tactile sensation over the contralateral body and head (Fig. 17–9).

The VPL receives ascending input from the medial lemniscus, and input to the VPM is from the trigeminothalamic tracts. Within VPL, medial lemniscal fibers from the contralateral cuneate nucleus terminate medial to those from the gracile nucleus. As a result, the representation of the lower limb and foot is lateral and of the upper limb is medial in VPL (see Fig. 17–9). The representation of an individual body part is organized as a C-shaped lamina. Tactile signals are also represented in other thalamic nuclei receiving lemniscal input, including the ventral posterior inferior nucleus and the pulvinar and lateral posterior group.

In addition to their somatotopic organization, the medial lemniscal fibers that terminate in the ventral posterior nucleus are segregated on the basis of their

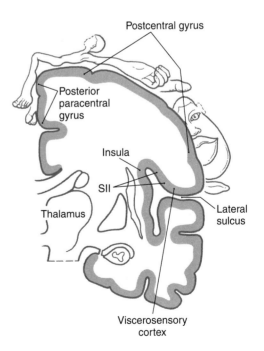

Figure 17–10. The homunculus (body representation) of the primary somatosensory (SI) cortex. (Adapted from Penfield and Rasmussen, 1968, with permission.)

and 17–9). This cortical region is bordered anteriorly by the central sulcus and posteriorly by the postcentral sulcus, and comprises the postcentral gyrus and the posterior paracentral gyrus (Fig. 17–11). The cortex contains a somatotopic representation of the body surface (a *homunculus*, or "little man"), which is laid out in a "foot-to-tongue" pattern along the medial-to-lateral axis (see Fig. 17–10). Body regions, such as the hand and the lips, with a high density of receptors have a disproportionately large amount of cortical tissue dedicated to their central representation. In contrast, regions, such as the back, with low receptor density have small cortical representations (see Figs. 17–3 and 17–10). Blood supply to the SI cortical areas is provided by the anterior and middle cerebral arteries. Vascular lesions involving the middle cerebral artery produce tactile loss over the contralateral upper body and face, and those involving the anterior cerebral artery affect the contralateral lower limb (see Fig. 17–9).

Histologically, the primary somatosensory cortex is subdivided into four distinct areas; from anterior to posterior, these are *Brodmann areas 3a, 3b, 1,* and *2* (see Fig. 17–11). Area 3a is located in the depths of the central sulcus and abuts area 4 (primary motor cortex). Areas 3b and 1 extend up the side of the sulcus onto the shoulder of the postcentral gyrus, whereas area 2 lies on the gyral surface and abuts area 5 (somatosensory association cortex).

Each of these four cytoarchitectural areas of the SI cortex receives submodality-specific inputs. Areas 3a and 2 are primarily targeted by neurons in the "shell" region of VPL. They receive proprioceptive inputs arising from muscle spindle afferents (mainly area 3a), Golgi tendon organs, and joint afferents (mainly area 2). These two areas are capable of processing kinesthetic information related to muscle length and tension, as well as static and transient joint position. Areas 3b and 1 are mainly targeted by neurons in the "core" region of the VPL. They receive cutaneous afferents from receptors such as Meissner corpuscles (RA) and Merkel cells (SA).

Damage to various parts of the somatosensory cortex results in characteristic types of sensory losses. Lesions involving area 1 produce a deficit in texture discrimination, whereas damage to area 2 results in loss of size and shape discrimination (*astereognosis*). Injury to area 3b has a more profound effect than that from damage to either area 1 or 2 alone, producing deficits in both texture and size and shape discrimination. This difference suggests that there is hierarchical processing of tactile information in SI cortex, with area 3b performing the initial processing and distributing the information to areas 1 and 2.

Additional Cortical Somatosensory Regions. The secondary somatosensory cortex (SII) lies deep in the inner face of the upper bank of the lateral sulcus (see Fig. 17–10). It, too, contains a somatotopically organized representation of the body surface. Inputs to SII cortex arise from the ipsilateral SI cortex, as well as from the ventral posterior inferior nucleus (VPI) of the thalamus, a triangle-shaped nucleus lying anterior (ventral) to VPL and VPM. This cortical area is also supplied primarily by the middle cerebral artery, so it cannot substitute functionally for SI following vascular compromise of this artery (see Fig. 17–9).

Posterior to area 2, additional parietal cortical regions also receive tactile inputs. These regions include area 5 and lateral portions of area 7 (7b). The anterior pulvinar and lateral posterior group, which receive some medial lemniscal input, project to areas 5 and 7 (see Fig. 17–11). In addition, they also receive input from primary somatosensory cortex. Lesions in the parietal association area can produce *agnosia*, in which contralateral body parts are lost from the personal body map. Sensation is not radically altered, but the limb is not dressed and is not recognized as part of the patient's own body.

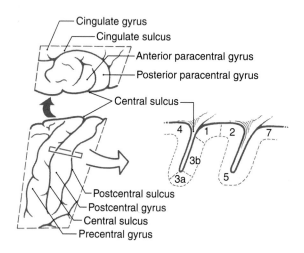

Figure 17–11. The primary somatosensory cortex (SI) of the parietal lobe. The views on the left show the location of SI in the postcentral gyrus (lateral view, bottom) and as it extends medially into the posterior paracentral gyrus (top view). See Figure 17–12 for an overview of the entire hemisphere. The cross section at right shows the subdivisions of the SI cortex into the four cytoarchitecturally distinct areas 3a, 3b, 1, and 2. Rostral to these is motor cortex (area 4), and caudal to them are association areas 5 and 7.

Trigeminal System

Overview. Most of the somatosensory information from craniofacial structures, including the oral and nasal cavities, is transmitted to the brainstem over the trigeminal nerve. The relay and central processing of input from trigeminal primary afferent neurons occur in a column of brainstem neurons that begins rostrally in the middle pons and extends caudally to overlap with the posterior horn of the cervical spinal cord. The primary afferent neurons of the trigeminal nerve, the brainstem nuclei, and pathways described in the following discussion are referred to as the trigeminal system. Like spinal cord somatosensory pathways, trigeminal pathways can be subdivided into those for sensations of pain and thermal sense (see Chapter 18) and those for tactile discrimination, flutter-vibration, and proprioception and kinesthesia.

Trigeminal Nerve. As its name implies, the trigeminal nerve has three peripheral divisions: *ophthalmic* (V_1), *maxillary* (V_2), and *mandibular* (V_3). A few nerve fibers in the *facial* (cranial nerve VII), the *glossopharyngeal* (cranial nerve IX), and the *vagus* (cranial nerve X) nerves transmit GSA innervation from a small cutaneous area around the ear. The peripheral distribution of these nerves delineates the facial "dermatomes" (see Chapter 18).

The cell bodies of trigeminal primary afferent neurons are located in the *trigeminal* (*gasserian* or *semilunar*) *ganglion* (Figs. 17–12 and 17–13) and in the mesencephalic trigeminal nucleus. The central processes of trigeminal ganglion cells form the large *sensory root* (*portio major*) of the trigeminal nerve as they enter the lateral aspect of the pons. Within the brainstem, central processes of most trigeminal ganglion cells bifurcate into ascending and descending branches before terminating on *second-order neurons* in the *brainstem trigeminal sensory nuclei.* The ascending branches terminate in the *principal sensory nucleus*, located in the pons, and the descending branches coalesce to form the *spinal* (*descending*) *tract of the trigeminal nerve.* The axons of this tract terminate throughout the rostrocaudal extent of the *spinal nucleus of the trigeminal nerve*, which lies just medial to the tract. Discussion in this chapter focuses on the more rostral components of the trigeminal system, including the *principal sensory nucleus* and the *trigeminal mesencephalic nucleus.* These nuclei and their connections are primarily involved in tactile discrimination, proprioception, and kinesthesia from the head. The role of more caudal components of the trigeminal system, which serve a primary role in pain and thermal sensations, is considered in Chapter 18.

Anterior and Posterior Trigeminothalamic Tracts

Peripheral Receptors. Tactile sensations originating in the head are transduced into nerve impulses by the same types of nerve terminals and specialized sensory receptor organs found in other parts of the body (see Fig. 17–1). However, owing to their association with structures unique to this region, some of these sensory endings serve specialized functions. For example, receptors in the periodontal ligament (the primarily collagenous connective tissue surrounding each tooth) are exquisitely sensitive to tooth displacement and bite force. A large number of encapsulated receptors, particularly Meissner corpuscles, are found beneath the surface of the lips and perioral skin. The precision of two-point tactile discrimination on the lips and perioral regions is comparable with that on the finger tips. Most of the primary afferent neurons concerned with perception of discriminative sensation from the face and oral cavity have large-diameter (e.g., Aβ) axons. Some of these ascend without branching, whereas others bifurcate before terminating in the *principal sensory nucleus.*

The *principal* (*chief*) *sensory nucleus* is situated in the middle pons at the rostral pole of the spinal trigeminal nucleus (see Figs. 17–12 and 17–13). The principal sensory nucleus can be divided into dorsomedial and ventrolateral regions. The dorsomedial division receives most of its primary afferent input from the oral cavity, and the ventrolateral division receives input from all three components of the trigeminal nerve. Thus, the somatotopic pattern of the principal sensory nucleus is inverted, with V_1 being ventral, V_3 dorsal, and V_2 sandwiched in between (see Fig. 17–12).

Second-order neurons in the principal sensory nucleus have relatively small receptive fields. In addition, they are subject to the same types of intranuclear modulations as are cells in the posterior column nuclei (e.g., lateral inhibition and descending modulation from the SI cortex), which serves to sharpen the contrast between adjacent receptive fields. Neurons in the principal sensory nucleus relay discriminative tactile information from the head to the *ventral posteromedial nucleus* (VPM). Neurons in the ventrolateral part of the principal sensory nucleus give rise to axons that project to the contralateral VPM along with fibers that originate in the more caudally located spinal trigeminal nucleus. This combined ascending projection forms the *trigeminal lemniscus*, or *anterior* (*ventral*) *trigeminothalamic tract* (see Figs. 17–12 and 17–13), which courses in close proximity to the medial lemniscus. Neurons in the dorsomedial division of the principal sensory nucleus project to the ipsilateral VPM by way of the *posterior* (*dorsal*) *trigeminothalamic tract* (see Fig. 17–12). This pathway ascends in the pontine tegmentum lateral to the periaqueductal gray in close association with the central tegmental tract. The afferent projections from the principal sensory nucleus terminate somatotopically within VPM so that the oral cavity is represented medially and the lateral facial structures are represented more laterally (see Figs. 17–9 and 17–12). Third-order thalamocortical neurons in VPM project via the posterior limb of the internal capsule to the laterally placed face area of SI in the postcentral gyrus (see Figs. 17–9 and 17–10). Perioral regions have the highest peripheral innervation density, and consequently the largest representation along the postcentral gyrus (see Fig. 17–10).

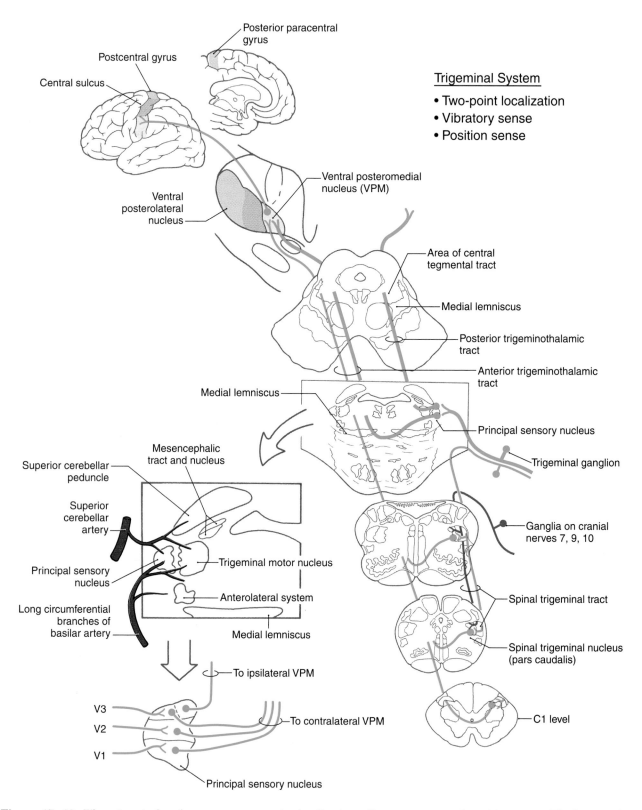

Figure 17–12. The trigeminal pathways carry two-point localization, vibratory sense, and position sense. Thick green lines represent large-diameter primary afferent fibers, and thin green lines represent smaller-diameter fibers conveying primarily thermal and nociceptive information. Note the inverted pattern of primary afferent fibers in the principal sensory nucleus.

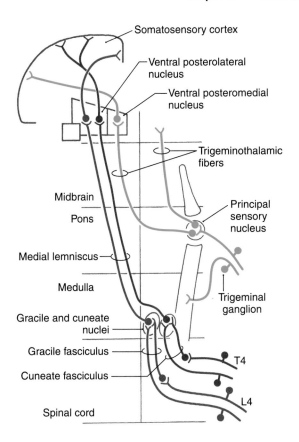

Figure 17–13. Summary of posterior column–medial lemniscal and trigeminothalamic pathways carrying discriminative touch, flutter-vibration, and position sense to the contralateral primary somatosensory cortex.

Proprioceptive endings (muscle spindles) in jaw muscles and some periodontal ligament receptors (modified Ruffini endings and free nerve endings) are innervated by primary afferent neurons located in the *trigeminal mesencephalic nucleus.* This brainstem nucleus consists of a slender column of pseudounipolar cells of neural crest origin that remain within the neural tube during development. Cells of the mesencephalic nucleus extend from the rostral pons to upper midbrain levels, where they form a thin band of neurons along the lateral edge of the periaqueductal gray. An important difference between the cell bodies of the mesencephalic nucleus and typical ganglion cells is that the former receive synaptic inputs from peptidergic and monoaminergic neurons in the brainstem. This synaptic influence on the neurons of the trigeminal mesencephalic nucleus provides a unique form of presynaptic modulation before central relay of the primary afferent information.

The processes of cells in the trigeminal *mesencephalic nucleus form the mesencephalic tract of the trigeminal nerve.* This tract is located directly adjacent to the mesencephalic nucleus (see Fig. 17–12; see also Fig. 17–16), and also extends rostrally, where it borders the midbrain aqueductal gray. The central processes of the

trigeminal mesencephalic neurons generally branch in the area posterior (dorsal) to the trigeminal motor nucleus to innervate cells of the motor nucleus. This input to the motor nucleus forms the afferent limb of the *myotatic jaw jerk reflex.* This clinically useful reflex consists of the processes of trigeminal mesencephalic nucleus neurons that innervate muscle spindles in jaw closing muscles and terminate monosynaptically on trigeminal motor neurons. In turn, the axons of trigeminal motor neurons innervate muscles (i.e., temporalis) that elevate the jaw. A gentle tap on the jaw activates the afferent fibers of this reflex and initiates a contraction of the homonymous muscle (e.g., temporalis) as well as its synergists (e.g., masseter). Trigeminal mesencephalic afferents from the periodontal ligament provide feedback to jaw muscle motor neurons during mastication to regulate bite force, although this input is not monosynaptic. In addition to providing collaterals to the trigeminal motor nucleus, a bundle of descending branches of mesencephalic tract fibers (called the Probst tract) distributes to the reticular formation, the spinal trigeminal nucleus, and the cerebellum. The central connections of the trigeminal mesencephalic nucleus are consistent with their broad participation in the coordination of oral motility patterns including mastication, swallowing, and speech.

Proprioceptive input from the mesencephalic nucleus is also provided to the principal sensory nucleus and the spinal trigeminal nucleus via collaterals of the Probst tract. Some proprioceptive receptors are innervated by trigeminal ganglion cells, such as those with receptors in the temporomandibular joint, the extraocular muscles, and some periodontal ligaments. Because most trigeminal ganglion axons bifurcate when they enter the brainstem, both the principal sensory and the spinal trigeminal nucleus receive proprioceptive input. Proprioceptive input to the spinal trigeminal nucleus is relayed to the cerebellum, the spinal cord, and the thalamus. However, the principal sensory nucleus receives a disproportionate share of large-diameter, heavily myelinated fibers and may be considered the trigeminal homologue of the posterior column nuclei. These pathways provide the substrate for cortical processing that permits the full hedonic appreciation of foods with different textural properties (oral stereognosis).

Receptive Field Properties of Cortical Neurons

Neurons located in cortical areas representing the body and head are organized into functional units called *cortical columns* (see Fig. 32–9). These are distributed from the pial surface to the cortical white matter. Each column contains neurons responsive to one submodality, and the cells in a column all have similar peripheral receptive field loci. Thalamocortical inputs terminate on stellate cells in layer IV and lower parts of layer III of the SI cortex. Axons of the stellate cells distribute information vertically to the pyramidal cells within individual columns.

The receptive field properties of cortical neurons are more complex than those at subcortical levels. Cortical neurons respond to a specific stimulus orientation (edges) and to specific textures. They are also capable of coding the velocity, speed, and direction of moving stimuli. At least three distinct populations of neurons receive proprioceptive inputs. The first consists of simple neurons that receive input from a single joint or muscle group. These rapidly adapting cells signal movement. The second group consists of postural neurons that signal the final position of a joint once the movement is completed. The third is made of neurons that receive inputs from several joints and muscle groups (multijoint) and signal complex joint-muscle interactions.

The functional properties of cortical neurons reflect the processing and integration of sensory information as it ascends from the posterior column and ventral posterior nuclei to the final processing station in the cortical columns. This sensory signal processing can include (1) convergence of afferent input, which increases receptive field size while decreasing resolution; (2) divergence of output signal, which allows relay cells to amplify the sensory signal and supply it to multiple targets; (3) facilitation; and (4) inhibition. These processes act in concert to enhance the signal-noise ratio in terms of both space and time.

In general, cortical neurons display larger receptive fields and more complex inhibitory surrounds than displayed by their subcortical inputs. For example, a tactile stimulus in the center of a receptive field results in amplification of the sensory signal and increased activity in a restricted population of cortical cells. Conversely, stimulation at the edge of the receptive field suppresses the activity in these neurons. This mechanism provides the circuits active in two-point discrimination.

Neuroimaging and Functional Localization

Neuroimaging techniques including functional magnetic resonance imaging (fMRI), regional cerebral blood flow (rCBF) studies, positron emission tomography (PET), and magnetoencephalography (MEG) have been used in recent clinical studies of the somatosensory pathway in humans. These techniques have elegantly demonstrated the functional organization of somatosensory areas activated by application of various tactile stimuli. For example, PET studies have identified cortical areas 3b and 1 as participating in the discrimination of moving stimuli, whereas cortical area 2 is activated when subjects palpate objects focusing on shape and curvature. Functional MRI studies have identified two areas of increased blood flow, suggesting a concomitant increase in cerebral cortical activity in response to air puff stimulation applied to various loci on the upper limb. One cortical area, located in the depth of the central sulcus, corresponds to area 3b. The other cortical locus of increased activity identified in these studies is posterior and lateral and corresponds to area 1. PET studies have also provided evidence that a tactile stimulus activates both primary (SI) and second somatosensory (SII) cortices.

Plasticity and Reorganization in the Primary Somatosensory Cortex

Brain injury, whether resulting from birth or other trauma, tumors, or stroke, can affect anyone and can be devastating. However, on closer examination, some individuals appear to "recover" lost functions, whereas others remain relatively unchanged. In general, the younger the person suffering the trauma, the more "recovery" is noted. To understand possible mechanisms for this apparent recovery we must first look at brain development. The brain of a child is quite malleable or plastic. It forms multiple, redundant neural connections linking various brain areas. Many of these connections will be retained, through usage and experience, whereas others will be "pruned" by programmed cell death (*apoptosis*) and other cellular mechanisms. These processes will continue for a finite period of time (a critical period), thus giving many brain regions the potential to function in a variety of ways. Children suffering brain trauma due to a birth injury may appear to be quite normal with respect to sensory, motor, and cognitive abilities. It is only after inspection of a brain scan (such as from an MRI study) that the abnormal brain anatomy resulting from the injury is appreciated. Apparently, the developing brain possesses the ability to reassign brain functions to other brain regions. This is commonly referred to as plasticity.

In contrast with that in children, the nervous system in the adult has passed beyond the critical periods of brain development and has become relatively nonmalleable. It has been a common view that most neural connections in the adult are stable and have lost much of their capacity to form new synapses. Evidence suggests, however, that the somatosensory cortex can undergo reorganization. An example of this phenomenon is the changes in the cortical map following limb or digit amputation. Normally (Fig. 17–14*A*), each digit has a sequential representation in the somatosensory cortex. When digits are amputated, there is a loss of input to the corresponding area(s) of the somatosensory cortex. The resultant reorganization usually involves cortical sensory neurons or areas that have lost their input as a result of the amputation. The cortical neurons or areas representing the missing body part now respond when skin regions adjacent to the ampu-

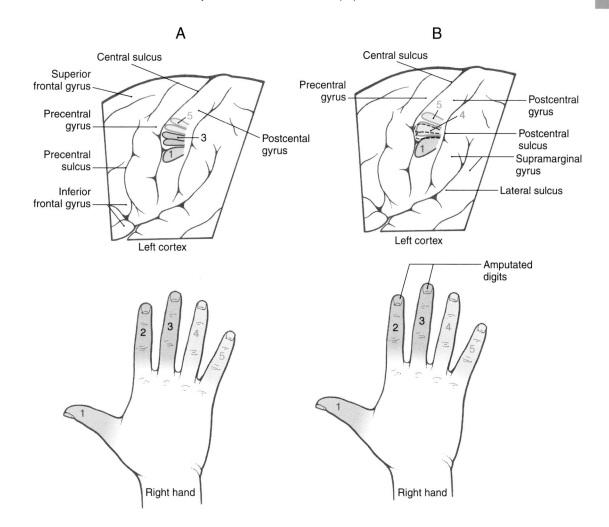

Figure 17–14. A representation of the digit region of the somatosensory cortex (*A*) and how this area is reorganized following amputation of the second and third digits (*B*).

tated body part are stimulated. There appears to be an expansion of cortical neurons representing body parts flanking the amputated digits into those cortical regions that previously had the map of the now-missing body parts (see Fig. 17–14*B*). Although many of these cortical changes are subtle, a recent study suggests that the time course of this reorganization can be quite rapid, beginning within 10 days after amputation. Thus, it appears that the adult brain can exhibit plastic changes and undergo reorganization in response to specific peripheral perturbations. Similar phenomena have been described by molecular biologic methods within hours after experimental induction by appropriate stimuli.

Nonconscious Proprioception: Spinocerebellar Pathways

Four spinocerebellar pathways transmit proprioceptive information and limited exteroceptive signals from cutaneous mechanoreceptors to the cerebellum (Fig.

17–15). These sensory signals include information about limb position, joint angles, and muscle tension and length. Input to the cerebellum plays an integral role in guiding cerebellar control of body muscle tone, movement, and posture.

Spinocerebellar tract axons terminate in the cerebellar nuclei and as mossy fibers in the vermis and paravermal region of the cerebellum. Sometimes these areas are collectively called the *spinocerebellum*. This afferent input to the cerebellum forms a pair of somatotopic representations of the body surface in the anterior lobe and the paravermis of the posterior lobe (see Fig. 17–15). Degeneration of the major spinocerebellar tracts occurs in diseases such as *Friedreich ataxia*. The result is *cerebellar ataxia*— lack of coordination during walking and other movements that occurs because the cerebellum is not receiving the sensory feedback necessary to regulate movement.

Posterior Spinocerebellar Tract. Proprioceptive afferents and a limited number of exteroceptive (cutaneous) afferents from the lower limb and lower trunk

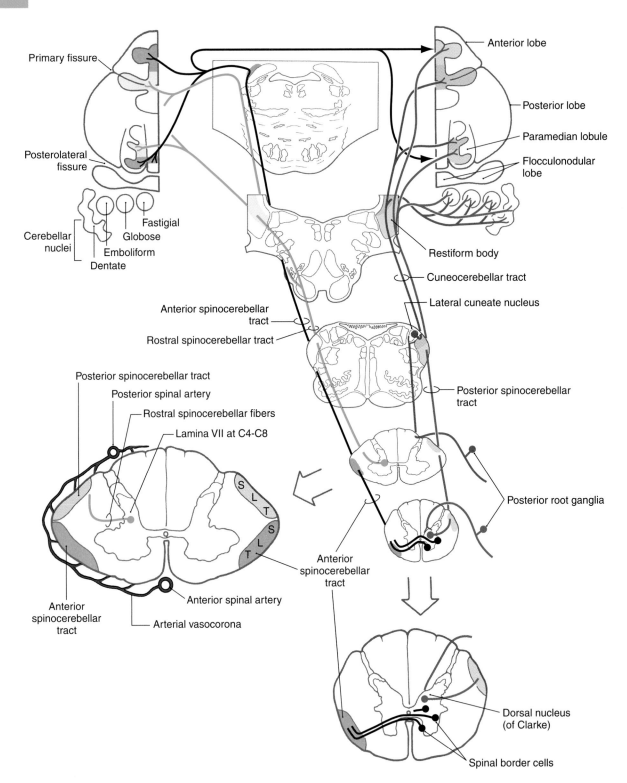

Figure 17–15. Organization of posterior, anterior, and rostral spinocerebellar tracts and of the cuneocerebellar tract.

travel in the posterior spinocerebellar tract to reach the ipsilateral cerebellar cortex (see Fig. 17–15). Posterior root fibers from the trunk and lower limb terminate on cells in the *dorsal nucleus of Clarke*, which is located in lamina VII of the intermediate zone in spinal segments T1 to L2. Primary afferent fibers from the spinal cord

levels caudal to L2 ascend in the posterior funiculus to reach this nucleus.

Group I muscle spindle and Golgi tendon organ afferents monosynaptically activate cells in the Clarke nucleus. The discharge rate of these posterior spinocerebellar tract cells shows a linear relationship to muscle length; therefore, their firing rate can encode muscle length as a frequency code. Group II and group III tactile fibers also terminate on other spinocerebellar cells in the Clarke nucleus. Axons from cells in the dorsal nucleus of Clarke traverse the ipsilateral lateral funiculus and collect on the surface of the spinal cord lateral to the corticospinal tract. These fibers ascend to reach the cerebellum via the restiform body.

Cuneocerebellar Tract. The cuneocerebellar tract is the upper limb equivalent of the posterior spinocerebellar tract (see Fig. 17–15). Posterior root fibers in spinal segments C2 to T4 carry muscle spindle and exteroceptive information in the ipsilateral cuneate fasciculus to the cuneate nucleus. In the lower medulla, proprioceptive primary afferent fibers terminate somatotopically in the *lateral cuneate nucleus*. Cells of the lateral cuneate nucleus project as *cuneocerebellar fibers* to the cerebellum via the restiform body. Exteroceptive input arising from the rostral end of the cuneate nucleus also ascends to the cerebellar cortex to terminate in the folia of the anterior lobe in lobule V.

Anterior Spinocerebellar Tract. This pathway relays information from group I afferents arising in the lower limb. The cells of origin of this pathway are located in lumbar segments L3 to L5. They are located in the lateral part of Rexed laminae V to VII, as well as along the anterolateral border of the anterior horn, where they are called *spinal border cells* (see Fig. 17–15). The axons of anterior spinocerebellar tract (ASCT) cells immediately *cross the midline* in the anterior white commissure and ascend in the lateral funiculus anterior to the posterior spinocerebellar tract. In the pons, these fibers turn posterolateral to enter the cerebellum via the *superior cerebellar peduncle*. Most fibers *recross* to terminate in the cerebellum ipsilateral to their side of origin. These ASCT fibers are distributed more laterally in the cerebellum than are those of the posterior tract. Cells giving rise to ASCT fibers are strongly influenced by descending projections of the reticulospinal and corticospinal pathways. Reticulospinal input inhibits ASCT cells, and the corticospinal input facilitates ASCT cells. Vestibulospinal and rubrospinal projections also monosynaptically excite ASCT cells.

Rostral Spinocerebellar Tract. This tract, the upper limb equivalent of the ASCT, arises from cell bodies located in lamina VII of the cervical enlargement (C4 to C8) (see Fig. 17–15). The efferent projections from these neurons ascend uncrossed in the lateral funiculus of the spinal cord. Although most of these axons enter the cerebellum via the restiform body, some travel in the superior cerebellar peduncle. The rostral spinocerebellar tract from the upper limb and the anterior spinocerebellar tract from the lower limb relay cutaneous

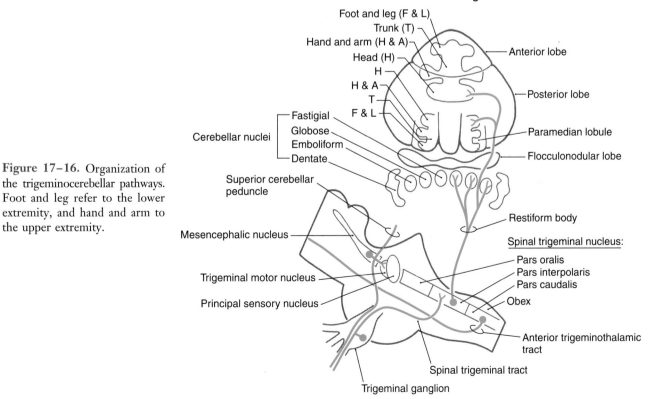

Figure 17–16. Organization of the trigeminocerebellar pathways. Foot and leg refer to the lower extremity, and hand and arm to the upper extremity.

Trigeminocerebellar Connections

Foot and leg (F & L)
Trunk (T)
Hand and arm (H & A)
Head (H)
H
H & A
T
F & L

Anterior lobe
Posterior lobe
Paramedian lobule
Flocculonodular lobe

Cerebellar nuclei
Fastigial
Globose
Emboliform
Dentate

Superior cerebellar peduncle

Restiform body

Spinal trigeminal nucleus:
Pars oralis
Pars interpolaris
Pars caudalis
Obex

Mesencephalic nucleus

Trigeminal motor nucleus
Principal sensory nucleus

Anterior trigeminothalamic tract

Spinal trigeminal tract
Trigeminal ganglion

tactile information from Meissner, Merkel, and Pacinian mechanoreceptors (group II and group III afferents) to the cerebellum.

Trigeminocerebellar Connections. The oral motor system requires continual feedback during mastication. As food is chewed, its texture and consistency are altered, changing the demands on jaw muscles. In addition, adaptation is required for long-range functional changes. For example, there are modifications in jaw motility patterns during the transition from suckling to chewing in the newborn and from natural dentition to the use of dentures. It is probable that proprioceptive information reaching the cerebellum from jaw muscle spindles, periodontal afferents, and the temporomandibular joint is involved in these processes.

Branches of the central processes of the mesencephalic trigeminal neurons are distributed to the cerebellar hemispheres and nuclei via the superior cerebellar peduncle (Fig. 17–16). Additional proprioceptive signals from the spinal trigeminal nucleus *pars interpolaris* and *pars candalis* enter the cerebellum by way of the restiform body. They contribute a head representation to the two somatotopic maps in the cerebellar cortex.

Sources and Additional Reading

Bodegard A, Geyer S, Naito E, Zilles K, Roland PE: Somatosensory areas in man activated by moving stimuli: Cytoarchitectural mapping and PET. Neuroreport 11:187–191, 2000.

Bodegard A, Ledberg A, Geyer S, Naito E, Zilles K, Roland PE: Object shape differences reflected by somatosensory cortical activation in human. J Neurosci 20:Rapid Communication 51, 2000.

Brodal A: Neurological Anatomy in Relation to Clinical Medicine, 3rd Ed. Oxford University Press, New York, 1981.

Burgess PR, Perl ER: Cutaneous mechanoreceptors and nociceptors. In A Iggo (ed): Handbook of Sensory Physiology, vol 2: Somatosensory System. Springer-Verlag, New York, 1973, pp 30–78.

Frot M, Mauguiere F: Timing and spatial distribution of somatosensory responses recorded in the upper bank of the sylvian fissure (SII area) in humans. Cerebral Cortex 9:854–863, 1999.

Johansson RS, Vallbo AB: Tactile sensory coding in the glabrous skin of the human hand. Trends Neurosci 6:27–32, 1983.

Johnsen-Berg H, Christensen V, Woolrich M, Matthew PM: Attention to touch modulates activity in both primary and second somatosensory areas. Neuroreport 11:1237–1241, 2000.

Jones EG: Cortical and subcortical contributions to activity-dependent plasticity in primate somatosensory cortex. Annu Rev Neurosci 23:1–37, 2000.

Kuwabara S, Mizobuchi K, Toma S, Nakajima Y, Ogawara K, Hattori T: "Tactile" sensory nerve potentials elicited by air-puff stimulation: A microneurographic study. Neurology 54:762–765, 2000.

Lenz FA, Dostrovsky JO, Tasker RR, Yamashiro K, Kwan HC, Murphy JT: Single-unit analysis of human ventral thalamic nuclear group: Somatosensory responses. J Neurophysiol 59:299–316, 1988.

Mountcastle VB: Neural mechanisms in somesthesis. In Mountcastle VB (ed): Medical Physiology, 14th Ed, vol I. CV Mosby, St Louis, 1980, pp 348–390.

Parent A: Carpenter's Human Neuroanatomy, 9th Ed. Williams & Wilkins, Baltimore, 1995.

Penfield W, Rasmussen T: The Cerebral Cortex of Man: A Clinical Study of Localization of Function. Hafner Publishing, New York, 1968 (Facsimile of 1950 ed).

Pizella V, Tecchio F, Romani GL, Rossini PM: Functional localization of the sensory hand area with respect to the motor central gyrus knob. Neuroreport 10:3809–3814, 1999.

Servos P, Zacks J, Rumelhart DE, Glover GH: Somatotopy of the human arm using fMRI. Neuroreport 9:605–609, 1998.

Vallbo AB, Olsson KA, Westberg K-G, Clarke FJ: Microstimulation of single tactile afferents from the human hand: sensory attributes related to unit type and properties of receptive fields. Brain 107:727–749, 1984.

Weiss T, Miltner WHR, Huonker R, Friedel R, Schmidt I, Taub E: Rapid functional plasticity of the somatosensory cortex after finger amputation. Exp Brain Res 134:199–203, 2000.

Chapter 18

The Somatosensory System II: Touch, Thermal Sense, and Pain

S. Warren, R. P. Yezierski, and N. F. Capra

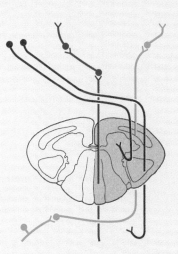

Overview 274

Anterolateral System 274
Overview
Receptors and Primary Neurons
Peripheral Sensitization and Primary Hyperalgesia
Central Sensitization and Secondary Hyperalgesia
Pain Receptors in Muscles, Joints, and Viscera
Central Pathways

Spinal Trigeminal Pathway: Anterior Trigeminothalamic Tract 285
Primary Neurons
Central Pathways

Imaging Studies of Pain in the Somatosensory Pathway 288

Pain Perception 288

Pain Perception in the Somatosensory Thalamus 288

Pain Transmission and Control 291

One crucial role of the somatosensory system is to supply the brain with information related to insults that could damage tissue. These signals ascend the neuraxis in a fiber bundle called the *anterolateral system* (ALS). Anyone who has used a hammer or hot skillet has had experience with this system. Hit your thumb with a hammer, and if you are lucky, only *high-threshold mechanoreceptors* that signal excess skin deformation will be activated. If you are unlucky, tissue is damaged and the result is pain (*nociception*). Specifically, *mechanonociceptors* have been stimulated. One common response is to gently rub the damaged area. This activates central nervous system (CNS) pathways that decrease the transmission of nociceptive signals and alter the perception of pain.

After the blow, damaged tissues release chemicals that activate another type of pain receptor, the *chemonociceptor*. These receptors may contribute to the mechanism underlying long-term pain and tenderness (*hyperalgesia*). Similarly, the temperature of a skillet is detected by *thermoreceptors* in the skin and transmitted through the ALS. If a burn is produced, the tissue damage is signaled by high-frequency firing of *thermonociceptors*. ALS activation can lead to a variety of responses, including withdrawal reflexes, the conscious perception of pain, emotional effects such as suffering, and behavioral changes aimed at avoiding the cause of the pain.

Overview

Nondiscriminative (*crude* or *poorly localized*) *touch, innocuous thermal,* and *pain* (*mechanical, chemical,* and *thermal*) sensations are conveyed by pathways that collectively make up the *ALS*. This system transmits signals originating in peripheral receptors to spinal cord and brainstem neurons (Fig. 18–1). These signals are then forwarded to thalamic nuclei and from there to the trunk and extremity representations in the primary somatosensory cortex. The *anterior trigeminothalamic pathway* (Fig. 18–1; see also Fig. 18–11) carries similar signals that originate from receptors in the head. These are relayed through brainstem and thalamic nuclei to the face area of the somatosensory cortex. The touch fibers of the ALS differ from those described for the posterior column–medial lemniscal system (PCML system) (see Chapter 17) in several ways: (1) they yield a generalized feeling of being touched but do not give precise localization, (2) their receptive fields are larger, and (3) they are smaller in diameter and more slowly conducting. Disruption of the ALS can produce symptoms ranging from reduced sensibility (*hypesthesia*), to numbness, tingling, and prickling (*paresthesia*), to a complete loss of sensibility (*anesthesia*).

Anterolateral System

Overview. The ALS is a composite bundle that includes *spinothalamic, spinomesencephalic, spinoreticular, spinobulbar,* and *spinohypothalamic fibers.* Spinothalamic fibers project directly from the spinal cord to the ventral posterolateral (VPL) nucleus, the posterior nuclear group, and intralaminar nuclei (central lateral and centromedian–parafascicularis nuclei) of the thalamus. Collaterals to the reticular formation arise from some of these axons. Spinomesencephalic axons project to the periaqueductal gray (PAG) and to the tectum; the latter are *spinotectal fibers.* Although spinoreticular fibers project to the reticular formation of the medulla, pons, and midbrain, collaterals may ascend to other targets such as the thalamus. Projections of less relevance to the somatosensory system, such as *spino-olivary fibers,* are grouped under the category of *spinobulbar fibers.* Spinohypothalamic fibers terminate in hypothalamic areas and nuclei, including some that give rise to hypothalamospinal axons.

Fibers classically described as composing the lateral spinothalamic tract were considered to carry *only* pain

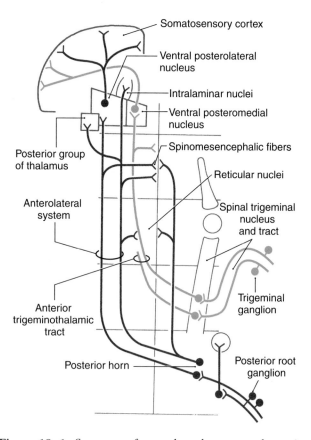

Figure 18–1. Summary of anterolateral system and anterior trigeminothalamic tract fibers conveying nondiscriminative tactile, thermal, and nociceptive inputs to the contralateral somatosensory cortex.

and thermal information, whereas the anterior spinothalamic tract was concerned *only* with nondiscriminative touch. This older view of separate tracts conveying separate types of information is not used in this chapter. Current thinking holds that all parts of the ALS carry all modalities (pain, temperature, and touch), but that there are direct and indirect routes. The former is the *neospinothalamic pathway* (spinal cord → thalamus), whereas the latter is the polysynaptic *paleospinothalamic pathway* (spinal cord → reticular formation → thalamus). Both of these pathways, plus other fibers as defined previously, collectively form the ALS.

Receptors and Primary Neurons. The receptors for nondiscriminative touch, innocuous thermal stimuli, and nociceptive stimuli are distributed in glabrous and hairy skin, as well as in deep tissues including joints and muscles (Table 18–1). Morphologically, these receptors are all *free (naked) nerve endings* (see Fig. 17–1); that is, they lack specialized receptor cells or encapsulations. Because of this lack, the basis for their submodality specificity is unclear. These submodalities are transduced by activation of peripheral branches of either *thinly myelinated Aδ fibers* or *unmyelinated C fibers*. The density of free nerve endings and the corresponding size of receptive fields vary over the body surface in the same way as for other cutaneous receptors (see Fig. 17–3), being highest on the hands and in the

perioral area. Regardless of size or location, however, each field is exquisitely sensitive to thermal, chemical, or mechanical stimuli.

Nociceptors (pain receptors) are found in cutaneous as well as in deep structures. Two major classes of cutaneous pain receptors have been identified. They are the Aδ *mechanical nociceptors* and the *C-polymodal nociceptors*. These receptors are found at the end of the peripheral processes of thinly myelinated (Aδ) or nonmyelinated (C) fibers (see Table 18–1). Aδ mechanical nociceptors respond to mechanical injury accompanied by tissue damage. C-polymodal nociceptors respond to mechanical, thermal, and chemical stimuli. In addition, other cutaneous receptors that respond to high threshold or noxious stimuli have been identified. They include receptors that respond to temperature changes (thermoreceptors or thermonociceptors) (see Table 18–1) and receptors that respond to chemicals, irritants, or algesic compounds (chemonociceptors) (see Table 18–1).

Nondiscriminative touch results from the stimulation of free nerve endings that act as non-noxious *high-threshold mechanoreceptors* (see Table 18–1). These receptors respond to any rough stimulus, including tapping, squeezing, rubbing, and stretching of the skin, that does not result in tissue damage (Fig. 18–2A). Nerve fibers associated with these receptors generally have no background activity when unstimulated and when stimulated they respond with a sustained discharge that signals stimulus duration (see Fig. 18–2A).

Non-nociceptive *thermoreceptors* fall into two classes: those activated by heat (35–45°C) and those activated by cold (17–35°C). They show a graded response to changes in ambient temperature (see Fig. 18–2B). With repeated stimulation, these receptors become sensitized and show a decreased threshold and larger response to the application of a stimulus. Levels of heat (>45°C) or cold (<17°C) that burn or freeze the skin produce high-frequency firing in both Aδ and C *thermonociceptors* (Fig. 18–2B).

Tissue damage causes the release of a number of chemical substances that activate a class of free nerve endings called *chemonociceptors* (see Table 18–1). Specifically, chemonociceptors are activated by the release of endogenous substances such as bradykinin, H$^+$ ions, and foreign irritants such as insect venoms. These free nerve endings are the peripheral processes of C fibers.

Peripheral Sensitization and Primary Hyperalgesia. Pain receptors, unlike Meissner corpuscles or Merkel cells, demonstrate a unique phenomenon called *sensitization*. These receptors become more sensitive (lower pain threshold) and thus more responsive (manifested by increases in firing rate) to noxious stimulation within their receptive fields. Although the mechanisms responsible for receptor sensitization are not completely known, chemicals released by the damaged skin

Table 18–1. Classification of Cutaneous Mechanical, Thermal, and Nociceptive Receptors Using Small-Diameter Fibers and Their Adequate Stimuli

Receptor	Stimulus
Cutaneous Mechanoreceptors	Respond to nondiscriminative tactile stimuli
Aδ and C fiber high-threshold mechanoreceptors	Pinch, rub, stretch, squeeze
Cutaneous Thermoreceptors	Respond to transient changes in temperature
Warm and cool thermoreceptors	Innocuous warm and cool stimuli
Cutaneous Nociceptors	Mediate cutaneous pain
Aδ mechanonociceptors	Mechanical tissue damage
C-polymodal nociceptors	Mechanical tissue damage, noxious thermal stimuli, algesic compounds
Other Cutaneous Nociceptors	Mediate cutaneous pain
C fiber mechanonociceptors	Mechanical tissue damage
Aδ, heat thermonociceptors	Noxious thermal stimuli, tissue damage (?)
Aδ, C fiber cold thermonociceptors	Noxious thermal stimuli, tissue damage (?)
C fiber chemonociceptors	Algesic compounds

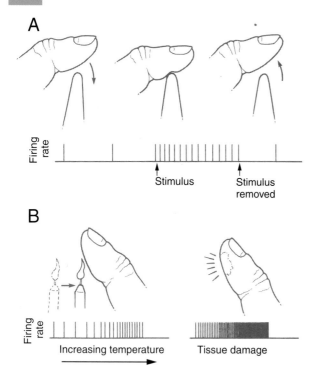

Figure 18–2. Fibers conveying information from high-threshold mechanoreceptors (*A*) respond to the application of a punctate stimulus. Thermoreceptors show a graded response to increases in temperature (*B, left*), whereas burns produced by prolonged thermal stimulation evoke high-frequency response in thermonociceptors (*B, right*).

or byproducts from plasma, or both, are thought to contribute to this phenomenon. As a result of this heightened sensitivity, the affected area is exquisitely sensitive to painful stimuli, and patients experience a sensory disturbance called *hyperalgesia* (exaggerated response to a painful stimuli). This condition can be differentiated into *primary hyperalgesia* and *secondary hyperalgesia*. *Primary hyperalgesia* occurs in the region of damaged skin and is probably the result of receptor sensitization. *Secondary hyperalgesia* occurs in the skin bordering the damaged tissue. Although receptor sensitization may contribute to secondary hyperalgesia, there is likely to be a central (e.g., spinal) component as well.

Central Sensitization and Secondary Hyperalgesia. Sensitization of peripheral nociceptors causes an increase in spontaneous activity in the Aδ and C fibers. The central processes of these fibers enter the posterior horn of the spinal cord, where they activate posterior horn neurons. Ongoing inputs from these injured peripheral nociceptors evoke a number of changes in the central processing of sensory information by the posterior horn neurons. These changes include a marked increase in the receptive field size of the posterior horn neuron (to include skin areas not involved in the initial injury), an increased response of the cells to

the application of suprathreshold stimuli, a decreased response to stimulus application in the receptive field, and activation of the cell by novel inputs (e.g., a light breeze). This phenomenon is known as *central sensitization*, and it represents *a potentiated state in which the system has been shifted from one functional level (normal) to another (sensitized) by a change in transcription*. In some persons, an innocuous stimulus, such as a gentle breeze or a light touch, can evoke pain sensation in the skin bordering the damaged tissue. The perception of an innocuous stimulus as painful is referred to as *allodynia* and can be the result of central sensitization.

Pain Receptors in Muscles, Joints, and Viscera. In addition to the cutaneous pain receptors, pain receptors in muscles, joints, and viscera have also been identified (Table 18–2). *Muscle pain* is mediated by receptors of both group III (thinly myelinated) and group IV (nonmyelinated) afferent fibers. Excessive stretch or contraction following strenuous exertion may activate the muscle pain receptors of group III fibers. Receptors of the group IV fibers may be activated in response to the release of algesic compounds after muscle injury or ischemia. *Joint pain*, including arthritis, may be caused by inflammation. This pain is mediated by receptors associated with group III and group

Table 18–2. Classification of Deep and Visceral Nociceptors and Their Adequate Stimuli

Receptor	Adequate Stimulus
Muscle Nociceptors	Mediate muscle pain
Group III afferent fibers	Bradykinin, 5-HT, K⁺ ions, prostaglandin (PGE₂)
Group IV afferent fibers	Bradykinin, 5-HT, K⁺ ions, prostaglandin (PGE₂)
Joint Nociceptors	Mediate joint pain: arthritis (?)
Group III afferent fibers	Inflammation (e.g., K⁺, bradykinin) kaolin, carrageenan
Group IV afferent fibers	Inflammation (e.g., K⁺, bradykinin) kaolin, carrageenan
Visceral Nociceptors	Mediate visceral pain
Heart	
Aδ and C afferent fibers	Prostaglandin (PGE₂), H⁺ ions, bradykinin, K⁺ ions, ischemia
Respiratory System	
Lung irritant receptors (Aδ fibers)	Irritant aerosols and gases, mechanical stimuli
J receptors (C fibers?)	Capsaicin, pulmonary congestion or edema, inhaled irritants
Gastrointestinal Tract	
Rapidly adapting mechanoreceptors, slowly adapting mechanoreceptors, chemoreceptors (Aδ and C fibers?)	Irritation of the mucosa, distention, powerful contraction, torsion, traction, bloat, cramping, appendicitis, impaction
Urogenital Tract	
C-polymodal nociceptors: testis	Intense mechanical stimuli, noxious heat, algesic chemicals

5-HT, 5-hydroxytryptamine.

IV fibers. *Visceral pain* is often described as being diffuse in nature and difficult to localize and is frequently referred to an overlying somatic body location. In addition, visceral pain usually involves autonomic reflexes. Visceral pain receptors located in the heart, respiratory structures, the gastrointestinal tract, and the urogenital tract are poorly identified (see Table 18–2). These receptors can be activated by intense mechanical stimuli including overdistention or traction, ischemia, and endogenous compounds, including bradykinin, prostaglandins, H^+ ions, and K^+ ions. Activation of these receptors produces pain.

Of the two primary afferent fiber types carrying nociceptive sensations, Aδ fibers have a slightly faster conduction velocity (5–30 m/sec) than that of C fibers. They carry well-localized sensations, which do not evoke an affective component to the sensory experience. A pinprick, used clinically to test ALS function, is one stimulus that activates Aδ fibers. On the other hand, C fibers are smaller and conduct more slowly (0.5–2 m/sec). They transmit poorly localized sensations that produce a noticeable affective component. For example, the dull, persistent ache that follows a muscle pull results from activation of C fibers. Both Aδ and C fibers are considerably smaller and conduct more slowly than fibers of the PCML system. Nerve blocks or anoxia preferentially affects large-diameter, heavily myelinated fibers and thus usually results in loss of discriminative tactile, vibratory, and postural sensations to varying degrees. Local anesthetics, such as lidocaine or bupivacaine, preferentially affect small-diameter Aδ and C fibers and thus result in loss of nociception (*analgesia*).

The cell bodies of C and Aδ fibers are generally small compared with other pseudounipolar neurons in the posterior root ganglion. The central processes of these cells enter the spinal cord via the lateral division of the posterior root (see Fig. 17–5). Many smaller fibers contain peptides such as substance P and calcitonin gene–related peptide (CGRP), which may serve as neurotransmitters. In addition to their normal trajectory into the posterior horn, a small number of C fibers enter the spinal cord through the anterior root of a spinal nerve. It is possible that these fibers provide a basis for the return of pain after *posterior rhizotomy*, a procedure in which posterior roots are sectioned in an attempt to alleviate intractable pain.

The strip of skin that is innervated by the peripheral cutaneous branches of a given spinal nerve is called a *dermatome* (Figs. 18–3 and 18–4). The central processes of these nerves that convey cutaneous input terminate in the posterior horn. It is clinically useful to examine dermatomes that have relationships to landmarks on the body; for example, C7 for the index finger, the T4–T5 border at the nipples, T10 at the navel, L1 along the pelvic rim, L5 for the big toe, and S4 and S5 for the genitalia and anus (Fig. 18–4).

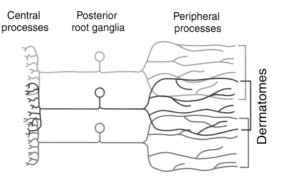

Figure 18–3. The dermatomes formed by the peripheral processes of adjacent spinal nerves overlap on the body surface (see also Fig. 18–4). The central processes of these fibers also overlap in their spinal distribution.

There is overlap between both the peripheral and central distribution of adjacent spinal nerves and consequently between their dermatomes (see Fig. 18–3). This overlap reduces the effects of injury to a single spinal root.

Shingles (*herpes zoster*) is a disease of viral etiology that is noteworthy for its dermatomal distribution (Fig. 18–5). Subsequent to a bout of chickenpox, viral DNA may infect and become latent in trigeminal and posterior root ganglion cells. The virus may reactivate periodically, producing infectious virions that travel down the peripheral processes of the neurons to produce a painful skin irritation in the dermatomal distribution of the ganglion (see Fig. 18–5). When an injury or disease process affects a series of nerve roots, the result is diminished sensibility (*hypesthesia*) over the dermatomes served by those roots. The borders of the hypesthetic region correspond to dermatomal boundaries. However, the most debilitating aspect of this disease is a poorly understood recurrence of pain. Known as *postherpetic neuralgia*, this condition is a neuropathic pain.

Clinically, it is important to test patients for intact ALS and PCML systems function. In testing the ALS, a single pin point applied to the skin should evoke a response (*perception of pain*) from the patient. Function of the intact PCML system is tested by the simultaneous application of two points spaced at measured intervals. As the points are moved closer, the ability to identify them as separate stimuli (*two-point discrimination*) decreases, and eventually disappears.

Central Pathways. As mentioned previously, Aδ and C fibers enter the spinal cord via the *lateral division* of the *posterior root entry zone*. The fibers enter the posterolateral fasciculus (*Lissauer's tract*) and bifurcate into ascending and descending branches (see Fig. 17–5). Some collaterals terminate on interneurons in the spinal gray matter. These connections participate in the circuits that mediate spinal reflexes such as the *flexor withdrawal reflex* (see Chapter 9).

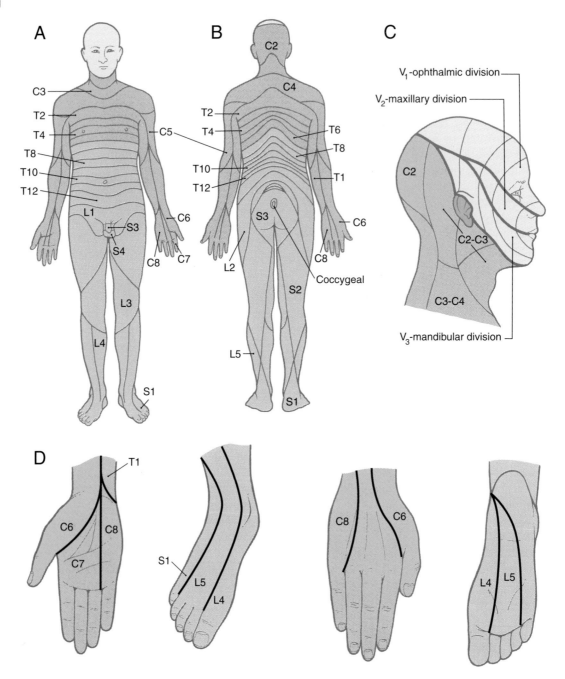

Figure 18–4. Dermatomal maps of the peripheral distribution of spinal nerves (*A* and *B*) and trigeminal nerve (*C*). *D*, Details of dermatomal maps on anterior and posterior surfaces of the hand and foot.

The functional properties of posterior horn neurons reflect the type of primary afferent fiber input received. These neurons are classified as *low-threshold* (non-nociceptive), *nociceptive-specific* (noxious), *wide dynamic range* (non-nociceptive and noxious), or *deep* on the basis of their responses to different stimulus modalities.

The central target of nociceptive primary afferent fibers includes laminae I, II, and V of the posterior horn (Fig. 18–6*A* and *B*). Aδ fibers target laminae I and V. Rexed's lamina I, the *posteromarginal nucleus* (or *zone*), receives mainly input from Aδ fibers (Fig. 18–7*A*). Neurons in this nucleus (or zone) project to

other spinal cord laminae, the brainstem reticular formation, and the thalamus (see Fig. 18–7*B*). Lamina II, the *substantia gelatinosa*, is divided into outer (IIo) and inner (IIi) layers. Input to IIo and IIi is derived primarily from C fiber primary afferents, and IIi also receives input from collaterals of non-nociceptive afferent fibers. Lamina II contains excitatory and inhibitory interneurons that project to other laminae of the posterior horn. In addition, some neurons in IIo are involved in relaying sensory information to supraspinal sites, including the thalamus (see Fig. 18–7*B*).

Neurons in laminae III and IV, the *nucleus proprius*

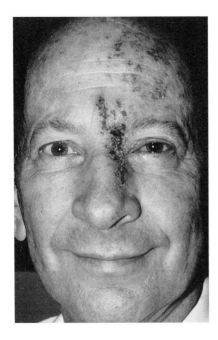

Figure 18–5. A case of herpes zoster (shingles) in an adult male. The ophthalmic division of the trigeminal nerve is involved; note that the lesions do not cross the midline. The border between the ophthalmic and maxillary divisions on the lateral aspect of the nose may very slightly between patients.

(also called the *posterior proper sensory nucleus*), receive non-noxious inputs from the periphery. Cells in these laminae project to deeper laminae of the spinal cord, to the posterior column nuclei, and to other supraspinal relay centers including the midbrain, thalamus, and hypothalamus (see Fig. 18–7*B*).

Lamina V neurons receive both noxious and non-noxious (nociceptive and non-nociceptive) inputs and project to the medullary and mesencephalic reticular formation, thalamus, and hypothalamus (see Fig. 18–7*B*). Neurons in deeper laminae of the spinal gray receive (directly and indirectly) noxious and non-noxious inputs and connect with neurons in other spinal cord levels (*propriospinal connections*).

As mentioned previously, the fibers of the ALS participate in both direct and indirect spinothalamic pathways (see Fig. 18–1). Most Aδ fibers participate in the *direct (neospinothalamic)* pathway, which carries nondiscriminative tactile, innocuous thermal, and nociceptive signals. When these Aδ fibers enter the posterolateral fasciculus and bifurcate, the main branch ascends three to five spinal levels to terminate on second-order neurons (tract cells) in lamina I of the posterior horn (see Fig. 18–17*A*). These tract cells, in turn, project to the thalamus. The great majority of their axons cross the midline of the spinal cord obliquely via the anterior

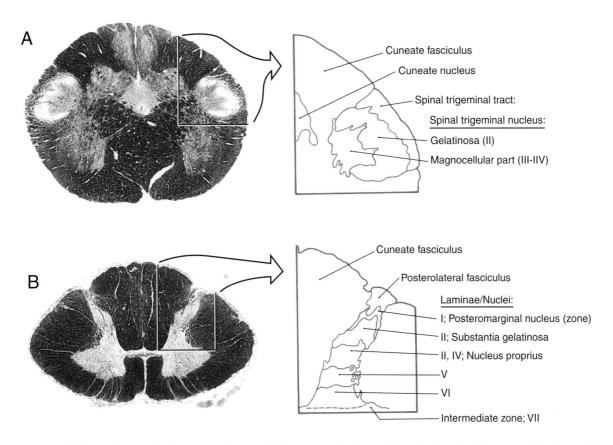

Figure 18–6. Weil (myelin)-stained section of the medullary posterior horn (*A*) at the level of the pyramidal decussation and the spinal posterior horn (*B*) at levels C7 to C8. The spinal laminae that correspond to medullary structures are labeled.

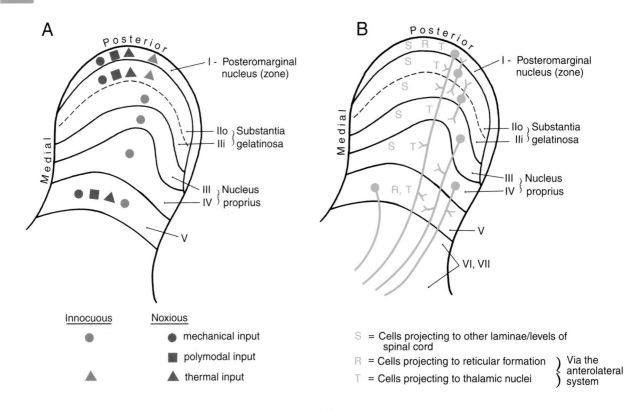

Figure 18–7. Summary of posterior horn laminae and their major sensory inputs (*A*) and major outputs (*B*).

(ventral) white commissure and ascend in the contralateral ALS. A few ascend in the *ipsilateral* ALS. The thalamic (third-order) neurons of these pathways are located mainly in the ventral posterolateral nucleus (VPL), the posterior nucleus, and the intralaminar nuclei.

The polysynaptic *indirect* (*paleospinothalamic*) component of the ALS relays noxious and innocuous mechanical and thermal information to the brainstem reticular formation. The input to this pathway originates chiefly from C fibers. Branches of these fibers ascend and descend by one or two levels in the posterolateral fasciculus to synapse on interneurons in laminae II and III (see Fig. 18–7B). These interneurons influence tract cells in laminae V to VIII, which send axons that cross obliquely through the anterior white commissure (over a distance of one to three segments) to join the contralateral ALS. These *spinoreticular fibers* terminate in the brainstem reticular formation, which in turn projects to the thalamus.

Fibers in the ALS are arranged *somatotopically* in the spinal cord. Axons from lower levels (coccygeal and sacral) of the body are found posterolaterally, whereas those from more rostral levels of the cord are added in an orderly anteromedial sequence (Fig. 18–8). Because of its location, the ALS does not receive blood supply from a single vessel. Instead, its blood supply originates from the *arterial vasocorona*, and via *sulcal branches* of the *anterior spinal artery* (see Fig. 18–8). Consequently,

occlusion of either of these vessels results in a *patchy loss of nociceptive, thermal*, and *touch sensations over the contralateral side of the body* beginning about two spinal segments below the lesion. In contrast, a *complete loss of these sensations* is seen in patients who have had an *anterolateral cordotomy* for relief of intractable pain.

The ALS may be involved in trauma or diseases of the spinal cord. For example, a hemisection of the spinal cord (as in the *Brown-Séquard syndrome*) results in a combination of sensory and motor losses. Sensory deficits include (1) *contralateral* loss of nociceptive and thermal sensations over the body below the level of the lesion (ALS damage), and (2) *ipsilateral* loss of discriminative tactile, vibratory, and position sense over the body below the level of the lesion (posterior column damage) (Fig. 18–9A). The motor loss is manifested as an ipsilateral paralysis of the leg or leg and arm, depending on the level of the hemisection (see Chapter 24). *Syringomyelia*, a condition in which there is cystic cavitation of central regions of the spinal gray matter, may impinge on the anterior white commissure and decussating ALS fibers (see Fig. 18–9B). When located at C3 to C4 levels of the spinal cord, this lesion produces *bilateral loss* of nondiscriminative tactile, nociceptive, and thermal sensations beginning several segments below the level where fibers are interrupted. The symptoms present as sensory losses in the configuration of a cape draped over the shoulders and extending down to nipple level.

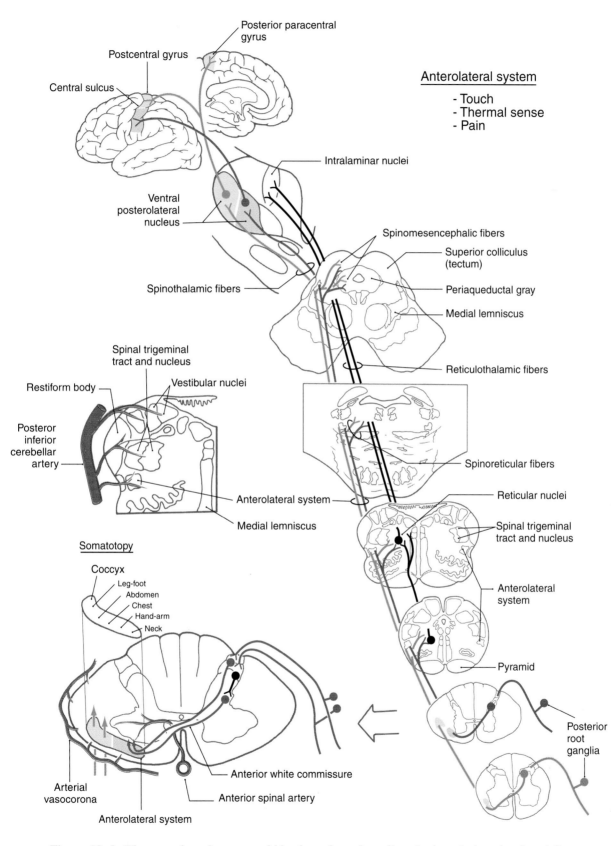

Figure 18–8. The anterolateral system and blood supply to these fibers in the spinal cord and medulla.

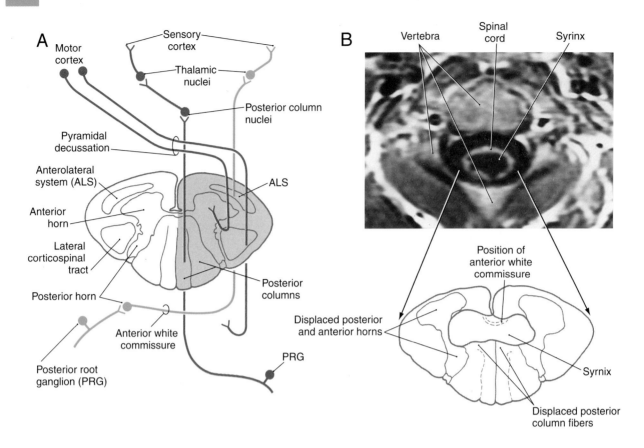

Figure 18–9. Major tracts interrupted in a spinal cord hemisection (*A*, Brown-Seqúard syndrome) that account for the characteristic sensory and motor losses. Magnetic resonance image of a cervical syringomyelia (*B*) with resultant expansion of the lesion into fibers of the anterior white commissure. In both *A* and *B*, the cross section of the spinal cord is shown in an orientation identical to that seen in the clinical setting.

In the medulla, ALS fibers retain their position near the anterolateral surface. They are located anterior to the spinal trigeminal nucleus and posterolateral to the interior olive, and remain separated from the PCML system as both course through the medulla and pons (see Fig. 18–8). Therefore, vascular lesions or tumors in the lower brainstem can affect discriminative touch and nociception differentially. At the pontomesencephalic junction, the medial lemniscus rotates posterolaterally so that ALS fibers are adjacent to its lateral margin. Thereafter, the PCML and ALS pathways course together to terminate in the thalamus (see Fig. 18–8).

As the ALS ascends through the medulla, it decreases in size because of the departure of the *spinoreticular axons*, which originate in laminae V to VIII and terminate in the reticular formation. The reticular formation also receives *collaterals from lamina I spinothalamic axons* (see Fig. 18–8). Several other pathways ascend in the ALS. For example, spinomesencephalic axons may terminate in the PAG or, as *spinotectal fibers*, in deep layers of the superior colliculus and anterior pretectum. Many tract cells with axons in the ALS project via collaterals to multiple targets as the primary axon ascends through the brainstem.

In addition to direct spinothalamic fibers, spinal cord neurons also project to brainstem targets that indirectly influence thalamic nuclei. Most notably, the reticular formation, which receives *spinoreticular fibers*, projects via *reticulothalamic fibers* to the intralaminar nuclei and posterior group (see Fig. 18–8). The intralaminar nuclei project to the striatum and wide areas of cerebral cortex, and subserve the alerting response to painful stimuli. Nuclei of the posterior thalamic group project to secondary somatosensory (SII) cortex, and the retroinsular cortex. These polysynaptic pathways may underlie the dull, poorly localized, but persistent painful sensations that are perceived with localized thalamic lesions.

Somatosensory information, including nociceptive input from posterior horn cells, also ascends directly to the hypothalamus via the *spinohypothalamic fibers* of the ALS. In addition, spinal input is indirectly conveyed to the hypothalamus by way of synaptic relays in the reticular formation and in the PAG (see Fig. 18–15). Through these ascending pathways, nociceptive information is transmitted to brain centers, such as the limbic system, that underlie emotional and autonomic responses to nociceptive stimuli.

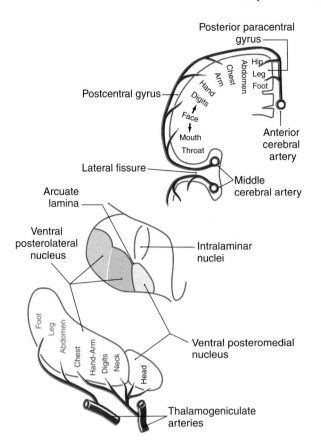

Figure 18–10. Somatotopic organization of, and blood supply to, the ventral posteromedial and posterolateral nuclei and the SI somatosensory cortex.

central gyrus (see Figs. 18–8 and 18–10). These thalamocortical fibers terminate primarily at the 3b/1 border on specific physiologic classes of SI neurons: *low-threshold non-nociceptive, nociceptive-specific,* and *wide dynamic range cells.* Loss of nociceptive and thermal sensations over the contralateral body and face can result from vascular compromise of either the middle (for trunk, upper extremity, and face) or the anterior (for the lower extremity) cerebral artery (see Fig. 18–10). Not only the sensation but also the ability to localize is lost.

Not all nociceptive information reaches the thalamus via the ALS. The *spinocervicothalamic pathway* is a supplemental multimodal pathway that carries discriminative innocuous tactile information as well as nociceptive signals (Fig. 18–11). This pathway begins with afferent fibers that terminate on second-order cells in laminae III and IV of the posterior horn. The axons of these second-order cells travel in the *ipsilateral lateral funiculus* to spinal levels C1 and C2, where they terminate on third-order neurons in the *lateral cervical nucleus.* The axons of these cells decussate at the level of the spinomedullary junction and ascend in the medial lemniscus (see Fig. 18–11). Like PCML axons, these *cervicothalamic axons* terminate in the VPL nucleus. This pathway is not essential for pain perception

Input to the VPL is somatotopically organized such that lower body areas are represented laterally, and upper body regions (exclusive of the head) are represented medially (Fig. 18–10; see also Fig. 18–9). Within the VPL, fibers of the ALS terminate on clusters of cells located in the periphery of the nucleus. Most of these cells are different from the ones targeted by PCML axons. However, some VPL neurons, called *multinodal cells*, receive input from both ALS and PCML pathways. The functional classes of cells found in VPL reflect the peripheral input received by tract cells of the spinal cord. These classes include, as in the spinal cord, *nociceptive-specific, wide dynamic range, low-threshold non-nociceptive,* and *deep neurons.*

Thalamocortical axons carrying nondiscriminative tactile, nociceptive, and thermal signals project via the posterior limb of the internal capsule to the somatosensory cortices (see Fig. 18–8). Fibers originating in the VPL project mainly to SI cortex (areas 3, 1, and 2), whereas those from the posterior nucleus terminate principally in the SII cortex. The somatotopy observed in the VPL is reflected in the cortex. Thalamocortical fibers from lateral areas of VPL project to the *posterior paracentral gyrus* (leg and foot), whereas progressively more medial parts of the VPL project in an orderly manner to sequentially more lateral areas of the *post-*

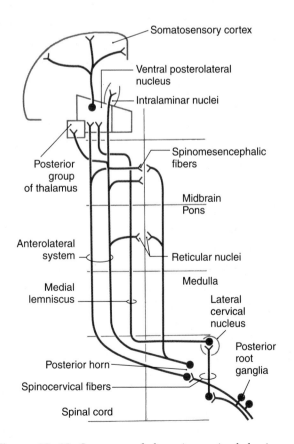

Figure 18–11. Summary of the spinocervicothalamic tract that carries innocuous discriminative tactile, thermal, and nociceptive sensations.

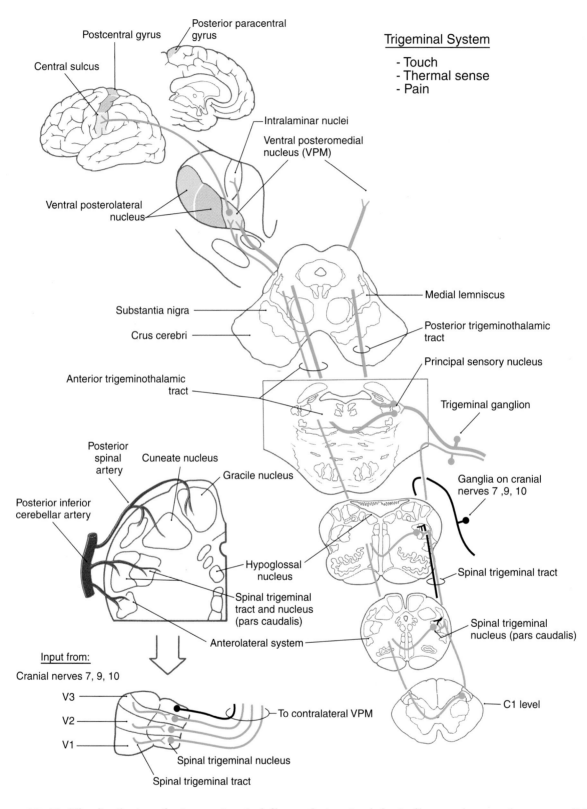

Figure 18–12. The distribution of primary trigeminal fibers, of trigeminothalamic fibers to the ventral posteromedial nucleus (VPM), and the blood supply to trigeminal structures in the medulla. The large-diameter fibers convey discriminative touch, vibratory, and proprioceptive input, whereas the smaller-diameter fibers constitute the pathway for nondiscriminative tactile, thermal, and nociceptive signals.

and is not especially prominent in humans. However, these fibers and the uncrossed axons in the ALS may be the basis for the retention of some nociceptive function after lesions involving the ALS, or for the return of pain perception after *anterolateral cordotomy*.

Spinal Trigeminal Pathway: Anterior Trigeminothalamic Tract

Primary Neurons. Cranial nerves V, VII, IX, and X serve the cutaneous receptors of the face, the oral cavity, and the dorsum of the head except for the area served by the cervical nerves (Fig. 18–12; see also Fig. 18–14). In addition to cutaneous structures, the trigeminal nerve also innervates deep tissues including the temporomandibular joint, the meninges, and the peridontium. The primary sensory fibers of these nerves have their cell bodies in the *trigeminal ganglion*, the *geniculate ganglion* of cranial nerve VII, and the *superior ganglia* of cranial nerves IX and X. Aδ and C fiber nociceptors are found throughout this region, and they are particularly prominent in the tooth pulp. Some of these fibers extend into the dentinal tubules, and carious lesions of the tooth expose these and other pulpal nerves to stimuli that result in dental pain. It is probable that the dull aching pain caused by pulp inflammation is the product of C fiber activity. Dental hypersensitivity, often characterized by sharp sensation, represents Aδ fiber activity. The cornea also receives a large number of nociceptive fibers. This corneal innervation forms the afferent limb of the *corneal (blink) reflex*. The meninges are also supplied by fibers of the trigeminal ganglion cells that terminate in the spinal trigeminal nucleus. These fibers are thought to be involved in the pain of migraine headache.

The central processes of small and large trigeminal ganglion cells are part of the *trigeminal sensory root*, which attaches to the pons (see Fig. 18–12). The bifurcating small-diameter axons course posteromedially into the pontine tegmentum, sending an ascending branch to the *principal sensory nucleus*. The descending branch of these fibers joins with numerous other unbranched small-diameter fibers to form a prominent fiber bundle in the posterolateral brainstem, the *spinal trigeminal tract*. Through the caudal pons and the rostral medulla, this tract is internal to the restiform body. However, in the lower medulla caudal to the obex, it forms a superficial landmark lateral to the cuneate tubercle, known as the *trigeminal tubercle (tuberculum cinereum)*. This landmark served as a useful reference point for surgeons, who discovered that sectioning the *spinal tract of the trigeminal nerve* at this level (*tractotomy*) provides substantial relief from facial pain on the operated side.

The spinal trigeminal tract extends from the middle pons to the second or third cervical spinal cord segment, where its fibers interdigitate with those of the posterolateral fasciculus (Lissauer tract) (see Figs. 18–6 and 18–12; see also Fig. 17–5). Fibers of the mandibular division lie dorsal to those of the maxillary division, which, in turn, are dorsal to the ophthalmic division; this arrangement creates an inverted hemiface representation (Fig. 18–13). The primary afferent neurons associated with cranial nerves VII, IX, and X have cell bodies in their respective ganglia, enter the medulla, and take a position dorsal to those of the mandibular division.

The peripheral distribution of the branches of the trigeminal nerve (V$_1$, V$_2$, and V$_3$) delineates the facial dermatomes (see Figs. 18–4 and 18–13). Unlike the spinal segmental dermatomes, which partially overlap, the *boundaries between adjacent facial dermatomes are sharply defined*. This segregation of trigeminal branches is maintained by their central processes in the spinal trigeminal tract. An unfortunate clinical condition that illustrates the divisional pattern of the trigeminal system is *herpes zoster*, or *shingles*. Patients with shingles have a characteristic rash that outlines the affected dermatome or spinal cord segment; the ophthalmic or maxillary division is usually affected, and the rash is unilateral (see Fig. 18–5).

Injury to trigeminal nerve fibers produces a *paresthesia* restricted to specific regions of the face. The pain of *tic douloureux (trigeminal neuralgia)* produces episodic "paroxysmal" pain usually restricted to the peripheral distribution of the maxillary or mandibular division on one side. Trigeminal neuralgia is further characterized by the presence of "trigger zones," which, upon the most gentle stimulation (such as a light breeze or a brush with a wisp of cotton), produce stabbing pain on one side of the face. The precise etiology of this condition remains enigmatic, but vascular compression of the trigeminal nerve root and the presence of microneuromas are likely causes.

Central Pathways. The *spinal trigeminal nucleus*, located medial to the spinal tract, is the site of termination for fibers of the spinal trigeminal tract (see Fig. 18–12). On the basis of cytoarchitecture, this nucleus is divided into a *pars caudalis*, a *pars interpolaris*, and a *pars oralis*. The caudal subnucleus (*pars caudalis*) (see Figs. 18–6 and 18–13) extends from C2 or C3 rostrally to the level of the obex. This part of the spinal nucleus shares many cytoarchitectural similarities with the posterior horn. For this reason, it has been termed the *medullary posterior horn* and has been divided into layers that correspond to Rexed spinal cord laminae (see Fig. 18–6). The *substantia gelatinosa* is largely continuous with lamina II of the spinal cord, and the *magnocellular* region is continuous with laminae III and IV. The pars caudalis and the posterior horn also show homology in the distribution of neurotransmitters. For example, substance P and calcitonin gene–related pep-

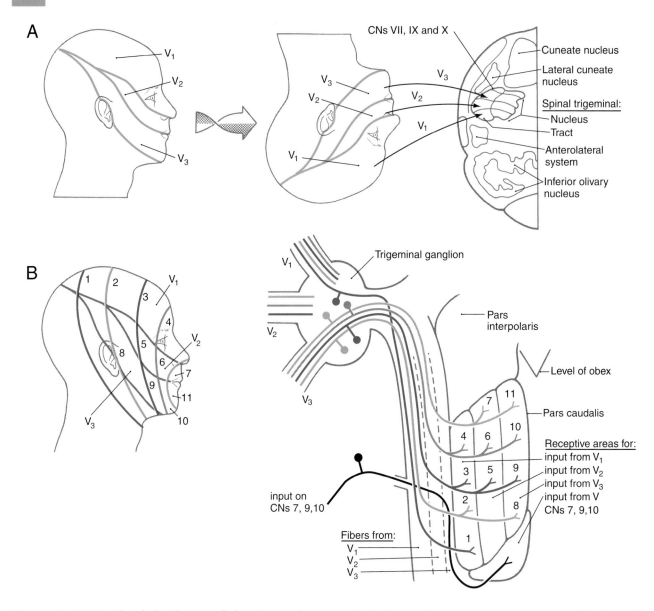

Figure 18–13. Peripheral distribution of the trigeminal nerve and the inverted somatotopic arrangement of the hemiface within the spinal trigeminal nucleus pars caudalis (*A*). Functional onion-skin pattern of facial pain is superimposed along the caudal to rostral axis of the pars caudalis (*B*). CN, cranial nerve.

tide (CGRP) are localized in nociceptive C fibers that terminate in both of these areas.

The pars caudalis plays an important role in the transmission of nondiscriminative touch, nociceptive, and thermal sensations. This role is reflected by the fact that central processes of Aδ and C fibers terminate somatotopically in this subnucleus. In addition to the somatotopy within the pars caudalis, an *onion-skin pattern of facial pain* representation is oriented along the rostrocaudal axis of the subnucleus (see Fig. 18–13*B*). The nociceptive fibers that innervate circumoral and intraoral zones (teeth, gums, and lips) terminate rostrally, close to the obex in the caudal pars interpolaris. Fibers innervating progressively more caudal and lateral regions of the face terminate caudally within the

spinal nucleus. Many second-order neurons in the subnucleus caudalis receive convergent input from small-diameter fibers that innervate cutaneous and deep tissues (jaw muscles and the temporomandibular joint). Convergence of information from different regions is thought to contribute to the referral of pain and may be involved in the manifestation of less well-understood clinical problems such as *temporomandibular disorders* and atypical facial pain.

At medullary levels, the posterior inferior cerebellar artery supplies the territory of the ALS fibers, as well as the spinal trigeminal nucleus and tract (see Figs. 18–8 and 18–12). Vascular lesions involving this vessel produce characteristic sensory symptoms collectively known as the *lateral medullary (Wallenberg) syndrome* or

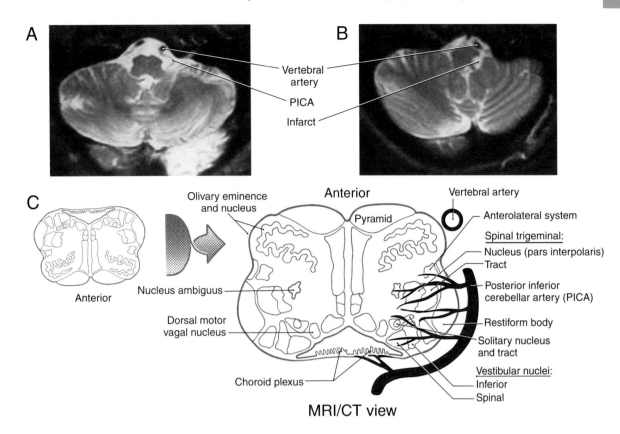

Figure 18–14. Lateral medullary (Wallenberg) syndrome. A normal magnetic resonance image (*A*) showing vertebral artery and posterior inferior cerebellar artery (PICA). An occlusion of PICA (*B*) resulting in an infarct of the posterolateral medulla. The lesioned area contains trigeminal structures, the anterolateral system, and other important nuclei and tracts (*C*). MRI/CT, magnetic resonance imaging/computed tomography.

the *posterior inferior cerebellar artery syndrome*. The sensory symptoms of this syndrome may include a contralateral loss of pain (*hemianalgesia*) and temperature (*hemithermoanesthesia*) sensibility over the body and ipsilateral loss of these modalities over the face. However, the extent of damage following posterior inferior cerebellar artery lesions shows remarkable variation, and the combination of symptoms is representative of the structures served by this artery (Fig. 18–14).

The interpolar subnucleus (*pars interpolaris*) is located between the level of the obex and the rostral pole of the hypoglossal (XII) nucleus. The most rostral subdivision is the oral subnucleus (*pars oralis*), which extends from the level of the rostral pole of the hypoglossal nucleus to the caudal end of the trigeminal motor nucleus (see Figs. 18–12 and 18–13). Some neurons in the pars interpolaris and the pars oralis contribute to ascending somatosensory pathways, whereas others project to the cerebellum (see Fig. 17–16). In addition to projection neurons, the spinal trigeminal nucleus, particularly the subnucleus oralis, contains many local circuit neurons involved in brainstem reflexes.

The axons of *second-order trigeminothalamic neurons* in the spinal trigeminal nucleus decussate, coalesce to form the *anterior trigeminothalamic tract*, and ascend through the brainstem just posterior to the medial lemniscus (see Fig. 18–12). These fibers terminate in the ventral posteromedial (VPM), the posterior, and the intralaminar nuclei of the thalamus. As noted in Chapter 17, this pathway also carries crossed fibers from the principal trigeminal nucleus. The principal nucleus fibers terminate in the core of VPM, whereas spinal nucleus fibers terminate in its periphery. At the pontomesencephalic junction, anterior trigeminothalamic fibers join with ALS fibers at the lateral margin of the medial lemniscus (see Figs. 18–8 and 18–12). Like ALS fibers, ascending anterior trigeminothalamic axons terminate in or give rise to collaterals that supply the reticular formation.

A particularly prominent target of some of these collaterals is the parabrachial nuclear complex. Located adjacent to the superior cerebellar peduncle (brachium conjunctivum), the parabrachial nuclei serve as an important relay for spinal and trigeminal pain fibers, as well as for ascending axons carrying visceral sensory information. In addition to regulating oral and facial reflexes, projections from the reticular formation terminate in the dorsal thalamus in the intralaminar nuclei and the medial region of the posterior nucleus. The intralaminar nuclei project widely to the striatum and cortex, especially frontal and somatosensory cortex. The medial region of the posterior nucleus projects to the head representation in secondary somatosensory cortex.

Imaging Studies of Pain in the Somatosensory Pathway

Imaging studies including electroencephalography (ECG), positron emission tomography (PET), functional magnetic resonance imaging (fMRI), and magnetoencephalography (MEG) have provided insight into the localization of brain regions responsible for the processing of pain signals. Electroencephalography recordings show patterns of increased brain activity following application of painful stimuli. This increased activity is especially prevalent in the somatosensory cortex and the frontal cortex.

When PET is used to identify changes in regional cerebral blood flow, painful stimuli are found to activate SI and SII cortices, as well as the anterior cingulate cortex, anterior insula, the supplemental motor area of motor cortex, and different thalamic nuclei. The cingulate cortex and the anterior insula are connected with nontraditional somatosensory brain regions including the limbic cortex. This widespread cortical activation may provide a morphologic basis for integrating the location of painful stimuli with memory and emotion. PET has also been used to study possible gender differences in pain perception to noxious heat stimuli. Similar increases in cortical activation in the posterior insula and anterior cingulate cortex, as well as in the cerebellar vermis, have been described for both sexes. However, increases in the contralateral prefrontal cortex activity, contralateral insula, and thalamus noted in female subjects suggest that pain perception and processing may be different in males and females.

Functional MRI techniques reveal increased thalamic and cortical activation in response to application of innocuous (tactile, cool, and warm) and nociceptive (cold and hot) stimuli. Regions of increased activity in response to innocuous stimuli included the contralateral thalamus (VPL), posterior insula, and bilateral SII cortex. There was no activation of SII in response to application of warm stimuli. Pain stimuli, however, activated the anterior insula cortex.

Using a CO_2 laser to activate Aδ and C nociceptors, magnetoencephalography (MEG) revealed that both the contralateral SI and bilateral SII cortices are involved in the processing of pain stimuli. These results support the view that SI provides a mechanism to code spatial, temporal, and intensity qualities of a painful stimulus. These various methods are providing a more complete identification of structures included in pain pathways and responsible for the processing of painful stimuli.

Pain Perception

Vascular compromise of middle or anterior cerebral arteries (see Fig. 18–10) produces a loss of sensibility (discriminative, nondiscriminative, thermal, and nociceptive) over contralateral regions of the body. Over time, however, appreciation of sensation returns. Pain sensations are first to return, followed by nondiscriminative tactile and thermal sensations. Discriminative tactile, vibratory, and proprioceptive sensations often fail to return to normal levels. If occlusion of the middle cerebral artery affects most of the postcentral gyrus, sensation begins returning first on the face and oral regions, then on the neck and trunk, and finally on the extremities and the distal parts of the limbs. This return of function indicates that other cortical areas may partially take over the appreciation of somatosensory stimuli, using the input they receive through non-lemniscal, non-ALS pathways.

At least some forms of somatosensory stimuli can be perceived at subcortical levels. In fact, electrical stimulation of the primary somatosensory cortex does not result in a complaint of pain, whereas thalamic stimulation may elicit paresthesia and sensations of dull pain and pressure. Furthermore, painful stimuli can be recognized and produce suffering without the presence of primary and secondary cortices, leading to the concept that *pain is perceived at subcortical levels*. However, damage to specific cortical regions eliminates the ability to precisely localize pain, suggesting that such localization is a function of somatosensory cortex and its lemniscal inputs (Fig. 18–15A).

A second dissociation can occur in pain pathways. Pain perception and its affective component, suffering, are served by separate brain regions. The neospinothalamic pathway to primary somatosensory cortex is involved in the localization of painful stimuli (see Fig. 18–15A). Paleospinothalamic pathways that access the hypothalamus and limbic system via the reticular formation and periaqueductal gray are involved in the suffering component of the pain experience (see Fig. 18–15B). This dissociation can be regulated pharmacologically, as some drugs eliminate suffering without affecting pain perception. Patients taking benzodiazepines report that the pain is still present, but that its unpleasant nature is diminished.

Pain Perception in the Somatosensory Thalamus

Results of stereotaxic surgery for treatment of chronic pain or movement disorders have provided tremendous

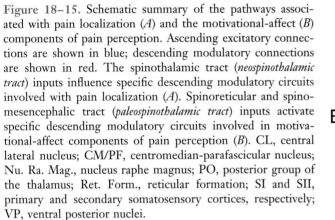
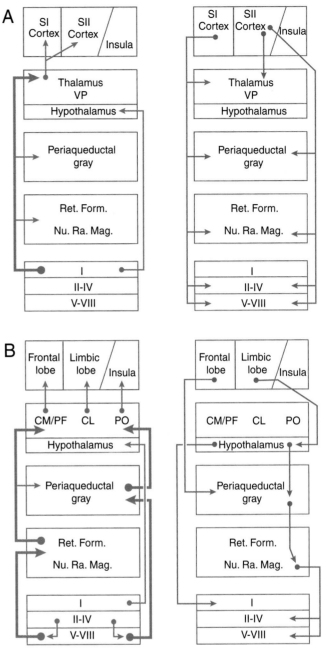

Figure 18–15. Schematic summary of the pathways associated with pain localization (*A*) and the motivational-affect (*B*) components of pain perception. Ascending excitatory connections are shown in blue; descending modulatory connections are shown in red. The spinothalamic tract (*neospinothalamic tract*) inputs influence specific descending modulatory circuits involved with pain localization (*A*). Spinoreticular and spino-mesencephalic tract (*paleospinothalamic tract*) inputs activate specific descending modulatory circuits involved in motivational-affect components of pain perception (*B*). CL, central lateral nucleus; CM/PF, centromedian-parafascicular nucleus; Nu. Ra. Mag., nucleus raphe magnus; PO, posterior group of the thalamus; Ret. Form., reticular formation; SI and SII, primary and secondary somatosensory cortices, respectively; VP, ventral posterior nuclei.

insights into the role of the thalamus in pain perception. Before the performance of such surgical procedures, physiologic identification of the desired target is undertaken in these patients. Single-neuron recording and microstimulation have demonstrated that neurons within the human VPM/VPL (collectively called ventrocaudal [Vc] nuclei by some neurosurgeons) are involved with processing of tactile, thermal, and pain signals. Patients report that microstimulation of Vc evokes sensations of touch, warmth, coolness, tingling, burning, or pain localized to specific body areas. These recordings also reveal that a population of thalamic cells activated by innocuous tactile stimuli are mixed with other neurons activated by mechanical and thermal stimuli in the painful range. Single-neuron recording has demonstrated that microstimulation in VPM/VPL (Vc) evokes the sensation of angina, suggesting that these nuclei play a role in pain localization regardless of its origin—that is, cutaneous or visceral.

The human VPL/VPM (Vc) can undergo changes (i.e., it exhibits plasticity) following *deafferentation*, which can occur directly as a result of damage to ascending pathways or secondarily as a result of removing sensory inputs (e.g., amputation). These changes may contribute to *chronic pain* or *phantom limb pain*. They involve the upregulation and downregulation of neurochemicals within the nucleus, changes in local circuitry, and changes in the functional state of Vc neurons. For example, in patients who have undergone leg amputation, single-neuron recordings reveal that the thalamic region formerly receiving input from the lower leg and foot responded to stimulation of the stump (thigh). These patients also described the presence of nonpainful tingling over the stump in response to microstimulation in this same area.

In an attempt to bring about relief for chronic or neuropathic pain, two therapies have been utilized: thalamic lesioning and deep brain stimulation. Lesions have been centered in either the lateral thalamus or the medial thalamus. Lateral thalamic lesions involve the somatosensory thalamus (VPL/VPM). Although producing some transient relief for pain, these lesions produce unwanted side effects including loss of cutaneous

and position sense in the affected limb, as well as impaired motor function. Lesions in the medial thalamus involve the centromedian-parafascicular (CM-PF) complex and the medial dorsal nucleus. These lesions produce longer-lasting relief with minimal side effects, which may include a transient confusion or somnolence, and no apparent sensory loss.

Deep brain electrical stimulation has been used for more than several decades in patients suffering from chronic pain or deafferentation pain. Stimulating electrodes centered in the somatosensory thalamus, the CM-PF complex, or the periventricular gray (PVG)–PAG activate neurons within their vicinity and thus may contribute to stimulus-induced analgesia. Cortical stimulation has also been shown to produce relief of chronic pain of neuropathic origin. An evaluation of different brain regions that may contribute to the stimulus-induced analgesia was carried out using PET. Following thalamic stimulation, increased regional cortical

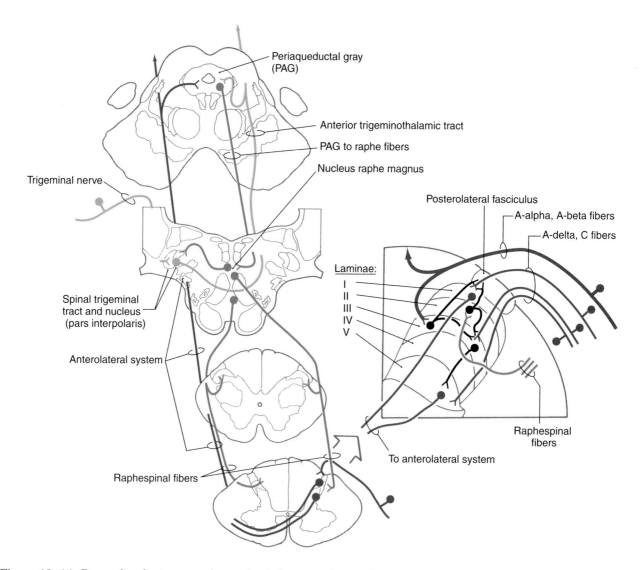

Figure 18–16. Descending brainstem pathways that influence and control pain transmission within the brainstem (for trigeminal pathways) and in the spinal cord (for anterolateral system projections).

blood flow was noted in the rostral insula, a region activated in studies of experimental pain, neuropathic pain, and warm and cool innocuous stimuli, as well as in the anterior insular cortex. These results suggest that stimulation of the somatosensory thalamus may activate a pain modulation circuit that involves thalamocortical thermal pathways.

Central or *thalamic pain* is a poorly understood sequela of natural or surgical lesions of structures involved in somatic sensibility. Central pain was originally observed with thalamic lesions, but it can occur with lesions of the ALS below the level of the thalamus.

Central pain syndrome can also result from vascular lesions. Patients surviving the *Wallenberg syndrome* (see Fig. 18–14) may ultimately develop central pain, suggesting some sparing of alternate or parallel pain pathways. In the *central pain syndrome*, the analgesia that initially results from the lesion is replaced after a period of weeks, months, or years by spontaneous *paresthesia*, *dysesthesia*, or unusual painful responses. *Allodynia*, pain resulting from a stimulus that does not normally evoke pain, and *hyperalgesia*, an increased response to a stimulus that is normally painful, are common neurologic signs associated with the central pain syndrome. Patients often characterize central pain as burning, aching, pricking, or lacerating and as occurring in paroxysms that vary in intensity and are poorly localized.

Central pain may last for years and is intractable to current analgesics. Pharmacologic agents, such as antidepressants and antiepileptic drugs, have been used with varying degrees of success to treat central pain. Although the etiology for this condition has not been elucidated, it is possible that this type of pain represents a deafferentation phenomenon; that is, it results from removal of primary afferent influence on central neurons. The time course and symptoms of central pain suggest that it may be due to the sprouting of inappropriate connections of non-nociceptive or nociceptive fibers, to increased excitability of central pain neurons, or to the removal of inhibitory influences on pain neurons.

Patients experiencing central pain may obtain temporary relief from *transcutaneous electrical nerve stimulation (TENS)* (electrical stimulation of nerves through the skin), from electrical stimulation of the posterior columns, or from chronic stimulation of the periaqueductal or periventricular gray regions by stereotaxically positioned electrodes (*deep brain stimulation*). These stimulation techniques are discussed later. Neuroablative surgical procedures that have been used in the treatment of central pain include anterolateral cordotomy, trigeminal tractotomy, lesions of the posterior root entry zone, thalamotomies, and cortical ablation. Unfortunately none of these procedures is successful in the long term.

Pain Transmission and Control

The relaying of information from the spinal cord to supraspinal centers is an important event in the higher-order processing of nociceptive sensory signals. On the basis of the localization of putative neurotransmitters and secondary messengers in the posterior horn, several candidates, such as *peptides* (calcitonin gene–related peptide, substance P), *glutamate*, and *nitric oxide*, may be involved in this process. These and other chemical agents underlie the central pharmacology of nociceptive transmission and are responsible for the varied qualities associated with central pain pathways. Pain can be classified as acute or chronic, fast or slow, dull or sharp, or burning, throbbing, or aching. Because pain is such a complex sensory experience, reduction or elimination of nociceptive sensations is of obvious clinical importance. Effective clinical approaches that can be used for the purpose of controlling pain include *pharmacologic intervention* and *stimulation-produced analgesia*.

The CNS has neural circuits designed to modulate pain transmission (Figs. 18–16 and 18–17; see also Fig. 18–15). These systems have components at all levels of the neuraxis. They are capable of controlling nociceptive neuron firing and are sensitive to opiates.

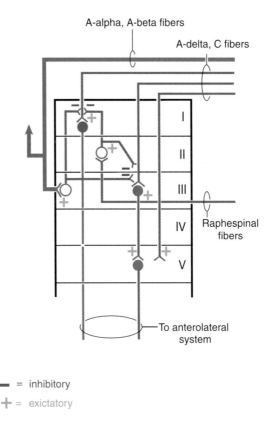

A-alpha, A-beta fibers

A-delta, C fibers

I

II

III

IV

V

Raphespinal fibers

To anterolateral system

▬ = inhibitory

✛ = exictatory

Figure 18–17. Posterior horn circuits that influence primary sensory fibers and ascending tract cells conveying nociceptive inputs. In addition to raphespinal fibers, some reticulospinal fibers also influence pain transmission in the posterior horn.

Central structures implicated in the *descending control* of nociceptive transmission include (1) the somatosensory cortex, frontal and limbic cortices, (2) the periventricular nucleus of the hypothalamus, (3) the PAG, and (4) raphe nuclei and adjacent medullary reticular formation. Descending pathways originating in these structures are activated by ascending afferent pain signals.

Cortical input may upregulate sites in the brainstem and hypothalamus that modulate the processing of nociceptive information in the brainstem and spinal cord (see Figs. 18–15 and 18–16). The *PVG* of the hypothalamus communicates with the *PAG* of the midbrain via an enkephalinergic pathway. Descending PAG fibers exert an excitatory influence on serotoninergic neurons in the medullary *nucleus raphe magnus*, both directly and through interneurons in the medullary reticular formation (see Fig. 18–16). This projection uses *serotonin, neurotensin, somatostatin,* and *glutamate. Raphespinal neurons* project, in turn, to the posterior horn and pars caudalis of the trigeminal nucleus (see Figs. 18–16 and 18–17). These serotoninergic axons terminate on enkephalinergic interneurons in laminae II and III, which act pre- and postsynaptically to suppress incoming activity in the pain fibers (see Figs. 18–16 and 18–17). In addition, the hypothalamus (via *hypothalamospinal fibers*) projects directly to the medullary and spinal cord posterior horns to act on incoming nociceptive signals (see Fig. 18–15). *Cholecystokinin* and substance P are among the putative neurotransmitters used by PAG projection neurons.

Stimulation-produced analgesia (SPA) relies on electrical stimulation of CNS structures to induce the release of endogenous chemicals, such as *enkephalin*, from cells in pain control circuits. As noted previously, endogenous opiates such as enkephalin inhibit pain transmission. Stimulation of PVG, the PAG, or the nucleus raphe magnus results in the release of enkephalin or monoamines and in analgesia. Systemic administration of pharmacologic opiates, such as *morphine*, excites periventricular and periaqueductal neurons, supplementing their natural activity. This increase in activity suppresses neurons in the spinal and medullary posterior horns that transmit painful information, producing analgesia. The direct delivery of opioids to the spinal cord (*epidural* anesthetic techniques) also is used to produce a powerful analgesia for surgical procedures and deliveries.

Current therapies for the control of pain transmission include transcutaneous electrical nerve stimulation and chronic stimulation of the posterior columns by implanted electrodes. Posterior column stimulation activates large-diameter myelinated fibers. Antidromic activation of these fibers discharges collaterals in the posterior horn (see Fig. 18–17). These collaterals stimulate the enkephalinergic interneurons in the posterior horn that inhibit the transmission of pain signals. This stimulation also provides long-term diminution of pain for reasons that are poorly understood. *Acupuncture-like stimulation* also may produce local analgesia by stimulating these fibers.

Sources and Additional Reading

Brodal A: Neurological Anatomy in Relation to Clinical Medicine, 3rd Ed. Oxford University Press, New York, 1981.

Burgess PR, Perl ER: Cutaneous mechanoreceptors and nociceptors. In Iggo A (ed): Handbook of Sensory Physiology, vol 2. Somatosensory System. Springer-Verlag, New York, 1973, pp 30–78.

Bushnell MC, Duncan GH, Hofbauer RK, Ha B, Chen JL, Carrier B: Pain perception: Is there a role for primary somatosensory cortex? Proc Natl Acad Sci USA 96:7705–7709, 1999.

Davis KD, Kwan CL, Crawley AP, Mikulis DJ: Functional MRI study of thalamic and cortical activations evoked by cutaneous heat, cold and tactile stimuli. J Neurophysiol 80:1533–1546, 1998.

Dubner R, Bennett GJ: Spinal and trigeminal mechanisms of nociception. Annu Rev Neurosci 6:381–418, 1983.

Dubner R, Sessle B, Storey A: The Neural Basis of Oral and Facial Function. Plenum Press, New York, 1978.

Duncan GH, Bushnell MC, Marchand S: Deep brain stimulation: A review of basic research and clinical studies. Pain 45:49–59, 1991.

Kiss ZHT, Dostrovsky JO, Tasker RR: Plasticity in human somatosensory thalamus as a result of deafferentation. Stereotact Funct Neurosurg 62:153–163, 1994.

Lenz FA, Dougherty PM: Pain processing in the human thalamus. In Steriade M, Jones EG, McCormick DA (eds): Thalamus, vol II. Elsevier, Oxford, 1997, pp. 617–651.

Light A: The Initial Processing of Pain and Its Descending Control: Spinal and Trigeminal Systems, vol 12. Pain and Headache. Karger, New York, 1992.

Mayer DJ, Liebeskind JC: Pain reduction by focal electrical stimulation of the brain: An anatomical and behavioral analysis. Brain Res 68:73–93, 1974.

Paulson PE, Minoshima S, Morrow TJ, Casey KL: Gender differences in pain perception and patterns of cerebral activation during noxious heat stimulation in humans. Pain 76:223–229, 1998.

Poggio GF, Mountcastle VB: A study of the functional contributions of the lemniscal and spinothalamic systems to somatic sensibility. Central nervous mechanisms in pain. Johns Hopkins Hosp Bull 106:266–316, 1960.

Talbot JD, Marrett S, Evans AC, Meyer E, Bushnell MC, Duncan GH: Multiple representations of pain in the human cerebral cortex. Science 251:1355–1358, 1991.

Wall PD, Melzack R: Textbook of Pain, 3rd Ed. Churchill Livingstone, Edinburgh, 1994.

Willis WD: The Pain System: The Neural Basis of Nociceptive Transmission in the Mammalian Nervous System, vol 8. Pain and Headache. Karger, New York, 1985.

Young RF: Effect of trigeminal tractotomy on dental sensation in humans. J Neurosurg 56:812–818, 1982.

Chapter 19

Viscerosensory Pathways

S. G. P. Hardy and J. P. Naftel

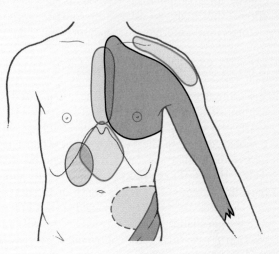

Viscerosensory Receptors   294

Viscerosensory Fibers   295

Ascending Pathway for Sympathetic Afferents   295
   Projections to Thalamus
   Projections to Reticular Formation
   Referred Pain
   Angina

Pathways for Parasympathetic Afferents   298
   Sacral Parasympathetic Afferents
   Cranial Parasympathetic Afferents
   Baroreceptor Reflex

Visceral Input to the Reticular Activating System   301

The somatosensory system conveys information from sensory receptors in the skin, joints, and skeletal muscles that allows one to perceive and respond to input from the external environment. Functioning in parallel with somatosensory pathways are fibers that convey information from visceral receptors. This input allows the body to make appropriate responses to changes in its internal environment.

Viscerosensory Receptors

Viscerosensory receptors may be categorized as nociceptors or physiologic receptors (Table 19–1). *Nociceptors* in the viscera are the free nerve endings of Aδ and C fibers located in the heart, respiratory structures, the gastrointestinal tract, and the urogenital tract (see Table 19–1). These receptors respond to stimuli that have the potential to damage tissue or to stimuli resulting from the presence of damaged tissue. For example, intense mechanical stimuli (such as overdistention or traction), ischemia, and endogenous compounds (including bradykinin, prostaglandins, and H$^+$ and K$^+$ ions) can activate these receptors and produce pain. These receptors signal changes in visceral structures that result from pathologic processes (such as myocardial ischemia or appendicitis) or from benign conditions such as gastrointestinal cramping or bloating. *Visceral pain* is often described as being diffuse in nature, difficult to localize, and is frequently referred to an overlying somatic body location.

Physiologic receptors are responsive to innocuous stimuli, and they monitor the functions of visceral structures on a continuing basis. These receptors also mediate normal visceral reflexes such as the baroreceptor reflex. Examples of physiologic receptors are (1) rapidly

adapting mechanoreceptors, (2) slowly adapting mechanoreceptors, and (3) various types of specialized receptors.

Rapidly adapting mechanoreceptors (see Table 19–1) signal the occurrence of dynamic events such as movement or sudden changes in pressure. This type of receptor is present in organs of the thoracic, abdominal, and pelvic cavities. In the thoracic cavity, it is represented by free nerve endings that exist in the epithelia of pulmonary airways. Because these nerve endings are sensitive to the presence of inhaled particles, they have been referred to as "cough receptors." Rapidly adapting mechanoreceptors in the abdominal and pelvic cavities vary greatly in size and location and may be either unencapsulated or encapsulated. The largest example of a rapidly adapting mechanoreceptor is the pacinian corpuscle.

Slowly adapting mechanoreceptors (see Table 19–1) signal the presence of stretch or tension within a visceral structure. These typically unencapsulated receptors are located in the smooth muscle layer of the pulmonary airways and in the smooth muscle layers of hollow abdominal and pelvic viscera. They are essential for the perception of a sense of fullness in certain viscera, such as the stomach or bladder.

Certain *specialized receptors* (see Table 19–1) are unique to the viscerosensory system. These include baroreceptors, chemoreceptors, osmoreceptors, and internal thermal receptors. *Baroreceptors* (Fig. 19–1*A*) are found in the walls of the aortic arch and carotid sinus and respond to rapid increases or decreases in blood pressure. For baroreceptors to effectively perform this task, blood pressure must be in the range of about 30 to 150 mm Hg. *Chemoreceptors* (see Fig. 19–1*B*) are found in structures called *carotid bodies* (located at the bifurca-

---

Table 19–1. Classification of Nociceptive Receptors and Physiologic Receptors in the Viscera and Their Adequate Stimuli

Viscerosensory Receptors	Adequate Stimulus(i)
Nociceptors	Mediate visceral pain
Heart	
Aδ and C afferent fibers	Prostaglandin (PGE$_2$), H$^+$ ions, bradykinin, K$^+$ ions, ischemia
Respiratory System	
Lung irritant receptors (Aδ fibers)	Irritant aerosols and gases, mechanical stimuli
J receptors (C fibers?)	Capsaicin, pulmonary congestion/edema, inhaled irritants
Gastrointestinal Tract	
Rapidly adapting mechanoreceptors, slowly adapting mechanoreceptors, chemoreceptors (Aδ and C fibers?)	Irritation of the mucosa, distention, powerful contraction, torsion, or traction, bloat, cramping, appendicitis, impaction
Urogenital Tract	
C-polymodal nociceptors	Intense mechanical stimuli, noxious heat and algesic chemicals
Physiologic Receptors	Monitor physiologic state of viscera; mediate visceral reflexes
Rapidly adapting visceral mechanoreceptors	Movement, sudden change in pressure
Slowly adapting visceral mechanoreceptors	Stretch/tension
Baroreceptors	Increase or decrease in blood pressure
Chemoreceptors	Changes in oxygen, carbon dioxide tension, H$^+$ ions
Osmoreceptors	Changes in blood osmolarity
Internal thermal receptors	Change in circulating blood temperature

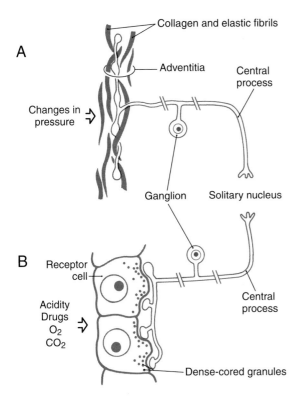

Figure 19–1. Diagrammatic representation of a baroreceptor (*A*) and a chemoreceptor (*B*). The baroreceptor is located in the adventitia in apposition to collagen and elastic fibrils. This receptor alters its firing rate in response to changes in blood pressure. The chemoreceptor is composed of a receptor cell and the numerous afferent endings it contacts. Changes in blood chemistry trigger responses in the receptor cell that result in an action potential in the subjacent afferent ending.

tion of the common carotid artery) and *aortic bodies* (located in the aortic arch) and are activated by changes in the composition of arterial blood. These changes include alterations in oxygen and carbon dioxide tension and in acidity. The presence of certain drugs such as nicotine or cyanide may also alter arterial blood gases.

In addition, specialized visceroreceptors also exist in the hypothalamus as *chemoreceptors, osmoreceptors,* and *internal thermal receptors.* These viscerosensory receptors are activated by changes in blood chemistry or osmolarity, or by changes in the temperature of blood circulating through the hypothalamus. Hypothalamic neurons that respond to these changes by altering their firing rates are considered to be the "receptor" cells.

Viscerosensory Fibers

Sympathetic and parasympathetic divisions of the autonomic (visceral motor) nervous system (see Chapter 29) have traditionally been considered to consist of only visceromotor (general visceral efferent [GVE]) fibers.

These fibers travel through sympathetic nerves (such as splanchnic and cardiac nerves) or through parasympathetic nerves (such as vagus and pelvic nerves). However, these *sympathetic and parasympathetic nerves also contain viscerosensory (general visceral afferent [GVA]) fibers* that serve many important functions. In this chapter, *the terms "sympathetic afferent" and "parasympathetic afferent" are used to describe viscerosensory fibers contained in sympathetic and parasympathetic nerves,* respectively. In addition to its conciseness, this usage complies with the terminology introduced by Langley, a pioneer in studies on the autonomic nervous system (see Cervero and Foreman).

Visceral afferents tend to predominate in parasympathetic nerves but are comparatively sparse in sympathetic nerves. For example, more than 80% of the fibers in the vagus nerve (a parasympathetic nerve) are viscerosensory, whereas less than 20% of the fibers in the greater splanchnic nerve (a sympathetic nerve) are visceral afferents. Most visceral afferents (90%; both sympathetic and parasympathetic) are either unmyelinated or thinly myelinated and, therefore, are slowly conducting fibers.

Information originating from *nociceptors* is conducted almost exclusively by sympathetic nerves. In contrast, input originating from *physiologic receptors* (innocuous input) travels primarily in parasympathetic nerves. Thus, there is a division of responsibility between parasympathetic and sympathetic nerves in terms of viscerosensory input.

Ascending Pathway for Sympathetic Afferents

Afferent fibers conveying nociceptive information from thoracic and abdominal viscera travel via the cardiac and splanchnic nerves (Fig. 19–2). For example, primary sensory fibers that originate from the stomach join the greater splanchnic nerve, enter the sympathetic trunk, and pass through a white ramus to join the spinal nerve. Nociceptive input from pelvic viscera such as the prostate and sigmoid colon is conveyed by viscerosensory fibers traveling through the hypogastric plexus and lumbar splanchnic nerves.

The cell bodies of origin of sympathetic afferent fibers are located in dorsal root ganglia at about levels T1 to L2 (see Fig. 19–2). The central processes of these fibers enter the spinal cord via the lateral division of the dorsal root. They may ascend or descend one or two spinal levels in the dorsolateral fasciculus before terminating in laminae I and V and/or laminae VII and VIII. Cells in laminae I and V project mainly to the contralateral side as part of the *anterolateral system,* whereas the neurons in laminae VII and VIII project bilaterally as *spinoreticular fibers.* In addition, some primary viscerosensory fibers terminate on preganglionic

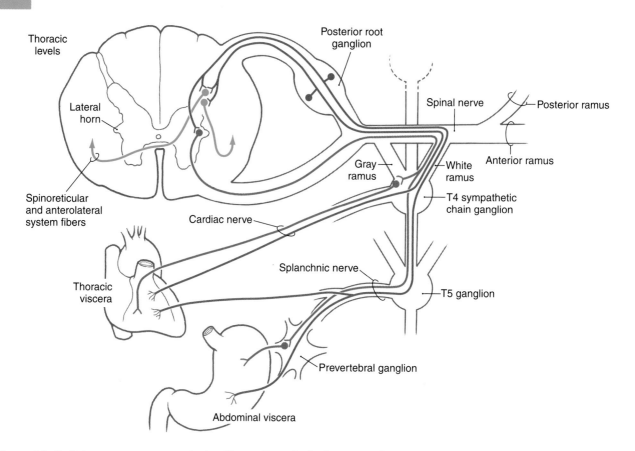

Figure 19–2. Primary sensory sympathetic afferent fibers (red) shown in relation to posterior horn tract cells (green) conveying visceral information to the thalamus and to general visceral efferent neurons (blue).

sympathetic cell bodies located in the intermediolateral cell column at spinal levels T1 to L2 (see Fig. 19–2). The axons of these latter cells, in turn, exit through the ventral root as GVE preganglionic sympathetic fibers.

In general, viscerosensory fibers that enter the spinal cord at a particular level originate from structures that receive GVE input from the same spinal level (see Fig. 19–2). For example, visceral afferent fibers from the stomach enter the spinal cord over the posterior roots of T5 to T9 and terminate in the same spinal segments that convey visceral efferent outflow to the stomach.

Projections to Thalamus. Some neurons located in laminae I and V receive nociceptive input from sympathetic afferent fibers and send their axons rostrally via two routes in the anterolateral system (ALS) (Fig. 19–3). Some fibers cross in the anterior white commissure and ascend in the ALS, whereas others ascend in this bundle on the ipsilateral side. These ALS fibers terminate in the ventral posterolateral nucleus (VPL), which, in turn, projects to the inferolateral part of the postcentral gyrus (the parietal operculum) and to the insular cortex (see Fig. 19–3). The location from which this visceral nociceptive information originated is encoded in these particular regions of the cerebral cortex. However, visceral pain is poorly localized (lacks

detailed point-to-point representation) because receptor density is low and receptive fields correspondingly large and because this input converges in the pathway. Consequently, it is not possible to tell whether pain is coming from the stomach or the duodenum; rather, it can be determined *only* that the pain is coming from the general area of the upper abdomen.

Projections to Reticular Formation. In addition to the direct path to thalamus and sensory cortex via the ALS and VPL, there are indirect routes via the reticular formation through which visceral nociceptive information can reach the cortex. The reticular formation receives spinoreticular inputs (mainly from laminae VII and VIII) and collaterals from the ALS (see Fig. 19–3). In turn, cells of the reticular formation project to progressively higher levels of the neuraxis, thus relaying viscerosensory information in a multisynaptic fashion to progressively higher levels of the brain. Neurons located in the reticular formation and in the periaqueductal gray ultimately project to the hypothalamus and to the intralaminar nuclei of the thalamus (see Fig. 19–3). These latter cell groups project to the cortex.

Reticulohypothalamic fibers travel via the *dorsal longitudinal fasciculus*, the *mammillary peduncle*, and the *medial forebrain bundle*. The first originates mainly from

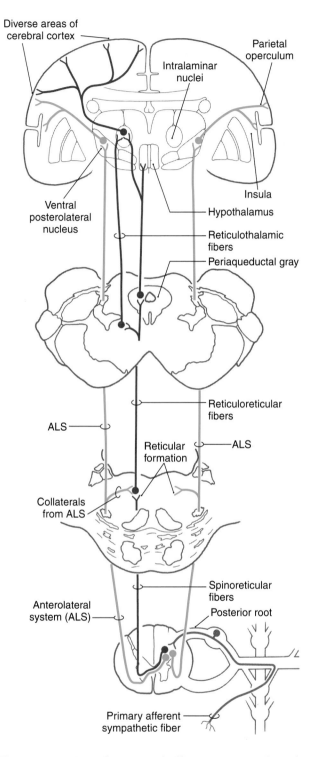

Figure 19-3. Ascending visceral afferent input travels in the anterolateral system (green) and through multisynaptic circuits via the reticular formation of the brainstem (gray). These fibers influence specific and diverse areas of the cerebral cortex.

tions, hypothalamic centers may be influenced by either system. For example, viscerosensory input resulting from distention of the bowel may result in increased heart rate or cutaneous flushing. On the other hand, somatosensory stimuli such as those associated with coitus or suckling may increase the release of the hypothalamic hormone oxytocin.

Referred Pain. Referred pain is the phenomenon whereby noxious stimuli that originate in a visceral structure, such as the heart or the stomach, are perceived by the patient as pain arising from a somatic portion of the body wall such as the skin, bones, or skeletal muscles (Fig. 19–4). Although such referral of pain may mask the true origin of the information, certain patterns of referred pain are clearly diagnostic of diseases in particular visceral locations. Visceral pain is transmitted by sympathetic sensory fibers and is typically referred to those somatic structures whose afferents enter the cord via the same dorsal roots.

The mechanism underlying referred pain is thought to involve a convergence of somatic and visceral afferent information onto pools of posterior horn neurons, the axons of which ascend to higher levels of the neuraxis (Fig. 19–5). Normally, a visceral nociceptive fiber (e.g., from the heart) synapses on a spinothalamic tract cell whose axon will travel to the VPL, and from there the information is relayed to visceral parts of the sensory cortex. Consequently, this sensory input is perceived as arising from deep within one of the body's cavities (e.g., the thoracic cavity) (see Fig. 19–5A). In some situations, however, collaterals of visceral afferent fibers may synapse on and excite posterior horn tract cells that usually transmit only somatosensory information (see Fig. 19–5B). In this case, the tract cell is

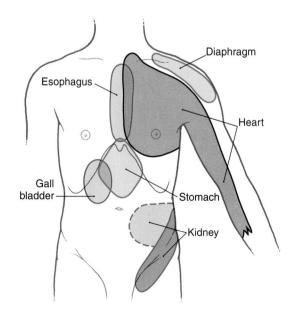

Figure 19-4. Superficial areas to which pain is commonly referred from the corresponding deep structures.

the periaqueductal gray, and the latter two originate mainly from the mesencephalic reticular formation. These midbrain centers receive both viscerosensory and somatosensory input and, through their projec-

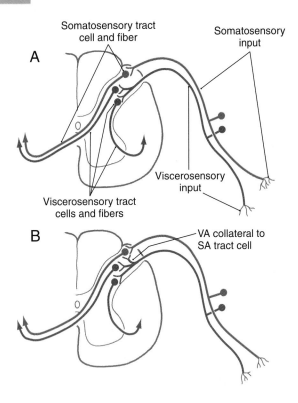

Figure 19-5. The circuits involved in referred pain. Somatic afferent (blue) and visceral afferent (red) fibers terminate on tract cells that convey their respective types of information to the thalamus (A). Collaterals from visceral afferent fibers (VA) may activate tract cells that usually convey somatic afferent (SA) data (B). The brain interprets this input as originating from the body wall.

activated by a visceral afferent collateral but sends information via the VPL to a part of the somatosensory cortex that represents the body wall (see Fig. 19-5B). Consequently, the pain is "referred" to (interpreted as coming from) the surface of the body (see Fig. 19-4) even though the stimulus actually originated from a visceral structure.

Angina. Referred pain may occur in conjunction with disease affecting any internal organ; however, it is frequently associated with diseases of the heart (see Fig. 19-4). The pain resulting from heart disease is termed *angina*. In about 80% of patients, angina is initially perceived as an unpleasant squeezing sensation originating from behind the sternum. This discomfort may also be perceived as pain radiating down the left arm or, more rarely, down both arms. On rare occasions, the pain has been reported to radiate bilaterally into the neck, jaw, and temporomandibular joints. The predilection of the pain for the left side of the chest, or extending down the left arm, reflects the predominance of myocardial disease in the left side of the heart. Consequently, nociception from the left side of the heart is referred to the left side of the body. Because angina is typically perceived as a pain of the

chest, including the sternum and pectoral muscles, it is frequently called *angina pectoris*.

The pathways involved in angina are shown in Figure 19-6. Afferent fibers from the heart enter the sympathetic trunk through either the *cervical cardiac* or *thoracic cardiac nerves*. The former join the sympathetic chain at superior, middle, and inferior cervical ganglia, whereas the latter nerves join the sympathetic ganglia associated with spinal nerves T1 to T5 (see Fig. 19-6). These primary viscerosensory fibers enter the spinal cord and terminate in laminae I and V of the posterior horn. These same spinal segments also receive cutaneous somatosensory input from dermatomes of the chest wall and arm (see Fig. 19-6). Tract cells in the posterior horn that receive primarily somatosensory input may also be activated, as noted previously, by collaterals of visceral afferent fibers from the heart. Consequently, the cerebral cortex interprets the pain as originating from the surface of the body (over the upper chest and/or arm) when actually the stimulus that has produced the painful input is located in a visceral structure (the heart).

Pathways for Parasympathetic Afferents

Sacral Parasympathetic Afferents. The *pelvic nerves* are parasympathetic and contain viscerosensory fibers passing to cord levels S2 to S4 and GVE preganglionic fibers originating from these levels. These primary sensory parasympathetic fibers pass through the pelvic nerves, enter the spinal nerves, and have their cell bodies in posterior root ganglia of S2 to S4 (Fig. 19-7). Rather than entering the posterior root, however, their central processes course into the spinal cord via the anterior root.

Once in the cord, these viscerosensory fibers terminate in the posterior horn and in the immediate vicinity of the visceral efferent preganglionic motor neurons. The posterior horn cells relay information on bladder or bowel distention (a sense of "fullness") to the VPL via the ALS and spinoreticular pathways described earlier and then, through this thalamic nucleus, to the insular and parietal opercular cortices (see Figs. 19-3 and 19-6). In this way, a full bladder is perceived and interpreted as such. In addition, ascending input from the pelvic viscera is also shunted into the hypothalamus for the initiation of *supraspinal autonomic reflexes*. Those GVE cell groups in S2 to S4 that receive viscerosensory afferents give rise to parasympathetic preganglionic axons that synapse on postganglionic cells located in pelvic viscera (see Fig. 19-7). This relationship between viscerosensory fibers and the GVE cell groups to which they project forms the basis for *spinal autonomic reflexes*.

Cranial Parasympathetic Afferents. Cranial nerves

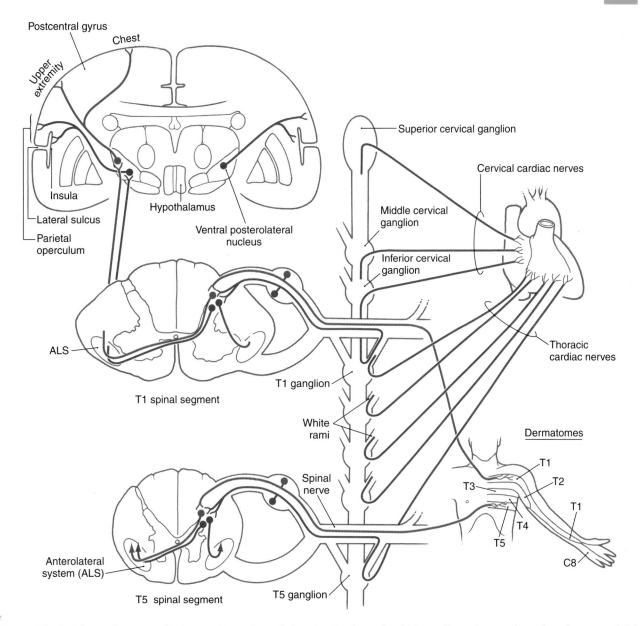

Figure 19–6. The pathways mediating cardiac pain and the circuits through which cardiac pain may be referred to superficial parts of the body wall.

III, VII, IX, and X contain fibers of GVE preganglionic parasympathetic motor neurons. However, only cranial nerves VII, IX, and X have sensory ganglia. Of these, only cranial nerves IX (the *glossopharyngeal*) and X (the *vagus*) have significant numbers of parasympathetic afferent fibers.

Visceral afferent fibers traveling in the glossopharyngeal nerve originate primarily from *chemoreceptors* of the carotid body and *baroreceptors* of the carotid sinus wall (Fig. 19–8). In addition, nociceptive and tactile input from the oropharynx (the general area of the palatine tonsil) is also conveyed on the ninth cranial nerve. These sensory fibers form the afferent limb of the gag reflex.

The *carotid body* is composed of specialized neural elements, *chemoreceptors*, and is innervated by viscerosensory branches of the glossopharyngeal nerve. Within the carotid body, these chemoreceptors are located in close proximity to a fenestrated capillary network. As a result, they are responsive to changes in arterial oxygen and carbon dioxide tension, to the acidity of the blood, and to drugs. Carotid *baroreceptors* are located in the carotid sinus wall and respond to rapid changes in arterial blood pressure (see Fig. 19–8). The *aortic arch* also contains chemoreceptors and baroreceptors that are similar in structure and function to those found in the carotid body and sinus. These specialized receptors, however, are innervated by aortic or cardiac branches of the vagus nerve.

Fibers of the vagus nerve transmit a wide variety of physiologic information from thoracic viscera and from all viscera of the abdominal cavity above the level of

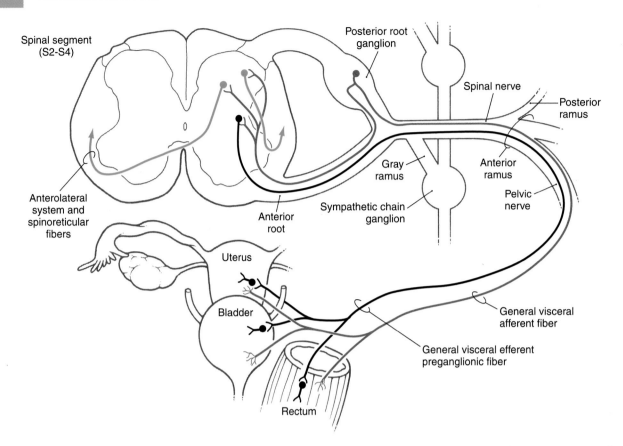

Figure 19–7. Primary sensory parasympathetic fibers (red) shown throughout their trajectory in relation to posterior horn tract cells (green), and general visceral efferent neurons (gray).

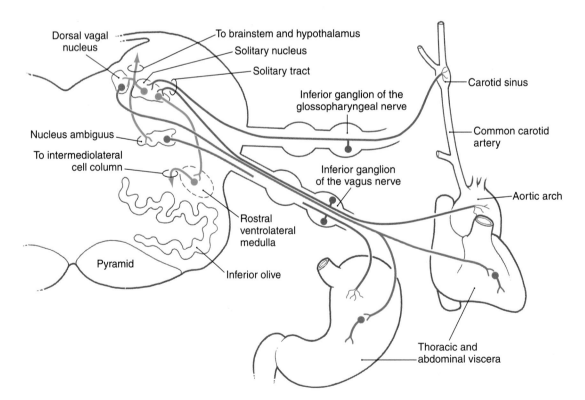

Figure 19–8. The nucleus of the solitary tract receives viscerosensory fibers via the glossopharyngeal (IX) and vagus (X) nerves and projects to a variety of other nuclei.

the splenic flexure of the large colon. These vagal fibers convey information regarding the functional status of these structures but are not responsible for conveying information on pain.

The peripheral viscerosensory fibers, traveling in the glossopharyngeal and vagus nerves, enter the skull through the jugular foramen. Within this foramen there is a superior and an inferior ganglion on each nerve (see Fig. 19–8). Cell bodies of primary viscerosensory neurons (GVA) are found within the inferior ganglion, whereas the superior ganglion contains the cell bodies of primary somatosensory neurons (general somatic afferent [GSA]).

The central processes of primary visceral fibers in cranial nerves IX and X enter the medulla, form the *solitary tract*, and synapse with neurons of the adjacent *solitary nucleus* (see Fig. 19–8). Some of these fibers use substance P and cholecystokinin as their transmitters. Second-order neurons in the solitary nucleus project to and influence a variety of neurons in the brainstem and hypothalamus. These targets include the dorsal vagal nucleus, the nucleus ambiguus, rostral areas of the anterolateral medulla, and the parabrachial nuclei. The *dorsal vagal nucleus* is the primary source of preganglionic parasympathetic neurons that project to thoracic and abdominal viscera. In addition, the *nucleus ambiguus* also contains some parasympathetic visceromotor cells (GVE, autonomic) that innervate cardiac ganglia. However, the targets of most nucleus ambiguus cells (these are special visceral efferent [SVE] motor neurons) are muscles of the larynx, pharynx, and esophagus.

Visceral efferent (GVE) motor neurons of the nucleus ambiguus and the dorsal vagal nucleus receive input from the solitary nucleus and project, via the vagus nerve, to parasympathetic ganglia of the heart. Activation of this pathway causes a decrease in heart rate and a corresponding decrease in blood pressure, a "vasodepressor" response. Conversely, neurons located in rostral parts of the anterolateral medulla receive solitary input and project to the spinal cord, where they influence the activity of preganglionic sympathetic motor neurons in the intermediolateral cell column. In doing so, this medullary center causes an increase in blood pressure and thereby serves a "vasopressor" function.

Baroreceptor Reflex. Projections from the solitary nucleus to the dorsal vagal nucleus, to cells associated with the nucleus ambiguus, and to rostral parts of the anterolateral medulla are essential to the normal operation of the *baroreceptor reflex* (see Fig. 19–8). In this reflex, afferent input from carotid and aortic baroreceptors enters the medulla on cranial nerves IX and X and terminates in the solitary nucleus. Increases in blood pressure cause the baroreceptors to increase their discharge frequency, whereas decreases in blood pressure result in a lower rate of baroreceptor dis-

charge. In this manner, blood pressure is continuously monitored and the resulting information is forwarded to the solitary nucleus. Within this nucleus, neurons projecting to the dorsal vagal nucleus and the nucleus ambiguus respond in a manner opposite to those neurons projecting to the rostral parts of the anterolateral medulla. For example, during a period of acute *hypertension*, solitary neurons excite "vasodepressor" cells of the dorsal vagal nucleus and nucleus ambiguus and inhibit "vasopressor" neurons of the rostral anterolateral medulla. The inhibitory solitary neurons may be GABAergic, whereas the excitatory neurons presumably use one of the excitatory amino acids. As a result of this dual influence from solitary neurons, blood pressure is lowered and the hypertension is diminished. Conversely, during a period of acute *hypotension*, projections from the solitary nucleus inhibit "vasodepressor" cells and excite "vasopressor" neurons, leading to an elevation of blood pressure. Consequently, blood pressure is elevated and the hypotension relieved.

In addition to projections from the solitary nucleus to ambiguus, dorsal vagal, and medullary nuclei, solitary neurons also project into the reticular formation. Consequently, visceral afferent information entering the medulla on the vagal and glossopharyngeal nerves may also influence the hypothalamus and diverse areas of the cerebral cortex. The solitary nucleus does not relay any appreciable amount of general visceral sensation to the dorsal thalamus. Consequently, most of the GVA afferent information conveyed by the glossopharyngeal and vagal nerves does not reach a level of consciousness. On the other hand, the solitary nucleus does relay taste information to the cerebral cortex via the ventral posteromedial nucleus.

Visceral Input to the Reticular Activating System

The reticular formation of the brainstem receives a wide range of inputs and projects to, among other targets, the intralaminar nuclei of the thalamus. These cell groups, in turn, send their axons to broad areas of the cerebral cortex, with the largest number terminating in the frontal lobe. These reticulothalamic and thalamocortical pathways "alert" or "activate" the cerebral cortex as a whole and constitute one important part of the *ascending reticular activating* system (ARAS) (see also Chapters 18 and 32).

As discussed previously, the reticular formation receives viscerosensory input via spinoreticular fibers and collaterals from the ALS. Some of these ascending fibers to the reticular formation convey nociceptive visceral afferent information originating from the gut on sympathetic afferent fibers. Other ascending fibers convey a sense of bladder (or bowel) fullness originating from pelvic viscera on parasympathetic afferents. Both types of viscerosensory input feed into the reticular formation and participate in the "arousal" of the cerebral cortex through the ARAS. For example, either sudden pain from the stomach or small intestine or the stimulus of a full bladder will excite the reticulothalamocortical circuit and wake a person from a deep sleep. The initial sensation is not one of specific information (full bladder, stomach pain) but rather the sense of just being awakened. However, once the cortex has been "alerted," the conscious/perceptive part of the brain takes over (recognizes the source of the arousal) and addresses the problem.

Sources and Additional Reading

Ammons WS: Cardiopulmonary sympathetic afferent excitation of lower thoracic spinoreticular and spinothalamic neurons. J Neurophysiol 64:1907–1916, 1990.

Bieger D, Hopkins DA: Viscerotopic representation of the upper alimentary tract in the medulla oblongata in the rat: the nucleus ambiguus. J Comp Neurol 262:546–562, 1987.

Brody MJ: Central nervous system mechanisms of arterial pressure regulation. Fed Proc 45:2700–2706, 1986.

Cechetto DF, Saper CB: Evidence for a viscerotopic sensory representation in the cortex and thalamus in the rat. J Comp Neurol 262:27–45, 1987.

Cervero F, Foreman RD: Sensory innervation of the viscera. In Loewy AD, Spyer KM (eds): Central Regulation of Autonomic Functions. Oxford University Press, New York, 1990, pp 104–125.

Coggeshall RE, Applebaum ML, Frazen M, Stubbs TB III, Sykes MT: Unmyelinated axons in human ventral roots, a possible explanation for the failure of dorsal rhizotomy to relieve pain. Brain 98:157–166, 1975.

Garrison DW, Chandler MJ, Foreman RD: Viscerosomatic convergence onto feline spinal neurons from esophagus, heart and somatic fields: effects of inflammation. Pain 49:373–382, 1992.

Procacci P, Zoppi M, Maresca M: Heart, vascular and haemopathic pain. In Wall PD, Melzack R (eds): Textbook of Pain. 4th ed. Churchill Livingstone, Edinburgh, 1999, pp 621–639.

Reis DJ, Granata AR, Joh TH, Ross CA, Ruggiero DA, Park DH: Brain stem catecholamine mechanisms in tonic and reflex control of blood pressure. Hypertension Suppl II 6:7–15, 1984.

Willis W: Visceral pain. In Brooks FP, Evers PW (eds): Nerves and the Gut. Charles B Slack, Thorofare, NJ, 1977, pp 350–364.

Chapter 20

The Visual System

J. B. Hutchins and J. J. Corbett

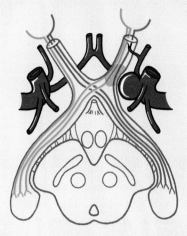

Overview 304

Anatomy of the Eye 304
Cornea
Chambers of the Eye
Iris
Lens
Uvea

The Neural Retina and Pigment Epithelium 305

Photoreceptor Cells 306
Rods
Cones
Macula and Fovea

Receptive Fields 309

Processing of Visual Input in the Retina 309
Outer and Inner Plexiform Layers
Horizontal Cells
Bipolar Cells
Amacrine Cells
Ganglion Cells

Retinal Projections 312
Retinogeniculate Projections
Optic Nerve, Chiasm, and Tract

Lateral Geniculate Nucleus 316
Magnocellular and Parvocellular Layers
Ipsilateral and Contralateral Layers

Optic Radiations 318

Primary Visual Cortex 319
Visual Cortical Columns

Abnormal Development of Visual Cortex 320

Other Visual Cortical Areas 321

Of all the senses, humans rely most on vision. Consequently, more is known about vision than about any of the other senses, and a great deal of basic science and clinical research effort has been spent examining its function. The visual system is enormously important in clinical diagnosis. About 40% of the fibers in the brain carry information related to visual function; therefore, damage to the brain is often reflected by some type of visual dysfunction.

Overview

Like other sensory systems, the visual system creates a location-coded (visuotopic) "map" of its sensory field (the visual world) that is preserved at all levels. Light information is received by the retinal photoreceptors, and the initial processing of the visual signal occurs in the retina. Although the retina projects to several diencephalon and midbrain structures, most retinal axons terminate in a thalamic relay nucleus, which in turn innervates a large region of cortex in the occipital lobe. From there, visual input is sent to several cortical regions, which govern much of the temporal and parietal lobes.

Anatomy of the Eye

The human eye is an optical instrument specialized for the collection of light and the initial processing of visual information. Its anatomy is considered only briefly here.

Cornea. The *cornea* provides a transparent protective coating for the optical structures of the eye (Fig. 20–1). Its lateral margin is continuous with the *conjunctiva*, a specialized epithelium covering the "white" (*sclera*) of the eye. Although the conjunctiva and sclera have a blood supply (most evident when the eyes are irritated), the central cornea is normally not vascularized. Several conditions, such as vascularization or the presence of opacities, may degrade the optical quality of this structure, requiring transplantation of a donor cornea.

Chambers of the Eye. Just behind the cornea is the fluid-filled *anterior chamber*, which is bounded posteriorly by the *iris* and the opening of the *pupil* (see Fig. 20–1). A second fluid-filled space, the *posterior chamber*, is bounded anteriorly by the iris and posteriorly by the lens and its encircling suspensory ligament (zonule fibers). Fluid is continuously produced by the epithelium over the *ciliary body* around the rim of the posterior chamber and flows through the pupillary opening into the anterior chamber. It then drains into a set of modified veins, the *canals of Schlemm*, that are located around the rim of the anterior chamber in the *angle* where the iris meets the cornea. Because the suspensory ligament encircling the lens consists of discrete strands, the fluid in the posterior chamber is in contact with the *vitreous body*, the gelatinous mass that fills the main space of the eyeball between the lens and the retina. Therefore, any condition that obstructs the drainage of fluid via the canals of Schlemm can lead to *glaucoma*, a buildup of pressure in the entire eyeball that may result in blindness.

Iris. The *iris* is a pigmented structure lying directly anterior to the lens (see Fig. 20–1). The connective tissue, or *stroma*, of the iris contains melanocytes that reflect or absorb light to give the iris its characteristic

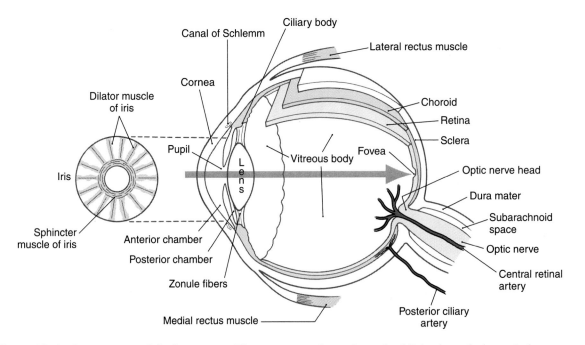

Figure 20–1. Cross section of the human eye. The gray arrow shows the path of light through the optical apparatus.

color. Also embedded in the stroma are the circumferentially organized *sphincter muscle* of the iris and the radially arranged *dilator muscle* (see Fig. 20–1).

The innervation of the *iris sphincter*, which closes the pupil, is parasympathetic. This pathway begins with preganglionic neurons whose cell bodies lie in the *Edinger-Westphal nucleus*, and whose axons terminate in the *ciliary ganglion*. Axons of postganglionic ciliary ganglion neurons, in turn, end as neuromuscular synapses on the sphincter muscle and release acetylcholine. When activated, this pathway results in a reduction in pupil diameter, or *miosis*.

The innervation of the *iris dilator*, which opens the pupil, is sympathetic. The pathway begins with preganglionic neurons whose cell bodies lie in the *intermediolateral cell column* of the spinal cord at upper thoracic levels, and whose axons terminate in the *superior cervical ganglion*. Axons of postganglionic superior cervical ganglion neurons, in turn, end as neuromuscular synapses on the dilator muscle and release norepinephrine. When activated, this pathway results in an increase in pupil diameter, or *mydriasis*. This phenomenon is a measure of the general state of the sympathetic tone. Anger, pain, or fear may result in an enlargement of the pupil in the absence of a change in lighting conditions. The *pupillary light reflex*, a contraction of the pupil in response to light, is used to assess the function of the nervous system at midbrain levels (see Chapter 28).

The circumference of the pupillary margin changes by a factor of 6. This proportional change in muscle length is greater than any other in the human body. To accomplish this change, acetylcholine is released onto *both* the sphincter and dilator muscles. The effect is to activate muscarinic receptors that *depolarize* sphincter muscle cells and cause contraction. Additionally, acetylcholine released by collaterals onto the dilator muscle mediates presynaptic inhibition of norepinephrine release and *blocks* dilator contraction. Thus, as the sphincter contracts, the dilator relaxes, strengthening the pupillary response to light.

Lens. The *lens* is a clear structure that focuses light on the *retina* (see Fig. 20–1). Mechanisms that change the curvature of the lens are discussed in detail in Chapter 28.

Beginning at about age 40, the lens begins to lose its elasticity, so that the shape it adopts when relaxed is more flattened than earlier in life. This change reduces the affected person's ability to focus on near objects, a condition called *presbyopia*. Reading or bifocal corrective lenses are prescribed to aid the patient in performing tasks requiring close, detailed vision.

Opacities in the lens, known as *cataracts*, are relatively common and can be seen as a cloudiness of the lens. Cataracts may be caused by *congenital defects* (e.g., secondary to maternal infection with *rubella*), persistent exposure to *ultraviolet light*, or poorly understood

mechanisms that occur in aging. Current therapy consists of replacement of the lens with an inert plastic prosthesis, restoring sight but with a concomitant loss of accommodation.

Uvea. The iris, ciliary body, and choroid make up the *vascular tunic* of the eye, also called the *uvea*. The *choroid* is a highly vascularized, pigmented tissue layer lying between the *retinal pigment epithelium* and the *sclera*, the tough outer coating of the eye. *Uveitis* is an inflammation of these structures, often secondary to eye injury.

The Neural Retina and Pigment Epithelium

The inner surface of the posterior aspect of the eye is covered by the *retina*, which is composed of the *neural retina* and the *retinal pigment epithelium* (Fig. 20–2). In describing the layers and cells of the retina, it is common to use the terms *inner* and *outer*. *Inner* refers to structures located toward the vitreous (i.e., the center of the eyeball), whereas *outer* is used in reference to structures located toward the pigment epithelium and choroid.

The *retinal pigment epithelium* is a continuous sheet of pigmented cuboidal cells bound together by tight junctions that block the flow of plasma or ions. Its functions are as follows: (1) it supplies the neural retina with nutrition in the form of glucose and essential ions; (2) it protects retinal photoreceptors from potentially damaging levels of light; and (3) it plays a key role in the maintenance of photoreceptor anatomy via phagocytosis.

The *neural retina* contains the photoreceptors and associated neurons of the eye and is specialized for sensing light and processing the resultant information. The *photoreceptors* absorb quanta of light (photons) and convert this input to an electrical signal. The signal is then processed by retinal neurons as discussed further on. Finally, the retinal neurons called *ganglion cells* send the processed signal to the brain via axons that travel in the optic nerve.

The contact between the neural retina and the pigment epithelium is the adult remnant of the ventricular space of the developing eyecup. As such, it is mechanically unstable. This instability is demonstrated in a *retinal detachment*, in which the neural retina tears away from the pigment epithelium. Because photoreceptors are metabolically dependent on their contact with pigment epithelial cells, a detached retina must be repaired to avoid further damage. The detached part of the neural retina is welded to the pigment epithelium using surgical procedures. Although this repair prevents an increase in the area of detachment, the detached portion of the retina does not regain function.

The neural retina has seven characteristic layers (see

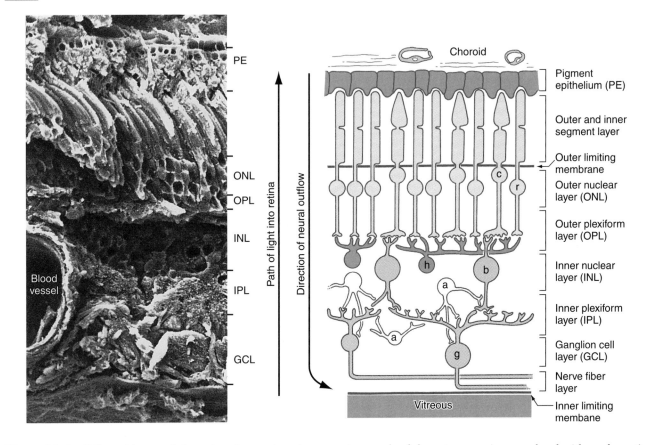

Figure 20–2. Cells and layers of the retina. A scanning electron micrograph of the primate retina correlated with a schematic drawing showing the main retinal cell types. Photoreceptors (rods, r, and cones, c) are shown in green. Horizontal cells (h, gray) and bipolar cells (b, blue) receive input from photoreceptors; the bipolar cells, in turn, synapse onto amacrine cells (a, white) and ganglion cells (g, red).

Fig. 20–2). From outer to inner they are (1) a layer containing the *photoreceptor cell outer and inner segments;* (2) an *outer nuclear layer* consisting of the nuclei of photoreceptor cells; (3) the *outer plexiform layer*, consisting of the synaptic connections of photoreceptors with second-order retinal cells; (4) the *inner nuclear layer*, containing somata of second-order and some third-order retinal cells; (5) the *inner plexiform layer*, another area of synaptic contact; (6) the *ganglion cell layer*, containing the cell bodies of the *ganglion cells;* and (7) the *nerve fiber layer* (or *optic fiber layer*), composed of the axons of the ganglion cells. These axons converge at the *optic disc* to form the *optic nerve*. Layers 2 through 7 are flanked by a pair of *limiting membranes*, which consist of glial cell processes joined by tight junctions. The *outer limiting membrane* is located between layers 1 and 2, and the *inner limiting membrane* is located between the nerve fiber layer and the vitreous.

The *photoreceptor outer segments* interdigitate with the melanin-filled processes of pigment epithelial cells (see Fig. 20–2). These processes are mobile, and they elongate into the pigmented layer when the light is bright (*photopic* conditions) and retract when the light is dim (*scotopic* conditions). This mechanism combines with

contractions of the iris to protect the retina from light conditions that would otherwise damage the photoreceptors. The iris, pigment epithelium, and circuitry of the retina all contribute to the eye's ability to resolve the visual world over a wide range of light conditions.

The blood supply of the neural retina arises from branches of the *ophthalmic artery:* the *central artery of the retina* and the *ciliary arteries*. The central artery branches out from the optic nerve head to serve inner portions of the neural retina. The ciliary arteries penetrate the sclera around the exit of the optic nerve and feed the *choriocapillaris* (a portion of the choroid), which, in turn, provides nutrients to the outer portions of the neural retina.

Photoreceptor Cells

The rods and cones of the retina are responsible for *photoreception*, the process by which photons are detected and the information is transduced into an electrochemical signal. There are two basic types of photoreceptors: *rods* and *cones* (Figs. 20–3 and 20–4). Both types have the same overall design. Light is detected and transduced in an *outer segment* that points toward

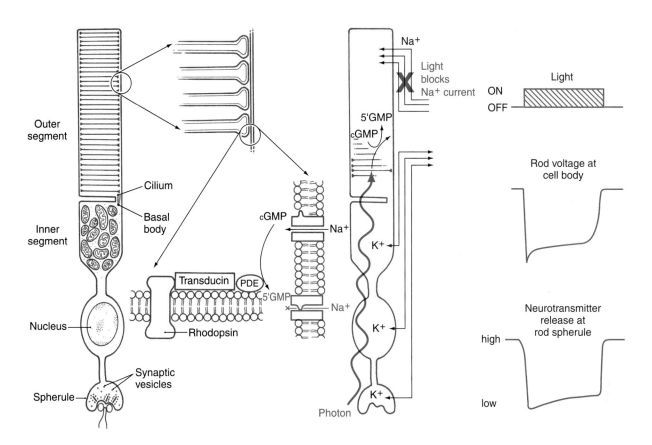

Figure 20–3. The rod photoreceptor and the physiologic and chemical changes that occur in response to light. Events associated with light are shown in red. cGMP, cyclic guanosine monophosphate; 5′GMP, 5′-guanosine monophosphate; PDE, phosphodiesterase.

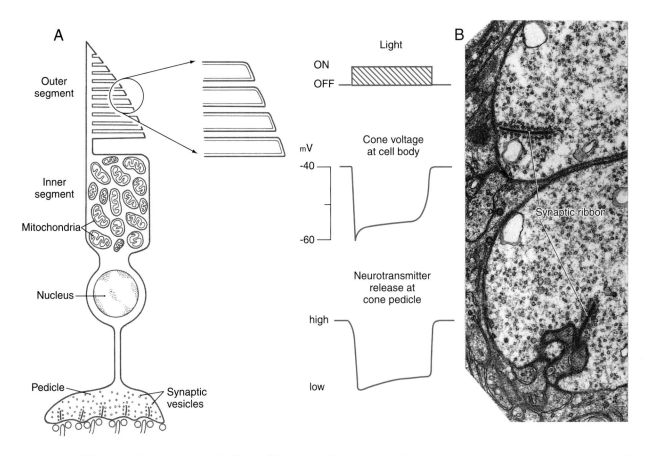

Figure 20–4. The cone photoreceptor *(A)*. Cones, like rods, reduce their levels of neurotransmitter release when stimulated by photons. Cones and rods are also distinguished by prominent electron-dense *synaptic ribbons* in their terminals *(B)*.

the pigment epithelium. A narrow stalk, the *cilium*, connects the outer segment to a second expanded region called the *inner segment*, which contains mitochondria and produces the energy that maintains the cell. The cilium contains nine pairs of microtubules emanating from a basal body located in the inner segment. The nucleus and perikaryon of the cell are found in the outer nuclear layer; finally, the cell terminates in the outer plexiform layer in an expansion that makes synaptic contacts with neurons. This synaptic expansion is called the *spherule* in rod cells and the *pedicle* in cone cells. Both rod and cone synaptic terminals contain a characteristic dark sheet of protein called the *synaptic ribbon*. This structure may act as a "conveyor belt," organizing vesicular release of transmitter.

Rods. Rod cells are named for the shape of their outer segment, which is a membrane-bound cylinder containing hundreds of tightly stacked membranous discs (see Fig. 20–3). The rod outer segment is a site of *transduction*. Photons travel through cells of the neural retina before striking the membranous discs of the rod outer segment. Molecules of *rhodopsin* within these membranes undergo a conformational change and along with transducin and phosphodiesterase (PDE) induce biochemical changes in the rod outer segment, which reduce levels of cyclic GMP (cGMP). In the dark, cGMP levels in the rod outer segment are high. Cyclic GMP mediates a *standing sodium current*. At rest, in the dark, sodium ions flow into the rod outer segment. This high resting level of sodium permeability results in a relatively high resting potential for rod cells, about −40 mV. These sodium channels of the outer segment membrane, which are normally open, close in response to increased calcium or a reduction in cGMP. This drives the membrane potential away from the sodium equilibrium potential and toward the potassium equilibrium potential, and the rod cell is *hyperpolarized* in response to a light stimulus (see Fig. 20–3). Note that photoreceptors are the only sensory neurons that hyperpolarize in response to the relevant stimulus.

The hyperpolarization of the rod outer segment propagates passively (i.e., without firing an action potential) through the perikaryon to the rod spherule. In the absence of light, the photoreceptor terminals constantly release the transmitter *glutamate* at these synapses. The arrival of a light-induced wave of hyperpolarization causes a transient *reduction* in this tonic release of glutamate. As explained further on, this event can *depolarize* some of the cells that receive synapses from photoreceptor terminals while *hyperpolarizing* others.

Rhodopsin molecules are capable of a huge but finite number of photoisomerization events. Rather than replace individual rhodopsin molecules, every morning the distal one tenth of the outer segment is broken off and phagocytosed by the pigment epithelium. Through this process of *rod shedding*, the outer segment is con-

stantly renewed. New discs are formed at the base of the outer segment and move outward so that the shed discs are replaced. In this way, the rod remains a constant length, and the outer segment is renewed about every 10 days.

Cones. Like rod outer segments, *cone outer segments* also consist of a membranous stack (see Fig. 20–4). Unlike in rods, however, these stacks of cone membranes are of constantly decreasing diameter (from cilium to tip), giving the cell its characteristic shape. Also, they are not enclosed within a second membrane but are open to the extracellular space adjacent to the pigment epithelium (see Fig. 20–4).

The process of transduction in cones is generally similar to that in rods. *Cone opsin* absorbs photons and undergoes a conformational change, resulting in a hyperpolarization of the cell membrane (see Fig. 20–4). This hyperpolarization propagates passively to the cone's synaptic ending, the *cone pedicle*, in the outer plexiform layer. Cone pedicles and rod spherules both contain synaptic ribbons surrounded by vesicles, but cone pedicles are larger (see Fig. 20–4). Serial-section electron microscopy has shown that the synaptic ribbons are actually a single extensive sheet of protein. Like rods, cones release the neurotransmitter *glutamate* tonically in the dark and respond to light with a decrease in glutamate release.

There are three types of cones, each tuned to a different light wavelength (Fig. 20–5). *L-cones* (red cones) are sensitive to long wavelengths, *M-cones* (green

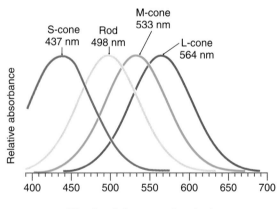

Figure 20–5. Absorption spectra of rods and the three types of cones. Because the three cone spectra are different but overlapping, any wavelength of light in the visual spectrum (bottom scale) will elicit a set of response intensities in the three types of cones that is different from the set elicited by any other wavelength. Therefore, any color in the visual spectrum can be uniquely encoded. The rod spectrum is shown for comparison even though rod input is not used in color recognition. Dim red light can be used to adapt humans to maximum rod sensitivity because red light (620 to 700 nm) is not absorbed by rods to any significant extent.

cones) to medium wavelengths, and *S-cones* (blue cones) to short wavelengths. Because any pure color represents a particular wavelength of light, each color will be represented by a unique combination of responses in the L-, M-, and S-cones.

If one of these cone types is absent because of a genetic defect in the corresponding opsin, the affected person will confuse certain colors that look different to visually normal people and is said to be *"color blind."*

The term is poorly chosen. It is better to think of this condition as "color confusion" because the patient can still see all colors of the visible spectrum; it is the ability to *distinguish* certain colors that has been lost. Because the genes for the L-cone (red-absorbing) and M-cone (green-absorbing) opsins are located on the X chromosome, color blindness is more common in men. Alteration of the gene for the S-cone (blue-sensitive) pigment, which is located on an autosome, is much rarer. The inability to detect a pure red is known as *protanopia*, and inability to detect green is known as *deuteranopia*.

Macula and Fovea. At the posterior pole of the eye is a yellowish spot, the *macula lutea*, the center of which is a depression called the *fovea centralis* (Fig. 20–6). Near the fovea, the inner retinal layers become thinner and eventually disappear so that, at the bottom of the foveal pit, only the outer nuclear layer and photoreceptor outer segments remain. This allows a maximum amount of light to reach the photoreceptors with optimal fidelity.

Most of the visual input that reaches the brain comes from the fovea. Cones, which are responsible for color vision, are the only type of photoreceptor present in the fovea. In contrast, rods, which are most sensitive at low levels of illumination, are not present in the fovea but are the predominant type of photoreceptor in the periphery of the retina. The visual world is a composite formed from a succession of foveal images carrying form and color information supplemented with input from the peripheral retina carrying motion information.

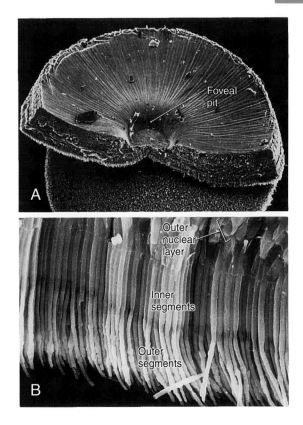

Figure 20–6. Scanning electron micrographs of the primate fovea centralis *(A)* and of inner and outer segments of photoreceptors *(B)*, mostly rods, in more peripheral areas of the retina. Only cones are present in the foveal pit. The surface striations *(A)* are ganglion cell axons en route to the optic nerve head. (Photographs courtesy of Dr. Bessie Borwein. From Borwein, 1983, with permission of Wiley-Liss, Inc.)

Receptive Fields

As with other sensory systems, the concept of a *receptive field* is key to an understanding of the visual system. Receptive fields in the visual system range from the very simple to quite complex.

The *receptive field* of any visually responsive cell is an exact location in the visual world (Fig. 20–7). Light originating in that location triggers a response in one or more retinal cells. This response can be either depolarization or hyperpolarization with a corresponding increase or decrease in the number of action potentials.

In the early stages of visual information processing, receptive fields have a characteristic *concentric center-surround* organization. The receptive field is roughly circular (see Fig. 20–7). Stimuli in the center of this circle tend to evoke one type of response (e.g., depolarization), whereas stimuli in the doughnut-shaped outer rim, or *annulus*, evoke the opposite response (e.g., hyperpolarization).

Processing of Visual Input in the Retina

Among retinal cells, only retinal ganglion cells have voltage-gated sodium channels on their axonal membranes. As a result, only ganglion cells use action potentials to carry information. So-called *calcium spikes* resulting from an increase in calcium permeability are seen in amacrine cells. All other retinal cells use only graded potentials to process information.

The receptive field properties of each retinal cell depend on the processing of information passing through the neurons between the photoreceptor and the retinal cell in question. For example, a bipolar

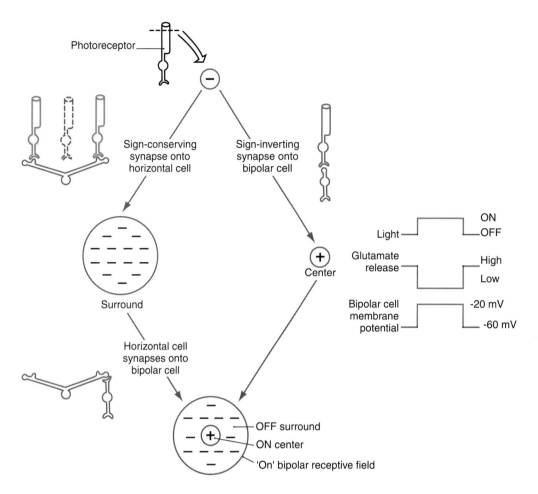

Figure 20–7. How center-surround receptive fields are built in the visual system. Inputs from both receptors and horizontal cells contribute to the characteristic center-surround receptive fields of bipolar cells. A "sign-conserving" synapse is one in which hyperpolarization in the presynaptic cell promotes hyperpolarization in the postsynaptic cell. A "sign-inverting" synapse is one in which hyperpolarization in the presynaptic cell promotes depolarization in the postsynaptic cell. The example illustrated is of an "on" bipolar cell.

cell's response is directly related to the activity of photoreceptors and horizontal cells (Fig. 20–8; see also Fig. 20–7). As with all sensory systems, the structural, electrical, and synaptic properties of the cell are reflected in receptive field properties.

Outer and Inner Plexiform Layers. Synaptic contacts in the retina are concentrated into the *outer* and *inner plexiform layers* (see Fig. 20–2).

The *outer plexiform layer* contains synapses among and between retinal photoreceptors, horizontal cells, and bipolar cells. Contacts between a single cone pedicle or rod spherule, a centrally placed postsynaptic bipolar cell process, and two laterally placed horizontal cell processes form a *triad*.

The *inner plexiform layer* contains synaptic contacts among and between bipolar, amacrine, and ganglion cells. In this layer "off" and "on" bipolar cells terminate, making synaptic contact with the corresponding type ganglion cell. Amacrine cells also synapse with ganglion cells, other amacrine cells, and bipolar cells.

Horizontal Cells. *Horizontal cells* consist of a cell body and its associated dendrites and an axon that courses parallel to the plane of the retina to nearby and distant photoreceptors (see Fig. 20–2). These cells receive glutaminergic input from photoreceptors, and, in turn, form GABAergic synaptic contacts on adjacent rods and cones. This arrangement allows horizontal cells to sharpen the edge of a receptive field by inhibiting surrounding photoreceptor cells.

Bipolar Cells. In their position between photoreceptor cells and ganglion cells, *bipolar cells* help to form a straight-through pathway for visual input (see Fig. 20–2). *Cone bipolar cells* and *rod bipolar cells* are differentiated based on their principal synaptic inputs. It is not surprising that cone bipolar cells predominate in the central retina and rod bipolar cells are most common in the retinal periphery.

Bipolar cells are the comparators, or edge-detectors, of the retina. With the horizontal cells, they compare the activity in each region of the visual field with that

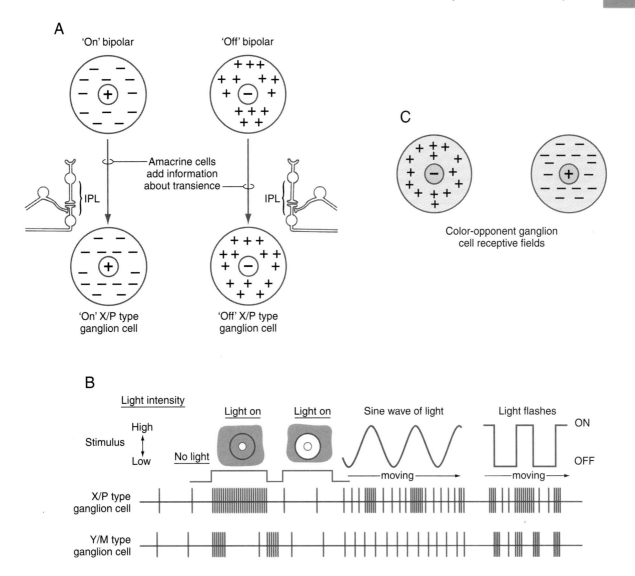

Figure 20–8. How ganglion cell receptive fields are built in the visual system. Both "on" and "off" bipolar cells contribute to the formation of receptive fields *(A)* in ganglion cells. Amacrine cells add information about transience (i.e., how long since the light has changed from on to off). X- or P-type ganglion cells respond linearly to the sine wave grating; the frequency of action potentials rises and falls in synchronization with the sine wave of light intensity used to stimulate them *(B)*. When light strikes both center and surround, there is no net change in ganglion cell activity in either cell type. On the other hand, Y- or M-type ganglion cells respond best to the changes between light onset and offset. X- or P-type ganglion cells are also responsive to color *(C)*. Two examples are shown, called R⁻G⁺ (for red inhibitory center and green excitatory surround) and R⁺G⁻. There are also G⁻R⁺ and G⁺R⁻ cells and blue-yellow combinations. IPL, inner plexiform layer.

in a nearby location. They are the first visual cells to exhibit the *center-surround receptive field* organization. In terms of physiologic response, there are two basic types of bipolar cells. "*On,*" or *depolarizing,* cells respond to a light stimulus in the receptive field center by depolarizing, whereas "*off,*" or *hyperpolarizing,* bipolar cells have the opposite center response (see Figs. 20–7 and 20–8).

Imagine a series of photons striking a photoreceptor outer segment. Recall that the photoreceptor is *hyperpolarized* in response to light input, so that release of its neurotransmitter, glutamate, is *decreased* in the presence of light.

In an "on" bipolar cell, glutamate must act through receptors to hyperpolarize the bipolar cell. Then, when the glutamate released constantly in the dark is removed by the presence of light, the "on" bipolar depolarizes. This effect can be confusing, because we are used to thinking of most glutamate receptors as excitatory, or "sign-conserving." These receptors must be of a different type from that found elsewhere— namely, inhibitory, or "sign-inverting." See Table 20–1.

In an "off" bipolar cell, the opposite type of glutamate receptor must be present on the postsynaptic membrane. In the dark, glutamate released tonically by

Table 20–1. Responses of Bipolar Cells to Stimuli

Condition	Bipolar Cell Response in Receptive Field Center	
	ON	**OFF**
Light	Depolarized	Hyperpolarized
No light	Hyperpolarized	Depolarized
Photoreceptor depolarization	Hyperpolarized	Depolarized
Photoreceptor hyperpolarization	Depolarized	Hyperpolarized
Increased transmitter release	Hyperpolarized	Depolarized
Decreased transmitter release	Depolarized	Hyperpolarized

the photoreceptor depolarizes the "off" bipolar cell. Then, when photons activate the photoreceptor, it hyperpolarizes and reduces its release of glutamate. The decrease in the amount of neurotransmitter causes a hyperpolarization of the postsynaptic "off" bipolar cell membrane. This effect could take place using the "conventional" type of glutamate receptor, which is "excitatory" or "sign-conserving."

Thus, the designation "sign-conserving" means that the electrical responses of the photoreceptor and bipolar cell are the same (hyperpolarization in one leads to hyperpolarization in the other; depolarization in one leads to depolarization in the other), and "sign-inverting" means that the electrical response is inverted (hyperpolarization in one leads to depolarization in the other, and vice versa). The terms "excitation" and "inhibition" lead to confusion in this context and therefore are avoided.

Amacrine Cells. These cells have a small soma, no obvious axon, and dendrites that are few but highly branched (see Fig. 20–2). Their cell bodies are usually found in the inner nuclear layer but may be displaced into the ganglion cell layer. Amacrine cells may contain two different transmitters, for example, GABA and acetylcholine or glycine and a neuropeptide.

Like horizontal cells, amacrine cells also have dendrites that extend over long distances, sampling and modifying bipolar cell output. Whereas horizontal cells sense change, amacrine cells *sense change in change*. For example, a fan blade rotating at constant speed alters the activity of horizontal cells as the dark and light areas of the fan blade stimulate them, but the amacrine cell network will not change its activity. However, if the fan is speeding up or slowing down, the amacrine cell network is maximally stimulated.

Ganglion Cells. *Ganglion cells* are the output cells of the retina (see Fig. 20–2). Their somata form the ganglion cell layer, and their axons converge on the *optic disc* and form the *optic nerve*. Ganglion cells are grouped in two ways: by size and by physiologic role. These classifications largely coincide. Like bipolar cells, ganglion cells have center-surround types of receptive fields.

The largest ganglion cells, called *alpha cells*, predominate in the peripheral retina and receive input mainly from rods. They have more extensive dendritic trees and thicker axons than those of other ganglion cell types. Physiologically, alpha cells correspond to the cell type called *Y* (or *M*). They participate little in color perception, in line with their largely rod input, and they show the "on" or "off" center-surround patterns of the bipolar cells with which they connect. These are called "M" because in humans and other primates, they consistently connect to other large cells, called *magnocellular* layers in the lateral geniculate nucleus, as we shall see later ("magnus" is Latin for large).

Medium-sized ganglion cells, *beta cells*, are found predominantly in the central retina and receive input mainly from cones. They correspond to the physiologic class *X* (or *P*). In keeping with their central location and small dendritic arbors, they have small receptive fields. They are responsive to color stimuli, and this fact gives the center-surround organization a new twist. The center responds to one color, and the surround responds maximally to the color opposite it on a color wheel (see Fig. 20–8*C*). For example, an X cell may have a yellow-responsive center and a blue-responsive surround. Still, the "*on-center*" and "*off-center*" categories remain. These cells are called "P" because in humans and other primates they consistently connect to other, smaller cells in *parvocellular* ("parvus" is Latin for small) layers in the lateral geniculate nucleus.

All ganglion cells left out of the preceding two categories are classified anatomically as *gamma*, *delta*, and *epsilon cells* and physiologically as *W cells*. By definition, these cells constitute a mixed bag; they tend to have smaller cell bodies and axons, and they show a variety of receptive field sizes and physiologic responses.

Retinal Projections

Retinal ganglion cells send axons to a variety of locations in the diencephalon and midbrain. Among the targets are the *suprachiasmatic nucleus*, a hypothalamic region that controls diurnal rhythms (see Chapter 30);

the *accessory optic* and *olivary pretectal nuclei*, which subserve the pupillary light reflex (see Chapter 28); and the *superior colliculus*, which helps to control eye movements (see Chapter 28) and mediates so-called visual reflexes. The superior colliculus, in turn, projects to the *pulvinar*, the largest nucleus of the thalamus. The pulvinar receives input from the superior colliculus, pretectum, and visual cortex (see later) and sends information to *visual association areas*.

Retinogeniculate Projections. Most retinal ganglion cells send their axons to the *lateral geniculate nucleus* by way of the optic nerve, chiasm, and tract. This connection is called the *retinogeniculate projection* (Fig. 20–9). In this pathway, an orderly map of visual space must be maintained. The receptive fields of photoreceptors (and of the ganglion cells to which they are connected) lie in a precise arrangement on the retinal surface. Adjacent points in the visual world are perceived by adjacent ganglion cells. This orderly representation of the visual world on the retina is called a *retinotopic map*.

The *visual field* is the part of the world seen by the patient with both eyes open and looking straight ahead (see Fig. 20–9A). It consists of a *binocular zone*—a broad central region seen by both eyes—and right and left *monocular zones* (or *monocular crescents*) seen only by the corresponding eye. In the clinical setting, it is common to test the visual function of the two eyes separately by covering first one eye and then the other. Consequently, visual field deficits are commonly illustrated as losses from the visual field of each eye (see, for example, Figs. 20–12 and 20–13). Each *visual field* is divided into nasal and temporal halves (*hemifields*), and each of these halves is divided into upper and lower parts (resulting in *quadrants*) (see Fig. 20–9A). Consequently, each visual field is composed of four quadrants.

A stream of photons can be thought of as a ray of light that enters the eye. Stray light rays are blocked by the pigmented cells of the iris; only light passing through the pupil reaches the retina. The light ray is bent (*refracted*) by the cornea and lens so that the image is focused on the retina. Light from the inferior visual world strikes the superior retina. Light from the right visual world (in the binocular zone) strikes the temporal retina of the left eye and the nasal retina of the right eye (see Fig. 20–9B). These patterns are essential to understanding normal vision and the defects in visual fields seen in patients with lesions in the visual pathways. The retinotopic map is maintained throughout the visual system.

Optic Nerve, Chiasm, and Tract. Axons of retinal ganglion cells conveying input from all areas of the retina converge at the *optic disc*, where they penetrate the choroid and sclera to form the *optic nerve*. Within the *nerve fiber layer* of the retina, ganglion cell axons

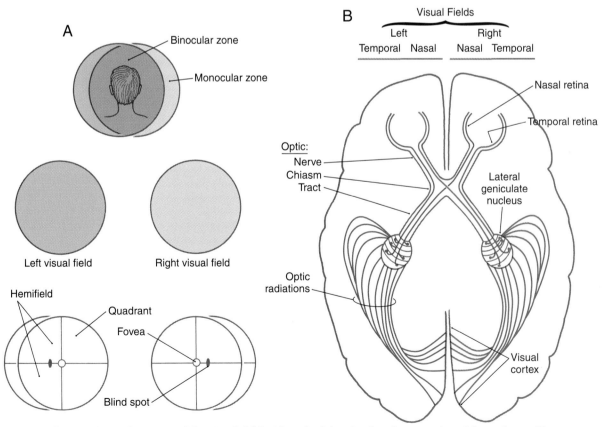

Figure 20–9. Overview of the visual fields *(A)* and of the visual pathway as viewed from above *(B)*.

are unmyelinated. However, as they pass through the sclera, they become ensheathed with myelin formed by oligodendrocytes. Because there are no photoreceptor cells in the optic disc (only ganglion cell axons), light striking this area is not perceived. Consequently, this part of the retina is commonly called the *blind spot* (Fig. 20–10*A* and *B*). Visual acuity is greatest at the fovea, but the peripheral retina has little form vision: fine details cannot be perceived in the peripheral retina because the "pixel density" (density of rod photoreceptors) is much lower.

The *optic nerve* extends from the caudal aspect of the eye to the *optic chiasm* (see Fig. 20–9*B*). This nerve is enclosed in a sleeve of dura and arachnoid mater that is continuous with the same layers around the brain. Thus, the subarachnoid space extends along the optic nerve, which is bathed in cerebrospinal fluid. For this reason, increases in intracranial pressure may be transmitted along the optic nerve(s) and can cause blockage of axoplasmic flow at the optic nerve head. This axoplasmic stasis results in swelling of the optic nerve head (*papilledema*) (see Fig. 20–10*D*). The resulting damage to the optic nerve may result in partial or complete loss of vision in that eye (similar to Fig. 20–11).

Terminal branches of the central retinal artery, a branch of the ophthalmic artery, issue from the optic disc and radiate over the retina. Examination of these vessels through an *ophthalmoscope* can help assess the health of the eye and the central nervous system (see Fig. 20–10*C* and *D*). Changes in the configuration of the retinal vessels or in the size or shape of the optic disc may indicate diseases of the retina, the vascular system, or the central nervous system.

Just rostrolateral to the pituitary stalk, the optic nerves come together to form the *optic chiasm*, from which the *optic tracts* diverge as they pass caudally. In the chiasm, the fibers from the *nasal* half of each retina (corresponding to the temporal hemifields) cross to enter the contralateral optic tract, whereas the fibers from the *temporal* half of each retina (corresponding to the nasal hemifields) remain on the same side and enter the ipsilateral optic tract. In this way, each half of the brain receives the fibers corresponding to the contralateral half of the visual world (Figs. 20–9 and 20–12).

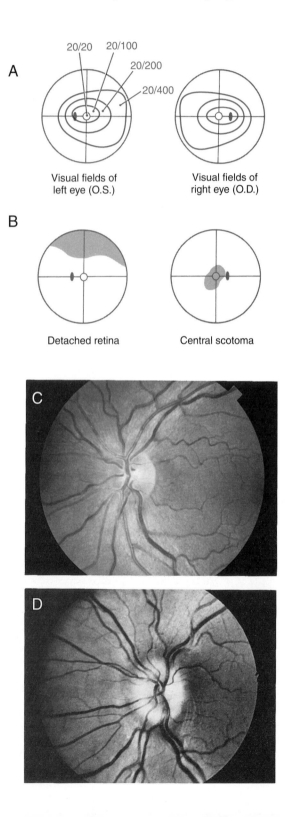

Figure 20–10. Contour diagram of visual acuity on the retinal surface *(A)*. The contour lines are isopters, lines of equal retinal sensitivity. Visual acuity is sharpest in the fovea (20/20) and drops precipitously in the outer parts of the retina (to 20/600). This drop correlates with a lower density of photoreceptors and ganglion cells in the peripheral regions of the retina. The standard abbreviations O.S. and O.D. stand for left eye (*oculus sinister*) and right eye (*oculus dexter*), respectively. Diagrams showing location of defects in the visual field *(B)*. A detached retina in the lower part of the eye results in an irregular defect in the upper visual field *(B, left)*, whereas an irregular lesion of the macula or compression of the optic nerve produces a central scotoma (area of reduced vision) in the center of the visual field *(B, right)*. Ophthalmoscopic view of the fundus of a normal right eye *(C)*. Increased intracranial pressure may produce a "choked disc" (papilledema), which is swelling of the optic nerve head visible through an ophthalmoscope *(D)*. Blood vessels emerge from the optic disc, the light area in the center of the photograph.

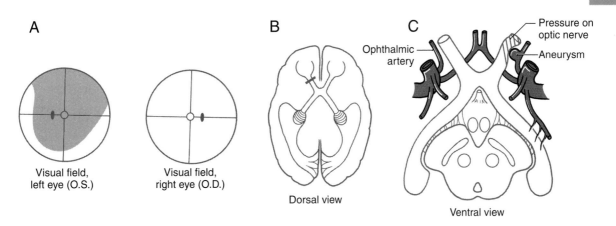

Figure 20–11. A visual field deficit, such as blindness in the left eye *(A)*, may result from a lesion of the left optic nerve *(B,* as seen from above). An aneurysm of the ophthalmic artery *(C,* seen from below) may cause damage to the optic nerve on that side.

Although many clinical events can affect the optic chiasm (and consequently vision), this structure is especially susceptible to tumors of the pituitary gland. Enlarging pituitary tumors that damage the crossing fibers in the midline of the chiasm will interrupt visual input from the temporal halves of both visual fields, resulting in a *bitemporal hemianopia* (see Fig. 20–12*D, E,* and *F*). A lesion that damages the lateral part of the chiasm

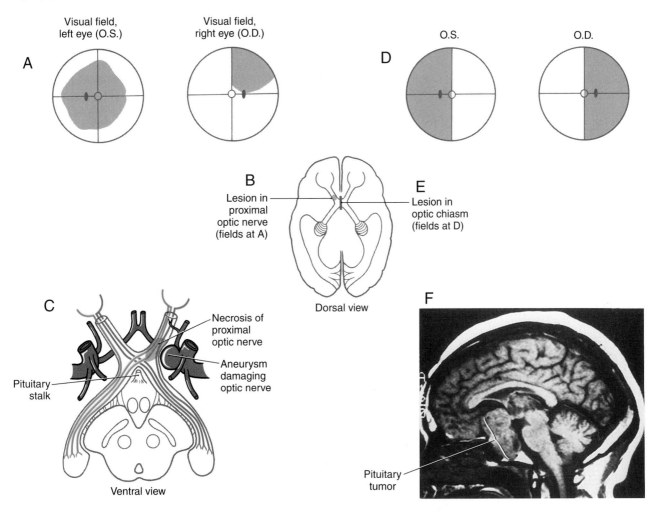

Figure 20–12. Visual field deficits *(A)* resulting from a lesion of the left optic nerve *(B* and *C)* at its junction with the optic chiasm (a *junctional lesion*). This is caused by destruction of fibers of the left nerve plus some of the crossed fibers from the lower nasal hemiretina on the right, producing an upper temporal field deficit on the right. Visual field deficits *(D,* bitemporal hemianopia) resulting from damage to the crossing fibers in the optic chiasm *(E)*. Pituitary tumors, such as the one shown in the magnetic resonance image *(F)*, are a common cause of such deficits.

may interrupt only fibers conveying information from the nasal visual field on the same side, although in practice this situation is quite rare. This deficit is called an *ipsilateral* (either *right* or *left*) *nasal hemianopia.*

Extending caudolateral from the chiasm, the axons of retinal ganglion cells continue as a compact bundle, the *optic tract.* This structure courses over the surface of the crus cerebri at its junction with the hemisphere and ends in the lateral geniculate nucleus of the diencephalon (see Fig. 15–4). Because the optic tract contains fibers conveying visual input from the ipsilateral nasal hemifield and the contralateral temporal hemifield (Fig. 20–13; see also Fig. 20–9), lesions of the optic tract result in a *contralateral* (*right* or *left*) *homonymous hemianopia.*

The optic chiasm receives blood from the small *anteromedial branches* of the anterior communicating artery and A₁ segment of the anterior cerebral artery. The optic nerve receives its blood supply from small branches of the ophthalmic artery traveling parallel to the nerve. As noted previously, the optic nerve head and retina are supplied by the central artery of the retina. The optic tract receives its main blood supply from the *anterior choroidal artery,* whereas the lateral geniculate nucleus is in the domain of the *thalamogeniculate artery,* a branch of the posterior cerebral artery (see Fig. 15–16).

Lateral Geniculate Nucleus

The *lateral geniculate nucleus* is located internal to an elevation on the caudoventral aspect of the diencephalon, the *lateral geniculate body* (Fig. 20–14A–C). The human lateral geniculate nucleus consists of six cellular layers with thin sheets of myelinated fibers sandwiched between them. The anterior (ventral) base of this nucleus is formed by the incoming *optic tract* fibers, whereas its posterior (dorsal) and lateral borders are formed by the outgoing *optic radiations.* The cell layers are numbered 1 through 6 from anterior to posterior. As explained in the next two sections, the layers can be grouped by both the type of ganglion cell input they receive and the side of the retina from which the input originates.

Magnocellular and Parvocellular Layers. Layers 1 and 2 of the lateral geniculate contain cells with large somata and are called the *magnocellular* layers. Layers 3 through 6 contain small cells and are therefore termed the *parvocellular* layers (see Fig. 20–14C–E). The subdivision of the lateral geniculate into magnocellular and parvocellular layers correlates with the subdivision of the retinal ganglion cells into Y and X (M and P) classes. The Y (M) fibers terminate in the magnocellular layers (layers 1 and 2), whereas the X (P) fibers terminate in the parvocellular layers (layers 3 through 6) (see Fig. 20–14E). (The M and P abbreviations are derived from the terms "magnocellular" and "parvocellular.") Recollect that the Y (M) ganglion cells draw their input mainly from rods and have larger receptive fields and thick, rapidly conducting axons. The X (P) ganglion cells receive input mainly from cones and have small receptive fields and slower-conducting axons; they arise mainly in the central retina and are responsible for high-acuity color vision. The ganglion cells of the remaining, mixed W class terminate on small cells scattered between the main layers.

Ipsilateral and Contralateral Layers. The ganglion cell axons that arise in the *temporal* retina remain uncrossed as they pass through the chiasm and terminate in layers 2, 3, and 5 of the *ipsilateral* lateral geniculate nucleus. On the other hand, the axons that arise in the *nasal* retina cross in the chiasm and terminate in layers 1, 4, and 6 of the *contralateral* lateral geniculate (see Fig. 20–9).

Ganglion cell axon terminals and relay cells on which they synapse are arranged so that the same point in visual space is represented six times, once for each layer of the lateral geniculate nucleus, and at the same medial-lateral point in each layer. The map progresses from the midline to the periphery in visual space as the layer runs from medial to lateral in the lateral geniculate nucleus. Layers also run rostral to caudal, repre-

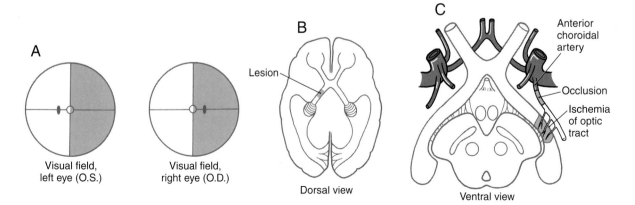

Figure 20–13. Visual field deficits (*A,* right homonymous hemianopia) resulting from a lesion of the left optic tract (*B,* seen from above). Interruption in the blood supply to the optic tract (*C,* seen from below) may produce such deficits.

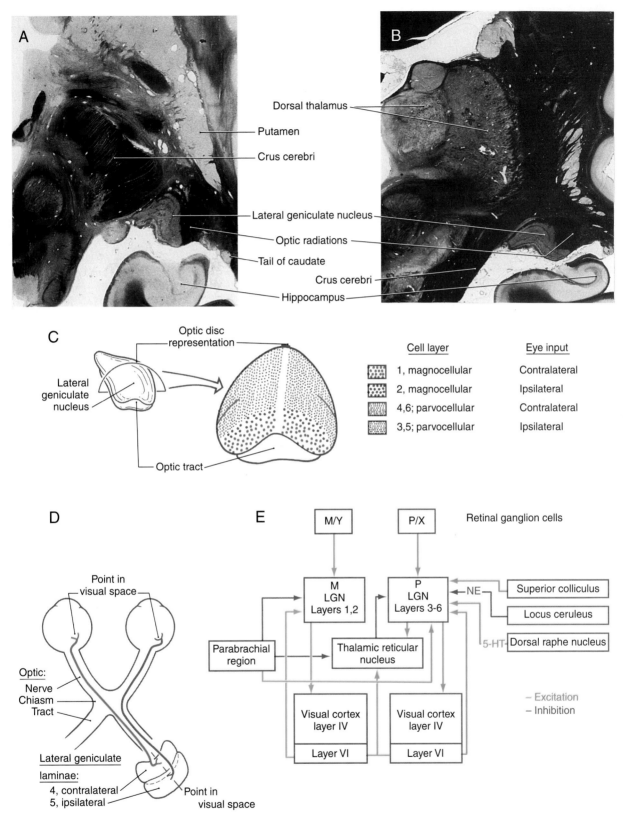

Figure 20–14. The lateral geniculate nucleus (LGN). The LGN has a characteristic layered structure in either nucleic acid–based stains or myelin stains (*A*, axial; *B*, coronal). Six layers can be distinguished. Note the absence of ipsilateral layers in the areas of the LGN representing the monocular crescents; only three contralateral layers are found here. A reconstruction of the human LGN is shown (*C*) (data from Hickey and Guillery, 1979). Note the "hole" in the LGN layers in the area representing the optic disc. Layers in the LGN are arranged so that ganglion cells receiving information from the same point in visual space innervate the LGN with axon terminals stacked one above the other (*D*). The LGN is not simply a relay; much information processing goes on there, as evidenced by the simplified wiring diagram (*E*) (data from Casagrande and Norton, 1990). 5-HT, 5-hydroxytryptamine (serotonin); NE, norepinephrine.

senting the inferior-superior axis. Consider the position of the blind spot. About halfway through the lateral geniculate nucleus rostrocaudally is the position of the horizontal midline. The optic nerve disc lies about 15 degrees nasal to the fovea. Because there are no photoreceptors at that point on the retina, there are no relay cells in the lateral geniculate nucleus representing that location. The optic nerve disc "representation" appears as a blank column extending through all six layers (see Fig. 20–14C). This column lies halfway in a mediolateral direction, not one sixth (15/90 degrees) of the way, because there are many more cells in the fovea and perifoveal region, and this part of the retina is expanded in the lateral geniculate nucleus. This enlargement of central visual areas is seen throughout the visual system.

Optic Radiations

Relay cells forming the layers of the lateral geniculate nucleus receive input from ganglion cells (as *retinogeniculate fibers*) and send their axons to the ipsilateral primary visual cortex as a large bundle of myelinated fibers, the *optic radiations* (Figs. 20–15, 20–16; see Fig. 20–9). The primary visual cortex (striate cortex) is located on the upper and lower banks of the *calcarine sulcus*. Consequently, the optic radiations are also called the *geniculostriate* (or *geniculocalcarine*) *pathway*.

The optic radiations can be divided into two main bundles, one serving the lower and one the upper quadrant of the contralateral hemifields (see Figs. 20–15 and 20–16). The fibers corresponding to the lower quadrant of the contralateral hemifields originate from the dorsomedial portion of the lateral geniculate nucleus, arch directly caudally to pass through the retrolenticular limb of the internal capsule, and synapse in the cortex of the superior bank of the calcarine sulcus,

on the cuneus. Consequently, a lesion in the upper portion of the optic radiations results in a *contralateral* (*right* or *left*) *inferior quadrantanopia*.

The fibers corresponding to the upper quadrant of the contralateral hemifields originate from the ventrolateral portion of the lateral geniculate nucleus. These fibers do not pass directly caudal to the visual cortex. Instead they arch rostrally, passing into the white matter of the temporal lobe, to form a broad U-turn (*Meyer's* or *Archambault's loop*) before passing caudally to synapse in the inferior bank of the calcarine sulcus, on the lingual gyrus (see Figs. 20–15 and 20–16). Damage to Meyer's loop in the temporal lobe, or to these fibers en route to the calcarine sulcus, results in a *contralateral* (*right* or *left*) *superior quadrantanopia* (see Fig. 20–16). Geniculostriate fibers conveying information from the macula (and fovea) originate from central regions of the lateral geniculate nucleus and pass to caudal portions of the visual cortex.

Lesions of the optic radiations may be small and result in a *quadrantanopia*. Lesions in the optic tracts and optic radiations are largely thought of in terms of congruity and incongruity. A deficit is called *congruous* when the visual field loss of one eye is superimposable on that of the other eye. The more anterior a lesion in the optic tract or radiations, the more likely it will be incongruous. Conversely, the closer a lesion is to the visual radiations and cortex, the more congruous it is likely to be.

The blood supply to the optic radiations is via branches of the *middle* and *posterior cerebral arteries* that penetrate deep into the white matter. In general, the more laterally located fibers of the optic radiations and the fibers of Meyer's loop are served by branches of the middle cerebral artery. The more medially located fibers and the visual cortex receive their blood supply from the posterior cerebral artery.

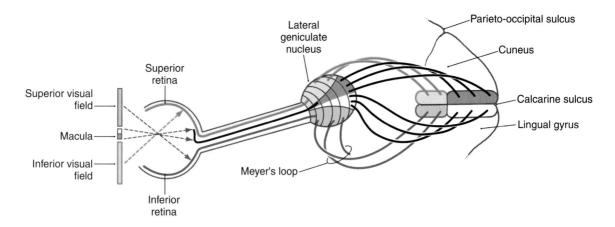

Figure 20–15. The retinogeniculate and geniculostriate pathways in the sagittal plane. Input from the *superior visual field* is received by the inferior retina and is relayed to the lower bank of the calcarine cortex. Similarly, input from the *inferior visual field* reaches the upper bank of the calcarine sulcus. Note the disproportionately large representation of the macula; the central 10 degrees of visual field space occupies about one-half of the visual cortex.

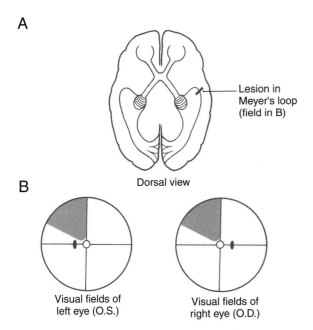

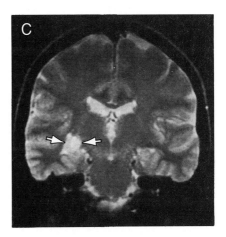

Figure 20–16. Visual field deficits (*B*, left superior homonymous quadrantanopia) resulting from lesions in the lower part of the optic radiations (*A*). Such a defect can be produced by lesions in the right Meyer loop, as shown in the magnetic resonance image (*C*).

Primary Visual Cortex

As mentioned previously, the primary visual cortex, which receives most of the axons from the lateral geniculate nuclei, lies on either bank of the calcarine sulcus in the occipital lobe. It is also called *area 17*, the *striate cortex*, or *V1*. The superior bank of the calcarine sulcus, on the *cuneus*, receives input from the inferior part of the contralateral hemifields, whereas the inferior bank of the sulcus, on the *lingual gyrus*, receives input from the superior part of the hemifields (see Fig. 20–15). Also, as mentioned previously, the *central part* of the visual field (i.e., the macula and fovea) is represented in the portion of the primary visual cortex closest to the occipital pole, and more peripheral regions are represented more rostrally on the cuneus and lingual gyrus (see Fig. 20–15). The central 10 degrees of the visual field occupies about one half of the visual cortex.

The six-layered neocortex of area 17 is characterized by a wide layer IV. This layer contains an extra band of myelinated fibers, the *stria of Gennari* (Fig. 20–17). This structure accounts for the name *striate cortex* and is indicative of the large geniculocalcarine input to this layer. In addition, layer VI is prominent and is the source of a cortical feedback projection to the lateral geniculate nucleus. The visual cortex is organized into an elaborate array of *cortical columns* (see later on), which extend perpendicularly from the pial surface to the white matter.

A large lesion of the visual cortex on one side (e.g., from occlusion of the calcarine artery) will result in a *contralateral* (*right* or *left*) *hemianopsia. Macular sparing* may result because caudal parts of the visual cortex can also be served by collateral branches of the middle cerebral artery.

Visual Cortical Columns. In the visual cortex, the center-surround receptive field organization found at previous levels is transformed. Layer IV of the cortex receives input from the lateral geniculate nucleus. This

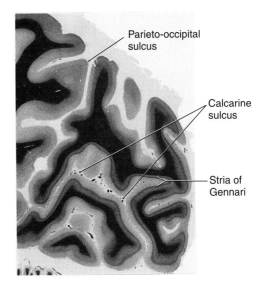

Figure 20–17. The characteristic appearance of the stria of Gennari in the primary visual cortex bordering on the calcarine sulcus. This stria consists of the geniculocalcarine fibers that project to layer IV of the visual cortex.

layer contains cells that respond best to bars or edges of light rather than to spots or rings. These cells are termed simple cells (Fig. 20–18). In general, on progressing toward the pial surface (i.e., toward layer I of cortex) or toward the white matter (i.e., toward layer VI of cortex), receptive field properties become even more complex. For this reason, some of the cells in these layers are termed complex cells. These cells respond vigorously to bars with a particular orientation. Unlike with simple cells, the location within the receptive field is not important to complex cells (see Fig. 20–18).

Cells lying directly above or below one another in visual cortex tend to respond to light stimuli in the same point in visual space. Thus, the *retinotopic order* seen at all levels of the visual system is preserved. However, there is an additional level of complexity: The simple cells that respond best to input from the right or left eye form narrow, parallel stripes called *ocular dominance columns* (see Fig. 20–18 *C* and *D*). If a special labeling method is used to mark the cells that

respond best to input from only one eye, the result is a characteristic "zebra-stripe" pattern that can be seen from the external surface of the cortex (see Fig. 20–18*C*).

Stripes called *orientation columns* cross the cortex at right angles to the ocular dominance columns (see Fig. 20–18*D*). These orientation columns contain cells that respond best to bars or edges of light with a particular orientation.

Abnormal Development of Visual Cortex

During development of the visual system, visually responsive cells compete for synaptic space on cortical cells. If both eyes receive the same visual information at the same time, this competition results in the devotion of equal numbers of layer IV visual cortical cells primarily to inputs from the right or the left eye (see Fig. 20–18*D*).

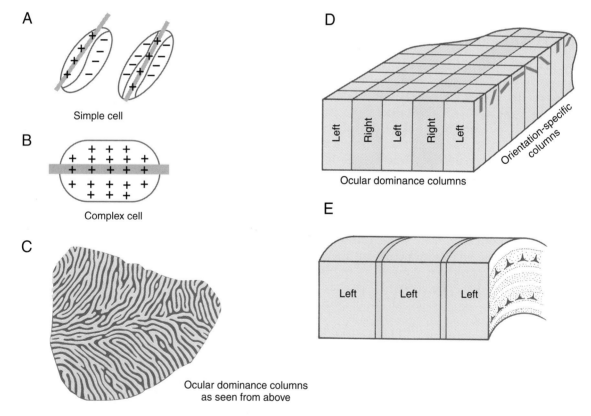

Figure 20–18. The organization of visual cortex. Receptive fields of visual cortical cells differ from earlier levels of the visual system. The center-surround organization is supplanted by cells that respond best to light in a particular orientation. Two such examples, of *simple (A)* and *complex (B)* cells, are shown. Complex cells are estimated to constitute 70% of the visual cortical cells. Although cells in visual cortex respond to the contralateral visual field, they are driven more strongly by either the ipsilateral or contralateral eye. These eye-specific stripes were reconstructed by LeVay, as shown in *C*. Seen in a cross section of visual cortex (*D*), eye-specific stripes are actually columns that alternate with orientation-specific columns. In amblyopia (*E*), the territory occupied by one eye becomes much larger than that driven by the opposite eye, and a functional blindness occurs in the nondominant eye. (*C*, courtesy of Simon LeVay. Modified from LeVay, 1975, with permission of Wiley-Liss, Inc.)

Two problems arise if the competition for cortical territory is disrupted. First, accurate perception of depth relies on a comparison, made by visual cortical cells, between the information arising from both eyes for the same point in visual space. If only one eye is capable of driving cortical cells, almost all depth perception is lost. This problem occurs in about 3% of the population.

Second, there is a *critical period* for an effective competition. During the critical period, competition results in the formation or loss of synaptic contacts between axons of lateral geniculate neurons and visual cortical cells. Roughly speaking, the number and position of these synapses are then translated into the number of action potentials generated in response to a particular visual stimulus. At some point, this competition is declared closed and a victor is named. The synaptic connections made during the competition phase become permanent, the lateral geniculate neurons that lost the competition are permanently shut out, and binocular vision cannot be regained. This condition is called *amblyopia*. Although the extent of the critical period in the human visual system is unknown, it probably does not extend beyond the age of 5 or 6 years.

Myopia (nearsightedness), *hyperopia* (farsightedness), *congenital cataracts*, and corneal abnormalities all may result in amblyopia if untreated. *Strabismus*, a deviation of one or both eyes, also may cause amblyopia; if the underlying deviation of the eyes cannot be treated, alternating a patch between both eyes prevents amblyopia.

Other Visual Cortical Areas

We have seen how the visual world is broken down into elements (dots, stripes, and so forth) for efficient processing of the visual image. How the visual image is reconstructed from its component parts, so that a complete perception of visual space emerges, is unclear. Neuroscientists have jokingly proposed the existence of a "grandmother cell" or "Aunt Tillie cell" that is responsible for remembering how the face of one's grandmother appears. Others argue that such properties are the responsibility of small groups of cells or even entire regions of the brain.

It is clear that a large part of the brain is devoted to the processing and perception of visual space (Fig. 20–19A and B). Areas 18 and 19, which surround area 17 in the occipital cortex, continue the general pattern of organization found in primary visual cortex. They receive inputs directly from area 17 and from the pulvinar.

Starting in area 18, the M and P pathways that originate in the retinal ganglion cells diverge. Up to this level, both of these pathways, or "streams," have been located in the same general region: M and P cells coexist in retina, lateral geniculate nucleus, and area 17 even though they process separate streams of information. This arrangement persists in the *V2 subregion* of area 18, but as the streams emerge from this subregion, they take different routes (see Fig. 20–19C). The *M stream* proceeds to a subregion of area 18 called *V3* and then to the medial temporal area (*V5*), and finally goes to the *posterior parietal area* (area 7a). Remember that the information carried by this stream originates largely in rod cells and in the peripheral portions of the retinas and that the receptive fields involved are large. Appropriately, this information is used in determining where relevant visual stimuli are located, and whether they are moving.

The *P stream* proceeds from the V2 subregion to the *V4* subregion of Brodmann area 19 and from there to the *inferior temporal cortex* (area 37). This stream, which originates mainly in cones and in the central area of the retina, codes for form and for color (see Fig. 20–19C). In fact, starting in the lateral geniculate nucleus, form and color information are carried by separate portions of the P stream. The portion that subserves form perception makes use of the small receptor fields and consequent high acuity of the P ganglion cells. The color-opponent receptive fields of these ganglion cells form the basis for color perception.

Stroke or trauma to higher-order visual processing areas may produce syndromes that seem odd to the casual observer, some of which are popularized in the Oliver Sacks book *The Man Who Mistook His Wife for a Hat*. For example, the process of perception seems to be anatomically distinct from the process of attaching meaning to what we see. Thus, an *apperceptive agnosia*, in which the patient cannot identify objects because of a perceptual deficit, is a separate entity from an *associative agnosia*, in which the patient can perceive the object, face, or photograph but cannot attach any meaning to it. This latter phenomenon was described by Teuber as "percepts stripped of their meaning."

These *agnosias* arise from lesions of the inferotemporal region in areas 18, 20, and 21, alone or in combination. In most people, the left hemisphere is dominant for speech. Thus, lesions in areas 18, 20, and 21 of the left (dominant) hemisphere typically result in *object agnosia*, in which the patient is unable to recognize (i.e., identify or name) real objects, although they are perceived. Lesions in these areas in the right (nondominant) hemisphere produce agnosia for drawings of objects. Smaller lesions bilaterally in these areas may produce *prosopagnosia*, inability to recognize faces. The patient can see the face but cannot interpret the object as a face.

Achromatopsia is the inability to recognize color. The color is perceived, but meaning cannot be attached to the perception. This condition may result from a lesion in the *fusiform gyrus* (also called the lateral *occipitotemporal gyrus*; see Fig. 16–6) in the dominant (left) hemisphere.

Balint syndrome results from lesions bilaterally in the parieto-occipital junctional region. It consists of

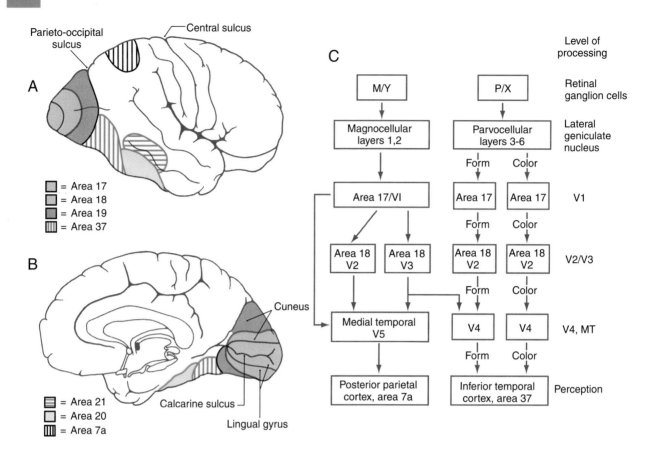

Figure 20–19. Overview of the cortical processing of visual input. Lateral *(A)* and medial *(B)* views of the cerebral cortex showing the cortical areas (using Brodmann's numbers) involved in the processing of visual input. The paths through which these areas interact to create a perceived image *(C)*. Area 18 is divided into V2 and V3 subregions on the basis of its cortical connections. The input from the magnocellular layers of the lateral geniculate nucleus, which originates in the X (P) retinal ganglion cells, is concerned mainly with detecting form and color. MT, medial temporal cortex.

loss of voluntary eye movements (reflex eye movements are preserved), optic ataxia (i.e., poor visual-motor co-ordination), and *asimultagnosia* (the inability to understand visual objects).

A similar dissociation of functions we normally think of as linked occurs in the phenomenon of *alexia without agraphia*. In this syndrome, affected persons can write but cannot read what they have written (or what anyone else has written). A lesion of the splenium of the corpus callosum, carrying visual information from one visual cortex to another, combined with damage to the adjacent occipital region can produce this syndrome, which usually (but not always) occurs in conjunction with a homonymous hemianopsia.

Sources and Additional Reading

Borwein B: Scanning electron microscopy of monkey foveal photoreceptors. Anat Rec 205:363–373, 1983.

Choisser B. Face Blind! Bill's Face Blindness (Prosopagnosia) Pages. Available at http://www.choisser.com/faceblind/

Curcio CA, Sloan KR, Kalina RE, Hendrickson AE: Human photoreceptor topography. J Comp Neurol 292:497–523, 1990.

Hubel DH: Eye, Brain and Vision. New York, Freeman, 1988.

Koretz JF, Handelman GH: How the human eye focuses. Sci Am 259:92–99, 1988.

LeVay S, Hubel DH, Wiesel TN: The pattern of ocular dominance columns in macaque visual cortex revealed by a reduced silver stain. J Comp Neurol 159:559–576, 1975.

Masland RH: Functional architecture of the retina. Sci Am 255:102–111, 1986.

Stryker MP: Is grandmother an oscillation? Nature 338:297–298, 1989.

Werblin FS: The control of sensitivity in the retina. Sci Am 228:70–79, 1973.

Zeki S: A Vision of the Brain. Boston, Blackwell Scientific Publications, 1993.

Chapter 21

The Auditory System

C. K. Henkel

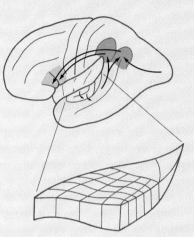

Overview 324

Properties of Sound Waves and Hearing 324

Processing of Sound: The Ear 325
External (Outer) Ear
Middle Ear
Inner Ear: Structure of the Cochlea
Mechanoelectrical Transduction
Tuning of the Cochlea
Primary Afferent Innervation and Function

An Overview of Central Auditory Pathways 329
Vascular Supply of the Auditory Brainstem and Cortex

Brainstem Auditory Nuclei and Pathways 331
Cochlear Nuclei
Superior Olivary Complex
Lateral Lemniscus and Its Nuclei
Inferior Colliculus
Medial Geniculate Nucleus

Auditory and Related Association Cortices 336

Descending Auditory Pathways 337
The Olivocochlear Bundle

Middle Ear Reflex 338

Acoustic Startle Reflex, Orientation, and Attention 339

Hearing is one of the most important senses. In combination with vision and the ability to speak, it contributes, in a significant way, to the quality of life. In our daily routine, we unconsciously sort out meaningful sounds from background noise, localize the source of sounds, and react (many times in a reflex mode) to unexpected sounds. About 12% of people in the general population experience a diminution or loss of hearing during their lifetime, which may represent a significant disability.

Overview

The auditory apparatus is adapted for receiving sound waves at the tympanic membrane and transmitting auditory signals to the central nervous system. Injury to elements of the peripheral apparatus, such as the ear ossicles, may result in *conductive deafness*. Alternatively, damage to the cochlea or the cochlear portion of the eighth cranial nerve may result in *sensorineural (nerve) deafness*. When central auditory pathways are injured, the apparent hearing dysfunction (*central deafness*) is usually combined with other signs and symptoms. Central lesions seldom result in complete deafness in one ear. To understand the neurophysiologic and audiologic methods used in assessing peripheral and central auditory disorders, it is essential to understand the structure and function of the cochlea and central auditory pathways.

Properties of Sound Waves and Hearing

Complex sounds are mixtures of pure tones that are either harmonically related, thus having *pitch*, or that are randomly related and therefore called *noise*. The cochlear apparatus is designed to analyze sounds by separating complex waveforms into their individual frequency components.

The *frequency* of audible sounds is measured in cycles per second, or hertz (Hz). A simple sine wave (Fig. 21–1) depicts the cyclic increase and decrease in the compression of air molecules that constitutes a pure tone. The time interval between two peaks is the *period*, the distance traveled is the *wavelength*, and the number of cycles per second is the *frequency*. The *intensity* is the peak-to-trough amplitude of force at the eardrum.

The normal frequency range for human hearing is 50 to 16,000 Hz. Most human speech takes place in the range of 100 to 8000 Hz, and the most sensitive part of the range is between 1000 and 3000 Hz. Exposure to loud noise can result in selective hearing loss for certain frequencies, and normal aging may reduce the range.

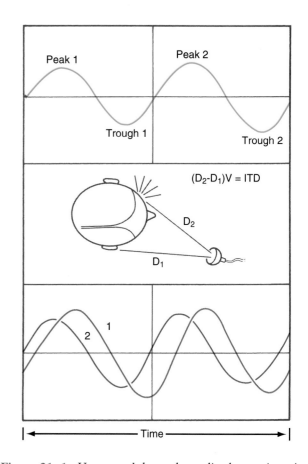

Figure 21–1. *Upper panel* shows the cyclic changes in a simple, pure tone. *Lower panel* shows that the arrival of a tone at right (D_1) ear and left (D_2) ear is affected by the distance traveled and shadowing effect of the head (*center panel*) when the source of the sound is displaced from the midline. The interaural time difference (ITD) is calculated by the equation $(D_2 - D_1)V = ITD$, where V is the speed of sound.

The hearing apparatus is exquisitely sensitive to sound *intensity* over an enormous *dynamic range*. Intensity of sound is related to the perception of loudness and is usually measured in units called *decibels* (dB). Intensity is also related to a measure of *sound pressure level* at the tympanic membrane. A sound that has 10 times the power of a just-audible sound is said to have 20 dB sound pressure level. Normal conversational levels of sound are about 50 dB. Sounds above 120 to 130 dB elicit pain, and permanent damage to the hearing apparatus is probable for exposure to repeated sounds above 150 dB (e.g., jet engine).

The brain derives the location of a sound by computing differences in the shape, timing, and intensity of the waveforms that reach each ear. The path of the sound is affected by the distance to the ears and by obstacles such as the head (see Fig. 21–1). Thus, *interaural time and intensity differences* are related to the angle between the direction in which the head is pointing and the direction of the sound source. *Interaural time differences are more important for localizing low-*

frequency sounds, whereas interaural intensity differences are more important for localizing high-frequency sounds.

Processing of Sound: The Ear

External (Outer) Ear. Sound waves are captured by the *external ear (pinna)* and channeled through the *external auditory meatus* to the *tympanic membrane* (Fig. 21–2A). Resonance features of the pinna and meatus enhance some frequencies more than others in a direction-dependent fashion. For example, sounds coming toward the back of the head are baffled compared with those coming toward the side of the head. *Monaural (single-ear) localization* depends on such cues, and accuracy in localizing sound is impaired by damage to the pinna.

Middle Ear. The *middle ear* or *tympanic cavity* is an air-filled space in the temporal bone that is interposed between the tympanic membrane and the inner ear structures (see Fig. 21–2A). Sounds are transmitted across the space from the tympanic membrane to the fluid-filled inner ear by a chain of three bony *ossicles:* the *malleus, incus,* and *stapes.* On one end of this chain, the arm of the malleus is attached to the tympanic membrane, and at the other end, the footplate of the stapes fits into the *oval window* of the membranous labyrinth of the inner ear. The three bones act as levers to reduce the magnitude of movements of the tympanic membrane while increasing their force at the oval window.

The mechanical stiffness of the ossicle chain acts to *compensate for the difference in impedance* between air and fluid environments (a function called *impedance matching*) so that there is optimal transfer of energy between the two media. Diseases such as *otosclerosis* and *otitis media* result in conductive hearing loss by affecting the efficiency of the ossicle movement. The stiffness of the ossicle chain can also be modified by two muscles of the middle ear, the tensor tympani and stapedius muscles (middle ear reflex).

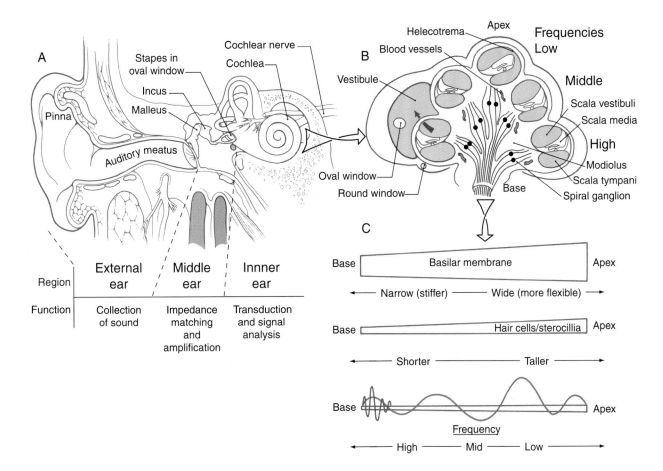

Figure 21–2. The path of auditory signals from the external ear, through the middle ear, and into the inner ear (*A*). The cochlea (*B*) is shown in cross section from apex to base. The basilar membrane (*C*) functions to separate waves of different frequencies within a sound. This membrane is narrow and stiff at its base and becomes wider and more flexible toward the apex, and the hair cell stereocilia increase correspondingly in height. These features "tune" the membrane so that each frequency of sound in the audible range will cause a wave in the basilar membrane that has its peak amplitude at a unique spot (near the base for high frequencies and near the apex for low frequencies). At this spot, the hair cells are excited most intensely, producing a peak in neural output.

Inner Ear: Structure of the Cochlea. The cochlea is named for its similarity to a conch shell (see Fig. 21–2B). The *membranous cochlea*, the coiled portion of the inner ear, is encased in the osseous cochlea and consists of three spiraling chambers. The cochlea makes approximately 2⅔ turns from base to apex. Uncoiled, it is about 34 mm long. The base of the cochlear spiral is connected to the saccule of the membranous labyrinth by the *ductus reuniens.*

The central chamber of the membranous cochlea is the *cochlear duct*, also called the *scala media* (see Fig. 21–2B). Above it, the *scala vestibuli* is positioned to communicate with the vestibule, the portion of the membranous inner ear between the oval window and the cochlea. Below, the *scala tympani* ends at the *round window*, which separates this space from the middle ear cavity. The scala vestibuli is continuous with the scala tympani through an opening at the apex of the cochlea called the *helicotrema* (see Fig. 21–2B). In cross section, the scala media is bounded by the *basilar membrane* below, the *vestibular or Reissner membrane* above, and the *stria vascularis* externally (Fig. 21–3; see also Fig. 21–2B). The screwlike bony core of the cochlea is the *modiolus.* A spiral osseous lamina extends outward from the modiolus to join the basilar membrane. The basilar membrane, in turn, is continuous laterally with the *spiral ligament.* The scala vestibuli and tympani are filled with *perilymph.* The *endolymph*, which fills the cochlear duct, is elaborated by the cells and rich capillary bed of the *stria vascularis* (see Fig. 21–3).

The *organ of Corti* is the specialized sensory epithelium resting on the basilar membrane (see Fig. 21–3). It comprises inner and outer hair cells, supporting cells, and the tectorial membrane. The inner hair cells are separated from the outer hair cells by the *tunnel of Corti* (see Fig. 21–3). This tunnel is formed by the filamentous arches of the *inner and outer pillar cells* and is filled with fluid.

Inner hair cells form a single line spiraling from base to apex, and the *outer hair cells* form three parallel lines that follow the same course (Fig. 21–4). In all, there are about 3500 inner and 12,000 outer hair cells in the cochlea. Projecting from the apical surface of each hair cell is a *hair bundle* consisting of 50 to 150 *stereocilia* arranged in curving rows (see Fig. 21–4). Each hair bundle is polarized so that the longest stereocilia are on the outer border (see Fig. 21–4), and the rows of stereocilia are linked by filamentous material at their tips.

The *tectorial membrane* is a gelatinous arm that extends outward over the sensory epithelium from the limbus of the osseous spiral lamina (see Fig. 21–4). The taller stereocilia in each hair bundle are in contact with or embedded in the tectorial membrane. Consequently, movement of the basilar membrane and the organ of Corti will bend the stereocilia against the tectorial membrane and cause a graded depolarization of the hair cells.

The bony modiolus, around which the cochlear duct turns, houses the *spiral ganglion* (see Figs. 21–2 and

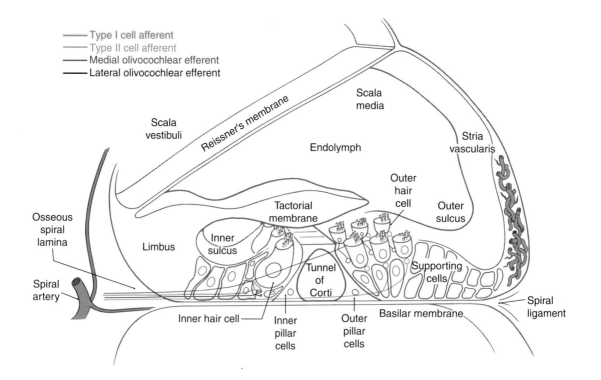

Figure 21–3. Cross section through a typical turn of the membranous cochlea.

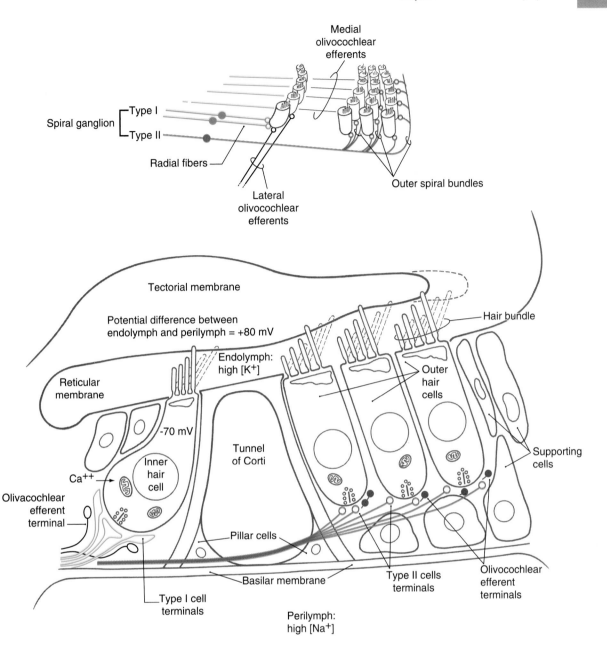

Figure 21–4. The structure and function of the organ of Corti (*lower*) and the relation of type I and type II afferent fibers to the spiraling ranks of inner and outer hair cells (*upper*). Note that the designation as lateral or medial olivocochlear efferents refers to their origin in the superior olive, not to their target in the organ of Corti.

21–3). At the edge of the osseous spiral lamina, the peripheral processes of the bipolar cells of this ganglion lose their myelin and pass through perforations to the basilar membrane, where they synapse on the base of the inner and outer hair cells (see Fig. 21–4). The central processes of the spiral ganglion cells form the *cochlear portion of the vestibulocochlear nerve (cranial nerve VIII). Efferent fibers* to the cochlea either spiral along the inner part of the basilar membrane to synapse on inner hair cells or travel radially across the tunnel of Corti to contact outer hair cells (see Fig. 21–4).

Mechanoelectrical Transduction. Inner hair cells are extremely sensitive transducers that convert the mechanical force applied to the hair bundle into an electrical signal (see Fig. 21–4). Endolymph, like extracellular fluid, has a high concentration of K^+. In contrast, perilymph, like cerebrospinal fluid, has a high concentration of Na^+. As indicated in Figure 21–4, the potential difference between the endolymph and the perilymph is +80 mV. This endolymphatic potential appears to be due to the selective secretion and absorption of ions by the stria vascularis. At the same time, ion pumps in the hair cell membrane produce a resting intracellular potential of about −70 mV (see Fig. 21–4).

As the basilar membrane moves up in response to fluid movement in the scala tympani, the taller stereocilia are displaced against the tectorial membrane. This causes ion channels at the tips of the stereocilia to open, allowing K+ flow along the electrical gradient to depolarize the cell (see Fig. 21–4). The large potential difference between the endolymph and the hair cell interior creates a force of 150 mV that drives K+ into the cell and that increases the range of the cell's graded electrical response to mechanical displacement. Damage to the stria vascularis results in loss of the endolymphatic potential and failure of mechanoelectrical transduction.

When a hair cell depolarizes, voltage-gated Ca²⁺ channels at the base of the cell open, and the resulting influx of Ca²⁺ causes synaptic vesicles to fuse to the cell membrane and release a neurotransmitter into the synaptic cleft between the hair cell and the cochlear nerve fibers (see Fig. 21–4). The transmitter causes depolarization of the afferent fiber, and an action potential is transmitted along the cochlear nerve fiber.

The stimulus-related changes in the electrical potential between the perilymph and the hair cells can be recorded anywhere in the cochlea. The potential varies synchronously with the sound stimulating the ear and is therefore referred to as the *cochlear microphonic*. This electrical record provides a clinically useful monitor of cochlear function.

Tuning of the Cochlea. The cochlea acts as a frequency filter to separate and analyze individual frequencies from complex sounds. These tuning properties result from anatomic and physiologic characteristics of hair cells and the basilar membrane (see Figs. 21–2 and 21–3).

The plunger-like motion of the stapes in the oval window compresses the perilymph. In the fluid medium of the cochlea, this pressure variation imparts motion to the basilar membrane, causing a wave to travel along it (see Fig. 21–2C). The basilar membrane is stiffest at its base and becomes progressively more flexible toward its tip. Therefore, any given frequency of sound (pure tone) will cause a wave in the basilar membrane that has its maximum displacement at a unique point along the membrane. For high tones, this point is close to the base of the cochlea, and for lower frequencies, it is more distal. The response of hair cells to the tone is strongest at the point of greatest displacement. Therefore, the position from base to apex along the spiral of the basilar membrane and organ of Corti is directly related to the frequency of the tone that will elicit a response. This relationship of frequency and cochlear position is the basis for the *place theory of cochlear tuning*. The *cochleotopic order*, and thus *tonotopic representation*, are highly conserved throughout the auditory pathways.

In patients with profound sensorineural hearing loss, some audible sensation may be regained with *cochlear implants* having a number of fine wire electrodes. Each wire is tuned to a broad frequency band from an electrical receiver, and the wires are implanted so that each stimulates nerve terminals at the appropriate tonotopic point along the cochlear spiral.

Primary Afferent Innervation and Function. The spiral ganglion is made up of two types of bipolar

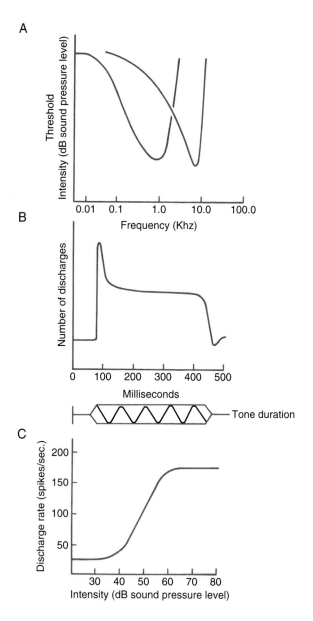

Figure 21–5. Typical response characteristics of type I cochlear afferent fibers. Frequency-tuning curves (*A*), post-stimulus time histogram of discharges through the duration of a tone burst at the characteristic frequency of a primary afferent fiber (*B*), and rate-intensity curve illustrating how a limited range of intensity can be coded in the response of primary afferent fibers to increasing intensity of a tone at its characteristic frequency (*C*).

sensory neurons. *Type I cells* make up 90% to 95% of the cells in the spiral ganglion and have radial branches that synapse with only one or two inner hair cells (see Fig. 21–4). As many as 20 or more type I radial fibers converge on each inner hair cell. As a result, type I cochlear nerve fibers respond to a narrow frequency range. In contrast, *type II ganglion cells* have widely distributed peripheral processes, which traverse the tunnel of Corti and synapse with over 10 outer hair cells (see Fig. 21–4). Thus, type II cochlear fibers are more sensitive to low-intensity sounds than are type I cells, but they may be less precisely tuned to frequency.

Frequency is coded in the cochlear nerve by the position of afferent fibers along the cochlear spiral. For loud sounds, each afferent fiber responds over a range of frequencies. As the intensity of the sound drops to near threshold, the frequency response range narrows. A *tuning curve* can be constructed that plots the threshold intensity for each frequency that will elicit a response (Fig. 21–5A). The *characteristic frequency* is the frequency at which the fiber has the lowest threshold. The discharge pattern of primary afferents over time to pure tone bursts is shown with post-stimulus histograms of the number of action potentials summed over many presentations (see Fig. 21–5B). Stimulus onset produces an initial high-frequency discharge followed by a lower sustained discharge level that is related to stimulus intensity. When the tone ends, the fiber drops back to a low, spontaneous discharge rate. For low-frequency fibers, the timing of each impulse is *phase-locked* with the stimulus cycle, so that the fiber output preserves the timing information of the signal.

The enormous dynamic range of the human ear to intensity cannot be coded in the response of single nerve fibers. *Intensity is coded both by the discharge rate of cochlear nerve fibers and by recruitment of activity in additional afferents as stimulus intensity increases.* The discharge rate for the cochlear nerve fibers increases proportionally with intensity over a range of about 40 dB sound pressure level and then plateaus (see Fig. 21–5C). At higher stimulus intensities, additional cochlear nerve fibers having sequentially greater thresholds are recruited.

An Overview of Central Auditory Pathways

In the major ascending auditory connections from cochlea to cortex, *the place code of the cochlea is, as a rule, strictly maintained* (Fig. 21–6). Within this tonotopic framework, projections connect similar frequency regions of successive nuclei. Information processing is, therefore, hierarchical with increasing complexity of feature extraction.

All fibers in the cochlear nerve synapse in the *cochlear nuclei.* As cochlear information ascends to the auditory cortex, information is distributed through multiple parallel pathways that ultimately converge in the inferior colliculus. The hierarchy of auditory nuclei involved in these parallel pathways includes the *cochlear nuclei, nuclei of the superior olive and trapezoid body, nuclei of the lateral lemniscus,* and *inferior colliculus.* Specific fiber bundles that convey this information from one level to the next are the *trapezoid body, acoustic stria,* and *lateral lemniscus.* From the midbrain, auditory information is conveyed from the inferior colliculus by its *brachium* to the *medial geniculate nucleus* of the thalamus and then through the *sublenticular limb of the internal capsule to the auditory cortex.* The hierarchy of auditory regions and the fiber bundles that connect one level to the next are summarized in Figures 21–6 and 21–10.

Although fibers conveying auditory input decussate at several levels, this information is routed in one of two orderly ways: (1) *monaural information* (information about sounds at a single ear) *is routed to the contralateral side* and (2) *binaural information* (information about differences between sounds at both ears) *is handled by central pathways that receive, compare, and transmit this input.* Binaural pathways perform the neural computation needed to localize brief sounds.

Unilateral damage to the cochlear nerve or cochlear nucleus results in monaural deafness. In contrast, unilateral damage at or above the superior olivary complex leaves intact routes from either ear that are conveyed through binaural pathways, so monaural deafness does not occur. The auditory decussations, particularly the trapezoid body, are functionally similar to the optic chiasm in the visual system and have been collectively referred to as a *functional acoustic chiasm.* Thus, central hearing dysfunction may result in inattention to stimuli on the contralateral side.

Vascular Supply of the Auditory Brainstem and Cortex. The blood supply to the cochlea and the auditory nuclei of the pons and medulla originates from the *basilar artery.* The *internal auditory (labyrinthine) artery,* usually a branch of the *anterior inferior cerebellar artery* (AICA), supplies the inner ear and the cochlear nuclei. Occlusion of the AICA will result in a monaural hearing loss. This lesion may also damage the emerging fibers of the facial nerve and the pontine gaze center, resulting in monaural deafness combined with ipsilateral facial paralysis and an inability to look toward the side of the lesion.

Vascular lesions higher in the ascending auditory system necessarily interrupt pathways conveying information from both ears. The superior olivary complex and lateral lemniscus are mainly supplied by *short circumferential branches of the basilar artery.* The *superior cerebellar* and *quadrigeminal arteries* supply the inferior

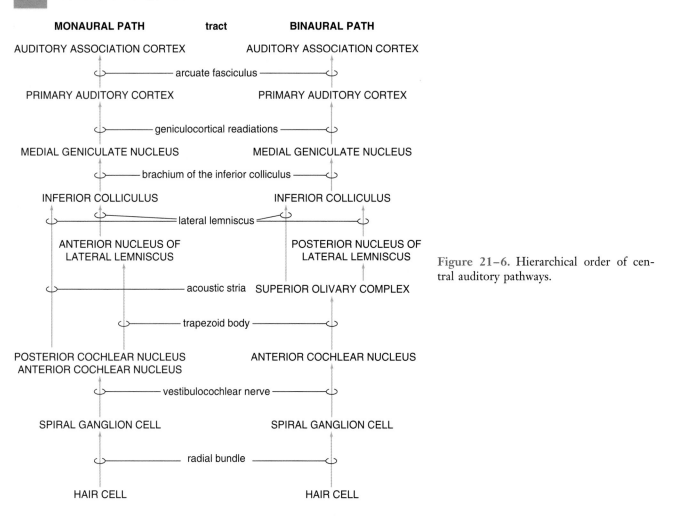

MONAURAL PATH   tract   **BINAURAL PATH**

AUDITORY ASSOCIATION CORTEX   AUDITORY ASSOCIATION CORTEX

───── arcuate fasciculus ─────

PRIMARY AUDITORY CORTEX   PRIMARY AUDITORY CORTEX

───── geniculocortical readiations ─────

MEDIAL GENICULATE NUCLEUS   MEDIAL GENICULATE NUCLEUS

───── brachium of the inferior colliculus ─────

INFERIOR COLLICULUS   INFERIOR COLLICULUS

───── lateral lemniscus ─────

ANTERIOR NUCLEUS OF   POSTERIOR NUCLEUS OF
LATERAL LEMNISCUS   LATERAL LEMNISCUS

───── acoustic stria   SUPERIOR OLIVARY COMPLEX

───── trapezoid body ─────

POSTERIOR COCHLEAR NUCLEUS   ANTERIOR COCHLEAR NUCLEUS
ANTERIOR COCHLEAR NUCLEUS

───── vestibulocochlear nerve ─────

SPIRAL GANGLION CELL   SPIRAL GANGLION CELL

───── radial bundle ─────

HAIR CELL   HAIR CELL

Figure 21–6. Hierarchical order of central auditory pathways.

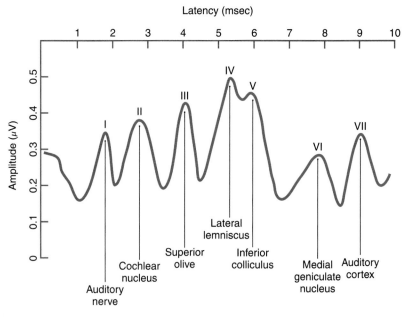

Figure 21–7. Stylized example of an auditory brainstem response (ABR) recording. Waves I to VII are labeled above, while below the sequence of auditory structures in which activity may be correlated with each wave is indicated. The vertical lines indicate the approximate correlation of wave and auditory structure, although more than one structure is likely to contribute to various waves. Levels and latency may vary in actual recordings.

colliculus, and the medial geniculate bodies lie in the vascular territory of the *thalamogeniculate arteries*. The blood supply to the primary auditory and association cortices is via branches of the M_2 segment of the *middle cerebral artery*.

When damage occurs to neural tissue from vascular insults, tumors, or demyelinating diseases such as multiple sclerosis, the effect on conduction time and activity levels in the auditory system can be utilized in clinical neurophysiology to assist in localization of the pathologic process. Brainstem auditory evoked responses, or, simply, ABRs, are average scalp potentials elicited by a train of clicks and recorded much as for an electroencephalogram. A pattern of seven waves occurs in the ABR, with peaks at regular latencies after the clicks that are correlated with activity levels of the

ascending auditory system. The activity from one or more auditory structures may be correlated with a specific wave, as summarized in Figure 21–7, and shifts in latency and amplitude of specific waves may be used to localize the lesion, assess hearing, or indicate swelling in response to neurosurgical procedures.

Brainstem Auditory Nuclei and Pathways

Cochlear Nuclei. The *posterior cochlear nucleus (dorsal cochlear nucleus)* and the *anterior cochlear nucleus (ventral cochlear nucleus)* are located lateral and posterior to the restiform body and are partially on the surface of the brainstem at the pontomedullary junction

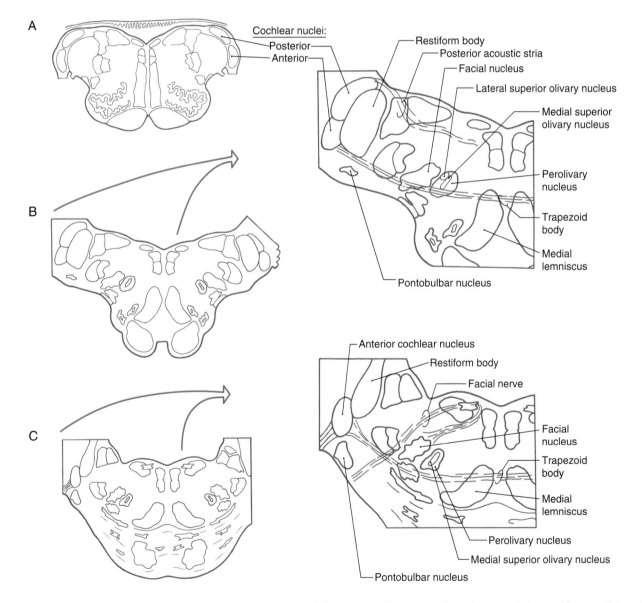

Figure 21–8. Levels of the rostral medulla (*A, B*) and caudal pons (*C*) illustrating the relations of the cochlear nuclei and superior olivary complex.

(Fig. 21–8*A*). The posterior cochlear nucleus drapes over the restiform body just inferior to the pontomedullary junction. At this level the posterior part of the anterior cochlear nucleus is small in proportion to the posterior cochlear nucleus (see Fig. 21–8*A* and *B*). The anterior cochlear nucleus extends rostral to the posterior cochlear nucleus (see Fig. 21–8*C*), where it may be covered by the flocculus and by caudal fascicles of the middle cerebellar peduncle.

All cochlear nerve fibers end in the cochlear nuclei on the ipsilateral side (Fig. 21–9*A*–*C*). As these fibers enter the brainstem at the cerebellopontine angle, they divide into ascending and descending bundles. Fibers in the ascending bundle synapse in the anterior part of the anterior cochlear nucleus, whereas fibers in the descending bundle synapse in the posterior part of the anterior cochlear nucleus and in the posterior cochlear nucleus.

In the cochlear nuclei, each afferent nerve fiber makes specialized synaptic contacts with several different cell types. Individual fibers and their synaptic contacts distribute along orderly rows so that the resulting order produces distinct tonotopic maps in each division (see Fig. 21–9*C*). The frequency-related lines are organized so that low frequencies are represented laterally and high frequencies medially (see Fig. 21–9*C*).

Specific cell types of the cochlear nuclei, in turn, give rise to parallel but separate ascending pathways in the auditory system that analyze and code different sound features while preserving frequency information. These projections are subdivided into pathways conveying monaural information to the inferior colliculus and those providing input to the superior olivary complex for binaural processing. Most fibers from the anterior cochlear nucleus course anterior to the restiform body part of the inferior cerebellar peduncle to form the *trapezoid body* (see Fig. 21–8*B* and *C*). Projections from the posterior cochlear nucleus and some from the anterior cochlear nucleus course posteriorly over the restiform body as the *posterior acoustic stria* and decussate in the pontine tegmentum before joining the lateral lemniscus.

Many of the cells in the posterior cochlear nucleus contribute to complex local circuits and are not easily correlated with distinct ascending channels. *Pyramidal cells* have fusiform cell bodies with apical and basal dendrites. A major output of the posterior cochlear nucleus is via a direct pyramidal cell projection to the contralateral inferior colliculus (Fig. 21–10).

The anterior cochlear nucleus is distinguished by the presence of anatomically and physiologically dis-

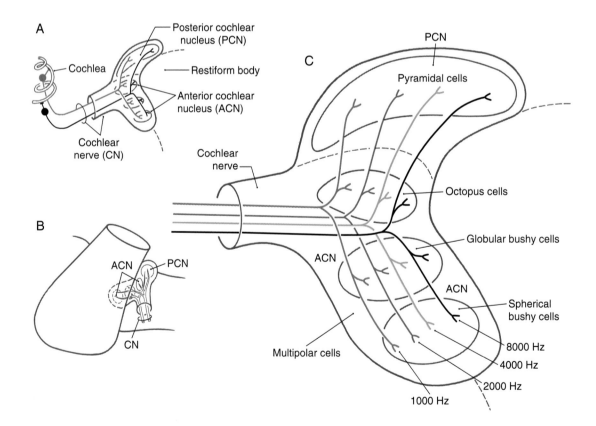

Figure 21–9. The posterior and anterior cochlear nuclei in cross section (*A*, *C*) and as viewed from the lateral aspect on the left (*B*). The anterior cochlear nucleus extends rostral to the posterior nucleus (*B*). The course, tonotopic organization of cochlear afferent fibers, and the locations of main cell groups are also indicated (*C*).

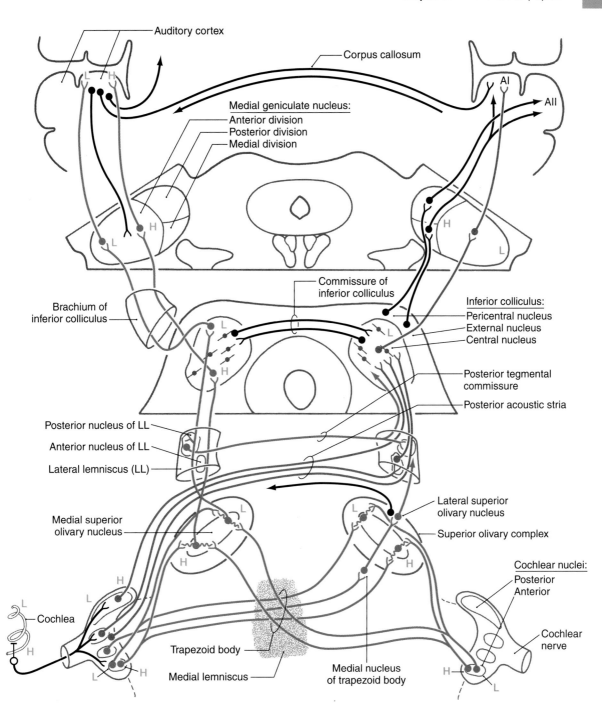

Figure 21–10. Ascending central auditory pathways. Monaural pathways are shown in red, binaural pathways in blue, and other connections in black. AI and AII, primary and secondary auditory cortices; H, high frequencies; L, low frequencies.

tinct output cell types (see Fig. 21–9). The anterior part of the anterior cochlear nucleus contains mainly *spherical* and *globular-shaped bushy cells*. There is almost a one-to-one relation between the synaptic endings of a cochlear nerve fiber and each bushy cell. Consequently, the temporal pattern of activity in bushy cells is remarkably similar to that of the primary afferents, and these cells show sustained responses to short

tones that convey information about the timing and phase of the tone. Axons of bushy cells travel in the trapezoid body. They are the central origin of ascending channels that process *binaural information*, which is useful for sound localization (see Figs. 21–6 and 21–10).

Multipolar cells in the anterior cochlear nucleus are sensitive to changes in sound pressure level. Accordingly, they convey information in a *direct monaural pathway*, primarily to the contralateral inferior colliculus (see Fig. 21–10) about the intensity of the sound.

Also in the posterior part of the anterior cochlear nucleus are *octopus cells*, which have long, relatively unbranched dendrites and integrate inputs from a wider array of cochlear afferents. Axons of octopus cells synapse mainly in the contralateral anterior nucleus of the lateral lemniscus, which, in turn, projects to the inferior colliculus. This *indirect monaural pathway* conveys information that is useful in analyzing brief components of speech sounds.

Superior Olivary Complex. The superior olivary complex is located near the facial motor nucleus in the caudal pons (see Figs. 21–8*B* and *C* and 21–10). It is the first site in the brainstem where information from both ears converges. This binaural processing is essential for accurate sound localization and the formation of a neural map of the contralateral auditory hemifield.

The *medial superior olivary nucleus* (MSO), which forms a distinct vertical bar within a diffuse group of *periolivary nuclei*, is the principal nucleus in the human superior olivary complex (see Figs. 21–8*B* and *C* and 21–10). The *lateral superior olivary nucleus* (LSO), located lateral to the MSO, is not distinct and contains fewer cells.

The *trapezoid body* is a bundle of myelinated fibers passing anterior to the superior olivary complex and intermingling with fibers of the medial lemniscus as it crosses the midline (see Figs. 21–8*B* and 21–10). Decussating fibers of the trapezoid body end in the contralateral superior olivary complex or ascend in the contralateral lateral lemniscus. The medial nucleus of the trapezoid body lies medial to the MSO. It receives projections from globular bushy cells in the contralateral anterior cochlear nucleus and gives rise to important local inhibitory circuits within the superior olive.

The topographic organization of the afferents to the MSO conserves the orderly representation of the cochlea and LSO (see Fig. 21–10). In the MSO, cells with low characteristic frequencies are found posteriorly, and those with higher characteristic frequencies are found more anteriorly. These are bipolar cells with medially and laterally directed dendrites (see Fig. 21–10). Axons in the trapezoid body arising from ipsilateral spherical bushy cells make excitatory synapses with the laterally directed dendrites. Axons from the contralateral spherical bushy cells decussating in the trapezoid body make excitatory synapses with the medially directed dendrites. The excitatory neurotransmitter is probably *glutamate* or *aspartate*. Local inhibitory circuits in the superior olivary complex use *glycine* as a neurotransmitter.

The pathways to the MSO from spherical bushy cells in the anterior cochlear nucleus are anatomically arranged so that signals from the contralateral ear arrive close enough in time to those from the ipsilateral ear to summate at each cell only for a sound from a specific point in the contralateral hemifield of space. This mechanism, referred to as *coincidence detection*, accounts for computation by the MSO of interaural time differences that constitute the basis for sound localization (Fig. 21–11). Coincidence of the signals occurs when the shorter neural conduction time in the path from the ipsilateral cochlear nucleus compensates for the shorter delay of the sound path to the contralateral ear (see Figs. 21–10 and 21–11).

Ascending projections from the MSO travel largely in the ipsilateral side in the lateral lemniscus and synapse in the *central nucleus* of the inferior colliculus (see

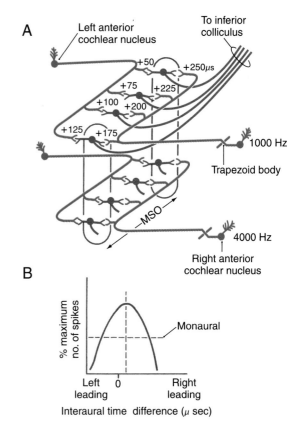

Figure 21–11. The mechanism by which the brain calculates interaural time afferences. Diagrammatic representation of how time delay lines are established in connections between spherical bushy cells in anterior cochlear nucleus and two medial superior olivary nucleus (MSO) isofrequency columns (*A*). Transmission times in microseconds for the neural signals are shown in the delay line that maps interaural time delay along the rostral-caudal axis of MSO. When the interaural delay is precisely the reciprocal of this delay (that is, when the arrival at the right ear precedes the arrival at the left ear by the same increment as the signal on the left precedes the signal on the right), then summation occurs as shown in the delay-response curve (*B*). The dotted horizontal line indicates the response level to a monaural stimulus.

Fig. 21–10). These fibers synapse in corresponding frequency regions (low to low, and so on) of the inferior colliculus (see Fig. 21–10). The LSO also contributes in a small way in humans to both of these pathways. Branches of these projections always end in the posterior nucleus of the lateral lemniscus (dorsal nucleus of the lateral lemniscus). This nucleus, in turn, projects to the contralateral inferior colliculus and constitutes an indirect binaural pathway from the superior olive to the inferior colliculus (see Fig. 21–10).

Detection of interaural intensity differences, which provide spatial cues for high-frequency stimuli caused by shadowing of sounds by the path from the contralateral side of the head, is also accomplished by summation of excitatory and inhibitory inputs to LSO neurons. Humans are capable of detecting small sound differences between the two ears for high-frequency signals, which serve as cues to the source of the auditory signal.

Lateral Lemniscus and Its Nuclei. The lateral lemniscus contains axons from second-order neurons in the cochlear nuclei, third-order neurons in the superior olive, and fourth-order neurons in the adjacent nuclei of the lateral lemniscus (see Fig. 21–10). *It is precisely this heterogeneous collection that prevents a simple correlation of nuclei or tracts with specific wave components of the ABRs that are widely used to assess clinically the level of brainstem function.* The interposition of synaptic delays in each of these components of the lateral lemniscus imparts temporal differences that contribute to at least the second, third, and fourth wave components of the evoked responses.

The larger *anterior nucleus of the lateral lemniscus* (*ventral nuleus of the lateral lemniscus*) consists of cells scattered among the ascending fibers of the lateral lemniscus (see Fig. 21–10). It extends from the rostral limit of the superior olive to just below the inferior colliculus. These cells project to the inferior colliculus, completing an *indirect monaural pathway* (see Fig. 21–10).

The smaller *posterior nucleus of the lateral lemniscus* (*dorsal nucleus of the lateral lemniscus*) is intercalated in the ascending fiber bundles of the lateral lemniscus just caudal to the inferior colliculus (see Fig. 21–10). This nucleus receives input mainly from the superior olivary complex. Ascending projections from the posterior nucleus of the lateral lemniscus decussate in the *posterior tegmental commissure*. These fibers terminate in the contralateral inferior colliculus and, to a lesser degree, in the contralateral posterior nucleus of the lateral lemniscus (see Fig. 21–10). This pathway is largely inhibitory, using gamma-aminobutyric acid (GABA) as the neurotransmitter. It conveys binaural information and inhibits activity from the opposite hemifield.

Inferior Colliculus. Virtually all ascending auditory pathways terminate in the inferior colliculus (see Fig.

21–10). The egg-shaped core of the inferior colliculus, the prominent *central nucleus*, is nested in a base of afferent fibers formed by fibers of the lateral lemniscus. These fibers are the major source of input to the inferior colliculus. In a shell around the central nucleus, other cells form the smaller *paracentral nuclei* (see Fig. 21–10). These are the *pericentral nucleus*, which lies posterior and is traversed by fibers from the commissure of the inferior colliculus, and the *external (lateral) nucleus*, which lies lateral and is intersected by fibers that form the *brachium of the inferior colliculus.*

The central nucleus integrates information from multiple hindbrain auditory sources and, in turn, projects to the anterior division of the medial geniculate nucleus (see Fig. 21–10). The central nucleus consists of parallel layers of cells with disc-shaped dendritic fields. Afferents from the lateral lemniscus course parallel to these dendritic fields, forming a series of *fibrodendritic laminae*. Ascending projections diverge and converge in a point-to-plane order in the central nucleus. As a result, each point along the cochlear spiral is represented in an *isofrequency lamina*. Functionally, cells in the central nucleus are narrowly tuned, with the lowest frequencies represented posterolaterally and higher frequencies anteromedially (see Fig. 21–10).

Many cells in the inferior colliculus respond to input from either ear. Among cells with low characteristic frequencies, many are sensitive to interaural time delays, and those with high characteristic frequencies are sensitive to interaural intensity differences. Thus, binaural responses of inferior collicular neurons resemble those of the superior olivary neurons, from which they receive a dominant binaural input. These responses are probably further modified by indirect binaural pathways from the posterior nucleus of the lateral lemniscus and by intrinsic circuits in the fibrodendritic laminae. Other cells in the fibrodendritic laminae of the central nucleus are monaural and are mainly excited only by the contralateral ear. Their responses resemble those of cells in the contralateral cochlear nucleus.

Cells in the *paracentral nuclei* are broadly tuned to frequency, and they habituate rapidly to repetitive stimuli. They receive input from the central nucleus and the cerebral cortex and nonauditory input from the spinal cord, posterior column nuclei, and superior colliculus. These nuclei project to the medial geniculate nucleus (see Fig. 21–10), superior colliculus, reticular formation, and precerebellar nuclei. Thus, the paracentral nuclei are probably involved in functions related to attention, multisensory integration, and auditory-motor reflexes (see Fig. 21–15).

Medial Geniculate Nucleus. The medial geniculate nucleus forms a small protuberance on the lower caudal surface of the thalamus between the lateral geniculate body and the pulvinar (see Fig. 21–10; see also Fig. 15–9). The *anterior division* of the medial geniculate nucleus receives afferents from the central nucleus of the inferior colliculus and projects to the primary auditory cortex. Isofrequency contours in the anterior division are arranged so that low frequencies are represented laterally and higher frequencies medially (see Fig. 21–10). As a result of collicular and thalamic integration, however, most cells are not reliably excited by simple tones and are probably involved in complex feature detection.

The *posterior division* receives input from the pericentral nucleus of the inferior colliculus and projects to secondary auditory cortex (see Fig. 21–10). These projections are also tonotopically arranged. More broadly tuned and sensitive to habituation, this pathway may convey information about moving or novel stimuli that direct auditory attention.

The *medial (magnocellular)* division receives afferents from the external nucleus of the inferior colliculus and projects to association areas of auditory cortex. It contains cells that are broadly tuned to auditory and other sensory stimuli, including vestibular and somesthetic inputs. The medial division projects to temporal and parietal association areas and to the amygdala, putamen, and pallidum. In view of the multisensory convergence that occurs in this pathway, it may be a part of the reticular activating system.

Auditory and Related Association Cortices

The *primary auditory cortex (AI)* is located in the *transverse gyri of Heschl* (Fig. 21–12; see also Fig. 21–10). Two transverse temporal gyri are buried in the lateral sylvian sulcus, covered by parts of the frontal and parietal opercula, and continuous with the superior temporal gyrus. Caudal to the transverse temporal gyri is a smooth area, the *planum temporale*, which is usually larger on the left side than on the right.

The primary auditory cortex (AI, Brodmann area 41) is located in the first (anterior) transverse temporal gyrus but may extend into the second (posterior) gyrus (see Fig. 21–12A). Cytoarchitecturally, area 41 encompasses the *granular cortex*, with its well-developed layer IV containing small granule cells and densely packed small pyramidal cells in layer VI (see Fig. 21–12B). Adjacent to the granular cortex in the second transverse gyrus and planum temporale is area 42, which constitutes the *secondary (AII) auditory* cortex (see Fig. 21–12A).

Area 41 is reciprocally connected with the anterior division, and area 42 with the posterior division, of the medial geniculate body (Fig. 21–13). Through the corpus callosum, each auditory cortical area is connected with the reciprocal areas in the other cerebral hemisphere. The tonotopic organization of constituent cells

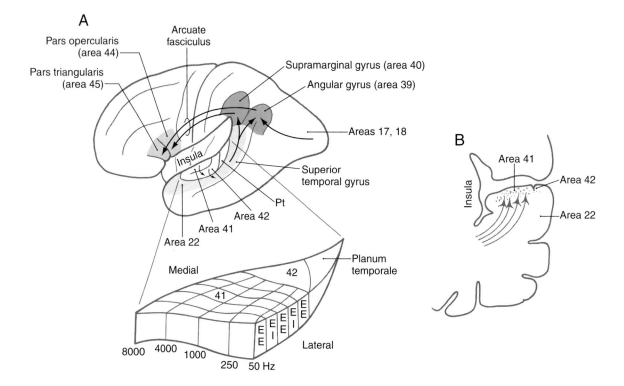

Figure 21–12. The organization of auditory cortical areas. Location and interconnections of auditory cortical areas (*A*) and of the granular cortex in area 41 (*B*), and the orthogonal isofrequency and binaural response columns in the primary auditory cortex (detail from *A*).

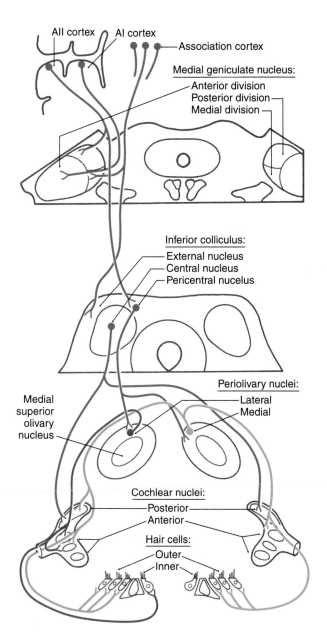

Figure 21–13. Descending auditory pathways that modulate sensory processing at central and peripheral auditory sites. The lateral olivocochlear efferents are shown in red and the medial olivocochlear efferents in green. AI and AII, primary and secondary auditory (cortices).

auditory area and is located mainly in the posterior portion of the superior temporal gyrus (see Fig. 21–12*A*). It is connected to the primary auditory cortex by the *arcuate fasciculus* (see Fig. 21–12*A*). Area 22 includes a part of the planum temporale and the posterior portion of the superior temporal gyrus. It receives connections from the primary auditory cortex, as well as visual and somesthetic information. This speech receptive area, known as the *Wernicke area*, may be as much as seven times larger on the left side than on the right. When this area is damaged by occlusion of branches of the middle cerebral artery, an *auditory aphasia* (*Wernicke aphasia*) results. In such cases, comprehension of speech sounds is impaired, but discrimination of nonverbal sounds is largely unaffected.

The higher association areas of auditory cortex also extend into the inferior parietal lobule (see Fig. 21–12*A*). This lobule is made up of the angular gyrus (area 39) and supramarginal gyrus (area 40). These two areas are important in aspects of language such as reading and writing and are sometimes included in the Wernicke area.

Brodmann areas 44 and 45 are known as the *Broca area* for expressive speech and language. They are located in the *pars opercularis* and *pars triangularis* of the *inferior frontal gyrus* (see Fig. 21–12*A*). The major pathway connecting these areas with the primary and association auditory cortex is the *arcuate fasciculus* (see Fig. 21–12). If areas 44 and 45 are damaged along with other motor cortices on the left side by a stroke involving branches of the middle cerebral artery, the result is *Broca aphasia*. In this disorder, speech is nonfluent, but comprehension of verbal and nonverbal sounds is largely unimpaired.

Descending Auditory Pathways

Descending projections make reciprocal connections throughout the auditory pathway. They form feedback loops that provide circuits to modulate information processing from the peripheral level to the cortex (see Fig. 21–13). For example, the auditory cortex projects to the medial geniculate nucleus and nuclei of the inferior colliculus. The inferior colliculus projects to the periolivary nuclei, which, in turn, send olivocochlear efferents to the cochlea. There are also descending projections from the periolivary nuclei to the cochlear nuclei.

The Olivocochlear Bundle. The *olivocochlear efferent system* arises from groups of cells in the periolivary nuclei of the superior olivary complex (see Fig. 21–13). These efferent systems travel as the *olivocochlear bundle* in the vestibular part of the vestibulocochlear nerve. *Lateral olivocochlear efferent* cells project to the ipsilateral inner hair cells, where they make axoaxonic synapses with type I spiral ganglion afferent

of the cortical layers and incoming afferent fibers form a series of orderly isofrequency columns that extend through the primary auditory cortex as long stripes (see Fig. 21–12). High frequencies are represented medially and low frequencies laterally. The series of stripes so formed have one subcomponent composed of cells excited by stimulation of both ears (EE) alternating with a subcomponent composed of cells excited by the contralateral ear and inhibited by the ipsilateral ear (EI).

The *auditory association cortex* surrounds the primary

fibers (see Figs. 21–4 and 21–13). *Medial olivocochlear efferent* cells have bilateral projections that terminate directly on outer hair cells (see Figs. 21–4 and 21–13).

Direct efferent feedback to outer hair cells, in particular, may influence cochlear mechanics and, consequently, the sensitivity and frequency selectivity of the cochlea. Efferent-induced changes in outer hair cell membrane potentials result in changes in the height of the cells and the stiffness of their stereocilia. These changes modulate basilar membrane motion and thereby influence cochlear function. The tight coupling of the basilar membrane to the tectorial membrane by the outer hair cells enables this efferent mechanism to feed energy back to the cochlea to amplify responses to specific tones. The cochlear amplifier effect is important in selectively tuning the cochlea to important sounds.

Middle Ear Reflex

The small striated muscles of the middle ear affect the mechanical impedance of the ossicular chain. These muscles are activated by the *middle ear reflex* (Fig. 21–14).

The *stapedius muscle* is innervated by *facial motor neu-*

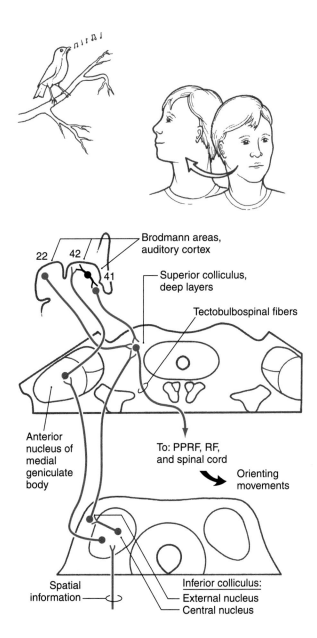

Figure 21–15. The pathways that subserve auditory-motor integration involved in simple orientation to a novel auditory stimulus. RF, reticular formation; PPRF, paramedian pontine reticular formation.

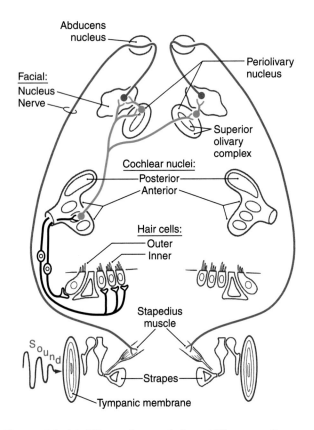

Figure 21–14. The pathway of the middle ear reflex arc. For simplicity, only the stapedius reflex is shown.

rons, and the *tensor tympani muscle* by *trigeminal motor neurons*. These motor neurons are intimately associated with the caudal end of the superior olivary complex, in the case of the stapedius muscle, and the rostal end of the superior olivary complex, in the case of the tensor tympani muscle. In these positions, auditory input via axons of neurons in the cochlear nuclei or the superior olivary complex provides the sensory limb of the reflex. The sensory pathways are bilateral, so that stimuli may be presented by earphones to one ear while the device to measure impedance is placed in the ear canal on the other side.

Acoustic Startle Reflex, Orientation, and Attention

Reflexive and learned responses to sound require sensory-motor integration. In addition to corticocortical interconnections for the dissemination of auditory information, there is also integration of auditory sensory input with motor pathways in the brainstem. Reticulospinal neurons in the region of the lateral lemniscus have dendrites that sample lemniscal activity and are involved in rapid *acoustic startle reflex* pathways. In addition, the *deep layers of the superior colliculus* receive auditory information from the inferior colliculus and auditory cortical areas (Fig. 21–15). The deep layers of the superior colliculus integrate auditory, visual, and somesthetic information and project to brainstem and cervical spinal cord nuclei via tectobulbospinal fibers, which are involved in controlling orientation of the head, eyes, and body to sound (see Fig. 21–15).

Sources and Additional Reading

Interesting www links related to the auditory system:

Promenade 'round the Cochlea. Available at www.iurc.montp.inserm.fr/cric/audition/english/start.htm

The Cochlea—graphic tour of the inner ear's machinery. Available at www.sissa.it/bp/Cochlea/index.html

Altschuler RA, Bobbin RP, Hoffman DW: Neurobiology of Hearing: The Cochlea. Raven Press, New York, 1986.
Altschuler RA, Bobbin RP, Clopton BM, Hoffman DW: Neurobiology of Hearing: The Central Auditory System. Raven Press, New York, 1991.
Gelfand SA: Hearing: An Introduction to Psychological and Physiological Acoustics. Marcel Dekker, New York, 1990.
Pickles JO: An Introduction to the Physiology of Hearing, 2nd Ed. Academic Press, London, 1988.
Webster D, Fay RR, Popper AN: Springer Handbook of Auditory Research, vol I. The Auditory Pathway: Neuroanatomy. Springer-Verlag, New York, 1992.
Yost WA: Fundamentals of Hearing: An Introduction. Academic Press, San Diego, 1994.

Chapter 22

The Vestibular System

J. D. Dickman

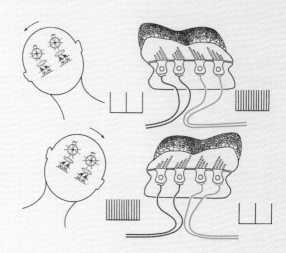

Overview   342

Peripheral Vestibular Labyrinth   342
Vestibular Receptor Organs
Membranous Labyrinth

Vestibular Sensory Receptors   345
Hair Cell Morphology
Hair Cell Transduction
Morphologic Polarization of Hair Cells

Semicircular Canals and Otolith Organs   347
Function of Semicircular Canals
Function of Otolith Organs

Vestibular Nuclei   349
Vestibular Afferent Inputs
Cerebellar Connections
Commissural Connections
Other Afferent Connections
Other Efferent Connections

Vestibulo-ocular Network   353
Rotational Vestibulo-ocular Reflex
Linear Vestibulo-ocular Reflex
Nystagmus

Vestibulospinal Network   354
Lateral Vestibulospinal Tract
Medial Vestibulospinal Tract

Vestibulo-Thalamo-Cortical Network   355
Vestibular Thalamus
Vestibular Cortex

Dizziness and Vertigo   356

Humans have the ability to control posture and movements of the body and eyes relative to the external environment. The *vestibular system* mediates these motor activities through a network of receptors and neural elements. This system integrates peripheral sensory information from vestibular, somatosensory, visceromotor, and visual receptors, as well as motor information from the cerebellum and cerebral cortex. Central processing of these inputs occurs rapidly, with the output of the vestibular system providing an appropriate signal to coordinate relevant movement reflexes. Although the vestibular system is considered to be a special sense, most vestibular activity is conducted at a subconscious level. However, in situations producing unusual or novel vestibular stimulation, such as rough air in a plane flight or wave motion on ships, vestibular perception becomes acute, with dizziness, vertigo, or nausea often resulting.

Overview

The vestibular system is an essential component in the production of motor responses that are crucial for daily function and survival. Throughout evolution, the highly conserved nature of the vestibular system is revealed through striking similarities in the anatomic organization of receptors and neuronal connections in fish, reptiles, birds, and mammals.

For the present discussion, the vestibular system can be divided into five components: (1) The *peripheral receptor apparatus* resides in the inner ear and is responsible for transducing head motion and position into neural information. (2) The *central vestibular nuclei* comprise a set of neurons in the brainstem that are responsible for receiving, integrating, and distributing information that controls motor activities such as eye and head movements, postural reflexes, and gravity-dependent autonomic reflexes and spatial orientation. (3) The *vestibulo-ocular network* arises from the vestibular nuclei and is involved in the control of eye movements. (4) The *vestibulospinal network* coordinates head movements, axial musculature, and postural reflexes. (5) The *vestibulo-thalamo-cortical network* is responsible for the conscious perception of movement and spatial orientation.

Peripheral Vestibular Labyrinth

The vestibular labyrinth contains specialized sensory receptors and is located lateral and posterior to the cochlea in the inner ear (Fig. 22–1). The vestibular labyrinth consists of five separate receptor structures, *three semicircular canals* and *two otolith organs*, which are contained in the petrous portion of the temporal bone. The labyrinth is actually composed of two distinct components. The *bony labyrinth* is a surrounding shell that contains and protects the sensitive underlying vestibular sensory structures (see Fig. 22–1). In humans,

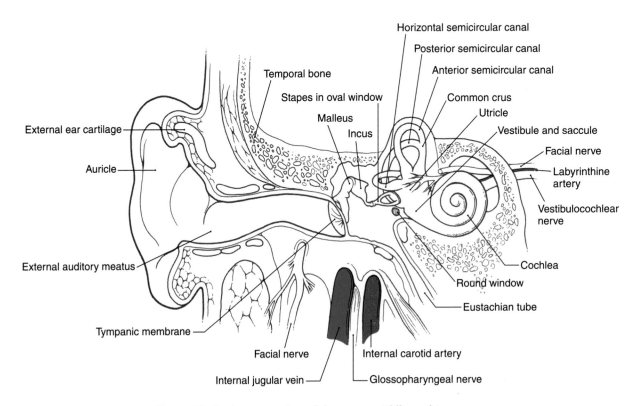

Figure 22–1. A cross section of the outer, middle, and inner ear.

the bony labyrinth can be visualized only on excision of the mastoid process. Inside the bony labyrinth is a closed, fluid-filled system, the *membranous labyrinth*, which consists of connecting tubes and prominences (Fig. 22–2). Vestibular receptors are located in specialized regions of the membranous labyrinth.

Between the membranous and bony labyrinth is a space containing fluid called *perilymph*, which is similar to cerebrospinal fluid. Perilymph has a high sodium content (150 mM) and a low potassium content (7 mM), and it bathes the vestibular portion of the eighth cranial nerve.

The membranous labyrinth is filled with a different type of fluid, called *endolymph*, which covers the specialized sensory receptors of both the vestibular and the auditory systems. Endolymph has a high concentration of potassium (150 mM) and a low concentration of sodium (16 mM). It is important to note the differences in these two fluids because both are involved in the normal functioning of the vestibular system. Disturbances in the distribution or ionic content of endolymph often lead to vestibular pathology.

Vestibular Receptor Organs. The five vestibular receptor organs in the inner ear complement each other in function. The semicircular canals (horizontal, anterior, and posterior) transduce rotational head movements (angular accelerations). The otolith organs (utricle and saccule) respond to translational head movements (linear accelerations) or to the orientation of the head relative to gravity. Each semicircular canal and otolith organ is spatially aligned to be most sensitive to movements in specific planes in three-dimensional space.

In humans, the horizontal semicircular canal and the utricle both lie in a plane that is slightly tilted anterodorsally relative to the naso-occipital plane (Fig. 22–3). When a person walks or runs, the head is normally declined (pitched downward) by approximately 30 degrees, so that the line of sight is directed a few meters in front of the feet. This orientation causes the plane of the horizontal canal and utricle to be parallel with the earth horizontal and perpendicular to gravity. The anterior and posterior semicircular canals and the saccule are arranged vertically in the head, orthogonal to the horizontal semicircular canal and utricle (see Fig. 22–3). The two vertical canals in each ear are positioned orthogonal to each other, whereas the plane of the anterior canal on one side of the head is coplanar with the plane of the contralateral posterior canal (see Fig. 22–3).

The receptor cells in each vestibular organ are innervated by primary afferent fibers that join with those from the cochlea to comprise the *vestibulocochlear (eighth) cranial nerve*. The cell bodies of these bipolar vestibular afferent neurons are in the vestibular ganglion (the Scarpa ganglion), which lies in the internal

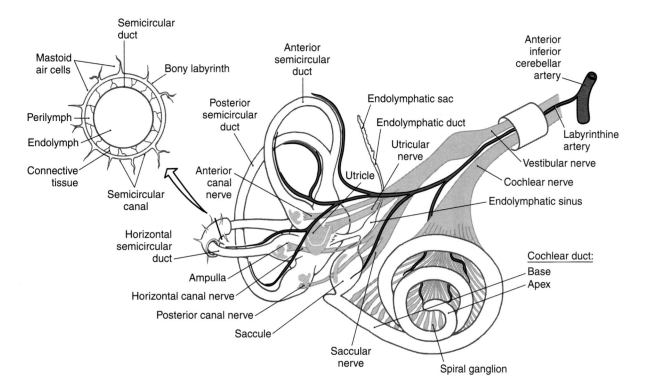

Figure 22–2. The membranous labyrinth and associated vessels and nerves. The approximate configuration of the receptor sites in the ampullae, utricle, and saccule are shown in green. The detail shows the relationship between bony and membranous labyrinths.

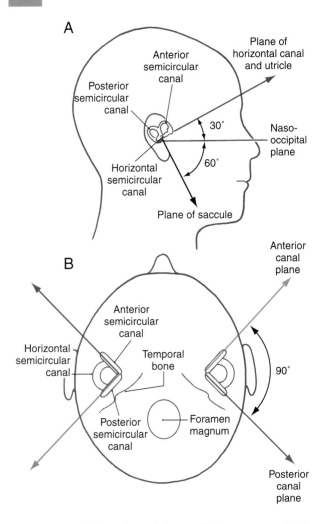

Figure 22–3. Orientation of the vestibular receptors. In the lateral view (*A*), the horizontal semicircular canal and the utricle lie in a plane that is tilted relative to the naso-occipital plane. In the axial view (*B*), the vertical semicircular canals lie at right angles to each other.

acoustic meatus (Fig. 22–4). The central processes of these bipolar cells enter the brainstem and terminate in the ipsilateral vestibular nuclei and cerebellum.

The blood supply to the labyrinth is primarily via the *labyrinthine artery*, usually a branch of the anterior inferior cerebellar artery. This vessel enters the temporal bone through the internal auditory meatus. Although not as important as the labyrinthine artery, the *stylomastoid artery* also provides branches to the labyrinth, mainly to the semicircular canals. An interruption of blood supply to the labyrinth will compromise vestibular (and cochlear) function, resulting in labyrinth-associated symptoms such as dizziness, nystagmus, and unstable gait.

Membranous Labyrinth. The membranous labyrinth is supported inside the bony labyrinth by connective tissue. The three *ducts of the semicircular canals* connect to the utricle, and each duct ends with a single prominent enlargement, the *ampulla* (see Fig. 22–2). Sensory receptors for the semicircular canals reside in a neuroepithelium at the base of each ampulla. The receptors in the utricle are oriented longitudinally along its base, and in the saccule they are oriented vertically along the medial wall (see Fig. 22–2). Endolymph in the labyrinth is drained into the endolymphatic sinus via small ducts. In turn, this sinus communicates through the *endolymphatic duct* with the *endolymphatic sac*, which is located adjacent to the dura mater (see Fig. 22–2). The saccule is also connected to the cochlea by the *ductus reuniens*.

The balance between the ionic contents of endolymph and perilymph is maintained by specialized secretory cells in the membranous labyrinth and the endolymphatic sac. In cases of advanced *Ménière disease*, there is disruption of normal endolymph volume resulting in *endolymphatic hydrops* (an abnormal distention

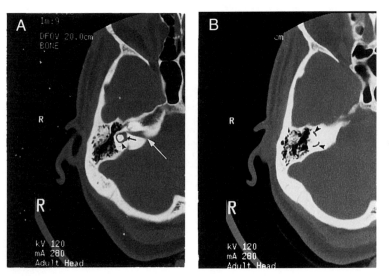

Figure 22–4. Computed tomography (CT) scans of the human temporal bone. The horizontal (*A*, *arrowhead*) and anterior and posterior (*B*, *arrowheads*) semicircular canals, utricle (*A*, *small arrow*), and internal acoustic canal (*A*, *large arrow*) are visible.

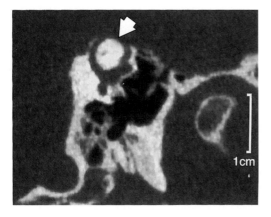

Figure 22–5. Computed tomographic image of the temporal bone projected into the plane of the left superior canal in a patient with superior canal dehiscence syndrome. The patient developed vertigo, oscillopsia, and eye movements in the plane of the left superior canal in response to loud noises and pressure in the left ear. A dehiscence is noted overlying the left superior canal (*arrowhead*).

of the membranous labyrinth). Symptoms of Ménière disease include severe vertigo, positional nystagmus, and nausea. Affected persons often suffer unpredictable attacks of auditory and vestibular symptoms, including vomiting, tinnitus (ringing in the ears), and a complete inability to make head movements or even stand passively. For patients with frequent debilitating attacks, the first course of treatment is often administration of a diuretic (e.g., hydrochlorothiazide) and a salt-restricted diet to reduce the hydrops. If a persistent Ménière symptom continues, a second treatment option is the implantation of a small tube or shunt into the abnormally swollen endolymphatic sac.

Occasionally, a condition may develop in which a portion of the temporal bone overlying either the anterior or the posterior semicircular canal thins so much that an opening (dehiscence) is created next to the dura (Fig. 22–5). In affected patients, the canal dehiscence exposes the normally closed bony labyrinth to the extradural space. Symptoms can include vertigo and oscillopsia in response to loud sounds (the Tulio phenomenon) or in response to maneuvers that change middle ear or intracranial pressure. The eye movements evoked by these stimuli (nystagmus) align with the plane of the dehiscent superior canal. Surgical closure of the defect by bone replacement is often performed.

Vestibular Sensory Receptors

Hair Cell Morphology. The sensory receptor cells in the vestibular system, like those in the auditory system, are called *hair cells* owing to the *stereocilia* that project from the apical surface of the cell (Fig. 22–6A). Each

hair cell contains 60 to 100 hexagonally arranged stereocilia and a single longer *kinocilium*. The stereocilia are oriented in rows of ascending height, with the tallest lying next to the lone kinocilium. The stereocilia arise from a region of dense actin, the *cuticular plate*, located at the apical end of the hair cell. The cuticular plate acts as an elastic spring to return the stereocilia to the normal upright position after bending. Each stereocilium is connected to its neighbor by small filaments.

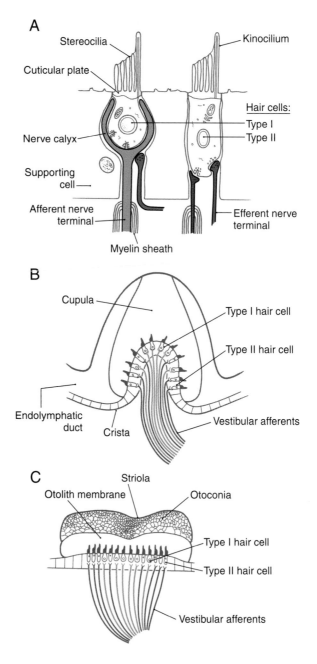

Figure 22–6. The receptor cells (*A*, type I and type II hair cells) of the vestibular system. The relation of these cells to the crista and cupula (*B*) in the ampullae and to the macula and otolith membrane (*C*) of the otolith organs is shown.

There are two types of hair cells, and they differ in their pattern of innervation by fibers of the eighth cranial nerve (see Fig. 22–6A). Type I hair cells are chalice-shaped and typically are surrounded by an afferent terminal that forms a *nerve calyx*. Type II hair cells are cylindrical and are innervated by simple synaptic boutons. Excitatory amino acids such as aspartate and glutamate are the neurotransmitters at the receptor cell–afferent fiber synapses. Both types of hair cells, or their afferents, receive synapses from *vestibular efferent fibers* that control the sensitivity of the receptor. These efferent fibers contain acetylcholine and calcitonin gene–related peptide (CGRP) as neurotransmitters. Efferent cell bodies are located in the brainstem just rostral to the vestibular nuclei and lateral to the abducens nucleus. They are activated by behaviorally arousing stimuli or by trigeminal stimulation.

Within each ampulla, the hair cells and their supporting cells lie embedded in a saddle-shaped neuroepithelial ridge, the *crista*, which extends across the base of the ampulla (see Fig. 22–6B). Type I hair cells are concentrated in central regions of the crista, and type II hair cells are more numerous in peripheral areas. Arising from the crista and completely enveloping the stereocilia of the hair cells is a gelatinous structure, the *cupula*. The cupula attaches to the roof and walls of the ampulla, forming a fluid-tight partition that has the same specific density as that of endolymph. Rotational head movements produce angular accelerations that cause the endolymph in the membranous ducts to be displaced, so that the cupula is pushed to one side or the other like the skin of a drum. These cupular movements displace the stereocilia (and kinocilium) of the hair cells in the same direction.

For the otolith organs, a structure analogous to the crista, the *macula*, contains the receptor hair cells (see Fig. 22–5C). The hair cell stereocilia of otolith organs extend into a gelatinous coating called the *otolith membrane*, which is covered by calcium carbonate crystals called *otoconia* (from Greek, "ear stones"). Otoconia are about three times as dense as the surrounding endolymph, and they are not displaced by normal endolymph movements. Instead, changes in head position relative to gravity, or linear accelerations (forward-backward, upward-downward) produce displacements of the otoconia, resulting in bending of the underlying hair cell stereocilia.

Hair Cell Transduction. The response of hair cells to deflection of their stereocilia is highly polarized (Figs. 22–7 and 22–8A). Movements of the stereocilia *toward the kinocilium* cause the hair cell membranes to *depolarize*, which results in an increased rate of firing in the vestibular afferent fibers. If the stereocilia are *deflected away from the kinocilium*, however, the hair cell is *hyperpolarized* and the afferent firing rate decreases.

The mechanisms underlying the depolarization and hyperpolarization of vestibular hair cells depend, re-

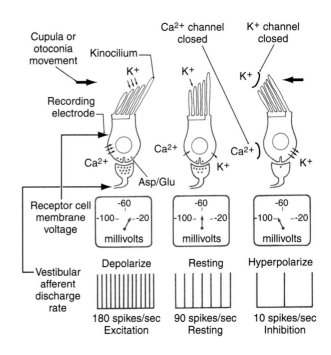

Figure 22–7. Physiologic responses of vestibular hair cells and their vestibular afferent fibers. Asp, aspartate; Glu, glutamate.

spectively, on the potassium-rich character of endolymph and the potassium-poor character of the perilymph that bathes the basal and lateral portions of the hair cells. Deflection of the stereocilia *toward* the kinocilium causes potassium channels in the apical portions of the stereocilia and kinocilium to open. K+ flows into the cell from the endolymph, depolarizing the cell membrane (see Fig. 22–7). This depolarization in turn causes voltage-gated calcium channels at the base of the hair cells to open, allowing Ca2+ to enter the cell. The influx of Ca2+ causes synaptic vesicles to release their transmitter (aspartate or glutamate) into the synaptic clefts, and the afferent fibers respond by undergoing depolarization and increasing their rate of firing. When the stimulus subsides, the stereocilia and kinocilium return to their resting position, allowing most calcium channels to close and voltage-gated potassium channels at the base of the cell to open. K+ efflux returns the hair cell membrane to its resting potential (see Fig. 22–7).

Deflection of the stereocilia *away* from the kinocilium causes potassium channels in the basolateral portions of the hair cell to open, allowing K+ to flow out from the cell into the interstitial space. The resulting hyperpolarization of the cell membrane decreases the rate at which the neurotransmitter is released by the hair cells and consequently, decreases the firing rate of afferent fibers.

Almost all vestibular primary afferent fibers have a moderate firing rate at rest (approximately 90 spikes per second). Therefore, it is likely that some hair cell calcium channels are open at all times, causing

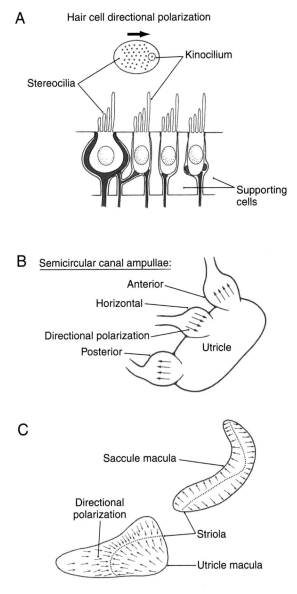

Figure 22–8. Morphologic polarization of vestibular receptor cells showing polarity of stereocilia and kinocilia (*A*), and the orientation of receptors in the ampullae (*B*) and maculae (*C*).

a slow, constant release of neurotransmitter. The ototoxic effects of some aminoglycoside antibiotics (e.g., streptomycin, gentamicin) may be due to direct reduction of the transduction currents of hair cells.

Morphologic Polarization of Hair Cells. Given that deflections of the stereocilia toward and away from the kinocilium cause opposing physiologic responses, it is clear that the directional orientation of the hair cells in the vestibular organs will play an essential role in signaling the direction of movement. On the cristae of the horizontal semicircular canal, the hair cells are all arranged with their kinocilium on the side closer to the utricle (see Fig. 22–8*B*). Thus, *movement of endolymph toward the ampulla in the horizontal canal* causes the

stereocilia to be deflected toward the kinocilium, resulting in depolarization of the hair cell. In the vertical semicircular canals, the hair cells are arranged with their kinocilia on the side farther from the utricle (closest to the endolymphatic duct). Thus the *hair cells of the vertical canals are hyperpolarized by movement of endolymph toward the ampulla* (ampullopetal movement) and are depolarized by movement away from the ampulla (ampullofugal movement).

In both the utricle and the saccule, the otolith membrane overlying the hair cells contains a small, curving depression, the *striola*, that roughly bisects the underlying macula (see Fig. 22–8*C*). Hair cells on the utricular macula are polarized so that the kinocilium is always on the side toward the striola (see Figs. 22–6*C* and 22–8*C*), which effectively splits the receptors into two morphologically opposed groups. In contrast, the kinocilia of saccular hair cells are oriented on the side away from the striola. Because the striola curves through the macula, otolith hair cells are polarized in *many different directions* (see Fig. 22–8*C*). In this way, utricular and saccular hair cells are directionally sensitive to a wide variety of head positions and linear movements.

Semicircular Canals and Otolith Organs

As stated previously, the vestibular receptors transduce *movement and position* stimuli into neural signals that are sent to the brain. The *semicircular canals are responsive to rotational acceleration* resulting from turns of the head or body. The *otolith organs are responsive to linear accelerations.* The most prominent linear acceleration on earth is the constant force of gravity. Linear motion, such as experienced during swinging on a swing or flying in an airplane through turbulence, couples with gravity to change the direction and amplitude of the resultant *gravitoinertial acceleration* (GIA). The GIA is sensed by the otolith organs and can be greatly reduced during space flight. Linear accelerations also occur in situations such as up-and-down motion during running or acceleration of an automobile. The otolith organs are also responsive to tilting of the head relative to gravity (pitch and roll movements). (Forward and backward tilting is called *pitch*; side-to-side tilting is called *roll*.)

Function of Semicircular Canals. The membranous semicircular duct can be thought of as a fluid-filled tube with a partition (the cupula) in the middle (see Fig. 22–6*B*). Because the utricles are located medially, as compared to the horizontal canals, on each side of the head, the hair cells of the complementary left and right semicircular canals are *oppositely polarized.* An example is seen in rotational head movements made in the horizontal plane (Fig. 22–9). When the head is stationary (no angular acceleration), the endolymph

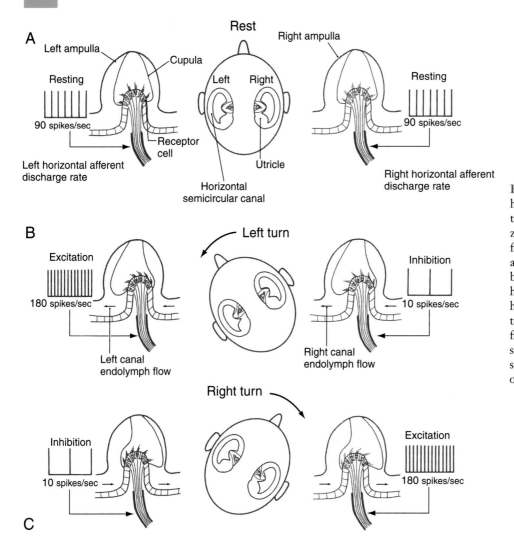

Figure 22–9. Response of the horizontal semicircular canals to head rotations in the horizontal plane. At rest (*A*), the firing rates of horizontal canal afferents are equivalent on both sides. With a leftward head turn (*B*) or a rightward head turn (*C*), there is receptor depolarization and afferent fiber excitation toward the side of the turn and corresponding inhibition on the opposite side.

and the cupula remain still, and the afferents from the two horizontal semicircular canals fire at the same (resting) rate (see Fig. 22–9*A*). When the head turns to the right or left, however, the horizontal semicircular ducts turn with it, but the endolymph lags owing to inertial forces and the viscous drag between the fluid and the duct wall. The lagging endolymph deflects the cupula, which in turn deflects the stereocilia of the hair cells. As Figure 22–9*B* shows, a leftward turn of the head causes the stereocilia in the left horizontal canal ampulla to be deflected toward their kinocilia, resulting in an increase in the discharge rate of the eighth nerve afferents on the left side. Simultaneously, the hair cells in the right horizontal canal ampulla are hyperpolarized, so their afferents show a decreased rate of firing. A rightward head turn produces the opposite pair of responses (see Fig. 22–9*C*).

It is important to realize that the left and right semicircular canals of each functional pair (such as the left and right horizontal canals) *always* respond oppositely to any head movement that affects them. This fact leads to the "push-pull" concept of vestibular func-

tion, which states *that directional sensitivity to head movement is coded by opposing receptor signals.* Because of commissural connections, neurons in the vestibular nuclei receive information from receptors on both sides of the head. These neurons act as *comparator units* that interpret head rotation on the basis of the relative discharge rates of left and right canal afferents. This pattern of connections also increases the sensitivity of the system, so that even small differences in the discharge rates of afferents from corresponding canal pairs (such as in slow head movements) can be perceived. During a leftward head turn, the comparator units receive impulses at a higher frequency from the left horizontal canal than from the right horizontal canal; the difference is interpreted as a left head turn. Similar conditions exist when the head is pitched or rolled so that the vertical semicircular canals are stimulated by rotational accelerations in their respective planes. However, in the case of the vertical canals, the opposing push-pull responses occur between the anterior semicircular canal in one ear and the posterior semicircular canal of the opposite ear (see Fig. 22–3).

Head trauma or disease can change the normal resting activity in eighth nerve afferent fibers. This change may be interpreted by the brain as turning, even though the head is stationary. For example, a lesion of the eighth nerve, such as that produced by a glomus tumor or acoustic neuroma (Fig. 22–10), may reduce the frequency of impulses in the ipsilateral afferent fibers or block their impulse transmission entirely. The comparator units of the vestibular nuclei will then consistently receive a higher impulse frequency from the intact side, which will be interpreted as a head turn away from the side of the lesion.

Function of Otolith Organs. The receptor hair cells in the maculae do not respond to head rotation but are sensitive to linear acceleration and tilt of the head (Fig. 22–11). When the head is moved with respect to gravity (*rolled* or *pitched*), the otoconia crystals are displaced because of their density with respect to the surrounding endolymph. This displacement shifts the underlying gelatinous coating on the maculae and produces stereocilia deflection in the hair cells. As in the responses of semicircular canal hair cells, otolith organ hair cells are either depolarized or hyperpolarized with stereocilia deflection toward or away from the kinocilium, respectively. However, hair cells on the maculae are oriented according to their position relative to the striola (see Fig. 22–8). Hair cells on one side of the striola will be depolarized, and hair cells on the other side of the striola will be hyperpolarized (see Fig. 22–11). Because the striola is curved, only certain groups of cells will be affected by a specific direction of head tilt or linear acceleration. Thus, movement is encoded by a macular map of directional space. The eighth nerve fibers maintain the directional signal because each afferent only innervates hair cells from a small region on the macular neuroepithelium.

Vestibular Nuclei

Neural information carried on vestibular afferent fibers is transmitted to the four vestibular nuclei, which lie in the rostral medulla and caudal pons (Fig. 22–12). The *superior vestibular nucleus* lies dorsally in the central pons and is bordered by the restiform body and the fourth ventricle (see Fig. 22–12B). The *medial vestibular nucleus* lies in the lateral floor of the fourth ventricle throughout most of its rostrocaudal extent (see Fig. 22–12B–E). The *lateral vestibular nucleus* lies lateral to the medial vestibular nucleus (see Fig. 22–12B and C) and contains some large neurons known as Deiters' cells. Located lateral to the medial vestibular nucleus, the *inferior* (or *descending*) vestibular nucleus extends through much of the medulla (see Fig. 22–12D–F).

The processing of positional and movement information for control of visual and postural reflexes largely takes place in the vestibular nuclei. Consequently, the major targets for efferents of the vestibular nuclei include the oculomotor nuclei, the vestibulocerebellum, the contralateral vestibular nuclei, the spinal cord, the reticular formation, and the thalamus. Each vestibular nucleus differs in its cytoarchitecture and its afferent and efferent connections.

Vestibular Afferent Inputs. Vestibular primary afferent fibers enter the brainstem at the pontomedullary junction. These fibers traverse the restiform body, then bifurcate into ascending and descending branches. Afferent fibers from the semicircular canals project primarily to the superior and medial vestibular nuclei, although lesser inputs also reach the lateral and inferior vestibular nuclei (Fig. 22–13). The otolith organs

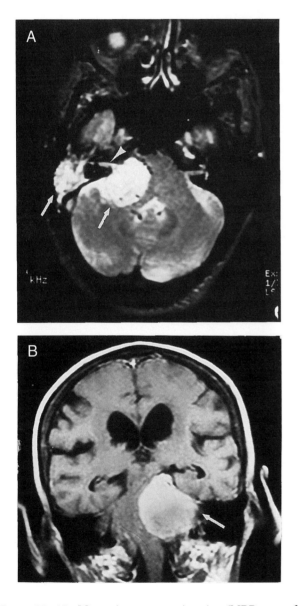

Figure 22–10. Magnetic resonance imaging (MRI) scan of a glomus tumor (*A*) and an acoustic neuroma (*B*) involving the vestibular nerve. Both patients complained of dizziness, nausea, and spatial disorientation.

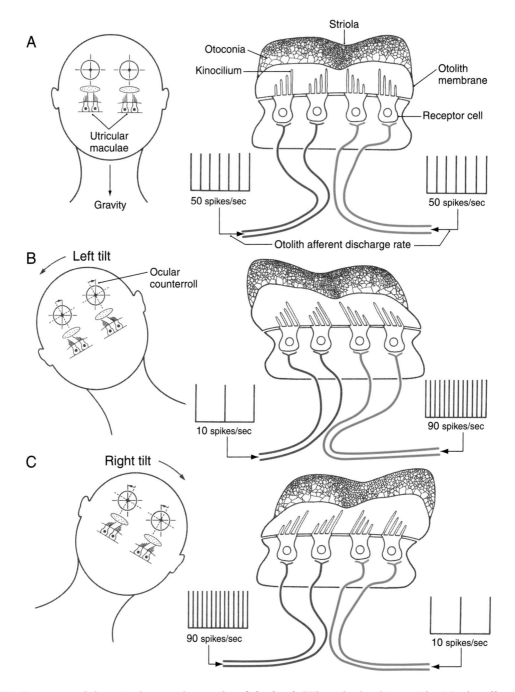

Figure 22–11. Responses of the utricular maculae to tilts of the head. When the head is upright (*A*), the afferent fibers have equivalent firing rates on both sides of the striola (red and green lines). With leftward tilt (*B*) or rightward tilt (*C*), hair cells and their innervating afferents are either excited or inhibited, depending on their position relative to the striola; the weight of the otoconia causes the stereocilia to be deflected. Hair cells on the "upslope" side of the striola increase their firing rate, and those on the "downslope" side decrease their firing rate.

project primarily to the lateral, medial, and inferior vestibular nuclei. Saccular afferents also project to cell group Y, which, in turn, excites neurons in the contralateral oculomotor nucleus and influences vertical eye movements.

The termination of vestibular afferent fibers on neurons of the vestibular nuclei is highly ordered. Individual central neurons in the superior and medial vestibu-

lar nuclei appear to receive information from otolith receptors and from one semicircular canal pair (either horizontal or vertical). Vestibular neurons in the lateral and inferior nuclei mostly receive information from several canal pairs and otolith receptors. As a result of their inputs, neurons in the vestibular nuclei show directional selectivity for particular head movements and can encode both the angular and linear components of

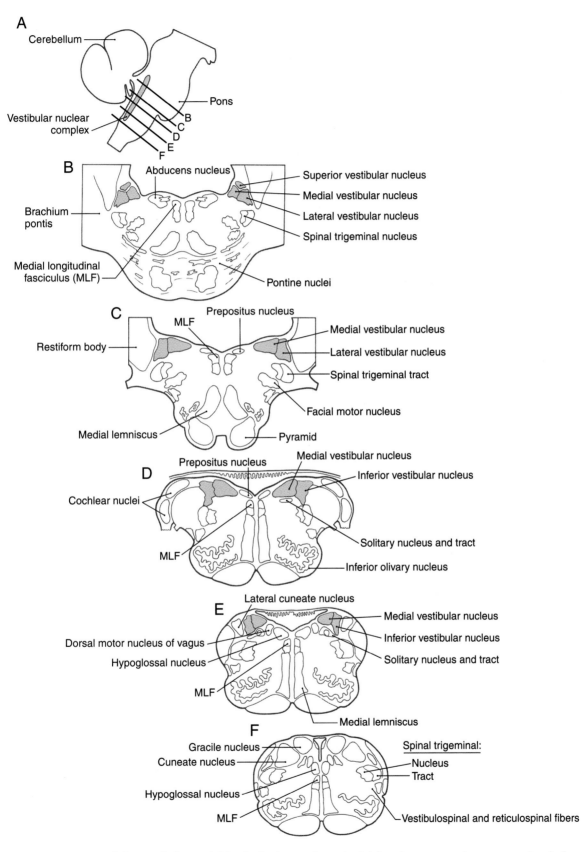

Figure 22–12. Location of the vestibular nuclei in the brainstem in sagittal (*A*) and representative cross-sectional planes (*B–F* from levels indicated in *A*).

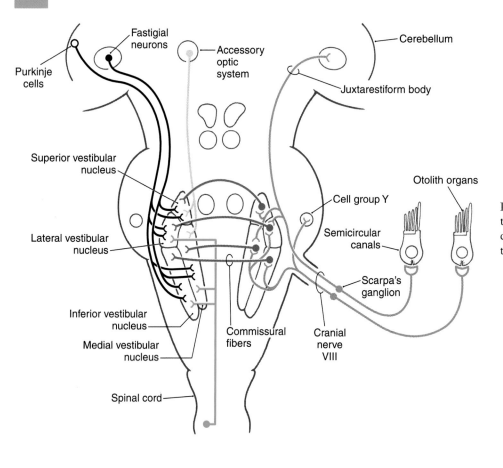

Figure 22–13. Afferents to the vestibular nuclei. Open cell bodies represent inhibitory projections.

head movements. These cells distribute information about both the direction and speed of the head movement, as well as the position of the head with respect to gravity, to many different regions of the brain.

Cerebellar Connections. The vestibular labyrinth is the only sensory organ in the body that sends direct primary afferent projections to the cerebellar cortex and nuclei (see Fig. 22–13). These *primary vestibulocerebellar* fibers course through the *juxtarestiform body*, the smaller medial part of the inferior cerebellar peduncle. Primary vestibulocerebellar fibers send collaterals to the dentate nucleus and terminate as mossy fibers in the nodulus, the uvula, and perhaps the flocculus. Neurons in all four vestibular nuclei also send axons to the cerebellum as *secondary vestibulocerebellar* projections. These axons end in the nodulus, uvula, flocculus, fastigial nucleus, and dentate nucleus.

The cerebellum forms reciprocal connections with the vestibular nuclei. This cerebellovestibular projection includes Purkinje cell axons (*cerebellar corticovestibular fibers*) from the nodulus, uvula, flocculus, and other areas of the cerebellar vermis. In addition, projections from the fastigial nucleus (*fastigiovestibular fibers*) also innervate the vestibular nuclei. Purkinje cells are GABAergic (γ-aminobutyric acid) and therefore inhibitory, whereas the fastigiovestibular fibers use glutamate or aspartate and are excitatory. These *vestibulocerebellar* and *cerebellovestibular* fibers all pass through the juxtarestiform body. The reciprocal connections between the cerebellum and the vestibular nuclei constitute important regulatory mechanisms for the control of eye movements, head movements, and posture.

Commissural Connections. Commissural *vestibulovestibular fibers* arise from all vestibular nuclei, but they appear to be most prominent from the superior and medial nuclei. Many of these fibers form reciprocal connections with the analogous contralateral nucleus. Most vestibulovestibular cells contain the inhibitory neurotransmitter GABA or glycine, although some may use the excitatory amino acids. These commissural fibers provide the pathways by which information from pairs of corresponding semicircular canals and otolith organs can be compared. Commissural fibers also play a major role in *vestibular compensation*, a process by which reflexes and postural control that are impaired as a result of unilateral loss of vestibular receptor function (through trauma or disease) are restored gradually by means of central adjustment.

Other Afferent Connections. *Spinovestibular fibers* arise from all levels of the spinal cord and provide proprioceptive input primarily to the medial and lateral vestibular nuclei. Information concerning the movement of the head through the visual world also reaches vestibular nuclei neurons through the *accessory optic system* (see Chapter 28). Finally, vestibular nuclear neurons receive input from the reticular formation, primarily from cells relaying information regarding proprioception.

Other Efferent Connections. Vestibular neurons also send efferent projections to the reticular formation, posterior (dorsal) pontine nuclei, and nucleus of the tractus solitarius, or solitary tract (NTS). The function of some of these efferent projections remains unknown. However, neurons arising from the medial and inferior vestibular nuclei project to the NTS, where it is believed that the vestibular function–mediated changes in breathing and circulation, which occur with changes in posture, are controlled. These compensatory vestibular visceromotor responses serve to stabilize respiration and blood pressure during normal body position changes relative to gravity and during locomotion. They may also be important for induction of motion sickness and emesis.

Vestibulo-ocular Network

It is often necessary to keep one's gaze fixed on an object of interest while the head is moving—as in reading a sign on a building while walking down the street. The vestibular system provides this capability by eliciting compensatory eye movements through a network of neural connections. These stabilizing eye movements, collectively known as the *vestibulo-ocular reflex*, are said to be *compensatory* because they are equal in magnitude and opposite in direction to the head movement perceived by the vestibular system. The vestibulo-ocular reflex occurs for any direction or speed of head movement, whether the movement is rotational, linear, or a combination of both. The reflex can also be suppressed at will if, for example, one wishes to focus on a moving target while turning the head in the same direction (as when watching an airplane or baseball move across the sky).

Rotational Vestibulo-ocular Reflex. There are three types of rotationally induced eye movements: *horizontal*, *vertical*, and *torsional*. Each of the six pairs of eye muscles (see Chapter 28) must be controlled in unison to produce the appropriate response. Thus, the vertical semicircular canals and the saccule are responsible for controlling vertical eye movements, whereas the horizontal canals and the utricle control horizontal eye movements. Torsional eye movements are controlled by the vertical semicircular canals and the utricle.

For purposes of example, only the horizontal vestibulo-ocular reflex is described here (Fig. 22–14). Primary afferents from the horizontal semicircular canals project to specific neurons in the medial and lateral vestibular nuclei. Most of these cells send excitatory signals through the *medial longitudinal fasciculus* to the *contralateral abducens nucleus*. Abducens motor neurons send impulses via the sixth cranial nerve to excite the *ipsilateral lateral rectus muscle*. At the same time, abdu-

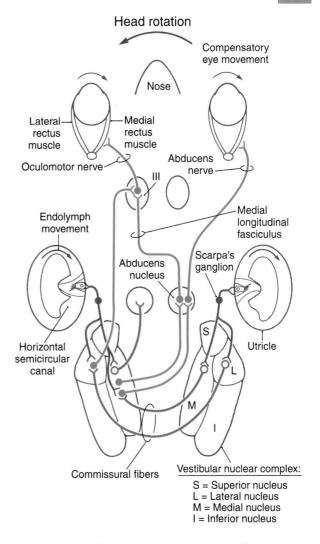

Figure 22–14. The connections subserving the horizontal vestibulo-ocular reflex. Open cell bodies represent inhibitory projections. III, oculomotor nucleus.

cens interneurons send excitatory signals to motor neurons in the *contralateral oculomotor nucleus*, which innervates the *medial rectus muscle*. A second population of vestibular neurons sends excitatory signals to the medial rectus subdivision of the ipsilateral oculomotor nucleus. A third group of vestibular neurons carry inhibitory signals to the ipsilateral abducens nucleus.

During a *leftward head turn*, excitatory signals from the left horizontal semicircular canal afferents increase the firing rate of neurons in the left vestibular nuclei neurons (see Fig. 22–14). At the same time, inhibitory signals from the right vestibular nuclei are decreased via commissural neurons. Neurons in the left vestibular nuclei then excite both the contralateral abducens motor neurons and interneurons, which, in turn, produce contraction in the right lateral rectus and the left medial rectus muscle (see Fig. 22–14). The resulting *rightward eye movement* keeps the object of interest on the fovea. Through matching bilateral connections, the left lateral rectus and right medial rectus eye muscles are inhibited.

A similar pattern of connections links the vertical semicircular canals with the motor neurons in the trochlear and oculomotor nuclei to control vertical and torsional responses (see also Chapter 28). The vertical vestibulo-ocular reflex originates primarily from neurons in the superior vestibular nucleus, although some medial vestibular nucleus neurons also participate.

Linear Vestibulo-ocular Reflex. During linear movements that do not involve head rotation, an appropriate vestibulo-ocular reflex also occurs. These reflexes depend on input from the otolith organ receptors and involve connections to the extraocular motor neuron pools that are similar to those described above for the rotational vestibulo-ocular reflex. For example, side-to-side head movements result in a horizontal eye movement in a direction opposite to the head movement. Vertical displacements of the body, such as occur during walking or running, elicit oppositely directed vertical eye movements to stabilize gaze. During roll tilts of the head, the compensatory eye movement is termed *counter-roll* and is actually a torsional eye movement (see Fig. 22–11).

Nystagmus. With large head rotations, such as with a 360-degree body turn, compensatory eye movements take another form (Fig. 22–15). Initially, the vestibulo-ocular reflex directs the eyes slowly in the direction opposite to the head motion. This movement is called

the *slow phase*. When the eye reaches the limit of how far it can turn in the orbit, it springs back rapidly to a central position, moving in the same direction as that of the head, also known as the *fast phase*. Another slow phase then begins. This combination of slow compensatory phases punctuated by fast return phases is called *nystagmus*. Nystagmus movements are named for the direction of the fast return phase—for example, as leftward-beating nystagmus or downward-beating nystagmus. Nystagmus takes many forms and is often observed clinically (see also Chapter 28). In cases of head injury with acute temporal bone fracture, the semicircular canals can be affected, producing rapid spontaneous nystagmus that can persist for hours or days.

Nystagmus can be used as a diagnostic indicator of vestibular system integrity. Typically, in patients complaining of dizziness or vertigo, the function of the vestibular labyrinth is assessed by administering a *caloric test*. Either warm (40°C) or cold (30°C) water is introduced into the external auditory canal. In normal persons, warm water induces nystagmus that beats toward the ear into which the water has been introduced, whereas cold water induces nystagmus that beats away from the ear into which the water has been introduced. (This relationship is encapsulated in the mnemonic COWS: *c*old water produces nystagmus beating to the *o*pposite side; *w*arm water produces nystagmus beating to the *s*ame side.) In normal persons, the two ears give equal responses. If there is a unilateral lesion in the vestibular pathway, however, nystagmus will be reduced or absent on the side of the lesion.

Vestibulospinal Network

The vestibular system influences muscle tone and produces reflexive postural adjustments of the head and body through two major descending pathways to the spinal cord, the *lateral vestibulospinal tract* and the *medial vestibulospinal tract* (Fig. 22–16). There is also a *reticulospinal pathway* that receives input from the vestibular system.

Lateral Vestibulospinal Tract. The lateral vestibulospinal tract (LVST) arises primarily from neurons in the lateral and inferior vestibular nuclei and projects to all levels of the ipsilateral spinal cord. This projection is topographically organized. Cells in anterorostral areas of the lateral nucleus project to the cervical cord, while cells in posterocaudal regions project to the lumbosacral cord. These vestibulospinal neurons receive substantial input from orthogonal semicircular canal pairs, the otolith organs, the vestibulocerebellum, and the fastigial nucleus, as well as proprioceptive inputs from the spinal cord.

Fibers of the LVST course through the lateral medulla dorsal to the inferior olivary complex (see Fig.

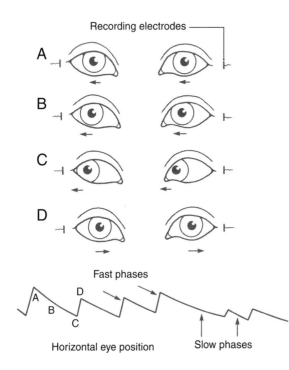

Figure 22–15. Vestibular nystagmus in the horizontal plane showing slow phase (*A–C*) and fast phase (*D*) eye movements.

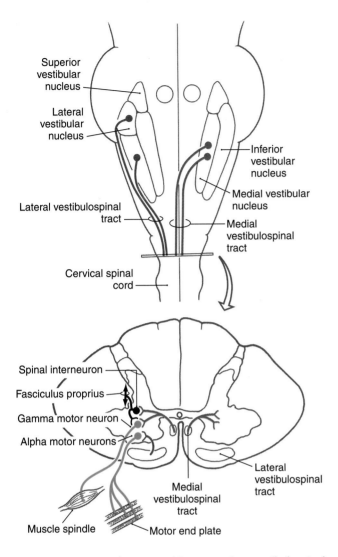

Figure 22–16. Pathways making up the vestibulospinal system.

through neurons in the medial vestibulospinal tract (MVST). These fibers originate primarily from the medial vestibular nucleus, although lesser projections arise from the inferior and lateral vestibular nuclei. Similar to LVST neurons, cells of the MVST receive input from vestibular receptors and the cerebellum, as well as somatosensory information from the spinal cord. Fibers of the MVST descend bilaterally through the medial longitudinal fasciculus to terminate in laminae VII to IX of the cervical spinal cord (see Fig. 22–16). These MVST fibers carry both excitatory and inhibitory signals, and they terminate on neck flexor and extensor motor neurons, as well as on propriospinal neurons.

The effects of vestibular–function induced responses can be seen in the *vestibulocollic* reflex, which is actually a series of responses that stabilize the head in space. If, for example, a person falls forward, MVST neurons will receive signals on downward linear acceleration from the saccule, signals on the changing head position relative to gravity from both the utricle and the saccule, and signals on forward rotational acceleration from the vertical semicircular canals. The MVST neurons process this information and transmit excitatory signals to the dorsal neck flexor muscles (splenius, biventer cervicus, and complexus muscles). At the same time, inhibitory signals are sent to the anterior neck extensor muscles. The result is a neck movement upward, opposite to the falling motion, to protect the head from impact.

Vestibulo-Thalamo-Cortical Network

Vestibular Thalamus. The cognitive perceptions of motion and spatial orientation arise through the convergence of information from the vestibular, visual, and somatosensory systems at the thalamocortical level. Neurons in the superior, lateral, and inferior vestibular nuclei project bilaterally to two thalamic areas (Fig. 22–17). The first is located in the *ventral posterolateral (VPL) nucleus and includes adjacent cells in the ventral posteroinferior (VPI) nucleus.* The second is the *posterior nuclear group,* located near the medial geniculate body. In humans, electrical stimulation of these areas produces sensations of movement and dizziness. Thalamic VPL and posterior nucleus neurons constitute separate, parallel pathways transmitting vestibular information from the brainstem to the cortex, because their connections with cortical areas are distinct.

Vestibular Cortex. Two cortical areas respond to vestibular stimulation (see Fig. 22–17). One region, *area 2v*, lies at the base of the intraparietal sulcus just posterior to the hand and mouth representations in the postcentral gyrus. Electrical stimulation of this area in humans produces sensations of moving, spinning, or

22–12), and then through the anterior funiculus of the cord (see Fig. 22–16) to terminate directly on alpha and gamma motor neurons and on interneurons in lamina VII to IX. Axons of many LVST neurons give off collaterals in different segments of the cord, thus ensuring that different muscle groups will be coordinated during postural control. The LVST neurons contain either acetylcholine or glutamate as a neurotransmitter and exert an excitatory influence on extensor muscle motor neurons. The coordinated actions of neurons that make up the LVST and provide postural stabilization are not completely understood. However, if a person begins tilting to the right, ipsilateral LVST fibers elicit extension of the left axial and limb musculature. Concurrently, right extensor muscles are inhibited. These actions stabilize the body's center of gravity and preserve upright posture.

Medial Vestibulospinal Tract. The action of vestibular stimulation on neck muscles arises primarily

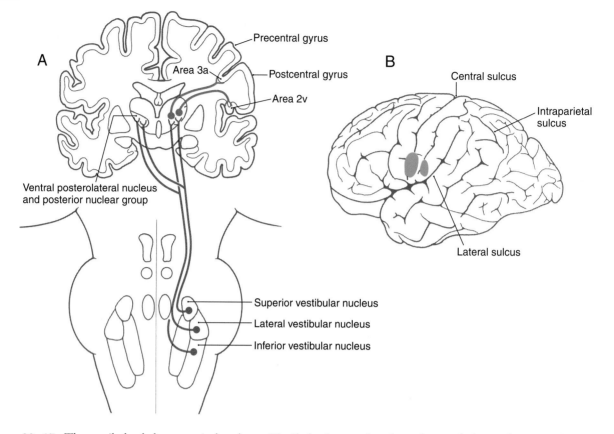

Figure 22–17. The vestibulo-thalamo-cortical pathway. Vestibular input arises from the vestibular nuclei as vestibulothalamic fibers and is relayed to the cortex as thalamocortical fibers (*A*). Areas 3a and 2v (*B*) are the main cortical regions that receive this input.

dizziness. Area 2v neurons respond to head movements and receive projections from the posterior thalamic nucleus. These 2v cells also receive visual and proprioceptive inputs. Area 2v is probably involved with motion perception and spatial orientation, because it has reciprocal connections with other parietal regions involved in similar functions (e.g., areas 5 and 7). Lesions of parietal cortical areas result in confusions in spatial awareness.

The second cortical area responding to vestibular stimulation, *area 3a*, lies at the base of the central sulcus, adjacent to the motor cortex (see Fig. 22–17) and receives input from VPL or VPI thalamic nuclear neurons. In addition, area 3a cells receive inputs from the somatosensory system. Because these cells project to area 4 of the motor cortex, it is believed that one of their functions is to integrate motor control of the head and body.

Dizziness and Vertigo

Dizziness is a *nonspecific* term that generally means a spatial disorientation that may or may not involve feelings of movement. Dizziness may be accompanied by nausea or postural instability. A large number of factors may produce a dizzy sensation, and many are not exclusively vestibular in origin.

Vertigo is a *specific* perception of body motion, often spinning or turning, experienced when no real motion is taking place. As children, we all learn to produce vertigo by whirling in place as fast as possible and then abruptly stopping. For a few moments, the world seems to be spinning in the opposite direction. Examination of the eyes during this phase will reveal a nystagmus that beats in the direction opposite to the original direction of rotation. Vertigo can also be elicited optokinetically if the visual surroundings are revolved while the body remains stationary. Many modern amusement games take advantage of this phenomenon to produce the sensation of motion.

One of the most common vestibular disorders observed clinically is *benign positional vertigo*. This condition is characterized by brief episodes of vertigo that coincide with particular changes in body position. Typically, episodes may be triggered by turning over in bed, getting up in the morning, bending over, or rising from a bent position. The pathophysiology of benign positional vertigo is not clearly understood, but posterior canal abnormalities are implicated. One possible explanation is that otoconial crystals from the utricle separate from the otolith membrane and become lodged in the cupula of the posterior canal (a condition called *cupulolithiasis*). The resulting increased density of the cupula produces abnormal cupula deflections when the head changes position relative to gravity.

Sources and Additional Readings

Baloh RW, Halmagyi GM: Disorders of the Vestibular System. Oxford University Press, New York, 1996.

Baloh RW, Honrubia V: Clinical Neurophysiology of the Vestibular System. FA Davis, Philadelphia, 1990.

Beitz AJ, Anderson JH: Neurochemistry of the Vestibular System. CRC Press, New York, 2000.

Dickman JD, Byer M, Hess BJ: Three-dimensional organization of vestibular related eye movements to rotational motion in pigeons. Vision Res 40:2831–2844, 2000.

Goldberg JM: The vestibular end organs: Morphological and physiological diversity of afferents. Curr Opin Neurobiol 1:229–235, 1991.

Highstein SM, Cohen B, Buttner-Ennever JA: New directions in vestibular research. Ann NY Acad Sci 781:1–739, 1996.

Highstein SM, McCrea RA: The anatomy of the vestibular nuclei. In Buttner-Ennever JA (ed): Reviews of Oculomotor Research, vol 2. Neuroanatomy of the Oculomotor System. Elsevier, Amsterdam, 1988.

Hudspeth AJ: How the ear's works work. Nature 341:397–404, 1989.

Minor LB, Solomon D, Zinreich JS, Zee DS: Sound- and/or pressure-induced vertigo due to bone dehiscence of the superior semicircular canal. Arch Otolaryngol Head Neck Surg 124:249–258, 1998.

Wilson VJ, McIvill Jones G: Mammalian Vestibular Physiology. Plenum Press, New York, 1979.

Yates BJ, Miller AD: Vestibular Autonomic Regulation. CRC Press, New York, 1996.

Chapter 23

Olfaction and Taste

R. D. Sweazey

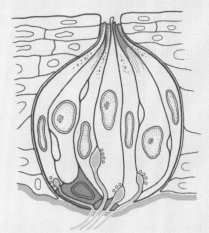

Overview   360

Olfactory Receptors   360

Olfactory Transduction   362

Central Olfactory Pathways   363
Olfactory Bulb
Olfactory Bulb Projections
Olfactory Cortex Projections

Taste Receptors   366

Distribution of Taste Receptors   367
Lingual Taste Buds
Extralingual Taste Buds

Taste Transduction   368

Peripheral Taste Pathways   369

Central Taste Pathways   370

The olfactory and taste systems sample the rich chemical environment that surrounds us. Information provided by these systems is intimately associated with the enjoyment of foods and beverages. When we refer to the taste of food, what we mean is a complex sensory experience correctly called *flavor*. Flavor perception results from a combination of the olfactory, taste, and somatosensory cues present in foods and beverages. *Olfaction* is the sensation of odors that results from the detection of odorous substances aerosolized in the environment. In contrast, taste (*gustation*) is the sensation evoked by stimulation of taste receptors located in the oropharyngeal cavity. The somatosensory system contributes to the experience of flavor by detecting irritating components in smells like ammonia or the "hot" in spicy food like peppers. In general, this refers to the activation of somatosensory endings by strong "aversive" chemical substances. Somatosensory cues include thermal, tactile, and the common chemical sense, and this information is relayed to the brain by branches of the trigeminal nerve that innervate oral and nasal mucosa.

Overview

For many mammals, smell is the principal means by which information about the environment is received. *Macrosmatic* animals have a well-developed sense of smell on which they rely for recognizing food, detecting predators and prey, and locating potential mates. In animals such as humans that are less dependent on smell (*microsmatic* animals), the olfactory system is less well developed. However, humans are still able to distinguish thousands of odors, many at extremely low concentrations. Through connections with cortical and limbic structures, the olfactory system plays a role in the pleasures associated with eating and with the many scents that make up our world.

In contrast to olfaction, the taste system exhibits a limited range of sensations. Traditionally, taste sensations have been divided into *sweet*, *salty*, *sour*, and *bitter*. In addition to these four basic tastes, a taste sensation termed *umami*, best exemplified by the taste of monosodium glutamate, may be important for identification of amino acids. Furthermore, recent evidence suggests that taste mechanisms for fats may also exist. Combinations of these different taste qualities account for much of our taste experience. Taste input, which originates from receptors in the oropharyngeal cavity, is important for determining the acceptance or rejection of foods. This information is relayed by neural pathways that underlie various ingestive and digestive functions.

Disorders of olfaction or taste may adversely affect the individual's quality of life. The intimate association between the chemical senses and ingestion means that chemosensory disorders impair the patient's ability to enjoy eating. In addition, these disorders can render the patient unable to detect hazards such as gas leaks or spoiled foods.

Olfactory Receptors

The receptors responsible for transduction of odor molecules are found in the *olfactory mucosa*. This portion of nasal mucosa is about 1 to 2 cm² in size and is located in the roof of the nasal cavity on the inferior surface of the cribriform plate and along the nasal septum and medial wall of the superior turbinate (Fig. 23–1). The olfactory mucosa is composed of a superficial acellular layer of *mucus* that covers the *olfactory epithelium* and underlying *lamina propria*. The olfactory epithelium is differentiated from the adjacent pinkish respiratory epithelium by its faint yellowish color and greater thickness. In humans the transition between olfactory and respiratory epithelia is gradual.

The olfactory epithelium is pseudostratified and contains three main cell types: *olfactory receptor neurons, supporting cells (sustentacular cells), and basal cells* (Fig. 23–2A and B). The small (5 μm) somata of bipolar olfactory receptor neurons are found in the basal two thirds of the epithelium. Each has a single thin apical dendrite and a basally located unmyelinated axon. The apical dendrite extends to the surface of the epithelium, where it terminates in a knoblike *olfactory vesicle* from which 10 to 30 nonmotile *cilia* arise and protrude into the overlying mucus layer (see Fig. 23–2C and D). These olfactory cilia contain receptors for odorant molecules.

The unmyelinated axon of an olfactory receptor neuron is about 0.2 μm in diameter, making it one of the smallest in the nervous system. These axons pass through the lamina propria and group together into bundles called *olfactory fila*, which collectively make up the *olfactory nerve* (cranial nerve I) (see Fig. 23–2A).

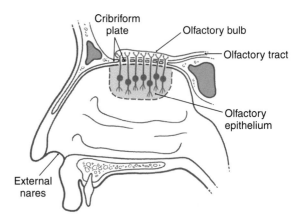

Figure 23–1. Sagittal section through the human nasal cavity showing the relationship of the olfactory epithelium and bulb to the cribriform plate.

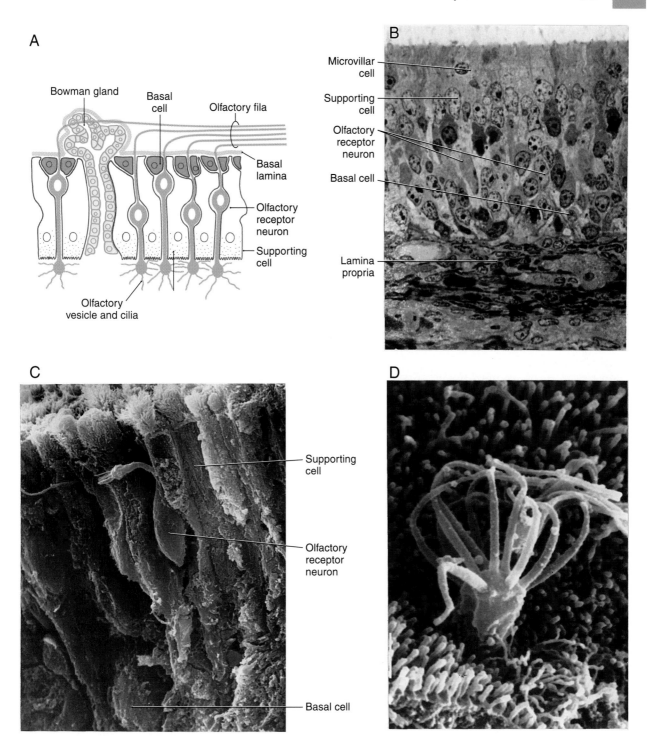

Figure 23–2. Schematic drawing of the olfactory epithelium (*A*). Light micrograph of the human olfactory epithelium and the underlying lamina propria (*B*). Scanning electron micrographs of the human olfactory epithelium showing its characteristic cell types (*C*) and the dendritic knob and cilia of a receptor neuron (*D*). (Photomicrographs courtesy of Dr. Richard M. Costanzo, Virginia Commonwealth University.)

The olfactory fila pass through the cribriform plate to terminate in the olfactory bulb.

Olfactory receptor cells are true neurons because they originate embryologically from the central nervous system. Olfactory receptor cells undergo continuous turnover, with an average life span between 30 and 60 days. They are replaced by receptors arising from undifferentiated basal cells by mitotic division (see Fig. 23–2*A* and *C*). Thus, basal cells are stem cells that give rise to the receptor cells.

The supporting cells are columnar and extend from the lamina propria to the surface of the epithelium, where they end in short microvilli that extend into the overlying mucus (see Fig. 23–2*A*, *C*, and *D*). Nuclei of the sustentacular cells are found near the surface of the epithelium. These cells provide mechanical support for the olfactory receptor cells (see Fig. 23–2*B*). In addition, they contribute secretions to the overlying mucus that may play a role in the binding or inactivation of odorant molecules.

A fourth and minor cell type, the *microvillar cell*, is found in the human olfactory epithelium (see Fig. 23–2*B*). These cells have an apical process that projects into the mucus and a basal process that extends to the lamina propria. Although their function is unknown, they may be a second type of receptor neuron.

The lamina propria contains bundles of olfactory axons, blood vessels, fibrous tissue, and numerous *Bowman glands* (see Fig. 23–2*A*). The serous secretions of the Bowman glands, combined with the secretions of the sustentacular cells, provide the mucus covering of the olfactory mucosa.

Disorders of smell are normally classified according to the type of loss experienced by the patient. The loss of smell (*anosmia*) or decreased sensitivity to odorants (*hyposmia*) is frequently associated with upper respiratory infections, sinus disease, and head trauma. Nasal and paranasal diseases (*rhinitis, sinusitis*) may block the access of odorants to the olfactory epithelium, but blockage is not the sole cause because even when blockage is absent, olfactory dysfunction may be present. Frequently, treatment with systemic anti-inflammatory drugs results in rapid attenuation of symptoms, suggesting that edema of the olfactory epithelium may be partly responsible for the olfactory dysfunction. In addition, the viruses associated with upper respiratory infections may permanently damage the olfactory epithelium. Although the process is not completely understood, it is thought that viral infection of an olfactory receptor cell may lead to cell death. Head trauma can produce olfactory deficits by damaging central olfactory pathways or olfactory receptor axons as they pass through the cribriform plate. In the case of head trauma, shearing movements of the olfactory bulb relative to the cribriform plate transect these thin axons. For example, anosmia or hyposmia is common in boxers.

Olfactory Transduction

Olfactory perception begins when volatile odor molecules are inhaled and contact the mucus layer that bathes the olfactory epithelium. This mucus is an aqueous solution of proteins and electrolytes. Odorants, particularly hydrophobic ones such as musk, cross the mucus by interacting with small, water-soluble proteins called *odorant-binding proteins*. These proteins are ubiquitous in the mucus layer.

After crossing the mucus, odor molecules bind to odorant receptors on the cilia of the olfactory receptor neurons, where transduction occurs (Fig. 23–3). The *odorant receptors* are membrane proteins belonging to a superfamily of G protein–coupled receptors. Binding of the odorant to the receptor leads to activation of a second-messenger pathway involving an *olfactory-specific G protein*, which, in turn, activates adenylyl cyclase to produce cyclic AMP (cAMP). The transient rise in ciliary cAMP opens a cyclic nucleotide–gated cation channel in the ciliary membrane, allowing cations to flow into the cell (see Fig. 23–3). The flow of cations into the cell results in a gradual depolarization (*generator potential*) that travels down the dendrite to the soma of the olfactory receptor neuron. A sufficiently large depolarization initiates an action potential that travels along the axon to the olfactory bulb.

There is also controversial evidence for another intracellular second-messenger pathway in olfactory transduction. This pathway, involving inositol 1,4,5-

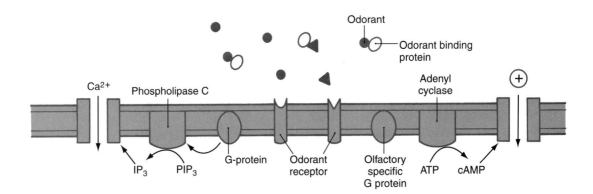

Figure 23–3. Pathways of olfaction transduction. Odorants are transported through the mucus by odorant-binding proteins. Binding of odorants to receptors on the olfactory cilia activates a second-messenger pathway involving either cyclic AMP (cAMP) or inositol 1,4,5-trisphosphate (IP_3). Both pathways lead to the opening of membrane cation channels and depolarization of the olfactory receptor neuron. PIP_3, phosphatidylinositol 4,5-bisphosphate.

trisphosphate (IP_3), is thought to act either separately or with the cAMP pathway. In this pathway, binding of odorant to the receptor activates a G protein that, in turn, activates phospholipase C to produce IP_3. The IP_3 opens a channel in the ciliary membrane that permits Ca^{2+} to enter the cell (see Fig. 23–3). Although some evidence exists for IP_3 transduction pathways, recent studies using cyclic nucleotide–gated ion channel knockout mice found that responses to odors using cAMP-mediated ion channels and odors thought to use IP_3-mediated channels were abolished in these animals. This finding suggests that the cAMP transduction mechanism may be the sole excitatory transduction mechanism in olfactory receptor cells.

Central Olfactory Pathways

Olfactory Bulb. The olfactory bulb, a forebrain structure, is located on the ventral surface of the frontal lobe in the olfactory sulcus and is attached to the rest of the brain by the *olfactory tract*. The olfactory tract is an inclusive structure that contains fibers of the *lateral olfactory tract*, cells of the *anterior olfactory nucleus*, and fibers of the *anterior limb of the anterior commissure*. The latter part of the olfactory tract is the route through which many centrifugal fibers reach the olfactory bulb (see Fig. 23–6).

The olfactory bulb consists of five well-defined layers of cells and fibers, which give it a laminated appearance. From superficial to deep these are the *olfactory nerve layer, glomerular layer, external plexiform layer, mitral cell layer,* and *granule cell layer* (Fig. 23–4).

The afferent projections from the olfactory epithelium form the *olfactory nerve layer* on the surface of the olfactory bulb. These axons terminate exclusively in structures called *olfactory glomeruli*, which are found in the glomerular layer of the bulb (Figs. 23–4 and 23–5). Their terminations, in general, preserve the topographic arrangements of the receptors.

Glomeruli are the most prominent feature of the olfactory bulb. The core of an olfactory glomerulus is made up of the axons of olfactory receptor neurons, which branch and synapse on the bushy endings of the *primary dendrites* (apical dendrites) of *mitral* and *tufted* cells (see Fig. 23–4). These two cells are functionally similar and together constitute the efferent neurons of the olfactory bulb. Adjacent to the glomerulus are small interneurons (juxtaglomerular cells), of which *periglomerular cells* are the principal type. This cell has short bushy dendrites that arborize extensively within a glomerulus and a short axon that distributes within a radius of about five glomeruli.

There is significant neural convergence at the level of the olfactory glomerulus; thousands of olfactory receptor neurons form excitatory (*glutaminergic, carnosine*) axodendritic synapses on mitral, tufted, and periglomerular cells (see Fig. 23–5). The other major synaptic connections within the glomerulus are reciprocal and serial dendrodendritic synapses between mitral or tufted cells and periglomerular cells. It appears that the synapses of mitral and tufted cells onto periglomerular cells are excitatory (*glutaminergic*), whereas those of periglomerular cells onto mitral and tufted cells are inhibitory (*GABAergic*).

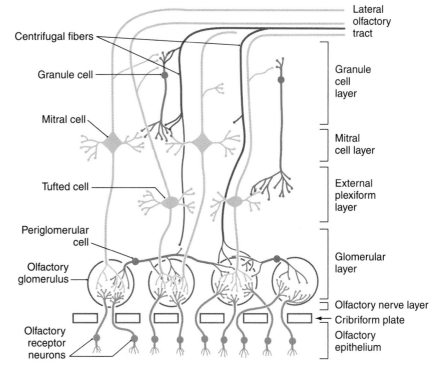

Figure 23–4. Schematic drawing of the olfactory bulb, showing the laminar organization, the major cell types, and the basic neuronal circuitry. Receptor neurons are shown in blue, interneurons in red, the efferent neurons of the bulb in green, and centrifugal fibers in black.

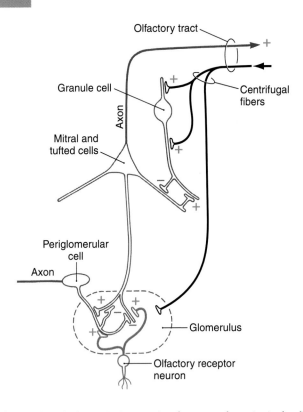

Figure 23–5. Synaptic interaction between the principal cell types of the olfactory epithelium and bulb. Excitatory synapses (+) are shown in green and inhibitory ones (−) in red. The action of centrifugal axons in the glomerulus is not clearly established.

ids, such as *glutamate*, are present in centrifugal fibers that arise in cortical structures.

The *external plexiform layer* is composed of the somata of tufted cells, along with the *primary* and *secondary dendrites* (basal dendrites) of tufted and mitral cells and the apical dendrites of *granule cells* (see Fig. 23–4). Within this layer, the apical dendrites of granule cells form reciprocal dendrodendritic *GABAergic* synapses with the secondary dendrites of tufted and mitral cells. These synapses modulate tufted and mitral cell output through lateral and feedback inhibition. Mitral and tufted cells, in turn, have excitatory (*glutaminergic*) synapses on granule cell dendrites (see Fig. 23–5).

The *mitral cell layer* is a thin layer containing the large somata of mitral cells. In addition, the axons of tufted cells, granule cell processes, and centrifugal fibers traverse this layer (see Fig. 23–4).

Internal to the mitral cell layer, the *granular cell layer* contains the cell bodies of *granule cells*, the principal interneuron of the olfactory bulb. This layer also contains primary and collateral axons of mitral and tufted cells and centrifugal afferents from the anterior olfactory nucleus, olfactory cortex, cells of the diagonal band, locus ceruleus, and raphe nucleus (see Fig. 23–4). Granule cells lack axons, their only output being via dendrodendritic GABAergic synapses with mitral and tufted cells. In addition, granule cells receive numerous synaptic inputs from both mitral and tufted cell axon collaterals and centrifugal afferent fibers. Granule cells presumably modulate olfactory bulb activity via an inhibitory feedback loop that shuts down the activity of the mitral and tufted neurons.

Olfactory Bulb Projections. Axons of mitral and tufted cells emerge from the caudal portion of the olfactory bulb to form the *lateral olfactory tract*. Although *glutamate* is the major neurotransmitter of these efferent fibers, *aspartate, dopamine,* and *substance P* also are used. These fibers course caudally to terminate in areas on the ventral surface of the telencephalon, which are broadly defined as the *olfactory cortex* (Fig. 23–6). The principal areas making up the olfactory

The glomerular layer also receives input from other central nervous system areas via *centrifugal afferents* that use a wide variety of neurotransmitters and neuromodulators (see Figs. 23–4 and 23–5). *Noradrenergic* centrifugal afferents from the *locus ceruleus* and *serotonergic* fibers from the *raphe nuclei* of the midbrain and rostral pons terminate in the glomeruli. Centrifugal fibers from the ipsilateral *anterior olfactory nucleus* and the *diagonal band* terminate in the periglomerular spaces, primarily on periglomerular cells. Excitatory amino ac-

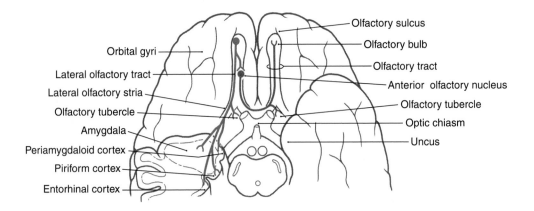

Figure 23–6. Major efferent projections of the olfactory bulb.

cortex are the *anterior olfactory nucleus, olfactory tubercle, piriform cortex, anterior cortical amygdaloid nucleus, periamygdaloid cortex,* and *lateral entorhinal cortex.* The olfactory cortex is an example of *paleocortex,* a phylogenetically older type of cortex that is less complex than the neocortex. Throughout most of its extent, olfactory cortex has three cell layers, as opposed to the six layers characteristic of neocortex. A unique aspect of the olfactory system is that the olfactory bulb projects directly to the cortex. In other sensory systems, information reaches the cortex after a relay in the thalamus.

Lateral olfactory tract axons send collaterals to the anterior olfactory nucleus, to other areas of olfactory cortex, and to subcortical limbic structures. The major targets of the anterior olfactory nucleus are the olfactory bulbs bilaterally and the contralateral anterior olfactory nucleus (Fig. 23–7; see also Fig. 23–6). The large numbers of interbulbar connections, via the anterior olfactory nucleus, suggest that interhemispheric processing of odors plays an important role in olfactory functions.

Axons of the lateral olfactory tract course caudally in the form of the *lateral olfactory stria* to terminate in the *olfactory tubercle* and the *piriform cortex* (see Fig. 23–6). The piriform cortex is a major component of the olfactory cortex. Fibers of the lateral olfactory tract also continue posteriorly to terminate in the *anterior cortical amygdaloid nucleus,* the *periamygdaloid cortex* (a part of the piriform cortex overlying the amygdala), and the *lateral entorhinal cortex.*

There is little evidence of a topographic projection from the olfactory bulb onto the various structures constituting the olfactory cortex. Mitral cells project to all areas of the olfactory cortex, whereas tufted cells terminate primarily in its anterior parts. However, each region of olfactory cortex is regarded as receiving input from all areas of the olfactory bulb.

Olfactory Cortex Projections. Cells of the olfactory cortex have reciprocal connections with other regions of the olfactory cortex (*intrinsic* or *associational connections*) and connections with regions outside the olfactory cortex (*extrinsic connections*) (see Fig. 23–7). Most intrinsic connections arise from the anterior olfactory nucleus, piriform cortex, and lateral entorhinal cortex. As a group, these associational fibers distribute to all areas of the olfactory cortex (see Fig. 23–7).

Extrinsic connections include extensive projections back to the olfactory bulb. These centrifugal fibers originate from most parts of the olfactory cortex, with the exception of the olfactory tubercle. As in other sensory systems, olfactory information is also relayed to the neocortex. This connection occurs through a direct projection from olfactory cortex to *orbitofrontal* and *ventral agranular insular cortices* or via a relay in the thalamus. The latter pathway originates from cells in the olfactory cortex that project to the *dorsomedial nucleus of the thalamus* (see Fig. 23–7). This neocortical representation of olfaction is important for discrimination and identification of odors, and lesions in this area, particularly the orbitofrontal cortex, result in the loss of these capabilities. Another noteworthy point is that the insular and orbitofrontal cortex also receive taste input. The medial orbitofrontal cortex appears to play an especially important role in integrating olfactory, taste, and other food-related cues that produce the experience of flavor.

In addition to neocortical projections, the olfactory cortex also sends fibers directly to the lateral hypothal-

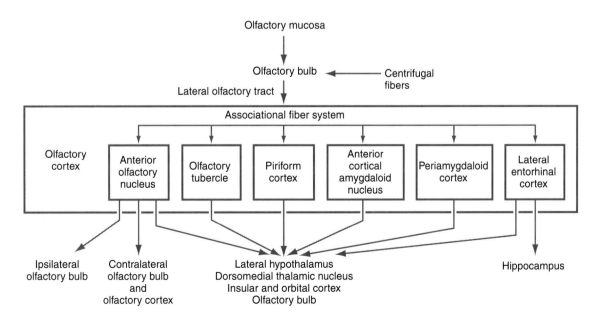

Figure 23–7. Major projections of the olfactory cortex.

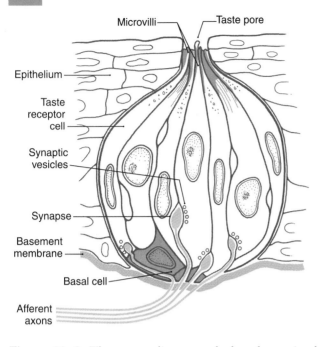

Figure 23–8. The mammalian taste bud and associated structures. (Adapted from Mistretta, 1989, with permission.)

Taste Receptors

The taste experiences of sweet, salty, sour, bitter, and umami result from an interaction between gustatory stimuli and receptor cells located in sensory organs called *taste buds*. Although found throughout the oropharyngeal cavity, taste buds are most obvious on the tongue, where they are ovoid structures with a constriction at their apical end.

Each taste bud contains 40 to 60 *taste receptor cells* that extend from a basal lamina to the surface of the epithelium. The apical ends of these receptor cells are covered with microvilli of variable lengths that extend into a *taste pore*. The pore forms a pocket to permit contact between the microvilli of the taste receptor cells and the external milieu (Figs. 23–8 and 23–9). Numerous junctional complexes located between the apices of receptor cells restrict access of stimuli to the microvilli where taste transduction occurs. The taste pore is filled with a protein-rich substance through which substances must pass to reach the receptor cell microvilli. Taste receptor cells undergo a continuous process of turnover, having a life span of 10 to 14 days. New taste cells are thought to arise from polygo-

amus and hippocampus. Those to the lateral hypothalamus arise primarily from the piriform cortex and anterior olfactory nucleus and are probably important for feeding behavior. The projection to the hippocampus arises from the entorhinal cortex and links olfactory input to centers concerned with learning and behavior.

The chief complaint of most patients with chemosensory disturbances is the loss or alteration of taste. However, clinical studies reveal that in all but a small number of patients, the dysfunction actually resides in the olfactory system. The reason for this discrepancy is that most people confuse taste with flavor.

Aged-related declines in olfactory function are quite common even in healthy persons. Typically, the loss occurs gradually, and the patient often fails to notice these changes. These gradual changes can affect the palatability of foods in the elderly. Olfactory dysfunction is also encountered in neurodegenerative diseases such as Alzheimer and Parkinson disease, or in Huntington chorea. These neurodegenerative diseases involve central olfactory pathways and produce marked reduction in the affected person's olfactory capabilities. Most notably, these olfactory deficits appear very early in the course of the disease and may be among its first manifestations.

Disorders of olfaction are also associated with epilepsy and various depressive and psychiatric disorders such as schizophrenia and Korsakoff psychosis. The patients frequently experience *parosmia* (dysosmia), a distortion in a smell experience or the perception of a smell when no odor is present.

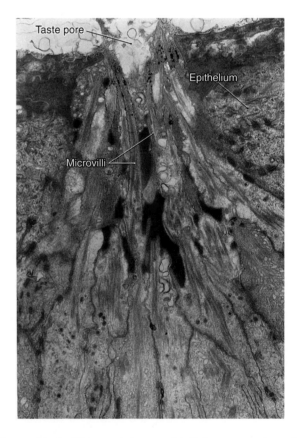

Figure 23–9. Electron micrograph of the pore region of a mouse taste bud in the circumvallate papilla. (Courtesy of Drs. F. Kinnamon and H. Linnen, University of Denver.)

nal *basal cells* located in basolateral areas of the taste bud. These cells are not involved in taste transduction.

Afferent fibers penetrate the basement membrane and then branch within the base of the taste bud (see Fig. 23–8). Each taste bud is typically innervated by more than one afferent fiber, and an individual fiber may innervate multiple taste buds. The taste afferent fibers form the postsynaptic element of a chemical synapse near the base of the taste receptor cell.

Distribution of Taste Receptors

Lingual Taste Buds. Taste buds are found, in variable numbers, on the human tongue, palate, pharynx, and larynx. On the tongue, taste buds are located exclusively in specialized structures called *papillae*, of which there are three types (Fig. 23–10A). Taste buds on the anterior two thirds of the tongue reside in mushroom-shaped *fungiform papillae* (see Fig. 23–10B).

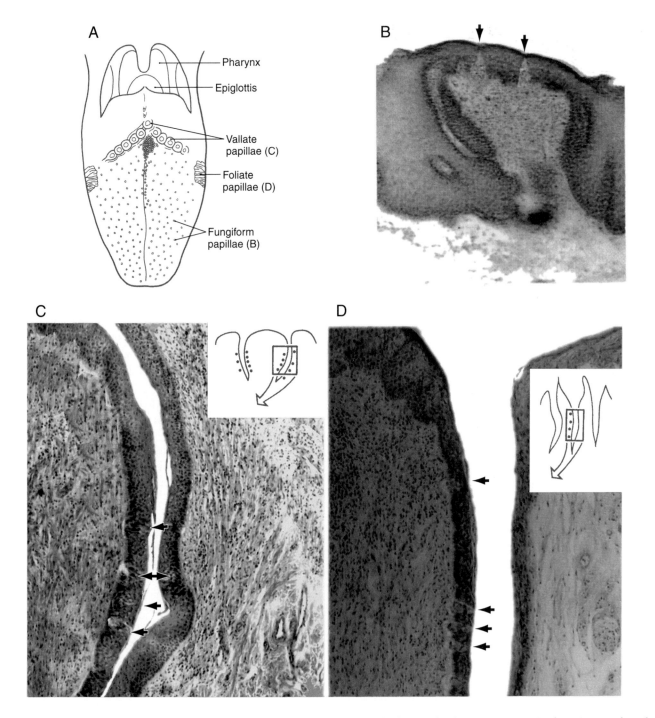

Figure 23–10. Distribution of papillae (*A*), and their associated taste buds, on the human tongue. Light micrographs of transverse sections through human fungiform (*B*), vallate (*C*), and foliate (*D*) papilla. Examples of taste buds are shown at the *arrows*. (Photomicrographs courtesy of Dr. I.J. Miller, Bowman Gray School of Medicine.)

These structures are scattered among the more numerous nongustatory *filiform papillae* distributed over the surface of the tongue. The size, shape, and number of fungiform papillae vary widely, and usually 2 to 4 taste buds are found in the dorsal epithelium of each. The *vallate (circumvallate) papillae* are located on the dorsal surface of the tongue at the junction of the oral and pharyngeal cavities (see Fig. 23–10A). There are from 8 to 12 vallate papillae, each of which is composed of a central papilla surrounded by a cleft containing taste buds in its epithelium (see Fig. 23–10C). A single *foliate papilla* on each side of the tongue appears as a series of clefts along the lateral margin of the tongue (see Fig. 23–10A). Each is composed of 2 to 9 clefts, 5 being the most common number. Taste buds in foliate papillae are also located in the epithelium that lines the clefts (see Fig. 23–10D).

Associated with both the vallate and foliate papillae are the *von Ebner* lingual salivary glands. These glands drain into the base of the clefts and influence their microenvironment. Taste stimulation of vallate and foliate papillae influences the secretions of the von Ebner glands via circuits located in the brainstem.

It was once believed that different regions of the tongue were specialized for the detection of particular taste qualities. It is now known that all taste qualities are detected in all regions of the tongue, although sensitivity to the different taste qualities and taste transduction mechanisms may vary by tongue region.

Extralingual Taste Buds. Additional taste buds are located on the human soft palate, oral and laryngeal pharynx, larynx, and upper esophagus. Extralingual taste buds are not located in papillae but rather are situated in the epithelium. Palatal taste buds are located at the juncture of the hard and soft palates and on the soft palate. Laryngeal taste buds are found on the laryngeal surface of the epiglottis and adjacent aryepiglottal folds. The number of extralingual taste buds is substantial, and they may contribute to the taste experience. Stimulation of some extralingual taste buds, particularly those near the larynx, elicits brainstem-mediated reflexes that prevent accidental aspiration of ingested materials.

Taste Transduction

As research has begun to clarify the mechanisms underlying taste transduction, it has become clear that individual taste qualities or even individual taste compounds utilize several transduction mechanisms. In general, taste transduction is initiated when soluble chemicals diffuse through the contents of the taste pore and interact with receptors located on the exposed apical microvilli of the receptor cells. Several different receptor types have recently been cloned, including a novel variant of the metabotropic glutamate receptor that functions as an umami taste receptor and G protein–coupled receptors that function as bitter receptors. The interaction of the chemical stimulus with the taste cell receptor results in either depolarization or hyperpolarization of the receptor cell microvilli. It is now generally accepted that depolarizing potentials of sufficient magnitude result in action potential generation within the taste receptor cells, which in turn produce an increase in intracellular Ca^{2+}, either by the release of Ca^{2+} from internal stores or by the activation of voltage-gated calcium channels located in the basolateral membrane of the taste receptor cells (Fig. 23–11). This Ca^{2+} release results in a release of chemical transmitters at the afferent synapse, which, in turn, leads to an action potential in the afferent fiber. Although several neurotransmitter candidates have been put forward, the transmitter at the synapses between taste receptor cells and primary afferent fibers remains unknown.

Transduction of stimuli leading to salty and perhaps some sour and bitter tastes appears to be the result of a direct interaction of these tastants with specific ion channels located in the apical membrane of the taste receptor cells. One mechanism for the transduction of sodium salts such as sodium chloride involves movement of Na^+ into the taste receptor cell through apically located amiloride-sensitive cation channels. Similar mechanisms have been proposed for K^+ salts (see Fig. 23–11). One pathway responsible for transduction of some sour and bitter stimuli is blockage of apical voltage-sensitive potassium channels. At the resting potential, there is a small outward K^+ current through the apical membranes of taste receptor cells. Protons provided by sour stimuli such as hydrochloric acid block this outward current, causing the cell to depolarize.

At least some sweet, sour, and bitter-tasting compounds are transduced by receptors that activate intracellular G protein–mediated second-messenger pathways (see Fig. 23–11). Binding of sweet-tasting compounds such as sucrose to apically located receptors stimulates an adenylyl cyclase–cAMP second-messenger pathway that closes basolateral potassium channels, leading to depolarization of the taste receptor cell. Additional mechanisms for sweet transduction have been proposed, and there is good evidence for multiple sweet receptor types. Some bitter compounds are thought to act through a different second-messenger pathway (IP_3) that releases Ca^{2+} from intracellular stores.

Evidence exists for several possible transduction pathways for amino acids. One pathway involves binding of amino acids by receptors that are directly coupled to cation channels having properties similar to those of the nicotinic acetylcholine receptor. Another receptor activates a G protein–dependent increase in

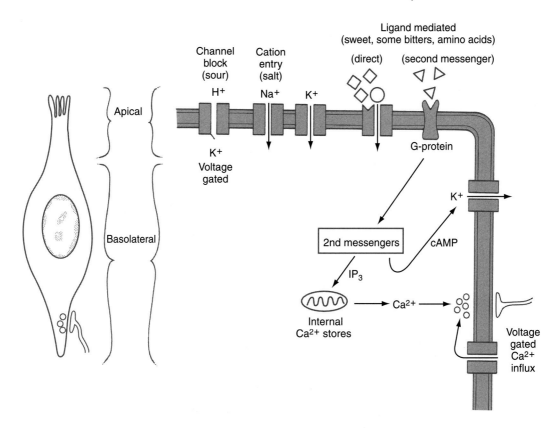

Figure 23–11. Some pathways of transduction in taste receptors. Some sour and bitter substances are transduced by the closing of apical voltage-sensitive potassium channels. The transduction of salts such as sodium chloride, involves the movement of ions (such as Na^+ and K^+) through amiloride-sensitive cation channels in the apical membrane. Sweet and some bitter compounds are thought to activate intracellular second-messenger pathways (cyclic AMP [cAMP], inositol 1,4,5-trisphosphate [IP_3]), which leads to activation of membrane channels in the basolateral membrane of the taste cell. (Adapted from Kinnamon, 1988, with permission.)

the second messengers cAMP and IP_3. For umami transduction, it is hypothesized that a glutamate receptor activates a G protein that stimulates phosphodiesterase, causing a reduction in intracellular cAMP and subsequent changes in receptor cell activity.

Peripheral Taste Pathways

The afferent fibers of first-order taste neurons (special visceral afferent [SVA]) innervating oropharyngeal taste buds travel in the *facial* (VII), *glossopharyngeal* (IX), and *vagus* (X) nerves (Fig. 23–12). The *chorda tympani* branch of the facial nerve innervates taste buds in the fungiform papillae on the anterior two thirds of the tongue and in the most anterior clefts of the foliate papillae. The *greater superficial petrosal nerve*, also a branch of the facial nerve, innervates taste buds on the soft palate. The cell bodies of facial nerve fibers subserving taste are located in the *geniculate ganglion*, and their central processes enter the brainstem at the pontomedullary junction in the *intermediate nerve*, which is actually a part of the facial nerve. These primary afferent taste fibers enter the *solitary tract*, travel caudally,

and terminate on cells of the surrounding *solitary nucleus* (see Fig. 23–12).

The chorda tympani branch leaves the facial nerve just distal to the geniculate ganglion. Consequently, lesions of the root of the seventh cranial nerve, or tumors in the internal auditory meatus (such as an acoustic neuroma), will result in loss of taste perception from the anterior two thirds of the tongue on the ipsilateral side. Accompanying this deficit are paralysis of the ipsilateral facial muscles, *hyperacusis* (paralysis of the stapedius muscle), and impaired secretion of the nasal and lacrimal glands and of submandibular and sublingual salivary glands. Damage just distal to the geniculate ganglion may or may not result in taste loss, depending on the origin of chorda tympani branches, but an ipsilateral facial paralysis will be seen.

Taste buds located in the vallate papillae and posterior clefts of the foliate papillae are innervated by the *lingual-tonsillar branch* of the glossopharyngeal nerve (cranial nerve IX). Those located on the epiglottis and esophagus are innervated by the *superior laryngeal nerve*, a branch of the vagus nerve (cranial nerve X) (see Fig. 23–12). Taste fibers in cranial

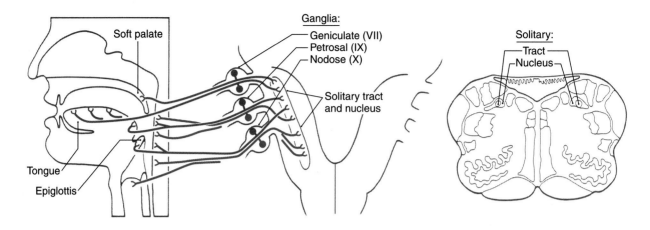

Figure 23–12. Summary of peripheral taste pathways. Special visceral afferent fibers for taste (in red) terminate in the rostral (gustatory) areas of the solitary nucleus, whereas general visceral afferent fibers (in blue) terminate in the caudal portion of the nucleus.

nerves IX and X have their cell bodies of origin in the *inferior ganglia* (*petrosal* and *nodose*, respectively) of these cranial nerves (see Fig. 23–12). The central processes of these fibers, like those of the facial nerve, enter the medulla, descend in the solitary tract, and terminate on neurons in the adjacent solitary nucleus (see Fig. 23–12).

Central Taste Pathways

The solitary nucleus is the principal visceral afferent nucleus of the brainstem. On the basis of functional characteristics, it is divided into a *rostral (gustatory) nucleus* and a *caudal (visceral* or *cardiorespiratory) nucleus.* Taste fibers traveling in cranial nerves VII, IX, and X

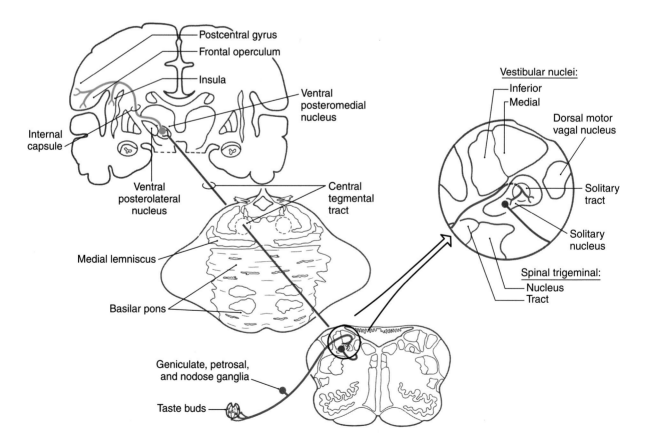

Figure 23–13. The ascending taste pathway to thalamus and cortex.

terminate primarily in the rostral portions of the solitary nucleus because this region contains most of the second-order neurons in the taste pathway (see Fig. 23–12). There is considerable overlap in the distribution of the terminals of the primary taste cranial afferents from these three nerves in the gustatory nucleus. General visceral afferent fibers of the vagus, and those that travel in the glossopharyngeal nerve, terminate in the caudal part of the solitary nucleus (see Fig. 23–12). These visceral fibers are involved in the central control of respiration, cardiac function, and certain aspects of swallowing.

Axons arising from second-order taste neurons in the gustatory nucleus ascend in association with the ipsilateral *central tegmental tract* and terminate in the *parvicellular division of the ventral posteromedial nucleus* of the thalamus (*VPMpc*), medial to the head representation (Fig. 23–13). Axons from these neurons in the VPMpc travel through the ipsilateral posterior limb of the internal capsule to terminate in the inner portion of the *frontal operculum* and *anterior insular cortex*, and in the rostral extension of *Brodmann area 3b* on the lateral convexity of the postcentral gyrus (see Fig. 23–13). This pathway (solitary nucleus → VPMpc → cortex) is responsible for the discriminative aspects of taste and, in contrast to other sensory pathways, is exclusively ipsilateral.

Physiologic studies in primates indicate that there is an additional region of the cortex that processes taste information. The *lateral posterior orbitofrontal cortex* receives inputs from primary taste cortex and acts as a site of integration for taste, olfactory, and visual cues associated with the ingestion of foods. Recent data suggest that cells in this area are involved in the appreciation of flavor, food reward, and the control of feeding. Taste-responsive cells have also been found in the primate amygdala and hypothalamus. These cells do not respond exclusively to taste, and their connecting pathways and role in taste-mediated behavior are not fully understood. Taste information is also relayed from cells in the solitary nucleus into medullary reflex connections that influence salivary secretion, mimetic responses, and swallowing.

The complete loss of taste (*ageusia*) is rarely encountered, in part because of the large numbers of nerves that relay taste information to the central nervous system. Rarely does a patient suffer from bilateral injury to all the nerves innervating the oropharyngeal region. More frequently a patient suffers from *hypogeusia*, decreased taste sensitivity, or *parageusia* (*dysgeusia*), distortions in the perception of a taste. One of the most frequent causes of taste disturbances is drug usage. A wide array of medications can affect taste function, the most frequent complaint being *dysgeusia*. Like olfactory disorders, taste disorders are associated with head trauma, viral infections, and various psychiatric disorders. Taste changes are one of the most frequent complaints of cancer patients undergoing radiation therapy and chemotherapy, and there is a progressive taste loss in diabetic patients.

Sources and Additional Reading

Anholt RRH: Molecular neurobiology of olfaction. Crit Rev Neurobiol 7:1–22, 1993.

Cowart BJ, Young IM, Feldman RS, Lowry LD: Clinical disorders of smell and taste. Occupational Med 12:465–483, 1997.

Getchell TV, Doty RL, Bartoshuk LM, Snow JB (eds): Smell and Taste in Health and Disease. Raven Press, New York, 1991.

Gold GH: Controversial issues in vertebrate olfactory transduction. Annu Rev Physiol 61:857–871, 1999.

Herness MS, Gilbertson TA: Cellular mechanisms of taste transduction. Annu Rev Physiol 61:873–900, 1999.

Kinnamon SC: Taste transduction: A diversity of mechanisms. Trends Neurosci 10:491–496, 1988.

Kinnamon SC: A plethora of taste receptors. Neuron 25:507–510, 2000.

Mistretta CM: Anatomy and neurophysiology of the taste system in aged animals. In Murphy C, Cain WS, Hegsted DM (eds): Nutrition and the Chemical Senses in Aging: Recent Advances and Current Research Needs. Ann NY Acad Sci 561:277–290, 1989.

Mori K: Membrane and synaptic properties of identified neurons in the olfactory bulb. Prog Neurobiol 29:275–320, 1987.

Mori K, Nagao H, Yoshihara Y: The olfactory bulb: Coding and processing of odor molecule information. Science 286:711–715, 1999.

Morrison EE, Costanzo RM: Morphology of the human olfactory epithelium. J Comp Neurol 297:1–13, 1990.

Norgren R: Gustatory system. In Paxinos G (ed): The Human Nervous System. Academic Press, San Diego, 1990, pp 845–861.

Scott JW, Wellis DP, Riggott MJ, Buonviso N: Functional organization of the main olfactory bulb. Microsc Res Tech 24:142–156, 1993.

Speielman AI: Chemosensory function and dysfunction. Crit Rev Oral Biol Med 9:267–291, 1998.

Zald DH, Pardo JV: Functional neuroimaging of the olfactory system in humans. Int J Psychophysiol 36:165–181, 2000.

Chapter 24

Motor System I: Peripheral Sensory, Brainstem, and Spinal Influence on Anterior Horn Neurons

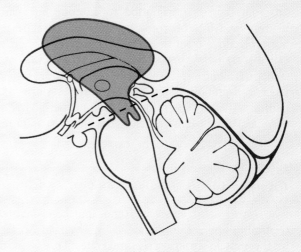

G. A. Mihailoff and D. E. Haines

Overview 374

Anterior (Ventral) Horn Motor Neurons 374
Types and Distribution
Neuromuscular Function
Motor Units
Size Principle

Peripheral Sensory Input to the Anterior (Ventral) Horn 376
Muscle Spindles
The Gamma Loop
Golgi Tendon Organ
Reflex Circuits

Brainstem and Spinal Systems: Anatomy and Function 379
Vestibulospinal Tracts
Reticulospinal Tracts
Rubrospinal Tract

Functional Role of Brainstem and Spinal Interactions 383
Decerebration
Posterior (Dorsal) Root Section
Cerebellar Anterior Lobe Section
Decortication

Spinal anterior (ventral) horn motor neurons whose axons innervate skeletal muscles are called *lower motor neurons*. These cells stimulate muscles to produce characteristic movements of a body part. The activity of these motor neurons is influenced from two sources. First, *peripheral sensory input* arrives via posterior (dorsal) roots and is transmitted to anterior horn motor neurons and interneurons. Second, extensive descending projections from the cerebral cortex and brainstem, called *supraspinal systems*, terminate at all levels of the spinal cord and are responsible for a mixture of excitatory and inhibitory effects on anterior horn motor neurons. This chapter focuses on the peripheral sensory and brainstem systems that influence anterior horn neurons.

Overview

The *lower motor neurons* of the spinal cord anterior (ventral) horn are topographically arranged according to the muscle groups they innervate. This is particularly evident in the cervical and lumbosacral enlargements, the levels of the spinal cord that innervate the musculature of the upper and lower limbs, respectively. Motor neurons that supply flexor muscles generally are more posteriorly located in the anterior horn than are extensor motor neurons. In addition, motor neurons that innervate paravertebral and proximal limb muscles are most medial, whereas those that innervate distal musculature are most lateral (Fig. 24–1). The anterior horn motor neurons receive sensory feedback from the muscles they control, as well as from synergist and antagonist muscles. The linkage of peripheral sensory input and anterior horn neurons forms the substrate for a number of spinal reflexes (see Figs. 9–8 to 9–10).

In addition to sensory feedback, the activity of lower motor neurons in the spinal cord is greatly influenced by descending projections from cells in the brainstem and cerebral cortex. These brainstem and cortical neurons are referred to as *upper motor neurons*, and, unlike lower motor neurons, they have no direct synaptic link with muscles. Because of their origin, these descending projections are also called *supraspinal systems*.

Anterior horn motor neurons represent the only direct link (the *final common path*) between the nervous system and skeletal muscle. As such, these neurons play a central role in the production of movement. The regulation of motor neuron activity by peripheral sensory input and descending brainstem influences is crucial to the performance of normal movement.

Anterior (Ventral) Horn Motor Neurons

Types and Distribution. There are two varieties of anterior (ventral) horn motor neurons, alpha and gamma, which are mingled in the anterior horn. *Alpha motor neurons* innervate the ordinary, working fibers of skeletal muscles called *extrafusal fibers*, and *gamma motor neurons* innervate a special type of skeletal muscle fiber, the *intrafusal fibers*, which are found only within *muscle spindles*. Recall that the anterior horn also contains small *interneurons* whose axons distribute locally within the spinal gray. Interneurons are numerous in the intermediate zone and anterior horn and are functionally quite essential in the regulation of alpha and gamma motor neurons. Their action on motor neurons may either be excitatory or inhibitory.

The axons of anterior horn motor neurons exit the spinal cord via the anterior roots and course distally in peripheral nerves. These fibers represent the *final common path* that links the nervous system and skeletal muscles. As the motor axon reaches the muscle it innervates, it loses its myelin sheath and forms a series of flattened boutons that indent the surface of a group of muscle fibers. This specialized type of synapse is called a *neuromuscular junction* or *motor end plate* (Fig. 24–2).

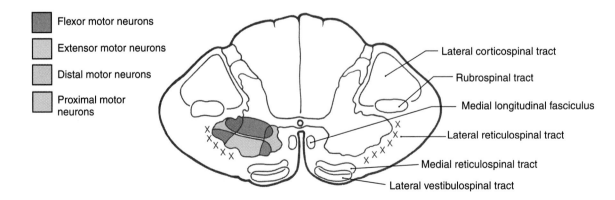

Figure 24–1. The locations of vestibulospinal and reticulospinal tracts at a representative cervical level of the spinal cord. Medial vestibulospinal fibers are located in the medial longitudinal fasciculus. The general positions of motor neuron pools are shown on the left.

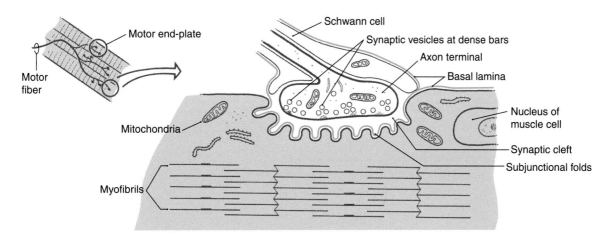

Figure 24–2. The structural elements characteristic of a motor end plate (neuromuscular junction).

Neuromuscular Function. Like synapses in the central nervous system, the junction between a motor axon and skeletal muscle fibers consists of presynaptic and postsynaptic components (see Fig. 24–2). The *presynaptic element*, the axon terminal, contains round, clear synaptic vesicles (filled with the neurotransmitter *acetylcholine*), mitochondria, and small patches of dense material around which the vesicles aggregate at the active site. The presynaptic element is separated from the postsynaptic element by an extracellular space called the *synaptic cleft*. The *postsynaptic membrane*, the specialized portion of the muscle cell plasma membrane subjacent to the axon terminal, exhibits a large number of folds that effectively increase the surface area of the muscle cell in contact with the axon terminal (see Fig. 24–2). These irregularities, called *subjunctional folds*, contain *nicotinic acetylcholine receptors* near their summit at the synaptic cleft. The receptors are linked to ligand-gated ion channels in the postsynaptic membrane. These channels mediate the ion flux that underlies the transmission of electrical impulses from nerve to muscle. Surrounding the entire muscle fiber and extending into the synaptic cleft is a *basal lamina* similar in composition to basement membranes present in other tissues.

When an action potential depolarizes the presynaptic element, there is an influx of calcium through voltage-gated membrane channels. Synaptic vesicles fuse with the presynaptic membrane at the active sites (which are marked by structures called *dense bars*) and release *acetylcholine* into the synaptic cleft. The transmitter binds to receptors on the postsynaptic membrane and opens specific ion channels. Ion flux then occurs, and a depolarizing potential called an *end plate potential* spreads over the surface of the muscle fiber. This potential triggers the release of Ca^{2+} (from the sarcoplasmic reticulum), which elicits the movement of actin and myosin filaments, resulting in muscle contraction. Synaptic transmission is terminated by an en-zyme called *acetylcholinesterase*, which is located in the matrix of the basal lamina in the depths of the postjunctional folds. This enzyme inactivates acetylcholine by detaching it from its receptor and hydrolyzing it to acetate and choline.

Motor Units. Each muscle fiber receives only one motor end plate, but the number of muscle fibers innervated by a single alpha motor neuron axon varies from a few to many. The aggregate of a motor neuron axon and *all* the muscle fibers it innervates is called a *motor unit* (Fig. 24–3). In general, as the need for fine control of a muscle increases, the size or innervation ratio of its motor unit decreases. That is, the number of muscle fibers innervated by a single axon decreases. The size of a motor unit is also related to the mass of the muscle and its speed of contraction. Small muscles that generate low levels of force typically have *small motor units* (10 to 100 muscle fibers per motor axon), whereas large, powerful muscles that generate high levels of force are usually innervated by *large motor units* (600 to 1000 muscle fibers per motor neuron axon).

Motor units can be divided into three categories on the basis of the physiologic properties of their muscle

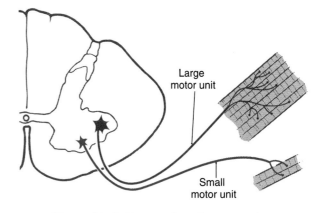

Large motor unit

Small motor unit

Figure 24–3. Large and small motor units.

and nerve components. *Type S* (slow) units have a relatively long contraction time and generate relatively low tension but are fatigue-resistant. They are sometimes referred to as *FR units* (for *fatigue-resistant*) and are innervated by small and slow-conducting alpha fibers. Muscles in which S units predominate are dark red and are called *slow twitch muscles*. A second type of motor unit generates large contractile force, exhibits a rapid contraction time, and is innervated by large alpha motor neurons with heavily myelinated, fast-conducting axons. However, these units fatigue quickly and are therefore referred to as *FF units* (for *fast-fatigable*). Muscles in which these units predominate are pale-colored and are called *fast twitch muscles*. *Type FR* units have a slightly slower contraction time than that of FF units and generate nearly as much force but are almost as fatigue-resistant as S units.

Muscles generally have a mixture of motor units, and the proportions vary according to the demands placed on the muscle. For example, the soleus muscle is a slow-twitch postural muscle containing mainly S-type units. The relatively slow conduction time of the small-diameter alpha axons serving these motor units is adequate for the demands of this muscle. By contrast, the gastrocnemius muscle is a dynamic, powerful muscle used in running and jumping. It is considered to be a fast-twitch muscle and contains mainly FF motor units innervated by large-diameter, fast-conducting axons.

Size Principle. The nervous system uses the size and functional properties of motor units as a means of grading the force of muscle contraction. When a group of motor neurons in the anterior horn is activated, those with the *smallest somata* are recruited first because they have the *lowest threshold* for synaptic activation and respond to the weakest input. They typically form small motor units and produce relatively weak movements. More intense input recruits successively *larger neurons* (and larger motor units), producing more powerful movements. This is called the *size principle*. Increased muscle force output can also be achieved by increasing the firing rate of motor units. Muscle contraction generally remains smooth because the component motor units are activated asynchronously (owing to the size principle), and this tends to "even out" the differences in motor unit properties.

Peripheral Sensory Input to the Anterior (Ventral) Horn

Muscle Spindles. Signals that transmit information from skeletal muscles into the nervous system enter the spinal cord via the posterior (dorsal) roots. For the most part, these signals are generated in specialized structures in muscles called *neuromuscular spindles (muscle spindles)*. The output of the muscle spindle signals a change in muscle length and the rate of change in muscle length.

A muscle spindle (Fig. 24–4) is a long, thin encapsulated structure that typically contains about seven skeletal striated *intrafusal muscle fibers*. Spindles range in length from 4 to 10 mm. The capsule of the spindle is attached to, and oriented in parallel with, the *extrafusal fibers* that constitute the bulk of the muscle.

There are two basic types of intrafusal fibers: *nuclear bag fibers* and *nuclear chain fibers* (see Fig. 24–4). Like other skeletal muscle cells, intrafusal fibers are multinucleated, and the arrangement of the nuclei is the most obvious structural feature distinguishing the two types. In both types, the nuclei occupy the central (*equatorial*) region of the cell. In *nuclear bag fibers*, the nuclei are clustered centrally, and the equatorial region is somewhat swollen. In *nuclear chain fibers*, the nuclei are arranged in a linear row, and the equatorial region is not obviously expanded. The contractile elements of both types of cell are located entirely in the distal (polar) regions of the cell. Because the ends of the cell are anchored, contraction of the intrafusal fibers causes the equatorial region to be stretched between the two polar regions.

The different types of intrafusal fiber perform different sensory functions. There are actually two sub-

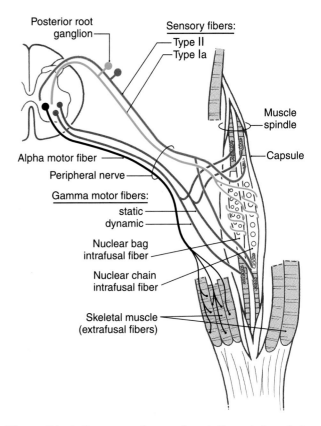

Figure 24–4. Structure of a muscle spindle and the relation of afferent and efferent nerve fibers to intrafusal and extrafusal muscle fibers.

types of nuclear bag fiber, which have different elastic properties and, correspondingly, different functions. One type, the *dynamic bag fiber*, is sensitive mainly to the rate of change in muscle length. The other, the *static bag fiber*, signals only a change in muscle length. Nuclear chain fibers, like static bag fibers, are mainly sensitive to changes in muscle length.

Intrafusal muscle fibers are associated with two types of sensory fibers, both of which are concentrated at the equatorial (noncontractile) region of infrafusal fibers. The *type Ia* fiber is heavily myelinated, has a conduction velocity of 80 to 120 m/sec, and is typically associated with nuclear bag fibers. The distal end of this fiber is wrapped around the central (noncontractile) region of an intrafusal muscle fiber. Because of this relationship, the type *Ia* afferent terminations are called *annulospiral endings*. These endings are, in effect, mechanoreceptors. Stretching of the central region of the intrafusal fiber will also stretch the sensory fiber and mechanically open ion channels in its membrane. If the induced ion flux raises the membrane potential above threshold, an action potential is initiated. The firing frequency is directly proportional to the degree to which the spindle is stretched.

The other type of muscle spindle sensory fiber, the *type II fiber*, is principally associated with nuclear chain fibers. Its connection with the equatorial region of the target infrafusal fiber has the form of a cluster of thin, radiating branches and is called a *secondary ending* or *flower-spray ending*. This sensory fiber is also activated by mechanical stretch, but it *codes only the event (change in muscle length), not the rate* of the stretch.

Each type of intrafusal fiber is also innervated by a gamma motor neuron. Dynamic nuclear bag fibers are associated with *dynamic gamma motor neurons*, whereas static nuclear bag fibers and nuclear chain fibers are innervated by *static gamma motor neurons*. When the gamma motor neuron is active, contractile elements at either pole of the intrafusal muscle fiber are activated,

resulting in increased stretch on its central region. This increases the frequency of action potentials that travel over the *Ia* sensory fibers. As explained further on, dynamic and static gamma motor neurons function to maintain spindle sensitivity and length, respectively.

The Gamma Loop. Muscle spindles play an essential role in movement and in the maintenance of muscle tone. Consider two situations: one in which a muscle—for example, the biceps brachii—is passively stretched, and another in which it contracts and shortens actively against a load.

A passive stretching of the biceps muscle, produced, for example, by tapping on its tendon, will elongate the muscle spindles. The stretching of the equatorial region of the nuclear bag fibers results in an increase in the firing rate of the Ia fibers (see Fig. 24-6A). These fibers enter the cervical spinal cord and form monosynaptic excitatory synapses with alpha motor neurons that innervate the biceps brachii (Figs. 24-5 and 24-6). This is the circuit that forms the basis of the monosynaptic tendon reflex explained in Chapter 9 (see Fig. 9-8).

The connection between the Ia fibers and the alpha motor neurons of a muscle also functions in a more complex mechanism called the *gamma loop*, which is crucial to the maintenance of stretch reflexes and muscle tone. In this mechanism, muscle contraction is produced by supraspinal activation of *gamma* rather than alpha motor neurons (see Fig. 24-5). Like alpha neurons, gamma motor neurons receive supraspinal input from the cerebral cortex and brainstem. In the gamma loop, this supraspinal input activates the gamma neurons so that the intrafusal muscle fibers contract. Because the contraction of an intrafusal fiber has the effect of stretching the equatorial region between the two shortened polar regions, it results in an increase in Ia fiber activity. In the spinal cord, this increase in Ia fiber discharge activates alpha motor neurons and extrafusal muscle fibers, resulting in muscle contraction

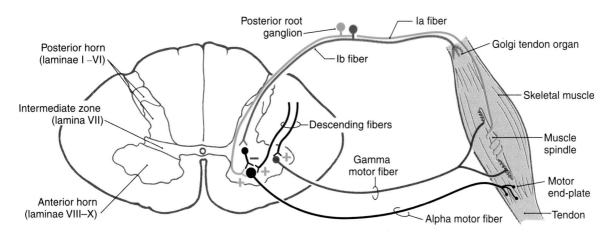

Figure 24-5. Circuits related to input and output mediated by Golgi tendon organs and to alpha-gamma coactivation.

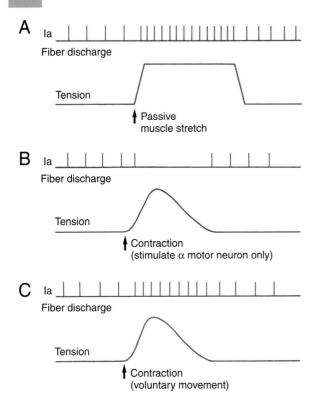

Figure 24–6. Responses shown by Ia fibers in three situations: passive stretching of the muscle (*A*), experimentally induced contraction of the muscle in which only the alpha motor neurons are stimulated (*B*), and a voluntary contraction of the muscle in which gamma and alpha motor neurons are coactivated (*C*). When only alpha neurons are activated (*B*), the intrafusal fibers stay relaxed while the surrounding muscle contracts, and Ia discharge ceases. In a normal contraction (*C*), the shortening of the spindle caused by extrafusal fiber contraction is coupled with the contraction of intrafusal fibers, and the Ia discharge persists.

(see Fig. 24–5). This circuit involving gamma motor neurons, Ia primary afferent fibers, and alpha motor neurons is called the *gamma loop*.

Now, consider the situation in which a muscle is contracting actively against a load. Because a muscle spindle is attached parallel to the adjacent extrafusal fibers, one might erroneously infer that its overall length is determined, in part, by the length of the surrounding muscle; when the muscle contracts, the spindle shortens. This is not so; if the intrafusal fibers remained passive during extrafusal muscle fiber contraction, the shortening of the spindle would relax the equatorial region of the intrafusal fibers, and the Ia fibers would cease firing (see Fig. 24–6*B*). This slack, inactive spindle would be useless for reporting muscle dynamics. In reality, the spindle retains its sensitivity (see Fig. 24–6*C*), and the Ia sensory fibers continue to fire during voluntary muscle contraction, because when the brain signals the alpha motor neurons to initiate muscle contraction, it sends parallel impulses to the gamma neurons to cause the intrafusal fibers to con-

tract. Therefore, when the extrafusal muscle fibers shorten, the intrafusal fibers also shorten because their gamma motor neurons are activated at the same time, and the equatorial regions of the intrafusal fibers remain under nearly constant tension. This phenomenon is called *alpha-gamma coactivation* (see Fig. 24–5).

Golgi Tendon Organ. Sensory feedback to the spinal anterior horn is also derived from the Golgi tendon organ. These structures are located in tendons near their junctions with muscle fibers and consist of networks of thin nerve fibers intertwined with the collagen fibers of the tendon (see Fig. 24–5). These nerve fibers, like the sensory fibers of muscle spindles, are mechanoreceptors. When force is applied to the tendon, the sensory fibers are stretched, which opens ion channels in the nerve fiber membrane. The fibers that lead from the tendon organs to the spinal cord are *type Ib* fibers. These fibers are large in diameter and heavily myelinated, with a conduction velocity of 70 to 110 m/sec. After entering the spinal cord, the type Ib fibers traverse the intermediate zone to reach the anterior horn, where they form *excitatory* synapses with interneurons. These interneurons in turn *inhibit* alpha motor neurons that innervate the muscle associated with the activated Golgi tendon organ. This action of the Golgi tendon organ is exactly opposite that of the muscle spindle; activation of the latter leads to excitation of alpha motor neurons innervating the muscle associated with the activated spindle.

Reflex Circuits. Afferent fibers from muscle spindles and Golgi tendon organs take part in a variety of reflex circuits that directly or indirectly influence the activity of anterior horn motor neurons. Several of the more prominent of these circuits are described in Chapter 9 (see Figs. 9–8 to 9–10); they are summarized only briefly here. As mentioned earlier, many type *Ia* spindle afferents form monosynaptic excitatory connections with alpha motor neurons that innervate the muscle from which the afferents originated. This circuity is the basis for the *tendon* or *stretch reflex* (see Fig. 9–8). At the same time, these *Ia* fibers activate *Ia interneurons* that inhibit motor neurons innervating *antagonist* muscles; this is called *reciprocal inhibition* (see Fig. 9–8). Incoming muscle afferents can also activate interneurons that project to the contralateral side of the spinal cord, as well as to propriospinal neurons that link the spinal segment at which the spindle afferents entered to more rostral or caudal spinal cord levels. Circuits of the first type, which convey cutaneous somatic inputs, form the basis for the *crossed extensor reflex* (see Fig. 9–10).

In general, the various spinal reflex pathways primarily target alpha motor neurons, although some may influence gamma motor neurons as well. For the most part, the activity of these basic spinal reflexes occurs in the background and is not under direct volitional control. However, other long loop reflexes involve ascending pathways that reach the cerebral cortex by way of a thalamic relay. These reflexes are subject to voluntary regulation, and their spinal effects can be enhanced or diminished on a volitional basis through various supraspinal systems.

Brainstem and Spinal Systems: Anatomy and Function

Of the several pathways that project to the spinal cord from the brainstem or cerebral cortex, four are particularly relevant to voluntary movement. All four originate from cell groups in the brainstem. Two of them, the *vestibulospinal* and *reticulospinal systems*, travel in the ventral funiculus of the spinal cord. The other two, the *rubrospinal* and *lateral corticospinal tracts*, travel in the lateral funiculus. The following sections focus on vestibulo-, reticulo-, and rubrospinal tracts.

Vestibulospinal Tracts. The vestibulospinal system comprises medial and lateral vestibulospinal tracts (Fig. 24–7; see also Fig. 24–1). The *medial vestibulospinal tract* is made up of axons that originate in the medial and inferior vestibular nuclei and *descend bilaterally* into

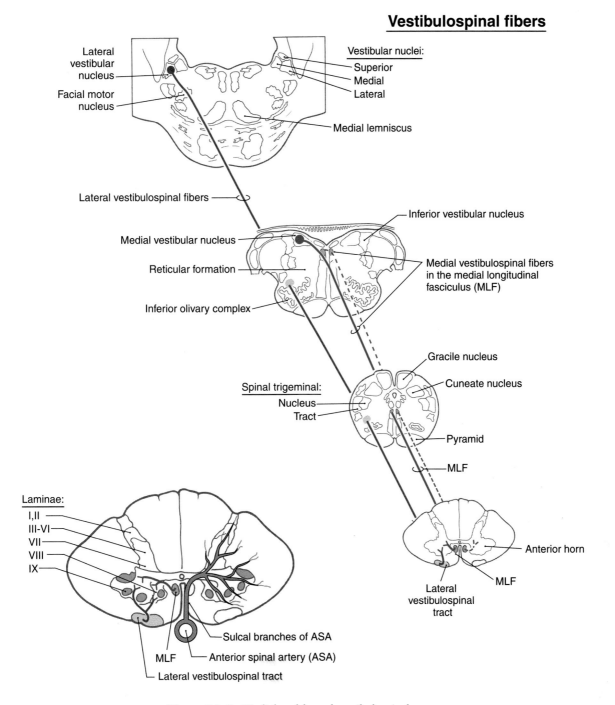

Figure 24–7. Medial and lateral vestibulospinal tracts.

the spinal cord as part of the medial longitudinal fasciculus. The *lateral vestibulospinal tract* is formed by axons that originate in cells of the lateral vestibular nucleus and *descend ipsilaterally* through the anterior portion of the brainstem to course in the anterior funiculus of the spinal cord.

The medial vestibulospinal tract projects only as far as cervical or upper thoracic spinal cord levels and influences motor neurons controlling neck musculature. The lateral vestibulospinal tract, in contrast, extends throughout the length of the cord. Cells in rostral portions of the lateral vestibular nucleus project to the cervical cord, cells in the midportion project to thoracic cord, and cells in the caudal part terminate in lumbosacral levels. The fibers of this tract terminate in the medial portions of laminae VII and VIII and *excite motor neurons that innervate paravertebral extensors and proximal limb extensors* (see Fig. 24–7). These muscles function to counteract the force of gravity and, therefore, are commonly called *antigravity muscles.* Through their effects on these extensor muscles, lateral vestibulospinal fibers function in the control of posture and balance. Evidence from experimental studies suggests that some vestibulospinal axons synapse directly on alpha motor neurons but that most exert their influence through spinal interneurons.

Activity in the lateral vestibulospinal tract is driven primarily by three ipsilateral inputs—two excitatory, and one inhibitory (Fig. 24–8). The two sources of excitatory input are the vestibular sensory apparatus and the cerebellar nuclei, mainly the fastigial nucleus. The inhibitory input consists of Purkinje cell axons from the cerebellar cortex.

The lateral vestibulospinal tract is the path by which input from the vestibular sensory apparatus is used to coordinate orientation of the head and body in space. Maintenance of body and limb posture is also influenced by extensive cerebellovestibular projections, which can be either excitatory or inhibitory. The cerebral cortex essentially has no direct projections to the vestibular nuclei; consequently, the vestibulospinal tract is not influenced by cortical mechanisms.

Reticulospinal Tracts. Cells at many levels of the reticular formation contribute to the reticulospinal system, and these fibers can be found in the lateral and anterior (ventral) funiculi throughout the spinal cord. Reticulospinal fibers participate in a wide variety of functions ranging from pain modulation to visceromotor activity. Most of the fibers involved in somatomotor function originate either from the oral and caudal pontine nuclei or from the gigantocellular reticular nucleus (Fig. 24–9). The fibers from the *oral* and *caudal pontine reticular nuclei* descend bilaterally, but with an ipsilateral predominance, in the anterior (ventral) funiculus. They constitute the *medial reticulospinal (or pontine reticulospinal) tract,* which runs the full length of the spinal cord. The fibers from the *gigantocellular re-*

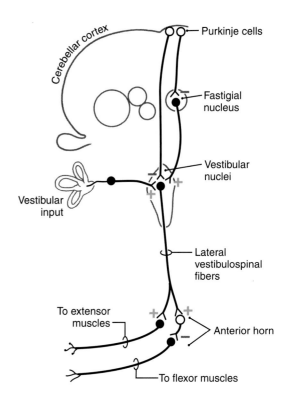

Figure 24–8. Summary of vestibular and cerebellar inputs to vestibular nuclei and the subsequent action of lateral vestibulospinal fibers on spinal motor neurons. Inhibitory neurons have open cell bodies.

ticular nucleus originate at medullary levels. Most of these *medullary reticulospinal fibers* remain ipsilateral and descend in the anterior funiculus, although a few decussate (see Fig. 24–9). Most take up a new position somewhat lateral and anterior to the anterior horn, where they are often called the *lateral reticulospinal tract.*

Like the vestibulospinal fibers, reticulospinal fibers terminate in anteromedial portions of laminae VII and VIII, where they influence *motor neurons supplying paravertebral and limb extensor musculature.* However, in contrast to the vestibulospinal tract, individual reticulospinal fibers commonly terminate at multiple spinal levels by means of collateral branches, and there is little evidence for monosynaptic contact with alpha motor neurons.

The reticulospinal system is activated by ipsilateral descending cortical projections (*corticoreticular fibers*) as well as ascending somatosensory systems (*spinoreticular fibers*), mainly those conveying nociceptive signals. Through influences on gamma motor neurons, the reticulospinal system is involved in the maintenance of posture and in the modulation of muscle tone. Pontine reticulospinal fibers tend to mediate excitatory effects, and medullary reticulospinal fibers usually produce inhibitory effects.

Rubrospinal Tract. In the midbrain, neurons in the

Rubrospinal and reticulospinal tracts

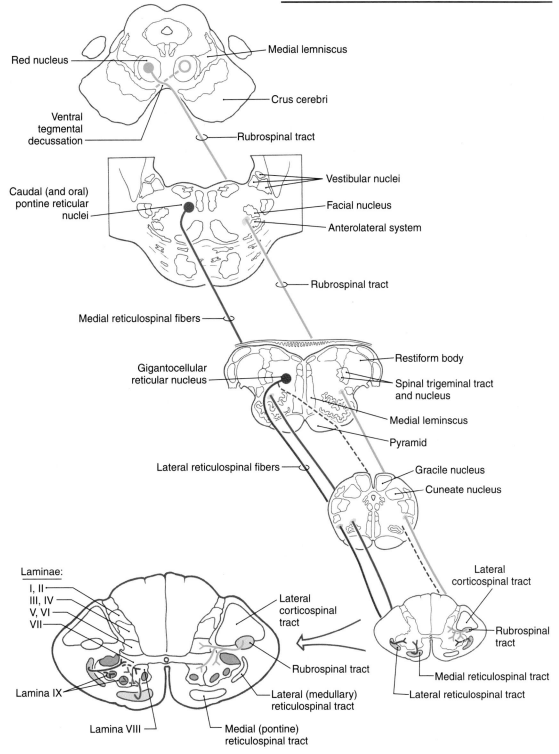

Figure 24–9. Rubrospinal and reticulospinal tracts.

red nucleus give rise to axons that cross the midline in the *anterior (ventral) tegmental decussation* (see Fig. 24–9). These fibers descend through the brainstem contralateral to their origin and enter the spinal cord anteriorly adjacent to the lateral, corticospinal tract. The red

nucleus consists of magnocellular and parvocellular subdivisions. In mammals that have been investigated, and probably also in humans, the magnocellular part gives rise to *rubrospinal fibers, and the parvocellular part gives rise to rubro-olivary fibers.* Each rubrospinal fiber terminates in a restricted area of the spinal cord; they do not innervate multiple cord levels by means of collaterals, as do reticulospinal fibers. In the spinal gray, rubrospinal fibers terminate in laminae V, VI, and VII. For the most part, they provide *excitatory influence to motor neurons innervating proximal limb flexors* (see Fig. 24–9).

The magnocellular portion of the red nucleus is relatively smaller in humans than in other mammals, and the rubrospinal tract is correspondingly small. In addition, relatively few rubrospinal axons extend caudal to the cervical enlargement, suggesting that this system is primarily involved with the upper extremity. Clinical findings in patients are consistent with this conclusion, indicating that the rubrospinal system exerts its control mainly over the upper extremity and has little influence over the lower extremity.

The rubrospinal system is influenced by the cerebral cortex and the cerebellar nuclei via *corticorubral* and *cerebellorubral* fibers, respectively. Precentral and premotor cortices project to the ipsilateral red nucleus, and the supplementary area contributes contralateral input. The latter pathways provide a route through which the cortex might influence flexor motor neurons and thus serve as a supplement to the corticospinal system. Connections that link the cerebellar nuclei, inferior olive, red nucleus, and rubrospinal tract may represent circuitry important for modifying motor performance or acquiring new motor skills.

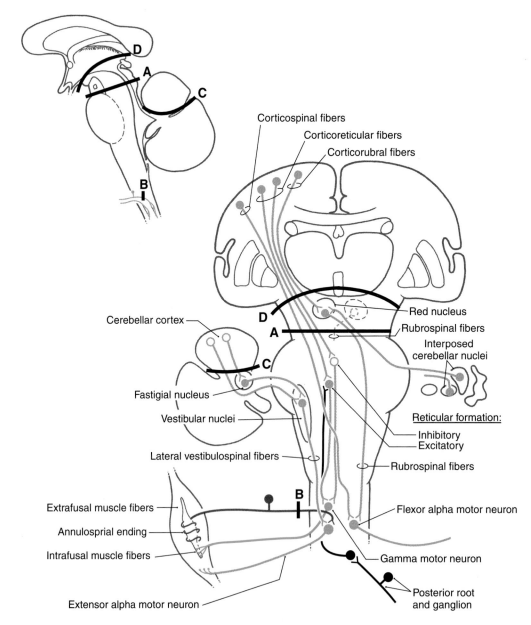

Figure 24–10. Locations of lesions and circuits involved in decerebrate rigidity (A), posterior root section in a decerebrate preparation (B), decerebellate rigidity (C), and decorticate posturing (D). Inhibitory neurons have open cell bodies. Ascending fibers in the anterolateral system (black fibers) are an excitatory input to excitatory cells of the reticular formation.

Functional Role of Brainstem and Spinal Interactions

Insight into the functional role of the brainstem and spinal systems has come from animal studies in which lesions have been created in specific locations of the brainstem. The resulting deficits mimic those of humans known to have or suspected of having damage in the same structures.

Decerebration. In the basic experiment, under deep anesthesia, the brainstem was completely transected bilaterally between the superior and the inferior colliculi (Fig. 24–10*A*). This procedure results in deficits that closely resemble those seen in humans after traumatic insults, vascular disease, or tumors in the midbrain. This lesion results in unopposed activity in extensor musculature in all four extremities, a condition called *decerebrate rigidity* (Fig. 24–11).

In this situation, all descending cortical systems are interrupted. These systems include the corticospinal tract, as well as the corticorubral and corticoreticular projections. In addition, the rubrospinal tract is transected, but the excitatory and inhibitory components of the reticular formation are intact. Also unaffected is the ascending somatosensory input to the reticular formation, most of which is directed to the excitatory elements of the reticulospinal system.

Posterior (Dorsal) Root Section. An important question that arose in relation to decerebration experiments was whether the extensor hypertonus was due to excessive activation of alpha or of gamma extensor motor neurons. To answer this question, the posterior (dorsal) root input from one extremity was interrupted in a decerebrate animal (see Fig. 24–10*B*). Immediately, the extensor hypertonus in that limb collapsed. What does this result indicate? Remember that supraspinal input can produce muscle contraction by two routes: by activation of alpha motor neurons or via the *gamma loop*. In the latter, the supraspinal input stimulates gamma motor neurons, leading to contraction of intrafusal fibers, and the resulting increase in Ia sensory input activates alpha motor neurons. Section of a posterior root eliminates the Ia input to the spinal cord and thus breaks the gamma loop. The collapse of extensor hypertonus caused by posterior root section therefore indicates that excitation of alpha motor neurons alone (in the absence of the gamma loop) is not sufficient to produce extensor hypertonus. Thus, decerebrate rigidity is due to excessive excitatory drive on extensor *gamma* motor neurons, coupled with diminished activation of all flexor motor neurons as a result of interruption of the corticospinal and corticorubrospinal system.

Having reached this conclusion, one may ask what

Decerebrate rigidity

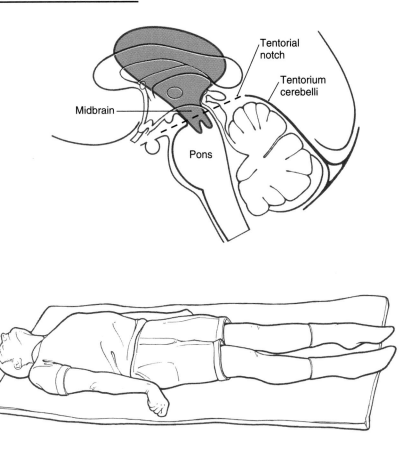

Figure 24–11. Decerebrate rigidity. A supratentorial lesion has extended through the tentorial notch. The patient's lower extremities are extended, with the toes pointed inward; the upper extremity is extended, with the fingers flexed and the forearms pronated; and the neck and head are extended. The rigidity may be so extreme that the patient's back is arched up off the bed. A patient may become decerebrate after a period of decorticate posturing (see Fig. 24–12).

Decorticate rigidity

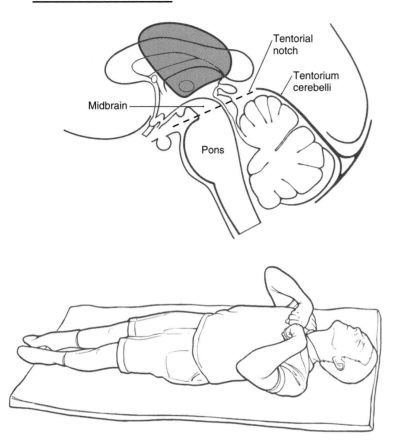

Figure 24–12. Decorticate rigidity. The lesion is in a supratentorial location. The lower extremities are extended, with the toes pointed slightly inward, and the upper extremities are flexed against the chest. The head is extended. A lesion in supratentorial location may produce decorticate rigidity or posturing that may proceed to decerebrate rigidity or posturing as the lesion expands inferiorly (caudally) through the tentorial notch (see Fig. 24–11).

produces the excessive activation of gamma motor neurons in decerebrate animals. Two supraspinal systems are intact in these experimental animals and could be the source of this excessive drive: the vestibulospinal and reticulospinal systems. The vestibular nuclei do not normally receive significant input from the cerebral cortex, however, and the major inputs they do receive—from the vestibular apparatus and from the cerebellar cortex and nuclei (see Fig. 24–8)—are not disrupted by decerebration. Therefore, the *vestibulospinal system* is not likely to be the culprit. By contrast, the *reticulospinal system* does receive important cortical input. The inhibitory portion of the reticulospinal system is driven largely from the cortex and becomes nonfunctional after decerebration. Although the excitatory elements have also lost their cortical input, they can still be activated by ascending sensory inputs not affected by the lesion. Thus, decerebration upsets the balance between excitatory and inhibitory reticulospinal influences on spinal neurons, creating a bias toward excitation. The conclusion is, therefore, that *extensor hypertonus in the decerebrate patient results from excessive excitatory input to extensor gamma motor neurons via reticulospinal fibers.* For this reason, decerebrate rigidity is sometimes called *gamma rigidity.*

Cerebellar Anterior Lobe Section. If extensor gamma motor neurons receive preferential input from the reticulospinal system, it is reasonable to ask if extensor alpha motor neurons are preferentially driven by vestibulospinal fibers. To investigate this point, the cerebellar anterior lobe was removed in an animal that was also rendered decerebrate (by midcollicular transection) (see Fig. 24–10C). Under these conditions the extensor hypertonus actually proved to be enhanced compared with that seen with decerebration alone, and the condition was called *decerebellate rigidity.* Section of the posterior roots from one extremity in such an animal produced only a slight decline in extensor rigidity of the limb.

Removal of the cerebellar anterior lobe cortex has two effects (see Fig. 24–10C). First, *it eliminates direct Purkinje cell inhibition of the vestibular nuclei,* resulting in enhanced output over the vestibulospinal tract. Second, Purkinje cell *inhibition of fastigial neurons is eliminated.* The resulting increase in fastigial excitatory output to the vestibular nuclei further augments the activity in the vestibulospinal tract. Therefore, the effect of cerebellar cortex removal is to substantially increase activity in the vestibulospinal system. Is this excessive drive delivered to the extensor alpha motor neurons, as hypothesized? Apparently it is, as is evident from the failure of the posterior root section to abolish the extensor hypertonus in this animal. The persistent hypertonus, therefore, cannot be due to activity in the gamma loop. Instead, the increased vestibulospinal output must reach *alpha* motor neurons directly. Consequently, *decerebellate extensor rigidity* is referred to as *alpha rigidity*.

Decortication. An extension of these experiments was undertaken to explain the neural substrate for the phenomenon called *decorticate posturing* or *decorticate rigidity* observed in humans (Fig. 24–12). In this situation, the patient presents with flexion of the upper extremities at the elbow combined with extensor hypertonus in the lower extremities.

In experimental animals, this condition can be mimicked by transecting the brainstem at a level *rostral to the superior colliculus* (see Fig. 24–10D). This lesion leaves the rubrospinal tract intact while eliminating the cortical input to the red nucleus. The rubrospinal system can still be activated because excitatory projections to the red nucleus from the cerebellar nuclei are unaf-

fected. The rubrospinal tract influences primarily *flexor muscles*, and most of this activity, particularly in humans, is limited to the upper extremity. Therefore, under these conditions, the lower extremities exhibit extensor hypertonus for the same reasons as in decerebration. The upper extremities do not exhibit extensor hypertonus but instead show an increase in flexor tone. This characteristic type of posturing is called *decorticate rigidity* (see Fig. 24–12).

These two conditions, *decerebration* and *decortication*, are frequently seen in patients (see Figs. 24–11 and 24–12), and knowledge of these symptoms and the underlying brain pathology is important in the diagnosis and clinical management of these patients. In some cases, the patient may be comatose and exhibit decorticate posturing that converts to decerebrate posturing. This is an ominous sign, as it suggests that the lesion has continued to progress and now involves more caudal portions of the brainstem. The patient's cardiovascular and respiratory control centers in the medulla may soon be compromised, necessitating prompt intervention.

Sources and Additional Reading

Boyd IA: The isolated mammalian muscle spindle. Trends Neurosci 3:258–265, 1980.

Brodal A: Neurological Anatomy in Relation to Clinical Medicine, 3rd Ed. Oxford University Press, New York, 1981, pp 148–211, 264–282.

Brooks VB (ed): Handbook of Physiology, section 1: *The Nervous System*, vol II. *Motor Control, Part 1*. American Physiological Society, Bethesda, 1981.

Desmedt JE (ed): Spinal and Supraspinal Mechanisms of Voluntary Motor Control and Locomotion, vol 8. *Progress in Clinical Neurophysiology*. Karger, Basel, 1980.

Mathews PBC: Evolving views on the internal operation and functional role of the muscle spindle. J Physiol (Lond) 320:1–30, 1981.

Shepherd GM: Neurobiology. Oxford University Press, New York, 1994.

Sherrington CS: Decerebrate rigidity and reflex coordination of movements. J Physiol (Lond) 22:319–332, 1898.

Taylor A, Prochazka A (eds): Muscle Receptors and Movement. Macmillan, London, 1981.

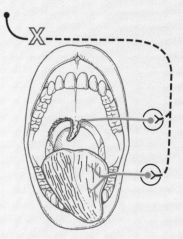

Chapter 25

Motor System II: Corticofugal Systems and the Control of Movement

G. A. Mihailoff and D. E. Haines

Overview 388

General Features of Motor Deficits 388
Lower Motor Neuron Signs
Upper Motor Neuron Signs
Spasticity

The Corticospinal System 389
Origin
Course
Termination

The Corticonuclear (Corticobulbar) System 394
Origin
Course
Termination

Other Corticofugal Systems 398
The Corticorubral System
The Corticoreticular System
The Corticopontine System

Motor Cortex and the Control of Movement 399
Primary Motor Cortex
Supplementary Motor Cortex
Premotor Cortex
Posterior Parietal Cortex
Cingulate Motor Cortex
Cerebellar and Pallidal Influences

Hierarchical Organization Versus Parallel Distributed Processing in the Motor System 402

Brushing your teeth seems like a simple voluntary movement. The neural basis for this action is, in fact, richly complex. For example, muscles in the upper limb are used cooperatively with jaw muscles, while neck and back muscles provide postural support. Sensory feedback from the teeth and gums is linked to muscle afferents conveying tension and proprioceptive signals from the forearm and hand. Even other, less obvious aspects, such as visual system input and memory of prior experience, are involved. This chapter focuses on elements of voluntary movement that are regulated by the cerebral cortex.

Overview

The control of voluntary movement is a complex, multifaceted process that involves many areas of the brain. One of the principal control sites is the cerebral cortex, specifically the *primary motor, premotor,* and *supplementary motor cortices* in the frontal lobe, along with portions of the parietal lobe. Although the frontal and parietal lobes have direct projections to the spinal cord, they also work cooperatively through the *primary motor cortex (upper motor neurons)* and the *corticospinal* and *corticonuclear (corticobulbar) systems* to influence the activity of ventral horn and cranial nerve motor neurons (*lower motor neurons*). The latter cells and their axons represent the *final common path* that links the central nervous system with skeletal muscles. Lesions that damage the descending cortical systems or lower motor neurons produce signs and symptoms of *upper* or *lower motor neuron disease*, respectively. These signs are among the most useful in diagnosing neurologic deficits related to the control of movement.

General Features of Motor Deficits

Lower Motor Neuron Signs. *Lower motor neurons* are those cells whose axons synapse directly on skeletal muscle. When these neurons or their axons are damaged, the innervated muscles will show some combination of the following signs: (1) *flaccid paralysis* followed eventually by *atrophy*, (2) *fibrillations* or *fasciculations* (involuntary contractions of one motor unit or a group of motor units), (3) *hypotonia* (decreased muscle tone), and (4) weakening or absence of tendon (stretch) reflexes (*hyporeflexia, areflexia*).

Upper Motor Neuron Signs. The term *upper motor neuron* is commonly used in reference to corticospinal cell bodies and their axons. Other neurons, such as rubrospinal or reticulospinal neurons, can also be included under the strict definition of this term. Corticospinal neurons are also called *pyramidal neurons* because their axons pass through the medullary pyramid. Therefore, *upper motor neuron signs* and *pyramidal tract signs* are often used synonymously. However, as described later in this chapter, these characteristic signs of "pyramidal tract damage" are in fact the result of injury to other descending motor systems in *combination* with damage to the pyramidal tract. For example, ischemic lesions of the internal capsule can potentially involve corticostriatal, corticothalamic, and corticoreticular fibers in addition to corticospinal axons.

Damage to upper motor neurons results in muscles that (1) are *initially weak* and flaccid but (2) eventually become *spastic*, (3) exhibit increased muscle tone (*hypertonia*), seen as an increase in resistance to passive movement of an extremity, and (4) show an increase in deep tendon reflexes (*hyperreflexia*), as may be seen in *clonus*. Upper motor neuron lesions usually *affect groups of muscles*, and certain pathologic reflexes and signs often appear. One of the most common is the *inverted plantar reflex*, also known as the *Babinski sign*. This is a dorsiflexion of the great toe in response to firm stroking of the lateral aspect of the sole of the foot with a blunt instrument. The response in the normal adult is plantar flexion of the great toe.

Spasticity. Muscles that are no longer under the influence of upper motor neurons exhibit spasticity. That is, when tested by the examiner, the affected muscles offer

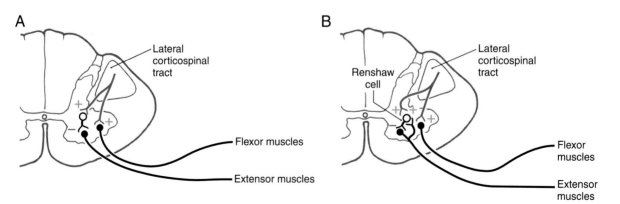

Figure 25–1. Connections by which descending fibers influence lower motor neurons. Direct influence and influence via an inhibitory interneuron *(A)*, and influence via a Renshaw cell, which participates in a recurrent inhibition circuit *(B)*.

an *increased resistance to passive movement or manipulation*. These effects are most pronounced in the antigravity muscles, which in humans include the proximal flexors in the upper extremity and extensors in the lower extremity. Also, the increased resistance to passive movement is velocity-dependent: *the more rapidly the examiner moves the affected extremity, the greater the resistance*. However, after a relatively brief period of applied force, the increased resistance totally collapses; this response is known as the *clasp-knife effect*.

Several hypotheses have been advanced to explain spasticity. One suggests that spasticity, with its associated hypertonia and hyperreflexia, is the result of excessive activation (or release from inhibitory control) of *dynamic gamma motor neurons*. This would lead to increased activity transmitted over type Ia muscle spindle afferents, resulting in increased excitatory tonic drive on the associated alpha motor neurons. Another suggestion is that spasticity may represent a *generalized failure of the descending cortical activation of spinal cord inhibitory interneurons*. For example, supraspinal fibers activate *type Ia* inhibitory interneurons that contact extensor motor neurons (Fig. 25–1*A*). If the upper motor neuron input to inhibitory interneurons were removed, the extensor motor neurons would be *released from inhibitory control*, and the result would be hypertonia and spasticity.

Descending cortical fibers also activate a type of interneuron called a *Renshaw cell* (see Fig. 25–1*B*). The Renshaw cell receives excitatory input from a lower motor neuron via an axon collateral, and in turn it inhibits the same lower motor neuron. It is also known, for example, that cortical fibers activating ankle flexors also contact Renshaw cells (as well as type Ia inhibitory interneurons) that inhibit ankle extensors (see Fig. 25–1*B*). This circuitry serves to prevent reflex stimulation of the extensors when flexors are active. Therefore, when the cortical fibers are lost (upper motor neuron lesion), the inhibition of antagonists is absent. The result is repetitive, alternating contraction of ankle flexors and extensors. Such a phenomenon is referred to as *clonus* and is often present in combination with spasticity and hyperreflexia.

The Corticospinal System

Before the specifics of corticospinal projections are considered, a general point about laterality should be made. The fibers that form the core of the corticospinal system cross the midline in the pyramidal decussation; this structure is commonly called the *motor decussation* because it is where corticospinal fibers cross. Corticospinal fibers arising in the left motor cortex therefore influence muscles on the right side of the body, and vice versa. Consequently, *lesions of corticospinal fibers rostral to the pyramidal (motor) decussation result in contralateral motor deficits, whereas lesions in the spinal*

cord result in ipsilateral deficits. An understanding of this concept of laterality is essential in the diagnosis of the neurologically impaired patient.

Origin. Neurons that give rise to *corticospinal* axons are located in deep portions of layer V of the cerebral cortex (Fig. 25–2*A*). A small number of these pyramidal neurons are especially large, with somata that may reach 100 μm or more in diameter. These neurons are called *Betz cells*, and at one time it was believed that they were the sole source of corticospinal axons. Now it is known that they account for only 1% to 2% of this fiber bundle.

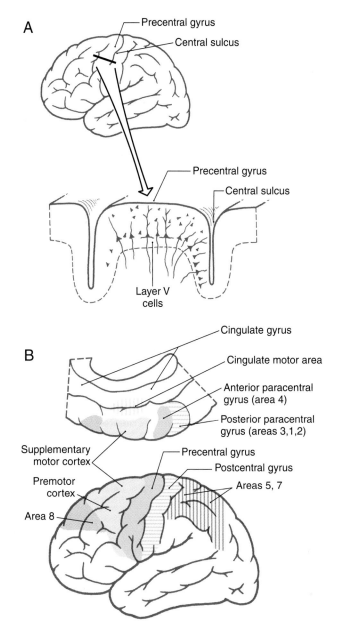

Figure 25–2. The motor-related areas of the cortex. A cross section through the precentral gyrus, showing pyramidal cells in layer V (*A*). The main areas of the cortex that give rise to corticospinal axons (*B*).

Corticospinal neurons are found primarily in six cortical locations (see Fig. 25–2B). The single largest concentration is in *Brodmann area 4*, which occupies the posterior portion of the *precentral gyrus* bordering on and extending into the depth of the *central sulcus.* This region is also called *MI*, the *primary motor cortex.* The *premotor* and *supplementary motor cortices*, which are located in *area 6*, also give rise to corticospinal axons. About two thirds of all corticospinal axons arise from neurons in the *frontal lobe*, with more than one half of this group coming from the *precentral* and *anterior paracentral gyri* (see Fig. 25–2B). The remaining one third arise from the *parietal lobe* and a few other regions. Included are cells in the *postcentral gyrus* (areas 3, 1, and 2), the *superior parietal lobule* (areas 5 and 7), and portions of the *cingulate gyrus.*

Within *MI*, corticospinal neurons are somatotopically organized in patterns that reflect their influence over specific muscles. The caricature thus created is called the *motor homunculus* (Fig. 25–3A). Neurons in medial *MI*, the anterior paracentral gyrus, project to lumbosacral cord levels to influence motor neurons which innervate muscles of the foot, leg, and thigh. Thoracic and cervical cord levels, which contain motor neurons innervating the trunk and upper extremity, receive input from neurons in the medial two thirds of the precentral gyrus. The musculature of the head, face, and oral cavity is influenced by neurons in the anterolateral (ventrolateral) part of the precentral gyrus (see Fig. 25–3A). These cells contribute to the *corticonuclear (corticobulbar) tract* that projects to cranial nerve motor nuclei. The disproportion in body part size in the *homunculus* reflects the density and distribution of corticospinal neurons devoted to the control of musculature in each particular region of the body (see Fig. 25–3A). Complete but less precise body representations are also found in other motor cortical regions. Thus, any single muscle may be influenced from multiple locations in the cerebral cortex.

The blood supply to MI arises from branches of the anterior and middle cerebral arteries (see Fig. 25–3B). The lower extremity area of MI is served by terminal branches of the A_2 segment of the anterior cerebral artery. Specifically, these branches arise from the *callosomarginal artery.* The trunk, upper extremity, and head areas of the motor cortex are supplied by branches of the M_4 segment of the middle cerebral artery, mainly its *rolandic* and *prerolandic branches* (see Figs. 8–5 and 8–6).

Lesions that involve only areas of motor cortex outside the MI usually do not result in paralysis, and the effects may dissipate over time. For example, vascular infarcts of the premotor cortex may produce an *apraxia.* This disorder involves difficulty in using the affected part of the body to perform voluntary actions, such as grasping a pencil, even though there is no obvious spasticity, paralysis, or altered tone in the mus-

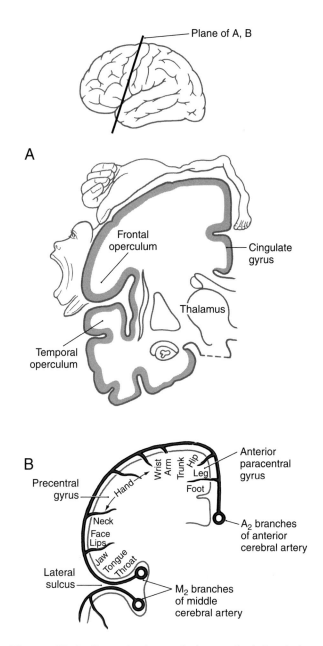

Figure 25–3. Coronal views of the cerebral hemisphere showing the somatotopy of the primary motor cortex *(A, B)* and the blood supply of the anterior paracentral lobule and the precentral gyrus *(B)*. (*A* is adapted from Penfield and Rasmussen, 1968, with permission.)

cles. For example, a premotor lesion may result in the inability to perform voluntary actions with the contralateral hand, although the strength and tone of the hand muscles are normal. Similarly, unilateral lesions in the supplementary motor cortex affect the ability to coordinate actions on the two sides of the body. The muscles, again, are normal. In contrast, lesions that affect both the primary motor cortex in combination with another motor cortical region usually result in spastic paralysis and hyperreflexia, signs characteristic of upper motor neuron lesions.

Course. The largest axons in the corticospinal tract are myelinated, range from 12 to 15 μm in diameter, and have conduction velocities of 70 m/sec but they make up less than 10% of the total corticospinal population. The remainder are less than 5 μm in diameter, and many are lightly myelinated or unmyelinated.

Corticospinal fibers pass through the corona radiata and converge to enter the *posterior limb of the internal capsule* (Fig. 25–4A). In this region, these fibers are somatotopically organized such that the axons that will terminate at the highest cord levels are located most rostrally and the axons that will terminate at progressively lower levels are located more caudally (see Fig. 25–4B).

Unlike lesions of the cortical gray matter, interruption of axons in the posterior limb of the internal capsule often results in catastrophic motor deficits. A

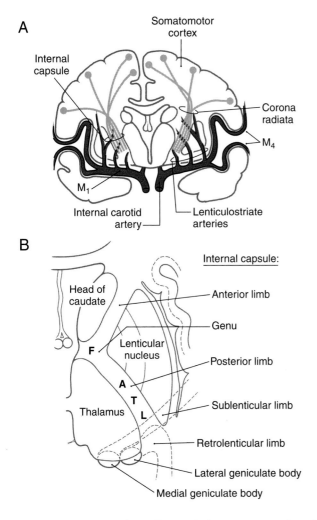

Figure 25–4. Descending fibers of the corticospinal and corticonuclear (corticobulbar) systems in the internal capsule in coronal *(A)* and axial *(B)* planes. The positions of fibers from face (F), arm-upper extremity (A), trunk (T), and leg-lower extremity (L) areas are shown in the internal capsule in the axial plane *(B)*.

common cause of lesions in this area is hemorrhage from *lenticulostriate branches* of the M$_1$ segment of the middle cerebral artery (see Fig. 25–4A). Motor symptoms of capsular infarcts appear in the contralateral upper and lower extremities and consist of weakness and transient flaccid paralysis of variable duration, which is followed by spastic paralysis (upper motor neuron signs) that typically never resolves. These symptoms appear because not only corticospinal fibers but also many other types of cortical axons are interrupted. Included are axons projecting to the neostriatum, thalamus, and brainstem, as well as thalamocortical axons involved in somatic sensation and vision. Damage to thalamocortical axons explains why *hemisensory loss or homonymous hemianopia may accompany the motor deficit*. It is important to note that deficits such as *spasticity*, *hypertonia*, and *hyperreflexia*, although commonly associated with pyramidal tract lesions, are in fact due to damage of other descending systems in *combination* with damage to corticospinal fibers.

As they pass caudally from the internal capsule, corticospinal fibers traverse the various divisions of the brainstem. In the midbrain they coalesce to form the middle third of the crus cerebri (Figs. 25–5A and 25–6). Within this part of the crus, fibers from forearm/hand areas of MI are located medially, whereas those from leg/foot areas are located laterally.

Fibers in the medial two thirds of the crus cerebri (frontopontine, corticonuclear [corticobulbar], and corticospinal) and the exiting rootlets of the oculomotor nerve fibers are served by *paramedian branches of P$_1$* and branches from the adjacent *posterior communicating artery* (see Fig. 25–6). Hemorrhage of these vessels will damage these groups of fibers, resulting in (1) *contralateral hemiparesis* of the arm and leg with spasticity and (2) *deviation of the ipsilateral eye* down and laterally, owing to the unopposed action of the superior oblique and lateral rectus muscles. *Direct* and *consensual light reflexes* and accommodation may also be lost in the eye on the side of the lesion. This condition is known as *superior alternating hemiplegia* because cranial nerve signs are seen on one side and corticospinal signs on the "alternate" side.

From the midbrain, corticospinal fibers continue into the basilar pons, where they make their way between the masses of neurons forming the basilar pontine nuclei (see Figs. 25–5B and 25–6). As corticospinal axons pass through the pontine gray, they give rise to collaterals that synapse on these neurons.

Corticospinal fibers in the basilar pons and the exiting fibers of the abducens nerve in the caudal pons are in the domain of the *paramedian branches of the basilar artery*. Occlusion or rupture of these vessels results in *hemiplegia* and *upper motor neuron* signs in the contralateral extremities. If the lesion extends caudally, it may involve *intra-axial* abducens fibers, resulting in lower motor neuron paralysis of the ipsilateral lateral rectus

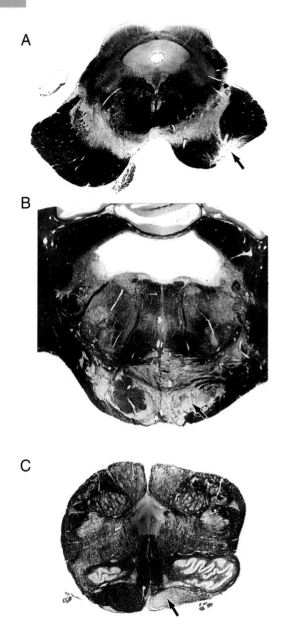

A

B

C

Figure 25–5. Degeneration of corticospinal fibers caused by an infarction in the posterior limb of the internal capsule. The degeneration serves as a marker to show the position of these fibers (at *arrows*) in the middle third of the crus cerebri *(A)*, the basilar pons *(B)*, and the pyramid of the medulla *(C)*.

nate at cervical levels tend to be located medially, whereas those projecting to lumbar and sacral levels are more lateral. Collaterals of these axons innervate the inferior olivary complex, posterior column nuclei, and various medullary reticular nuclei.

The pyramid, the laterally adjacent exiting fibers of the hypoglossal nerve, and the medial lemniscus receive their blood supply through penetrating branches of the *anterior spinal artery* (see Fig. 25–6). Occlusion of these branches results in a *contralateral hemiparesis of the extremities* (with spasticity) and an *ipsilateral flaccid paralysis of the tongue*. When protruded, the tongue deviates toward the side of the lesion (the flaccid side). This combination of symptoms is called *inferior alternating hemiplegia*. Because branches of the anterior spinal artery also serve the medial lemniscus, an inferior alternating hemiplegia may be accompanied by a *contralateral loss of two-point discrimination and vibration sense*.

At the medullospinal junction, some 85% to 90% of the corticospinal fibers cross the midline as the *pyramidal (motor) decussation* (see Fig. 25–6). Fibers that cross in the rostral and caudal portions of the pyramidal decussation originate, respectively, from the upper and lower extremity areas of MI cortex. This arrangement explains why small vascular lesions in the decussation (which is also served by branches of the *anterior spinal artery*) may result in bilateral weakness or paralysis of either the upper or lower extremities. The decussating fibers extend into the lateral funiculus to form the *lateral corticospinal tract*. The corticospinal axons that do not cross in the decussation continue into the ipsilateral anterior funiculus of the spinal cord as the *anterior corticospinal tract* (see Fig. 25–6). Damage to this tract is of little clinical significance.

Termination. Fibers of the lateral corticospinal tract are topographically organized (see Fig. 25–6). Axons terminating in cervical cord levels are most medial in this tract, whereas those distributing to lumbosacral levels are most lateral. This pattern means that, as the medially located fibers enter and terminate in the spinal gray, the adjacent, more lateral fascicles shift medially.

Corticospinal fibers that arise from the frontal lobe terminate primarily in the intermediate zone and anterior horn (laminae VII to IX), whereas fibers arising from the parietal lobe terminate in the base of the posterior horn (laminae IV to VI). As might be expected, most fibers terminate in the spinal cord enlargements that serve the extremities; about 55% terminate in the cervical enlargement, and about 25% in the lumbosacral enlargement. The remaining fibers terminate at thoracic levels. As mentioned earlier, some corticospinal fibers terminate, via their collateral branches, at multiple levels. However, the influence exerted by any single axon or its collaterals depends on the *number* of synapses it forms and the locus of the synaptic contacts on the postsynaptic neuron. Thus, a given

muscle (Fig. 25–7). This combination of deficits is called *middle alternating hemiplegia*. The paramedian branches of the basilar artery may penetrate deep into the pons and also serve the medial lemniscus (see Fig. 25–7). In such cases, damage to these vessels can produce not only the motor deficits described above but also *contralateral loss of vibratory sense and two-point tactile discrimination*.

In the medulla, corticospinal fibers aggregate on the anterior (ventral) surface of the brainstem, where they course within the medullary pyramids (see Figs. 25–5C and 25–6). Within the pyramid, fibers that will termi-

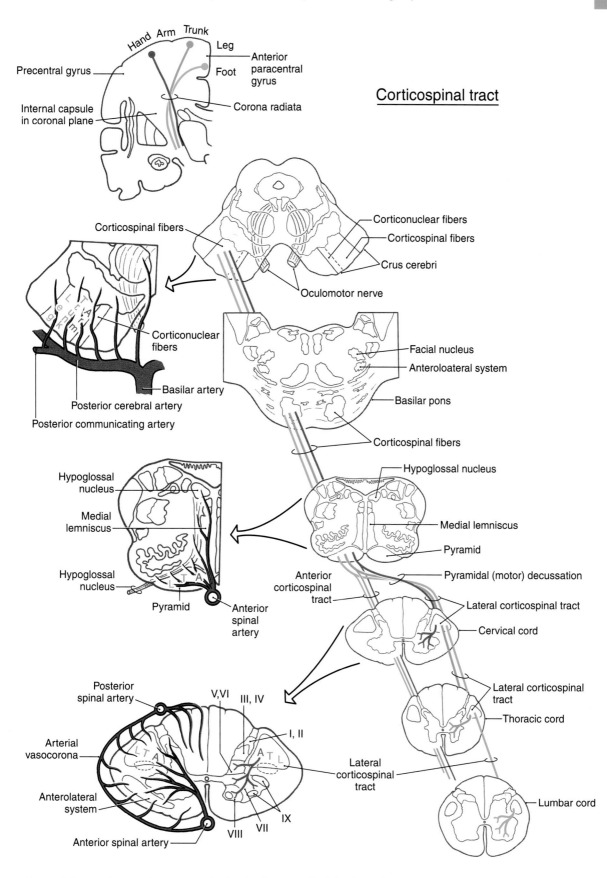

Corticospinal tract

Figure 25–6. The corticospinal system with details showing the blood supply to these fibers in the midbrain, medulla, and spinal cord.

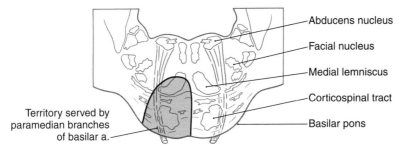

Abducens nucleus

Facial nucleus

Medial lemniscus

Corticospinal tract

Basilar pons

Territory served by paramedian branches of basilar a.

Figure 25–7. The blood supply to corticospinal fibers, the abducens fibers, and parts of the medial lemniscus in the caudal pons. The motor deficits seen with this lesion constitute a middle alternating hemiplegia.

corticospinal axon may have a powerful action on some spinal cord neurons and only a weak influence on others.

At their level of termination, particularly in the cord enlargements, corticospinal fibers synapse primarily on *interneurons* in laminae V to VII. In animals capable of dexterous finger movements, such as monkeys and humans, some corticospinal fibers terminate among clusters of lamina IX alpha motor neurons, which innervate distal flexor muscles. However, most corticospinal fibers, at least in nonhuman primates, synapse with excitatory and inhibitory interneurons, which, in turn, influence flexor and extensor motor neurons, respectively.

Interruption of lateral corticospinal axons in the *upper cervical cord* (C1, C2) results in spastic hemiplegia involving the ipsilateral upper and lower extremities (Fig. 25–8). Common upper motor neuron signs such as *hypertonia, hyperreflexia,* and the *Babinski sign* will be present ipsilateral to the lesion. If the lesion is sufficiently large, the innervation of the diaphragm (from C3 to C5 via the phrenic nerve) may be disrupted, necessitating the use of a respirator.

A lesion of the *cervical enlargement* results in a different pattern of motor deficits (see Fig. 25–8). If the damage involves only the lateral funiculus, then the ipsilateral upper and lower extremities will exhibit typical upper motor neuron signs. If, however, the C6 to C8 anterior horn and the lateral funiculus are both included in the lesion, then *lower motor neuron signs will appear in the upper extremity ipsilaterally, whereas upper motor neuron signs will be seen in the ipsilateral lower extremity.* When anterior horn motor neurons or their axons are damaged, the affected muscles exhibit lower motor neuron signs, despite the fact that supraspinal systems providing input to these cells may also have been interrupted.

At *lumbosacral levels,* injury to the spinal cord frequently affects motor neurons in the anterior horn as well as descending supraspinal fibers (see Fig. 25–8). Characteristically, affected patients exhibit lower motor neuron signs in the ipsilateral lower extremity if both the corticospinal fibers and anterior horn motor neurons are damaged.

The blood supply to the lateral corticospinal tract is derived from *penetrating branches of the arterial vasocorona* and *sulcal branches of the anterior spinal artery* (see Fig. 25–6). The former serve fibers in lateral parts of the tract, and the latter serve the medially located fibers. Hyperextension of the neck may result in injury to the cord or in occlusion of the sulcal arteries (*central cord syndrome*); either can result in bilateral hemiparesis of the upper extremities secondary to vascular infarcts involving medial regions of both lateral corticospinal tracts. In addition, affected patients exhibit both urinary retention and a bilateral, patchy loss of pain and temperature sensations below the lesion.

A functional hemisection of the spinal cord, such as may be caused by a tumor or by trauma, results in a characteristic set of deficits known as the *Brown-Séquard syndrome* (Fig. 25–9; see also Fig. 18–8). These deficits begin about two levels below the lesion and consist of (1) *ipsilateral* loss of two-point discrimination and vibration (from damage to the dorsal columns), (2) *contralateral* loss of pain and thermal sensation (from damage to the anterolateral system), and (3) an *ipsilateral* paresis or paralysis (from damage to the corticospinal tract). The paralysis involves the upper and lower extremities or only the lower, depending on the level of the injury. Also if the lesion is large enough that it involves several spinal cord levels, damage to a sufficient number of primary afferent fibers entering the cord may result in a narrow band of complete anesthesia on the side ipsilateral to the lesion in dermatomes corresponding to the damaged cord segments.

The Corticonuclear (Corticobulbar) System

Origin. Organized in parallel with the corticospinal system is the *corticonuclear (corticobulbar) system* (Fig. 25–10). As defined here, the corticonuclear (cortico-

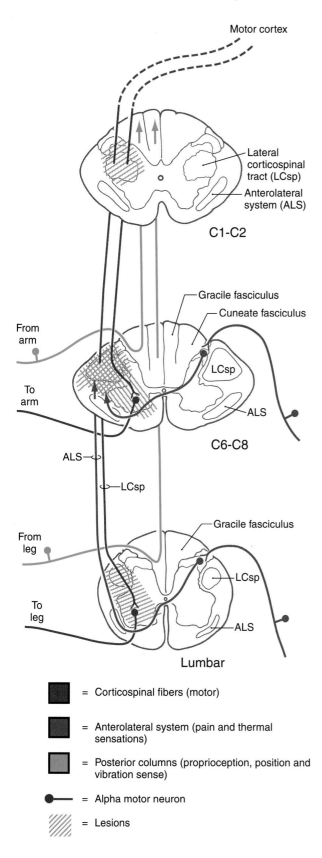

Figure 25–8. Examples of spinal cord lesions at C1 to C2 (involving only corticospinal fibers), at C6 to C8 (involving corticospinal fibers only, or these plus the anterior horn), and at lumbar levels (involving corticospinal fibers and the anterior horn).

bulbar) system consists of the cortical neurons that influence the movements of striated muscles innervated by the motor nuclei of cranial nerves V, VII, and XII, by the nucleus ambiguus (cranial nerves IX and X), and by the accessory nucleus.

The term "corticobulbar" was historically used to describe all cortical projections to cranial nerve nuclei of the brainstem. The suffix "bulbar" comes from "bulb," an obsolete term for the medulla oblongata. The international committee (see Preface) charged with establishing anatomic terminology studied the inherent problems with this term; for example, how can there be "corticobulbar" projections to cranial nerve nuclei that are not in the "bulb"? After detailed consideration by the committee and with the publication of the new terminology in 1998, the term "corticobulbar" was replaced with terms that accurately describe these connections. These terms are as follows: *fibrae corticonucleares bulbi* for cortical projections to cranial nerve nuclei in the bulb/medulla (medullary corticonuclear fibers), *fibrae corticonucleares pontis* for cortical projections to cranial nerve nuclei of the pons (pontine corticonuclear fibers), and *fibrae corticonucleares mesencephali* for similar projections to the midbrain (mesencephalic corticonuclear fibers). Here these terms are shortened to simply *corticonuclear*, and this new and preferred term is used synonymously with the replaced term "corticobulbar."

Some corticonuclear (corticobulbar) axons project directly to cranial motor neurons, but most terminate on reticular formation interneurons immediately adjacent to the cranial nerve nuclei. The corticonuclear system originates for the most part from the face and head area of the precentral gyrus (see Fig. 25–10). Because most of the musculature innervated by cranial nerves is located in the facial region, this area of MI is typically called *face motor cortex*.

The oculomotor, trochlear, and abducens nuclei do not receive direct input from the face motor cortex. Instead, voluntary control of eye movement is mediated via cortical projections from *frontal* and *parietal motor eye fields* to eye movement (gaze) control centers in the midbrain and pons. These centers in the *midbrain reticular formation* and the *paramedian pontine reticular formation*, in turn, relay input from the cortical eye fields to somatic motor neurons in the nuclei of cranial nerves III, IV, and VI (see Chapter 28 for details). Although the cortex of each hemisphere influences these nuclei bilaterally, the resulting eye movements are conjugate and are toward the side contralateral to the cortex from which the input originated. Because these cortical axons are often several synapses removed from the actual cranial nerve motor neurons, they are not considered part of the corticonuclear system as defined in this text.

Course. Corticonuclear (corticobulbar) axons that originate from cells in layer V of the face motor cortex

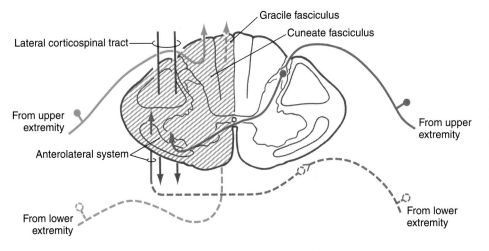

Figure 25–9. Hemisection of the spinal cord (Brown-Séquard syndrome).

funnel into the *genu of the internal capsule*, where they are positioned just anterior to corticospinal fibers (see Fig. 25–4). Corticonuclear fibers continue into the crus cerebri, where they are located medial to corticospinal fibers traveling to cervical cord levels (see Figs. 25–6 and 25–10). From here, these axons descend into the pons and medulla in association with corticospinal fibers. Corticonuclear fibers in the genu and in the crus cerebri receive their blood supply from lenticulostriate arteries and from the paramedian branches of the basilar bifurcation, respectively.

Termination. As corticonuclear (corticobulbar) fibers pass through the basilar pons, branches arch dorsally into the pontine tegmentum to terminate in the area of the trigeminal and facial motor nuclei (see Fig. 25–10). The fibers to the trigeminal motor nuclei terminate on interneurons adjacent to the nuclei. The corticonuclear system sends nearly equal numbers of fibers to the left and right trigeminal motor nuclei. Likewise, nearly equal numbers of fibers are sent to the left and right facial motor nuclei. Whereas the muscles of facial expression in the upper half of the face are controlled about equally from both hemispheres, *muscles in the lower half of the face are influenced primarily from the contralateral hemisphere.* Consequently, a lesion of corticonuclear fibers rostral to the facial motor nucleus results in drooping of muscles at the corner of the mouth and on the lower portion of the face *on the side opposite the lesion* (Fig. 25–11*A*). This deficit is called *central facial paralysis (central seven).* In contrast, a lesion of the root of the facial nerve will result in a flaccid paralysis of facial muscles of *upper and lower portions of the face on the ipsilateral side* (see Fig. 25–11*B*); this deficit is called *Bell (facial) palsy.*

At mid-medullary levels, corticonuclear (corticobulbar) fibers pass dorsally to reach the ambiguus and hypoglossal nuclei (see Fig. 25–10). Projections to nucleus ambiguus motor neurons are generally bilateral. However, the motor neurons that innervate muscular parts of the soft palate and uvula receive mainly con-

tralateral input. Consequently, a lesion of corticonuclear fibers to the nucleus ambiguus may produce a weakness in the affected muscles and *result in failure of the soft palate to elevate on the contralateral side and in deviation of the uvula toward the side of the lesion* (Fig. 25–12).

In the case of the hypoglossal nuclei, although corticonuclear (corticobulbar) fibers in general distribute bilaterally, those *motor neurons that innervate the genioglossus muscles receive primarily contralateral corticonuclear input.* Each genioglossus muscle pulls its half of the tongue anteriorly and slightly medially. When the two muscles function together and symmetrically, the tongue protrudes straight out of the mouth. However, a lesion of corticonuclear fibers to the hypoglossal nucleus will cause the tongue to *deviate toward the weak side when protruded, because of the unopposed pull of the intact muscle* (see Fig. 25–12). In this example the tongue will deviate *toward the side opposite the lesion* (see Fig. 25–12), and the patient may have other symptoms characteristic of a lesion of the genu of the internal capsule, such as a central seven or a deviation of the uvula. In contrast, an injury to the hypoglossal nerve will result in *deviation of the tongue toward the side of the lesion* (Fig. 25–13*A* and *B*) and ultimately lead to the appearance of lower motor neuron signs, such as muscle atrophy and a flaccid paralysis. If, however, the lesion is in the medulla and involves the root of the hypoglossal nerve, the patient may experience an ipsilateral deviation of the tongue along with a contralateral hemiparesis (corticospinal fiber involvement) and a contralateral loss of posterior column modalities (medial lemniscus involvement). Hypoglossal root fibers, corticospinal fibers, and the medial lemniscus have a common blood supply in the medulla from the anterior spinal artery.

The final contingent of corticonuclear (corticobulbar) fibers innervates the spinal portion of cranial nerve XI (*accessory nucleus*) (see Fig. 25–10). These fibers continue into the upper cervical spinal cord along

Corticonuclear (corticobulbar) Fibers

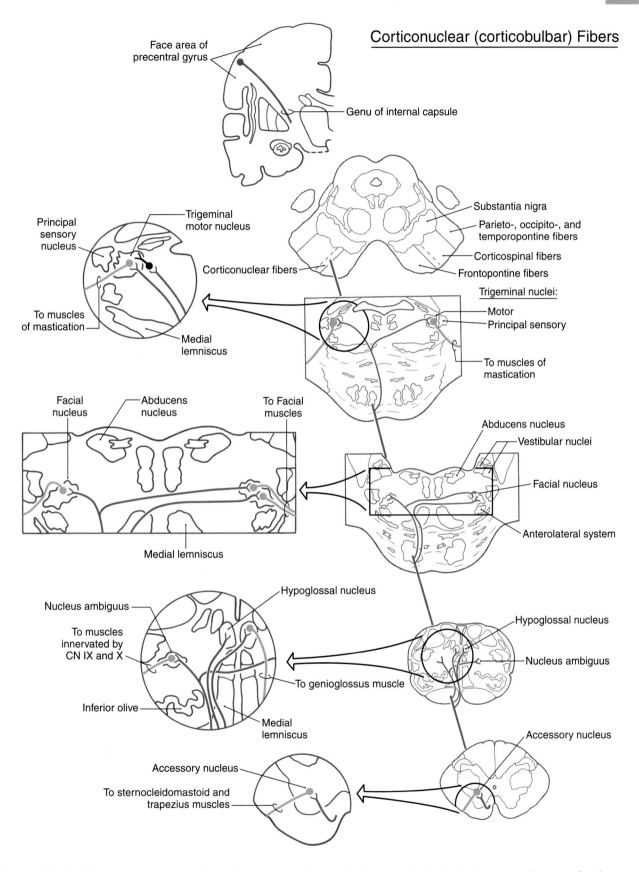

Figure 25–10. The corticonuclear (corticobulbar) system with details shown at the levels of the trigeminal motor, facial motor, hypoglossal and ambiguus, and accessory nuclei. CN, cranial nerve.

with corticospinal fibers. Clinical observations in patients with cortical or internal capsule lesions reveal that the sternocleidomastoid and trapezius muscles (targets of accessory motor neurons) are affected mainly on the side of the lesion. The patient is unable to shrug that shoulder or to turn the head away from the side of the lesion. This finding suggests that *corticonuclear fibers distribute primarily to the ipsilateral accessory nucleus.*

Because the corticospinal and corticonuclear (corticobulbar) systems course adjacent to each other, it is common for lesions of the internal capsule or midbrain to affect both fiber bundles. For example, lenticulostriate arteries serve the genu and most of the posterior limb of the internal capsule (see Fig. 25–4A). Consequently, hemorrhage of these vessels on the *left side* results in (1) a *right spastic hemiparesis* of the extremities (corticospinal damage), (2) a *central facial paralysis* on the right, (3) a *deviation of the uvula to the left* on phonation, and (4) a *deviation of the tongue to the right* when protruded. The latter three deficits are the result of damage to corticonuclear fibers. Effects on the tra-

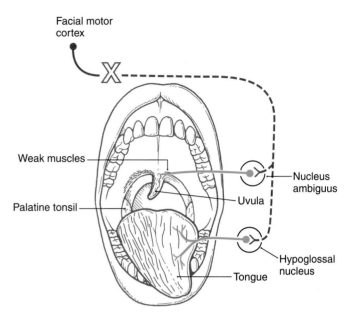

Figure 25–12. Deviation of the uvula and of the tongue following a lesion of corticonuclear (corticobulbar) fibers to the nucleus ambiguus and hypoglossal nucleus, respectively. The large X indicates the side of the lesion. Note that the tongue deviates to the side opposite the lesion, whereas the uvula deviates toward the side of the lesion. Compare the deviation of the tongue with that in Figure 25–13.

pezius and sternocleidomastoid muscles are variable and are absent in some patients. Lesions at lower brainstem levels may produce corticospinal and corticonuclear signs in combination with signs and symptoms of cranial nerve root injury. For example, a lesion in the territory of the paramedian branches of the basilar artery will damage corticospinal fibers, corticonuclear fibers to relatively caudal targets (nucleus ambiguus, hypoglossal nucleus), and the exiting rootlets of the abducens nerve (see Fig. 25–7). A patient with such a lesion presents with (1) a *contralateral spastic hemiparesis* of the extremities, (2) *ipsilateral deviation of the uvula* on phonation, (3) *deviation of the tongue* (on protrusion) *to the contralateral side*, and (4) a *medial rotation of the ipsilateral eye* caused by a flaccid paralysis of the lateral rectus muscle and the unopposed action of the medial rectus. Comparison of corticospinal with corticonuclear deficits is essential to localizing in neurologically compromised patients.

Other Corticofugal Systems

The Corticorubral System. Cortical projections to the red nucleus arise primarily from areas 4 and 6, and to a lesser extent from areas 5 and 7. Both large and small neurons in the red nucleus receive ipsilateral corticorubral input. Although pyramidal tract neurons provide some collaterals to the red nucleus, many cor-

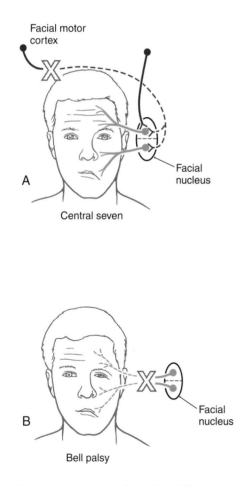

Figure 25–11. Appearance of the face following a lesion of corticonuclear (corticobulbar) fibers to the facial nucleus *(A)* versus a lesion of the root of the facial nerve *(B)*. The large X indicates the location of the lesion.

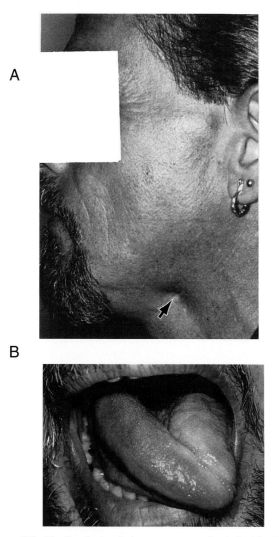

Figure 25–13. Paralysis of the tongue on the left side. Removal of a lymph node from the left side of the neck (*A, arrow*) inadvertently resulted in damage to peripheral fibers of the hypoglossal nerve on that side. The tongue deviates to the left (side of the lesion) on protrusion (*B*).

ticorubral axons are not collaterals of pyramidal tract fibers. In general, the corticorubral-rubrospinal projection is topographically organized. For example, the arm region of the MI cortex projects to cells of the red nucleus that, in turn, send their axons to contralateral cervical levels of the spinal cord (see Fig. 24–10). Because the rubrospinal system primarily influences flexor musculature, this pathway may supplement the function of the corticospinal tract. It is known from experimental studies that section of corticospinal fibers in the medullary pyramid leaves the animal still able to walk, climb, and pick up food but unable to perform fine, dexterous movements with its digits. This finding suggests that the *corticorubrospinal system may partially compensate for the loss of the corticospinal tract.*

The red nucleus also receives input from the contralateral interposed nuclei of the cerebellum (see Chapter

27). Consequently, a fairly small population of brainstem upper motor neurons are capable of integrating signals from motor-related areas of the cerebral cortex and from the cerebellum. Input from the interposed nuclei is excitatory, and this projection may be part of a circuit specialized for rapid control or adjustment of movements based on sensory processing by the cerebellum.

The Corticoreticular System. The pontine and medullary nuclei that give rise to the reticulospinal tracts receive cortical input from the premotor cortex and, to a lesser extent, from the supplementary motor cortex. Because reticulospinal systems influence extensor muscles, including the paravertebral extensors as well as those of the limbs, the *corticoreticulospinal system* may provide the cortex with the means to influence extensor musculature in parallel with its regulation of flexors (see Fig. 24–9). It should also be noted that the cerebellar nuclei project to the motor-related areas of the reticular formation, thus providing for a cerebellar influence on extensor musculature.

The Corticopontine System. Axons from nearly all regions of the cerebral cortex contribute to the corticopontine projection, and this pathway is particularly well developed in the human brain. Although most of these fibers originate from motor-related areas, nonmotor regions in the frontal lobe and in parietal, temporal, and occipital association cortices also contribute fibers. Corticopontine axons descend through the internal capsule (see Fig. 16–11) and continue into medial and lateral parts of the crus cerebri. Frontopontine fibers are located medially, and parieto-, occipito-, and temporopontine fibers are laterally placed (see Fig. 13–8). These corticopontine projections synapse in the ipsilateral basilar pontine nuclei. Although most neurons in the pontine nuclei send their axons into the contralateral cerebellum via the middle cerebellar peduncle, there is a notable ipsilateral projection.

Although little is known about the function of this vast system, it certainly must be involved in some aspects of motor control. However, recent studies in humans have shown that the cerebellum is also active during mental problem solving and internal (nonvocal) language functions. This finding implies that the corticopontine system is an important route of communication between the cerebral cortex and the cerebellum, structures that have no direct connections in the mature brain.

Motor Cortex and the Control of Movement

The classical view of voluntary movement control is that the various motor-related areas of the cerebral cortex function hierarchically. At one time, the primary motor cortex was thought to form the apex of this hierarchy, the output of the other cortical areas being funneled through it. Recent findings suggest that the motor-related cortical areas outside MI and their respective descending projections carry out the tasks involved in planning and executing a movement *in parallel* with MI and its projections.

Primary Motor Cortex. Recall that the primary motor cortex (MI) is organized into a detailed somatotopic map of the body (see Fig. 25–2) and that many corticospinal fibers originate from the pyramidal neurons in its layer V. What is this area's contribution to the control of movement?

Like other cortical areas, such as the striate and primary somatosensory cortices, MI is organized into a series of modules or *vertical columns.* Microstimulation in MI can result in discrete movements of individual muscles. For example, stimulation in a vertical column in the hand area of *MI* may evoke flexion of a digit (Fig. 25–14). Neurons in the same columnar array receive somatosensory feedback from the patch of skin on the volar (glabrous) side of the digit, which is the area that would come in contact with a surface when the digit is flexed, as to grasp an object. These connections are part of *long-latency reflex* circuits. The sensory information reaches MI from the ascending

sensory systems indirectly via synapses in the thalamus and the primary sensory cortex (SI). The inference here is that *motor cortical neurons are informed of the result of their output.*

Furthermore, studies have shown that certain muscles, particularly distal muscles of the upper extremity, are regulated from more than one cortical location. Conversely, microstimulation at one cortical locus can activate more than one muscle. Thus, the classical view that a discrete, somatotopically organized projection emanating from area 4 is primarily responsible for the control of individual muscles may be an oversimplification.

The activity of corticospinal neurons in MI can also be modified during a movement. In experiments involving nonhuman primates, microelectrodes placed in layer V identified single corticospinal neurons by their response to *antidromic stimulation* of a medullary pyramid. These monkeys were trained to make wrist flexion or extension movements in a situation in which the movement was either assisted or impeded by a weight attached to the wrist by a pulley system. Under these conditions, corticospinal neurons were found to be *active slightly in advance of the movement*, and they did not simply code for flexion or extension but rather for *the amount of force required to make the movement.* For example, when the weight was arranged to oppose the movement, increases in the amount of weight were matched by increases in the activity of corticospinal neurons.

Other populations of MI cortical neurons encode *movement direction.* When a monkey is trained to move a handle toward one of several targets arranged concentrically around a central starting location, the activity of individual cortical neurons varies with the direction of the required movement. That is, some neurons fire rapidly for a movement in one direction but are silent for a movement in the opposite direction. This directional tuning is rather broad, however, with most neurons firing with movement in a preferred direction but exhibiting less vigorous activity in relation to movements in other directions.

Supplementary Motor Cortex. The supplementary motor cortex occupies the portion of Brodmann area 6 that lies rostral to MI near the convexity and extends onto the medial wall of the hemisphere rostral to the paracentral gyri (see Fig. 25–2). It contains a map of the body musculature that is complete, although less precisely organized than that of MI. It receives input from the parietal lobe and projects to MI and directly to the reticular formation and spinal cord.

Stimulation of the supplementary cortex can evoke movements. In contrast to the single-muscle movements evoked by MI stimulation, however, these movements involve sequences or groups of muscles and orient the body or limbs in space. In addition, stimuli of higher intensities are required than for MI, and bilat-

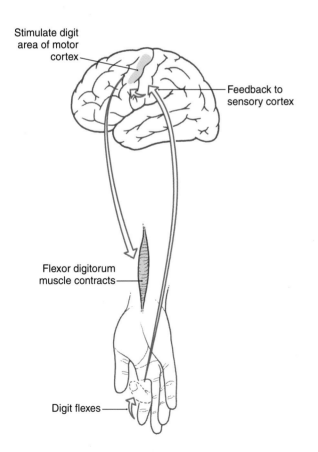

Figure 25–14. Schematic representation of a long-latency reflex.

eral movements of the hands or upper extremities are often produced.

Experiments performed in humans illustrate the functional role of the supplementary motor cortex. By using a device similar to a computed tomography scanner and injecting small amounts of a solution containing radioactive xenon into the vascular system, it is possible to measure small increases in local blood flow (as enhanced radioactivity) in brain regions where neuronal activity is increased. When a volunteer made a series of random finger flexion movements, an increase in neural activity was observed only over the hand region of MI (Fig. 25–15A). The subject was then asked to make flexion movements with several fingers of the same hand, but in a specific sequence. Activity was increased both in the *supplementary motor cortex* and in the hand region of MI (see Fig. 25–15B). Finally, when the subject was asked to mentally rehearse the sequence of finger movements without actually moving the fingers, increased activity was restricted to the supplementary cortex, and the hand region of *MI* was

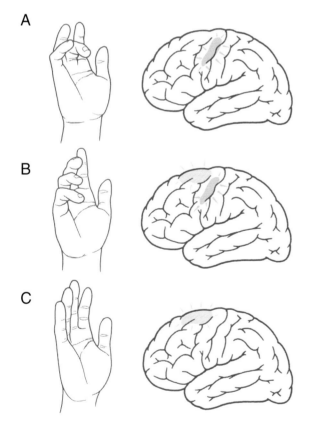

Figure 25–15. Random movements made without prior planning and in no particular order *(A)* result in increased activity in only the hand area of motor cortex. When the movement is planned and executed in a specific sequence *(B)*, both motor and supplementary cortices are active. When the movement is mentally planned and rehearsed but never executed *(C)*, only the supplementary cortex is activated.

silent (see Fig. 25–15C). These findings indicate that the supplementary motor cortex is involved in *organizing* or *planning* the sequence of muscle activation required to make a movement, whereas the primary motor cortex functions mainly to *execute* the movement.

Premotor Cortex. The premotor cortex occupies the portion of area 6 lying just rostral to the anterolateral part of MI (see Fig. 25–2). Like the supplementary motor cortex, this region contains a somatotopic representation of the body musculature that is complete, although less precisely organized than that of MI. The premotor cortex receives considerable input from sensory areas of the parietal cortex and projects to MI, the spinal cord, and the reticular formation. The reticular formation gives rise to reticulospinal fibers, which, in turn, influence spinal motor neurons that innervate paravertebral and proximal limb musculature.

On the basis of these connections, it was suggested that premotor cortex, like the supplementary cortex, is involved in the *preparation to move*. That is, it organizes those postural adjustments that are required to make a movement. To test this concept, monkeys were trained to move one hand to a specific target location that differed from trial to trial. The monkey was first given a cue signaling which target to reach for and then a "go" signal to actually make the movement. Recordings of *cell activity* revealed that premotor neurons were active only *during the interval between presentation of the cue and the "go" signal*. The premotor cortex is most active in directing the control of proximal limb muscles that are used to position the arm for movement tasks or, more generally, to orient the body for movement.

Posterior Parietal Cortex. The motor regions of the posterior parietal cortex comprise Brodmann areas 5 and 7, which largely occupy the superior parietal lobule (see Fig. 25–2). These areas carry out some of the "background computations" necessary for making movements in space. To organize such a movement, it is necessary to collate input from a variety of sensory systems to create a map of space and to compute a trajectory by which a body part can reach its target. Area 5 receives extensive projections from somatosensory cortex and input from the vestibular system, whereas area 7 processes visual information related to the location of objects in space. Both areas project primarily to supplementary and premotor cortices and have few spinal or brainstem targets.

Experiments in monkeys provide the best insight into the function of areas 5 and 7. In area 5, *arm projection neurons* are active only when the monkey reaches for a specific object of interest. They are not active when the same arm movement is made but the object is not present. In area 7, many different types of neurons are present. Neurons of one of these types, the *eye-hand coordination neurons*, are vigorously active only when the eyes fixate a target and the hand reaches for that target.

Cingulate Motor Cortex. Two aggregates of corticospinal neurons are associated with the cingulate gyrus (see Fig. 25–2). One occupies the anterior (ventral) bank of the cingulate sulcus, and the other, located slightly more caudally, occupies the posterior (dorsal) and anterior (ventral) banks of the cingulate sulcus. Each is topographically organized with respect to its spinal cord projections, and each also projects to primary motor cortex. Little is known about the functional role of these areas, other than that stimulation in either area produces motor effects. Because of their proximity to limbic cortex, these motor neurons may be involved in movements that have an intense motivational or emotional component.

Cerebellar and Pallidal Influences. The basal nuclei and cerebellum play essential roles in the control of movement through their interaction with motor-related areas of the cerebral cortex. Although these pathways are detailed in Chapters 26 and 27, their general relationships are summarized here. The cerebellar nuclei and the globus pallidus project primarily to spatially segregated regions in the ventral anterior, ventral lateral, and oral parts of the ventral posterolateral nuclei of the dorsal thalamus. These so-called motor areas of the thalamus give rise to thalamocortical projections to MI and the supplementary motor area. The thalamic areas that receive input from the globus pallidus project mainly to the supplementary motor cortex, and the thalamic areas that receive input from the cerebellum

project to MI. The degree to which these two lines of communication are separate is not clear. Therefore, it is important to realize that signals transmitted through the corticospinal and corticonuclear systems can be modified by outputs that reach the cortex from the cerebellum, basal nuclei, and thalamus.

Hierarchical Organization Versus Parallel Distributed Processing in the Motor System

Until recently, it was generally thought that the control of voluntary movement could be satisfactorily described by the hierarchical scheme shown in Figure 25–16. In this plan, lower motor neurons and their associated interneurons are influenced by (1) segmental peripheral sensory feedback circuits, (2) descending brainstem-spinal systems modulated by the cerebral cortex, and (3) the corticospinal system.

The motor areas in the cortex are organized such that the output of the corticospinal system is regulated and modulated by higher-order motor cortices. When a voluntary movement is desired, a plan for the movement is organized by the combined efforts of the higher-order motor areas and then transmitted to MI. The primary motor cortex then executes the plan by communicating with the spinal motor apparatus either directly or indirectly via brainstem-spinal systems (see

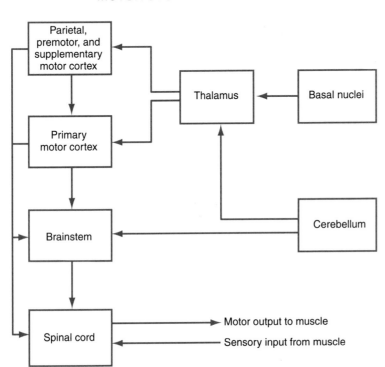

MOTOR SYSTEM HIERARCHY

Figure 25–16. The hierarchical organization of the motor system. Recent evidence suggests that the direct links from the various brainstem and cortical centers to the spinal cord, which operate as a series of systems, may be more important than was previously thought, and that hierarchical organization may be less important.

Fig. 25–16). However, a re-examination of the origins of the corticospinal tract indicates that more of these fibers than had been previously thought originate from cells located *outside MI*. In addition, circuits linking the basal nuclei, thalamus, and motor cortical regions are segregated to some extent from pathways linking the cerebellum, thalamus, and motor cortical areas.

These observations have led to an evolving hypothesis that motor system control is achieved by a series of *parallel systems* formed by somatotopically organized, descending cortical projections that link the various motor-related areas of cortex more directly with spinal

motor circuits. Included here are spinal projections originating from the so-called higher-order motor regions such as the premotor and supplementary motor areas. The concept is that each descending cortical pathway contributes its own element or series of elements to movement control. This idea is supported by the fact that stimulation in different cortical locations can elicit movements involving the same muscles or muscle groups, but the characteristics of the movement differ somewhat for each site of stimulation. However, additional information is needed before this hypothesis can be validated.

Sources and Additional Reading

Asanuma H, Rosen I: Topographical organization of cortical efferent zones projecting to distal forelimb muscles in the monkey. Exp Brain Res 14:243–256, 1972.

Asanuma H: The Motor Cortex. Raven Press, New York, 1988.

Brodal A: Neurological Anatomy in Relation to Clinical Medicine, 3rd Ed. Oxford University Press, New York, 1981.

Evarts EV: Relation of pyramidal tract activity to force exerted during voluntary movement. J Neurophysiol 31:14–27, 1968.

Georgopoulos AP: Higher order motor control. Annu Rev Neurosci 14:361–377, 1991.

Humphrey DR: On the cortical control of visually directed reaching: contributions by nonprecentral motor areas. In Talbot RE, Humphrey DR (eds): Posture and Movement. Raven Press, New York, 1979, pp 51–112.

Humphrey DR: Representation of movements and muscles within the primate precentral motor cortex: Historical and current perspectives. Fed Proc 45:2687–2699, 1986.

Penfield W, Rasmussen T: The Cerebral Cortex of Man: A Clinical Study of Localization of Function. Hafner Publishing, New York, 1968 (facsimile of 1950 ed).

Roland PE, Larsen B, Lassen NA, Skinholf E: Supplementary motor area and other cortical areas in organization of voluntary movements in man. J Neurophysiol 43:118–136, 1980.

Terminologia Anatomica: International Anatomical Terminology. Federative International Committee on Anatomical Terminology. Thieme, Stuttgart, 1998.

Chapter 26

The Basal Nuclei

T. P. Ma

Overview 406

Components of the Basal Nuclei 407
The Striatal Complex
The Pallidal Complex
The Subthalamic Nucleus
The Nigral Complex
The Parabrachial Pontine Reticular
Formation
Ventral Basal Nuclei

**Direct and Indirect Pathways of Basal Nuclear
Activity 412**
Disinhibition as the Primary Mode of Basal Nuclear Function

**Parallel Circuits of Information Flow Through the Basal
Nuclei 413**
The Motor Loop

Behavioral Functions of the Basal Nuclei 415
Hypokinetic Disturbances
Hyperkinetic Disturbances
Integrated Function of the Basal Nuclei

Etiology of Basal Nuclear Related Disorders 419
Huntington Disease
Parkinson Disease
Wilson Disease
Sydenham Chorea
Tardive Dyskinesia

Voluntary movement is essential to the well-being of living animals. Such behaviors are accomplished by signals that direct the actions of individual muscles. Although these signals originate in the cerebral cortex, they are modulated by a variety of subcortical structures. One such group of structures is the basal nuclei (ganglia) and their functionally associated cell groups. Classically, motor systems have been divided into "pyramidal" and "extrapyramidal" on the basis of whether the pathway is mediated by corticofugal neurons (pyramidal) or by the basal nuclei, cerebellum, or descending brainstem pathways (extrapyramidal). However, this distinction is overly simplistic, if not inaccurate. Consequently, it is not used here. The basal nuclei are involved in a wide variety of motor and affective behaviors, in sensorimotor integration, and in cognitive functions.

Overview

These nuclei are traditionally called the *basal ganglia* rather than the *basal nuclei*, even though "ganglia" is usually reserved for groups of nerve cell bodies in the peripheral nervous system. The official term *basal nuclei* is used throughout this chapter, although the unofficial term *basal ganglia* is commonly used in other sources.

For practical purposes they may be considered interchangeable.

The *basal nuclei* consist of cell groups embedded in the cerebral hemisphere. Although not classified as nuclei of the basal nuclei in a strict sense, the *subthalamic nucleus, substantia nigra,* and *pedunculopontine tegmental nucleus* are integral parts of the pathways passing through these forebrain cell groups. Collectively, the basal nuclei and their associated nuclei function primarily as components in a series of parallel circuits from the cerebral cortex through the basal nuclei to the thalamus and then back to the cerebral cortex.

Four fundamental concepts are crucial to understanding the basal nuclei. First, damage to or disorders of the basal nuclei result in disruption of movements, and may also cause significant deficits in other neural functions such as cognition, perception, and mentation. Second, the basal nuclei are anatomically and functionally segregated into parallel circuits that process different types of behaviorally significant information. Third, the basal nuclei function primarily through *disinhibition* (release from inhibition). Fourth, diseases of the basal nuclei can be described as disruptions of the neurochemical interactions between elements of the basal nuclei. These neurochemical relationships rely not simply on the neurotransmitters involved but also on the characteristics of the transmitter receptors, on the loca-

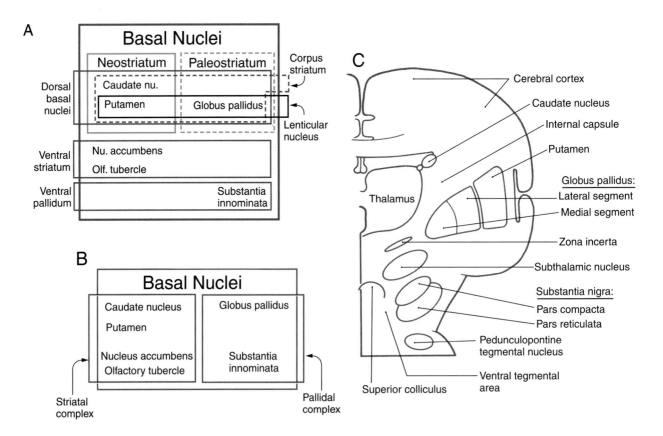

Figure 26–1. A series of stacked boxes (*A*) illustrating which nuclei form the various parts of the basal nuclei and how these groups are used in this chapter (*B*). A standard drawing of the basal nuclei (*C*) used throughout this chapter.

tions of the synapses, and on other inputs received by these cells.

In summary, the basal nuclei integrate and modulate cortical information along multiple parallel channels. These channels affect behavior indirectly by feedback to the cerebral cortex and directly by providing information to subcortical centers that influence movements. Disruption of these channels by stroke or disease results in dysfunctions characteristic of the circuits damaged.

Components of the Basal Nuclei

The basal nuclei are typically divided into dorsal (posterior) and ventral (anterior) divisions. The dorsal basal nuclei include the *caudate* and *putamen* (together constituting the *neostriatum*) and the *globus pallidus* (constituting the *paleostriatum*) (Fig. 26–1A). Associated with the dorsal basal nuclei are the *substantia nigra*, the *subthalamic nucleus*, and the *parabrachial pontine reticular formation* (containing the *pedunculopontine tegmental nucleus*). The ventral basal nuclei are located inferior to the anterior commissure and include the *substantia innominata*, *nucleus basalis of Meynert*, *nucleus accumbens*, and *olfactory tubercle*. This ventral region is intimately associated with portions of the amygdala and ventral tegmental area. For the purposes of this chapter, the basal nuclei are regarded as making up two complexes: the striatal complex and the pallidal complex (see Fig. 26–1B).

The telencephalic regions of the basal nuclei are supplied by the *medial striate artery, lenticulostriate branches* of the M_1 segment of the middle cerebral artery, and the *anterior choroidal artery* (Fig. 26–2A and B). The diencephalic and mesencephalic regions are supplied by the posteromedial branches of the P_1 segment of the posterior cerebral artery and branches of the posterior communicating artery. Diseases of these vessels may result in various behavioral or motor deficits, depending on which vessel and region is affected.

The Striatal Complex. The *striatal complex* is a functional unit composed of the *neostriatum* and *ventral striatum* (Fig. 26–3A and B). The neostriatum consists of the *caudate nucleus* and *putamen*. These two nuclei have the same embryological origin and similar connections. Although fused rostroventrally, they are separated throughout most of their extents by fibers of the internal capsule. The ventral striatum is composed of the *nucleus accumbens* and the *olfactory tubercle* (see Figs. 26–1A and B and 26–3A). The nucleus accumbens is located rostroventrally in the hemisphere, subjacent to where the putamen is continuous with the head of the caudate. It is internal to part of the anterior perforated substance. Portions of the olfactory tubercle are considered part of the ventral striatum because of functional, cytoarchitectural, and chemoarchitectural similarities. The olfactory tubercle is unique in that it has striatal characteristics yet receives primary olfactory information.

A defining characteristic of the striatal complex, *striosomes* (also called *patches*) are particularly prominent in the head of the caudate (see Fig. 26–4). Striosomes are acetylcholinesterase-poor regions within the striatal complex. They contain large amounts of one or more

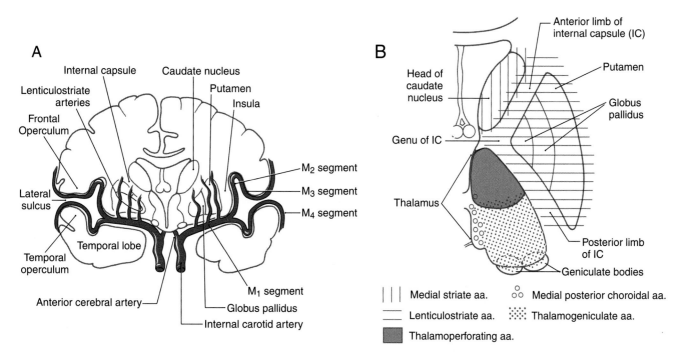

Figure 26–2. Blood supply to the basal nuclei in coronal (*A*) and axial (*B*) planes.

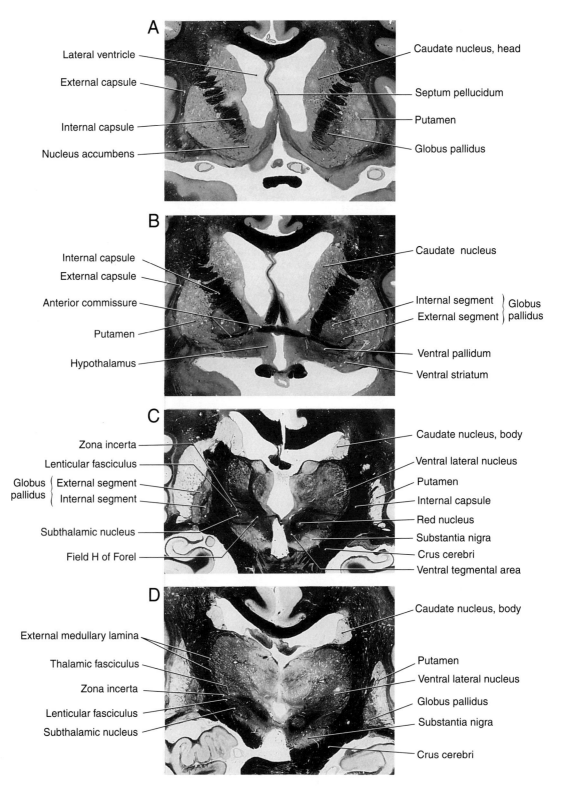

Figure 26–3. Cross sections of the human brain from rostral (*A*) to caudal (*D*) showing the basal nuclei and related structures. Myelin stain.

neuropeptides and one or more types of opiate receptors. The remainder of the striatal complex, called the *matrix*, contains high concentrations of acetylcholinesterase and therefore stains darkly when tissue is histochemically reacted to reveal this enzyme.

The largest afferent projections to the striatum are from the cerebral cortex (*corticostriatal fibers*) (Fig. 26–5A). Other afferents are from the thalamus (*thalamostriatal fibers*), substantia nigra (*nigrostriatal fibers*), and parabrachial pontine reticular formation (*pedunculopon-*

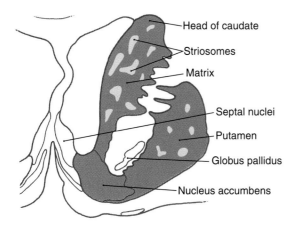

Figure 26-4. Schematic representation of striosomes and matrix in the striatal complex.

tostriatal fibers) (see Fig. 26–5A). The efferent projections of the striatum reach primarily the pallidum (striatopallidal fibers) and the nigral complex (striatonigral fibers), and to a small degree the subthalamic nucleus.

Most of the neurons in the neostriatum are called medium spiny neurons, so named because of their medium-sized cell bodies and the large numbers of spines on their dendrites (Fig. 26–6). Most medium spiny cells have dendritic fields that are restricted to the region in which the cell bodies are located. That is, medium spiny neurons in striosomes have dendrites that usually ramify within striosomes; medium spiny neurons within the matrix primarily ramify within the matrix. Both the afferent and efferent connections of neurons within the striosomes and matrix also differ significantly. Adjacent neurons across a striosome-matrix border may receive cortical projections from widely different areas. Axon collaterals of neurons within striosomes remain within striosomes, whereas those from the matrix remain within the matrix. Even when both compartments project to the same nucleus, they differ in the region to which they project, thus resulting in different functional effects.

Medium spiny neurons fire few action potentials spontaneously and thus require activation by their afferent fibers. These cells use the inhibitory neurotransmitter gamma-aminobutyric acid (GABA) and may also contain neuroactive peptides such as substance P and enkephalin. Thus, when medium spiny neurons are activated, they subserve both direct inhibitory and neuromodulatory functions at their targets. These cells are the only efferent neurons of the neostriatum.

Also found in the neostriatum are large, acetylcholine-containing local circuit neurons that modulate local activity within the neostriatum. Huntington disease is characterized by progressive loss of medium spiny neurons and acetylcholine-containing neurons throughout the striatal complex.

The Pallidal Complex. The pallidal complex is composed of the globus pallidus and the ventral pallidum. The latter is largely synonymous with the substantia innominata (see Fig. 26–3B and C). The pallidal complex contains primarily GABAergic neurons with high rates of spontaneous activity. Consequently, these cells tonically inhibit their targets.

The globus pallidus is divided into medial (internal) and lateral (external) segments by a sheet of white matter (the medullary lamina) (see Fig. 26–3C). The substantia innominata is located anterior (ventral) to the anterior commissure and internal to the anterior perforated substance. One important cell group in the substantia innominata is the basal nucleus of Meynert. This nucleus has large acetylcholine-containing neurons, which are lost in Alzheimer disease. However, this disease is not considered a basal nuclear disorder because acetylcholine-containing cells in the cerebral cortex, hippocampus, and septum are also lost in Alzheimer disease patients. This disease is further characterized by other biochemical and pathologic features such as senile plaques.

The two divisions of the globus pallidus are reciprocally connected (pallidopallidal fibers) (see Fig. 26–5B) but subserve different functions. The main afferent input to the pallidum is from the striatal complex. Medium spiny neurons from the striatum that project to the medial segment and substantia nigra use GABA and substance P; those that project to the lateral segment use GABA and enkephalin (see Fig. 26–5B).

The medial division is composed of the medial segment of the globus pallidus. It subserves the direct basal nuclear pathway (described below) and projects primarily to the thalamus (pallidothalamic fibers) (see Fig. 26–5B). These fibers exit the globus pallidus as two bundles: the ansa lenticularis and the lenticular fasciculus (Fig. 26–7). The ansa lenticularis originates from lateral portions of the medial segment and loops around the posterior limb of the internal capsule to enter field H of Forel. The lenticular fasciculus, on the other hand, originates in the posteromedial portion of the medial segment. These fibers traverse the internal capsule as small groups of axons, merge to form the lenticular fasciculus between the zona incerta and subthalamic nucleus, and then enter field H of Forel. In Forel's field, the ansa lenticularis and lenticular fasciculus join the thalamic fasciculus, which courses posterior (dorsal) to the zona incerta (see Fig. 26–7). These fibers ultimately terminate in ventral anterior, ventral lateral, and centromedian nuclei of the thalamus. The medial division of the pallidal complex is a principal efferent nucleus of the basal nuclei.

The lateral division is composed of the external (or lateral) segment of the globus pallidus and the ventral pallidum. This division subserves the indirect basal nuclear pathway (see below). These nuclei receive a large input from the striatal complex (striatopallidal fibers) and

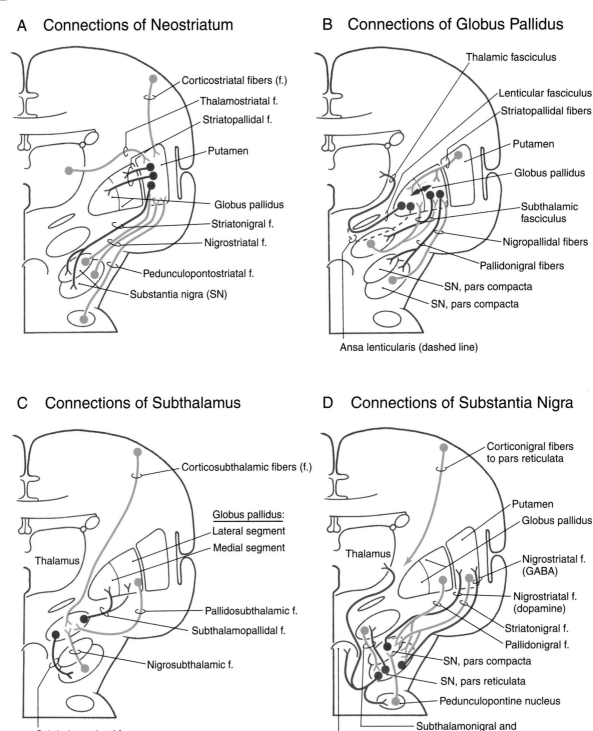

A   Connections of Neostriatum

Corticostriatal fibers (f.)
Thalamostriatal f.
Striatopallidal f.
Putamen
Globus pallidus
Striatonigral f.
Nigrostriatal f.
Pedunculopontostriatal f.
Substantia nigra (SN)

B   Connections of Globus Pallidus

Thalamic fasciculus
Lenticular fasciculus
Striatopallidal fibers
Putamen
Globus pallidus
Subthalamic fasciculus
Nigropallidal fibers
Pallidonigral fibers
SN, pars compacta
SN, pars compacta
Ansa lenticularis (dashed line)

C   Connections of Subthalamus

Corticosubthalamic fibers (f.)
Globus pallidus:
Lateral segment
Medial segment
Thalamus
Pallidosubthalamic f.
Subthalamopallidal f.
Nigrosubthalamic f.
Subthalamonigral f.

D   Connections of Substantia Nigra

Corticonigral fibers to pars reticulata
Putamen
Globus pallidus
Nigrostriatal f. (GABA)
Nigrostriatal f. (dopamine)
Striatonigral f.
Pallidonigral f.
SN, pars compacta
SN, pars reticulata
Pedunculopontine nucleus
Subthalamonigral and nigrosubthalamic f.
Thalamus
Superior colliculus

Figure 26–5. Schematic representations of the afferent (in green) and efferent (in red) connections of the neostriatum (*A*) and subthalamus (*C*) and of the afferent (in green) and efferent (in red and blue) connections of the globus pallidus (*B*) and substantia nigra (*D*). The *double-headed arrow* in *B* represents pallidopallidal fibers.

small projections from the subthalamic nucleus (*subthalamopallidal fibers*) and the substantia nigra pars reticulata (*nigropallidal fibers*). They project strongly to the subthalamic nucleus (*pallidosubthalamic fibers*) and are

also connected with the substantia nigra (*pallidonigral fibers*) (see Fig. 26–5B).

The Subthalamic Nucleus. The subthalamic nucleus is a lens-shaped cell group that makes up the largest

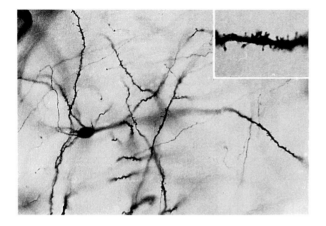

Figure 26–6. Medium spiny neuron from the primate neostriatum. The detail shows the characteristic appearance of dendritic spines on these cells. (Photos courtesy of Dr. José Rafols.)

part of the ventral thalamus. It is immediately anterior (ventral) to the zona incerta and rostral to the substantia nigra (see Fig. 26–3C and D). It receives projections from the lateral pallidal division (*pallidosubthalamic fibers*), cerebral cortex (*corticosubthalamic fibers*), nigral complex (*nigrosubthalamic fibers*), and parabrachial pontine reticular formation. The subthalamic nucleus projects to both pallidal divisions (*subthalamopallidal fibers*) and to the substantia nigra (*subthalamonigral fibers*) (see Fig. 26–5C). These connections, especially the subthalamopallidal projections to the medial globus palli-

dus, are an essential part of the indirect pathway underlying basal nuclear function.

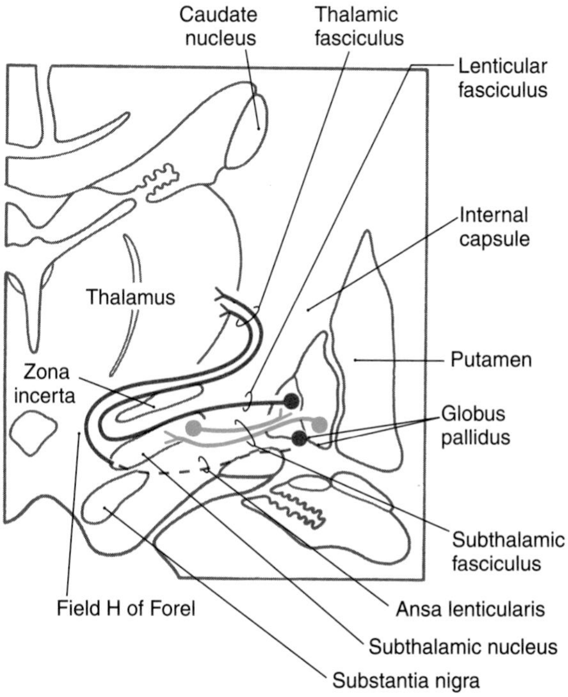

Subthalamic neurons use the excitatory neurotransmitter glutamate. Most of the time, subthalamic cells are inactive because of the constant inhibition by cells of the external pallidal segment. However, if this inhibition is removed, subthalamic neurons have a high level of activity. This activity is mediated in part by a large corticosubthalamic projection.

The Nigral Complex. The nigral complex is composed of the *substantia nigra* and the *ventral (anterior) tegmental area* (see Fig. 26–3C and D). The substantia nigra is divided into a cell-dense portion (*pars compacta*) and a reticulated portion (referred to here as the *pars reticulata*, although it can be divided into a *pars reticulata* and a *pars lateralis*). The pars reticulata is located at and within the medial edge of the descending corticofugal fibers that form the crus cerebri. The pars compacta and the adjacent ventral (anterior) tegmental area appear to subserve similar functions and to have a similar chemoarchitectural organization. The major afferents to the nigral complex are from the striatal and pallidal complexes. The nigral complex also receives cortical (*corticonigral*), subthalamic (*subthalamonigral*), and pedunculopontine fibers (see Fig. 26–5D).

The pars compacta includes a large number of neuromelanin-containing cells, whose dark color gives the nucleus its name (*substantia nigra*, "dark substance"). Neurons in the pars compacta use the neurotransmitter dopamine and project primarily to the neostriatum as *nigrostriatal fibers*. The dopamine released by these cells may excite or inhibit striatal neurons, depending on the *type of receptor* on the postsynaptic membrane.

The pars reticulata is formed by loose aggregations of medium- to large-sized GABAergic neurons that are indistinguishable from those of the medial pallidum. Neurons in the pars reticulata have axons with an extensive system of collaterals; consequently, they may project to and inhibit one or more target structures. These targets include the neostriatum (*nigrostriatal fibers*), thalamus (*nigrothalamic fibers*), superior colliculus (*nigrotectal fibers*), and the parabrachial pontine reticular formation. These cells have a high rate of discharge and tonically inhibit their targets. Projections of pars reticulata neurons represent an important pathway by which the basal nuclei influence other motor centers.

The pars compacta and the pars reticulata are interconnected. Dendrites of dopaminergic pars compacta neurons extend into the pars reticulata, where they release free dopamine (by a nonvesicular mechanism). The level of dopamine modulates the resting membrane potential of pars reticula cells, making them either more or less likely to discharge, depending on the subtype of dopamine receptor they possess. In turn, pars reticulata neurons have axon collaterals that ramify extensively in the pars compacta and form GABAergic synapses. Collectively, these interactions form modulat-

Labels (Figure 26–7): Caudate nucleus · Thalamic fasciculus · Lenticular fasciculus · Internal capsule · Putamen · Globus pallidus · Subthalamic fasciculus · Subthalamic nucleus · Substantia nigra · Ansa lenticularis · Field H of Forel · Zona incerta · Thalamus

Figure 26–7. Schematic representation of pallidal connections to the thalamus and with the subthalamic nucleus.

ory loops between neurons of the pars compacta and the pars reticulata.

These modulatory loops are influenced by the segregated output of the striatal striosome-matrix system. Neurons in the striosomes project predominantly onto the cells in the pars compacta whereas the neurons in the matrix project predominantly onto the cells in the pars reticulata. Thus, the striatonigral projection from the striosomes directly inhibits the dopaminergic nigrostriatal neurons and the projection from the matrix inhibits the GABAergic neurons in the pars reticulata. Consequently, inhibition of the pars reticulata is reduced and its targets, such as the pars compacta, are released from inhibition and may become more active.

The ventral (anterior) tegmental area is located medial to the substantia nigra. It contains large numbers of dopaminergic neurons and forms connections with the ventral striatum, the amygdala, and other limbic system structures. Cells of the ventral (anterior) tegmental area project to, and terminate on, striatal neurons that have postsynaptic D_2 (dopamine) receptors. In schizophrenia there is an increase in number and in sensitivity of these receptors. Neuroleptic drugs help to control schizophrenia by blocking (down-regulating) these D_2 receptors.

The Parabrachial Pontine Reticular Formation. Nuclei in the region of the parabrachial pontine reticular formation, primarily the pedunculopontine tegmental nucleus, are intimately connected with all portions of the basal nuclei and their associated nuclei. For example, GABAergic substantia nigra pars reticulata neurons project onto cells of the pedunculopontine tegmental nucleus. In return, acetylcholine-containing pedunculopontine tegmental neurons project to the substantia nigra pars compacta. In addition, the parabrachial pontine nuclei are connected with motor centers in the brainstem, which project to the spinal cord via the descending spinal pathways. Thus, these nuclei serve as an efferent pathway for the basal nuclei. It has been suggested that damage to the connections between the pedunculopontine nucleus and the basal nuclei may partially account for motor deficits, such as tremor or chorea, in some patients.

Ventral Basal Nuclei. The ventral (anterior) basal nuclear pathways are similar to those of the dorsal (posterior) basal nuclei. The pathways originate in the allocortex (limbic-related cortical areas) and the orbital and medial prefrontal cortices. The corticostriatal pathway terminates in the ventral striatum (nucleus accumbens and portions of the olfactory tubercle). Striatopallidal neurons then project to the ventral pallidum (substantia innominata). Then the pallidothalamic neurons project to the mediodorsal nucleus of the thalamus, whose neurons then project to the cerebral cortex. As with the dorsal basal nuclei, distinct regions of each nucleus in these pathways are connected with each other. Thus, there is a clear loop through the ventral basal

nuclei for information originating from different areas of the cortex, as is discussed later on.

Numerous functions are ascribed to the ventral basal nuclei. Generally, they can be grouped as natural or biologic rewards, such as feeding, drinking, sex, exploration, and appetitive learning. Consequently, diseases that can be associated with imbalances of "rewards," such as drug and alcohol abuse, can be related to the ventral basal nuclei. Similarly, other psychiatric disorders such as schizophrenia and certain forms of affective disorders are also linked to the ventral basal nuclei. Thus, it is not surprising that the pharmacologic treatments of many psychiatric disorders have their primary effect in the ventral basal nuclear pathways. Although the most evident symptoms of classic basal nuclear disorders are those of motor disturbances, almost every patient will exhibit psychiatric complications owing to the involvement of the ventral basal nuclei.

Direct and Indirect Pathways of Basal Nuclear Activity

Pathways though the basal nuclei consist of parallel circuits that share certain features. The basic circuit is divided into *direct* and *indirect* pathways that have opposing actions on targets of the basal nuclei (Fig. 26–

Direct and Indirect Pathways

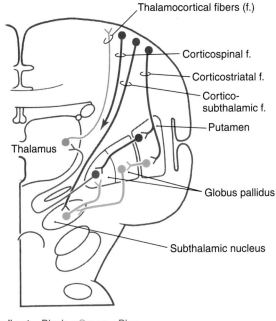

Indirect = Black → Green → Blue
Direct = Black → Red − Blue

Figure 26–8. Schematic representation of direct and indirect pathways through the basal nuclei.

8). As a general concept, the *direct pathway facilitates* a flow of information through the thalamus, and the *indirect pathway inhibits* this flow. These pathways create a balance in the inhibitory outflow of the basal nuclei and function by modulating the extent of this inhibition on target nuclei.

The *direct pathway* (Fig. 26–9A) begins as an excitatory, glutamatergic projection from the cerebral cortex to the striatal complex. Striatal neurons inhibit cells in the internal (or medial) segment of the globus pallidus and in the substantia nigra pars reticulata. These *striatopallidal* and *striatonigral fibers* use GABA and substance P. Cells of the internal segment of the globus pallidus (as *pallidothalamic fibers*) and of the substantia nigra pars reticulata (as *nigrothalamic fibers*) project to thalamic neurons. These fibers have a high rate of spontaneous activity and thus tonically inhibit target thalamic neurons. Inhibition of these pallidal and nigral projections by striatal cells decreases the inhibitory inputs to thalamocortical neurons (*thalamic disinhibition*). The *net effect of the direct pathway is to increase the activity of the thalamus and the consequent excitation of the cerebral cortex* (see Fig. 26–9B).

The *indirect pathway* includes a loop through the globus pallidus and subthalamic nucleus (see Fig. 26–9C). *Striatopallidal* neurons involved in this pathway contain GABA and enkephalin. They project into the lateral pallidal segment, which, in turn, sends *pallidosubthalamic fibers* into the subthalamic nucleus. These pallidosubthalamic fibers are GABAergic, have high spontaneous firing rates, and tonically inhibit subthalamic cells. Inhibition of these fibers by the neostriatum releases these subthalamic cells from their tonically inhibited state (*subthalamic disinhibition*). These subthalamic neurons have spontaneous activity and are also influenced by an excitatory *corticosubthalamic* projection. Together, these inputs increase the firing rates of glutamatergic *subthalamopallidal fibers* to the medial pallidal segment. As a consequence, the firing rate of inhibitory *pallidothalamic fibers* is increased, with a resultant decrease in the activity of thalamocortical neurons. The *net effect of the indirect pathway is to decrease activity of the thalamus and, consequently, decrease activity of the cerebral cortex* (see Fig. 26–9D).

Disinhibition as the Primary Mode of Basal Nuclear Function. The function of the direct pathway is to release the thalamus from its pallidal inhibition. This release is accomplished by striatopallidal inhibition of pallidothalamic neurons. In the indirect pathway, the subthalamic nucleus is released from inhibition by the lateral pallidal segment so that it can excite the inhibitory pallidothalamic cells. This mechanism that *releases cells from inhibition* is called *disinhibition*. Balance between *thalamic disinhibition* by the direct pathway and *subthalamic disinhibition* by the indirect pathway results in normal basal nuclear function. Behavioral deficits that accompany basal nuclear disorders

can ultimately be traced to imbalances between the direct and indirect pathways.

Although the balance between the direct and indirect pathways determines the net outflow of the basal nuclei, other elements of the basal nuclei are also modulated by disinhibitory mechanisms. For example, in the earlier description of the differential projection from the striosomes and matrix to the substantia nigra, the striosomal pathway to the pars compacta is a direct inhibitory pathway, whereas the pathway from the matrix, which inhibits the pars reticulata and releases the pars compacta nigrostriatal neurons, functions to *disinhibit* the pars compacta neurons.

Parallel Circuits of Information Flow Through the Basal Nuclei

Information flow through the basal nuclei is separated into five distinct parallel circuits. They are the *motor loop, oculomotor loop, dorsolateral prefrontal loop, lateral orbitofrontal loop,* and *limbic loop*. The *motor loop* is involved in somatosensory and somatomotor control (Fig. 26–10). The *oculomotor loop* is primarily related to the control of orientation and gaze. The *dorsolateral prefrontal* and *lateral orbitofrontal loops* are related to cognitive processes. The *limbic loop* is concerned with emotional and visceral functions. All of these circuits have both the direct and indirect pathways described previously.

The five circuits originate from functionally distinct regions of the cerebral cortex, pass through distinct regions of each basal nuclear component, modulate different areas of the thalamus, and return to functionally distinct cortical regions (see Fig. 26–10). That is, each circuit can be considered an independent channel that processes information from one functional type of cortex by way of its own areas of the basal nuclei and thalamus and returns to the appropriate functionally related part of cortex.

Anatomic studies in nonhuman primates have demonstrated that each loop projects to a restricted portion of each nucleus. Therefore, there is an *anatomic* as well as a functional separation of the basal nuclear circuits. This separation is called the *closed component* of the circuitry. However, at every stage of each circuit, the information is modulated and integrated with input from other centers by intrinsic basal nuclear connections. Thus, although the circuits are anatomically and functionally distinct, their activities are modulated by the other modalities of the basal nuclei. This integration is called the *open component* of basal nuclear circuits.

The Motor Loop. Because the most obvious symptoms of basal nuclear disorders are those associated with the motor system, it is important to understand the motor loop. Although this loop contains both di-

A Direct Pathway

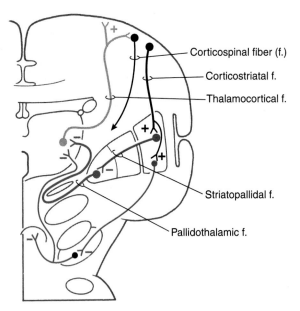

Corticospinal fiber (f.)

Corticostriatal f.

Thalamocortical f.

Striatopallidal f.

Pallidothalamic f.

B Firing Patterns of Neurons

Corticostriatal neurons

Striatopallidal neuron

Pallidothalamic neuron

Thalamocortical neuron

Corticospinal, corticobulbar neurons

C Indirect Pathway

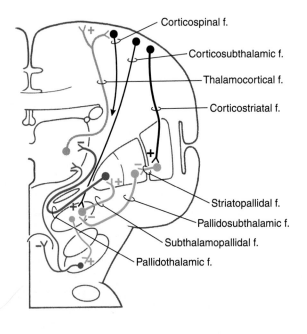

Corticospinal f.

Corticosubthalamic f.

Thalamocortical f.

Corticostriatal f.

Striatopallidal f.

Pallidosubthalamic f.

Subthalamopallidal f.

Pallidothalamic f.

D Firing Patterns of Neurons

Corticostriatal, corticosubthalamic neurons

Striatopallidal neuron

Pallidosubthalamic neuron

Subthalamopallidal neuron

Pallidothalamic neuron

Thalamocortical neuron

Corticospinal, corticobulbar neurons

Figure 26–9. The direct and indirect pathways (*A, C*) and the corresponding firing patterns of their neurons (*B, D*). The color of each fiber correlates with the color of the firing pattern (action potentials) of that neuron (*B* and *D*). The bold fibers in *A* and *C* represent the primary pathway in each example. +, excitatory synapse; −, inhibitory synapse.

Motor Loop

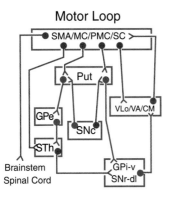

Figure 26–10. Flow diagram of the direct and indirect pathways of the motor loop through the basal nuclei as discussed in the text. CM, centromedian nucleus; GPe, external segments of globus pallidus; GPi-v, internal segment of globus pallidus, ventral portion; MC, motor cortex; PMC, premotor cortex; Put, putamen; SC, somatosensory cortex; SMA, supplementary motor area; SNc, substantia nigra pars compacta; SNr-dl, substantia nigra pars reticulata, dorsolateral portion; STh, subthalamic nucleus; VA, ventral anterior nucleus; VLo, ventral lateral nucleus, oral part.

rect and indirect pathways, which are associated with specific parts of each nucleus, only the direct pathway of the motor loop is described here (see Fig. 26–10).

The motor loop originates mainly in supplementary motor area (SMA), primary motor (MC), and premotor (PMC) cortices (see Fig. 26–10). These corticostriatal projections terminate in the putamen (Put), which also receives projections from somatosensory cortex (SC). Efferents from the putamen terminate in specific areas of the internal segment of the globus pallidus (GPi-v) and the substantia nigra pars reticulata (SNr-dl). These two regions project to the oral part of the ventral lateral nucleus (VLo), the ventral anterior nucleus (VA), and the centromedian nucleus (CM) of the thalamus. In turn, the VLo and VA nuclei project to the supplementary motor cortex; the VA nucleus, to premotor cortex; and the VLo and CM nuclei, to motor cortex. Although the primary projections of the globus pallidus and substantia nigra are to the thalamus, they also project to the superior colliculus and brainstem reticular formation. In this way, this basal nuclear circuit affects both cortical motor efferents and brainstem motor centers.

Comparable circuits also exist in each of the other four loops passing through the basal nuclei. Although the pathways through the other basal nuclear loops are not described here in detail, it is important to note that investigators are giving increasing attention to the deficits resulting from damage to those other pathways. These deficits include those related to emotions, cognition, eye movements, and mentation.

Behavioral Functions of the Basal Nuclei

The best-understood functions of the basal nuclei are associated with the motor systems, in particular, the somatomotor (motor loop) and visuomotor (oculomotor loop) systems. The role of the dorsolateral prefrontal, lateral orbitofrontal, and limbic loops in causing the cognitive and associative disturbances associated with basal nuclei syndromes is less well described. Lesions in the basal nuclei resulting from stroke or other disease processes lead to significant changes in the motor behavior of the patient. Examination of the pathways damaged in these cases reveals how disruption of the basal nuclei can lead to seemingly opposite effects in different patients. For example, movements can be either reduced (*hypokinetic disturbances*) or increased (*hyperkinetic disturbances*).

Hypokinetic Disturbances. The two major types of hypokinetic disturbances seen in patients with basal nuclear disorders are *akinesia* and *bradykinesia* (see Fig. 26–15). Akinesia is an impairment in the initiation of movement; bradykinesia is a reduction in velocity and amplitude of movement. Both disturbances are characteristic in patients with Parkinson disease.

Akinesia, the impaired ability to initiate voluntary movements, may be due to disruption of the ability to plan a movement or to guide a movement to some desired position. Experimental studies in nonhuman primates reveal that many basal nuclear neurons are most active during the *planning phase of a movement* or when the subject is making an *internally guided movement*. The latter is a movement to a location at which no particular stimulus is present. Thus, *patients with akinesia have a generalized disruption of the role of the basal nuclei in planning and generating programmed movements.*

Bradykinesia, the reduction in velocity and amplitude of movements, is due to disruption of the balance between the outflows of the direct and indirect pathways to the thalamus. The result is an increase in the activation of the antagonist muscles. Thus, the observed abnormalities are due to an inappropriate activation of the antagonistic muscles, and not necessarily to an overall decrease in muscular activity.

Hypokinetic disorders can be considered as lesions of the neostriatum (Fig. 26–11*A* and *B*). These lesions result in the loss of inhibitory connections between the neostriatum and internal segment of the globus pallidus. Thus, tonically active pallidothalamic neurons continuously inhibit their thalamic targets. The thalamus is not disinhibited, so there is a decreased flow of information through the thalamus to the cerebral cortex. This decrease, in turn, causes a decrease in the activity of the appropriate corticospinal and other corticofugal neurons. In addition, most of the connections

A Neostriatum Lesion (Direct Pathway)

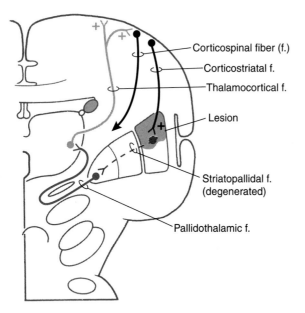

- Corticospinal fiber (f.)
- Corticostriatal f.
- Thalamocortical f.
- Lesion
- Striatopallidal f. (degenerated)
- Pallidothalamic f.

B Altered Firing Patterns

Corticostriatal neuron

Striatopallidal neuron

Pallidothalamic neuron

Thalamocortical neuron

Corticospinal, corticobulbar neurons

C Subthalamic Lesion (Indirect Pathway)

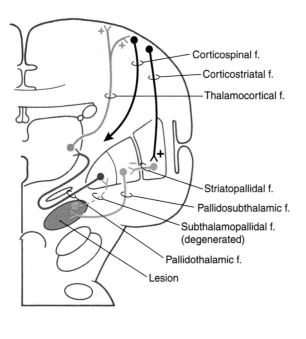

- Corticospinal f.
- Corticostriatal f.
- Thalamocortical f.
- Striatopallidal f.
- Pallidosubthalamic f.
- Subthalamopallidal f. (degenerated)
- Pallidothalamic f.
- Lesion

D Altered Firing Patterns

Corticostriatal, corticosubthalamic neurons

Striatopallidal neuron

Pallidosubthalamic neuron

Subthalamopallidal neuron

Pallidothalamic neuron

Thalamocortical neuron

Corticospinal, corticobulbar neurons

Figure 26–11. Schematic representation of a lesion in the neostriatum (*A*) and a lesion of the subthalamic nucleus (*C*), and the corresponding alterations of neuronal firing patterns of all fibers involved in each pathway (*B*, *D*). The color of each fiber in *A* and *C* correlates with the colors of the altered firing patterns in *B* and *D*. +, excitatory synapses; −, inhibitory synapse.

that subserve the indirect pathway remain intact. Therefore, when those connections are activated, as, for example, by a larger-than-normal burst in cortical neurons, subthalamopallidal cells excite pallidothalamic neurons, which results in increased inhibition of the thalamus. The combination of a *lack of disinhibition of the thalamus* by the direct pathway and an *increased inhibition of the thalamus* by the indirect pathway significantly *decreases the level of appropriate activity in the cerebral cortex and increases the level of inappropriate cortical activity.* Therefore, the patient becomes less able to execute the appropriate movements. Experiments in nonhuman primates have shown that bradykinetic animals have levels of neuronal activity that are consistent with this view.

Hyperkinetic Disturbances. Hyperkinetic disturbances take the form of dyskinesias. The three most common forms are *ballismus, choreiform movements,* and *athetoid movements. Ballismus* is most typically seen as *hemiballismus* because it usually occurs on one side. It consists of uncontrolled flinging (ballistic) movements of an upper or lower extremity. This motor disorder is most commonly seen in patients with vascular lesions localized to the contralateral subthalamic nucleus (see Fig. 26–11C and D). *Choreiform movements,* which are present in Huntington disease and sometimes in treated Parkinson disease, are generalized irregular dancelike movements of the limbs. Similar movements may occur in oral and facial musculature. In Huntington disease, there is an initial selective loss of the medium spiny cells in the striatum, which project to the lateral pallidum, and of acetylcholine-containing neurons in the striatal complex. It is thus likely that the neurons specifically associated with the indirect pathway are lost.

Finally, *athetoid movements (athetosis)* are a continuous writhing of distal portions of the extremity (Fig. 26–12). Athetosis generally presents as slow, sinuous movements more obvious in the upper extremities and hands (see Fig. 26–12) and face, although any muscle group may be affected. However, athetoid movements may also be seen as a range of abnormal movements. When they are more brisk and resemble chorea, the term *choreoathetosis* may be appropriate. When the movements are more intense and sustained, they may resemble dystonia, and the term *athetotic dystonia* may be used.

These hyperkinetic disturbances can most easily be explained by the disruption of the indirect pathway through the motor loop, resulting from the loss of excitatory subthalamopallidal neurons (see Fig. 26–11C and D). The balance between excitation of pallidothalamic neurons (by the direct pathway) and their inhibition (by the indirect pathway) is skewed. The result is a decrease in the net amount of inhibition of thalamic cells, which results in more activity in the cerebral cortex.

Integrated Function of the Basal Nuclei. By combining the direct and indirect pathways, it is possible to gain a better appreciation of how the basal nuclei affect their targets through a balance between activation by the direct pathway and inactivation by the indirect pathway, as seen in Figure 26–13A and B. The activity in the cerebral cortex activates the striatopallidal neurons in both the direct (red) and indirect (green) pathways. Activation of the striatopallidal neurons in the direct pathway will inhibit pallidothalamic neurons (blue in Fig. 26–13B1). This pathway may be considered to be the initiator of the pause in pallidothalamic activity. At the same time, activation of the striatopallidal neurons in the indirect pathway inhibits the pallidosubthalamic neurons (see Fig. 26–13B2). This inhibition releases the subthalamic neurons, which then excite the pallidothalamic neurons. This excitation by the subthalamopallidal neurons is what reactivates the pallidothalamic neurons that were inhibited through the direct pathway. Thus, careful balance of the activity between the two pathways modulates the amount of time during which the thalamocortical neurons are activated.

In addition to the activities of these pathways, the dopaminergic loop from the neostriatum to the substantia nigra and back to the neostriatum is also active. Activity of striatonigral projections originating in the striosomes will inhibit the dopaminergic nigrostriatal pathway (see Fig. 26–13B3). Subsequent reactivation of the nigral neurons is due in part to a number of factors, including disinhibition of the pars compacta by means of the striatal projection originating from the matrix. The dopaminergic neurons project back onto striatal neurons. The effect of the dopamine depends on the receptor on the postsynaptic striatal neuron. D_1 receptors are found on striatal neurons in the "direct" pathway. Stimulation of the D_1 receptors causes excitation of the striatopallidal neurons in the direct pathway. This excitation accounts for the slightly longer train of action potentials seen in Figure 26–13B. The neurons belonging to the "indirect" pathway have D_2

Figure 26–12. Athetoid movements (athetosis) of the upper extremity.

Direct and Indirect Pathways
(Including the Substantia Nigra)

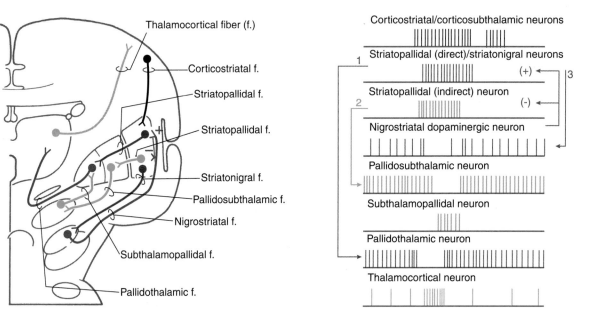

A   Connections

B   Firing Patterns of Neurons

Loss of Dopaminergic Nigral Connections

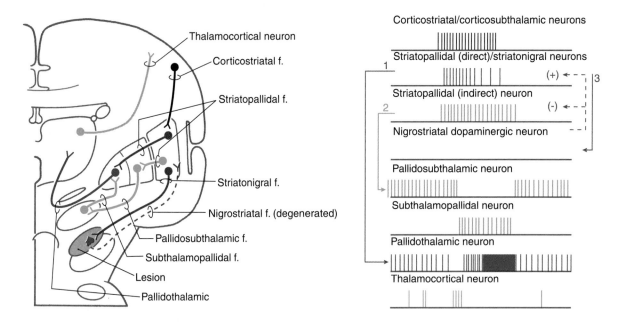

C   Connections

D   Altered Firing Patterns

Figure 26-13. The role of the direct and indirect pathways and the influence of dopamine. Fibers and their corresponding firing patterns for the intact pathway are shown in *A* and *B*. The flow of signals through the direct (1), indirect (2), and a striatonigral-nigrostriatal (3) pathways are shown by their respective arrows. Lesions of the pars compacta (*C*), as in Parkinson disease, is characterized by the loss of dopaminergic neurons and results in the altered firing patterns of neurons in the entire pathway (*D*). The colors of the fibers correlate with the colors of the action potentials. +, excitatory synapses; −, inhibitory synapses.

receptors. When dopamine binds to the D_2 receptor, the neurons are inhibited. This inhibition accounts for the shorter train of action potentials for indirect striatopallidal neurons in the indirect pathway.

The modulation between the direct and indirect pathways by dopamine is significantly affected if dopamine is absent. For example, in Parkinson disease, the loss of melanin-containing dopaminergic neurons in the nigral complex has the net effect of decreasing thalamocortical neuronal activity (see Fig. 26–13C and D). The striatopallidal activity of the direct pathway initiates the pause (see Fig. 26–13D1). However, the activity is less sustained owing to the loss of excitation from the substantia nigra. The activity of the striatopallidal neurons in the indirect pathway is not inhibited by the dopamine and thus is of a somewhat longer duration. This longer duration of activity results in a longer-lasting inhibition of the pallidosubthalamic neurons (see Fig. 26–13D2), which increases the activity of the subthalamopallidal neurons and results in increased activation of the pallidothalamic neurons. Consequently, not only is the pause in pallidothalamic activity shortened, but the inhibition of the thalamus by these neurons is enhanced. The net result is an increased inhibition of the thalamocortical neurons with decreased activity in the cerebral cortex.

Etiology of Basal Nuclear Related Disorders

The hallmark of basal nuclear disorders is a change in the neurochemical environment within the striatal complex. Careful examination of patients with basal nuclear syndromes reveals that the classic motor signs and symptoms constitute only one characteristic of these disorders and that associative memory and limbic dysfunctions also occur. Recent clinical and basic science studies of the chemistry of basal nuclear disorders have provided significant insight into their etiology and treatment.

Huntington Disease. As originally described, Huntington disease is a progressive, untreatable disorder in which patients lose their ability to function and experience increasing dementia; death occurs 10 to 15 years after onset. In the United States, the incidence of Huntington disease is approximately 1 in every 10,000 individuals. In the early stages, this disease is characterized by absentmindedness, irritability, depression, clumsiness, and sudden falls. Gradually, choreiform movements increase until the patient is bedridden. Cognitive functions and speech progressively deteriorate. The later stages of this disease are characterized by severe dementia. One salient feature in the progressive memory loss is a difficulty in integrating newly acquired memory into useful information for planning movements. In addition, a large proportion of these patients develop psychiatric disorders such as ma-

jor affective disorder, schizophrenia, and other behavioral disturbances.

Postmortem examination reveals decrease in the size of the striatal complex caused by loss of about 90% of all striatal neurons and extensive astrocytosis. In particular, medium spiny cells, which project to the lateral pallidum, and the large acetylcholine-containing local circuit cells are lost. This disease can be demonstrated on magnetic resonance images (MRI) as a flattening of the head of the caudate nucleus (Fig. 26–14).

Huntington disease is a genetic disorder inherited in autosomal dominant fashion, and any person who inherits the gene will develop the disease. It is most prevalent among persons of European descent. This disease is known to be due to a mutation on the short arm of chromosome 4. At the molecular level, the mutated gene has dozens of copies of the DNA sequence CAG (cytosine-adenine-guanine), which codes for glutamate. These CAG repeats result in the addition of glutamate residues inside the gene coding for the pro-

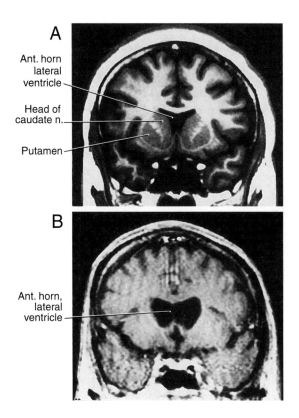

Figure 26–14. Coronal magnetic resonance image obtained through the frontal lobe, and head of the caudate nucleus, of a normal person (A) and of a patient with Huntington disease (B). The head of the caudate normally forms a prominent bulge into the anterior horn of the lateral ventricle (A, inversion recovery image). Profound cell loss in the neostriatum in Huntington disease greatly diminishes the size of the caudate and renders the lateral wall of the ventricle flat (B, T1-weighted image). The slightly wavy appearance of the image in B is the result of movement (tremor) while the scan was being done.

tein *huntingtin*. Patients with larger numbers of repeats have a more severe illness with an earlier onset. Thus, there are now reliable markers for the disease.

Much effort has been devoted to understanding the mechanism underlying the death of striatal neurons in this disease. In the very earliest stages of the disease, which may be months or years before the onset of clinical features, there is a diminution of glucose metabolism in the neostriatum of some patients. *Glutamate excitotoxicity* is thought to be primarily responsible for this process. Normally, cortical axons release glutamate as their neurotransmitter in the caudate and putamen. Glutamate binds with its receptor on the medium spiny neurons and opens the receptor channel to an influx of ions. This action depolarizes the membrane, resulting in an excitatory postsynaptic potential. As glutamate dissociates from the receptor, it is cleared from the extracellular space by uptake by astrocytes. In Huntington disease, an unknown mechanism causes glutamate to persist at one type of receptor, the *N*-methyl-D-aspartate (NMDA) receptor, which opens calcium ion channels. The resulting excessive influx of calcium causes an increase in intracellular calcium, which triggers a cascade that leads to cell death. Glutamate excitotoxicity is also thought to be the primary cause of localized neuronal death following acute brain injury, such as stroke (see Chapter 2).

Parkinson Disease. Parkinson disease also is a progressive, debilitating disorder. It affects over 500,000 Americans and there are about 50,000 new cases each year. It is the third most common neurologic disorder. Parkinson disease usually affects persons over 55 years of age. Some familial groups with Parkinson disease have been described.

This disorder is characterized by a progressive onset of movement and affective disturbances (Fig. 26–15). The movement disorders include *tremor at rest, cogwheel (gamma) rigidity* (increased muscular tone), *akinesia, bradykinesia, disturbances of eye movements,* and *loss of postural reflexes.* A classic picture of a patient with Parkinson disease is a person sitting or standing with *pill-rolling tremor,* a *blank stare* (reptilian or decreased blink), a *flexed posture,* and a *paucity of movement* (Fig. 26–15). When the patient starts to move there is a shuffling start, as if the feet were stuck in place (also called a *festinating gait*), followed by nearly normal gait.

At autopsy, the nigral complex is found to be devoid of dopaminergic neurons. Although there is degeneration of serotoninergic and noradrenergic pathways, it is this specific loss of dopamine that results in the observed symptoms. The standard treatment aims at replacing the lost dopamine. Because dopamine itself will not cross the blood-brain barrier, patients are given L-3,4-hydroxyphenylalanine (L-dopa), which will. This agent is now combined with a second drug, carbidopa, which does not cross the blood-brain barrier but has the effect of inhibiting peripheral uptake of L-dopa and thus increasing the amount of L-dopa available to brain

Figure 26–15. Parkinsonian tremor (resting tremor) and characteristic posturing. The patient may have difficulty initiating a movement (akinesia), or once movement is initiated, it may be slow and lack spontaneity (bradykinesia).

tissue. Persons receiving this combination therapy, along with other drugs such as dopamine agonists and monoamine oxidase inhibitors, show significant reductions in signs and symptoms, although progression of the disease is not arrested. Why L-dopa works is not clear. The method by which L-dopa is converted to dopamine in the brain of Parkinson disease patients is not known, as these persons have very little tyrosine hydroxylase, the enzyme that is necessary for this catabolism. Moreover, the dopamine is not localized to specific nerve terminals or to the basal nuclei. Thus, it appears that dopamine simply needs to be in the neural environment of the striatal complex to reduce the signs and symptoms of the patients.

Insight into one possible mechanism of Parkinson disease was inadvertently realized by a group of illicit drug manufacturers whose heroin was contaminated by a compound called 1-methyl-4-phenyl-1,2,3,6-tetrahydropyridine (MPTP). Persons who used this drug developed symptoms exactly like those of patients with Parkinson disease, but at an age (early twenties) when this disease is normally never manifested. Autopsy revealed a profound loss of dopaminergic neurons in the substantia nigra pars compacta.

Studies on the mechanism of action of MPTP in animal models showed that MPTP is converted into an active form (MPP$^+$) before it causes a loss of dopaminergic cells. This metabolic pathway requires monoamine oxidase (MAO). Thus, it was hypothesized that the use of MAO inhibitors may affect the progression of Parkinson disease. In fact, clinical trials have now shown that the drug L-deprenyl, an MAO-B inhibitor, slows the progression of Parkinson disease and increases the levels of dopamine in the brain. The increase in dopamine may result both from protection of neurons against toxicity and from blocking of the degradation pathway for dopamine, which requires the MAO enzyme.

In recent years, there has been an increasing number of surgical treatments for Parkinson disease. The most common type of surgical intervention is ablative surgery in which either the ventral intermediate nucleus of the thalamus (thalamotomy) or the posterolateral part of the internal segment of the globus pallidus (pallidotomy) is lesioned. Another surgical approach involves introduction of stimulation electrodes directed at the thalamus, globus pallidus, or subthalamic nuclei. Each of these surgical approaches appears to be effective for treating one set of symptoms. However, none of these approaches completely eliminates the disease or prevents its progression.

A controversial method of treatment for Parkinson disease patients involves the use of human embryonic or autologous transplants. Tissues that produce dopamine, such as substantia nigra (embryonic) and the adrenal cortex (autologous), are obtained and separated into cell suspensions. These are then injected into the lateral ventricles of the patient. The idea is that these cells will adhere to the walls of the ventricle and produce dopamine. The dopamine will then diffuse into the nearby cerebral cortex and basal nuclei. This procedure has now been performed in humans as well as in animals. In the clinical trials to date, only a small number of patients have benefited from this approach.

Wilson Disease. Wilson disease, also known as *hepatolenticular degeneration*, is also related to the basal nuclei. This genetic disorder inherited as an autosomal recessive trait, is due to a mutation on the long arm of chromosome 13. Onset of the disorder is typically between 11 and 25 years of age. Wilson disease is a disorder in copper metabolism that results in accumulation of the metal in the liver, resulting in small necrotic lesions leading to cirrhotic nodules and progressive liver damage. Signs of liver damage may precede the onset of neurologic abnormalities by several years. Another metabolic feature, probably due to damage of the tubules in the kidney, is *aminoaciduria*. In the eye, copper accumulates in the periphery of the cornea; these deposits are responsible for the Kayser-Fleischer ring (Fig. 26–16), which may appear yellow to green to brown. Degeneration of the putamen, often forming small cavities, is the predominant pathoanatomic feature in the brain. However, other regions of the brain, including the frontal lobe of the cerebral cortex, may also show similar changes. This degeneration is due to a loss of neurons, axonal degeneration, and increasing numbers of protoplasmic astrocytes.

As in other basal nuclear disorders, many patients

Figure 26–16. Deposition of copper (Kayser-Fleischer ring, *arrows*) in the periphery of the cornea as seen in hepatolenticular degeneration (Wilson disease).

with Wilson disease will develop psychiatric symptoms. However, the motor disturbances are often the most evident signs and include *tremor, dysarthria,* diminished dexterity, unsteady gait, and *rigidity.* The most common form of tremor is known as *asterixis,* a "wing-beating" tremor. Affected patients do not have a tremor at rest. However, after the arms are extended, a beating movement develops that can be restricted to the wrists or can result in the arms being thrown up and down in a wide arc. Treatment is essential, and the goal is to decrease the amount of copper within the body, thus limiting its toxic effects. Prognosis for patients who have completed the first few years of treatment is very good, with diminution of neurologic signs within 5 to 6 months from the onset of therapy.

Sydenham Chorea. This is a childhood autoimmune disease that is infrequently seen. This disease typically affects children between the ages of 5 and 15 years. It is a consequence of rheumatic fever, which is caused by infection with group A beta-hemolytic streptococci. The disease is self-limited and is rarely fatal; fatalities are usually attributed to consequences of rheumatic fever. The chorea may not appear until 6 months or longer after infection and typically lasts for 3 to 6 weeks. Patients present with rapid, irregular, aimless movements of the limbs, face, and trunk. These movements are more flowing and "restless" than those in Huntington disease patients. In addition, patients with Sydenham chorea have some muscular weakness and hypotonia. Other signs and symptoms may include irritability, emotional lability, obsessive-compulsive behaviors, attention deficit, and anxiety. Fortunately, this is a benign disease, and most patients experience complete recovery from the symptoms. However, about one third of patients may have recurrences of signs and symptoms after several months or even years.

Tardive Dyskinesia. *Tardive dyskinesia* is a basal nuclear disorder that is iatrogenic in nature—that is, caused by medical intervention for another disease. This condition is caused by chronic treatment with neuroleptic medications such as the phenothiazines (e.g., chlorpromazine, thioridazine) and butyrophenones (e.g., haloperidol). The manifestation of this condition is uncontrolled involuntary movements, particularly of the face and tongue, and cogwheel rigidity. These abnormalities may be temporary or permanent.

The action of these neuroleptic drugs is to block dopaminergic transmission throughout the brain. The primary target cells are those in the ventral tegmental area that form the mesolimbic dopaminergic pathway. Prolonged treatment with neuroleptic drugs leads to a hypersensitivity at the D_3 dopamine receptor, which causes an imbalance in the nigrostriatal influence on the basal nuclear motor loop and ultimately results in movement disorders.

Sources and Additional Reading

Alexander GE, DeLong MR, Strick PL: Parallel organization of functionally segregated circuits linking basal ganglia and cortex. Annu Rev Neurosci 9:357–381, 1986.

Gerfen CR: The neostriatal mosaic: Multiple levels of compartmental organization in the basal ganglia. Annu Rev Neurosci 15:285–320, 1992.

Goldman-Rakic PS (ed): Basal ganglia research (Special Issue). Trends Neurosci 13:241–308, 1990.

Graybiel AM, Aosaki T, Flaherty AM, Kimura M: The basal ganglia and adaptive motor control. Science 265:1826–1831, 1994.

Haber SN, McFarland NR: The concept of the ventral striatum in nonhuman primates. Ann N Y Acad Sci 877:33–48, 1999.

Middleton FA, Strick PL: Anatomical evidence for cerebellar and basal ganglia involvement in higher cognitive function. Science 266:458–461, 1994.

Parent A, Hazrati L-N: Functional anatomy of the basal ganglia. I. The cortico-basal ganglia–thalamo-cortical loop. Brain Res Rev 20:91–127, 1995.

Parent A, Hazrati L-N: Functional anatomy of the basal ganglia. II. The place of the subthalamic nucleus and external pallidum in basal ganglia circuitry. Brain Res Rev 20:128–154, 1995.

Pedro BM, Pilowsky LS, Costa DC, Hemsley DR, Ell PJ, Verhoeff NP, Kerwin RW, Gray NS: Stereotypy, schizophrenia and dopamine D_2 receptor binding in the basal ganglia. Psychol Med 24:423–429, 1994.

Rowland LP (ed): Merritt's Neurology, 10th ed. Lippincott Williams & Wilkins, Philadelphia, 2000.

Shultz W: Predictive reward signal of dopamine neurons. J Neurophysiol 80:1–27, 1998.

Chapter 27

The Cerebellum

*D. E. Haines, G. A. Mihailoff, and
J. R. Bloedel*

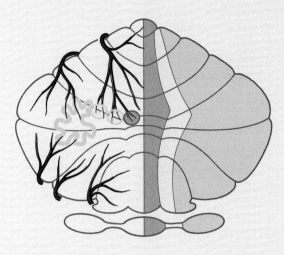

Overview 424

Basic Structural Features 424
Cerebellar Peduncles
Cerebellar Lobes, Lobules, and Zones
Cerebellar Nuclei
Blood Supply to Cerebellar Structures

Cerebellar Cortex 428
Purkinje Cell Layer
Granule Cell Layer
Molecular Layer
Cerebellar Afferent Fibers
Topographic Localization
Synaptic Interactions in the Cerebellar Cortex

Functional Cerebellar Modules 435
Vestibulocerebellar Module
Vestibulocerebellar Dysfunction
Vestibular Connections of the Vermis
Spinocerebellar Module
Pontocerebellar Module
Pontocerebellar Dysfunction

Cerebellar Influence on Visceromotor Functions 442

The Cerebellum and Motor Learning 443

As indicated by its relative size (about 10% of the weight of the central nervous system), the cerebellum is important in brain function. However, it executes these responsibilities in unique ways. First, it receives extensive sensory input, but it is not involved in sensory discrimination or interpretation. Second, although it profoundly influences motor function, resection of relatively large portions of the cerebellar cortex does not result in lasting paralysis. Third, the cerebellum is not critical for most cognitive functions, but it may play a role in motor learning and higher mental function.

Overview

The cerebellum is composed of a highly convoluted *cerebellar cortex* and a core of white matter containing the *cerebellar nuclei*. This structure is anchored to the brainstem via the *cerebellar peduncles*. The cerebellum is located posterior (dorsal) to the brainstem, inferior to the tentorium cerebelli, and internal to the occipital bone. The cerebellum has a *superior surface* apposed to the tentorium and a convex *inferior surface* that abuts the inner surface of the occipital bone.

The cerebellum receives input from many areas of the neuraxis and influences motor performance through connections with the dorsal thalamus and, ultimately, the motor cortices. Lesions of these pathways result in characteristic motor dysfunctions, which may involve either proximal (axial) or distal musculature. These deficits are actually the result of altered activity in the motor cortex and its descending brainstem and spinal projections, which influence lower motor neurons of the spinal cord.

Basic Structural Features

Cerebellar Peduncles. The cerebellum is connected to the brainstem by three pairs of *cerebellar peduncles* (Fig. 27–1A and B). The *inferior cerebellar peduncle* is composed of a *restiform body* and a *juxtarestiform body*. The former is the large ridge on the dorsolateral aspect of the medulla rostral to the level of the obex. This bundle contains mainly fibers that arise in the spinal cord or medulla. The juxtarestiform body is located in the wall of the fourth ventricle. Its fibers form reciprocal connections between the cerebellum and vestibular structures.

The exiting fibers of the trigeminal nerve represent the boundary between the basilar pons and the *middle cerebellar peduncle* (*brachium pontis*) (see Fig. 27–1A and B). This large peduncle conveys fibers from the nuclei of the basilar pons into the cerebellum.

The *superior cerebellar peduncle* (*brachium conjunctivum*) sweeps rostrally out of the cerebellum and courses into the midbrain just caudal to the exit of the trochlear nerve (see Fig. 27–1A and B). This bundle contains predominantly *cerebellar efferent fibers* that originate from neurons of the cerebellar nuclei and distribute to the diencephalon and brainstem.

Cerebellar Lobes, Lobules, and Zones. At the most general level, it is common to divide the cerebellum into a narrow midline *vermis* and expansive lateral *hemispheres* (Figs. 27–2 and 27–3). The cerebellum is further divided into *anterior*, *posterior*, and *flocculonodular lobes* by the *primary* and *posterolateral fissures*, respectively. The lobes of the cerebellum are composed of yet smaller divisions called *lobules* (Fig. 27–4; see also Fig. 27–2). The lobules of the vermis are identified by Roman numerals I to X; the corresponding lateral

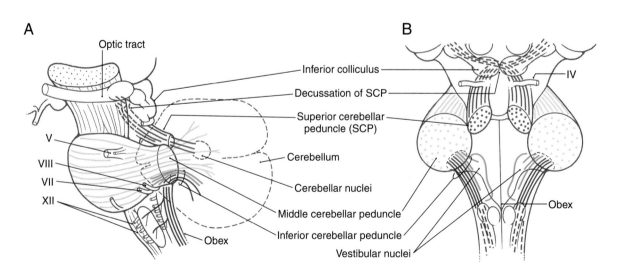

Figure 27–1. Lateral (*A*) and posterior (*B*) views of the brainstem showing the inferior (dark green + red), middle (light green), and superior (blue) cerebellar peduncles. The inferior peduncle is composed of juxtarestiform (dark green) and restiform (red) bodies. Cranial nerves are identified by Roman numerals.

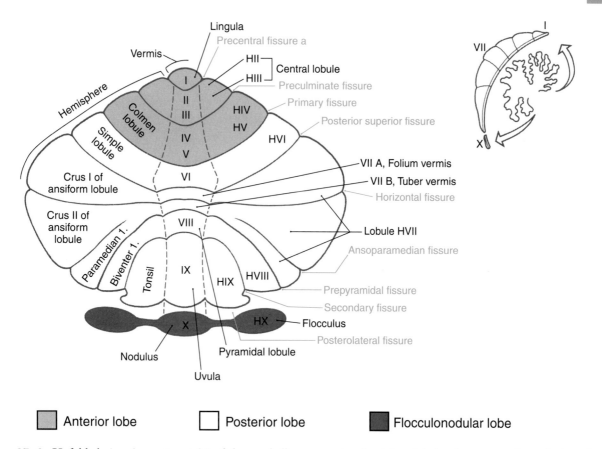

Figure 27–2. Unfolded view (see *upper right*) of the cerebellar cortex showing lobes, lobules (by name and number), and main fissures (printed in blue). The lobules of the hemisphere are designated by the prefix "H," to show which lobule of the hemisphere (H) is continuous with its corresponding (designated by the Roman numeral) vermal lobule. l., lobule.

(hemisphere) portion of each vermis lobule is identified by the same Roman numeral but with the prefix "H" (see Figs. 27–2 and 27–4). Vermis lobules II to X have hemisphere portions HII to HX (see Fig. 27–4); lobule I does not have a hemisphere part in humans. The anterior lobe comprises *lobules I to V and HII to HV*, and the posterior lobe, *lobules VI to IX and HVI to HIX*. The flocculonodular lobe consists of the nodulus (lobule X) and its hemisphere counterpart, the flocculus (lobule HX). In turn, each lobule consists of a series of individual ridges of cortex called *folia* (singular, *folium*).

The individual folia are continuous from one hemisphere to the other, across the midline (see Fig. 27–3A). This pattern, obvious on the superior cerebellar surface, is disrupted on the inferior surface by the enlargement of the lateral parts of the cerebellum and consequent infolding of the midline area (see Fig. 27–3B).

Superimposed on the lobes and lobules of the cerebellum are rostrocaudally oriented cortical zones that are defined on the basis of their connections. There are three principal zones on each side: the *medial (vermal)*, *intermediate (paravermal)*, and *lateral (hemisphere)* zones (see Fig. 27–4). These three larger zones can be

subdivided further, on the basis of their connections, into nine smaller zones. In general, these zone patterns are the basis for the modules discussed later in this chapter. The clinical deficits that result from a cerebellar lesion depend mainly on which of the three principal zones is involved; consequently, the three-zone terminology is used in this chapter.

The *medial (vermal) zone* is a narrow strip of cortex adjacent to the midline that extends throughout anterior and posterior lobes and includes the nodulus (see Figs. 27–3A and B and 27–4). This zone is widest in lobule VI and tapers rostrally and caudally. The *intermediate (paravermal) zone* lies adjacent to the medial zone and extends throughout anterior and posterior lobes but has little representation in the flocculonodular lobe (see Fig. 27–4). The *lateral (hemisphere) zone* occupies by far the largest part of the cerebellar cortex. It includes large portions of anterior and posterior lobes and the flocculus (see Figs. 27–2 and 27–4).

Cerebellar Nuclei. The four pairs of cerebellar nuclei are located within the white matter core of the cerebellum and are accessed easily by fibers traveling to and from the overlying cortex (Fig. 27–5; see also Fig. 27–4). The *fastigial (medial cerebellar) nucleus* lies immediately adjacent to the midline and is functionally re-

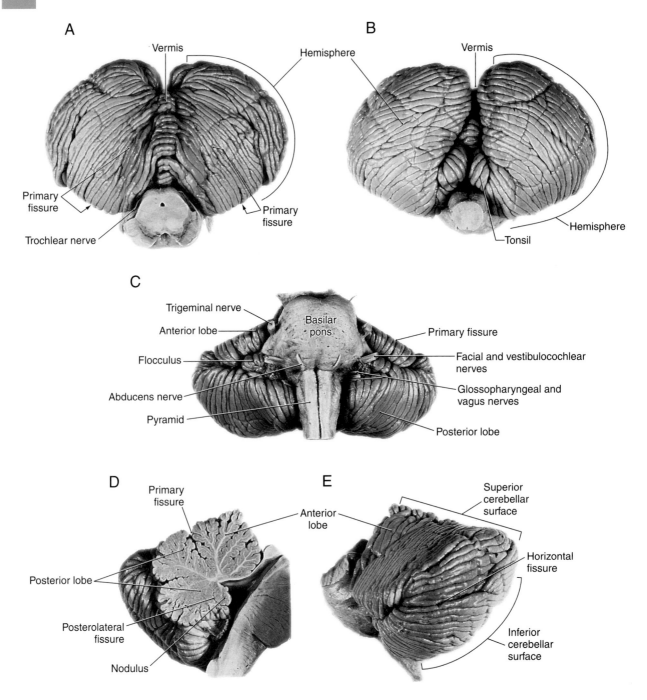

Figure 27–3. Rostral (*A*, superior surface), caudal (*B*, inferior surface), anterior or ventral (*C*, with brainstem intact), midsagittal (*D*), and lateral (*E*) views of the cerebellum. Only the major structures and relationships are indicated.

lated to the overlying medial zone of the cerebellar cortex. Lateral to the fastigial nucleus are the two interposed nuclei: the *globose* (*posterior interposed*) *nucleus* and the *emboliform* (*anterior interposed*) *nucleus*. These nuclei are functionally related to the intermediate zone of the cortex. Lateral to the emboliform nucleus is the *dentate* (*lateral cerebellar*) *nucleus*, which appears as a large, undulating sheet of cells shaped like a partially crumpled paper bag. Its opening, the *hilus*, is directed

anteromedially (see Fig. 27–5). This nucleus is functionally related to the lateral zone of the cortex; its large size correlates with the large size of this cortical zone.

Most of the signals that leave the cerebellum do so via axons that arise in the cerebellar nuclei; the remainder travel on fibers that originate in the cerebellar cortex. Collectively, the former constitute *cerebellar efferent projections* (or *fibers*). These axons originate from

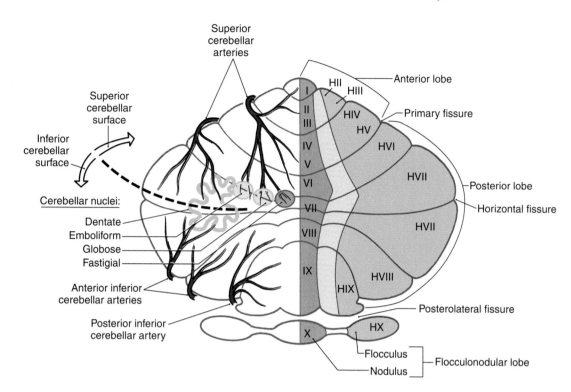

Figure 27–4. An unfolded view of the cerebellum showing medial (gray), intermediate (green), and lateral (blue) zones on the right and the nuclei with which each zone is functionally associated on the left in the corresponding color. The general areas of cortex and nuclei served by the cerebellar arteries are also indicated on the left. Roman numerals indicate lobules of the vermis; numerals preceded by H indicate the corresponding lobules of the hemisphere.

cells in the cerebellar nuclei and use one of the excitatory neurotransmitters *glutamate* or *aspartate* and thus function to activate their targets. The fastigial nuclei generally project bilaterally to the brainstem through the *juxtarestiform bodies*. Fibers originating in the dentate, emboliform, and globose nuclei exit the cerebellum via the *superior cerebellar peduncle* and cross in its decussation.

Some neurons in each cerebellar nucleus send axons or axon collaterals into the overlying cortex, where they terminate in the granular layer as mossy fibers. These axons are called *nucleocortical fibers*, and they exert an excitatory influence on the cerebellar cortex.

Blood Supply to Cerebellar Structures. The blood supply to the cerebellar cortex, nuclei, and peduncles is via the *posterior inferior artery* (PICA), *anterior inferior artery* (AICA), and *superior cerebellar artery* (Fig. 27–6; see also Fig. 27–4). The PICA originates from the vertebral artery and supplies the posterolateral medulla (including the restiform body), the choroid plexus of

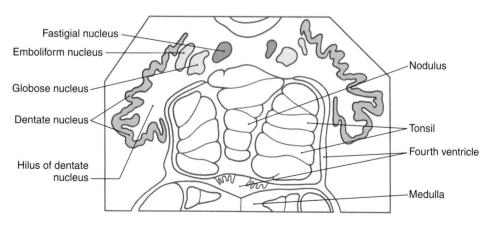

Figure 27–5. The cerebellar nuclei in cross section, drawn from a slide. The color coding of each nucleus corresponds to its appropriate zone in Figure 27–4.

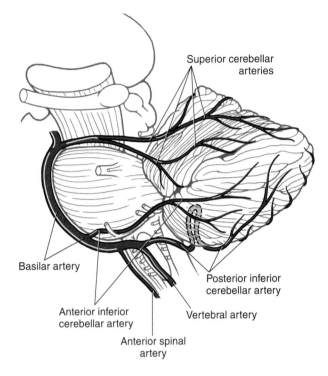

Figure 27–6. Origin and course of arteries serving the cerebellum as seen in lateral aspect.

the fourth ventricle, and caudomedial regions of the inferior cerebellar surface (including the vermis) (see Figs. 27–4 and 27–6). Caudal parts of the middle cerebellar peduncle and caudolateral portions of the inferior cerebellar surface are served by the AICA. This vessel may also supply caudal parts of the dentate nucleus (see Fig. 27–4). The entire superior surface of the cerebellum, most of the cerebellar nuclei, the rostral parts of the middle cerebellar peduncle, and the superior cerebellar peduncle are served by the superior cerebellar artery (see Fig. 27–6).

Cerebellar Cortex

Histologically, each folium of the cerebellum has a superficial cellular layer, the *cerebellar cortex*, and a core of myelinated fibers traveling to (*afferent*) or from (*efferent*) the overlying cortex. The cortex consists of a *Purkinje cell layer* insinuated between a cell-dense inner region immediately adjacent to the white matter core, the *granule cell layer*, and an outer pale and relatively cell-sparse portion, the *molecular layer* (Figs. 27–7 and 27–8).

Purkinje Cell Layer. The large (40 to 65 μm in diameter) somata of Purkinje cells form a single layer at the interface of the granular and molecular layers (see Fig. 27–7). Each Purkinje cell gives rise to an elaborate dendritic tree that radiates into the molecular layer. This dendritic tree is shaped like a fan with its wide flattened side oriented perpendicular to the long

axis of the folium (see Figs. 27–8 and 27–9). The "trunk" of the tree is a single primary dendrite, which gives rise to several secondary dendrites, which in turn branch into many tertiary dendrites. Synaptic contacts are formed mainly on short *terminal branchlets* that emerge primarily from the secondary and tertiary dendrites (see Figs. 27–8 and 27–9). There are two types of branchlets. *Smooth branchlets* emerge from secondary and tertiary dendrites, whereas *spiny branchlets* (*gemmules*) arise mainly from tertiary dendrites.

Purkinje cells are the only efferent neurons of the cerebellar cortex. Axons of Purkinje cells arise from the basal aspect of the cell body and may give rise to recurrent collaterals. The former processes traverse the granular layer and the subcortical white matter to eventually terminate in either the cerebellar or the vestibular nuclei. Purkinje cells projecting into the cerebellar nuclei (as *cerebellar corticonuclear fibers*) arise from all areas of the cortex, whereas those projecting into the vestibular nuclei (as *cerebellar corticovestibular fibers*) originate only from parts of the vermis and the flocculonodular lobe. Purkinje cells release gamma-aminobutyric acid (GABA) at their synaptic terminals and inhibit target neurons in the cerebellar and vestibular nuclei.

Granule Cell Layer. The most numerous neuron of the granule cell layer is the small (5 to 8 μm in diameter) *granule cell* (see Figs. 27–7, 27–8, and 27–9C). The dendrites of these cells form clawlike endings (*dendritic digits*) that ramify in the vicinity of the cell body. Their axons ascend into the molecular layer, where they bifurcate to form *parallel fibers*. As indicated by their name, *parallel fibers run parallel to the long axis of the folium*. Consequently, these fibers pass through the fanlike dendritic trees of the Purkinje cell and, when doing so, make synaptic contacts with spiny branchlets (see Figs. 27–8 and 27–9B and F). They also synapse with the cells intrinsic to the molecular layer (discussed further on). The distance spanned by the parallel fibers of a granule cell ranges from about 0.3 to 5.0 mm, and the number of Purkinje and other cells contacted varies accordingly. Granule cells use *glutamate* (or perhaps *aspartate*) as their neurotransmitter and thus have an excitatory effect on their target cells. In fact, the *granule cells are the only excitatory neurons of the cerebellar cortex*. All the others, as we shall see, are inhibitory.

The second cell type of the granular layer is the *Golgi cell*. The soma of this neuron is larger (at 18 to 25 μm in diameter) than that of the granule cell and is usually found in the granular layer adjacent to the Purkinje cells (see Figs. 27–8 and 27–9D). Dendrites of Golgi cells branch in the granular layer but extend primarily into the molecular layer without regard to plane of orientation. Axons of Golgi neurons branch in the granular layer and form synaptic contacts on granule cell dendrites (see Fig. 27–8). The Golgi cell utilizes GABA as a neurotransmitter and is an inhibitory interneuron in the cerebellar cortex.

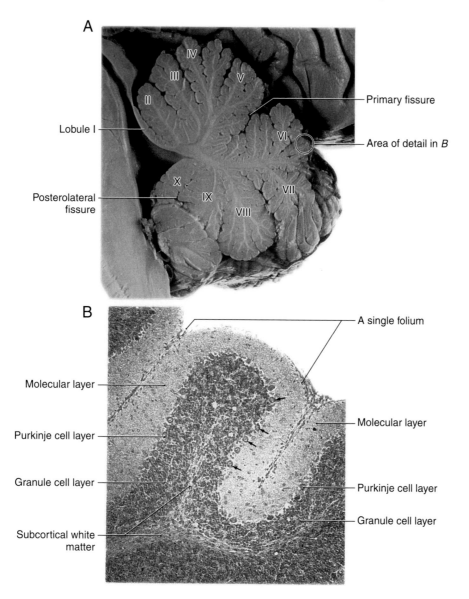

A

Lobule I

Posterolateral
fissure

Primary fissure

Area of detail in *B*

B

Molecular layer

Purkinje cell layer

Granule cell layer

Subcortical white
matter

A single folium

Molecular layer

Purkinje cell layer

Granule cell layer

Figure 27–7. Midsagittal view (*A*) of the cerebellum showing the main fissures and lobules and a detail (*B*) of the cortex showing the cortical layers as seen in a neutral red–stained section. Arrowheads point to representative Purkinje cell bodies.

The *cerebellar glomerulus* is a synaptic complex found in the neuropil of the granular layer. The glomerulus includes granule cell and Golgi cell dendrites and Golgi cell axons, and a specialized synaptic segment (*the mossy fiber rosette*) of a *mossy fiber*, one type of cerebellar afferent axon (see Figs. 27–8 and 27–9*C–E*). The mossy fiber rosette is centrally located and forms synapses with several granule cell dendrites. Golgi cell axons contact granule cell dendrites in the glomerulus and the entire complex is encapsulated by glial cell processes.

Molecular Layer. The molecular layer has considerably fewer cell bodies than the granule cell layer (see Fig. 27–7) but has proportionately more cell processes. These processes include *parallel fibers, Purkinje cell dendrites, Golgi cell dendrites, climbing fibers* (see later on), and the processes of cells intrinsic to the molecular layer (see Fig. 27–8).

The intrinsic cell types of the molecular layer are *stellate cells* and *basket cells* (see Fig. 27–8). Stellate cells are usually found in outer regions of the molecular layer and are frequently referred to as *superficial* or *outer stellate cells.* The somata of basket cells are located immediately above the Purkinje cell layer. The basket cell axon travels in the sagittal plane and gives rise to descending branches that form elaborate "baskets" around the Purkinje cell body. This cell derives its name from this characteristic feature.

In general, the dendritic and axonal plexuses of basket and stellate cells are *oriented primarily in the sagittal plane,* much like those of Purkinje cell dendrites (see Fig. 27–8). Although basket and stellate cells are similar in general shape, the *extent* of the dendritic and axonal fields is much larger in basket cells than in stellate cells. Consequently, basket cells may influence

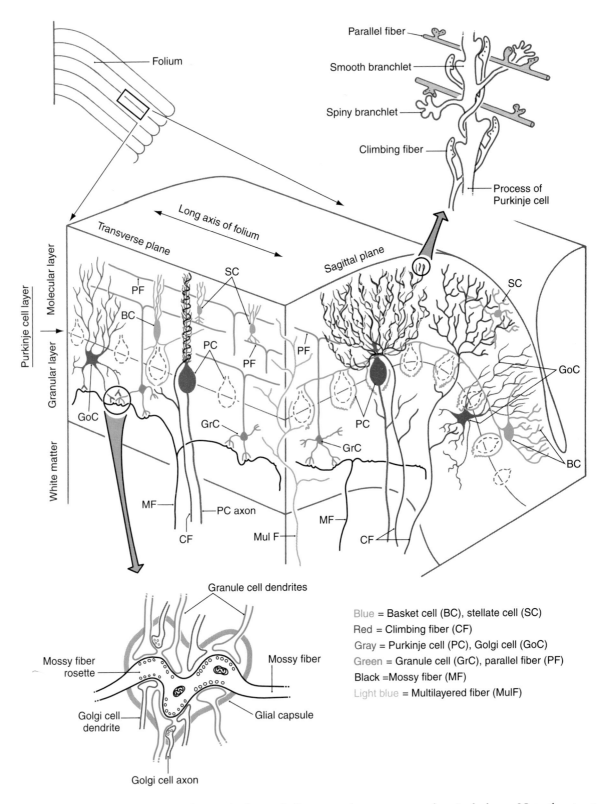

Figure 27–8. Cell types and synaptic relations in the cerebellar cortex in transverse and sagittal planes. Note the structure of the cerebellar glomerulus (*lower left*) and the interaction of parallel and climbing fibers (*upper right*) with the dendritic processes of Purkinje cells. Compare with Figure 27–9.

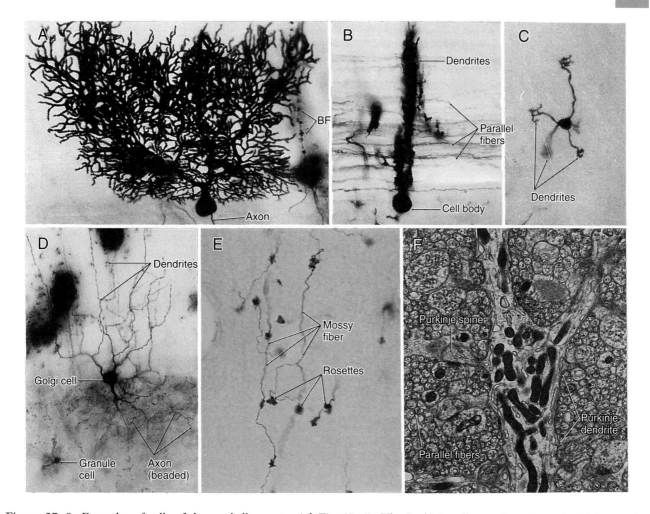

Figure 27-9. Examples of cells of the cerebellar cortex (cf. Fig. 27-8). The Purkinje cells are shown in sagittal (*A;* note the beaded appearance of the dendrites) and transverse (*B;* note the many parallel fibers) planes. Granule cell (*C, D*) dendrites end as a cluster of short, clawlike processes. Dendrites of Golgi cells (*D*) branch into molecular and granular layers, whereas their axons (*D,* beaded structures) ramify in only the granular layer. Mossy fibers (*E*) branch profusely and have many rosettes; synaptic contacts between mossy fibers and granule cell dendrites take place at the rosette in the cerebellar glomerulus (see Fig. 27-8). At the ultrastructural level, the Purkinje cell dendrite is surrounded by the numerous small profiles of parallel fibers. BF, Bergmann fiber, a type of glial cell process. (*A-D,* Courtesy of Dr. José Rafols, Wayne State University.)

a large number of Purkinje cells, mainly in the sagittal plane, whereas stellate cells influence a much smaller population, but also in the sagittal plane. Stellate and basket cells receive excitatory inputs from parallel fibers.

Basket and stellate cells are GABAergic and inhibit their target neurons. Although these cells influence several targets in the molecular layer, for our purposes the Purkinje cell is the most important. Similarly, Purkinje and Golgi cells are also inhibitory, thus making *the granule cell the only neuron in the cerebellar cortex whose output evokes an excitatory response.*

Cerebellar Afferent Fibers. The afferent fibers to the cerebellar cortex are grouped into three types on the basis of the *morphology and connections of their terminals in the cortex.* These three types of cortical terminals arise as *mossy fibers, climbing fibers,* and *multilayered* (monoaminergic) *fibers.*

The cerebellar afferent axons that end as *mossy fibers* originate from cell bodies in the cerebellar nuclei (*nucleocortical fibers*) and from a variety of other nuclei in the spinal cord, medulla, and pons (Table 27-1). En route to the cerebellar cortex, most of these cerebellar afferent fibers send collaterals into a cerebellar nucleus. In the granular layer, *mossy fibers* branch profusely, and their large terminals contact other cells at irregular intervals (the *mossy fiber rosette*). The rosette, which is the central element of the cerebellar glomerulus, gives the fiber a mossy appearance (see Figs. 27-8 and 27-9E). Single mossy fibers may form up to 50 rosettes, and each rosette may participate in synaptic contacts with up to 10 to 15 granule cells in a cerebellar glomerulus. In addition, a mossy fiber may branch and distribute to more than one folium. Mossy fibers utilize *glutamate* as their neurotransmitter and are excitatory to granule cell and Golgi cell dendrites in the cerebel-

Table 27–1.   Synopsis of Selected Afferent and Efferent Fibers of the Cerebellum as Contained in, or Associated with, the Cerebellar Peduncles

	Laterality	
Inferior Cerebellar Peduncle		
Restiform Body		
Dorsal spinocerebellar fibers	−	Mossy fibers
Cuneocerebellar fibers	−	Mossy fibers
Olivocerebellar fibers	X	Climbing fibers
Reticulocerebellar fibers (from reticulotegmental pontine nucleus)	X,−	Mossy fibers
Reticulocerebellar fibers (from lateral reticular nucleus)	−,(X)	Mossy fibers
Reticulocerebellar fibers (from paramedian reticular nucleus)	−,(X)	Mossy fibers
Trigeminocerebellar fibers	−	Mossy fibers
Raphecerebellar fibers	−,(X)	Multilayered fibers
Juxtarestiform Body		
Vestibulocerebellar fibers (primary and secondary)	−	Mossy fibers
Cerebellar corticovestibular fibers	−	Vestibular nuclei
Fastigiovestibular fibers	−,X	Vestibular nuclei
Fastigioreticular fibers	−,X	Reticular nuclei
Fastigio-olivary fibers	X	Caudal parts of accessory olivary nuclei
Fastigiospinal fibers	X	Spinal cord
Middle Cerebellar Peduncle		
Pontocerebellar fibers	X,(−)	Mossy fibers
Raphecerebellar fibers	−,(X)	Multilayered fibers
Superior Cerebellar Peduncle		
Ventral spinocerebellar fibers	X,−	Mossy fibers
Rostral spinocerebellar fibers	X,−	Mossy fibers
Ceruleocerebellar fibers	−,X	Multilayered fibers
Hypothalamocerebellar fibers	−,X	Multilayered fibers
Raphecerebellar fibers	−	Multilayered fibers
Cerebellar efferent fibers		
Dentatothalamic	X	Dorsal thalamus
Dentatorubral	X	Red nucleus
Dentatoreticular	X	Reticular nuclei
Dentatopontine	X	Pontine nuclei
Dentato-olivary	X	Principal olivary nucleus
Dentatohypothalamic	X	Hypothalamus
Interpositothalamic	X	Dorsal thalamus
Interpositorubral	X	Red nucleus
Interpositoreticular	X	Reticular nuclei
Interposito-olivary	X	Rostral parts of accessory olivary nuclei
Interpositohypothalamic	X	Hypothalamus
Interpositospinal	X	Spinal cord

−, uncrossed; X, crossed; (−), some uncrossed; (X), some crossed.

lar glomerulus and to cerebellar nuclear neurons on which their collaterals terminate.

The inferior olivary nuclei are the only source of cerebellar afferent axons that end as *climbing fibers* in the cerebellar cortex (see Table 27–1). Olivocerebellar fibers send collaterals to the appropriate cerebellar nucleus. The climbing fibers then terminate in the molecular layer by entwining, ivy-like, up the dendritic trees of Purkinje dendrites (see Fig. 27–8). Each Purkinje cell is innervated by a single climbing fiber, but olivocerebellar axons may branch to serve several Purkinje cells. Climbing fibers use the neurotransmitter *aspartate*, and they excite Purkinje cells and cerebellar nuclear neurons.

Multilayered fibers (monoaminergic or peptide-containing) originate from cells of the locus ceruleus (*noradrenergic*), the raphe nuclei (*serotoninergic*), the hypo-thalamus (some are *histaminergic*), and other select locations. These fibers enter the cerebellum via the cerebellar peduncles and, in the case of some hypothalamocerebellar fibers, by passing through the periventricular gray and then into the cerebellum. En route to the cerebellar cortex, many of these fibers send collaterals to the cerebellar nuclei. In the cortex, these axons branch diffusely and terminate in molecular and granular layers, where they may influence all major cell types (see Fig. 27–8). In general, these fibers modulate the output of the cerebellar cortex through two mechanisms. First, they decrease the spontaneous discharge rates of Purkinje cells. Second, both directly and via interneurons, multilayered fibers alter the responsiveness of Purkinje cells to excitation by climbing fibers and by the mossy fiber–granule cell projection.

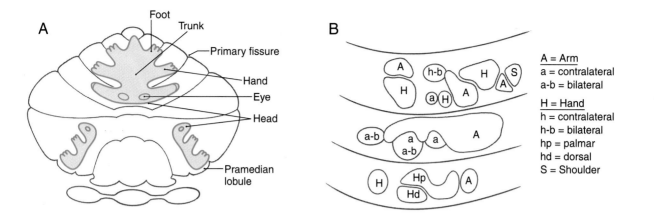

Figure 27–10. Somatotopy in the cerebellar cortex (*A*) and a summary representation of fractured somatotopy in the paramedian lobule (*B*) of a primate. In the somatotopic map, body areas were originally thought to be continuous (*A*), but more recent studies suggest that discontinuous body parts (or areas) may be represented in immediately adjacent cortical regions (*B*). (*B* is adapted from Welker et al, 1988, with permission.)

Topographic Localization. The cerebellum receives input from a wide range of sources. Some input that originates in the periphery is conveyed via spinocerebellar and vestibulocerebellar pathways that project *directly* to the cerebellum. Other afferent information is *indirect*, having passed through multiple central pathways before entering the cerebellum. For example, responses can be recorded in selected regions of the cerebellar cortex following stimuli that activate visual, auditory, or sensorimotor areas of the cerebral cortex in primates (Fig. 27–10*A*). These pathways involve a cerebropontine-pontocerebellar connection. Visual and auditory cortices project to cells of the basilar pons, which, in turn, provide a mossy fiber input to areas of the cerebellar cortex where the eye and ear are represented (see Fig. 27–10*A*). Similarly, the sensorimotor cortex, also via projections through the basilar pons, influences cerebellar cortical regions containing representations of the body.

At a finer level of resolution, experimental studies in mammals have shown that body parts are not represented continuously over a large area of cerebellar cortex but instead are broken into smaller, discontinuous patches. In this pattern, a small area of cortex that receives sensory input from the arm (via mossy fiber–granule cell connections) may be located adjacent to an area that receives input from a noncontiguous region of the same upper extremity (see Fig. 27–10*B*). In addition, each body part is represented in several locations. This pattern of spatial representation is referred to as *fractured somatotopy*.

Synaptic Interactions in the Cerebellar Cortex. In general, cerebellar function involves an ongoing series of *excitatory inputs to the cerebellar nuclei*, delivered through the collaterals of cerebellar afferent fibers, the effects of which are modulated by the *inhibitory action of Purkinje cell axons (corticonuclear fibers)* descending from the overlying cortex (Fig. 27–11). These synaptic interactions continuously modify the efferent signals generated by cerebellar nuclear neurons.

Climbing fibers synapse directly on Purkinje cells, whereas mossy fibers act through granule cells. Because

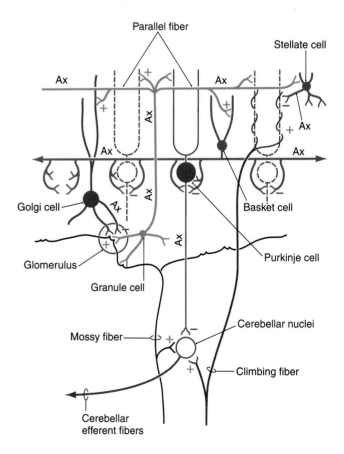

Figure 27–11. A diagrammatic representation of synaptic interactions in the cerebellar cortex. +, excitatory synapses; −, inhibitory synapses; Ax, axon.

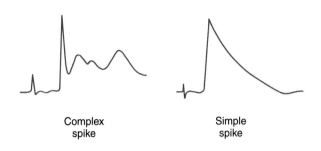

| Complex | Simple |
| spike | spike |

Figure 27–12. Complex and simple spikes of Purkinje cells, as recorded intracellularly following excitation by climbing and mossy fibers, respectively.

a *single climbing fiber* makes numerous synaptic contacts on a *single Purkinje cell*, its influence over that cell is substantial. Consequently, the Purkinje cell response to input from one climbing fiber is represented by a complicated waveform called a complex spike (Fig. 27–12). These spikes are unique and are the result of the combined action of multiple excitatory climbing fiber synapses formed throughout the Purkinje cell dendritic tree. In contrast, each Purkinje cell receives excitatory input from *many granule cells* via their parallel fibers. Summation, both spatial and temporal, of parallel fiber input is responsible for the generation of a single Purkinje cell response, the *simple spike* (see Fig. 27–12). At any given moment in time, there is an ongoing background level of simple spike activity in the cerebellum.

This level can be modulated by phasic increases or decreases in afferent inputs to the mossy fiber–granule cell–parallel fiber system. Collectively, mossy fibers have a powerful influence over Purkinje cells, and this influence may be modulated by climbing fibers through mechanisms not fully understood.

Let us review the mossy fiber–granule cell connection and its effects (see Figs. 27–8 and 27–11). In the cerebellar glomerulus, mossy fibers excite granule cell and Golgi cell dendrites. The Golgi cell axon, in turn, synapses on and inhibits granule cell dendrites in the glomerulus. Thus, the Golgi cell provides feedback inhibition to granule cells previously excited by mossy fiber activity. The granule cell axon enters the molecular layer, branches into parallel fibers, and excites Purkinje, stellate, basket, and Golgi cells (see Fig. 27–11). At a basic level, mossy fiber input leads to excitation of Purkinje cells via parallel fibers, and the GABAergic Purkinje cells respond by inhibiting the cerebellar nuclei.

The synaptic interactions within the cerebellar cortex are described in the following simplified model, which considers the cytoarchitectural and electrophysiologic properties of cerebellar cortical neurons. The inhibitory Purkinje cell outflow is modulated, in part, by the feed-forward inhibition resulting from stellate and basket cell activation (Fig. 27–13; see also Fig. 27–11). Parallel fibers excite a specific population of

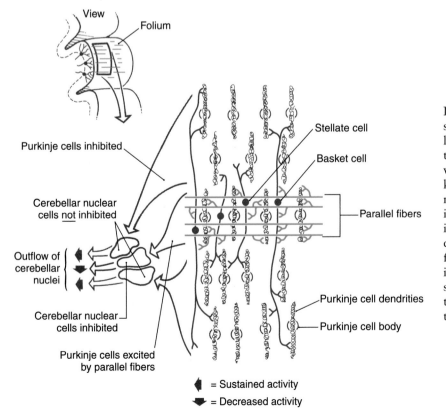

Figure 27–13. A diagrammatic representation of a portion of a cerebellar folium as viewed from the surface. Activation of a bundle of parallel fibers (green) will lead to activation of a row of Purkinje cells (red) located within their domain. Simultaneously, the Purkinje fibers in the flanking zones (gray) will be inhibited by the action of basket and stellate cells, which are also activated by parallel fibers. These populations of activated and inhibited Purkinje cells will cause, respectively, inhibition (red) and disinhibition (gray) of cerebellar nuclear cells via the corticonuclear pathway.

Purkinje cells, as well as stellate and basket cells located within their domain (see Fig. 27–13). The latter two GABAergic interneurons, in turn, inhibit Purkinje cells located adjacent to (via stellate synapses), or at some distance from (basket synapses), the row of activated parallel fibers. When a narrow bundle of parallel fibers is activated, under certain experimental conditions, basket and stellate cells can define a central row, or beam, of excited Purkinje cells. Within the beam, Purkinje cells are activated by parallel fibers and, in turn, inhibit cells in the cerebellar or vestibular nuclei (see Fig. 27–13). Purkinje cells on either side of the activated row are inhibited by stellate and basket axons and, consequently, do not inhibit their target neurons in the cerebellar (or vestibular) nuclei (see Fig. 27–13). These target cells are removed from their normal (background) inhibitory influence; that is, they are disinhibited.

Synaptic interactions between cells of the cerebellar cortex contribute to the activity of cerebellar nuclear neurons. The unique structural and functional properties of the cortex provide circuits for *the temporal and spatial processing of information* that contributes substantially to the cerebellar capacity to coordinate movement. The exact nature of these interactions has not been fully determined. However, the cytoarchitecture and synaptology within the cerebellum suggest one hypothetical model. According to this model, *the excitatory outflow from each cerebellar nucleus varies dynamically* in response to the combined effects of (1) the excitatory input from cerebellar afferent collaterals and (2) the inhibitory influence mediated by Purkinje cells.

Functional Cerebellar Modules

It is convenient to think of the cerebellum as being arranged into *compartments* or *modules*. Each module consists of (1) an area of cortex (usually a cortical zone), (2) a white matter core that contains afferent and efferent fibers to and from that cortical area, and (3) a nucleus (or nuclei) that is functionally related to the overlying cortical area. A cerebellar cortical zone and its corresponding nucleus (nuclei) and white matter constitute a module.

Vestibulocerebellar Module. The flocculonodular lobe and adjacent portions of vermal lobule IX receive afferents from the ipsilateral vestibular ganglion (*primary vestibulocerebellar fibers*) and vestibular nuclei (*secondary vestibulocerebellar fibers*). Therefore, these cortical areas are commonly called the *vestibulocerebellum*. Along with the *fastigial nucleus*, they form the *vestibulocerebellar* module (Fig. 27–14). Because this is phylogenetically the oldest part of the cerebellum, the vestibulocerebellum is sometimes called the archicerebellum (Greek *arche*, "beginning"). However, this latter term is not commonly used.

Vestibulocerebellar fibers access the flocculonodular cortex and fastigial nucleus via the juxtarestiform body and convey information concerning the position of the head and body in space, as well as information useful in orienting the eyes during movements. This information is supplemented by inputs carried on *olivocerebellar fibers* from the contralateral olivary nuclei and on *pontocerebellar fibers* (only to the flocculus) from the contralateral basilar pons. The latter pathways convey *indirect* inputs from nuclei of the diencephalon and brainstem, which are concerned with a broad spectrum of information regarding visual processing and eye movements (see also Chapter 28).

The outflow of the vestibulocerebellar module consists of *cerebellar corticovestibular fibers* from the flocculonodular lobe, *cerebellar corticonuclear fibers* from the nodulus to the fastigial nucleus, and efferent fibers arising in the *fastigial nucleus* (see Fig. 27–14). Purkinje cells in the flocculonodular cortex project, via the juxtarestiform body, directly to the ipsilateral vestibular nuclei (*cerebellar corticovestibular fibers*). Other Purkinje cells in the nodulus project into caudal regions of the fastigial nucleus as *cerebellar corticonuclear fibers*. Both of these projections are inhibitory (GABAergic) pathways. Fastigial neurons provide bilateral excitatory inputs to the vestibular and reticular nuclei (see Fig. 27–14). On the ipsilateral side, these axons pass directly through the juxtarestiform body. Fibers passing to the contralateral side cross in the cerebellar white matter and form the *uncinate fasciculus* as they loop over the superior cerebellar peduncle. These crossed fibers enter the vestibular complex via the juxtarestiform body.

The brainstem targets of cerebellar corticovestibular fibers and of fastigial efferent fibers, the vestibular and reticular nuclei, are the sources of vestibulospinal and reticulospinal tracts, respectively. The action of cerebellar corticovestibular fibers on the vestibular nuclei is inhibitory, whereas the action of fastigial efferents on the vestibular and reticular nuclei is excitatory.

Vestibulocerebellar Dysfunction. The vestibulocerebellum influences posture, balance, and equilibrium through vestibulospinal and reticulospinal projections to extensor motor neurons that influence axial and proximal limb muscles. The vestibular nuclei also bilaterally innervate the motor nuclei of cranial nerves III, IV, and VI through fibers that ascend in the medial longitudinal fasciculus.

Damage to the flocculonodular lobe, or to midline structures such as the nodulus and fastigial nucleus, will result in an unsteady lurching gait (*truncal ataxia*) that resembles that seen with drunkenness. This instability is also manifested as exaggerated movements of the legs and a *tendency to fall* to the side, forward, or backward. The patient may stand with feet farther apart than usual (*wide-based stance*) in an effort to maintain balance. Patients with such lesions are *unable to walk in tandem* (heel-to-toe) or to walk on their heels

Vestibulocerebellum

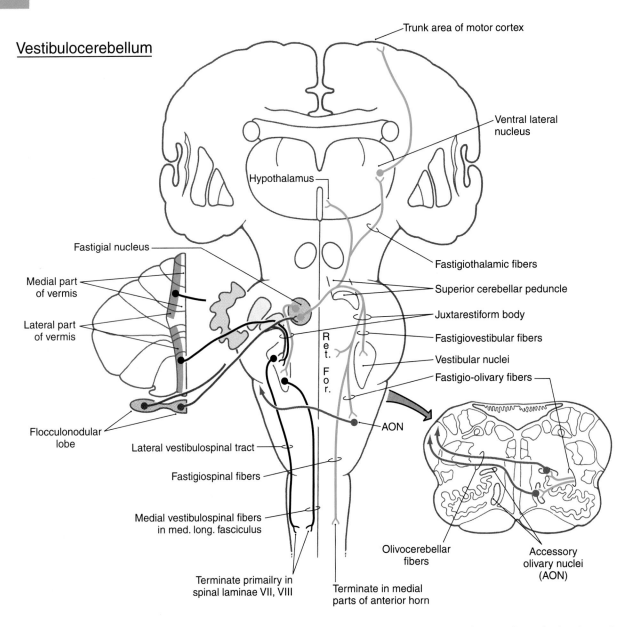

Figure 27–14. Projections of the vestibulocerebellum and of the lateral part of the medial zone through the fastigial and vestibular nuclei. med. long., medial longitudinal; Ret. For., reticular formation.

or toes. Midline lesions may also result in a tremor of the axial body or head called *titubation*. This tremor can range in amplitude from barely noticeable to so powerful that the patient is unable to sit or stand unsupported. *Nystagmus* is frequently seen, and deficits in pursuit eye movements are also common. In addition, the patient's head may *tilt* or turn to one side, the direction being unrelated to the laterality of the lesion. **Vestibular Connections of the Vermis.** In addition to the nodulus (of the flocculonodular lobe), most lobules of the vermal zone also have vestibular connections (see Fig. 27–14). For example, lateral portions of the vermal cortex receive secondary vestibulocerebellar fibers and project to the ipsilateral vestibular nuclei. Like the nodular cortex of the vestibulocerebellar mod-

ule, the medial portions of the vermal cortex send cerebellar corticonuclear fibers into the ipsilateral fastigial nucleus (Fig. 27–15). Consequently, the *fastigial nucleus* links vestibulocerebellar cortex *and* portions of the vermal cortex with the vestibular and reticular nuclei of the brainstem. In this respect, the vermal cortex and fastigial nucleus share the task of influencing axial musculature along with vestibulocerebellar and spinocerebellar modules.

Spinocerebellar Module. The vermal and intermediate zones receive input mainly via the *posterior* and *anterior spinocerebellar tracts* and, from the upper extremity, through *cuneocerebellar fibers*. Owing to this predominant input, these zones are collectively called the *spinocerebellum* (sometimes the *paleocerebellum*, al-

Spinocerebellum - Vermal Zone

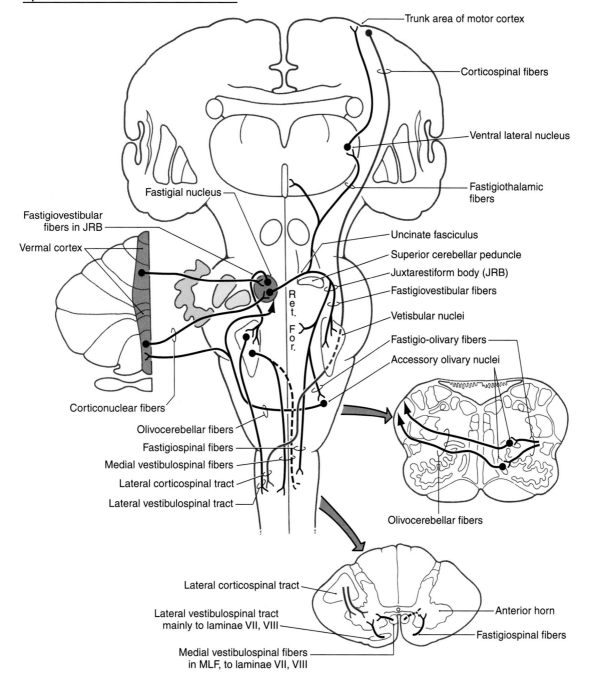

Figure 27–15. Projections of the spinocerebellum (vermal zone) through the fastigial and vestibular nuclei. MLF, medial longitudinal fasciculus; Ret. For., reticular formation.

though this term is not frequently used) (Fig. 27–16; see also Fig. 27–15). Posterior spinocerebellar and cuneocerebellar fibers enter the cerebellum via the restiform body, whereas anterior spinocerebellar fibers course into the cerebellum in association with the superior cerebellar peduncle. Fibers that enter the vermal zone send collaterals into the *fastigial nucleus*, and those

passing into the intermediate zone send branches into the *emboliform* and *globose nuclei*.

The output of the spinocerebellum is focused primarily on the control of axial musculature through vermal cortex and fastigial efferents, and on the control of limb musculature through efferents of the globose and emboliform nuclei. Posterior spinocerebellar and cuneocerebellar fibers inform

Spinocerebellum - Intermediate Zone

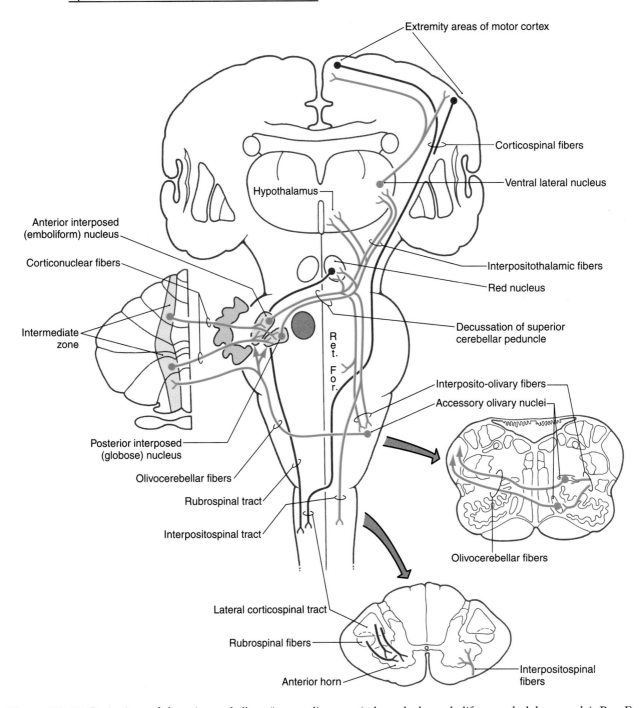

Figure 27–16. Projections of the spinocerebellum (intermediate zone) through the emboliform and globose nuclei. Ret. For., reticular formation.

the cerebellum of limb position and movement. This information is processed in the cerebellum and, through connections with the motor cortex via the thalamus, influences movements of the extremities and muscle tone. Cells in the spinal cord that give rise to anterior spinocerebellar fibers receive primary sensory inputs and are also under the influence of descending reticulospinal and corticospinal fibers. In this respect, anterior spinocerebellar fibers provide afferent signals *and* feedback to the cerebellum regarding motor circuits in the spinal cord.

There are additional inputs to the spinocerebellar

cortex. These arise in the contralateral accessory olivary nuclei (*olivocerebellar fibers*), the vestibular nuclei (*secondary vestibulocerebellar fibers*), the contralateral pontine nuclei (*pontocerebellar fibers*), and the reticular nuclei (*reticulocerebellar fibers*). These afferent axons also send collaterals into the fastigial and interposed nuclei.

The outflow of the spinocerebellar module consists of *cerebellar corticonuclear fibers* from vermal and intermediate cortex to *fastigial, emboliform,* and *globose nuclei* and of *cerebellar efferent axons* arising in these nuclei (see Figs. 27–15 and 27–16). Corticonuclear fibers project in a topographic sequence into their respective nuclei on the ipsilateral side. For example, fibers from anterior parts of the vermis enter rostral portions of the fastigial nucleus, whereas those of the posterior vermis project caudally in the same nucleus. In general, this pattern is repeated between the intermediate zone and the emboliform and globose nuclei.

As indicated previously, the *fastigial nucleus* projects bilaterally to vestibular and reticular nuclei, which, through their spinal projections, influence axial muscles. The fastigial nucleus also projects to (1) the contralateral *medial accessory olivary* nucleus, from which it receives input; (2) the medial areas of the anterior horn in upper levels of the spinal cord as *fastigiospinal fibers,* and (3) the ventral lateral nucleus of the thalamus, which, in turn, projects to trunk regions of the motor cortex (see Fig. 27–15).

Axons from the *globose* and *emboliform nuclei* exit the cerebellum via the superior cerebellar peduncle and cross in its decussation (see Fig. 27–16). From this point, some of these cerebellar efferent fibers course rostrally to terminate in the magnocellular part of the red nucleus (*cerebellorubral fibers*) and in the ventral lateral nucleus of the thalamus (*cerebellothalamic fibers*). These particular thalamic neurons project mainly to areas of the primary motor cortex. The red nucleus, via *rubrospinal fibers,* and the motor cortex, through *corticospinal fibers,* influence motor neurons in the contralateral spinal cord that control distal limb musculature (see Fig. 27–16). Other globose and emboliform efferents travel caudally to terminate in the reticular formation (*cerebelloreticular fibers*) and in the inferior olivary complex (*cerebello-olivary fibers*). Reticular cells influence spinal motor neurons and project back to the spinocerebellum as *reticulocerebellar fibers.* The globose and emboliform nuclei also receive *olivocerebellar fibers* from the accessory olivary nuclei to which they project (see Fig. 27–16).

Damage to spinocerebellar structures is frequently the result of extension from more medially or laterally located lesions. Consequently, the clinical picture is dominated by deficits characteristic of these medial or lateral regions. The lateral regions are more frequently affected (see below).

Pontocerebellar Module. The large lateral zone receives significant input from the *basilar pontine nuclei* via a primarily crossed *pontocerebellar fiber* projection. These fibers enter the cerebellum via the middle cerebellar peduncle. Because of this predominant source of afferent fibers, the lateral zone is called the *pontocerebellum* (see Fig. 27–16). Because the pontine nuclei receive a major projection from the ipsilateral cerebral cortex (as *corticopontine fibers*), this lateral zone is sometimes called the *cerebrocerebellum* (or *neocerebellum*). However, the term *pontocerebellum* is more appropriate and is in parallel with vestibulocerebellum and spinocerebellum. Afferent fibers that enter the lateral zone also send collaterals into the *dentate nucleus* (Fig. 27–17).

The pontocerebellum functions in the planning and control of precise dexterous movements of the extremities, particularly in the arm, forearm, and hand, and in the timing of these movements. Through its connections to motor cortical areas, the dentate nucleus is capable of modulating activity in cortical neurons that project to the contralateral spinal cord.

Another important source of afferents to the pontocerebellum is the *principal inferior olivary nucleus* (see Fig. 27–17). These *olivocerebellar fibers* are exclusively crossed. They enter the cerebellum via the restiform body, send collaterals to the dentate nucleus, and end in the molecular layer as climbing fibers.

The outflow of the pontocerebellar module consists of *cerebellar corticonuclear fibers* from the lateral zone to the dentate nucleus and *cerebellar efferent fibers* originating in the dentate nucleus (see Fig. 27–17). As for other cerebellar regions, corticonuclear fibers of the lateral zone are topographically organized; rostral and caudal areas of the zone project to the corresponding portions of the dentate nucleus.

Neurons of the *dentate nucleus* send their axons out of the cerebellum via the superior cerebellar peduncle and through its decussation (see Fig. 27–17). The fibers that pass rostrally project mainly to the parvocellular part of the red nucleus (*dentatorubral fibers*) and to the intralaminar and ventral lateral nuclei of the thalamus (*dentatothalamic fibers*). Some neurons of the red nucleus project, as part of the central tegmental tract, to the ipsilateral inferior olivary complex (*rubro-olivary fibers*). At the same time, cells of the ventral lateral thalamic nucleus project to wide areas of the motor and premotor cortices. The motor cortex, in turn, projects to the contralateral spinal cord (as *corticospinal fibers*) to influence motor neurons that innervate distal limb musculature (see Fig. 27–17). Descending crossed projections from the dentate nucleus pass mainly to the principal olivary nucleus (*dentato-olivary fibers*) and, in limited numbers, to the reticular and basilar pontine nuclei. *Olivocerebellar fibers* arising in the principal nucleus cross the midline and distribute to the cortex of the lateral zone and to the dentate nucleus. There is also feedback to the dentate nucleus via pontocerebellar and reticulocerebellar fibers.

Pontocerebellum

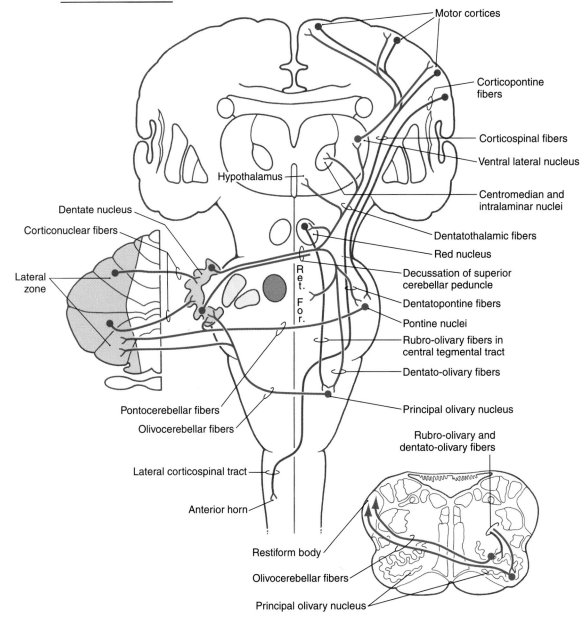

Figure 27–17. Projections of the pontocerebellum (lateral zone) through the dentate nucleus. Ret. For., reticular formation.

The relationship between the dentate nucleus and movement has been explored in experiments in monkeys. A cooling probe implanted in the cerebellar white matter adjacent to the dentate nucleus halts most of the electrical activity in dentate neurons. This significant reduction in electrical signals temporarily disconnects the dentate nucleus from its targets without permanently destroying the nucleus. Reversing the cooling results in a resumption of normal neuronal activity. After dentate cooling, the electrical activity of neurons in the primary motor cortex (MI) that signals the initiation of a movement (in response to a visual stimulus) was delayed, as was the execution of the movement it-

self. This observation indicates that dentate projections, which reach the cortex via the ventral lateral thalamic nucleus, are essential for the initial activation of MI corticospinal neurons at the beginning of a movement.

As a result of the delay of excitatory output from the motor cortex, there are corresponding delays in muscle contraction. For example, the *initial activation* of an agonist muscle (biceps brachii) to a load is slowed and its overall contraction time is longer. Similarly, the activation of the antagonist muscle (triceps brachii) that occurs when the load is removed is also delayed. This observation indicates that the reciprocal pattern of activation in agonists and antagonists that accompanies some movements is dramatically disrupted. Thus, cerebellar output is involved in *timing* muscle activation (and inactivation), as well as influencing the duration of muscle contraction.

Pontocerebellar Dysfunction. Before the consequence of lesions affecting the pontocerebellar module is considered, two important points merit emphasis. First, damage that involves *only* the cerebellar cortex rarely results in permanent motor deficits. However, damage to the cortex plus nuclei or to only the nuclei results in a wide range of motor problems. Second, *lesions of the cerebellar hemisphere result in motor deficits on the ipsilateral side of the body because the motor expression of cerebellar injury is mediated primarily through corticospinal and rubrospinal pathways.* In brief, the right lateral and interposed nuclei influence the left motor cortex and red nucleus, which, in turn, project to the right side of the spinal cord. Thus, a lesion in the cerebellum on the right results in deficits on the right side of the body. The exception is a midline lesion, which produces bilateral deficits restricted to axial/truncal parts of the body. Lesions that involve the cerebellar hemisphere frequently affect portions of the lateral and intermediate modules. It is common for these disorders to be categorized as disorders of the *lateral* (or *hemisphere*) *zone* or as *neocerebellar disorders.*

In general, lesions of the lateral cerebellum result in a deterioration of coordinated movement, which is sometimes referred to as a *decomposition of movement* (or *dyssynergia*). This deficit consists of the breakdown of movement into its individual component parts. There may also be a decrease in muscle tone (*hypotonia*) and in deep tendon reflexes. *Ataxia* involving the extremities generally is also seen in patients with lateral cerebellar lesions. Because of ataxia of the lower extremity, these patients may also have an *unsteady gait* and a tendency to lean or fall to the side of the lesion.

Dysmetria (also called *past-pointing*) is apparent in patients when they attempt to point accurately or rapidly to moving or stationary targets. The patient may reach past the target (*hypermetria*) or fall short of the target (*hypometria*).

Tremor is a consistent finding in patients with lateral cerebellar lesions. A *kinetic tremor,* commonly called an *intention tremor,* is evident when the patient performs a voluntary movement and is most obvious as the endpoint, or target, is approached. This deficit is commonly seen when the patient extends the arm and then attempts to touch the index finger to the nose (Fig. 27–18). At rest there is little or no tremor, but as the finger approaches the nose, the tremor is markedly accentuated. This finding is opposite to that seen in patients with Parkinson disease, whose tremor is evident at rest (*resting tremor*) but largely diminishes during a voluntary movement. Patients with cerebellar lesions may also demonstrate a *static tremor.* This tremor is manifested when the patient stands with the arms

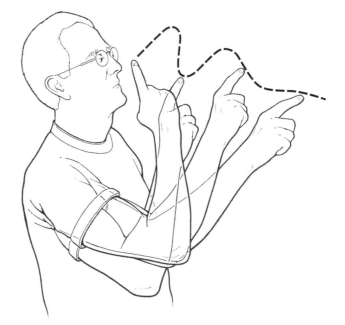

Figure 27–18. Intention tremor. Note that as the patient moves his finger closer to the target (his nose), the tremor becomes worse. In other words, as the patient "intends" to make a precise movement, the tremor gets progressively worse as the target is approached; this sequence of events is an easy way to remember this deficit.

extended (muscles contracted against gravity). There is rhythmic movement of the shoulders that also involves the upper extremities.

Awkward performance of rapid alternating movements, such as supinating and pronating the hand against the thigh, is called *dysdiadochokinesia* (Fig. 27–19). The patient may also be unable to perform rhythmic movements. This deficit is demonstrated by asking the patient to rapidly tap the table three times with the index finger, pause two seconds, tap three times, and so on. For patients with lateral cerebellar lesions, this task will be difficult or impossible.

Other lateral zone deficits include rebound phenomena, dysarthria, and ocular motor dysfunction. The *rebound phenomenon* (or *impaired check*) is an inability of agonist and antagonist muscles to adapt to rapid changes in load. For example, if the patient is asked to push against the physician's hand and the hand is then unexpectedly removed, the patient's arm will overshoot beyond the point where it would normally halt. Tremors or oscillations may also be evident as the arm returns to its starting point. Patients with *dysarthria* have slurred, garbled speech that may also be alternately slow or staccato in nature (*scanning speech*). This is a motor problem (not an aphasia), because the patient is still able to use words and grammar correctly. Characteristic ocular motor dysfunctions seen in patients with

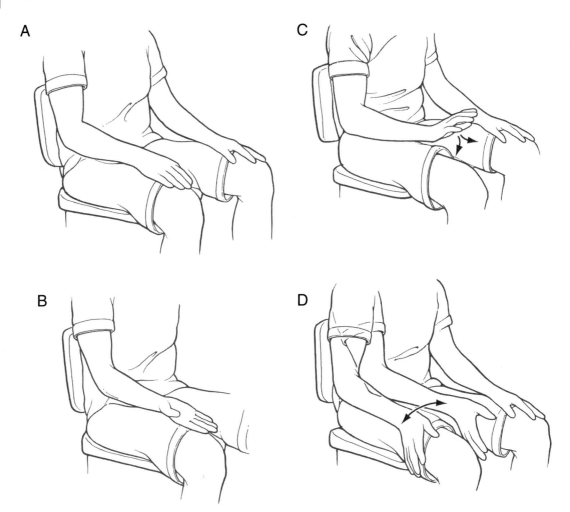

Figure 27–19. Difficulty performing rapid alternating movements: dysdiadochokinesia. Normally the patient can rapidly and precisely pronate (*A*) and supinate (*B*) his hand, but with a cerebellar lesion the patient cannot perform this task. This affected hand flails about the target (*C, D*) and is unable to make the movement smoothly or otherwise. If the patient has a lesion in the right side of the cerebellum, the right hand will be affected.

lateral cerebellar lesions are *nystagmus* and abnormalities of target-directed eye movements. Nystagmus most commonly presents as abnormal horizontal eye movements that consist of a slow conjugate movement away from the side of the lesion. This abnormality is opposite to that seen in lesions of the vestibular receptors, primary sensory fibers, and nuclei. In other cases, the velocity of these conjugate movements may be the same in both directions (*pendular nystagmus*). Abnormal target-directed eye movements may present as an inability to follow a slowly moving target (disturbed pursuit movements) or difficulty in maintaining fixation on a stationary target.

Cerebellar Influence on Visceromotor Functions

The cerebellum receives input from the *solitary* and *dorsal motor vagal nuclei* and from a number of nuclei in the *hypothalamus*. These areas are directly involved in

the control and modulation of a variety of visceromotor functions. The hypothalamus, in addition to having direct connections with the cerebellar cortex and nuclei, also projects to brainstem and spinal nuclei that are involved in the regulation of visceral functions.

Neurons in several hypothalamic areas and nuclei project to the cerebellar cortex and nuclei (*hypothalamocerebellar fibers*). The cerebellar nuclei, in turn, send a primarily crossed projection to the hypothalamus (*cerebellohypothalamic fibers*) via the superior cerebellar peduncle (see Figs. 27–15 and 27–17). Through these reciprocal connections, the cerebellum may receive visceral input and influence neurons that control visceral functions.

Visceral deficits related to cerebellar lesions are rarely reported, for two reasons. First, the somatomotor deficits seen in such cases are overwhelmingly diagnostic, and there is no need to look further. Second, cerebellar lesions may cause an increase in intracranial pressure with resultant pressure on the medulla. Consequently, it is difficult to separate visceral deficits that relate to the cerebellar lesion from those that relate to pressure on the medulla.

In some situations, however, visceral deficits can be related directly to the cerebellar lesion. For example, deficits of this type occurred in a patient who had an occlusion of branches of the left superior cerebellar artery with resultant damage to the cerebellar nuclei on that side. There was no indication of increased intracranial pressure. This patient had characteristic somatomotor tremors in the left arm and leg *during attempted movement* on that side and showed two distinct visceral responses *that occurred concurrent with the somatomotor tremor*, but not before or after. First, the patient's pupils dilated *during* the tremor. Second, the skin on the patient's face flushed and felt warm to the touch *during* the tremor. Immediately upon cessation of the attempted movement (and resulting tremor) of the left hand, the patient fanned his face with his right hand and complained, "It's hot in here." In other cases, small lesions restricted to the fastigial nucleus have been noted to result in decreases in heart rate and blood pressure. These observations suggest that the cerebellum is actively involved in the regulation of visceral function concurrent with its classically recognized influence in the somatomotor sphere.

The Cerebellum and Motor Learning

The cerebellum is unequivocally involved in the learning of a variety of relatively simple reflexive motor behaviors. It is extremely difficult to produce specific modifications in certain reflexes without this structure. Such modifications include the adaptation of the vestibulo-ocular reflex and the classic (pavlovian) conditioning of reflexes evoked by aversive stimuli, such as the eyeblink and withdrawal reflexes.

Although the cerebellum is believed to be involved in the acquisition of voluntary, complex motor skills, its role in this process is not well understood. Initial studies demonstrated that normal subjects performed better after practicing a new task than did patients with cerebellar damage. This finding was initially ascribed to the presence of a learning deficit in these patients. However, recent experiments that examined the *rate* at which a new motor skill is acquired revealed that cerebellar patients can learn to perform new movements even though the *quality* of the performance is affected by their deficit. These studies, together with animal experiments showing that volitional limb movements can be acquired during inactivation of the cerebellar nuclei, confirm that new (and complex) motor tasks can be learned despite impairment of cerebellar function.

Under normal conditions, the cerebellum plays an important role in the *acquisition* component of motor learning. For example, in animals, complex movements learned during the inactivation of the cerebellar nuclei are more variable. In other studies, imaging of the living brain has revealed that specific regions in the cerebellum are activated *during* the learning of novel movements. In addition, the acquisition of complex behaviors in animals is associated with increased modulation of cells in the cerebellar nuclei when the animal first learns to perform the task correctly and consistently on successive trials. Together, these data support the view that the cerebellum is actively involved in the acquisition of novel movements.

The participation of the cerebellum in the storage of memory engrams required for the recall of previously learned movements remains an area of active investigation. It has been proposed that the climbing fiber input to the dendrites of Purkinje cells can induce plastic changes that modify the responsiveness of these neurons to specific inputs mediated by parallel fibers. However, the relevance of this mechanism to normal physiologic conditions continues to be debated. There is considerably stronger evidence that the cerebellar nuclei may be a storage site in certain types of motor learning such as the classically conditioned eyeblink reflex.

There is evidence that memory storage sites also exist outside the cerebellum. For example, it is known that modification of the vestibulo-ocular reflex (see Chapter 22) involves persistent changes in synaptic transmission in the vestibular nuclei. However, some scientists contend that such modifications also occur within the cerebellum when this reflex is changed.

Studies that examined the basis for the classic conditioning of the eyeblink and withdrawal reflexes have implicated the cerebellum as the critical storage site for the plastic changes established during this process. This argument is not yet supported by a satisfactory body of evidence. Some researchers contend that sites outside the cerebellum are involved in this storage process. However, it is unclear which locations are most important, because the cerebellar nuclei are so critical for the *performance* of these reflexes once they are learned. As a consequence, learning may appear to be deficient only because of the difficulty in performing the desired movement.

The issue of memory storage for volitional movements is clearer. The cerebellum is not essential for the retention of previously learned complex motor behaviors. The patterns of movement required to perform complex volitional motor tasks can be recalled during the inactivation of the ipsilateral dentate and interposed nuclei.

In summary, certain reflexes cannot be modified in the absence of the cerebellum, and lesions of this structure impair the *performance* of certain previously learned behaviors. This structure plays an important role in the *acquisition of several motor behaviors including skilled volitional movements*, although the nature of its contribution has not been characterized. The cerebellum's participation in memory storage for certain reflex behaviors is unclear, although it is not considered to be an essential storage site for engrams related to complex volitional movements.

Sources and Additional Reading

Bloedel JR, Dichgans J, Precht W: Cerebellar Functions. Springer Verlag, New York, 1985.

Brooks VB, Thach WT: Cerebellar control of posture. In Brooks VB (ed): Handbook of Physiology, Section 1: The Nervous System, vol II: Motor Control, Part 2. American Physiological Society, Bethesda, Md, 1981.

Dietrichs E, Haines DE, Roste GK, Roste LS: Hypothalamocerebellar and cerebellohypothalamic projections—circuits for regulating nonsomatic cerebellar activity. Histol Histopathol 9:603–614, 1994.

Duvernoy HM: The Human Brain Stem and Cerebellum: Surface, Structure, Vascularization, and Three-Dimensional Sectional Anatomy with MRI. Springer Verlag, Vienna, 1995.

Gerrits NM, Ruigrok TJH, de Zeeuw CI: Cerebellar modules: Molecules, morphology and function. Prog Brain Res 124:1–330, 2000.

Gilman S, Bloedel JR, Lechtenberg R: Disorders of the Cerebellum. Contemporary Neurology Series, vol 21. FA Davis, Philadelphia, 1981.

Haines DE, Patrick GW, Satrulee P: Organization of cerebellar corticonuclear fiber systems. Exp Brain Res Suppl 6:320–371, 1982.

Haines DE, Dietrichs E, Mihailoff GA, McDonald EF: The cerebellar-hypothalamic axis: Basic circuits and clinical observations. In Schmahmann JD (ed): The Cerebellum and Cognition. International Review of Neurobiology, Vol 41:83–107, Academic Press, San Diego, 1997.

Hore J, Flament D: Evidence that a disordered servo-like mechanism contributes to tremor in movements during cerebellar dysfunction. J Neurophysiol 56:123–136, 1986.

Ito M: The Cerebellum and Neural Control. Raven Press, New York, 1984.

Larsell O, Jansen J: The Comparative Anatomy and Histology of the Cerebellum: The Human Cerebellum, Cerebellar Connections, and Cerebellar Cortex. University of Minnesota Press, Minneapolis, 1972.

Mihailoff GA: Identification of pontocerebellar axon collateral synaptic boutons in the rat cerebellar nuclei. Brain Res 648:313–318, 1994.

Ojakangas CI, Ebner TJ: Purkinje cell complex and simple spike changes during a voluntary arm movement learning task in the monkey. J Neurophysiol 68:2222–2236, 1992.

Robertson L: Organization of climbing fiber representation in the anterior lobe. In King JS (ed): New Concepts in Cerebellar Neurobiology. Alan R. Liss, New York, 1987, pp 281–320.

Thach WT, Goodkin HP, Keating JG: The cerebellum and the adaptive coordination of movement. Annu Rev Neurosci 15:403–442, 1992.

Welker W, Blair C, Shambes GM: Somatosensory projections to cerebellar granule cell layer of giant bushbaby, *Galago crassicaudatus*. Brain Behav Evol 31:150–160, 1988.

Chapter 28

Visual Motor Systems

P. J. May and J. J. Corbett

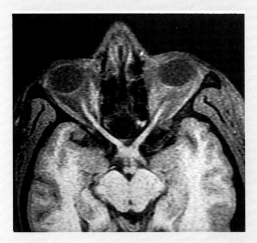

Overview 446

Peripheral Structures 446
Extraocular Muscles
Intraocular Eye Muscles
Eyelid

Central Structures 449
Oculomotor Nucleus
Edinger-Westphal Nucleus
Trochlear Nucleus
Abducens Nucleus
Abducens Internuclear Neurons
Sympathetic Supply to the Orbit

Targeting Movements 453
Saccades
Extraocular Muscle Motor Neurons
Horizontal and Vertical Gaze Centers
Supranuclear Control
Smooth Pursuit
Vergence Movements and the Near Triad

Reflex Movements 458
Optokinetic Eye Movements
Pupillary Light Reflex
Blinking and Other Lid Movements

All animals use their sensory organs to scan the environment in search of information. Often these organs are actively oriented toward relevant targets. This *orienting behavior* is exhibited by creatures from honey bees to human beings. Movement of the eyes, for example, allows closer inspection of the environment as needed. However, human eyes have a fovea, a small portion of the central retina that has exquisite visual sensitivity. Accurately directing the fovea to targets of interest represents a crucial orienting behavior in humans. Extraocular muscles orient our highly mobile eyes, and the *oculomotor* (or ocular motor) system controls these muscles. It is one of several *visual motor* systems that support the function of visual sensation.

Overview

The oculomotor system includes *gaze systems* that redirect the eyes to each new target. There are three basic types of "targeting" movements: (1) *saccades*, rapid movements that direct the eyes to each new target; (2) *smooth pursuit*, slower movements that allow the eyes to follow moving targets; and (3) *vergence movements*, which adjust for target distance by changing the angle between the eyes. Vergence is coupled with changes in the curvature of the *lens* and the size of the *pupil* that focus the target image on the fovea. Saccades and smooth pursuit are *conjugate* movements, in which the eyes move in the same direction, often with accompanying movements of the head and body. Vergence movements are *disconjugate* (see Table 28–1).

Visual motor systems also mediate a set of reflex actions. *Compensatory reflexes* keep the eyes on target despite body movements. Sensory inputs from the vestibular and visual systems tell the brain that the body is in motion. During movement, the *vestibulo-ocular reflex* compensates for acceleration, which is sensed by the vestibular labyrinth, whereas the *optokinetic reflex* compensates for velocity, which is indicated by movement of the whole visual field (see Table 28–1). Visual motor systems also compensate for the amount of light falling on the retina. The *pupillary light reflex* maintains the level of retinal illumination within the working range of the photopigments in the photoreceptor cells (rods or cones). Finally, the *blink reflex* protects the eye.

Disturbances of the visual motor systems are common and often produce the first symptoms recognized by a patient. Understanding these ocular signs provides for timely and effective diagnosis. For instance, *strabismus* is a defect in which the eyes are misaligned. If this defect is left untreated, the brain reacts to the constant diplopia (double vision) by ignoring the input from one eye and failing to focus it (*amblyopia*) and, eventually, failing even to orient it. However, amblyopia can be avoided through early treatment of strabismus.

Peripheral Structures

Extraocular Muscles. The eye is moved in the orbit by six extraocular muscles (Fig. 28–1). These muscles produce movements in the horizontal plane (left and right) around a vertical axis, movements in the vertical plane (up and down) around a horizontal axis, and torsional movements (clockwise and counterclockwise) around an axis running through the center of the pupil to the fovea. There are two pair of rectus muscles, with the members of each antagonistic pair arranged

Table 28–1.   Summary of Eye Movement Characteristics

Class	Function	Conjugate	Speed	Latency	Clinical Point(s)
Foveation	Hold fovea on target	—	None Only very small oscillations	—	Look for drifts, especially holding eccentric gaze
Saccade	Acquisition of new target	Yes	100–700 degrees/sec Amplitude-dependent	200 msec	Sedative-sensitive Test: Look between points
Smooth pursuit	Follow a moving target	Yes	Variable <100 degrees/sec Target speed–dependent	100 msec	Test: Follow a slow-moving target
Vergence	Direct fovea at targets of different distances	No	~15 degrees/sec Disconjugate saccades faster	150 msec	Test: Move target toward and away from nose
Vestibular (VOR)	Compensate for initial head movement	Yes	Variable <500 degrees/sec Head acceleration–dependent	<15 msec Lasts 45 sec	Dizziness and nausea Test: Caloric (irrigate ear)
Optokinetic (OKR)	Compensate for continued head movement	Yes	Variable <100 degrees/sec Retinal slip–dependent	50–100 msec	Test: Steady movement of a striped drum or cloth
Nystagmic fast phase	Recenter eye during continuous VOR or OKR	Yes	500 degrees/sec	—	Nystagmus: Alternating fast and slow phase movements

OKR, optokinetic reflex; VOR, vestibulo-ocular reflex.

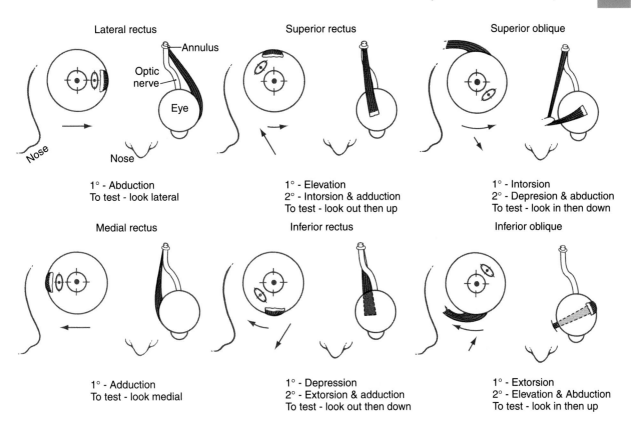

Figure 28–1. The six extraocular muscles of the left eye in frontal (*left*) and dorsal (*right*) views. The primary (*long arrow*) and secondary (*short arrow*) actions of each muscle and the position of the pupil after the movement (*red oval*) are indicated on the frontal view. These muscles originate at the common tendon (annulus of Zinn), with the exception of the inferior rectus, which originates from the orbit's nasal wall. Rectus muscles insert in front of the equator of the globe, and the oblique muscles insert behind it. Secondary actions are important for clinical evaluation of muscle function. In these cases, the eye is first rotated to align its axis with that of the muscle; for example, for the inferior oblique, the eye is first adducted and then elevated.

opposite one another on the globe, and a single pair of oblique muscles, which also act as antagonists. For horizontal eye movements, the *medial rectus muscle* rotates the eye toward the nose (*adduction*) and the *lateral rectus muscle* rotates the eye toward the temple (*abduction*). For vertical eye movements, the primary action of the *superior rectus* is to rotate the eye upward (*elevation*), and the primary action of the *inferior rectus* is to rotate the eye downward (*depression*). The direction of pull of the *superior oblique muscle* is modified because its tendon passes through a loop of connective tissue, the *trochlea*, on the medial wall of the bony orbit. Its insertion on the globe is caudal to that of the superior rectus. As a consequence, the actions of the superior oblique are *intorsion*, depression, and abduction. Conversely, the *inferior oblique muscle* actions are *extorsion* and elevation, as well as abduction.

The extraocular muscles are supplied by three cranial nerves. The *abducens nerve* (cranial nerve VI) supplies the lateral rectus muscle, the *trochlear nerve* (cranial nerve IV) supplies the superior oblique muscle, and the rest are innervated by the *oculomotor nerve* (cranial nerve III). The extraocular muscles are striated and contain fibers adapted to produce extremely high contraction velocities and nearly constant tension. The muscles run through connective tissue sheaths that control their pulling directions. Each muscles has an inner *global* layer and an outer *orbital* layer. The thick global layer inserts onto the sclera of the eye, whereas the thin orbital layer inserts into the connective tissue sheath to regulate the muscle's pulling direction. The trigeminal nerve carries sensory information from the extraocular muscles. This proprioceptive signal appears to be critical for normal development of *stereoscopic vision* (perception of three-dimensional space) and may have other roles. Visual feedback is the primary source of information on eye movement accuracy.

Intraocular Muscles. The eyes contain three intrinsic smooth muscles (Fig. 28–2). The *ciliary muscle* changes the curvature of the lens in order to bring visual targets into focus on the retina. The *sphincter* (or *constrictor*) *pupillae muscle* and *dilator pupillae muscle* control the size of the pupils in an antagonistic fashion, to regulate the amount of light entering the eyes and the depth of field.

The ciliary muscle, found in the *ciliary body*, is con-

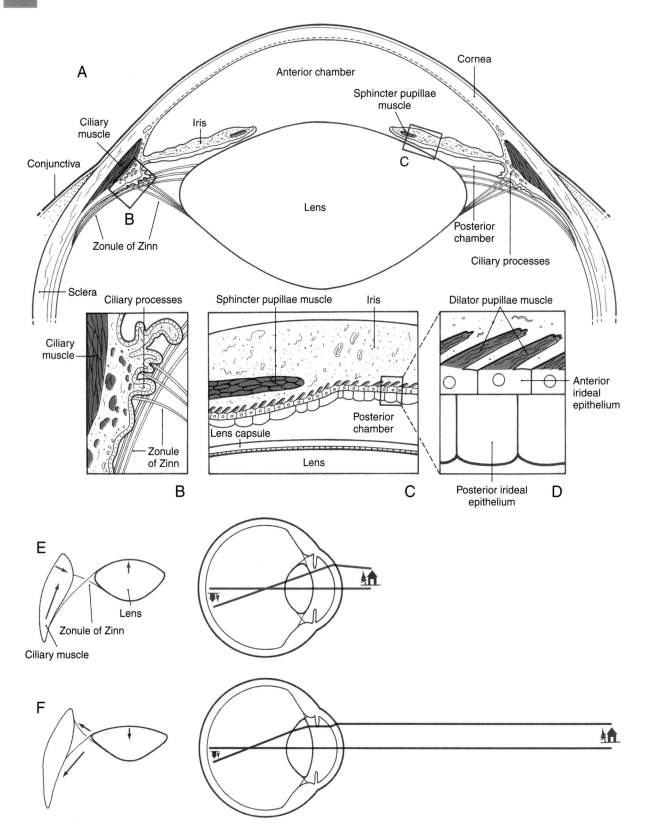

Figure 28–2. The anterior segment of the eye showing the intrinsic eye muscles and optical components (*A*). The zonule of Zinn (suspensory ligament) extends from the lens capsule and inserts into the epithelium covering the ciliary muscle (*B*). The pupillary sphincter and dilator muscles are separate muscles in the iris (*C*). The latter is actually made up of the myoid processes of the anterior iridial epithelium, as shown in *D*. The mechanism (*left*) and effects (*right*) of accommodation (*E, F*) are shown. On looking at a nearby target (*E*), the ciliary muscle contracts (*arrows*), releasing tension in the zonule and allowing the anterior surface of the lens to round up (*arrow*) owing to its own elasticity. On looking at a distant target (*F*), the ciliary muscle relaxes, and the unopposed tension in the zonule (*arrows*) flattens the lens (*arrow*).

nected to the lens by the *suspensory ligaments (zonule of Zinn)*. These fine connective tissue threads resemble the spokes of a bicycle wheel. The action of the ciliary muscle changes the shape of the lens (via the zonule) to adjust its refractive state and focus the image on the retina (see Fig. 28–2*E* and *F*). These changes are termed *lens accommodation*. Constriction of the ciliary muscle is produced by activation of cholinergic parasympathetic fibers from the ciliary ganglion. With age, the lens grows less elastic, so that the actions of the ciliary muscle have less effect on refraction. This loss of accommodation produces a blurring of near vision termed *presbyopia*. *Myopia*, a loss of distant acuity, generally appears at an early age and may be due to genetic or environmental factors.

The sphincter pupillae is a ring-shaped muscle that lies along the pupillary margin (see Fig. 28–2). It contracts in response to activation of cholinergic parasympathetic fibers from the ciliary ganglion, to constrict the pupil (*miosis*). The dilator muscle is radially arranged, so that its action folds the iris and draws open the pupil (*mydriasis*). The dilator is activated by adrenergic sympathetic postganglionic fibers from the superior cervical ganglion.

Eyelid. The eyelid is controlled by the *levator palpebrae superioris*, the *orbicularis oculi*, and the tarsal or *Müller muscles* (Fig. 28–3). The levator palpebrae is supplied

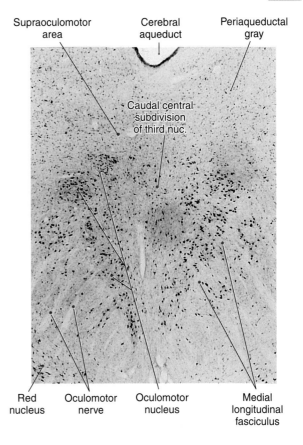

Figure 28–4. The human oculomotor nucleus and adjacent structures in a Nissl-stained cross section showing neuron cell bodies.

by cranial nerve III. It originates with and travels parallel to the superior rectus but continues forward to insert into the upper lid. The levator holds the eyelid up when the eyes are open, and it functions in concert with the superior rectus, increasing the elevation of the lids when the eyes look up. The orbicularis oculi, supplied by cranial nerve VII, closes the eyes by depressing the upper lid and elevating the lower lid. The tarsal muscles are small smooth muscles at the edge of the bony orbit. They are supplied by postganglionic sympathetic fibers and help keep the lids open.

Central Structures

Oculomotor Nucleus. The oculomotor nucleus (Fig. 28–4) lies near the midline in the ventral periaqueductal gray of the rostral midbrain. Beneath it lie the fibers of the *medial longitudinal fasciculus*, many of which synapse within the oculomotor nucleus. Axons from motor neurons in the oculomotor nucleus pass either medial to or through the red nucleus and exit the midbrain just medial to the crus cerebri. These structures are supplied by paramedian branches of the basilar artery and the proximal part of the posterior cerebral artery (P_1 segment). Consequently, vascular le-

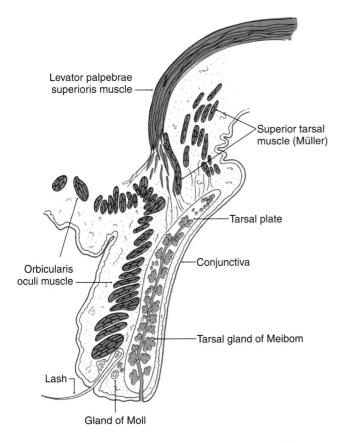

Figure 28–3. The structure of the upper eyelid.

sions in this region produce oculomotor deficits in combination with other symptoms (Table 28–2).

The third cranial nerve passes along the wall of the cavernous sinus and enters the ipsilateral orbit by way of the superior orbital fissure. Branches supply the superior rectus and levator palpebrae muscles, the inferior rectus and inferior oblique muscles, and the medial rectus muscle. The motor neurons supplying each of these individual muscles form rostrocaudally oriented columns within the nucleus (Fig. 28–5). The motor neurons supplying the levator palpebrae superioris muscle form a separate dorsal midline subnucleus called the *caudal central subdivision.*

Edinger-Westphal Nucleus. The nucleus of *Edinger and Westphal* contains the cholinergic, preganglionic parasympathetic motor neurons that control lens accommodation and pupillary constriction. In humans, the Edinger-Westphal nuclei form a pair of cell columns that lie near the midline, posterior (dorsal) to the oculomotor nucleus (see Fig. 28–5). The preganglionic fibers travel with the ipsilateral oculomotor nerve and synapse in the *ciliary ganglion.* Cholinergic, postganglionic motor neurons send their axons to the globe via the *short ciliary nerves.* These motor neurons supply the ciliary muscle and pupillary constrictor, with the great

majority supplying the former. Thus, in Weber syndrome (see Table 28–2) or other lesions of the oculomotor nerve, loss of the preganglionic fibers may result in ipsilateral *mydriasis* (dilation of the pupil) and paralysis of accommodation.

Trochlear Nucleus. The trochlear nucleus is a small, ovoid group of neurons nestled in the medial longitudinal fasciculus of the caudal midbrain (see Fig. 28–6). These motor neurons supply the contralateral superior oblique muscle. The axons forming the trochlear nerve arch dorsally and caudally around the periaqueductal gray, cross the midline in the anterior medullary vellum, and exit the dorsal surface of the brainstem at the base of the inferior colliculus. This nerve then courses laterally and then anteriorly, hugging the surface of the midbrain. It passes rostrally along the wall of the cavernous sinus, before entering the orbit via the superior orbital fissure, to innervate the superior oblique muscle.

Abducens Nucleus. The abducens nucleus is a spherical cell group located in the facial colliculus, adjacent to the internal genu of the facial nerve in the caudal pons (Fig. 28–7; see also Fig. 12–10). Abducens fibers pass slightly caudally and then anteriorly (ventrally) to exit near the midline at the pontomedullary junction.

Table 28–2. **Summary of Lesions of the Exiting Oculomotor Nerve**

Structures Involved	Deficits

Weber Syndrome

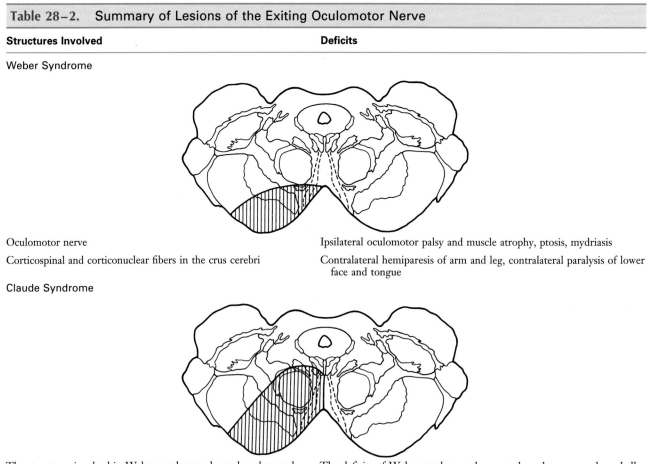

| Oculomotor nerve | Ipsilateral oculomotor palsy and muscle atrophy, ptosis, mydriasis |
| Corticospinal and corticonuclear fibers in the crus cerebri | Contralateral hemiparesis of arm and leg, contralateral paralysis of lower face and tongue |

Claude Syndrome

| The structures involved in Weber syndrome plus red nucleus and cerebellothalamic fibers | The deficits of Weber syndrome plus contralateral tremor and cerebellar ataxia |

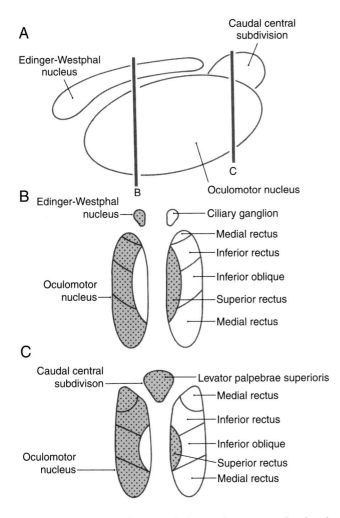

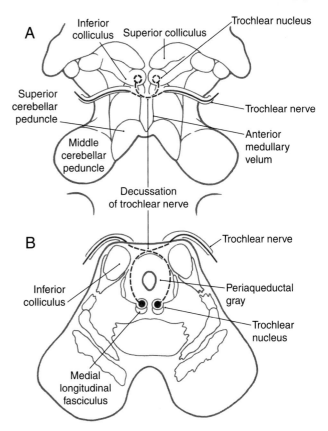

Figure 28-5. Lateral view of the oculomotor and related nuclei (*A*). The motor neuron pools in the nucleus are shown in frontal sections through anterior (*B*) and posterior (*C*) parts of the nucleus. The names of the major nuclei are indicated on the left in *B* and *C*, and the targets of each subdivision are indicated on the right in *B* and *C*. As indicated by the stippled and clear areas, these motor neurons project ipsilaterally, with the exception of the contralaterally projecting superior rectus motor neurons and the bilaterally distributed levator motor neurons. The axons of contralateral motor neurons cross immediately to join the ipsilateral oculomotor nerve. Thus, the oculomotor nerve projects entirely to ipsilateral muscles.

Figure 28-6. The trochlear nucleus and nerve shown in a dorsal view of the midbrain (*A*) and in a coronal section through the nucleus (*B*).

En route, they pass adjacent to the medial lemniscus and corticospinal tract. Owing to this relationship, loss of paramedian branches of the basilar artery compromises both the corticospinal tract and the exiting abducens fibers, resulting in a *contralateral hemiplegia* and paralysis of abduction in the *ipsilateral* eye (*Foville syndrome*). This pairing of motor symptoms is called an *alternating hemiplegia* and can also occur in association with cranial nerves III and XII. After exiting the brainstem, the abducens nerve (cranial nerve VI) enters the dura and ascends the clivus. It then passes beneath the petroclinoid ligament (the Dorello canal) to enter and pass through the cavernous sinus. Cranial nerves III,

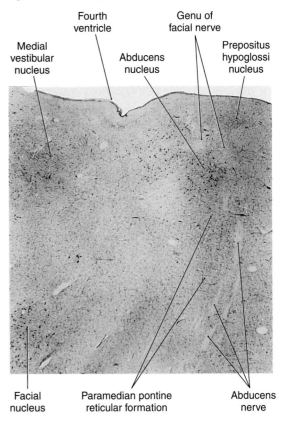

Figure 28-7. Human abducens nucleus and adjacent structures in Nissl-stained cross section showing neuron cell bodies.

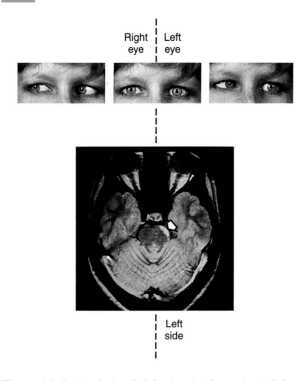

Right | Left
eye | eye

Left
side

Figure 28–8. Paralysis of abduction in the patient's left eye on looking to the left. The computed tomography scan shows demyelination of the pons at the level of the exiting left abducens nerve (*arrow*).

IV, and VI, and branches of cranial nerve V may be damaged in this sinus, singly or in combination, by pituitary tumors or carotid aneurysms. Cranial nerve VI supplies the lateral rectus muscle, so damage to this nerve in isolation will paralyze abduction ipsilaterally (Fig. 28–8).

Abducens Internuclear Neurons. Conjugate eye movement is achieved in the horizontal plane by a pathway from the abducens nucleus to the contralateral medial rectus subdivision of the oculomotor nucleus (see Fig. 28–12). Inputs to the abducens nucleus not only supply motor neurons but also drive *abducens internuclear neurons* found in this nucleus. Abducens internuclear neurons transmit signals to the medial rectus motor neurons on the contralateral side via the medial longitudinal fasciculus (MLF). In this way the lateral rectus of one eye works in tandem with the medial rectus of the other eye. Lesions in the MLF *between the abducens and oculomotor nuclei* damage the abducens internuclear neuron axons, producing *internuclear ophthalmoplegia*. In this situation, the lateral rectus contracts appropriately during horizontal eye movements, but the medial rectus for the opposite eye does not. However, because the oculomotor nerve and nucleus are intact, no deficits are present on convergence. The presence of internuclear neurons also explains why the symptoms of abducens nerve lesions differ from those of abducens nucleus lesions. In the latter, the action of the contralateral medial rectus muscle is impaired dur-

ing conjugate horizontal movements, in addition to the expected paralysis of the ipsilateral lateral rectus muscle.

Sympathetic Supply to the Orbit. The sympathetic innervation of the orbital contents is from the ipsilateral *superior cervical ganglion* (Fig. 28–9). These adrenergic postganglionic fibers enter the cranium on the internal carotid artery. In the cavernous sinus, they run briefly with cranial nerve VI and then join cranial nerves III and V. Sympathetic fibers traveling with the levator branch of cranial nerve III supply the superior tarsal muscle. Fibers traveling with the trigeminal nasociliary nerve exit as the *long ciliary nerves* to supply the eye, including the dilator pupillae muscle.

The cholinergic preganglionic motor neurons that supply the superior cervical ganglion are located at spinal cord levels T1 to T3. Their axons enter and ascend in the sympathetic trunk, to terminate in the ipsilateral superior cervical ganglion. Damage along this long path produces the *Horner syndrome*. The cardinal symptoms of this loss of sympathetic input to the

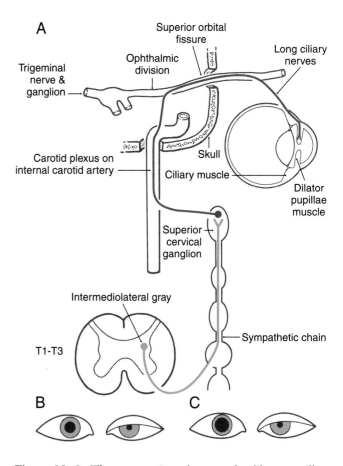

Figure 28–9. The autonomic pathway to the dilator pupillae muscle (*A*). A patient with a presumed Horner syndrome (disruption of this pathway) presents with a partial *ptosis* and *miosis* (*B*). Application of cocaine drops to both eyes establishes that the *anisocoria* is not due to natural or pharmacologic causes (*C*). The normal eye dilates in response to blockade of norepinephrine reuptake, but the deafferented eye shows no change.

head are partial *ptosis* (a drooping lid resulting from relaxation of the superior tarsal muscle), *miosis* (constriction of the pupil that occurs because the action of the sphincter is no longer opposed by the action of the dilator), and *anhidrosis* (loss of facial sweating). The Horner syndrome may also result from interruption of pathways linking the hypothalamus and brainstem to preganglionic motor neurons in the thoracic cord.

Targeting Movements

Saccades. To obtain detailed information about the visual world, the eyes make a series of very rapid movements (200 to 700 degrees/sec) from one point to another, stopping briefly at each point to allow detailed foveal inspection. These rapid conjugate eye movements are *saccades*, and the places at which the detailed visual inspection occurs are *fixation points*. During the saccade, the visual system suppresses incoming visual input. Consequently, one is not aware of these movements. Other processes in visual association cortex provide *visual constancy* by weaving together the information obtained at each fixation into a seamless view of the visual world.

Extraocular Muscle Motor Neurons. Extraocular muscle motor neurons have a characteristic *burst-tonic* firing pattern for saccadic eye movements. For example, before a leftward horizontal movement (Fig. 28–10*A* and *B*), left lateral and right medial rectus motor neurons show an initial burst of action potentials, which is then reduced to a sustained tonic level of activity. At the new eye position, the firing rate is slower than during the initial burst but higher than the rate for the previous eye position (see Fig. 28–10*B* and *C*).

These two parts of the motor neuron and target muscle response are referred to as a *pulse* and *step* of activity. The pulse, the initial burst of motor neuron activity, directs the phasic portion of the movement, producing the muscle contraction necessary to overcome the viscosity of the orbit and send the globe toward the target. The step in the activity supports the tonic action of the muscle, which is required to maintain the eye at its new position. The activity of the antagonists is silenced for the saccade and then resumes at a lower rate. However, when the eyes look to the right (see Fig. 28–10*C* and *D*), motor neurons for the left medial rectus and right lateral rectus muscles are activated. The brainstem distribution of activated motor neurons defines which muscles are activated and hence the direction of movement. The number of cells activated and their firing rates define the speed and distance (metrics) of the movement.

Horizontal and Vertical Gaze Centers. The brainstem circuitry that controls saccades is subdivided into systems controlling horizontal and vertical eye movement. The pontine reticular formation near the midline, which contains neurons that project to the extraocular motor nuclei, is sometimes called the *paramedian pontine reticular formation* (PPRF) or the *horizontal gaze center* (Figs. 28–11 and 28–12). The PPRF occupies portions of the *oral* and *caudal pontine reticular nuclei*. Cells of this premotor region show activity related to horizontal saccades, and lesions in this region produce *horizontal gaze palsies*. PPRF cells that project to extraocular motor neurons include *excitatory burst neurons* (EBNs), found rostral to the sixth nucleus, and *inhibitory burst neurons* (IBNs), found caudal to the sixth nucleus. Both of these cell types have phasic activity patterns; that is, they produce a burst of action poten-

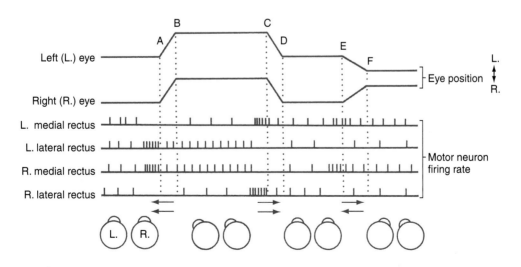

Figure 28–10. Relationship between eye movements and motor neuron firing patterns for horizontal saccades and vergence movements. The top two traces illustrate the changes in eye position. The bottom four traces show idealized firing patterns for motor neurons during horizontal movements. The cartoons at the bottom indicate that the eyes make a saccade to the left (*A*, *B*) and then to the right (*C*, *D*) and, finally, a convergent movement (*E*, *F*).

from supranuclear structures, including the superior colliculus and the frontal eye field, and processed through interneurons within the PPRF. In addition, the burst of action potentials produced by EBNs and IBNs is gated by inhibition from cells found in the midline of the pontine tegmentum (see Figs. 28–11 and 28–12). These inhibitory neurons are called *omnipause cells* because they fire spontaneously during fixation but are silent during a saccadic eye movement in any direction. Thus, the burst of action potentials in PPRF neurons is due, in part, to release from inhibition. PPRF neurons directly contribute to the pulse of motor neuron activity that produces a saccade but not

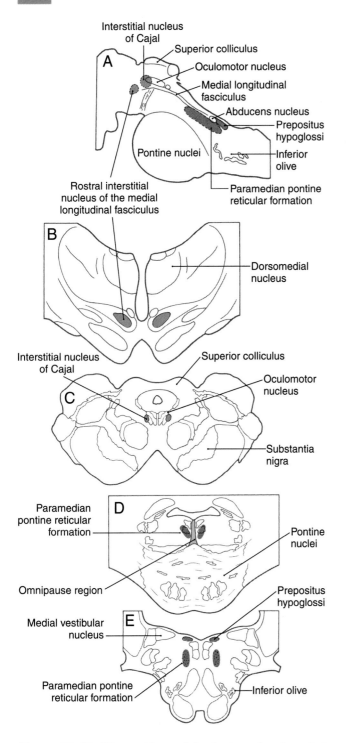

Figure 28–11. The positions of the horizontal and vertical gaze centers in the sagittal (*A*) and frontal (*B–E*) planes.

tials that slightly precedes the activity of the motor neurons (see Fig. 28–12). When gaze shifts to the right, the EBNs activate the abducens nucleus neurons on the right side, as the IBNs suppress the abducens nucleus neurons on the left side. If this inhibition of antagonists does not occur, eye movements are slowed and undershoot.

The pattern of activity in PPRF neurons is a product of the signal (coded in the cell firing pattern) sent

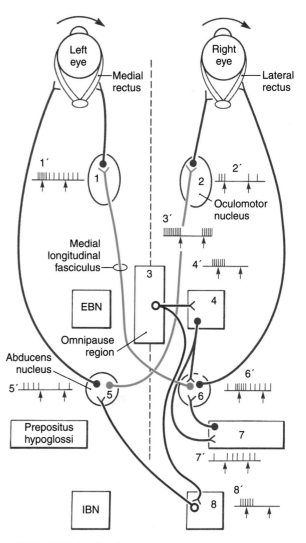

EBN = Excitatory burst neuron area

IBN = Inhibitory burst neuron area

Figure 28–12. Pathway for horizontal saccades. Inhibitory circuits are indicated by open circles (red neurons). Idealized firing patterns (1′ etc.) for a *saccade to the right* are indicated for each nucleus (numbers). The beginning and the end of the saccade are indicated by *arrows* on the firing patterns. Cells indicated by green neurons are abducens internuclear neurons.

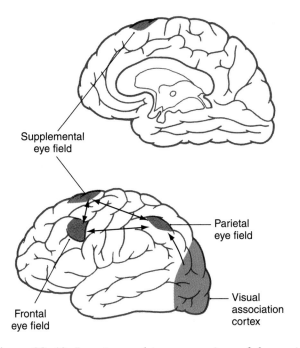

Figure 28–13. Locations and interconnections of the cortical eye fields on medial (*upper*) and lateral (*lower*) views of the cerebral cortex.

to the step change in activity that maintains eye position. The "where the eye is going" signal produced by the PPRF is transformed (integrated) into a "keep the eye in that position" signal by another brainstem structure. The likely source of this tonic position signal for horizontal saccades is the *nucleus prepositus hypoglossi* (see Figs. 28–11 and 28–12).

Vertical gaze palsies are often encountered with lesions of the midbrain-diencephalon junction. The *vertical gaze center* is located in the *rostral interstitial nucleus* of the *medial longitudinal fasciculus (riMLF)*, found at the rostral end of the MLF (see Fig. 28–11). This region receives supranuclear input from the superior colliculus and frontal eye field, as well as input from omnipause cells. It also contains burst neurons. These neurons provide the phasic signal for the saccade-related pulse of activity present in vertical gaze motor neurons. The tonic signal for the step in vertical gaze motor neuron activity is provided by neurons in the *interstitial nucleus of Cajal* (see Fig. 28–11). Because the superior and inferior rectus muscles act in pairs during vertical eye movements, the projections of the riMLF and interstitial nucleus of Cajal on both sides of the brainstem must work in concert. (Unilateral activation produces torsion.) The connections between the two interstitial nuclei of Cajal pass through the posterior commissure. Pinealomas pressing on the commissure produce vertical gaze deficits. Oblique saccades are produced by the vertical and horizontal gaze centers working in concert.

Saccadic eye movements are often accompanied by orienting movements of the head. Cells in the PPRF,

riMLF, and interstitial nucleus of Cajal and in the adjacent reticular formation are responsible for combined head and eye movements. These areas receive collicular and cortical projections, and they project to the extraocular motor neurons and to the cervical spinal cord as *reticulospinal* and *interstitiospinal* fibers. The superior colliculus also projects to the cervical spinal cord, but the tectospinal portion of the tectoreticulospinal system is very small.

Supranuclear Control. For each saccade, the central nervous system must determine the position of the next target of interest and transform this position, which is coded in a sensory map, into the appropriate pattern of motor neuron activity. The areas of the brain that direct saccadic eye movements include the *cortical eye fields* and the *superior colliculus*. In each area, stimulation will initiate contralaterally directed saccades, and single cell recordings reveal neuronal activity before saccades occur. Specifically, stimulation of a location in which cells are active before a 20-degree saccade to the left will produce a 20-degree leftward saccade.

The *frontal eye field* (area 8 of Brodmann) is located rostral to motor cortex (Figs. 28–13 and 28–14). It is

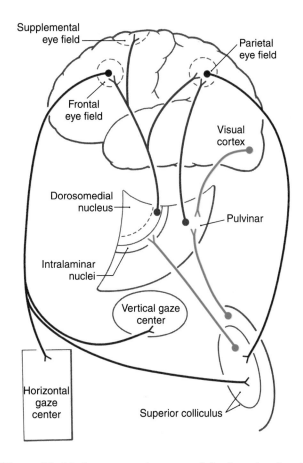

Figure 28–14. Inputs to and outputs of the frontal and parietal eye fields. The *dashed line* in the dorsomedial nucleus indicates the paralamellar subdivision. The pathway to the horizontal gaze center is crossed.

apprised of the location of targets via input from visual association cortex and from basal nuclei and thalamic relays (paralamellar dorsomedial nucleus and intralaminar nuclei) (see Fig. 28–14). The frontal eye field influences eye movements through projections to the vertical and horizontal gaze centers and the superior colliculus. Additional cortical regions influencing saccades include the *supplemental eye field* and the *parietal eye field* in the lateral intraparietal cortex (area 7 of Brodmann) (see Fig. 28–13). They have features similar to those of the frontal eye field but are less directly connected to the brainstem saccade circuits. These three cortical eye fields are reciprocally connected, and all three project to the superior colliculus. Perhaps for this reason, loss of any one of these four structures produces few visual motor symptoms.

The *superior colliculus (optic tectum)* is a layered structure found in the roof of the midbrain (Fig. 28–15). This area of the midbrain tectum receives its blood supply from the *quadrigeminal artery*, a branch of the posterior cerebral artery. The superficial layer of the superior colliculus is visual sensory. It is a target of retinal axons with Y- and W-type physiologic characteristics and projects to the dorsal lateral geniculate and pulvinar nuclei. In contrast, the intermediate layer is visual motor. It is the source of the *crossed tectoreticulospinal system* (predorsal bundle) that runs ventral to the MLF and terminates in the vertical and horizontal gaze centers (Fig. 28–16).

As for the frontal eye field, the major inputs to the intermediate layers of the superior colliculus arise from parietal association cortex and basal nuclei circuits (see Fig. 28–16). GABAergic nigrotectal cells in the substantia nigra pars lateralis and pars reticulata are spontaneously active but cease firing before saccadic eye movements. Tectal saccade-related activity is partly the result of release from this nigral inhibition. Basal nuclei diseases produce eye movement disorders; for example, patients with Parkinson disease have a dearth of spontaneous eye movements because of unmodulated activity in the nigrotectal pathway.

The superior colliculus and the frontal eye field differ in the types of saccades they control. The frontal eye field is important for voluntary and memory-guided eye movements, and the superior colliculus directs reflexive orienting movements. With loss of either structure, few deficits remain after recovery because the remaining structure compensates for the loss, but loss of both produces profound visuomotor impairment.

Smooth Pursuit. The eyes also make conjugate movements that allow the foveas to follow a moving target (Fig. 28–17). Usually, *smooth pursuit* eye movements are used to follow slow-moving, predictable targets (30 degrees/sec or less). They are, however, capable of following targets at speeds up to 100 degrees/sec. The lateral parietal and midtemporal cortices contain neurons that are sensitive to the speed and direction of a

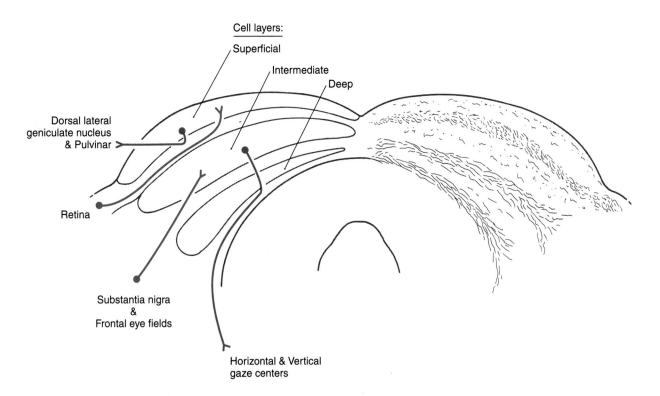

Figure 28–15. The superior colliculus in frontal section. The *right* side shows the layering seen in myelin stains, and the major inputs and outputs are indicated on the *left.*

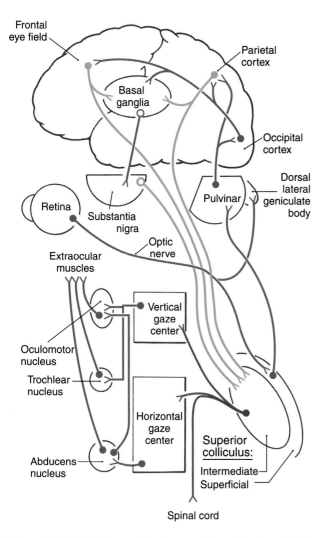

Figure 28–16. Pathways for the superior colliculus. Inhibitory circuits are indicated by *open circles*. The tectoreticulospinal pathway (red) to the horizontal gaze center and spinal cord is crossed.

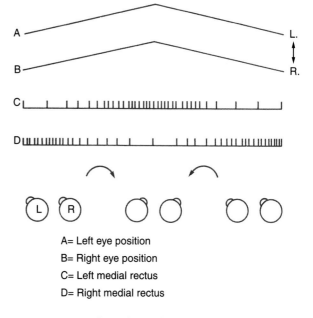

A= Left eye position
B= Right eye position
C= Left medial rectus
D= Right medial rectus

Figure 28–17. The relationship between eye movements and motor neuron firing for smooth pursuit. The eyes change position as they follow a slow-moving target from the left to the right and back to the left (*A, B*). Idealized and graded firing patterns in left (*C*) and right (*B*) medial rectus motor neurons.

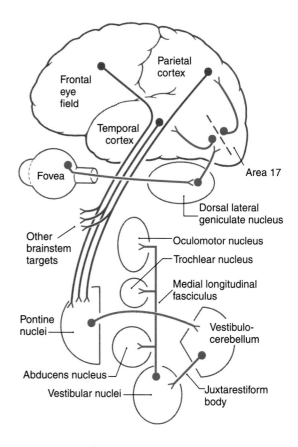

Figure 28–18. Pathways for smooth pursuit eye movements. The retinogeniculostriate pathway is the source of relevant visual sensory input to parietal and temporal association cortices.

target moving across the retina (Fig. 28–18). This input determines the speed and direction of the pursuit eye movements needed to keep the foveas on target. Neurons that display pursuit-related motor activity are found in a portion of the frontal eye field. These three cortical regions project to the *flocculus* and *paraflocculus* of the cerebellum, by way of a synaptic relay in the *posterolateral (dorsolateral) pons*. This portion of the cerebellum, in turn, provides input to the *vestibular nuclei*. Although located in a "sensory" nucleus, vestibular smooth pursuit cells fire with respect to the position and velocity of the eyes, not the visual sensory input. They are, in fact, *premotor neurons*, which project to the third, fourth, and sixth cranial nerve nuclei.

Smooth pursuit premotor neurons fire in a graded manner, depending on the degree and rate of eye excursion. This firing pattern is similar to that displayed by their motor neuron targets during smooth pursuit movements (see Fig. 28–17). They do not have the pulse-and-step form seen with saccades. Presumably, the cerebellum plays a role in precisely determining the rate of movement and predicting target trajectory. Floccular lesions do, in fact, produce deficits in smooth pursuit movements.

Vergence Movements and the Near Triad. Foveation involves directing the eyes toward targets in three-dimensional space. To look from a distant target to a closer one, three changes are made in the eyes. First, the eyes converge by the simultaneous activation of both medial rectus muscles to point both foveas at the closer target (see Fig. 28–10E and F). This is a disconjugate movement because the eyes move in opposite directions. Second, the curvature of the lens is increased, producing an increase in refractive power, to focus the closer target on the fovea (see Fig. 28–2). Third, the pupil is constricted, thereby increasing the depth of field of the eye. The combination of these three actions is termed the *near triad* or *near response*. The opposite effects (*divergence*, flattening of the lens, and pupillary dilation) occur when the gaze is shifted from a closer target to one farther away.

As the term "near triad" suggests, the three actions are generally yoked. Although vergence and accommodation can be dissociated under special conditions (closing one eye and bringing the target straight at the open eye), under normal conditions the vergence angle is used by the brain to adjust the accommodation of the lens. Other cues used to direct the near response include *retinal disparity* and focus. Cells in visual cortex with binocular visual fields are activated when the retinal images have a specific degree of disparity (the difference in the points on each retina where an image falls). This difference is an appropriate signal for the control of vergence. Lack of focus, or blurring of the image, supplies a feedback signal for adjusting the lens until *blur* is minimized.

Vergence movements are generally slower than saccades, and the burst of motor neuron activity related to the initial pulse in muscle activity is less evident (see Fig. 28–10E and F). Premotor neurons whose activity correlates with the near response are found in a *midbrain near response region* located in the *supraoculomotor area* (SOA) (Fig. 28–19). Cells in the SOA project to the medial and lateral rectus motor neurons and also to the preganglionic motor neurons in the Edinger-Westphal nucleus that control the lens and pupil. Although the near response pathways to the SOA remain undetermined, it is likely that the cerebellum influences vergence and accommodation through projections to this region.

Although combined vergence and saccade movements occur and when combined have a similar time course, the saccadic and vergence systems can act independently. For instance, following a lesion in the paramedian pontine reticular formation, horizontal saccades

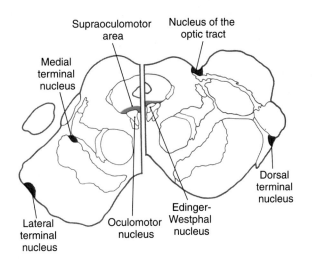

Figure 28–19. Nuclei of the accessory optic system and the supraoculomotor area in caudal (*left*) and rostral (*right*) coronal sections through the rostral midbrain.

are disrupted but vergence movements in the horizontal plane are unaltered.

Reflex Movements

The body is equipped with compensatory systems that keep the eyes directed at a target despite external perturbations of the body or head. The vestibular system specializes in sensing the acceleration that usually occurs at the beginning of a movement, and it also senses gravity. Activation of the vestibular labyrinths elicits a series of compensatory eye movements commonly termed the *vestibulo-ocular reflexes* (VOR). The pathways subserving these reflexes are discussed in Chapter 22. In contrast, the *optokinetic system* compensates for continuous-velocity movements. This section covers the optokinetic system as well as the pupillary and blink reflexes.

Optokinetic Eye Movements. As we move in the world, or move our heads, the entire visual scene moves across the retina. This whole-field movement of the visual scene is called *retinal slip*. Under these conditions, the eyes automatically move in a compensatory fashion to stabilize the image on the retina. For example, if the body is rotating to the left, the visual world will seem to move to the right, and rightward compensatory eye movements match the apparent movement of the visual scene (Fig. 28–20). These *optokinetic movements* are generally slow and match the *velocity* of retinal slip. As with smooth pursuit, they are produced by graded increases and decreases in the tonic firing rate of the appropriate motor neurons (see Fig. 28–20). When the eyes approach the limit of their rotation, a quick saccade brings them back to their primary position and another slow following movement begins.

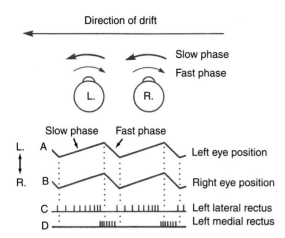

Figure 28-20. The relationship between eye movements and motor neuron firing during optokinetic nystagmus (OKN). The *long arrow* indicates the drift of the visual scene. The *thick arrow* shows the slow optokinetic movement of the eye as it follows the visual scene, and the *thin arrow* indicates the fast saccadic resetting movement. These changes in eye position are illustrated in *A* and *B*. Idealized motor neuron firing patterns are shown in *C* and *D*. This would be rightward OKN.

This set of alternating *slow* and *fast* (saccadic) *phases* of movement is called *optokinetic nystagmus* (OKN) (see Fig. 28–20). The direction of OKN is specified by the fast phase direction. A series of stripes can be moved in front of a subject to elicit nystagmus and test the *optokinetic reflex*.

The afferent limb of this reflex begins with stimulation of wide-field retinal ganglion cells that are sensitive to slow movements of the whole receptive field (Fig. 28–21). The receptive fields of these retinal cells are tuned to directions of movement that are comparable to the spatial planes in which the vestibular labyrinths are oriented. Axons of these retinal ganglion cells terminate in a series of small nuclei along the incoming optic tract, termed the *accessory optic system* (AOS) (see Fig. 28–19). These nuclei consist of the *nucleus of the optic tract* and the *medial, lateral,* and *dorsal accessory optic nuclei*. Each of these four nuclei contains cells that are activated by retinal slip in specific directions; for example, the nucleus of the optic tract cells is sensitive to temporal-to-nasal movements. The AOS nuclei also receive input from visual sensory association cortex, presumably from neurons of the smooth pursuit system (see Fig. 28–21). In humans, the optokinetic reflex comes into play only when the broad field movement cells in the accessory optic system indicate retinal slip and the pursuit system subserved by the geniculo-cortical pathways notes equivalent foveal target movement. Otherwise, the pursuit system overrides the optokinetic system.

The AOS nuclei project to the portions of the *nu-cleus reticularis tegmenti pontis* and the *inferior olive* that supply the vestibulocerebellum, and to the vestibular nuclei (see Fig. 28–21). In addition to vestibulo-ocular premotor neurons, the latter nuclei contain optokinetic neurons that influence extraocular motor neurons. In fact, lesions in the vestibular nuclei produce severe deficits in both reflexes. The pathways through the cerebellum are involved in adapting the gains of the optokinetic and vestibulo-ocular reflexes so that they combine to produce appropriate eye movements.

Pupillary Light Reflex. In addition to the changes in pupillary size that result from the near response, the pupil also responds to the amount of ambient light. The actions of the iris in each eye are yoked, so that light directed into one eye results in pupillary constriction in both the illuminated eye (*direct response*) and the opposite eye (*consensual response*). The pupil changes diameter to maintain luminance of the retina in the optimal range of the receptor photopigments. Many of the structures involved in the pupillary light reflex pathway are clearly visible in a magnetic resonance image oriented specifically to the long axis of the optic nerve (Fig. 28–22).

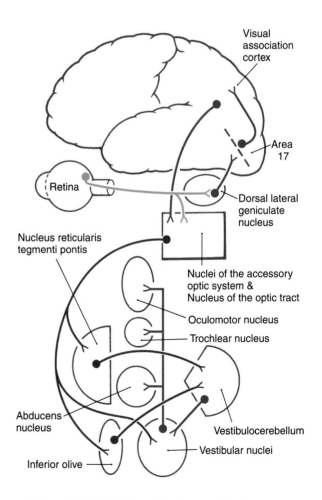

Figure 28-21. Pathway for the optokinetic system.

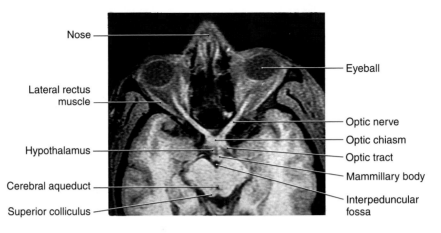

Figure 28–22. Magnetic resonance image (T1-weighted image) of optic structures (nerve, chiasm, and tract) in relation to the hypothalamus and midbrain. Compare with Figure 28–23.

The pupillary light reflex is a four-neuron arc (Fig. 28–23; see also Fig. 28–22). Retinal ganglion cells with broad receptive fields that respond in a linear fashion to luminance levels project via the optic nerve and tract to the midbrain. The decussation of approximately half of these fibers in the chiasm is one of the structural features responsible for the consensual response. The retinal axons terminate in the pretectum within the *olivary pretectal nucleus*, which, in turn, projects bilaterally to the Edinger-Westphal nucleus, with the decussating fibers crossing in the posterior commissure. Parasympathetic, preganglionic fibers from the Edinger-Westphal nucleus exit with the oculomotor nerve and terminate in the ciliary ganglion. The cholinergic postganglionic fibers reach the iris, where they excite the pupillary constrictor muscle. Damage to these postganglionic fibers produces a *tonic* dilated pupil (the *Adie syndrome*) in which the constrictor muscle is supersensitive to cholinergic drugs. The pupillary light reflex is a useful diagnostic tool for testing brainstem and cranial nerve function (see Figs. 28–22 and 28–23). Lesions may result in loss of the direct or consensual pupillary responses or in uneven pupil size (*anisocoria*). A dilated, unresponsive (fixed) pupil (or pupils) in the unconscious victim of head trauma is a grave sign. For example, it may indicate that a space-occupying lesion has forced the parahippocampal gyrus or uncus over the edge of the tentorium (*uncal herniation*), compressing the third cranial nerve. The pupillary fibers are superficially located in the oculomotor nerve and are particularly sensitive to pressure. Their loss may indicate that compression of the brainstem is imminent.

The effects of lesions in the sympathetic pathways (Horner syndrome) have already been discussed. Another syndrome, called *Argyll Robertson pupil*, is found in cases of *tabes dorsalis* (central nervous system syphilis). Affected patients show small pupils with very weak

or absent pupillary light reflexes bilaterally, but there is no loss of visual acuity, and the pupils do constrict in the near response. This sparing indicates that the afferent and efferent limbs of the pupillary light reflex must be intact. Consequently, it is assumed that bilateral degeneration in either the olivary pretectal nuclei or the pathways connecting them to the Edinger-Westphal nuclei must be the source of the pupillary dysfunction.

Pupil size also reflects visceromotor tone. Greater levels of excitement, including desire, result in dilation of the pupil via sympathetic activation. This fact was known to Elizabethan women, who used tincture of belladonna to dilate their eyes for cosmetic purposes. Today, cholinergic blockers are used to dilate the pupils for ophthalmologic examination.

Blinking and Other Lid Movements. The delicate structures of the eye are protected by the eyelids. Blinks, some of which occur in response to somatosensory stimulation, ensure protection for the eye. The *blink reflex* is used to assess trigeminal sensory and facial nerve function, as well as the integrity of the lid pathways through the lateral pons. Trigeminal nerve fibers with free nerve endings in the cornea or lid have central processes terminating in the spinal portion of the trigeminal sensory nucleus (Fig. 28–24). Second-order trigeminal neurons project directly and indirectly to the facial nucleus, where they excite orbicularis oculi motor neurons, which produces lid closure. In addition, an inhibitory pathway suppresses the activity of antagonist levator palpebrae motor neurons in the oculomotor nucleus.

Blinks also occur at regular intervals (averaging 12 blinks/minute), as automatically triggered movements that spread the tear film over the cornea. Although the precise origin of these rhythmic blinks is unknown, the constant dispersal of the tear film prevents corneal lesions and scarring. *Blepharospasm* is a disorder in this

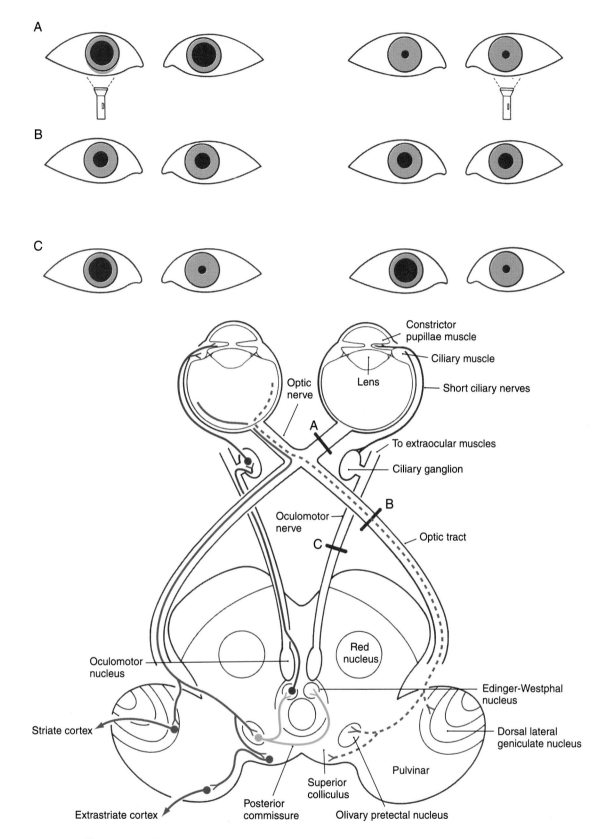

Figure 28–23. Pupillary light reflex pathways. If the optic nerve is partially damaged (*A*), shining a light into that eye will produce a diminished direct and consensual response (*left*); but both will be present when the undamaged side is illuminated (*right*). This is termed a *relative afferent pupillary defect*. A total lesion at *A* would produce a blind eye, which would induce neither a direct nor a consensual response when illuminated. If the lesion occurs in the optic tract (*B*) or pretectum, neither response is lost. Although the reflexes may be weaker, this is *not* easily discerned clinically. However, a large lesion in the posterior (dorsal) midbrain (e.g., pinealoma) would weaken pupillary responses bilaterally. If the lesion occurs in the oculomotor nucleus or nerve (*C*), both direct and consensual responses will be lost in the eye on the lesion side, but they will be present in the other eye.

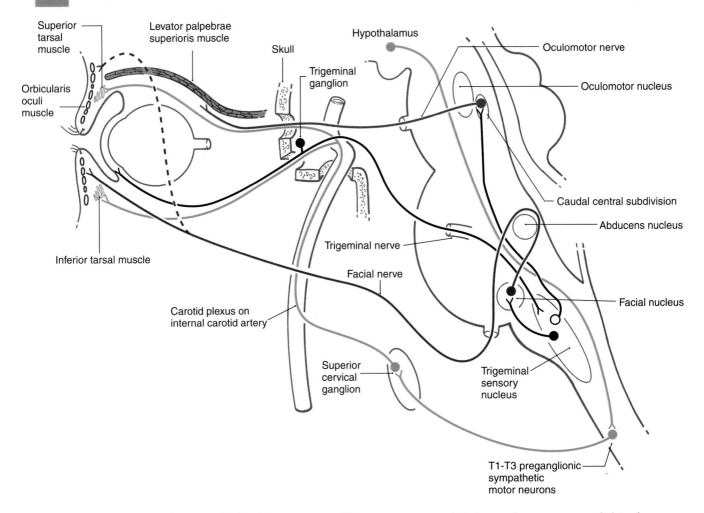

Figure 28–24. Pathways for control of eyelid movements. The motor nerves and their muscle targets are coded in the same color. Inhibitory projection is indicated by the *open circle*.

rhythmic behavior that results in bouts of high-frequency blinking, whereas parkinsonism produces a decreased blink rate. To keep out of the line of vision, the lids also move with the eyes during vertical eye movements. These movements are produced by actions of the levator palpebrae muscle, which works in concert with the superior rectus. Levator motor neurons receive inputs from cells in and near the vertical gaze centers. The orbicularis oculi muscle is not involved. Consequently, *Bell palsy*, in which the facial nerve is damaged along its peripheral course, results in a loss of the blink reflex on the affected side, but no ptosis or loss of vertical gaze–related lid movements. Lesions of the oculomotor nerve produce just the opposite results.

The tarsal muscles help to keep the lids open, as indicated by the partial ptosis present in the Horner syndrome. Their sympathetic innervation suggests that they regulate lid position with respect to emotional state. For example, high sympathetic tone produces widely opened eyes. Relaxation of the tarsal muscles leads to the feeling of "heavy lids," which signals the general tone of the autonomic system, as the brain prepares to rest.

Sources and Additional Reading

Brandt T: Vertigo: Its Multisensory Syndromes, 2nd ed. Springer-Verlag, London, 1999, p 503.

Büttner-Ennever JA (ed): Neuroanatomy of the oculomotor system. Rev Oculomot Res 2, Elsevier, Amsterdam, 1988, p 489

Clark RA, Miller JM, Demer JL: Three-dimensional location of human rectus pulleys by path inflections in secondary gaze position. Invest Ophthalmol Vis Sci 41:3787–3797, 2000.

Evinger C: A brainstem reflex in the blink of an eye. News Physiolog Sci 10:147–153, 1995.

Hall WC, May PJ: The anatomical basis for sensorimotor transformations in the superior colliculus. In Neff WD (ed): Contributions to Sensory Physiology, vol 8. Academic Press, San Diego, 1984, pp 1–40.

Huerta MF, Halting JK: The mammalian superior colliculus: Studies of its morphology and connections. In Vanegas H (ed): Comparative Neurology of the Optic Tectum. Plenum Press, New York, 1984, pp 678–773.

Keller EL, Heinen SJ: Generation of smooth-pursuit eye movements: Neuronal mechanism and pathways. Neurosci Res 11:79–107, 1991.

Leigh RJ, Zee DS: The Neurology of Eye Movements, vol 55 in series Contemporary Neurology. FA Davis, Philadelphia, 1999.

Loewenfeld IE: The Pupil. Anatomy, Physiology and Clinical Applications. Iowa State University Press, Ames, 1993.

Mays LE: Neural control of vergence eye movements: Convergence and divergence neurons in midbrain. J Neurophysiol 51:1091–1108, 1984.

Schiller PH, True SC, Conway JL: Deficits in eye movements following frontal eye-field and superior colliculus ablations. J Neurophysiol 44:1175–1189, 1980.

Schlag J, Schlag-Rey M: Evidence for a supplementary eye field. J Neurophysiol 57:179–200, 1987.

Scudder CA, Fuchs AF, Langer TP: Characteristics and functional identification of inhibitory burst neurons in the trained monkey. J Neurophysiol 59:1430–1454, 1988.

Sparks DL: Functional properties of neurons in the monkey superior colliculus: Coupling of neural activity with saccade onset. Brain Res 156:1–16, 1978.

Wurtz RH, Goldberg ME (eds): The Neurobiology of Saccadic Eye Movements. Rev Oculomot Res 3. Elsevier, Amsterdam, 1989, 429.

Chapter 29

Visceral Motor Pathways

J. P. Naftel and S. G. P. Hardy

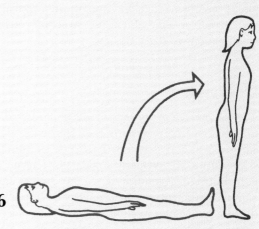

Overview   466

Organization of the Visceral Motor System   466
Targets of Visceral Motor Outflow
General Features of Peripheral Visceral Motor Outflow

Development   467
Preganglionic Visceral Motor Neurons
Postganglionic Visceral Motor Neurons

Sympathetic Division   468
Sympathetic Preganglionic Neurons
Sympathetic Ganglia
Internal Organization of Sympathetic Ganglia
Functional and Chemical Coding
Receptor Types in Sympathetic Targets
Causalgia

Parasympathetic Division   473
Preganglionic and Postganglionic Neurons
Parasympathetic Outflow Pathways
Functional and Chemical Coding
Receptor Types in Parasympathetic Targets

Enteric Nervous System   475

Regulation of Visceral Motor Outflow   475
Major Central Nervous System Components
Cardiovascular System
Urinary Bladder and Micturition

The primary function of the visceral motor system is the regulation of cardiovascular, respiratory, digestive, urinary, and reproductive organs. These organs are the main effectors of *homeostasis*, the maintenance of a stable internal environment against perturbing influences, both external and internal. In general, visceral motor neurons innervate smooth and cardiac muscles and glandular epithelium, or structures made up of combinations of these tissues.

Overview

The *visceral motor (autonomic) system* ensures that tissues of the body receive appropriate nutrients, electrolytes, and oxygen, and that functions such as osmolarity and temperature are properly regulated. The nervous system contributes significantly to the control and coordination of homeostatic mechanisms in response to continually changing requirements. Two overlapping control systems influence visceral effectors. One is *humoral (endocrine)*. Hormonal responses tend to develop slowly, but the effects are prolonged. The other system is *neural (autonomic)*. Visceral motor responses tend to be immediate, but their effects are short term.

The endocrine and autonomic systems are interdependent. They are both under the control of widely distributed central nervous system (CNS) structures, which generate commands after integrating inputs from a wide variety of sources. Thus, visceral motor output is influenced by emotional status, as well as by sensory signals reporting conditions inside and outside the body.

The visceral motor system has two major subdivisions, *sympathetic* and *parasympathetic*. In addition, neurons located in the wall of the alimentary canal form a somewhat autonomous component called the *enteric nervous system*. This is sometimes regarded as a third subdivision of the autonomic system. Within each of these components are populations of chemically coded, target-specific neurons.

Organization of the Visceral Motor System

Targets of Visceral Motor Outflow. The autonomic system provides neural control of *smooth muscle, cardiac muscle*, and *glandular secretory cells*. The *sympathetic* and *parasympathetic divisions* have overlapping and generally antagonistic influences on those viscera located in body cavities and on some structures of the head, such as the iris (Table 29–1). There are also visceral targets in the body wall and limbs. These are found in skeletal muscle (blood vessels) and in the skin (blood vessels, sweat glands, and arrector pili muscles). Visceral structures of the body wall and extremities are generally regulated

Table 29–1. Comparison of Effects of Sympathetic and Parasympathetic Activity on Some Visceral Functions

Physiologic Process	Sympathetic Stimulation	Parasympathetic Stimulation
Eye		
Pupil diameter	+	−
Lens refraction	0	+
Palpebral fissure width	+	0
Tear flow	0	+
Salivary gland flow	−	+
Skin		
Piloerection	+	0
Sweating	+	0
Blood flow	−	0
Skeletal muscle blood flow	±	0
Cardiovascular system		
Cardiac output	+	−
Total peripheral resistance	+	0
Bronchial diameter	+	−
Gut		
Peristalsis	−	+
Secretion	−	+
Sphincter tone	+	−
Blood flow	−	+
Liver glycogenolysis	+	0
Pancreatic insulin secretion	−	+
Pancreatic glucagon secretion	+	+
Urinary bladder detrusor tone	−	+
Urethra sphincter tone	+	±
Penile or clitoral erection	0	+
Ejaculation	+	0

+, positive effect; −, negative effect; 0, no effect; ±, variable effect.

by the sympathetic division alone. The sympathetic outflow thus has a global distribution in that it innervates visceral structures in all parts of the body, whereas the parasympathetic outflow serves only targets in the head and body cavities (see Table 29–1).

General Features of Peripheral Visceral Motor Outflow. There are similarities and differences between the neural control of skeletal muscle and visceral effectors such as smooth muscle (Fig. 29–1). As described in Chapter 24, lower motor neurons (alpha motor neurons) function as the final common pathway linking the CNS to skeletal muscle fibers (see Fig. 29–1A). Similarly, sympathetic and parasympathetic outflows serve as the final, but often dual, common neural pathway from the CNS to visceral effectors. However, unlike the somatic motor system, the peripheral visceral motor pathway consists of two neurons (see Fig. 29–1B and C). The first, the *preganglionic neuron*, has its cell body in either the brainstem or the spinal cord. Its axon projects as a thinly myelinated *preganglionic fiber* to an autonomic ganglion. The second, the *postganglionic neuron*, has its cell body in the ganglion and sends an unmyelinated axon (*postganglionic fiber*) to visceral effector cells such as smooth muscle. In general, parasympathetic ganglia are close to the effector tissue, and sympathetic ganglia are close to the CNS. Conse-

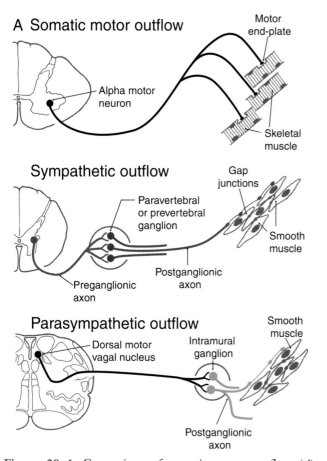

A Somatic motor outflow

Motor end-plate
Alpha motor neuron
Skeletal muscle

Sympathetic outflow

Gap junctions
Paravertebral or prevertebral ganglion
Smooth muscle
Postganglionic axon
Preganglionic axon

Parasympathetic outflow

Smooth muscle
Dorsal motor vagal nucleus
Intramural ganglion
Postganglionic axon

Figure 29–1. Comparison of somatic motor outflow (*A*) with sympathetic (*B*) and parasympathetic (*C*) outflow.

quently, *parasympathetic pathways typically have long preganglionic fibers and short postganglionic fibers, whereas sympathetic pathways more often have short preganglionic fibers and long postganglionic fibers.*

Visceral motor neurons and their targets are not organized into discrete motor units like those of the somatic motor system. Recall that an alpha motor neuron makes synaptic contacts with a definite group of skeletal muscle fibers over which it has exclusive control. In contrast, the terminal branches of a postganglionic visceral motor axon typically have a series of swellings containing neurotransmitter vesicles along their length, giving them a beaded (varicose) appearance (see Fig. 29–1*B* and *C*). The neurotransmitters released from these terminals may act on effector cells at a distance of up to 100 μm. Moreover, unlike skeletal muscle fibers, cardiac muscle fibers and the smooth muscle cells of some organs are electrically coupled by *gap junctions.* Because of this, neurochemical signaling to a few cells is sufficient to regulate a large group of cells that act as a unit. The main features of sympathetic and parasympathetic divisions are summarized in Table 29–2.

Development

Preganglionic Visceral Motor Neurons. Cell bodies of these neurons are located in nuclei or cell columns embryologically derived from the *general visceral efferent* cell column. This column arises from neuroblasts in

---

Table 29–2.   Comparison of the Sympathetic and Parasympathetic Divisions of Autonomic Outflow

Feature	Sympathetic (Thoracolumbar)	Parasympathetic (Craniosacral)
Location of preganglionic cell bodies	Spinal segments T1 to L2, mainly intermediolateral cell column	Spinal segments S2 to S4, intermediate gray; general visceral efferent nuclei of cranial nerves III, VII, IX, X
Location of preganglionic fibers	White rami T1 to L2, sympathetic trunks, splanchnic nerves	Pelvic nerves, cranial nerves III, VII, IX, X
Location of postganglionic cell bodies	Paravertebral ganglia, prevertebral ganglia (celiac, aorticorenal, superior mesenteric, inferior mesenteric)	Ganglion cell clusters in walls of viscera, cranial nerve autonomic ganglia (ciliary—III; pterygopalatine and submandibular—VII; otic—IX)
Location of postganglionic nerve fibers	Fibers to structures of body wall and limbs in gray rami and spinal nerves, plexuses associated with arteries supplying visceral structures of the head and body cavities	Within the viscera of body cavities; short nerves or plexuses extending from cranial ganglia to target organs; often accompany trigeminal nerve branches in head
Target effectors	Smooth muscle, cardiac muscle, and secretory cells throughout body	Mostly viscera of the head and the thoracic, abdominal, and pelvic cavities
Primary neurotransmitter of preganglionic neurons	Acetylcholine	Acetylcholine
Primary neurotransmitter of postganglionic neurons	Norepinephrine; cells supplying sweat glands use acetylcholine	Acetylcholine
Neuropeptides of postganglionic neurons	Neuropeptide Y and others	Vasoactive intestinal polypeptide and others
General physiologic effects	Mobilization of resources for intensive activity	Promotion of restorative processes

the dorsal part of the basal (motor) plate of the brainstem and spinal cord portions of the neural tube.

Postganglionic Visceral Motor Neurons. Cell bodies of these multipolar neurons are located in autonomic ganglia, which may be either well-defined, encapsulated structures, such as the superior cervical ganglion, or clusters of somata found in nerve plexuses or in the walls and capsules of visceral organs. Like most primary sensory neurons, autonomic ganglion cells are derived from *neural crest cells* that migrate to appropriate locations during development. *Congenital megacolon,* or *Hirschsprung syndrome,* results from failure of enteric neuronal precursor cells to migrate into the wall of the lower gut. As a result, the affected segment of the colon is paralyzed in a constricted state, with consequent distention of the proximal, normally innervated intestine.

Neurotrophins are a family of proteins, each of which regulates development of specific populations of neurons. The existence of these proteins was established when *nerve growth factor* (NGF) was identified as a target tissue–derived messenger molecule essential for the survival and development of sympathetic postganglionic neurons (Fig. 29–2). The pathologic changes in animals deprived of NGF during development are similar to those associated with familial dysautonomia (Riley-Day syndrome), an autosomal recessive disease. However, the genes for NGF and its receptor have now been eliminated as sites of the gene defect in familial dysautonomia. Nevertheless, the rapidly growing understanding of the functions of neurotrophins carries great potential for developing strategies to treat injury and disease of specific neuronal populations in both the CNS and the peripheral nervous system (PNS).

Sympathetic Division

Sympathetic Preganglionic Neurons. Although the sympathetic outflow influences visceral targets throughout the body, sympathetic preganglionic neurons are found only in spinal cord segments T1 through L2 (sometimes in C8 and L3). These cell bodies are located in Rexed lamina VII, primarily in the *intermediolateral nucleus* (*cell column*) of the lateral horn. The axons of these preganglionic cells exit the spinal cord in the *ventral root* and enter the *sympathetic trunk* via the *white communicating ramus.* Once in the sympathetic chain, a preganglionic fiber may (1) synapse at that level, with the postganglionic fiber joining that spinal nerve; (2) ascend or descend in the sympathetic chain to synapse on postganglionic neurons whose axons either join spinal nerves or project to targets in the thoracic cavity or head; or (3) pass through the chain ganglion as a preganglionic fiber to form part of a *splanchnic nerve* (Fig. 29–3A–C).

The sympathetic outflow for the entire body originates from thoracic and upper lumbar spinal cord segments. Although the segmental pattern of innervation is not straightforward, there is a general viscerotopic organization (Figs. 29–4 and 29–5). Neurons of the *superior, middle,* and *inferior cervical ganglia* receive input, via the sympathetic trunk, from preganglionic neurons of the upper thoracic spinal segments. Lower lumbar and sacral ganglia are supplied by neurons of

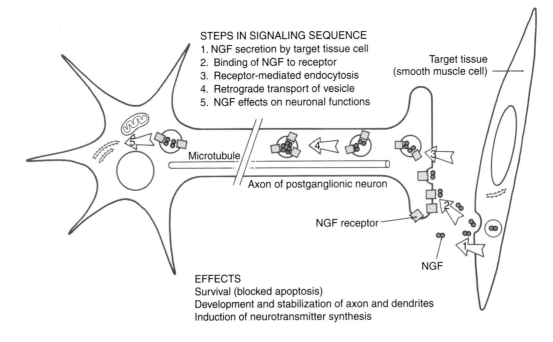

STEPS IN SIGNALING SEQUENCE
1. NGF secretion by target tissue cell
2. Binding of NGF to receptor
3. Receptor-mediated endocytosis
4. Retrograde transport of vesicle
5. NGF effects on neuronal functions

Target tissue
(smooth muscle cell)

Microtubule

Axon of postganglionic neuron

NGF receptor

NGF

EFFECTS
Survival (blocked apoptosis)
Development and stabilization of axon and dendrites
Induction of neurotransmitter synthesis

Figure 29–2. Mechanism by which the neurotrophin *nerve growth factor* (NGF) regulates the development and function of sympathetic ganglion neurons.

the lower thoracic and upper lumbar spinal segments. The ganglia between these regions are supplied by their corresponding spinal levels. Thus, visceral targets in the head, neck, and upper extremity, as well as the

viscera of the thoracic cavity, are served by preganglionic sympathetic neurons in the upper thoracic segments. The main abdominal viscera and other targets in the trunk are served by the central and lower thoracic spinal cord segments, whereas the pelvic viscera, the lower trunk, and the lower extremity are served by the lower thoracic and upper lumbar spinal cord segments.

Preganglionic sympathetic neurons also target the adrenal gland (see Fig. 29–5). Chromaffin cells of the adrenal medulla are related to sympathetic ganglion neurons in both derivation (neural crest) and function. These cells secrete catecholamines (mostly epinephrine) into the bloodstream in response to signals from preganglionic neurons. Thus, the sympathetic system, through this endocrine pathway, regulates functions of cells that are not directly contacted by nerve terminals. **Sympathetic Ganglia.** Cell bodies of sympathetic postganglionic neurons are generally grouped into discrete ganglia that are located at some distance from the target tissue. Most of them make up the *sympathetic chain (paravertebral) ganglia* and the *prevertebral ganglia* associated with the abdominal aorta or its large branches (the celiac, aorticorenal, superior mesenteric, and inferior mesenteric ganglia; see Fig. 29–5). In addition, small clusters of cell bodies are also scattered among nerve fibers in communicating rami, the sympathetic trunk, and peripheral plexuses.

The *sympathetic chain* extends along the full length of the vertebral column, but the number of ganglia does not exactly match the number of spinal nerves. Generally, there are 3 cervical, 10 to 11 thoracic, 3 to 5 lumbar, and 3 to 5 sacral sympathetic ganglia on each side, and a single *coccygeal ganglion (ganglion impar)* where the two chains meet caudally (see Figs. 29–4 and 29–5). Typically, the inferior cervical ganglion and the first thoracic ganglion fuse to form the *stellate ganglion*.

The sympathetic chain ganglia are connected to spinal nerves via *white* and *gray communicating rami*. The former appear white because they contain myelinated *preganglionic fibers*, and the latter appear gray because they are composed of unmyelinated *postganglionic fibers* that serve the limbs and body wall (see Fig. 29–3). Consequently, only spinal nerves T1 to L2 have white

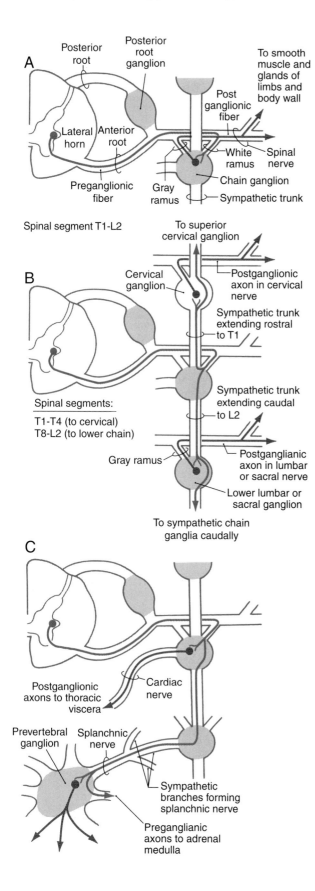

Figure 29–3. The types of routes that can be taken by peripheral sympathetic pathways. A preganglionic fiber may either (*A*) terminate in the chain ganglion at its level of origin, (*B*) ascend or descend in the chain to terminate in different ganglia, or (*C*) traverse the ganglion to enter a splanchnic nerve and terminate in a prevertebral ganglion. Postganglionic neurons that originate in a sympathetic chain ganglion may exit via either the gray ramus and spinal nerve (*A* and *B*) or directly via a nerve such as the cardiac nerve shown in *C*.

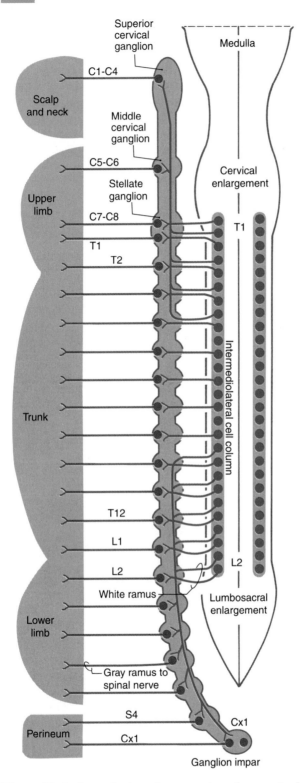

Figure 29–4. Sympathetic pathways to visceral targets in the body wall, limbs, and scalp and neck. Postganglionic sympathetic fibers that distribute to targets of the head are shown in Figure 29–5.

Sympathetic postganglionic neurons in paravertebral ganglia send their axons in two general directions. First, some postganglionic fibers join the spinal nerve via a gray ramus. These fibers distribute to blood vessels, sweat glands, and arrector pili muscles in the body wall and extremities (see Figs. 29–3*A* and *B* and 29–4). These structures receive little or no parasympathetic innervation, so they are exceptions to the dual arrangement of visceral innervation. Second, some of the postganglionic fibers that arise from cervical and upper thoracic sympathetic chain ganglia form *cervical* and *thoracic cardiac nerves* and *pulmonary nerves* that emerge directly from the ganglia (see Figs. 29–3*C* and 29–5). These fibers innervate the vascular smooth muscle of the esophagus, heart, and lung; the glandular epithelium of respiratory structures; the smooth muscles of the esophagus; and cardiac muscle. Axons of these sympathetic neurons mingle with parasympathetic fibers of the vagus nerve to form the autonomic plexuses of the thorax.

The largest of the paravertebral (sympathetic chain) ganglia is the *superior cervical ganglion*. Postganglionic fibers from these cells innervate blood vessels and cutaneous targets of the face and scalp and neck of the territories supplied by the first four cervical nerves (see Figs. 29–4 and 29–5). The superior cervical ganglion also innervates the salivary glands, nasal glands, lacrimal gland, and structures of the eye such as the pupillary dilator muscle and the superior and inferior tarsal muscles (see Fig. 29–5). Accordingly, a constellation of signs and symptoms results from interruption (central or peripheral) of the sympathetic pathway through the superior cervical ganglion (Fig. 29–6). These include constriction of the pupil (*miosis*) caused by the unopposed action of the parasympathetically innervated pupillary constrictor, drooping of the upper eyelid (*ptosis*) resulting from paralysis of the superior tarsal muscle (of Müller), *flushing of the face* from loss of sympathetically mediated vascular tone, and diminished or absent sweating (*anhidrosis*) on the face. Collectively, these signs and symptoms are known as *Horner syndrome* (see Fig. 29–6).

The *prevertebral sympathetic ganglia* are found in visceral motor plexuses associated with the abdominal aorta and its major branches, and they receive input via the splanchnic nerves (see Figs. 29–3*C* and 29–5). In general, the postganglionic fibers arising from each of these ganglia supply the same visceral targets as the corresponding branch of the aorta. Thus, the *celiac ganglion* is located at the origin of the celiac artery and supplies postganglionic fibers to the spleen and to viscera derived from the embryonic foregut. The *aorticorenal ganglion*, associated with the renal arteries, contains cell bodies of neurons that innervate the blood vessels of the kidneys. The *superior mesenteric ganglion* provides postganglionic fibers that supply the territory of the superior mesenteric artery (derivatives of the midgut). *Inferior mesenteric ganglion* neurons project to

rami (and contain preganglionic axons), whereas every spinal nerve is connected to the sympathetic trunk by a gray ramus that conveys postganglionic axons (see Fig. 29–4).

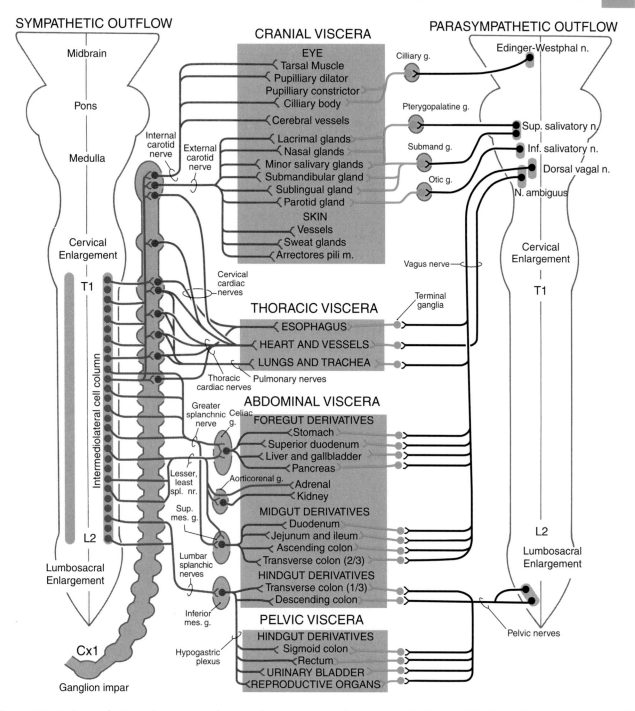

Figure 29–5. Sympathetic and parasympathetic pathways to visceral targets in the head and body cavities. g., ganglion; inf., inferior; m., muscle; mes., mesenteric; n., nucleus; nr., nerve; spl., splanchnic; submand., submandibular; sup., superior.

hindgut derivatives and to the urinary bladder, urethra, and reproductive organs. All abdominal aortic plexuses contain sympathetic fibers mixed with parasympathetic preganglionic fibers of either vagal or sacral origin.

Internal Organization of Sympathetic Ganglia. Axons of preganglionic sympathetic neurons branch in the periphery and synapse on many postganglionic neurons; the output of the preganglionic neurons is thus widely *divergent* (Fig. 29–7; see also Fig. 29–1). The number of postganglionic neurons exceeds preganglionic neurons by over a hundred-fold. Each postganglionic neuron, however, receives synaptic input from a number of preganglionic neurons, so there is also considerable *convergence* within sympathetic ganglia.

Sympathetic ganglia are commonly referred to as "relay ganglia," implying they are sites of simple signal transduction between preganglionic and postganglionic neurons. However, it is clear that more complicated signal processing takes place in prevertebral ganglia. In

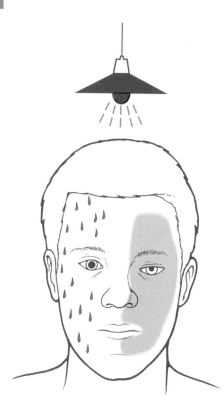

Figure 29–6. Features of the Horner syndrome. Note the lack of sweating on the affected side in response to radiant heat.

addition to receiving diverging and converging input from functionally coded preganglionic neurons, postganglionic neurons are influenced by a variety of other sources. These sources include synaptic inputs from collaterals of general visceral afferent fibers and from local neurons (see Fig. 29–7). These connections indicate a high degree of integration in prevertebral ganglia.

Functional and Chemical Coding. Conditions of extreme excitement or exertion bring about a comprehensive ("en masse") activation of sympathetic outflow, with widespread effects. These effects include increases in heart rate, blood pressure, blood flow to skeletal muscles, blood glucose level, sweating, and pupil diameter. Concurrently, there are decreases in gut motility, digestive gland secretion, and blood flow to abdominal viscera and skin (see Table 29–1). This constellation of effects has led to the concept that the sympathetic system acts in a global, nonselective manner. However, in less extreme conditions, there is ongoing, selective control of function-specific and target-specific subpopulations of preganglionic and postganglionic neurons. Such selectivity can be found, for example, among cell groups that control vascular tone. For instance, sympathetic pathways to blood vessels in the skin are primarily influenced by temperature, whereas sympathetic outflow to vessels in skeletal muscle responds mainly to changes in blood pressure signaled by

baroreceptors. Also, stabilization of blood flow to the head during movement from a reclining to a standing position is a function of the sympathetic division. This adaption to posture requires rapid changes in vascular tone that must vary according to the region of the body.

Just as components of the sympathetic system can be regulated independently, there is evidence that distinct populations of preganglionic and postganglionic neurons exist. For example, even though preganglionic neurons are all cholinergic, some also express one or more neuropeptides, such as substance P and enkephalins. In addition, distinct populations of preganglionic neurons have been identified on the basis of (1) location of the cell body in the visceral motor cell groups of the spinal cord, (2) morphology of the dendritic tree, and (3) specific target cell type among ganglion cells.

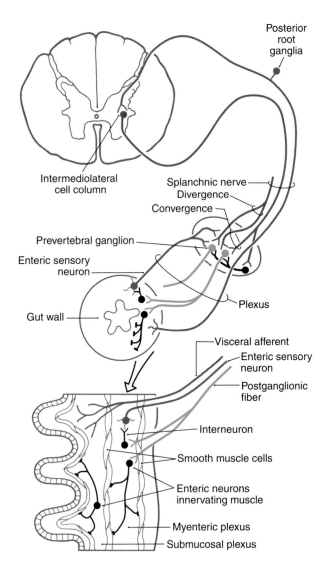

Figure 29–7. Sources of synaptic input to postganglionic neurons in prevertebral ganglia, and the organization of the enteric nervous system. The lower drawing is a detail of the gut wall.

More is known about the functional significance of chemical coding in postganglionic sympathetic neurons. Although most of these cells use *norepinephrine* as a transmitter, some are *cholinergic*. The latter neurons provide secretomotor innervation to most sweat glands and possibly innervation to other targets such as arrector pili muscles and arterioles of skeletal muscle. In addition, postganglionic cells express a variety of neuropeptides, some of which have been linked to specific functional populations of cells (Fig. 29–8) (see Table 29–2). Most prevalent among sympathetic peptides is *neuropeptide Y*, which is released along with norepinephrine by vasoconstrictor postganglionic fibers. This peptide has multiple effects at the adrenergic ending. These effects include stimulation of vascular smooth muscle contraction, potentiation of epinephrine effects, and, paradoxically, inhibition of norepinephrine release.

Receptor Types in Sympathetic Targets. The effect of a neurotransmitter on a target cell is determined by the nature of the target cell receptor and the particular signal transduction mechanism to which it is linked. Thus, the effects of norepinephrine, the main neurotransmitter of most postganglionic neurons, and of epinephrine, the main hormone of the adrenal medulla, vary among different target cells according to the type or types of adrenergic receptor they express (subclasses of α- and β-adrenergic receptors). For example, α_1 receptors on vascular smooth muscle cells mediate vasoconstriction, whereas activation of β_2 receptors results in relaxation. Increases in heart rate and cardiac output are mediated by β_1 receptors on cardiac muscle. Epinephrine is a more potent ligand than norepinephrine at most α- and β-adrenergic receptors. Consequently, epinephrine is administered to counteract symptoms of anaphylactic shock, including bronchospasm, edema, congestion of mucous membranes, and cardiovascular collapse. As mentioned, some tissues receive cholinergic sympathetic innervation (see Fig.

29–8). For example, stimulation of sweat gland secretion is mediated by muscarinic acetylcholine receptors.

Causalgia. In special circumstances, sympathetic activation can become linked to pain. Causalgia (complex regional pain syndrome type II) is a syndrome that can result from partial injury to a peripheral nerve, typically a nerve serving an extremity. Signs and symptoms include spontaneous burning pain, hypersensitivity of the skin, pain triggered by loud noises or strong emotions, sweating and reduced temperature of the limb, mottling of the skin, and swelling of the extremity. A striking feature of causalgia is that symptoms can often be alleviated by sympathectomy or otherwise blocking sympathetic function. Thus, a prevailing theory of the etiology of the pain associated with the syndrome is that sympathetic postganglionic neurons coursing in the injured nerve develop abnormal connections to nociceptive dorsal root ganglion neurons. This pathologic process could occur either in the tangle of regenerated nerve fibers that form a neuroma or within the sensory ganglion. The nociceptive neurons, in turn, may develop abnormal responsiveness to adrenergic stimulation.

Parasympathetic Division

Preganglionic and Postganglionic Neurons. Compared with the sympathetic division, the parasympathetic division is more restricted in its distribution. The cell bodies of parasympathetic preganglionic neurons are located in either sacral segments S2 to S4 or in the nuclei that provide the *general visceral efferent* (GVE) fibers that travel in cranial nerves III, VII, IX, and X (Table 29–3) (see Fig. 29–5). The parasympathetic division of the visceral motor system is accordingly called the *craniosacral* system, as distinct from the sympathetic division, which is *thoracolumbar*.

The cell bodies of postganglionic parasympathetic neurons supplying cranial structures are located in dis-

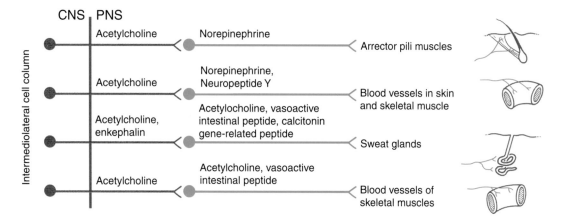

Figure 29–8. Chemical coding of sympathetic preganglionic and postganglionic neurons. Note that some details are based on animal studies and have not been confirmed in humans. CNS, central nervous system; PNS, peripheral nervous system.

Table 29–3.   Peripheral Pathways of Parasympathetic Outflow

Cranial Nerve	Location of Preganglionic Cell Bodies	Course of Preganglionic Fibers	Location of Postganglionic Cell Bodies	Target Tissue(s)	Effect on Target
Oculomotor	Midbrain: Edinger-Westphal nucleus	With cranial nerve III	Ciliary ganglion	Ciliary body, pupillary constrictor	Ciliary muscle contraction; contraction of pupillary sphincter
Facial	Pons: superior salivatory nucleus	Nervus intermedius, greater petrosal nerve to pterygopalatine ganglion or chorda tympani to submandibular ganglion	Pterygopalatine ganglion and submandibular ganglion	Lacrimal gland, nasal glands, submandibular and sublingual glands	↑ secretion
Glossopharyngeal	Medulla: inferior salivatory nucleus	Tympanic branch of cranial nerve IX, tympanic plexus, lesser petrosal nerve	Otic ganglion	Parotid gland	↑ secretion
Vagus	Medulla: dorsal motor vagal nucleus and nucleus ambiguus*	Various branches of cranial nerve X	Terminal ganglia in or on wall of target organ	Heart and great vessels, respiratory system, esophagus, foregut and midgut derivatives	↓ heart rate; bronchial constriction; ↑ blood flow to gut; ↑ peristalsis and secretion
Sacral splanchnic	S2 to S4 of spinal cord: intermediate gray matter	Pelvic nerve	Terminal ganglia in or on wall of target organ	Hindgut derivatives, reproductive organs, urinary bladder	

*Some general visceral efferent preganglionic parasympathetic cells that innervate the heart are found in this nucleus, although its main function is to provide special visceral efferent fibers that distribute on cranial nerves IX and X.

↑, increase; ↓, decrease.

crete ganglia, and, in general, their axons travel distally with branches of the trigeminal nerve. Cell bodies of parasympathetic postganglionic neurons supplying viscera of the body cavities are not grouped into macroscopic ganglia. Rather, these cells are scattered within nerve plexuses of the target organ or in the wall of the gut (*terminal* or *intermural ganglia*), where they intermingle with neurons of the *enteric nervous system*.

Parasympathetic Outflow Pathways. The visceral motor component of the oculomotor nerve arises from the *Edinger-Westphal nucleus*. These preganglionic fibers terminate in the ciliary ganglion. Axons of postganglionic cells of the ciliary ganglion innervate the sphincter muscle of the iris (for pupillary constriction) and the ciliary muscle (for near vision accommodation) (see Table 29–3 and Fig. 29–5).

The GVE preganglionic parasympathetic fibers of the *facial nerve* originate in the *superior salivatory nucleus*. Some investigators distinguish a separate lacrimal nucleus that targets the lacrimal gland. Preganglionic GVE axons from the superior salivatory nucleus exit the brainstem in the *intermediate nerve*, which is classically considered a part of the facial nerve. Some of these fibers course via the greater petrosal nerve to terminate in the *pterygopalatine ganglion*, which supplies the lacrimal gland and nasal and palatal mucous glands.

Other preganglionic fibers travel via the chorda tympani to the *submandibular ganglion*, which innervates the submandibular and sublingual salivary glands (see Table 29–3 and Fig. 29–5).

The *glossopharyngeal nerve* contains preganglionic parasympathetic fibers that originate in the *inferior salivatory nucleus*. These fibers take a tortuous course, via the tympanic nerve and plexus, to form the lesser petrosal nerve, which ends in the *otic ganglion*. Postganglionic fibers from the otic ganglion join the auriculotemporal nerve to reach the parotid gland (see Table 29–3 and Fig. 29–5).

The visceral motor component of the *vagus nerve* provides parasympathetic innervation to organs of the thoracic and abdominal cavities. Preganglionic GVE fibers of the vagus nerve originate in the *dorsal motor vagal nucleus*. In addition, a part of the *nucleus ambiguus* contains GVE preganglionic cells, whose axons travel with the vagus to innervate the heart. It should be emphasized, however, that the main outflow from the nucleus ambiguus consists of SVE fibers to the glossopharyngeal and vagus nerves. Preganglionic fibers of the vagus terminate on postganglionic neurons located in the walls of viscera of the thorax and abdomen (see Table 29–3 and Fig. 29–5). Thus, postganglionic neurons of the vagus nerve are not aggregated into discrete ganglia, as they are for cranial nerves III, VII, and IX.

The *sacral component of the parasympathetic division* innervates the lower digestive tract (beginning at about the left colic flexure) and the urinary bladder, urethra, and reproductive organs. Preganglionic neurons of the *sacral parasympathetic nucleus* occupy a position at sacral levels S2 to S4 comparable to the intermediolateral cell column at thoracic levels (see Fig. 29–10). Preganglionic fibers exit the spinal cord via ventral roots and form the *pelvic nerves (nervi erigentes)*. These nerves mingle with sympathetic fibers of the inferior hypogastric plexuses to form the pelvic visceral plexus lateral to the rectum, bladder, and uterus (see Table 29–3).

Functional and Chemical Coding. Preganglionic parasympathetic neurons, like preganglionic sympathetic neurons, use *acetylcholine* as their main neurotransmitter. Postganglionic parasympathetic neurons are also cholinergic. Both preganglionic and postganglionic parasympathetic neurons release molecules in addition to the principal transmitter at their terminals. These include neuropeptides, most prominently vasoactive intestinal peptide, which act as modulators of the postsynaptic response to the main transmitter.

Receptor Types in Parasympathetic Targets. Nicotinic receptors are limited to skeletal muscle and cholinergic synapses in autonomic ganglia and the CNS. Muscarinic cholinergic receptors appear to be the only class involved in the response of smooth muscle, cardiac muscle, and glandular cells to acetylcholine. The nature of the response depends on which type of muscarinic receptor (M_1, M_2, and so on) is expressed. For example, parasympathetic stimulation of gastric acid secretion is mediated by M_1 muscarinic receptors, whereas M_2 receptors mediate parasympathetic depression of heart rate and of contraction of cardiac muscle.

The cholinergic receptor blocker atropine has effects that are clinically useful in certain circumstances. These effects include pupillary dilation, relaxation of bronchiolar muscle, and reduction of peristalsis and secretion in the stomach.

Enteric Nervous System

It has been estimated that the digestive tract contains about as many nerve cells as the spinal cord. This enormous population of enteric neurons is concentrated mostly in the *myenteric* and *submucosal plexuses* of the gut wall. Neurons of the gut exhibit a wide range of structural, chemical, and physiologic properties, and they form an elaborate system of neural connections (see Fig. 29–7).

The *enteric nervous system* is influenced by inputs of the sympathetic and parasympathetic divisions, and some of these enteric cells may be regarded as postganglionic parasympathetic neurons. However, the enteric nervous system is largely self-sufficient in its regulation of digestive tract activities. This regulation is a result of intrinsic neuronal circuits involving both sensory and motor neurons, as well as interneurons within the myenteric and submucosal plexuses (see Fig. 29–7). For example, although smooth muscle cells undergo spontaneous contraction, there are elaborate neural circuits that coordinate waves of relaxation and contraction (peristalsis). These intrinsic neuronal circuits include mechanoreceptors, which signal stretch, and burst-type oscillator cells that spontaneously generate a chain of action potentials. Noradrenergic cells excited by these signals inhibit smooth muscle contraction. In addition to intrinsic circuits that control peristalsis, other circuits regulate blood flow and secretion in response to the contents of the lumen.

Enteric neurons express a large variety of neuropeptides, such as vasoactive intestinal peptide, neuropeptide Y, and cholecystokinin. Many of these peptides were first identified in the gut but have since been found in other central and peripheral neurons.

Regulation of Visceral Motor Outflow

Sensory input occurs at every level of the visceral motor pathway, including prevertebral postganglionic neurons, preganglionic neurons, and a wide variety of CNS structures that project, either directly or indirectly, to preganglionic neurons. Many different kinds of sensory information are integrated by a series of CNS structures collectively termed the *central autonomic network* (CAN), which generates coordinated signals to the visceral motor, endocrine, and somatic motor outflow pathways. Although visceral motor activities are generally beyond conscious control, emotional status and mental activity clearly influence visceral structures. Accordingly, the CAN integrates input from higher CNS centers involved in cognition and complex behavioral functions.

Major Central Nervous System Components. The preganglionic neurons of the autonomic motor pathways are influenced by cells in various brainstem and forebrain areas. The *hypothalamus* is the highest integrator of autonomic and endocrine functions. It directly regulates the secretory activity of the anterior and posterior pituitary and has reciprocal connections with the solitary nucleus and other components of the CAN in the forebrain and brainstem. Some hypothalamic nuclei project directly to preganglionic visceral motor neurons in the dorsal vagal nucleus, nucleus ambiguus, and intermediolateral cell column. The organization and functions of the hypothalamus are considered in Chapter 30.

Because of its diverse connections, the solitary nucleus is the most important brainstem structure coordinating autonomic functions. It receives general and special visceral sensory input, and it projects to vagal motor neurons, to salivatory and reticular nuclei, and to populations of brainstem neurons, which in turn project to sympathetic preganglionic neurons. The solitary nucleus also has reciprocal connections with other components of the CAN.

Other cell groups that are important in autonomic regulation reside in the reticular formation of the brainstem. These neurons are not always restricted to specific nuclei, and they are therefore sometimes designated by their relative positions. For example, cells in the *rostral ventrolateral medulla* project to the intermediolateral cell column, particularly to preganglionic neurons involved in cardiovascular regulation. This area is called the *vasopressor center* because stimulation results in increased peripheral vascular resistance and increased cardiac output. Areas such as this have been designated *centers*—for example, the respiration center, micturition center, and vomiting center. Although this terminology is convenient, it should be understood that these are not well-defined anatomic entities, but are components of widely distributed neural networks.

The importance of the supraspinal control of autonomic function is illustrated by some of the deficits associated with spinal cord injuries at higher levels (T6 or above). Initially, the interruption of descending reticulospinal and hypothalamospinal fibers that regulate sympathetic preganglionic neurons in the intermediolateral cell column is manifested as an overall reduction in sympathetic activity. Thus, clinical signs include lowered blood pressure, orthostatic hypotension, and reduced heart rate (bradycardia). With time, hyperactivity of sympathetic reflexes (termed *autonomic dysreflexia*) develops, probably as a result of denervation hypersensitivity of sympathetic neurons and target tissues. Signs and symptoms include hypertension, urinary retention, piloerection, profuse sweating, and reduction of blood flow to peripheral tissues in response to any of a wide variety of noxious stimuli below the level of the spinal cord injury.

Cardiovascular System. The function of the cardiovascular system is influenced by mental activity, emotional state, posture, muscular exertion, visceral activity, body temperature, and concentrations of blood gases and electrolytes. In addition to mechanisms that regulate blood pressure, there is precise neural control of blood flow to specific organs and regions of the body.

The *baroreceptor reflex* (Fig. 29–9) functions to buffer blood pressure against a sudden change in posture. Failure of this reflex results in *orthostatic hypotension*, a severe drop in blood pressure when the patient assumes an upright position. Primary visceral sensory neurons of the glossopharyngeal and vagus nerves convey signals from mechanoreceptors in the carotid and aortic sinuses centrally, where they terminate in the solitary nucleus (see Fig. 19–8). Projections of solitary

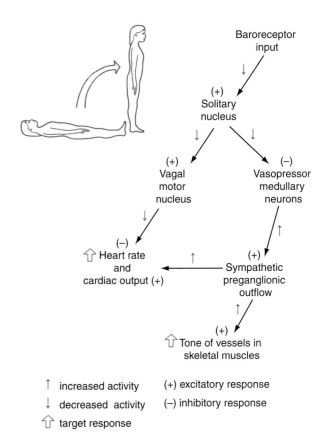

↑   increased activity   (+) excitatory response
↓   decreased activity   (−) inhibitory response
⇧   target response

Figure 29–9. Pathways of the baroreceptor reflex. A sudden movement to an upright position results in modification of neuronal activity (red) in these pathways to produce the compensatory changes in the cardiovascular system.

neurons influence the tonic activity of parasympathetic (vagal) output to the heart and sympathetic output to the heart and peripheral vessels.

When a reclining individual stands, there is a rapid reduction in baroreceptor discharge that results in a *decrease* in signals from the solitary nucleus to two brainstem targets (see Fig. 29–9). The first of these targets, the vagal preganglionic parasympathetic neurons that suppress heart rate and cardiac output, receive an excitatory drive from neurons of the solitary nucleus. Thus, a reduction of baroreceptor discharge results in a release of the heart from this inhibitory parasympathetic drive. The second target of solitary cells consists of *vasopressor neurons* in the *rostral ventrolateral medulla*. The neurons in this region have intrinsic pacemaker features and receive inhibitory inputs from the solitary nucleus. When these rostral neurons are released from the inhibitory drive of solitary neurons, the result is increased sympathetic outflow (see Fig. 29–9). This input is mediated via a major descending projection from these rostral medullary cells to the sympathetic preganglionic neurons in the intermediolateral cell column. Reduced inhibition of this excitatory projection results in increased cardiac output and increased resistance in vascular beds of skeletal muscle and abdominal visceral organs, but not of the skin, heart, and brain. Some of this solitary input to vasopressor neurons may be relayed through vasodepressor cell groups in the caudal ventrolateral medulla. Thus, when *a person moves from a reclining to standing posture, the resultant pooling of blood in the lower half of the body is quickly countered by increased vascular tone and increased cardiac output.* Without this reflex, movement to a standing position results in dizziness or fainting because of decreased blood flow to the brain. This manifestation of orthostatic hypotension is a serious consequence of many forms of autonomic dysfunction.

The *chemoreceptor reflex* maintains homeostasis of blood gas composition by adjusting respiration, cardiac output, and peripheral blood flow. Decreased PO_2 and increased PCO_2, detected by receptors in the carotid and aortic bodies, are signaled by glossopharyngeal and vagal afferents that terminate in the solitary nucleus. Within the medulla, the reflex pathway for cardiovascular effects parallels that for the baroreceptor reflex. A decrease in blood PO_2 activates this reflex and promotes increased heart rate and vascular tone. These changes result in a decreased blood flow to skeletal muscles and viscera, whereas blood flow to the brain is maintained. Thus, proportionately more oxygenated blood is available to the brain than to skeletal muscle and viscera. The resulting conservation of oxygen preserves vital functioning of the CNS.

The cardiovascular component of the chemoreceptor reflex is closely coordinated with respiration, a somatic motor function coordinated by other neurons of the brainstem reticular formation. For example, if breathing is suspended (as in diving), heart rate is slowed (*bradycardia*) rather than accelerated (*tachycardia*). **Urinary Bladder and Micturition.** Emptying of the urinary bladder, *micturition*, is brought about by contraction of smooth muscle of the bladder wall (*detrusor muscle*) and relaxation of skeletal muscle of the *external urethral sphincter* (Fig. 29–10). Contraction of the detrusor is mediated by parasympathetic outflow. Preganglionic neurons from the sacral cord innervate postganglionic neurons in the bladder wall (see Fig. 29–10). The bladder wall also has a sympathetic innervation. Its influence is mainly inhibitory on both the detrusor muscle and the parasympathetic postganglionic neurons in the bladder wall (see Fig. 29–10). The external urethral sphincter, which is subject to both reflex and

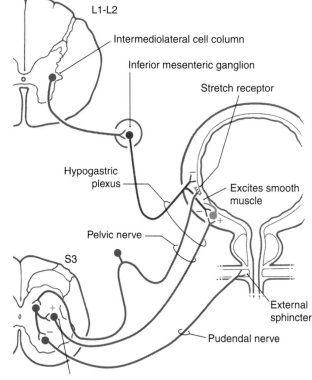

Figure 29–10. Neural pathways mediating control of the urinary bladder.

voluntary control, is supplied by alpha motor neurons in segments S3 and S4.

During periods of urine storage, activity of bladder afferent neurons is low. The low activity of the sensory neurons results in (1) low activity of parasympathetic excitatory innervation to the detrusor, (2) tonic activity of sympathetic neurons that inhibit both the parasympathetic ganglion cells in the bladder wall and the detrusor muscle directly, and (3) tonic activity of sacral somatic motor neurons mediating constriction of the external sphincter. As urine accumulates, pressure on the bladder wall activates tension receptors until bladder afferent activity rises to a threshold level. This increased activity of bladder afferents induces micturition by way of both spinal and brainstem reflexes that result in inhibition of sympathetic outflow, activation of parasympathetic outflow, and inhibition of somatic motor neurons supplying external sphincter muscle.

Sources and Additional Reading

Appenzeller O: The Autonomic Nervous System, 4th Ed. Elsevier, Amsterdam, 1990.

Bannister R, Mathias CJ (eds): Autonomic Failure. A Textbook of Clinical Disorders of the Autonomic Nervous System, 3rd ed. Oxford University Press, Oxford, 1992.

Baron R, Levine JD, Fields HL: Causalgia and reflex sympathetic dystrophy: Does the sympathetic nervous system contribute to the generation of pain? Muscle Nerve 22:678–695, 1999.

Benarroch EE: Neuropeptides in the sympathetic system: Presence, plasticity, modulation, and implications. Ann Neurol 36:6–13, 1994.

Brodal P: The Central Nervous System: Structure and Function. Oxford University Press, Oxford, 1998.

Gabella G: Structure of the Autonomic Nervous System. Chapman and Hall, London/John Wiley & Sons, New York, 1976.

Jänig W, Schmidt RF (eds): Reflex Sympathetic Dystrophy. VHC, Weinheim Germany/VHC Publishers, New York, 1990.

Jänig W, Stanton-Hicks M (eds): Reflex Sympathetic Dystrophy: A Reappraisal. IASP Press, Seattle, 1996.

Loewy AD, Spyer KM (eds): Central Regulation of Autonomic Functions. Oxford University Press, New York, 1990.

Low PA (ed): Clinical Autonomic Disorders. Evaluation and Management. Little, Brown, Boston, 1992.

Pick J: The Autonomic Nervous System. Morphological, Comparative, Clinical and Surgical Aspects. JB Lippincott, Philadelphia, 1970.

Shephard GM: Neurobiology, 3rd ed. Oxford University Press, New York, 1994.

Teasell RW, Arnold JM, Krassioukov A, Delaney GA: Cardiovascular consequences of loss of supraspinal control of the sympathetic nervous system after spinal cord injury. Arch Phys Med Rehabil 81:506–516, 2000.

Chapter 30

The Hypothalamus

*S. G. P. Hardy, R. B. Chronister,
and A. D. Parent*

Overview   480

Boundaries of the Hypothalamus   480

Divisions of the Hypothalamus   480
  Preoptic Area
  Lateral Zone
  Medial Zone
  Periventricular Zone

Blood Supply of the Hypothalamus   484

Hypothalamic Afferent Fibers   484
  Fornix
  Medial Forebrain Bundle
  Amygdalohypothalamic Fibers
  Other Afferent Fibers

Hypothalamic Efferent Fibers   486
  Ascending Projections
  Descending Projections

Intrinsic Hypothalamic Connections   487
  Supraopticohypophysial Tract
  Tuberoinfundibular Tract

Pituitary Tumors   488
  Secreting Tumors
  Cushing Disease
  Prolactin-Secreting Tumors
  Gonadotrope Tumors

Regional Functions of the Hypothalamus   491
  Caudolateral Hypothalamus
  Rostromedial Hypothalamus

Hypothalamic Reflexes   491
  Baroreceptor Reflex
  Temperature Regulation Reflex
  Water Balance Reflex

One of the most rostral cell groups to influence visceral function, and the one that has direct input to all other visceral nuclei in the neuraxis, is the hypothalamus. In addition to its role in regulating visceromotor functions, the hypothalamus also influences neural circuits that modify behavior.

Overview

The hypothalamus is the part of the diencephalon involved in the central control of visceral functions (through the *visceromotor* and *endocrine systems*) and affective or emotional behavior (via the *limbic system*) (Fig. 30–1). Although its primary role is in the maintenance of *homeostasis*, the hypothalamus partially regulates numerous functions including water and electrolyte balance, food intake, temperature, blood pressure, possibly the sleep-waking mechanism, circadian rhythmicity, and general body metabolism. The hypothalamus (at about 4 g) is dwarfed in size by the rest of the brain (weighing approximately 1400 g). However, it is perhaps the most important 4 g in the entire body. In short, the hypothalamus influences our responses to both the internal and external environments (see Fig. 30–1) and is necessary for life.

Boundaries of the Hypothalamus

The rostral boundary of the hypothalamus is the *lamina terminalis*, a thin membrane that extends ventrally from the anterior commissure to the rostral edge of the optic chiasm (Fig. 30–2A). The lamina terminalis separates the hypothalamus from the more anterior septal nuclei. Posteriorly (dorsally), the hypothalamus is bounded by the *hypothalamic sulcus*, a shallow groove that separates the hypothalamus from the dorsal thalamus (Fig. 30–2A, C). The lateral boundary of the hypothalamus is formed rostrally by the substantia innominata and caudally by the medial edge of the posterior limb of the *internal capsule* (see Fig. 30–2B and C; see also Fig. 15–6). Medially, the hypothalamus is bordered by the inferior portion of the *third ventricle*. Caudally, the hypothalamus is not sharply demarcated,

merging instead into the *midbrain tegmentum* and the *periaqueductal gray*.

Divisions of the Hypothalamus

The hypothalamus can be divided into the *preoptic area* and the *lateral, medial,* and *periventricular zones* (Fig. 30–3). The preoptic area is a transition region that extends rostrally, by passing laterally to the lamina terminalis, to form a continuation with structures in the basal forebrain. Three zones are located caudal to the preoptic area. The thin periventricular zone is the most medial and is subjacent to the ependymal cells that line the third ventricle. The medial zone is located lateral to the periventricular zone, and a line drawn from the post-commissural fornix to the mammillothalamic tract separates it from the lateral zone (see Fig. 30–3).

Preoptic Area. The preoptic area, although functionally a part of the hypothalamus (and diencephalon) is embryologically derived from the telencephalon. This area is composed primarily of the medial and lateral preoptic nuclei (see Fig. 30–3). The *medial preoptic nucleus* contains neurons that manufacture gonadotropin-releasing hormone (GnRH). GnRH is transported along the *tuberoinfundibular tract* to capillaries of the hypophysial portal system and thence to the anterior lobe of the pituitary gland (Fig. 30–4), where it causes the release of gonadotropins (luteinizing hormone and follicle-stimulating hormone). Because gonadotropin release is continuous in males and cyclical in females, the medial preoptic nucleus of males tends to be more active and consequently larger than that of females. Accordingly, the medial preoptic nucleus is often referred to as the *sexually dimorphic* nucleus of the preoptic area. The medial preoptic nucleus also influences behaviors that are related to eating, reproductive activities, and locomotion. The *lateral preoptic nucleus* is located immediately rostral to the lateral hypothalamic zone (see Fig. 30–3). The function of this nucleus is not fully established. However, through its connections with the ventral pallidum, it may function in part in locomotor regulation. Some investigators consider the nuclei of the preoptic area to be part of the supraoptic region of the medial hypothalamic zone.

Lateral Zone. The *lateral zone* (see Fig. 30–3) contains a large bundle of axons collectively called the *medial forebrain bundle* (see Fig. 30–3; see also Fig. 30–7). This diffuse bundle of fibers traverses the lateral hypothalamic zone and interconnects the hypothalamus with rostral areas such as the septal nuclei and with caudal regions such as the brainstem reticular formation.

The *lateral hypothalamic zone* comprises a large, diffuse population of neurons commonly called the *lateral*

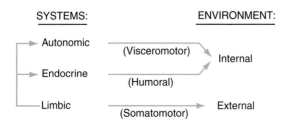

SYSTEMS:

ENVIRONMENT:

Autonomic

Endocrine

Limbic

(Visceromotor)

(Humoral)

(Somatomotor)

Internal

External

Figure 30–1. The interrelationships among the autonomic, endocrine, and limbic systems. All three systems are under the control of the hypothalamus.

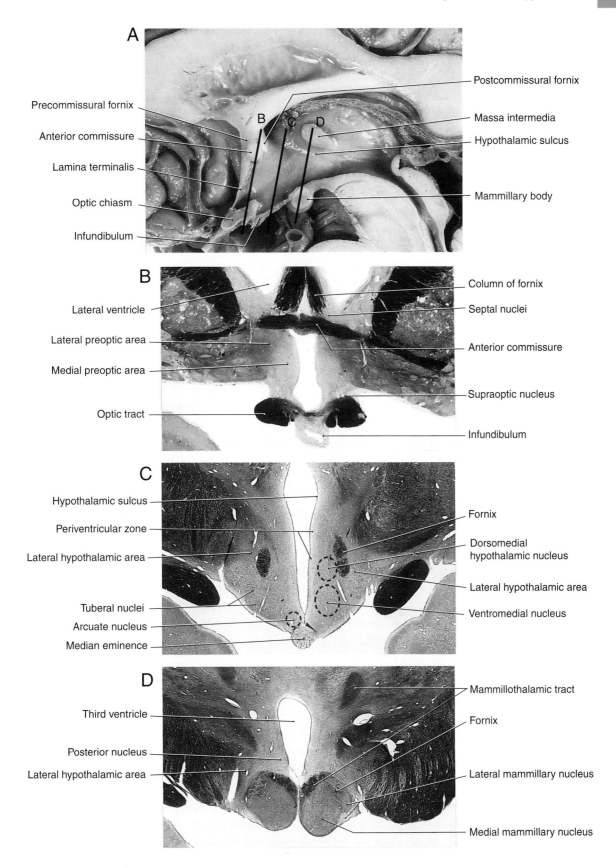

Figure 30–2. Midsagittal view (*A*) of the brain emphasizing hypothalamic structures. Cross sections of the hypothalamus through preoptic (*B*), tuberal (*C*), and mammillary (*D*) regions. The myelin-stained sections in *B*, *C*, and *D* correspond to the comparably labeled lines in *A*.

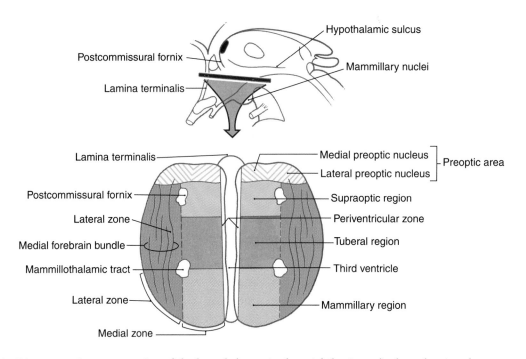

Figure 30–3. Diagrammatic representation of the hypothalamus in the axial (horizontal) plane showing the zones and regions.

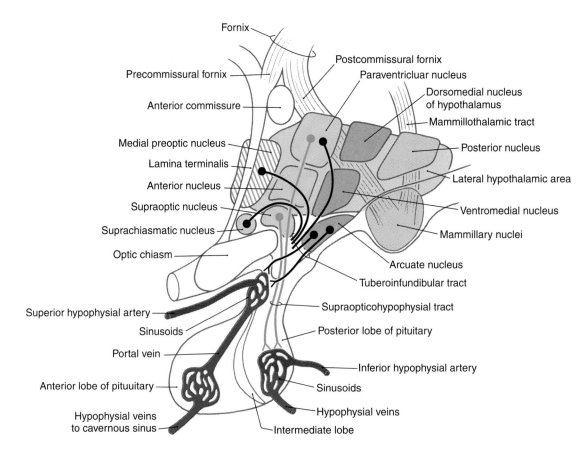

Figure 30–4. Midsagittal view of the hypothalamus emphasizing the nuclei, which contribute to the tuberoinfundibular and supraopticohypophysial tracts, the hypophysial portal system, and the general relations of the fornix and mammillothalamic tract.

hypothalamic area as well as smaller condensations of cells located in its anterior (ventral) portions. The latter cell groups are the *lateral hypothalamic nucleus* and the *tuberal nuclei*. The *lateral hypothalamic nucleus* is a loose aggregation of relatively large cells that extends throughout the rostrocaudal extent of the lateral hypothalamic zone. This nucleus constitutes a "feeding center." Stimulation of this nucleus in laboratory animals promotes feeding behavior; destruction of it causes feeding behavior to attenuate and the animal loses weight. The *tuberal nuclei* consist of small clusters of neurons, each containing small, pale, multipolar cells. Some tuberal neurons project into the tuberoinfundibular tract and therefore may convey releasing hormones to the hypophysial portal system. Others send a histaminergic input to the cerebellum that may be involved in the regulation of motor activity.

Medial Zone. The *medial zone* is a cell-rich region composed of many individual nuclei (see Figs. 30–3 and 30–4). It is divided into three regions: the *supraoptic (chiasmatic) region*, the *tuberal region*, and the *mammillary region* (see Fig. 30–3). The *supraoptic region* is located posterior (dorsal) to the optic chiasm. The *tuberal region* is the widest part of the hypothalamus and corresponds, in general, to the position of the *tuber cinereum*. The *mammillary region*, the most posterior of the three, corresponds to the location of the mammillary bodies.

The *supraoptic region* contains four nuclei: the *supraoptic, paraventricular, suprachiasmatic*, and *anterior nuclei* (see Fig. 30–4). Neurons of the *supraoptic* and *paraventricular nuclei* contain oxytocin and antidiuretic hormone (ADH) (i.e., vasopressin) and transmit these substances to the posterior pituitary by way of the *supraopticohypophysial tract* for release into the circulatory system (see Fig. 30–4). The functions of these hormones are discussed later in this chapter. The *suprachiasmatic nucleus* receives direct input from the retina and can influence other hypothalamic structures such as the medial preoptic nucleus. It is believed that the suprachiasmatic nucleus may mediate *circadian rhythms*, these being the hormonal fluctuations that are secondary to light-darkness cycles. The *anterior nucleus* is located immediately caudal to the preoptic area. Although this nucleus participates in a wide range of visceral and somatic functions, many of its neurons are involved in the maintenance of body temperature.

The *tuberal region* contains three nuclei: the *ventromedial, dorsomedial*, and *arcuate nuclei* (see Figs. 30–2*C* and 30–4). The *ventromedial nucleus*, one of the largest and best-defined of the hypothalamic nuclei, is considered to be a "satiety center." If this nucleus is stimulated in the laboratory, the experimental animal will not engage in feeding behavior. Conversely, a lesion to this nucleus causes the animal to eat excessively and gain weight. The *dorsomedial nucleus*, located immediately posterior (dorsal) to the ventromedial nucleus,

subserves a function relating to emotion or, at least, to emotional behavior. In laboratory animals, stimulation of the dorsomedial nucleus results in unusually aggressive behavior, which lasts only so long as the stimulation is present. This phenomenon, known as *sham rage*, can also be elicited by the stimulation of other hypothalamic and extra-hypothalamic sites. The *arcuate nucleus* is the primary location of neurons that contain releasing hormones. These substances are transmitted to the anterior pituitary by way of the tuberoinfundibular tract and hypophysial portal system, whereupon they influence the release of various pituitary hormones (see Fig. 30–4).

The *mammillary region* contains four nuclei: the *medial, intermediate*, and *lateral mammillary* and the *posterior hypothalamic nuclei* (see Figs. 30–2*A* and *D* and 30–4). The *medial mammillary nucleus* is large and especially well developed in the human. It represents the primary termination point for the axons of the postcommissural fornix, which originate primarily from the subiculum of the hippocampal complex. The medial mammillary nucleus is also the source of axons that are directed to the anterior nucleus of the dorsal thalamus as the *mammillothalamic tract* (see Fig. 30–4; see also Fig. 30–6). The latter pathway represents an important part of the limbic system. The much smaller *intermediate* and *lateral mammillary nuclei* are located lateral to the medial mammillary nucleus. The lateral mammillary nucleus receives input from the medial aspects of the midbrain reticular formation by way of the *mammillary peduncle* (see Fig. 30–7).

Insight into the function of the mammillary nuclei comes from experimental and clinical observations. For example, lesions of the mammillary bodies tend to impede the retention of newly acquired memory, so that an immediate memory or a *short-term memory* is not processed into *long-term memory*. A patient with a mammillary lesion has no difficulty in remembering events occurring months or years prior to the lesion. Memory for events occurring *after* the lesion is, however, limited to the short term (a period of minutes), and long-term memories are not established. As a result of this *anterograde amnesia*, affected patients typically have severe difficulties learning new tasks and transforming these experiences into long-term memory. These specific memory deficits are characteristic of the *Korsakoff syndrome*, a condition that is caused by thiamine deficiency and is typically associated with chronic alcoholism. The memory deficits in this syndrome are caused by progressive degeneration in the mammillary bodies and in functionally related brain structures, such as the hippocampal complex and the dorsomedial thalamic nucleus.

Patients with the Korsakoff syndrome may have difficulty in understanding written material and in conducting meaningful conversations because they tend to forget what was just read or said. An interesting feature of this syndrome is the patient's tendency to *confabulate*, that is, to string together fragmentary memories from various events into a synthesized memory of an "event" that never occurred.

The *posterior hypothalamic nucleus* merges imperceptibly with the midbrain periaqueductal gray. Accordingly, this nucleus is associated with the same myriad of emotional, cardiovascular, and analgesic functions that have been attributed to the periaqueductal gray.

Periventricular Zone. The periventricular zone (see Fig. 30–3), not to be confused with the paraventricular nucleus, is a very thin region composed of small cell bodies lying medial to the medial zone and immediately subjacent to the ependymal cells of the third ventricle. Many neurons of the periventricular zone synthesize releasing hormones. These neurons project by way of the *tuberoinfundibular tract* to the hypophysial portal system and thus influence the release of various hormones by the anterior pituitary. Consequently, many of the neurons in the periventricular zone serve a function similar to that of neurons located in the arcuate nucleus.

Blood Supply of the Hypothalamus

The hypothalamus and some immediately adjacent structures are served by small perforating arteries that arise from the *circle of Willis* (Fig. 30–5). Those branches from the anterior communicating artery and the A_1 segment of the anterior cerebral artery constitute the *anteromedial group* of perforating arteries (see also Fig. 8–10). In general, these vessels serve the nuclei of the preoptic area and supraoptic region, the septal nuclei, and rostral portions of the lateral hypothalamic area. A few perforating arteries may also arise from the bifurcation of the internal carotid artery.

The small perforating arteries that originate from the posterior communicating artery and the P_1 segment of the posterior cerebral artery constitute the *posteromedial group* (see Fig. 30–5; see also Fig. 8–10). These vessels serve primarily the nuclei of the tuberal and mammillary regions. Branches arising from the rostral portion of the posterior communicating artery distribute to the former region, whereas the latter region is served by branches of the caudal parts of the posterior communicating artery and of P_1. In addition, the posteromedial group also sends branches into the middle and caudal parts of the lateral hypothalamic area. The large *thalamoperforating arteries* usually arise from P_1. Although these vessels distribute mainly to rostral areas of the dorsal thalamus, they do give rise to some small branches that enter the posterior hypothalamus.

The *hypophysial arteries* arise from the internal carotid artery. The *inferior branches* (or rami) originate from the cavernous part of this large vessel, whereas the *superior branches* (see Fig. 30–4) are from the cerebral (supraclinoid) part of the internal carotid. Although small, these are important vessels.

Hypothalamic Afferent Fibers

The hypothalamus is connected to diverse sites, including the hippocampus, amygdala, brainstem tegmentum, various thalamic nuclei, septal nuclei, and even neocortical areas such as the infralimbic and cingulate cortex. With very few exceptions, these connections are reciprocal. The following are the most important input-output relationships of the hypothalamus.

Fornix. The *fornix* arises from neurons in the *subiculum* and the *hippocampus* (two components of the *hippocampal complex*) and is the largest single input to the hypothalamus (Fig. 30–6; see also Fig. 30–4). As the fornix approaches the anterior commissure, it divides into a small *precommissural bundle*, derived largely from the hippocampus, and a large *postcommissural bundle*,

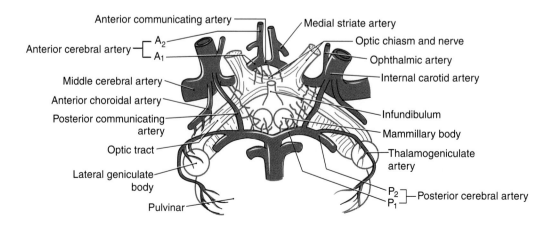

Figure 30–5. The blood supply to the hypothalamus.

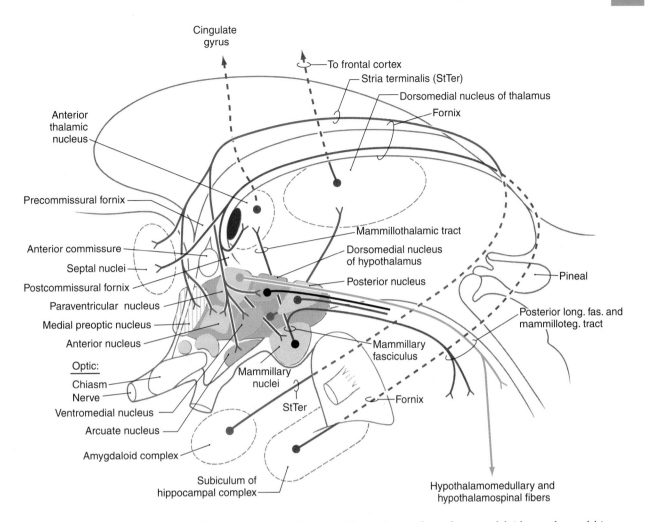

Figure 30–6. Midsagittal view of the hypothalamus emphasizing afferent inputs from the amygdaloid complex and hippocampus. Also shown are the origins of descending fibers to the brainstem and spinal cord. long. fas., longitudinal fasciculus; nuc., nucleus.

arising mainly from the subiculum. The former passes to septal and preoptic nuclei and to the anterior hypothalamic region, whereas the latter projects primarily to the medial mammillary nucleus, with lesser inputs to the anterior thalamic nucleus and lateral hypothalamus (Fig. 30–7).

Medial Forebrain Bundle. The *medial forebrain bundle* is a diffuse, composite structure containing mainly fibers that course rostrocaudally through the lateral hypothalamic zone. It contains ascending and descending fibers that interconnect the septal nuclei (including the nucleus accumbens septi), hypothalamus, and midbrain tegmentum (see Fig. 30–7).

Amygdalohypothalamic Fibers. Two major afferent fiber systems project to the hypothalamus from the amygdaloid complex: the stria terminalis, which is phylogenetically older, and a newer composite bundle called the ventral amygdalofugal pathway (see Fig. 30–7). The *stria terminalis* originates from the corticomedial portion of the amygdala and terminates in the

septal nuclei, preoptic area, and medial hypothalamic zone (see Fig. 30–7). This bundle, which accompanies the terminal vein at the juncture of the caudate nucleus and the thalamus, follows the arched configuration of the caudate nucleus and thus takes a path quite similar to that of the fornix (Fig. 30–6). The *ventral amygdalofugal pathway* originates from the basolateral portion of the amygdaloid complex and passes rostromedially beneath the lentiform nucleus and through the area of the substantia innominata to enter the hypothalamus (see Fig. 30–7). The axons in this pathway terminate primarily in the lateral hypothalamic zone and the septal and preoptic nuclei. This bundle also contains fibers that pass from the hypothalamus to the amygdala.

Other Afferent Fibers. The *mammillary peduncle* is a diffuse bundle of fibers that originates from medial portions of the midbrain reticular formation and terminates primarily in the lateral mammillary nucleus. Some of these axons enter the medial forebrain bundle and project to the septal nuclei. The dorsomedial nu-

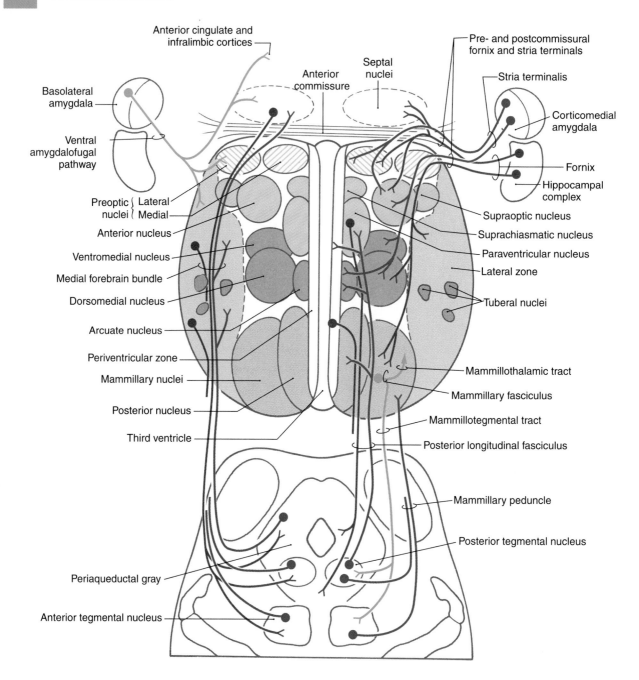

Figure 30–7. Diagrammatic representation of hypothalamic connections as seen in the axial (horizontal) plane. Note the distribution of fibers of the ventral amygdalofugal pathway, the precommissural and postcommissural fornix, and the stria terminalis.

cleus of the thalamus gives rise to *thalamohypothalamic fibers* that course anteriorly (ventrally) to enter the lateral hypothalamus. The only direct neocortical projection to the hypothalamus originates from the prefrontal cortex. This *corticohypothalamic fiber* projection is sparse and terminates primarily in the lateral hypothalamic area. As mentioned earlier, the suprachiasmatic nucleus receives a projection from the retina that is involved in behavioral rhythms and light-dark cycles. These *retinohypothalamic fibers* arise as direct axons from the optic chiasm or as collaterals of retinogeniculate fibers.

Hypothalamic Efferent Fibers

The various subdivisions of the hypothalamus have diffuse projections to numerous sites throughout the neuraxis. Only the major projections are summarized here. A useful generalization is that most structures projecting to the hypothalamus receive a reciprocal input from the hypothalamus. For example, the hypothalamus receives an amygdalohypothalamic projection and gives rise to hypothalamoamygdaloid fibers. For ease of discussion, the efferents are divided into those to forebrain structures (ascending) and those to brainstem and spinal targets (descending).

Ascending Projections. The *mammillary fasciculus* originates as a well-defined bundle from the medial mammillary nucleus (see Figs. 30–6 and 30–7). It passes posteriorly (dorsally) for a short distance then bifurcates into the *mammillothalamic tract* and the *mammillotegmental tract* (see Fig. 30–6). The former projects to the anterior nucleus of the thalamus and is an important part of the circuit of Papez (see Chapter 31). The latter (discussed further on) turns caudally and distributes to the tegmental nuclei of the midbrain reticular formation, thus reciprocating the mammillary peduncle. *Hypothalamothalamic fibers* arise mainly from the lateral preoptic area and project to the dorsomedial nucleus of the thalamus. The hypothalamus also projects to the amygdaloid nucleus (*hypothalamoamygdaloid fibers*) via the stria terminalis and the ventral amygdalofugal pathway. These fibers originate from various hypothalamic nuclei and project primarily to corticomedial nuclei of the amygdaloid complex.

Descending Projections. There are four main descending projections from the hypothalamus. These are *hypothalamospinal* and *hypothalamomedullary* fibers, the *posterior (dorsal) longitudinal fasciculus*, and the *mammillotegmental tract* (see Figs. 30–6 and 30–7).

Hypothalamospinal and *hypothalamomedullary* fibers arise mainly from the paraventricular nucleus, although some originate from the dorsomedial and lateral hypothalamic nuclei (see Fig. 30–6). These fibers descend through the periaqueductal gray and adjacent reticular formation of the midbrain and rostral pons, and then shift to an anterolateral (ventrolateral) position in the medulla. *Hypothalamomedullary fibers* terminate in the solitary nucleus, dorsal vagal motor nucleus, nucleus ambiguus, and other nuclei of the ventrolateral medulla. *Hypothalamospinal fibers* traverse the anterolateral (ventrolateral) medulla and lateral funiculus of the spinal cord to terminate on neurons of the intermediolateral cell column (general visceral efferent preganglionic cells). Hypothalamomedullary and hypothalamospinal fibers form an essential, *direct* link between the hypothalamus and autonomic nuclei of the medulla and spinal cord. Lesions in the ventrolateral medulla may disrupt these fibers. Although the functional effect of disrupting hypothalamomedullary fibers is not well understood, injury to hypothalamospinal fibers results in a loss of sympathetic outflow to the ipsilateral face and head (causing a *Horner syndrome*) and to the ipsilateral side of the body.

The *posterior (dorsal) longitudinal fasciculus* originates from nuclei of the medial hypothalamic zone, whereas fibers of the *mammillotegmental tract* arise from the medial mammillary nucleus (see Figs. 30–6 and 30–7). Fibers of these pathways descend through, and largely terminate in, the periaqueductal gray. The mammillotegmental tract, being more anteriorly (ventrally)

located, also terminates on neurons of the *posterior (dorsal)* and *anterior (ventral) tegmental nuclei* situated in the periaqueductal gray of the caudal midbrain (see Fig. 30–7). Some neurons of the periaqueductal gray function, at least in part, in relaying information to visceral areas of the brainstem, such as the solitary and dorsal motor vagal nuclei. Consequently, the posterior longitudinal fasciculus and mammillotegmental tract are generally viewed as relatively short tracts that *indirectly* influence the autonomic nuclei of the brainstem.

Intrinsic Hypothalamic Connections

The pathways that interconnect the many nuclei of the hypothalamus are numerous and complex. Only two especially important ones are considered here: the *supraopticohypophysial tract* and the *tuberoinfundibular tract*. Both of these tracts link the hypothalamus to the pituitary (see Fig. 30–4).

Supraopticohypophysial Tract. Two hormones are released by the posterior pituitary: *oxytocin* and *antidiuretic hormone* (ADH, vasopressin). These hormones are synthesized in large (magnocellular) neurons of the *supraoptic* and *paraventricular nuclei*. They are transported to the posterior pituitary in the axons of these neurons, which form the *supraopticohypohysial tract*. In the posterior pituitary, they are stored in specialized axon terminals, sometimes called *Herring bodies*, which release them in response to the arrival of action potentials from the nerve cell body. The activity of these hypothalamic neurons—and thus the release of the hormones—is regulated in response to appropriate stimuli. Once released, the hormones enter a capillary plexus in the posterior pituitary and are conveyed to the general circulation by hypophysial veins.

Neurons containing oxytocin release this hormone during coitus, nipple suckling, and periods in which there is an increased level of estrogen. The release of oxytocin induces the contraction of smooth muscle in the uterus and of the myoepithelial cells in the mammary gland. The effect of oxytocin on the uterus is critical during and after childbirth. A synthetic form of oxytocin (*Pitocin*) is often administered to hasten labor and delivery. After the baby is born, oxytocin continues to be important. During nursing, for example, the baby's suckling causes oxytocin to be released from the posterior pituitary, and oxytocin in turn causes the myoepithelial cells of the milk glands to contract, expelling milk. The oxytocin released during nursing also has beneficial effects on the postpartum uterus. Specifically, the contractions of uterine muscles caused by oxytocin help this organ to gradually regain its original form and size.

Neurons containing ADH are influenced primarily by fluctuations in the osmolarity of the blood. The relative concentration of sodium chloride in blood plasma is normally about 300 milliosmoles. This osmolarity is largely a function of how much water is retained within the body. It is important to note that, in the process of maintaining fluid balance homeostasis, small deviations from normal blood osmolarity occur throughout each day. These deviations serve as stimuli that influence the release of ADH from the posterior pituitary. Because these stimuli occur frequently, the neurons containing ADH (unlike those containing oxytocin) are tonically active. Thus, small amounts of ADH are released numerous times each day.

When the blood osmolarity is high, the release of ADH from the posterior pituitary is facilitated. Upon entering the systemic circulation, ADH has a primary effect on the kidneys. Specifically, ADH causes the collecting tubules to increase their resorption of water from the developing urine, thereby returning water to the circulatory system. The additional water serves to dilute the blood, causing the blood osmolarity to be decreased. Consequently, however, the urine becomes more concentrated. As a result, urine output is diminished, and the urine that is produced has a darker color.

When the blood osmolarity is low, the release of ADH from the posterior pituitary is inhibited. Consequently, the amount of ADH in the systemic circulation will be diminished. In response, the collecting tubules of the kidneys decrease their resorption of water from the developing urine. Consequently, water remains in the urine and is not returned to the circulatory system. The effect of this renal conservation of water is an increase in the concentration of the blood, causing the blood osmolarity to be increased. Accordingly, there is also an increased output of pale-colored (dilute) urine.

Lesions of the supraoptic or paraventricular nucleus or of the supraopticohypophysial tract produce a syndrome known as *diabetes insipidus*, which is characterized by *polyuria* (increased urination) and *polydipsia* (increased consumption of water). This condition is due to a deficit of circulating ADH. It is of interest that ethanol causes a decrease in the release of ADH from the posterior pituitary. This is the reason why consumption of alcoholic beverages tends to cause copious urination and consequent dehydration and thirst.

The main function of ADH (vasopressin) is to assist in the maintenance of normal blood osmolarity and blood pressure. Normally, ADH increases blood pressure by increasing blood volume. However, ADH at high levels will cause contraction of vascular smooth muscle and may also result in increased blood pressure. In this regard, the release of ADH from the posterior pituitary often occurs in those situations in which an increase of blood pressure would be beneficial. For example, the hypotension that occurs in conjunction with *hypovolemia* (decreased blood volume) represents a stimulus that promotes the release of ADH. As a result, arterial constriction takes place and blood pressure is elevated. Accordingly, the hypotensive state is partially alleviated.

Tuberoinfundibular Tract. Most of the input to the pituitary through the tuberoinfundibular tract comes from small (parvicellular) neurons located in the arcuate nucleus and the *periventricular zone*. Neurons of the paraventricular, suprachiasmatic, tuberal, and medial preoptic nuclei also contribute to this tract (see Fig. 30–4). These axons convey various *releasing hormones* to the median eminence (the most inferior aspect of the tuberal area) and to the infundibulum of the pituitary gland. The substances are then released into a *primary plexus* of fenestrated capillaries (sinusoids), from which they are carried by *portal veins* to a *secondary plexus* of fenestrated capillaries in the pituitary (see Fig. 30–4). The releasing hormones of the hypothalamus include thyrotropin-releasing hormone, growth hormone–releasing hormone, growth hormone release inhibition hormone (somatostatin), corticotropin-releasing hormone, gonadotropin-releasing hormone, and prolactin-releasing hormone.

In the anterior lobe, the hypothalamic hormones regulate the functioning of hormone-producing adenohypophysial cells. The hormones of the adenohypophysis include *growth hormone* (primarily affecting the development of the musculoskeletal system), *gonadotropins* (affecting the ovary and testis), *corticotropin* (affecting the cortex of the adrenal gland), *thyrotropin* (affecting the thyroid gland), and *prolactin* (affecting milk production). Hormones leave the anterior pituitary via hypophysial veins and are distributed in the systemic circulation.

Pituitary Tumors

The pituitary gland is anatomically and functionally linked to the hypothalamus. Indeed, it is through the pituitary that many hypothalamic functions are expressed. Hormones and releasing hormones manufactured in the hypothalamus are transported via the supraopticohypophysial and tuberoinfundibular tracts and are released into the hypophysial portal system. Small lesions within the hypothalamus can block the manufacture and transportation of these substances, thereby adversely affecting pituitary functions. Similarly, tumors (*adenomas*) occurring within the pituitary can easily encroach on the neighboring hypothalamus and thus jeopardize its functions. It is therefore appropriate to briefly consider pituitary tumors at this point.

Although visual deficits are not a specific topic of this chapter (see Chapter 20), it should be noted that such deficits are frequently experienced by patients with pituitary tumors. These deficits may reflect damage to the optic nerve immediately rostral to the chi-

Figure 30–8. Gigantism resulting from overproduction of growth hormone. The normal adult male (*right*) is 6 feet 2.5 inches in height. The patient (*left*) was 7 feet 6 inches in height, wore size 17EEE shoes, and had a 12-inch span from the tip of his little finger to the tip of his thumb when his hand was spread.

asm, to the chiasm itself (uncrossed fibers or decussating fibers), or to the optic tract immediately caudal to the chiasm.

Tumors occurring within the pituitary gland account for 12% of primary brain tumors. In autopsy studies, the overall occurrence of incidental (undiagnosed) pituitary adenomas has been as high as 22.5% and as low as 3.2%, depending upon the thickness of sectioned pituitary glands. These tumors, occurring most frequently in young adults, are generally noncancerous. Incidental adenomas occur with increasing frequency in the fifth, sixth, and seventh decades of life.

Pituitary tumors can be classified according to their secretory characteristics, size, or biologic invasiveness. Secreting tumors produce an excess of one or more pituitary hormones; prolactin is the most commonly affected of these hormones. A second classification of these tumors is by size; *microadenomas* are tumors less than 1 cm in greatest dimension, whereas tumors greater than 1 cm across are referred to as *macroadenomas*. A third classification of these tumors is according to invasiveness; invasive tumors may erode and extend into the dura mater and even the sphenoid bone.

Nonsecreting pituitary tumors do not secrete hormones and as a result are often undiagnosed until they grow to a considerable size and exert pressure upon nearby structures (e.g., the optic chiasm and hypothalamus). Patients with these large pituitary tumors have

symptoms that often include visual disturbances (in 60% to 70% of patients) and headaches.

Secreting Tumors. In the clinical setting, secreting tumors are also commonly referred to as *hormonally active tumors* or *hypersecreting tumors*. The clinical manifestations of a secreting tumor are the effects of the biologic activity of the specific pituitary hormone that is overproduced. In cases of excessive production of *growth hormone*, patients experience uncontrolled growth in height referred to as *gigantism* (Fig. 30–8), if the growth occurs before closure of the epiphyseal plates of the long bones. These patients may have large muscles, but they are actually rather weak because these muscles contain excessive amounts of connective tissue rather than muscle fibers.

On the other hand, the overproduction of growth hormone after the growth plates have closed results in a condition termed *acromegaly*, referring to the enlargement of the digits of the patient (Fig. 30–9A). Patients with acromegaly have typical facial changes (see Fig. 30–9B), with elongation of the face, malocclusion of the jaw, and gaps in the lower dentition. Other changes include frontal and mastoid sinus bulges, a bulbous

A
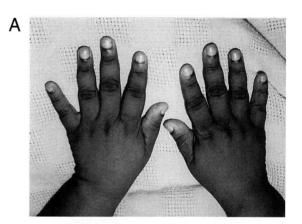

B

Figure 30–9. A female patient with acromegaly. Note the characteristic appearance of the enlarged hands (*A*) and the facial features (*B*).

nose, thickened lips, and very large hands and feet (see Fig. 30–9A and B). Associated with these external manifestations are excessive cortical thickening of bone, cardiac failure secondary to heart enlargement (*cardiomegaly*), as well as hypertension, and diabetes mellitus.

A much rarer form of pituitary tumor can result from the overproduction of *thyrotropin hormone*. The result may be either hypothyroidism or hyperthyroidism. Frequently, the patients suffer from abnormal cardiovascular function, as well as tremor. With the progression of these tumors, symptoms such as headaches, visual disturbances, and parasellar cranial nerve (i.e., cranial nerves III, IV, and VI) dysfunction may be recognized. The cranial nerve dysfunction generally involves the abducens nerve(s) first, and then the oculomotor and trochlear nerves, as the tumor enlarges in a lateral direction.

Cushing Disease. An overproduction of *corticotropin* leads to a form of hyperadrenalism known as *Cushing disease* (Fig. 30–10). In this instance, the excessive adrenocorticotropic hormone from the pituitary gland results in excessive adrenal cortisol secretion with the resultant classic clinical features of Cushing disease. In general, affected patients have central truncal obesity, including moonlike facies, facial hirsutism, and a dorsal cervical hump, commonly called a buffalo hump, which results from enlargement of the fat pad in this area (see Fig. 30–10A). The centropedal obesity has a hallmark finding of violaceous striae (stretch marks). These striae are clearly different from those seen in preg-

nancy or in excessive weight gain. Striae in Cushing disease (see Fig. 30–10B) are purple to violet, whereas striae in other situations are usually white. These patients also suffer from hyperpigmentation (see Fig. 30–10C), easy bruising, hypertension, osteopenia, and emotional lability.

Prolactin-Secreting Tumors. An overproduction of *prolactin* in women results in the syndrome of *galactorrhea* (*milk production*) (Fig. 30–11) and *amenorrhea* (*absence of menstrual periods*). *Hyperprolactinemia* is seen physiologically as part of a normal pregnancy. There are, however, other causes of hyperprolactinemia that require differentiation from either tumor or pregnancy (e.g., hypothyroidism, drug use). In men, hyperprolactinemia may be indicated by decreased libido, impotency, or infertility. This type of hypersecretory tumor is the most common pathologic cause of infertility.

Gonadotrope Tumors. These tumors consist of cells that produce either excessive luteinizing hormone (LH) or follicle-stimulating hormone (FSH). Excessive secretion of FSH causes no known symptoms in either men or postmenopausal women. Excessive LH secretion has been reported to cause premature pubertal changes (*precocious puberty*) in males and possibly disruption of the ovarian cyclicity in females. In most instances, however, gonadotrope adenomas come to clinical attention because of their mass effect, with visual impairment, headaches, and occasionally diplopia (*double vision*) caused by optic nerve compression due to lateral extension of the tumor.

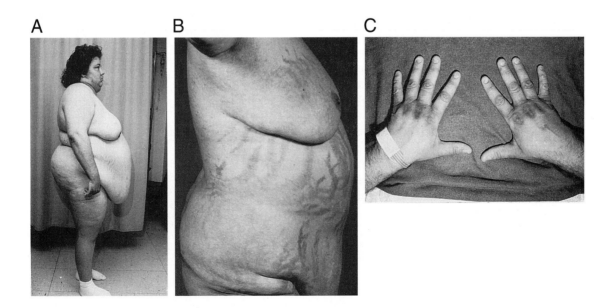

Figure 30–10. Cushing disease resulting from an overproduction of corticotropin. Note the characteristic central (truncal) obesity, the moonlike face, and the cervical hump (*A*). The striae in Cushing disease (*B*) can be very prominent and are purplish in color. Hyperpigmentation (*C*), here seen as dark oval areas over the knuckles, is also a characteristic feature.

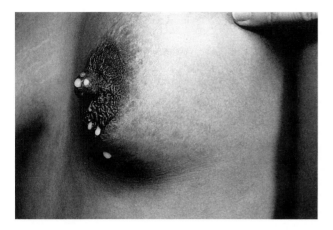

Figure 30–11. Milk production in a nonpregnant woman resulting from a prolactinoma.

Regional Functions of the Hypothalamus

Because of the interrelationships existing among the autonomic, endocrine, and limbic systems (see Fig. 30–1), and because of the small size of the hypothalamus, it is difficult to assign a specific function to each of its individual nuclei. Some hypothalamic nuclei participate in functions that are shared with other nuclei within the same vicinity. Consequently, these latter nuclei may participate in functions that are not uniquely their own. Instead, they may act in concert to affect functions that are attributable to hypothalamic *regions*, rather than specific nuclei.

Although the functions of some hypothalamic nuclei are largely known, the functions of others are poorly understood. Because of this and the fact that many hypothalamic nuclei participate in regional functions, some investigators believe that the hypothalamus should not be described as containing functional "centers," such as a "feeding center" or a "satiety center." Furthermore, because of the functional interrelationships that exist among certain nuclei within given hypothalamic regions, there is merit in thinking of the hypothalamus in terms of regional functions. Clinical observations are consistent with this view.

On the basis of clinical observations and experimental data, it is appropriate to divide the hypothalamus into two areas that share similar but opposing functions. These regions are the caudolateral and rostromedial areas of the hypothalamus (see Fig. 30–3). In general, the *caudolateral area* consists of the lateral hypothalamic zone and the mammillary region, and the *rostromedial area* consists of the supraoptic region and much of the tuberal regions (see Fig. 30–7).

Caudolateral Hypothalamus. *Activation* (stimulation) of the caudolateral hypothalamus produces behavioral manifestations that are generally associated with anxi-

ety. These include (1) increased activity of the *sympathetic* division of the visceromotor system, (2) increased *aggressive* behavior (see Chapter 31), (3) increased *hunger*, and (4) *increased body temperature* (resulting from cutaneous vasoconstriction and shivering).

A *lesion* in the caudolateral hypothalamus typically results in manifestations opposite to those caused by stimulation. For example, damage in this area results in the inhibition of sympathetic activities and the reduction of body temperature.

Rostromedial Hypothalamus. *Activation* (stimulation) of the rostromedial hypothalamus produces behavioral manifestations that are generally associated with contentment. These include (1) increased activity of the *parasympathetic* division of the visceromotor system, (2) increased *passive* behavior, (3) increased *satiety*, and (4) *decreased body temperature* (owing to cutaneous vasodilation and sweating).

The phenomenon of sweating is something of an oddity. Even though sweat glands are innervated by sympathetic nerve fibers, sweating is compatible with parasympathetic function in that it helps keep us cool. That the sympathetic terminals innervating most sweat glands are cholinergic, like the parasympathetic terminals innervating viscera, and that sweating can be elicited from the rostromedial hypothalamus support this observation.

Lesions in the rostromedial hypothalamus usually elicit behaviors opposite to those described for stimulation. For example, a lesion in this area results in the inhibition of parasympathetic activities and an increase in body temperature.

Hypothalamic Reflexes

All the vital functions of the hypothalamus, including the maintenance of blood pressure, body temperature, and water balance, are controlled through reflexes and are *typically* not subject to conscious control. However, through meditation or other means, some people can learn to alter certain hypothalamic responses. For example, *biofeedback* training enables some people to alter blood pressure and body temperature, which are generally under hypothalamic control. The neural mechanisms underlying these feats are largely unknown.

The internal environment is controlled partly through hypothalamic reflexes, which are typically mediated by the autonomic or endocrine systems (see Fig. 30–1). Three examples are described in the following sections.

Baroreceptor Reflex. The baroreceptor reflex (which is discussed in more detail in Chapter 19 and is illustrated in Fig. 19–8) regulates blood pressure in response to input from baroreceptors in the aortic arch and carotid sinus. These receptors are called extrinsic because they are outside the central nervous system. They sense variations in blood pressure and transmit the information to neurons in the solitary nucleus of the medulla. Neurons of the solitary nucleus project to and activate cells in the dorsal vagal nucleus, which in turn project to the terminal ganglia of the heart and influence heart rate. A blood pressure level above normal activates the solitary nucleus, leading to a decrease in heart rate and force of cardiac contraction, and consequently a decrease in blood pressure; a blood pressure level below normal has the opposite effect.

The hypothalamus is capable of influencing the baroreceptor reflex via a somewhat more complex pathway. The solitary nucleus contains cells that transmit baroreceptor information to the paraventricular, dorsomedial, and lateral hypothalamic nuclei. In turn, neurons in these hypothalamic areas project to the dorsal vagal nucleus of the medulla. By this route, the hypothalamus can powerfully modulate the responsiveness of the baroreceptor reflex.

Temperature Regulation Reflex. The reflex that maintains a constant body temperature depends on input from specialized temperature-sensing neurons in the hypothalamus (called *intrinsic receptors* because they are inside the central nervous system). When temperature of the blood reaching the hypothalamus rises above normal, these neurons stimulate regions in the *rostral hypothalamus* that are responsible for activating physiologic mechanisms for *heat dissipation*—sweating and cutaneous vasodilation. These effects are mediated by autonomic pathways. Conversely, when the blood temperature is below normal, the temperature sensors stimulate regions in the *caudal hypothalamus* that activate mechanisms for *heat conservation* (cutaneous vasoconstriction, mediated by autonomic pathways) and for *heat production* (shivering, mediated by reticulospinal pathways).

Water Balance Reflex. As noted previously, the volume and osmolarity of the blood are kept constant by reflex mechanisms. Unlike the baroreceptor and temperature regulation reflexes, which are entirely neural, the water balance reflex is *neurohumoral;* its efferent limb consists of a hormonal signal carried by ADH. In brief, the reflex works as follows. The osmolarity of the blood is monitored by specialized osmolarity-sensitive neurons located in the anterior hypothalamus near the preoptic and paraventricular nuclei. The output from these receptors influences the release of ADH by ADH-producing neurons in the supraoptic and paraventricular nuclei. When blood osmolarity is too high, more ADH is released, and water resorption from the collecting tubules of the kidney is increased. When blood osmolarity is too low, the release of ADH is inhibited, and renal water resorption is reduced. Accordingly, water is not resorbed into the blood but stays in the urine.

Sources and Additional Reading

Haymaker W, Anderson E, Nauta WJH: The Hypothalamus. Charles C Thomas, Springfield, Ill, 1969.

Koizumi K, Kollai M, Oomura Y, Yamashita H, Wayner MJ (eds): The hypothalamus: Selected topics. Brain Res Bull 20:651–902, 1988.

Renaud LP: A neurophysiological approach to the identification, connections and pharmacology of the hypothalamic tuberoinfundibular system. Neuroendocrinology 33:186–191, 1981.

Saper CB, Lowey AD, Swanson LW, Cowan WM: Direct hypothalamo-autonomic connections. Brain Res 117:305–312, 1976.

Swanson LW, Sawchenko PE: Hypothalamic integration: Organization of the paraventricular and supraoptic nuclei. Annu Rev Neurosci 6:269–324, 1983.

Ter Horst GJ, de Boer P, Luiten PGM, van Willigen JD: Ascending projections from the solitary tract nucleus to the hypothalamus. A *Phaseolus vulgaris* lectin tracing study in the rat. Neuroscience 31: 785–797, 1989.

van der Kooy D, Koda LY, McGinty JF, Gerfin CR, Bloom FE: The organization of projections from the cortex, amygdala, and hypothalamus to the nucleus of the solitary tract in rat. J Comp Neurol 224:1–24, 1984.

Chapter 31

The Limbic System

R. B. Chronister and S. G. P. Hardy

Overview 494

Cytoarchitectural Definitions of the Limbic Cortex 494

Early Functional Concepts 494

Blood Supply to the Limbic System 496

The Hippocampal Formation 496
Structure
Afferent Fibers
Efferent Fibers
The Complete Circuit of Papez
Dysfunctions and Korsakoff Syndrome

Long-Term Potentiation and Memory 500

The Amygdaloid Complex 500
Structure
Afferent Fibers
Efferent Fibers
Klüver-Bucy Syndrome

The Septal Region 503

The Nucleus Accumbens 504

The Limbic System and Emotions 504

The Limbic System and Cognitive Function 504

Many complex brain systems are organized in a way that allows their function to be readily deduced. For example, even though the connections of the somatosensory pathways with the brainstem, thalamus, and cortex are complex, each component plays a fairly clear-cut role. The processing of somatosensory information is generally well understood. In contrast, some systems are interconnected in such a way that a given function may be carried out by several components acting in cooperation, and a given component may participate in several functions. The *limbic system* is a case in point. This system comprises structures that receive inputs from diverse areas of the neuraxis and participate in complicated and interrelated behaviors such as memory, learning, and social interactions. Thus, lesions involving the limbic system generally result in a wide range of deficits.

Overview

The concept of a *"limbic system"* actually encompasses two levels of structural and functional organization. The *first level* consists of the cortical structures on the most medial edge (the limbus) of the hemisphere; these collectively form the *limbic lobe* (Fig. 31–1*A*). Beginning just anterior to the lamina terminalis and proceeding caudally, these are the *subcallosal area*, containing the parolfactory and paraterminal gyri; the *cingulate gyrus*; the *isthmus of the cingulate gyrus*; the *parahippocumpal gyrus*; and the *uncus* (Fig. 31–2). The limbic lobe also includes the *hippocampal formation*.

In 1878, Broca noted that the limbic lobe, which is present in all mammals, represents a relatively large part of the cerebral cortex in phylogenetically lower forms, and he postulated that it might be related to olfaction. Because of this latter point, the term *rhinencephalon* ("smell-brain") was later coined, and used interchangeably with *limbic lobe*. However, it is now known that the limbic lobe has little olfactory function in humans. Thus, the term "rhinencephalon" is antiquated and has largely disappeared from use.

The *second level* includes structures of the limbic lobe plus a variety of subcortical nuclei and tracts that collectively form the *limbic system* (see Fig. 31–1*B*). The subcortical nuclei of the limbic system include, among others, the *septal nuclei* and *nucleus accumbens* (*nucleus accumbens septi*); various nuclei of the hypothalamus, especially those associated with the *mammillary body*; the nuclei of the *amygdaloid complex* and adjacent *substantia innominata*; and parts of the dorsal thalamus, particularly the *anterior* and *dorsomedial nuclei*. Additional structures connected with the limbic system include the *habenular nuclei*, *ventral tegmental area*, and *periaqueductal gray*. Furthermore, the *prefrontal cortex* is considered by some investigators to be an important component of the limbic system, primarily because of its potential influence on various other cortical and subcortical parts of the limbic system. Cortical targets of the prefrontal cortex include the cingulate gyrus, whereas the hypothalamus, dorsal thalamus, amygdaloid complex, and nuclei of the midbrain represent subcortical targets.

The main efferent fiber bundles of the limbic system are the *fornix* (primarily efferents of the hippocampus and subiculum), the *stria terminalis* and *ventral amygdalofugal pathway* (both are mainly efferents of the amygdaloid complex), and the *mammillothalamic tract* (efferents of the medial mammillary nucleus (see Fig. 31–1*B*). A few additional nuclei and smaller tracts will be introduced as connections and functions of the limbic system are described.

Cytoarchitectural Definitions of the Limbic Cortex

The human cerebral cortex can be divided into several areas on the basis of the number of cell layers present. Most of the cerebral cortex (more than 90%) has six cell layers and is called the *neocortex* or *neopallium* (*isocortex*). Examples of neocortex include the primary sensory, motor, and association cortices. The cortical regions that have less than six layers are structurally and functionally associated with the limbic system or with olfaction and are classified as *allocortex*. Those structures that comprise three to five cellular layers are called the *paleocortex* (*paleopallium* or *periallocortex*) and are represented by the cortex of the parahippocampal gyrus (the *entorhinal cortex*), the uncus (the *piriform cortex*), and the cortex overlying the termination of the lateral olfactory stria (lateral olfacrory gyrus) (see Fig. 31–2). The lateral olfactory stria is directly rostromedial to the piriform cortex. Structures having only three cellular layers are classified as *archicortex* (*archipallium* or *allocortex*) and are represented by the dentate gyrus and hippocampus.

The separation between neocortex and allocortex is never sharp but instead consists of transitional areas where one cortical region blends into the next. Such areas are represented by caudal parts of the orbitofrontal cortex, the temporal pole, parts of the insula, and portions of the parahippocampal and cingulate gyri. They are especially important because they funnel input from association areas of the neocortex into the allocortex.

Early Functional Concepts

In the late 1930s, two pivotal observations were made that formed the basis for the concept of a limbic system. First, based largely on the morphology of the brain, a circuit for the elaboration of emotion was pro-

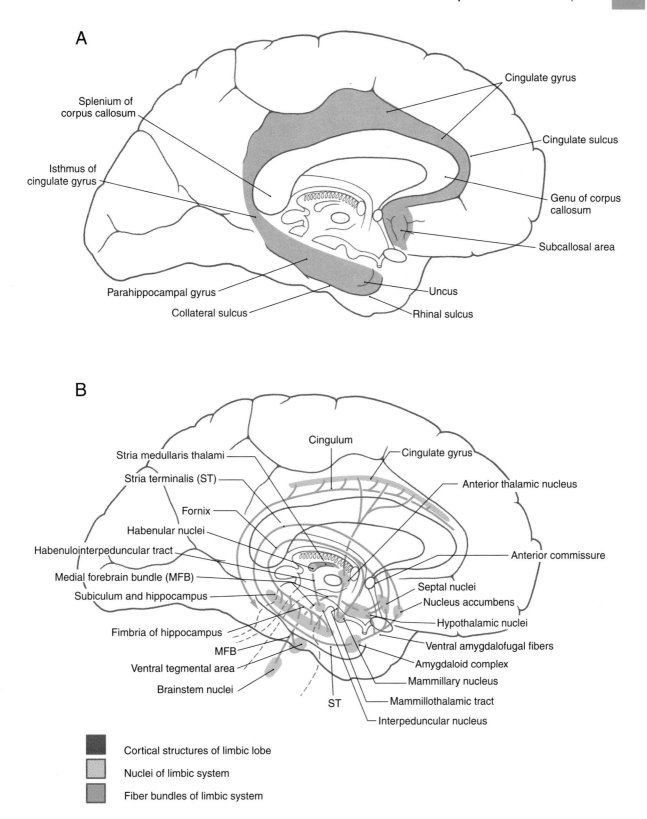

Figure 31–1. Cortical structures forming the limbic lobe *(A)* and the subcortical structures *(B,* nuclei in red, fiber bundles in blue) which, with the lobe, represent the main components of the limbic system.

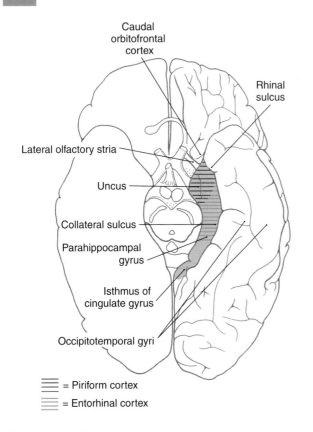

Caudal
orbitofrontal
cortex

Rhinal
sulcus

Lateral olfactory stria

Uncus

Collateral sulcus

Parahippocampal
gyrus

Isthmus of
cingulate gyrus

Occipitotemporal gyri

☰ = Piriform cortex

☰ = Entorhinal cortex

Figure 31–2. Cortical structures of the limbic lobe as seen on an anterior (ventral) view of the hemisphere.

posed. This pathway is now called the *Papez circuit* in recognition of James Papez, who initially described its components. This model, although surprisingly simple, has proved to be quite important in understanding limbic function. The circuit suggested that emotion, mediated through the hypothalamus, is controlled and modulated by fibers from the fornix. Specifically, the cortical control of emotional activity is presumed to originate from cingulate and hippocampal regions. These cortical influences are ultimately conveyed to the mammillary body of the hypothalamus via the fornix. In turn, the medial mammillary nucleus, via the mammillothalamic tract, projects to the anterior nucleus of the thalamus, and this cell group sends axons to the cingulate gyrus. This was the first time a specific anatomic substrate was proposed for a phenomenon as complex as emotion.

The second pivotal observation was that bilateral removal of large parts of the temporal lobe in monkeys resulted in a constellation of dysfunctions that came to be known as the *Klüver-Bucy syndrome*. This syndrome can also be caused in humans by temporal lobe injuries that involve primarily the amygdaloid complex. Its nature and significance are discussed later in the chapter.

Blood Supply to the Limbic System

The blood supply to the limbic system originates from several sources. The main vessels that serve much of the limbic system are the *anterior* and *posterior cerebral arteries*, the *anterior choroidal artery*, and branches arising from the *circle of Willis*.

The subcallosal area and rostral parts of the cingulate gyrus are supplied by branches of the anterior cerebral artery as it loops around the genu of the corpus callosum (see Fig. 8–3). Most of the cingulate gyrus and its isthmus receives its blood supply via the *pericallosal artery*, a branch of the anterior cerebral artery. *Temporal branches* of the posterior cerebral artery supply the parahippocampal gyrus. Although the uncus may receive some small branches from the posterior cerebral artery, it is served primarily by *uncal arteries*, which are branches of the M_1 segment of the middle cerebral artery (see Fig. 8–2).

The anterior choroidal artery usually originates from the internal carotid artery and follows the general trajectory of the optic tract. En route it sends branches into the choroidal fissure of the temporal horn of the lateral ventricle. This vessel serves the choroid plexus of the temporal horn, the hippocampal formation, parts of the amygdaloid complex, and adjacent structures such as the tail of the caudate nucleus, the stria terminalis, and the sublenticular and retrolenticular limbs of the internal capsule.

Vessels serving hypothalamic nuclei that are functionally associated with the limbic system originate from the circle of Willis. In general, rostral areas of the hypothalamus are served by branches from the anterior communicating artery and anterior cerebral artery, and posterior areas, by branches from the posterior communicating artery and proximal posterior cerebral artery (see Fig. 8–13). The anterior nucleus of the thalamus, an important synaptic station in the limbic system, is supplied by *thalamoperforating arteries* that arise from the P_1 segment of the posterior cerebral artery.

The Hippocampal Formation

The *hippocampal formation* is composed of the *subiculum, hippocampus* (also called the hippocampus proper or horn of Ammon), and the *dentate gyrus* (see Fig. 31–4), all of which constitute the allocortex of Brodmann. The subiculum is laterally continuous with the cortex of the parahippocampal gyrus and area of the periallocortex. Medially the edge of the hippocampal formation is formed by the dentate gyrus and the fimbria of the hippocampus.

Developmentally, the hippocampal formation originates dorsally and migrates into its ventral and medial position in the temporal lobe. During this migration, small remnants of the hippocampal formation remain behind to form the *medial* and *lateral longitudinal striae* and their associated gray matter, the *indusium griseum* (Fig. 31–3). These structures are quite small in the human brain and extend rostrally along the dorsal aspect of the corpus callosum into the subcallosal area.

Structure. The *subiculum* of the hippocampal formation is the transitional area between the three-layered hippocampus (archicortex or allocortex) and the five-layered entorhinal cortex (paleocortex or periallocortex) of the parahippocampal gyrus (Fig. 31–4). This transitional zone, although small, can be divided into a prosubiculum, subiculum proper, presubiculum, and parasubiculum. These areas are essential for the flow of information into the hippocampal formation.

The dentate gyrus and the hippocampus are each composed of three layers, which are characteristic of archicortex (see Fig. 31–4). The external layer is called the *molecular layer* and contains afferent axons and den-

drites of cells intrinsic to each structure. The middle layer, called the *granule cell layer* in the dentate gyrus and the *pyramidal layer* in the hippocampus, contains the efferent neurons of each structure (see Fig. 31–4). These layers are named according to the shape of the cell body of the principal type of neuron found therein. The dendrites of granule and pyramidal cells radiate into the molecular layer. The inner layer, called the *polymorphic layer* (also called the stratum oriens in the hippocampus), contains the axons of pyramidal and granule cells, a few intrinsic neurons, and many glial elements. In addition, the polymorphic layer of the hippocampus contains the elaborate basal dendrites of some larger pyramidal somata that are located in the pyramidal layer. These are called *double pyramid cells* because they have dendrites extending into both molecular and polymorphic layers (see Fig. 31–4). The innermost part of the hippocampus borders on the wall of the lateral ventricle and is a layer of myelinated axons arising from cell bodies located in the subiculum and hippocampus. This layer, called the *alveus*, is continuous with the *fimbria of the hippocampus*, which, in turn, becomes the *fornix* (see Fig. 31–4).

The hippocampus can be divided into four regions on the basis of a variety of cytoarchitectural criteria (see Fig. 31–4). These areas are designated CA1 to CA4, where *CA* stands for cornu ammonis (horn of Ammon). Area CA1 (a parvocellular region that can be separated into two cell layers in humans) is located at the subiculum-hippocampal interface. Area CA2 (a mixed zone) and area CA3 (a magnocellular zone) are located within the hippocampus. Area CA4 is located at the junction of the hippocampus with the dentate gyrus, but within the hilus of the dentate gyrus.

Afferent Fibers. The major input to the hippocampal formation is from cells of the entorhinal cortex via a diffuse projection called the *perforant pathway* (Fig. 31–5; see also Fig. 31–4). Most fibers of the perforant pathway terminate in the molecular layer of the dentate gyrus, although a few terminate in the subiculum and hippocampus. Granule cells in the dentate gyrus project into the molecular layer of the CA3 region of the hippocampus. CA3 neurons project into CA1 of the hippocampus, which, in turn, provides an input to the subiculum. In addition, the subiculum also receives a modest projection from the amygdaloid complex. Although the fornix is mainly an efferent path from the hippocampus, it also conveys cholinergic septohippocampal projections to the hippocampal formation and entorhinal cortex.

Efferent Fibers. The outflow of the hippocampal formation originates primarily from cells of the subiculum and, to a lesser degree, from pyramidal cells of the hippocampus (see Figs. 31–4 and 31–5). In both cases, axons of these neurons enter the alveus, coalesce to form the fimbria of the hippocampus, and then continue as the fornix. These glutaminergic fibers traverse

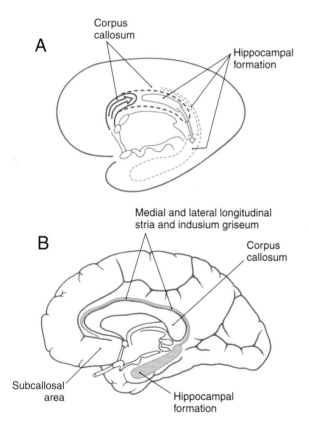

Figure 31–3. Developmental relationships between the corpus callosum (in red) and the hippocampal formation (in green). As the corpus callosum expands caudally from the general area of the anterior commissure *(A)*, the hippocampal primordium migrates into the temporal lobe. In the adult brain, remnants of the hippocampal formation left behind during development are located dorsal to the corpus callosum *(B)*.

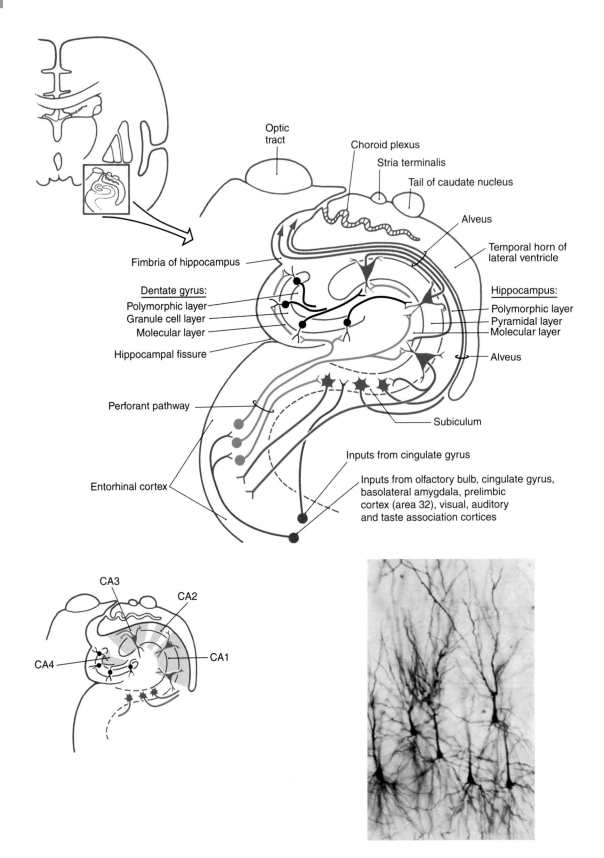

Figure 31-4. The basic structure of the hippocampal formation and its relation to adjacent structures. The cell types of the dentate gyrus, hippocampus, and subiculum are shown diagrammatically. The general locations of fields C1 to C4 are shown on the lower left; a Golgi stain of double pyramid cells is shown on the lower right. (Golgi stain courtesy of Dr. José Rafols, Wayne State University.)

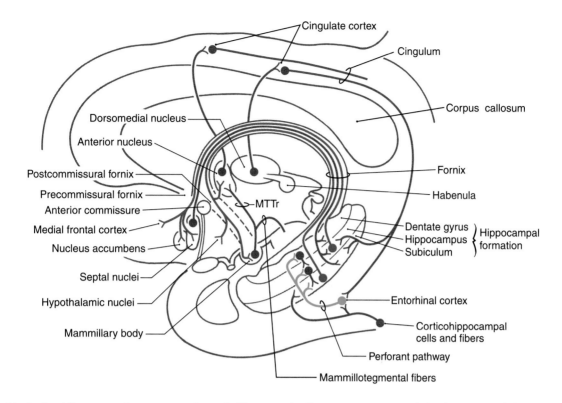

Figure 31–5. Semidiagrammatic representation of afferent and efferent connections of the hippocampal formation. MTTr, mammillothalamic tract.

the entire extent of the fornix, although some cross the midline in the *hippocampal decussation* just anterior (ventral) to the splenium of the corpus callosum.

At the level of the anterior commissure, the fornix divides into postcommissural and precommissural parts (see Fig. 31–5). The fibers originating in the subiculum mainly form the *postcommissural fornix*. Most of these fibers terminate in the *medial mammillary nucleus,* although some enter the ventromedial nucleus of the hypothalamus and the anterior nucleus of the dorsal thalamus (see Fig. 31–5). The *precommissural fornix* is composed of fibers arising primarily in the hippocampus. These fibers are somewhat diffusely organized and distribute to the septal nuclei, the medial areas of the frontal cortex, the preoptic and anterior nuclei of the hypothalamus, and the nucleus accumbens (see Fig. 31–5).

The Complete Circuit of Papez. As noted previously, the initial segment of the *Papez circuit* is a projection primarily from the subiculum, to the medial mammillary nucleus via the postcommissural fornix. The circuit is completed by the following connections (see Fig. 31–5): (1) a *mammillothalamic tract* that connects the *medial mammillary nucleus* to the *anterior nucleus* of the thalamus; (2) *thalamocortical fibers* from the anterior nucleus to broad expanses of the cortex of the *cingulate gyrus;* and (3) a projection from the cingulate

cortex, via the cingulum, to the entorhinal cortex and also directly to the subiculum and hippocampus. The subiculum returns information to the mammillary body.

Other areas of the cerebral cortex are recruited into the various functions associated with the Papez circuit largely through connections of the cingulate gyrus. For example, the cingulate receives input from premotor and prefrontal areas and from visual, auditory, and somatosensory association cortices. In turn, the cingulate cortex not only is a major source of afferent fibers to the hippocampal formation but also projects to most cortical areas from which it receives input. The cingulate gyrus thus is not only an integral part of the Papez circuit but also an important conduit through which a wide range of information can reach the limbic system.

Dysfunctions and Korsakoff Syndrome. The basic function of the hippocampal formation appears to be the consolidation of long-term memories from immediate and short-term memories—a point first observed in 1900 by Bechterew but essentially ignored until the 1950s. *Immediate memory* and *short-term memory* refer to types of memory that persist for seconds and minutes, respectively. Normally, these memories can be incorporated into long-term memory, which can be recalled days, months, or years later. However, in persons with hippocampal lesions, this conversion is not accomplished. Although patients may be able to perform a task for seconds or minutes, if distracted from the task they are unable to return to it. In other words, they do not "remember" what they were doing; the short-term experience is not incorporated into long-term memory. The redundancy and feedback in the hippocampus are ideal for this imprinting of memory.

One condition in which loss of memory and cognitive function is particularly obvious is *Alzheimer disease*. This disease is characterized, in part, by the presence of neurofibrillary tangles, neuritic plaques, and neuronal loss in specific brain regions. The subiculum and entorhinal cortices are among the first sites in which these abnormalities appear. As a result, the relay of information through the hippocampal formation is markedly impeded. It is believed that this damage is at least partially responsible for the memory deficits characteristic of Alzheimer disease.

As mentioned previously, the *Korsakoff syndrome* (Korsakoff psychosis) is a condition that is caused by prolonged thiamine deficiency and is typically seen in chronic alcoholics. The thiamine deficiency causes a characteristic pattern of degeneration in the brain. Typically, the mammillary bodies are involved, with some incursion into the dorsomedial nucleus of the thalamus and the columns of the fornix. There is also a loss of neurons in the hippocampal formation. These patients show a defect in short-term memory and consequently also in long-term memory for events occurring since the onset of disease. They may appear demented, and they are prone to *confabulation;* that is, they tend to string together fragments of memory from several different events to form a synthetic "memory" of an event that never occurred. In some chronic alcoholics, the memory loss and general confusion are accompanied by gaze palsies and ataxia, which occur secondary to cerebellar damage. When these deficits accompany profound memory losses and learning difficulties, the condition is called *Wernicke-Korsakoff syndrome.*

Bilateral damage to the hippocampal formation sometimes occurs in victims of heart attack or near-drownings as a result of transient cerebral ischemia. The part of the hippocampal formation most vulnerable to anoxia during an ischemic episode is the CA1 area. The CA1-subiculum interface region is referred to as *Sommer's sector* in pathologic conditions. Affected patients retain their long-term memories from the time before the ischemic event, but they are unable to retain memories for events occurring after the ischemic episode. Consequently, they also have difficulty learning new skills because the new information is not retained (remembered) long enough to become a long-term memory. Damage to Sommer's sector is also related to epilepsy. It is known that malfunction of the hippocampus is indeed related to the genesis of epileptic activity.

Bilateral lesions of the anterior part of the cingulate gyrus greatly diminish the emotional responses of the patient and may result in *akinetic mutism.* This is a state in which the patient is immobile, mute, and unresponsive but not in a coma. Other patients with cingulate damage may be alert but have no idea of who they are. Patients may also be unable to recall the order in which past events occurred.

Long-Term Potentiation and Memory

The process of *long-term potentiation* at individual synapses is the probable mechanism that underlies the consolidation of short-term into long-term memory. When several synapses are present on a single cell, the input from these synapses is *integrated;* that is, the small potential changes (epsp and ipsp, see Chapter 3) are added together. In long-term potentiation, one synapse fires in a particular temporal pattern, (such as bursts or trains of action potentials). This synaptic activity increases the likelihood that the target cells will be activated by that synapse and other synapses. This increased likelihood may be due to an increased probability that transmitter will be released from the presynaptic cell, or to an increased response in the postsynaptic cell to the same amount of neurotransmitter, or to both. Long-term potentiation has been demonstrated at terminals of the perforant pathway in the dentate gyrus and at the synapses of CA3 pyramidal cells on CA1 cells. These connections use the neurotransmitter glutamate.

According to one current model, the release of glutamate causes a change in the biochemistry of the *N*-methyl-D-aspartate (NMDA)-type glutamate receptors of the hippocampal cells, allowing an increased number of calcium ions to enter the cell. The calcium influx causes a second postsynaptic biochemical change: The gaseous neuromodulator nitric oxide is released and diffuses back to the presynaptic terminal. It acts on the presynaptic terminal to permanently increase the release of glutamate. At this type of synapse, *a brief, sustained increase in current synaptic activity increases the probability that future synaptic activity will take place.* That is, the more the circuit is activated, the easier it is to activate. This mechanism causes stimuli and responses to be paired in the process we call "memory." The increased probability of activation lasts in isolated preparations for hours; it cannot be measured in human brains but may be permanent.

The Amygdaloid Complex

Structure. The amygdaloid complex is an almond-shaped group of cells in the rostromedial part of the temporal lobe internal to the uncus (see Fig. 31–1*A* and *B*). It is immediately rostral to the hippocampal formation and the anterior end of the temporal horn of the lateral ventricle. The amygdaloid complex is composed of a number of nuclei. For our purposes, these nuclei can be grouped into a larger *basolateral group* and a smaller *corticomedial group* (including the *central nucleus*). The latter group is more closely related to olfaction, whereas the former has extensive interconnections with cortical structures.

Afferent Fibers. The basolateral cell groups of the amygdala receive inputs from the dorsal thalamus, the prefrontal cortex, the cingulate and parahippocampal gyri, the temporal lobe and insular cortex, and the subiculum (Fig. 31–6A). These fibers supply a wide range of somatosensory, visual, and visceral information to the amygdaloid complex.

The corticomedial cell group receives olfactory input, fibers from the hypothalamus (ventromedial nucleus, lateral hypothalamic area), and fibers from the dorsomedial and medial nuclei of the dorsal thalamus (see Fig. 31–6A). In addition, this cell group, particularly its central nucleus, receives ascending input from nuclei in the brainstem known to be involved in visceral functions. Among others, these include the parabrachial nuclei, the solitary nucleus, and portions of the periaqueductal gray.

Efferent Fibers. The two major efferent pathways of the amygdaloid complex are the stria terminalis and the ventral amygdalofugal pathway (see Figs. 31–1B and 31–6B). The *stria terminalis* is a small fiber bundle that arises primarily from cells of the corticomedial group. Through most of its course, this bundle lies in the groove between the caudate nucleus and the dorsal thalamus, where it is accompanied by the terminal vein. It is associated along its length with discontinuous aggregations of cells, which collectively are called the *bed nucleus of the stria terminalis*. This tract distributes to various nuclei of the *hypothalamus* (the preoptic nuclei, ventromedial nucleus, anterior nucleus, and lateral hypothalamic area), to the *nucleus accumbens* and the *septal nuclei*, and to the rostral areas of the caudate nucleus and putamen (see Fig. 31–6B).

The *ventral amygdalofugal pathway* is the major efferent fiber bundle of the amygdaloid complex. These fibers arise from the basolateral cell group and the central nucleus of the corticomedial cell group and follow two general trajectories (see Fig. 31–6B). Axons primarily from the basolateral cells pass medially through the substantia innominata (in which some of these fibers terminate) to eventually synapse in the hypothalamus and septal nuclei. The substantia innominata gives rise to a diffuse cholinergic projection to the cerebral cortex. It is probable that these fibers play a role in the activation of the cerebral cortex in response to behaviorally significant stimuli. In addition, cells of the basolateral group also project diffusely to frontal and prefrontal, cingulate, insular, and inferior temporal cortices (see Fig. 31–6B). Other fibers, mainly from the central nucleus, turn caudally and descend diffusely in the brainstem to terminate in visceral (dorsal motor vagal) nuclei, raphe nuclei (magnus, obscurus, pallidus), and other areas such as the locus ceruleus, parabrachial nuclei, and periaqueductal gray (see Fig. 31–6B). As noted previously, most of those brainstem areas that receive input from the amygdala project back to this structure.

Another route by which hippocampal and amygdaloid efferents influence the brainstem is through the *stria medullaris thalami*. This bundle conveys fibers from the septal nuclei (targets of amygdaloid and hippocampal inputs) to the *habenular nuclei* (Fig. 31–7A). The latter cell groups, in turn, give rise to the *habenulointerpeduncular tract*, which projects to the interpeduncular nucleus and other midbrain sites, including the ventral tegmental area and periaqueductal gray.

Klüver-Bucy Syndrome. As mentioned previously, bilateral temporal lobe lesions that largely abolish the amygdaloid complex cause a set of behavioral changes called the *Klüver-Bucy syndrome*. These deficits were initially described in a series of animal experiments, but they have also been seen in patients as a result of either trauma to the temporal lobe or temporal lobe surgery for epilepsy. Damage to the amygdaloid complex frequently involves portions of adjacent structures and of the surrounding white matter, and these incursions into other structures may contribute to the clinical picture. Damage to the amygdala and the hippocampus results in a greater memory deficit than the deficit noted with damage to either one alone.

The Klüver-Bucy syndrome is characterized by the following deficits. First, the patient is no longer able to recognize objects by sight (*visual agnosia*) and may also exhibit *tactile* and *auditory agnosia*. Second, there is the tendency to examine objects excessively by mouth (*hyperorality*) or to smell them. Even a harmful object such as a lit match may be examined by being brought to the lips or touched with the tongue. Third, the patient may have a compulsion to intensively explore the immediate environment (*hypermetamorphosis*) and to overreact to visual stimuli. Fourth, *placidity* is characteristically seen. The animal or the patient may no longer show fear or anger, even when such a reaction is appropriate. Fifth, the subject may eat in excessive amounts (*hyperphagia*), even when not hungry, or may eat objects that are not food, or food that is inappropriate to the species. For example, a monkey may eat raw meat, or a patient may eat leaves. Sixth, there is a striking augmentation in sexual behavior (*hypersexuality*). In humans, this takes the form of suggestive behavior, talk, and vague, ill-conceived attempts at sexual contact. In addition to these predictable deficits, these patients may also experience *amnesia*, *dementia*, or *aphasia*, depending on the extent of the lesion of the temporal lobe.

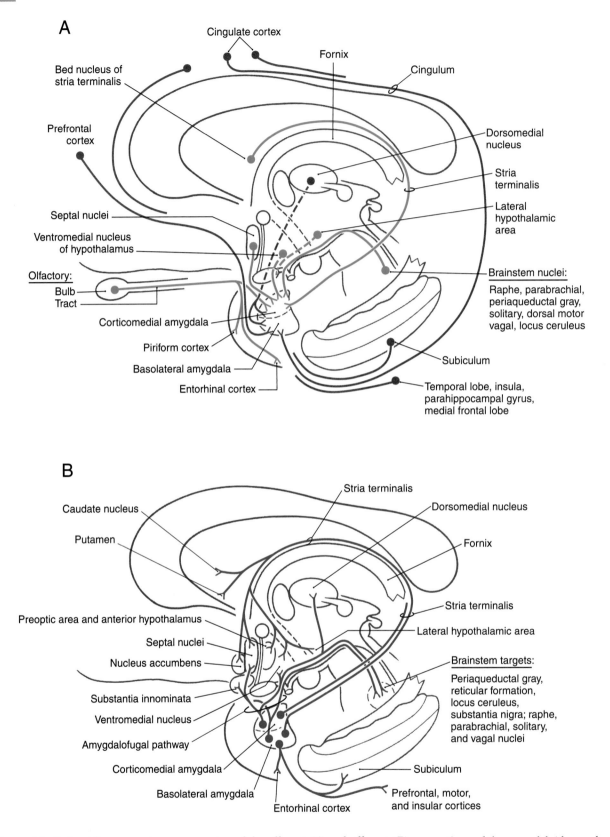

A

Cingulate cortex

Fornix

Cingulum

Bed nucleus of stria terminalis

Dorsomedial nucleus

Prefrontal cortex

Stria terminalis

Lateral hypothalamic area

Septal nuclei

Ventromedial nucleus of hypothalamus

Olfactory:
Bulb
Tract

Brainstem nuclei:
Raphe, parabrachial, periaqueductal gray, solitary, dorsal motor vagal, locus ceruleus

Corticomedial amygdala

Piriform cortex

Basolateral amygdala

Subiculum

Entorhinal cortex

Temporal lobe, insula, parahippocampal gyrus, medial frontal lobe

B

Stria terminalis

Dorsomedial nucleus

Caudate nucleus

Fornix

Putamen

Stria terminalis

Preoptic area and anterior hypothalamus

Lateral hypothalamic area

Septal nuclei

Nucleus accumbens

Brainstem targets:
Periaqueductal gray, reticular formation, locus ceruleus, substantia nigra; raphe, parabrachial, solitary, and vagal nuclei

Substantia innominata

Ventromedial nucleus

Amygdalofugal pathway

Corticomedial amygdala

Subiculum

Basolateral amygdala

Prefrontal, motor, and insular cortices

Entorhinal cortex

Figure 31–6. Semidiagrammatic representation of the afferent *(A)* and efferent *(B)* connections of the amygdaloid complex.

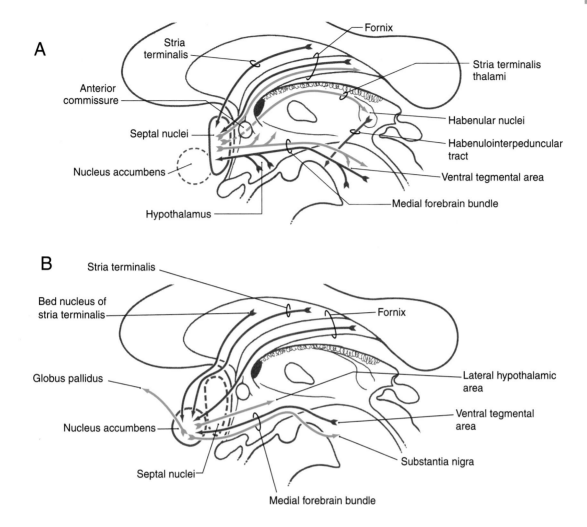

Figure 31–7. Summary of the main afferent (red arrows) and efferent (green arrows) connections of the septal nuclei *(A)* and of the nucleus accumbens *(B)*.

The Septal Region

The septal region (septal nuclei), excluding the nucleus accumbens, is a small area just rostral to the anterior commissure and in the medial wall of the hemisphere (see Figs. 31–1*B* and 31–7*A*). These nuclei extend into the base of the septum pellucidum. Despite their relatively small size, the septal nuclei have been implicated in a myriad of functions in animal models on the basis of the patterns of their inputs and outputs. In contrast, there is little clinical information regarding their function in humans. Rage behavior has been seen in a small group of patients with midline infarcts in this area.

The principal afferent pathways to the septal nuclei include fibers from the hippocampus (via the fornix), the amygdaloid complex (via the stria terminalis and ventral amygdalofugal pathways), and the ventral tegmental area of the midbrain (see Fig. 31–7*A*). Fibers

also originate from the preoptic, anterior, and paraventricular hypothalamic nuclei and from the lateral hypothalamic area. Many of the fibers in the stria terminalis and fornix also send branches into the nucleus accumbens.

The main efferent projections from the septal nuclei (see Fig. 31–7*A*) are septohippocampal fibers (in the fornix), projections to the habenular nuclei, the medial thalamic nuclei (via the stria medullaris thalami), and the ventral tegmental area (via the medial forebrain bundle). The preoptic, anterior, and ventromedial nuclei and the lateral hypothalamic areas also receive input from the septal nuclei.

The *medial forebrain bundle* is a diffuse group of fibers that courses rostrocaudally through the lateral hypothalamic area (see Fig. 31–7*A*). This bundle is complex in that it conveys ascending inputs into the hypothalamus and through this area into the septal region. It also is a major conduit through which the septal nuclei and portions of the hypothalamus communicate with the brainstem (see Fig. 30–7). The dopamine-containing fibers in this area are thought to be related to perceptions of pleasure or drive reduction.

The Nucleus Accumbens

The nucleus accumbens (nucleus accumbens septi) is located in the rostral and ventral forebrain where the head of the caudate nucleus and the putamen are continuous (see Fig. 26–3). These cells receive input from the amygdaloid complex (primarily via the ventral amygdalofugal pathway), from the hippocampal formation (through the precommissural fornix), and from cells of the bed nucleus of the stria terminalis (see Fig. 31–7B). The ventral tegmental area also gives rise to ascending fibers that enter the nucleus accumbens via the medial forebrain bundle. In addition, amygdalofugal fibers traversing the stria terminalis also enter the nucleus accumbens.

Cells within the nucleus accumbens have receptors for a variety of neurotransmitters, including endogenous opiates. Studies in animal models indicate that the nucleus accumbens may play an important role in behaviors related to addiction.

Efferent projections of the nucleus accumbens include fibers to the hypothalamus, nuclei of the brainstem, and the globus pallidus (see Fig. 31–7B). Nucleus accumbens fibers to the latter target represent an important route through which the limbic system may access the motor system.

The Limbic System and Emotions

In recent years, the term *limbic system* has been used mainly in reference to emotion-related areas of the brain and the pathways that interconnect them. These areas are generally composed of sites that function to alter the emotions. These sites, which are often interspersed in a given region of the brain, are frequently called either *aversion centers* or *gratification centers*. If an aversion center is stimulated, the person will experience fear or sorrow. On the other hand, stimulation of a gratification center will result in pleasure. Functional interconnections between aversion and gratifications centers probably contribute to emotional stability.

Although most limbic structures contain both gratification and aversion centers, in some structures one or the other type of center seems to predominate. For example, the hippocampus and amygdala have an abundance of aversion centers, whereas the nucleus accumbens contains an abundance of gratification centers. Consequently, stimulation of the amygdala may elicit fear, whereas stimulation of the nucleus accumbens results in feelings of joy or pleasure.

The emotion-related deficits resulting from small lesions in the limbic system are difficult to predict. However, the effects of relatively large lesions are more stereotypic. They typically result in the flattening of emotions, as reflected by the fact that emotional extremes (joy and anxiety) are reduced. This phenomenon, presumably resulting from the loss of both aversion and gratification centers, commonly results from large lesions in the amygdala, hippocampus, fornix, or cingulate or prefrontal cortices.

The Limbic System and Cognitive Function

There is a trend to look at the limbic system as a set of structures that influence not only emotion per se but also cognitive functions. The area on which there is the most agreement in this regard is memory. Other influences, however, are mediated through the nucleus basalis and are undoubtedly related to the control of cortical excitability. The full contribution of this upstream control is related to the transfer of information from limbic structures to the cerebral cortex.

Sources and Additional Reading

Hodges JR, Patterson K: Is semantic memory consistently impaired early in the course of Alzheimer's disease? Neuroanatomical and diagnostic implications. Neuropsychologia 33:441–459, 1995.

Isaacson RL: The Limbic System. Plenum Press, New York, 1974.

Kalivas PW, Barnes CD (eds): Limbic Motor Circuits and Neuropsychiatry. CRC Press, Boca Raton, Fl, 1993.

Kötter R, Meyer N: The limbic system: A review of its empirical foundation. Behav Brain Res 52:105–127, 1992.

Masliah E, Mallory M, Hansen L, DeTeresa R, Alford M, Terry R: Synaptic and neuritic alterations during the progression of Alzheimer's disease. Neurosci Lett 174:67–72, 1994.

Nauta WJH: Hippocampal projections and related neural pathways to the midbrain in the cat. Brain 81:319–340, 1958.

Reep R: Relationship between prefrontal and limbic cortex: a comparative anatomical review. Brain Behav Evol 25:5–80, 1984.

Sandner G, Oberling P, Silveira MC, Di Scala G, Rocha B, Bagri A, Depoortere R: What brain structures are active during emotions? Effects of brain stimulation elicited aversion on c-fos immunoreactivity and behavior. Behav Brain Res 58:9–18, 1993.

Squire LR: Memory and Brain. Oxford University Press, New York, 1987.

Zola-Morgan S, Squire LR, Amaral DG: Human amnesia and the medial temporal region: enduring memory impairment following a bilateral lesion limited to field CA1 of the hippocampus. J Neurosci 6:2950–2967, 1986.

Chapter 32

The Cerebral Cortex

J. C. Lynch

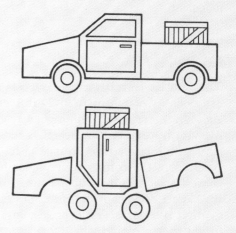

Overview 506

Histology of the Cerebral Cortex 506

Neuron Types in the Cerebral Cortex 508
Pyramidal Cells
Local Circuit Neurons

Laminar Organization 510
Intrinsic Circuitry of the Cerebral Cortex
Cytoarchitecture

Columnar Organization 511

Synopsis of Thalamocortical Relationships 513

Blood Supply to the Cerebral Cortex 514

Higher Cortical Functions 515
The Dominant Hemisphere and Language
Parietal Association Cortex: Space and Attention
Prefrontal Cortex and "Plans for Future Operation"

The cerebral cortex is the organ of thought. More than any other part of the nervous system, the cerebral cortex is the site of the intellectual functions that make us human and that make each of us a unique individual. These intellectual functions include the ability to use language and logic and to exercise imagination and judgment.

Overview

The cerebral cortex is a dense aggregation of neuron cell bodies that ranges from 2 to 4 mm in thickness and forms the surface of each cerebral hemisphere. The total area of the cerebral cortex is about 2500 cm², a little larger than a single page of a newspaper. Neurons in the cortex receive input from many subcortical structures by way of the thalamus and also from other regions of the cortex via association fibers. Cortical neurons, in turn, project to a wide range of neural structures, including other areas of the cerebral cortex, the thalamus, the basal nuclei, the cerebellum via the pontine nuclei, many of the brainstem nuclei, and the spinal cord.

The cerebral cortex is divided into distinct functional areas, some of which are devoted to the processing of incoming sensory information, others to the organization of motor activity, and still others primarily to what are considered "higher intellectual functions." These functions include memory, judgment, the planning of complex activities, the processing of language, mathematical calculations, and the construction of an internal image of an individual's surroundings. This chapter considers (1) the basic internal organization of the cerebral cortex at the cellular level, (2) the parceling of the cortex into distinct subregions on the basis of cellular organization and neural connections, and (3) the functional properties of some higher-order association cortical regions.

Histology of the Cerebral Cortex

The gray matter of the cerebral cortex is composed of neuron cell bodies of variable sizes and shapes, intermixed with myelinated and unmyelinated fibers (Figs. 32–1 and 32–2A). These cell bodies may be visualized with stains that bind to the rough endoplasmic reticulum (Nissl substance). Such stains leave the axons and dendrites almost invisible. Substances that bind to the lipoprotein of the myelin sheath surrounding some axons will make the myelinated portion of the fibers

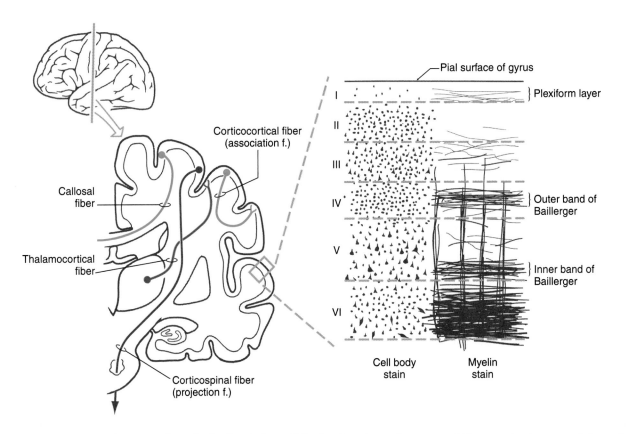

Figure 32–1. A coronal section through the hemisphere (*left*) showing the major types of fibers projecting to and from the cerebral cortex. The representation on the *right* shows layers (I to VI) of the cerebral cortex as they appear after staining for cell bodies or the myelin sheath.

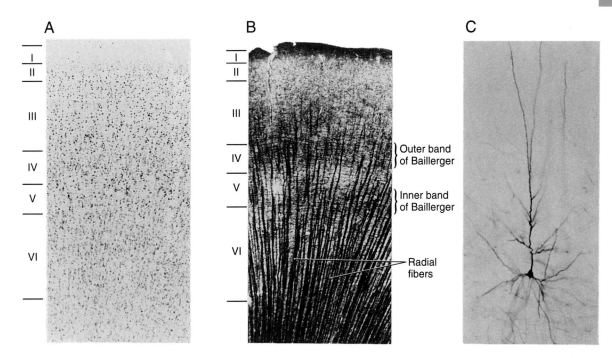

Figure 32–2. Nissl (*A*) and myelin (*B*) stains of adjacent sections of the human cerebral cortex, and a Golgi impregnation (*C*) of a pyramidal neuron in the primate neocortex. (*A* and *B* courtesy of Drs. Grayzna Rajkowska and Patricia Goldman-Rakic, University of Mississippi Medical Center and Yale University; *C* courtesy of Dr. José Rafols, Wayne State University.)

visible (see Figs. 32–1 and 32–2*B*). Yet another way of looking at cortical cells is to immerse small blocks of tissue in dilute silver salts, which precipitate on the membranes of the entire neuron. This reaction causes the cell body, its dendrites, and portions of the axon to become visible (see Fig. 32–2*C*); this technique is called the *Golgi method*. The basic connections of a given region of cortex include *projection fibers* to subcortical structures, *callosal fibers* to cortex in the opposite hemisphere, *association fibers* to cortex in the same hemisphere, and *thalamocortical fibers*, which provide virtually all of the neural input to the cortex that originates in noncortical structures (see Fig. 32–1).

The pattern of distribution of neuron cell bodies is, in general, called *cytoarchitecture*. Specifically, the cytoarchitecture of the cerebral cortex is characterized by layers. Most of the cerebral cortex has six distinct layers of neurons and is classified as *neocortex*. Two regions of the cerebral cortex have fewer than six layers. The first contains only three layers, is classified as *archicortex*, and includes the hippocampal formation. The second contains from three to five layers, is classified as *paleocortex*, and includes the olfactory sensory area and the nearby entorhinal and periamygdaloid cortices. The following discussion concentrates primarily on the neocortex.

The neuronal layers in the neocortex are designated by Roman numerals, beginning at the pial surface (see Fig. 32–1). Layer I, the *molecular layer*, contains very few neuron cell bodies and consists primarily of axons

running parallel (horizontal) to the surface of the cortex. Layer II, the *external granular layer*, is composed of a mixture of small neurons called *granule cells* and slightly larger neurons, which are called *pyramidal cells* based on the shape of their cell body. Layer III, the *external pyramidal layer*, contains primarily small to medium-sized pyramidal cells, along with some neurons of other types. Layer IV, the *internal granular layer*, consists almost exclusively of *smooth (aspiny) stellate (star-like) neurons* and *spiny stellate neurons*, both of which have sometimes been categorized as "granule cells." Layer V, the *internal pyramidal layer*, consists predominantly of medium to large pyramidal cells. Layer VI, the *multiform layer*, contains an assortment of neuron types including some with *pyramidal* and *fusiform* cell bodies.

Two features of the myelinated fibers in the neocortex are noteworthy. First, there are prominent plexuses of horizontally running myelinated fibers in layers IV and V. These are called the *outer* and *inner bands of Baillerger*, respectively (see Fig. 32–1). In the primary visual cortex, bordering on the calcarine sulcus, the outer band of Baillerger is greatly expanded. This band can be seen with the naked eye in fresh and stained sections and is called the *stria (line) of Gennari* (see Fig. 20–17). Second, in most regions of the neocortex, there are many radially oriented bundles of axons passing between the subcortical white matter and various parts of the cortex or between inner and outer cortical layers (see Fig. 32–2*B*).

Neuron Types in the Cerebral Cortex

Pyramidal Cells. The most common type of neuron in the cerebral cortex is the pyramidal cell (Fig. 32–3A; see also Fig. 32–2C). Pyramidal cells are found in all layers of the cortex with the exception of the molecular layer (layer I), and they are the predominant cell type in layers II, III, and V (Fig. 32–4). Pyramidal cells are characterized by (1) a roughly triangular cell body; (2) a single large *apical dendrite* that arises from the apex of the cell body and usually extends toward the molecular layer, giving off branches along the way; (3) an array of *basal dendrites* that run in a predominantly horizontal direction; and (4) an axon that originates from the base of the soma, leaves the cortex, and passes through the white matter.

The cell bodies of most pyramidal neurons range in size from 10 to 50 μm in height. The largest, called *giant pyramidal cells of Betz* or *Betz cells*, are found almost exclusively in the primary motor cortex of the precentral gyrus. Their somata may reach 100 μm in height. Betz cells are most common in the region of motor cortex that projects to the anterior horn of the lumbar spinal cord and hence are concerned with the control of leg movement. These cells are so large that they can be distinguished with the naked eye in Nissl-stained sections of the human brain.

Both apical and basal dendrites of pyramidal cells are characterized by membrane specializations called *dendritic spines*. These spines are small outgrowths from the dendrite that give the impression of thorns on a rose bush (see Fig. 32–3A–C). The vast majority of synaptic contacts received by a pyramidal cell are located on dendritic spines rather than directly on the dendrite shaft or on the cell body.

Pyramidal neurons represent virtually the only output pathway for the cerebral cortex. Almost all other cell types in the cortex are local circuit neurons that exert their influence within their own immediate vicinity. Axons of pyramidal cells may terminate in another region of the cortex in the same hemisphere (*association fibers*), decussate in the corpus callosum to terminate in the cerebral cortex of the opposite hemisphere (*callosal fibers*), or course through the white matter to any of the numerous subcortical targets in the forebrain, brainstem, or spinal cord (*projection fibers*).

Pyramidal cells display a laminar organization, with the cell bodies in a given layer projecting to specific neural targets (see Fig. 32–4). In general, pyramidal neurons in layers II and III give rise to association and callosal fibers. Pyramidal cells in layer V project to many subcortical structures, including the spinal cord, as projection fibers. The neurons in layer VI send their axons to a variety of locations, including thalamic nuclei and other regions of cortex. Within the cortex, axons of pyramidal cells send off an extensive and relatively dense array of *axon collaterals*. These collaterals

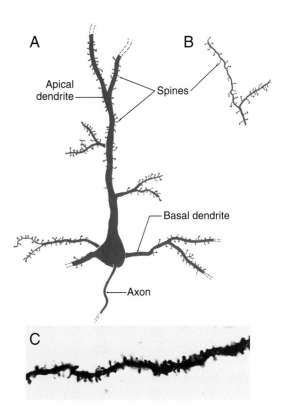

Figure 32–3. Examples of spines on basal and apical dendrites (*A* and *C*) and on the terminal ramifications of apical dendrites (*B*).

terminate in all cortical layers and extend through a horizontal area covering several millimeters around the cell body (Fig. 32–5).

Local Circuit Neurons. As mentioned previously, all the various nonpyramidal neurons of the cerebral cortex function as cortical *interneurons;* that is, their axons do not leave the immediate region of the cell body. These cells are often referred to as *local circuit neurons* or *intrinsic cortical neurons.*

Santiago Ramón y Cajal, working at the turn of the century, described a rich variety of intrinsic cortical neurons. However, by the 1950s it had become customary to refer to virtually all intrinsic cortical neurons as *stellate cells*, even though many were not actually star-shaped. Now the pendulum has swung in the other direction, and a number of distinct morphologic types are recognized. Some of the more important of these are illustrated in Figure 32–4: spiny and aspiny stellate cells, basket cells, and chandelier cells.

Three types of intrinsic neurons receive thalamocortical axon terminals in layer IV: the *small spiny cells,* the *aspiny stellate cells,* and dendrites of the *large basket cells.* Of these, the spiny cells are believed to be excitatory, whereas basket cells and aspiny stellate cells use the neurotransmitter gamma-aminobutyric acid (GABA) and are thus considered to be inhibitory interneurons. Most other intrinsic neurons are presumed to be inhibitory. On the other hand, pyramidal neurons

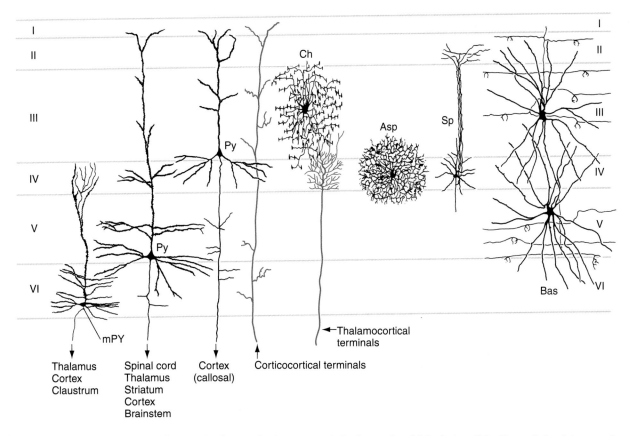

Figure 32–4. Representative cell types in the cerebral cortex and the layers in which their cell bodies and dendrites are found. Dendrites of pyramidal cells (Py) of layers II, III, and V extend into layer I, whereas those of modified pyramidal cells (mPy) in layer VI extend only to about layer IV. Chandelier cells (Ch) are restricted almost entirely to layer III. The somata of aspiny and spiny stellate neurons (Asp, Sp) are in layer IV, although their processes extend into other layers. Basket cells (Bas) have processes that collectively extend into all cortical layers from cell bodies located mainly in layers III and V. (Adapted from Hendry and Jones, 1981, and Jones, 1984, with permission.)

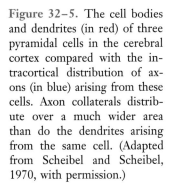

Figure 32–5. The cell bodies and dendrites (in red) of three pyramidal cells in the cerebral cortex compared with the intracortical distribution of axons (in blue) arising from these cells. Axon collaterals distribute over a much wider area than do the dendrites arising from the same cell. (Adapted from Scheibel and Scheibel, 1970, with permission.)

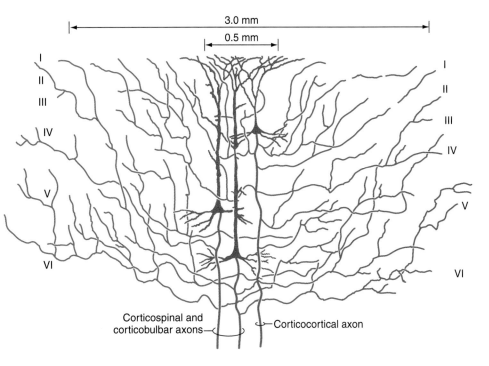

are uniformly associated with excitatory neurotransmitters, glutamate in particular.

Laminar Organization

Intrinsic Circuitry of the Cerebral Cortex. The *basic framework* of the internal circuit diagram of small regions of the cerebral cortex is well understood. In contrast, the *details* of this circuitry are only partially known and are in fact so complex as to defy the construction of a detailed circuit diagram like those used to represent a computer's electronic hardware. For example, a single axon may branch repeatedly and contact hundreds of other neurons. A single neuron may also receive synaptic contacts from thousands of other neurons. Within a small volume of cortex, there may be *millions* of neurons.

The basic framework of cortical circuitry consists of afferent fibers, local circuits for the processing of this afferent information, and efferent fibers that convey the processed information to another site (Fig. 32–6). Thalamocortical axons terminate primarily in layer IV and to a lesser extent in layers III and VI. In layer IV, they terminate on excitatory and inhibitory interneu-

rons as well as on dendrites from neurons in other layers (see Fig. 32–4). The axons of interneurons, in turn, may end on dendrites of pyramidal cells or of other interneurons. The local processing of information culminates in connections to pyramidal cells, which carry the information to other cortical or subcortical regions. A copy of the information also goes to neurons in the immediate vicinity via *axon collaterals* (see Fig. 32–5).

The general pattern of termination of *corticocortical* axons is quite different from that of thalamocortical axons. Corticocortical axons branch repeatedly and make synaptic contacts on neurons in *all* layers of the cortex (see Fig. 32–4).

The cerebral cortex receives a third set of inputs, called *diffuse inputs*, which consists of fibers that branch extensively and end diffusely over a wide area of cortex without respect for cytoarchitectural boundaries (see Fig. 32–6). These inputs arise from a variety of sources, including certain *nonspecific nuclei of the thalamus* (for example, the ventral anterior, central lateral, and midline nuclei), the *locus ceruleus*, and the *basal nucleus (of Meynert)*. These structures are generally concerned with regulating overall levels of cortical excit-

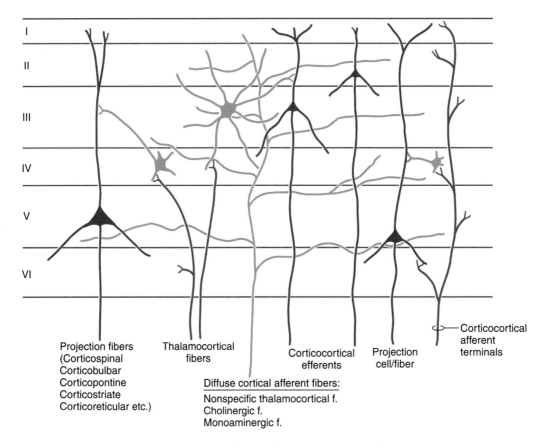

Figure 32–6. Basic circuits in the cerebral cortex. Afferent fibers are shown in blue and gray, interneurons in green, and efferent fibers in red. Thalamocortical fibers terminate primarily in layer IV, whereas corticocortical fibers and diffuse cortical afferents synapse in all layers. Pyramidal cells in the outer layers give rise to corticocortical projections, and those in layer V project to a wide range of subcortical targets.

ability and the associated phenomena of arousal, sleep, and wakefulness.

Cytoarchitecture. The cytoarchitecture of cortex differs from one area to another in ways that are related to function (Fig. 32–7). In primary sensory cortex, layer IV, the major *input* layer of cortex, is especially thick, whereas layer V, the major *projection layer*, is narrow and indistinct. Cortex with this pattern is called *heterotypical granular cortex.* In primary motor cortex, the pattern is reversed: layer IV is almost invisible, and layer V is very thick, seeming to merge directly with layer III. Thus, the *projection* layer is prominent, and the *input* layer is small. Cortex of this type is called *heterotypical agranular cortex.* In most other areas of the neocortex, including the association cortices, the six layers are all clearly represented and are of roughly equal thickness. This type of cortex is called *homotypical.*

The cerebral cortex has been subdivided on the basis of cytoarchitectural differences by many different investigators. The most famous of these, Korbinian Brodmann, worked in the early part of the twentieth century. He identified 47 distinct areas (Fig. 32–8), and his numbering scheme is still in common use today in both research and clinical settings. For example, the primary visual cortex is Brodmann area 17, and the primary motor cortex is area 4. In most instances, Brodmann cytoarchitectural areas are coextensive with cortical regions that have specific functional character-

istics. This numbering scheme is used throughout the chapter.

Columnar Organization

A second, vertical pattern of organization is superimposed on the horizontal layered pattern described above. Unlike the cortical layers, this vertical pattern is not immediately obvious in histologic sections stained for neuron cell bodies (Nissl stains). However, when Golgi-stained material is studied, it is clear that neurons are often grouped together so that their cell bodies, axons, and apical dendrites form clusters that are oriented at right angles to the surface of the cortex.

Mountcastle was the first to demonstrate physiologically the existence of a vertical ("columnar") organization in the cerebral cortex by recording the activity of hundreds of individual neurons in the primary somatosensory cortex of cats and monkeys. Within an area of cortex a few millimeters in diameter, all neurons had overlapping or adjacent receptive fields. For example, in one cortical region, all neurons might have receptive fields on a finger, whereas in a nearby region, the neurons might have receptive fields on the wrist. Within a cortical region in which all neurons had about the same receptive field, the neurons responded to different *sensory submodalities.* Some neurons were activated by light touch on the skin, others by joint

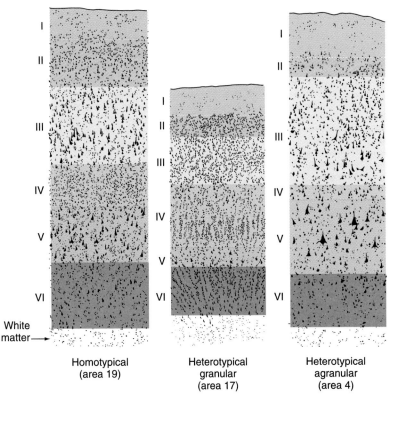

Figure 32–7. Typical cytoarchitectural patterns for homotypical, heterotypical granular, and heterotypical agranular regions. (Adapted from Campbell AW: Histological Studies on the Localization of Cerebral Function. Cambridge University Press, Cambridge, England, 1905.)

White matter→

Homotypical (area 19)

Heterotypical granular (area 17)

Heterotypical agranular (area 4)

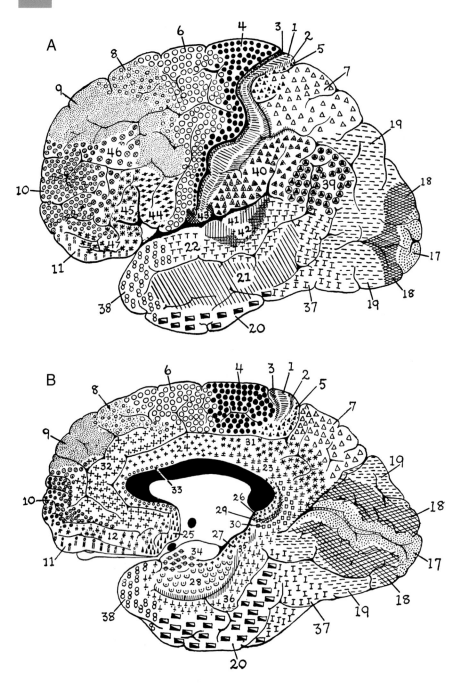

Figure 32–8. Cytoarchitectural map showing Brodmann areas on the lateral (*A*) and medial (*B*) surfaces of the hemisphere. (Modified after Brodmann K, from Carpenter MB, Sutin J: Human Neuroanatomy. Williams & Wilkins, Baltimore, 1983, with permission.)

rotation, and still others by strong pressure on deep tissue. However, when a microelectrode was inserted at right angles to the surface of the cortex, all the neurons encountered were activated by only one of these submodalities (Fig. 32–9*B*). In contrast, when a microelectrode was moved parallel or obliquely relative to the surface of the cortex, it encountered neurons of different submodalities as it moved from one functionally related group of neurons to another (see Fig. 32–9*A*).

The basis of columnar organization in primary sensory cortices is selective input from relay nuclei of the thalamus. Obviously, if all of the cells in one column respond to maintained pressure on the skin while the

cells in an adjacent column respond to joint position, the signals from the respective sensory receptors must have been continuously segregated all the way from the periphery through the posterior column nuclei and the ventrobasal complex of the thalamus to terminate in the cortex.

The anatomic basis of the *columnar organization* of the cortex is understood in the greatest detail in the visual cortex. In this region, at least three types of regularly repeating features are superimposed on the laminar patterns of neurons: the stimulus orientation columns, the ocular dominance columns, and the cytochrome oxidase–rich blobs. These features are discussed in detail in Chapter 20.

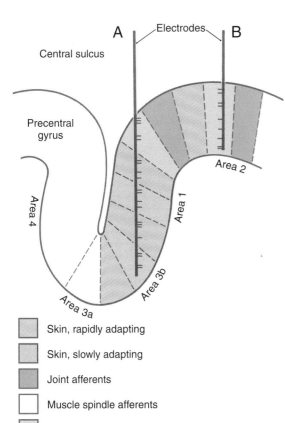

Skin, rapidly adapting

Skin, slowly adapting

Joint afferents

Muscle spindle afferents

Receptors in deep tissue

Figure 32–9. Diagrammatic section through the precentral and postcentral gyri showing the organization of columns in the somatosensory cortex. The columns are shown as colored compartments oriented, in general, perpendicular to the surface of the cortex. An electrode (at A) passing parallel to the surface of the cortex will pass through several columns with resultant recordings of the several modalities represented by the types of afferent information arriving at each column. An electrode passing through one column (at B) passing perpendicular to the surface of the cortex penetrates only a single column. Therefore, it records activity related to the single submodality received by that column.

The ocular dominance columns in the visual cortex provide a clear example of the role of thalamic input in columnar organization. Neurons in the layers of the lateral geniculate nucleus that receive input from the right eye send their axons to layer IV of the right-eye–dominant columns (Fig. 32–10). Here, the axons terminate predominantly on spiny and aspiny stellate cells, which, in turn, project to pyramidal cells. Collaterals of pyramidal cell axons provide one pathway by which neural signals can spread from one column to influence activity in adjacent columns (see Fig. 32–5). This influence may be either excitatory via direct connections or inhibitory via interneurons. The right eye, therefore, has a direct and strong influence on neurons in right-eye–dominant columns (R_L in Fig. 32–10) and an indirect and weaker influence on neurons in the adjacent left-eye–dominant columns (L_R in Fig. 32–10).

Connections between one region of the cortex and another, through either association fibers or callosal fibers, may also be arranged in a columnar pattern. For example, axons that originate in the inferior parietal lobule terminate in multiple columns in the ipsilateral and contralateral cingulate cortices. Columns of corticocortical axon terminals that originate in different functional regions may either overlap each other or interdigitate with each other.

Synopsis of Thalamocortical Relationships

The details of thalamocortical projections are described in the chapters devoted to specific systems. At this juncture, however, it is appropriate to briefly review what areas of the cortex are functionally related to which of the thalamic nuclei (Fig. 32–11).

The cortex of the frontal lobe encompasses Brodmann areas 4, 6, 8 to 12, 32, and 44 to 47 (see Fig. 32–8). The primary somatomotor cortex (area 4) and the premotor and supplementary motor cortices (area 6) receive input mainly from the ventral lateral nucleus of the thalamus and subserve important motor func-

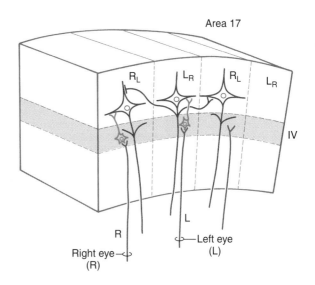

Figure 32–10. Functional columnar organization of sensory cortices, using the visual cortex (ocular dominance columns) as an example. Axons from the layers of the lateral geniculate nucleus related to the right (R) and left (L) eyes terminate in alternating columns. In right eye–dominant columns (R_L), cortical neurons are influenced predominantly by visual stimulation of the right retina, although they are also influenced to a lesser degree by stimulation of the left retina. The reverse is true for left eye–dominant columns (L_R). The neurons in each column influence the adjacent columns (dominant for the other eye) via axon collaterals of pyramidal cells or through the action of cortical interneurons. In general, other sensory cortices are organized similarly in regard to their sensory inputs.

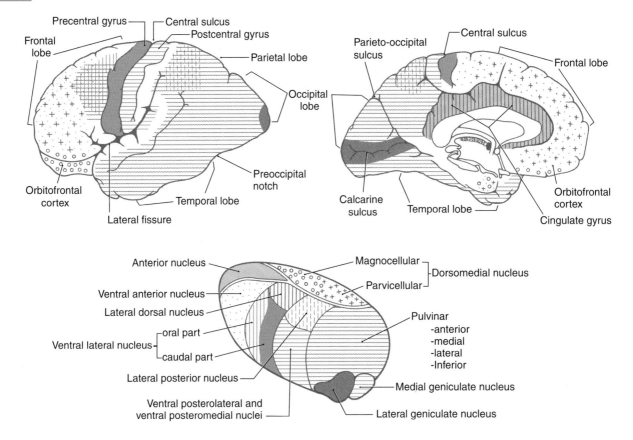

Figure 32–11. Relationships of the thalamic nuclei to the cerebral cortex as revealed by the patterns of thalamocortical connections. Each thalamic nucleus is pattern-coded or color-coded to match its target area in the cerebral cortex.

tions. Lateral, medial, and orbital aspects of the frontal lobe receive thalamocortical fibers mainly from the dorsomedial and anterior nuclei of the thalamus (see Fig. 32–11). These latter cortical areas, through a variety of direct and indirect connections, relate primarily to functions of the limbic system. Of particular note are the pars orbitalis and pars triangularis of the inferior frontal gyrus, damage to which results in Broca aphasia (discussed below).

Areas 3, 1, 2, 5, 7, 39, 40, and 43 are located in the parietal lobe (see Fig. 32–8). The primary somatosensory cortex (areas 3, 1, and 2) receives inputs from the ventral posterolateral and ventral posteromedial nuclei. These thalamic nuclei receive a full range of somatosensory input through synaptic relays in the spinal cord and brainstem and transmit this information to the cerebral cortex. The inferior parietal lobule comprises, in general, areas 39 and 40. Along with area 22, these areas are the cortical regions associated with Wernicke aphasia (discussed below).

The occipital and temporal lobes encompass areas 17 to 22, 36 to 38, and 41 and 42 (see Fig. 32–8). These areas of the cortex, plus portions of the parietal lobe, have extensive connections with the pulvinar nucleus of the thalamus and are involved in the processing of visual and auditory information at several differ-

ent functional levels (see Fig. 32–11). Located in this geographic area are the primary sensory cortices for vision and hearing. Area 17, on the banks of the calcarine sulcus, is the primary visual cortex; areas 41 and 42, in the depth of the lateral fissure in the transverse temporal gyri, constitute the primary auditory cortex (see Fig. 32–8). These cortical areas receive input from the lateral and medial geniculate nuclei of the thalamus, respectively.

The limbic lobe, which forms the most medial edge of the hemisphere, contains areas 23 to 31 and 33 to 35. The cingulate cortex receives fibers primarily from the anterior nucleus of the thalamus, but also from the lateral dorsal nucleus (see Fig. 32–11). Other regions of the limbic lobe have some connections with the dorsomedial nucleus. However, many of the subcortical targets of the parahippocampal and uncal cortices are structures such as the hippocampal formation. This area, in turn, projects to a variety of thalamic and basal forebrain targets.

Blood Supply to the Cerebral Cortex

The blood supply to the cerebral cortex and to subcortical structures of the telencephalon, including the internal capsule, is discussed in Chapters 8 and 16. This section summarizes the general nature of these patterns.

The cerebral cortex is served, in toto, by the *anterior, middle,* and *posterior cerebral arteries.* The anterior and middle cerebral arteries are the terminal branches of the internal carotid artery and the posterior cerebral artery is formed by the bifurcation of the basilar artery (see Figs. 8–3 and 8–8).

The anterior cerebral artery is joined to its counterpart just anterior to the optic chiasm by the anterior communicating artery. Proximal to the anterior communicating artery, the anterior cerebral artery gives rise to small branches that serve rostral portions of the hypothalamus. Cortical branches of the anterior cerebral artery distribute to the medial surface of the hemisphere caudally to about the position of the parieto-occipital fissure. The distal portions of these branches arch over the edge of the hemisphere (from its medial to its lateral surface) for a short distance (see Figs. 8–3, 8–9, and 16–9). Located in the domain of the branches of this major vessel are the foot, lower extremity, and hip areas of the primary somatomotor and primary somatosensory cortices.

The middle cerebral artery passes laterally from its origin and branches, in general, into superior and inferior trunks (these are M_2 branches) over the insular cortex. Terminal branches of the superior and inferior trunks serve the cortex on the lateral surface of the hemisphere above and below the lateral fissure, respectively (see Figs. 8–6 and 16–9). In addition to lateral portions of the frontal cortex, parietal and temporal association cortices are served by branches of the middle cerebral artery. Also located in the distribution area of this vessel are the trunk, upper extremity, and head regions of the primary somatomotor and primary somatosensory cortices (via branches of the superior trunk) and the primary auditory cortex (inferior trunk).

The cortex forming the inferior surface of the temporal lobe and the medial aspect of the occipital lobe is served by branches of the posterior cerebral artery (see Figs. 8–3, 8–8, and 16–9). As with the anterior cerebral artery, the terminal rami of the posterior cerebral artery loop over the edge of the inferior and medial surface of the hemisphere to serve small portions of the lateral aspect of the hemisphere. The posterior cerebral artery serves large expanses of visual association cortex and some of the cortical structures associated with the limbic system. In addition, the calcarine artery supplies the primary visual cortex located on and in the banks of the calcarine sulcus.

The wedge-shaped areas of overlap between the distal branches of the anterior and middle cerebral arteries and the posterior and middle cerebral arteries form what are called *border zones* (see Fig. 8–9). These areas are particularly susceptible to hypoperfusion during episodes of systemic hypotension. Such events may result in *watershed infarcts.*

Higher Cortical Functions

The cerebral cortex is generally considered to be the seat of *higher intellectual functions,* those faculties of thought that have reached their most complex levels in humans. Although other brain structures, including the thalamus, corpus striatum, claustrum, and cerebellum, contribute to these functions, the *multimodal association cortex* is closely linked to the most complex intellectual functions, such as logical analysis, judgment, language, and imagination.

The cerebral cortex can be divided into four general functional categories: *sensory, motor, unimodal association cortex,* and *multimodal association cortex* (Fig. 32–12). The primary sensory areas, except that for olfaction, receive thalamocortical fibers from diencephalic relay nuclei that are functionally related to each modality. For example, the ventral posterior complex of the thalamus projects to primary somatosensory cortex (Brodmann areas 3, 1, and 2) in the postcentral gyrus. Similarly, the lateral geniculate nucleus projects to primary visual cortex (area 17) in the banks of the calcarine sulcus, and the medial geniculate nucleus projects to primary auditory cortex in the transverse temporal gyri (areas 41 and 42).

Adjacent to each primary sensory area is a region of cortex that is devoted to a higher level of information processing relevant to that specific sensory modality. These areas are called *unimodal association cortices* (see Fig. 32–12). For example, *visual unimodal association cortex* (areas 18, 19, 20, 21, and 37) occupies all of the occipital lobe outside area 17 (the primary visual sensory area), as well as much of the inferior gyrus of the temporal lobe. Within these visual association areas, the basic elements of visual sensation are molded into an overall perception of the visual world. Similarly, *somatosensory association cortex* lies just posterior to the postcentral gyrus in area 5, and the *auditory association cortex* is in the superior temporal gyrus (area 22) next to primary auditory cortex. In all of these examples, the primary sensory cortices (i.e., areas 3, 1, 2; 17; and 41 and 42) receive input from their respective thalamic relay nuclei. In turn, the primary sensory areas project, via corticocortical fibers, to their corresponding association cortices (i.e., areas 3, 1, 2 to area 5; area 17 to areas 18 and 19; and areas 41 and 42 to area 22).

The remaining portions of the cerebral cortex that are not motor in function are classified as *multimodal association cortex* (see Fig. 32–12). These areas receive information from several different sensory modalities and create for us a complete experience of our surroundings. Multimodal association areas are critical to our ability to communicate using language, to use reason to extrapolate future events on the basis of present experience, to make complex and long-range plans, and to imagine and create things that have never existed. An example of long-range planning is going to college so you can go to medical school so you can do a

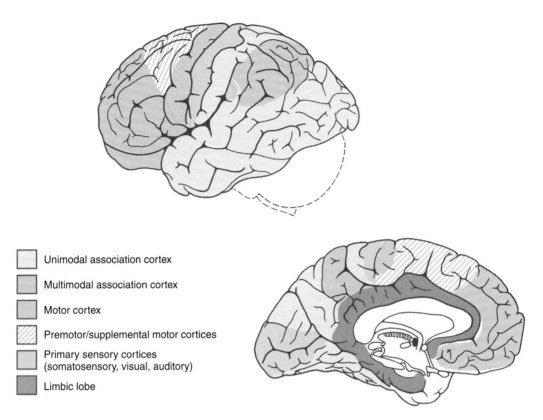

Figure 32-12. Primary motor and sensory (blue), unimodal association (green), and multimodal association (red) areas of the cerebral cortex are shown on lateral (*upper*) and medial (*lower*) surfaces of the hemisphere.

Unimodal association cortex

Multimodal association cortex

Motor cortex

Premotor/supplemental motor cortices

Primary sensory cortices
(somatosensory, visual, auditory)

Limbic lobe

residency and become a physician. This chapter concentrates on the cortical areas responsible for three of these higher functions: language, the appreciation of space, and the planning of behavior.

The Dominant Hemisphere and Language. The cerebral hemisphere that controls language is called the *dominant hemisphere*. In the vast majority of people, language functions are processed in the *left hemisphere*. As evidence of this left brain dominance, brain lesions that adversely affect language are found in the left hemisphere in about 95% of cases. Almost all right-handed individuals and about half of left-handed individuals are *left cerebral dominant*. It follows that the right cerebral hemisphere, in most of the general population, is the *nondominant hemisphere*.

Language is the faculty of communication using symbols organized by a system of grammar to describe things and events and to express ideas. In humans, the senses of vision and audition are closely linked to language, but language itself transcends any particular sensory system. Helen Keller was blind and deaf but used language eloquently to communicate very complex and subtle ideas. Language ability can be impaired selectively, with little or no change in the senses of vision or hearing, by brain damage in either the parietal-temporal junction or the frontal lobe. This impairment, termed *aphasia*, is a *disturbance of the comprehen-*sion and formulation of language, not a disorder of hearing, vision, or motor control.

The two classic types of aphasia are *Broca aphasia* and *Wernicke aphasia*. Broca aphasia, also termed *expressive aphasia* or *nonfluent aphasia*, consists of a loss of the ability to speak fluently. Lesions that produce this deficit are located in the inferior frontal gyrus of the *left hemisphere*, primarily in Brodmann areas 44 and 45 (Fig. 32–13). Wernicke aphasia is primarily a defect of the *comprehension* rather than the *expression* of language. This deficit is seen after injury to the supramarginal and angular gyri (areas 37, 39, and 40) and the posterior part of the superior temporal gyrus (area 22) in the *left hemisphere* (see Fig. 32–13).

Patients with the most severe form of *Broca aphasia* are unable to speak (*mutism*), although they are able to swallow and breathe normally and make guttural sounds. Their problem is not one of paralysis of the vocal apparatus. Rather, it is a difficulty in turning a concept or thought into a sequence of meaningful sounds. In less severe cases, or in patients in the recovery process, limited speech is possible. Short, habitual phrases such as "hi," "fine, thank you," and "yes" and "no" are the first to come back. However, speech is slow and labored, enunciation is poor, and nonessential words are commonly omitted (*telegraphic speech*). Affected persons typically have as much difficulty with

Figure 32–13. Cortical areas that mediate the processing of language. Lesions in the pars orbitalis and pars triangularis of the inferior frontal lobe will result in Broca aphasia, whereas damage in the supramarginal and angular gyri and adjacent superior temporal gyrus will result in Wernicke aphasia.

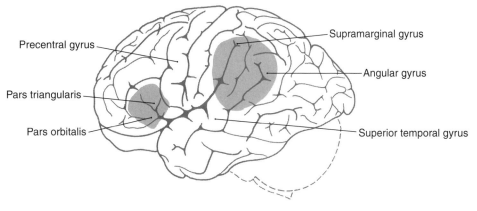

Precentral gyrus

Pars triangularis

Pars orbitalis

Supramarginal gyrus

Angular gyrus

Superior temporal gyrus

writing (*agraphia*) as with speaking. Although the patient is able to understand spoken or written language and can communicate verbally to some degree, the extremely laborious nature of the process of communication causes considerable frustration. Under particular emotional stress, patients may use inappropriate or vulgar words or phrases to express their distress.

The most common causes of Broca aphasia are tumors and occlusions of frontal M_4 branches of the middle cerebral artery. Mild aphasia without other deficits indicates that the damage affects only cortical areas. However, full-blown Broca aphasia indicates that the damage extends beyond the Broca area of the cortex to include insular cortex and subjacent white matter. Patients typically have *contralateral motor signs and symptoms*, such as weakness (*paresis*) of the lower part of the face, lateral deviation of the tongue when protruded, and weakness of the arm. Aphasia plus these motor problems suggests an occlusion of branches from the proximal parts of the middle cerebral artery (M_1), including the lenticulostriate arteries, which serve the internal capsule (see Fig. 16–10).

The second major type of aphasia is *Wernicke aphasia* (also described as *receptive* or *fluent* aphasia). Patients with severe Wernicke aphasia (1) are unable to understand what is said to them; (2) are unable to read (*alexia*); (3) are unable to write comprehensible language (*agraphia*); and (4) display *fluent paraphasic speech*. Paraphasic speech refers to the ability of patients to produce clear, fluent, melodic speech at a normal or even faster than normal rate. The content of the speech, however, may be unintelligible because of frequent errors of word choice, inappropriate use of words, or use of made-up nonsense words. An example of this type of speech is "we went to drive in the bridge for red pymarids were crooking the lawn browsers." Such speech is sometimes called "word salad." In

less severe cases, *paraphasias* frequently occur. For example, in trying to say "the cat has claws," the patient may use an incorrect but similar-sounding word ("the cat has clads"—a *literal paraphasia*) or a word that seems appropriate to the patient but is incorrect ("the cat has tires"—a *verbal paraphasia*). A surprising finding is that patients with Wernicke aphasia are much less aware of the extent of their disability than patients with Broca aphasia, and they are usually less frustrated and depressed about it. In contrast, patients with Broca aphasia are completely conscious of their communication problems and are often exceedingly frustrated or despairing.

Wernicke aphasia may result from occlusion of temporal and parietal M_4 branches of the middle cerebral artery. In addition, hemorrhage into the thalamus (or tumors in the thalamus) may produce Wernicke aphasia by extending laterally and caudally to invade the subcortical white matter. If this damage impinges on the Meyer loop and interrupts the optic radiations (see Chapter 20), a contralateral homonymous hemianopia may accompany the patient's other disabilities.

The severity and duration of the aphasia depend on the severity of the associated brain damage. In mild cases, only one or two symptoms may be discernible, and those may resolve quickly. A major stroke or severe traumatic injury, however, may produce a full-blown set of signs and symptoms that will never completely disappear.

Other, less common types of aphasia have also been described. These include *conduction aphasia*, which results from interruption of the connections linking the Broca and Wernicke areas. In this disorder, comprehension is normal and expression is fluent, but the patient has difficulty translating what someone has said to him or her into an appropriate reply. A more profound disorder is *global aphasia*, which occurs when occlusion of the left internal carotid or the most proximal portion of the middle cerebral artery (M_1 segment) produces damage that encroaches on both the Broca and the Wernicke areas, and the loss of language is virtually complete.

Several additional points should be mentioned. First, damage to the basal nuclei, particularly to the head of the caudate on the *left side*, has been associated with language disorders similar to Wernicke aphasia. Second, although we have referred so far to spoken and written language (i.e., *verbal language*), aphasia can also affect *nonverbal* language. A deaf person who uses American Sign Language can lose the ability to use or understand sign language after focal brain damage in the left hemisphere. Third, although most aspects of language are processed in the left hemisphere, some features are influenced by lesions in the nondominant parietal lobe. In particular, a patient with a right parietal lesion may have difficulty appreciating the *prosody* of speech. This term refers to the variations in vocal inflections, emotional content, and melody that may alter the meaning of a spoken sentence, as in: "George is here." versus "*George* is here!" versus "George is *here?*" versus "George *is* here!"

Parietal Association Cortex: Space and Attention. A completely different set of intellectual functions is mediated in the parietal association cortex of the nondominant hemisphere. Although the segregation of functions between the two parietal lobes is not complete, the parietal association cortex is nevertheless the most highly *lateralized* in the brain, with language functions concentrated in the left hemisphere and spatial relationships and related selective attention concentrated in the right hemisphere.

Much of our knowledge of the functional properties of different regions of the cerebral cortex has been gained from neurologic case studies of patients with cortical damage produced by stroke or head trauma. In this respect, the two great wars of the first half of the twentieth century led, inadvertently, to great progress in our understanding of the effects of brain injuries. One of the most striking symptoms of damage to *right parietal association cortex* (nondominant) is a defect of *attention*, in which the patient seems to be completely unaware of objects and events in the left half of his or her surrounding space. This symptom is termed *contralateral neglect* (Fig. 32–14*A* and *B*).

In its milder forms, contralateral neglect may simply be a tendency to ignore things on the left side of the patient's surroundings. For example, the patient may be asked to read a short passage and check off each word in the process. As the patient reads, words on the left side of the passage are progressively ignored, and only those on the right part are perceived (see Fig. 32–14*A*). Another way to demonstrate contralateral neglect is to draw a circle and ask the patient to draw in the numbers of a clock face. Typically, the patient with right parietal damage will put all of the numbers (1 to

12) on the right side of the circle (the side ipsilateral to the lesion), completely ignoring the left (contralateral) side of the circle (see Fig. 32–14*B*). A patient with contralateral neglect may not be aware of people standing to the left, may bump into large stationary objects on the left, and may not respond to sounds or words coming from the left. In extreme cases, the patient may not even recognize the left side of his or her own body (*asomatognosia*). For example, the patient may ignore the left side when dressing or grooming (*dressing apraxia*) or, if in a hospital, may even demand that the staff get this "other person" (the left side of his or her own body) out of the bed.

Another characteristic group of symptoms of right parietal lobe lesions concerns the ability to function successfully within the *spatial surroundings*. For example, the affected person may be unable to describe his route between home and work, draw a floor plan of his house (see Fig. 32–14*C*), or find a soft drink machine that is just down the hall. In extreme cases, the patient may not be able to navigate successfully from the bed to a chair that is just across the room and in full view.

Yet another difficulty is an inability to successfully manipulate objects in space. The patient may be unable to duplicate a simple block construction while looking at a model (see Fig. 32–14*D*). This difficulty is termed *constructional apraxia*; it is related not to visual acuity or to fine motor control but rather to an inability to internalize and duplicate the spatial relationships of the individual parts of the model. In addition, disorders of *affect* are common, including a reduced ability to understand and appreciate humor, a loss of the ability to appreciate the *prosody* of speech, and often an inappropriate cheerfulness and lack of concern for, or even awareness of, the implications of the illness. This lack of concern may be noted even when the deficit is as serious as total left hemiplegia.

Prefrontal Cortex and "Plans for Future Operation." The other major region of multimodal association cortex is the large expanse anterior to the primary motor and premotor cortices, the *prefrontal association cortex*. This region has historically been connected with some of the most distinctly human intellectual traits, such as *judgment*, *foresight*, a sense of *purpose*, a sense of *responsibility*, and a sense of *social propriety*.

One of the earliest accounts of the effect of brain injury on higher intellectual functions described a series of events that began on September 13, 1848. A crew of railroad construction workers was blasting a right-of-way through the rugged granite mountains of Vermont. The well-liked young foreman of the crew, Phineas Gage, was in charge of placing a black powder charge in a deep hole drilled in the rock, adding a fuse, covering the powder with sand, and finally tamping the sand and powder down firmly with an iron rod before lighting the fuse and running for cover. On this day, something apparently distracted Gage, and he began to

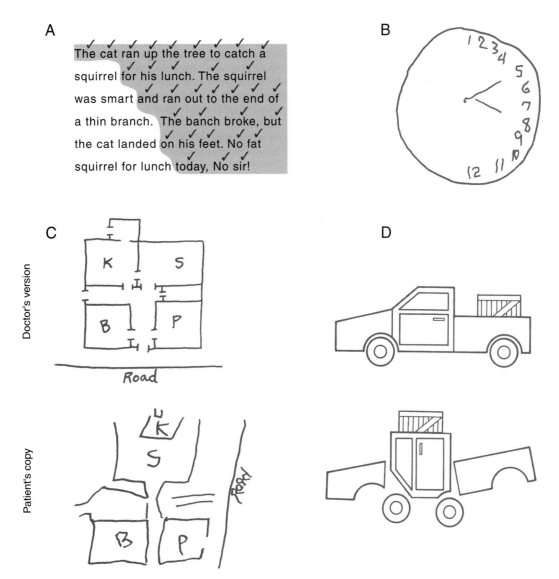

A

The cat ran up the tree to catch a squirrel for his lunch. The squirrel was smart and ran out to the end of a thin branch. The banch broke, but the cat landed on his feet. No fat squirrel for lunch today, No sir!

B

C

Doctor's version

Road

Patient's copy

Road

D

Figure 32–14. Signs (*A–D*) of damage to the nondominant parietal association cortex. See text for details.

tamp down a charge before the sand had been added. The iron rod struck the granite wall of the hole and a spark ignited the powder. The 3.5-ft-long, 13-lb rod was propelled out of the hole like a giant bullet.

The rod struck Gage just beneath the left eye and exited through the top of his head, destroying most of his prefrontal cortex. Amazingly, Gage was not killed instantly, and even more incredibly, he survived the inevitable serious wound infection that followed. Eventually he recovered his health, or at least the physical portion of it. Mentally, however, he was changed forever. Although he did not suffer paralysis, language disorders, or memory loss, his personality was radically altered. John Harlow, one of the doctors who attended Gage, perceived the importance of this case with respect to the localization of intellectual functions in the brain. In an article describing the injury and Gage's persisting intellectual symptoms, Harlow said:

His physical health is good, and I am inclined to say that he has recovered. . . . The equilibrium or balance, so to speak, between his intellectual faculties and animal propensities seems to have been destroyed. He is fitful, irreverent, indulging at times in the grossest profanity (which was not previously his custom), manifesting but little deference for his fellows, impatient of restraint or advice when it conflicts with his desires, at times pertinaciously obstinate, yet capricious and vacillating, devising many plans for future operation, which are no sooner arranged than they are abandoned. . . . In this regard his mind was radically changed, so decidedly that his friends and acquaintances said that he was "no longer Gage."

This passage, written more than 130 years ago, provides an accurate and insightful description of the major symptoms associated with destruction of prefrontal cortex. Patients with significant bilateral damage to the prefrontal cortex have a constellation of deficits that can be summarized as follows. First, they are *highly distractible*, turning from one activity to another according to the novelty of a new stimulus rather than according to a plan. This deficit is sometimes described as a *lack of consistency of purpose*. Second, these persons have a *lack of foresight*. They are not able to anticipate or predict future events on the basis of past events or present conditions. Third, they may be *unusually stubborn* in the face of advice with which they do not agree, and they may also *perseverate* in the performance of a task. Fourth, the patient with prefrontal damage displays a profound *lack of ambition*, a loss of the *sense of responsibility*, and a loss of a *sense of social propriety*. The first and third symptoms (*distractibility* versus *perseveration*) are obviously in conflict. It is impossible to predict which will dominate at a given moment, but *both* exemplify the affected person's loss of the ability to govern his own actions and life according to a *plan*. He is instead imprisoned in a chaotic world, his actions governed by randomly changing whims.

It was this set of symptoms that prompted the Portuguese neurosurgeon Egas Moniz to develop the prefrontal lobotomy procedure in the late 1930s to treat a range of severe, intractable mental problems. At that time, mental hospitals ("insane asylums") all over the world contained many patients who were so immobilized by anxiety that they could not even take care of their own bodily needs. They were warehoused under reprehensible conditions. The discovery that a neurosurgical procedure could alleviate the anxiety to the extent that the patients could lead a somewhat more normal existence (albeit still within the confines of a mental institution) was hailed as a great breakthrough. In these desperate patients, the symptoms as described above seemed a justifiable price to pay for freedom from the crushing anxiety that had immobilized them. Unfortunately, by the late 1940s and early 1950s, the procedure had acquired a popularity out of all proportion to its actual benefits, and it was widely misapplied (as in the movie "One Flew Over the Cuckoo's Nest"). The discovery of tranquilizers in the late 1950s provided a more effective method of treatment, having fewer undesirable side effects, and prefrontal lobotomy was rapidly abandoned as a method of treatment.

Sources and Additional Reading

Blakemore C: Mechanics of the Mind. Cambridge University Press, Cambridge, 1977.

Damasio AR: Aphasia. N Engl J Med 326:531–538, 1992.

Damasio H, Grabowski T, Frank R, Galaburda AM, Damasio AR: The return of Phineas Gage: Clues about the brain from the skull of a famous patient. Science 264:1102–1105, 1994.

Harlow JM: Recovery from the passage of an iron bar through the head. Pub Mass Med Soc 2:327–347, 1868.

Hendry SHC, Jones EG: Sizes and distributions of intrinsic neurons incorporating tritiated GABA in monkey sensory-motor cortex. J Neurosci 1:390–408, 1981.

Hubel DH, Wiesel TN: Functional architecture of macaque monkey visual cortex. Proc R Soc Lond B Biol Sci 198:1–59, 1977.

Jones EG: Varieties and distribution of non-pyramidal cells in the somatic sensory cortex of the squirrel monkey. J Comp Neurol 160:205–268, 1975.

Jones EG: Laminar distribution of cortical efferent cells. Cerebral Cortex, vol 1. Plenum Press, New York, 1984, pp 521–553.

Lynch JC: Parietal association cortex. Encyclopedia of Neuroscience, vol 2. Birkhauser, Boston, 1987, pp 925–926.

Lynch JC: Columnar organization of the cerebral cortex (cortical columns). Neuroscience Year (Supplement to Encyclopedia of Neuroscience). Birkhauser, Boston, 1989, pp 37–40.

Mountcastle VB: Modality and topographic properties of single neurons of cat's somatic sensory cortex. J Neurophysiol 20:408–434, 1957.

Peters A, Jones EG: Cerebral Cortex, vols 1–14. Plenum Press, New York, 1984–1999.

Rowland LP: Merritt's Neurology, 10th ed. Lippincott Williams & Wilkins, Baltimore, 2000.

Scheibel ME, Scheibel AB: Elementary processes in selected thalamic and cortical subsystems—the structural substrates. The Neurosciences, Second Study Programs, vol 2. Rockefeller University Press, New York, 1970, pp 443–457.

Valenstein ES: Great and Desperate Cures. Basic Books, New York, 1986.

Victor M, Ropper AH: Adams and Victor's Principles of Neurology, 7th ed. McGraw-Hill, New York, 2001.

Chapter 33

The Neurologic Examination

M. E. Santiago and J. J. Corbett

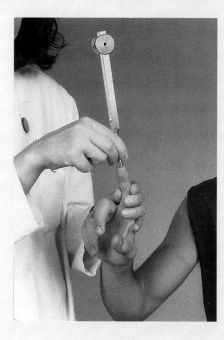

Overview 522

The Mental Status Examination 522

Cranial Nerve Function Testing 522
Cranial Nerve I
Cranial Nerve II
Cranial Nerves III, IV, and VI
Cranial Nerve V
Cranial Nerve VII
Cranial Nerve VIII
Cranial Nerves IX and X
Cranial Nerve XI
Cranial Nerve XII

The Motor Examination 530
Muscle Stretch Reflexes
Cerebellar Testing

The Sensory Examination 536

In many respects this chapter is a prologue to the experience of working directly with the patient. Now that many aspects of functional systems neurobiology have been mastered, the opportunity to apply this knowledge is at hand. Performing the neurologic examination is an excellent example of how basic neuroscience can apply directly to events (both normal and abnormal) encountered in the clinical setting. After all, the neurologically compromised patient is simply a normal person whose nervous system is not functioning properly.

Overview

No other branch of medicine lends itself so well to the correlation of the signs and symptoms of disease with structure and function as does neurology. The neurologic diagnosis of the impaired patient is a deductive process and is reached by a synthesis of all of the details from the history, the examination, and laboratory studies. The neurologic examination is divided into four main segments: mental status, cranial nerves, motor and cerebellar, and sensory.

Figure 33–1 shows a sample set of tools necessary to perform a routine neurologic examination: visual acuity card and eye occluder, ophthalmoscope, dilating eye drops, a flashlight, test tube with coffee to assess smell, disposable tongue blade, safety pin, tissue paper, and cotton-tipped applicator. Tuning forks, measuring tape, and a reflex hammer should also be included, as well as a quarter or a wooden cube for sensation testing.

The Mental Status Examination

The mental status examination starts first with an assessment of the level of consciousness of the patient.

Orientation to time, place, and person should also be documented. Memory of past events and short-term memory, as well as the ability to calculate, are also evaluated at this time. This basic examination is known as the Folstein Mini-Mental Status test (Fig. 33–2). Special tests of parietal lobe function include drawing a clock face (Fig. 33–3), bisecting a line, and copying a picture of a daisy or drawing a set of intersecting pentagons.

Speech disorders such as *dysarthria* are detectable in ordinary conversation and result from defects of articulation of the words secondary to tongue (cranial nerve XII), palate (cranial nerves IX and X), lips (cranial nerve VII), or pharyngeal muscle weakness or incoordination. Evidence of a speech disorder is usually pursued by asking the patient to repeat a difficult phrase like "Methodist Episcopal" or to repeat the sounds "puh-tuh-kuh" rapidly.

Language is the ability to use and understand written and spoken speech and is a function of the cortical, thalamic, and basal nuclei language circuits located in the dominant cerebral hemisphere. Language is assessed by asking the patient to repeat words or phrases ("no ifs, ands, or buts"), to name simple objects (watch, finger, pen), to follow commands (touch your left shoulder, close your eyes, point to the ceiling), and to write a sentence and read it aloud.

Language abnormalities are called aphasias. There are two major types: *nonfluent aphasia* and *fluent aphasia*. In nonfluent aphasia the patient has difficulty with verbal self-expression, producing the words only with great effort, but is able to understand and follow commands appropriately. Nonfluent aphasia is also called *expressive aphasia*. In fluent aphasia the patient has normal or even increased production of words, sometimes in long sentences with normal prosody (rhythm of speech); well-articulated but frequent neologisms (a series of meaningless words) give these sentences no content or meaning. In fluent aphasia, also called *receptive aphasia*, neither the patient nor the examiner is able to understand the meaning of the patient's speech.

Cranial Nerve Function Testing

Cranial Nerve I. The *olfactory nerve (cranial nerve I)* is rarely tested, because of the deleterious effects of smoking and sinus disease on the sense of smell in the general population. The nerve can be unilaterally damaged by trauma or a tumor of the skull base in the olfactory groove such as an olfactory groove meningioma (see Fig. 7–7). Total loss of the ability to smell (*anosmia*) is always associated with the inability to taste food (*ageusia*) as well—a familiar example being the unappealing taste of food associated with the nasal congestion of a head cold. *Dysageusia* is an unappealing or altered sense of taste, and *parosmia* is an altered or perverted perception of odors.

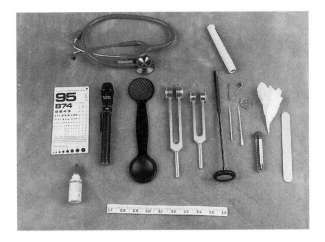

Figure 33–1. Instruments used to conduct a general neurologic examination.

Subject _____ , _____ _____/_____/_____ Examiner _____
                                                           yr   mo   d

Normal or DX _____ Age _____ Edu _____ M F 3MS _____ MMS _____
                                                    yrs   yrs   100   30

3MS MMS

5	

DATE AND PLACE OF BIRTH

Date: year _____, month _____, day _____
Place: town _____, state _____

3	3

REGISTRATION (No. of presentations: _____)

SHIRT, BROWN, HONESTY
(or: SHOES, BLACK, MODESTY)
(or: SOCKS, BLUE, CHARITY)

7	5

MENTAL REVERSAL

5 to 1
　　Accurate　　　　　　　　　　　2
　　1 or 2 errors/misses　　　　　0 1
DLROW
　　　　　　　　0 1 2 3 4 5

9	3

FIRST RECALL

Spontaneous recall　　　　　　　3
After "Something to wear"　　　　2
"SHOES, SHIRT, SOCKS"　　　　0 1

Spontaneous recall　　　　　　　3
After "A color"　　　　　　　　　2
"BLUE, BLACK, BROWN"　　　　0 1

Spontaneous recall　　　　　　　3
After "A good personal quality"　2
"HONESTY, CHARITY, MODESTY"　0 1

15	5

TEMPORAL ORIENTATION

Year
　　Accurate　　　　　　　　　　8
　　Missed by 1 year　　　　　　4
　　Missed by 2–5 years　　　　0 2
Season
　　Accurate or within 1 month　0 1
Month
　　Accurate or within 5 days　　2
　　Missed by 1 month　　　　0 1
Day of month
　　Accurate　　　　　　　　　　3
　　Missed by 1 or 2 days　　　　2
　　Missed by 3–5 days　　　　0 1
Day of week
　　Accurate　　　　　　　　　0 1

5	5

SPATIAL ORIENTATION

State　　　　　　　　　　　　0 2
County　　　　　　　　　　　0 1
City (town)　　　　　　　　　0 1
HOSPITAL/OFFICE BUILDING/HOME?　0 1

5	2

NAMING　(MMS: Pencil ___, Watch ___)

Forehead ___, Chin ___, Shoulder ___
Elbow ___, Knuckle ___

10	

FOUR-LEGGED ANIMALS (30 seconds) 1 point each

6	

SIMILARITIES

Arm-Leg
　　Body part; limb; etc.　　　　　　2
　　Less correct answer　　　　　0 1
Laughing-Crying
　　Feeling; emotion　　　　　　　2
　　Other correct answer　　　　0 1
Eating-Sleeping
　　Essential for life　　　　　　　2
　　Other correct answer　　　　0 1

5	1

REPETITION

"I WOULD LIKE TO GO HOME/OUT."　　2
1 or 2 missed/wrong words　　　　0 1
"NO IFS ___ ANDS ___ OR BUTS ___"

3	1

READ AND OBEY "CLOSE YOUR EYES"

Obeys without prompting　　　　　3
Obeys after prompting　　　　　　2
Reads aloud only　　　　　　　0 1
(spontaneously or by request)

5	1

WRITING (1 minute)

(I) WOULD LIKE TO GO HOME/OUT.
(MMS: Spontaneous sentence: 0 1)

10	1

COPYING TWO PENTAGONS (1 minute)

	Each Pentagon	
5 approximately equal sides	4	4
5 unequal (>2:1) sides	3	3
Other enclosed figure	2	2
2 or more lines	0 1	0 1
	Intersection	
4 corners	2	
Not-4-corner enclosure	0	1

3	3

THREE-STAGE COMMAND

___ TAKE THIS PAPER WITH YOUR
　　LEFT/RIGHT HAND
___ FOLD IT IN HALF, AND
___ HAND IT BACK TO ME

9

SECOND RECALL

(Something to wear)　　　　0 1 2 3
(Color)　　　　　　　　　　0 1 2 3
(Good personal quality)　　　0 1 2 3

Figure 33–2. The Folstein Mini-Mental examination.

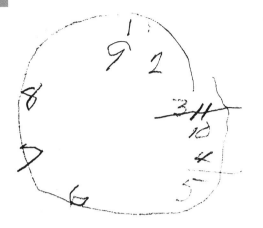

Figure 33–3. A clock face drawn by a patient with a parietal lobe lesion.

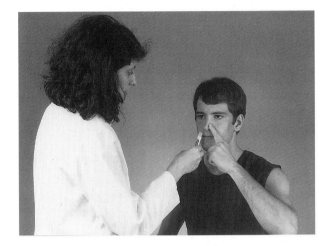

Figure 33–4. Testing the sense of smell (olfaction). The patient presses one nostril closed, and the open nostril is exposed to an aromatic substance.

Olfactory stimuli should be nontrigeminal—that is, it should not tickle or irritate the inside of the nose (as does ammonia, for example), which is innervated by the trigeminal nerve. Commonly used substances are vanilla, coffee, and perfumed soap. With the patient's eyes closed, occlude one nostril and bring a vial of the substance near the open nostril (Fig. 33–4). Ask the patient whether he or she smells something or not. The sensing of odor is more important than its identification. The process is then repeated for the other nostril.

Cranial Nerve II. The *optic nerve (cranial nerve II)* is tested by measuring the visual acuity and assessing the extent of peripheral vision by examining visual fields and by inspecting the retina and the optic nerve head using the ophthalmoscope. Visual acuity (also called visual resolution) is tested separately for each eye and should be recorded using the patient's best spectacle correction and a handheld visual acuity chart or a Snellen chart at 20 feet (Fig. 33–5). The number beside each line of letters indicates the number of feet at which the letters can be read by a person who has normal vision; thus, normally, the letters in the line designated 20 can be read at 20 feet, and the visual acuity is recorded as 20/20.

Examination of the visual fields is an important part of the ophthalmologic and neurologic examination. This procedure provides information about the entire visual pathway from the optic nerve to the occipital cortex. Because lesions interrupting various parts of the pathway cause specific types of defects in the visual field, it is frequently possible to determine the location of the lesion (see Chapter 20 for examples). There are several different methods for evaluation of the visual field. The commonest method used by most neurologists at the bedside is the *confrontation visual field examination.* The examiner faces the patient being examined. The patient should cover one eye with the palm of the hand, or with an eye occluder, and fixate the gaze of

the eye to be examined on the examiner's nose. Then the examiner presents a stimulus in each of the four quadrants—upper and lower nasal and upper and lower temporal—of the visual field; finger movement,

Figure 33–5. The Snellen chart.

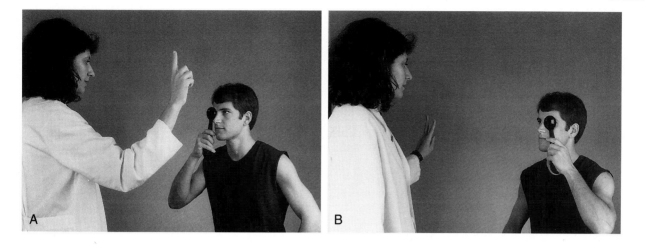

Figure 33–6. Visual field examination by confrontation. One eye is covered *(A)*, and all visual quadrants are tested for that eye. The procedure is repeated for the other eye *(B)*.

rapid finger counting, or hand comparison may be used for this purpose (Fig. 33–6*A* and *B*).

A *scotoma* (Greek for "spot") is a defect of the visual field surrounded by normal vision. The *blind spot* is a physiologic scotoma that represents the position of the optic disc within the visual field (the optic disc has no rods, cones, or ganglion cells; see Chapter 20). Loss of vision in one half of the field in one eye is called *hemianopsia*, and loss of vision in corresponding halves of the visual fields of both eyes is called *right* or *left homonymous hemianopsia* depending on which visual fields are lost. Loss of vision in different halves of the visual fields of both eyes is called *bitemporal visual field loss*.

The appearance of the optic nerve head or the optic disc is examined with an ophthalmoscope (Fig. 33–7), while the patient looks at a distant object. To examine

the patient's right eye, the examiner holds the ophthalmoscope with the right hand and uses his or her own right eye; this technique is reversed for examination of the patient's left eye. With the ophthalmoscope dial set on zero, the pupillary red reflex (the point at which the retinal reflex is seen "glowing" in the pupil) is located from a distance of about 2 or 3 feet (see Fig. 33–7*A*). The examiner slowly approaches the patient's eye as if viewing the eye through a keyhole. At the same time, plus or minus lenses, as needed, are dialed on the ophthalmoscope to focus on the patient's retina. The *optic disc* is located by directing the ophthalmoscope slightly toward the nasal side of the patient's retina (see Fig. 33–7*B*). The appearance of the optic disc is important. Normally it is round or slightly oval in shape and of a yellowish-red color, with clearly defined margins (Fig. 33–8*A*). Veins are darker in color and

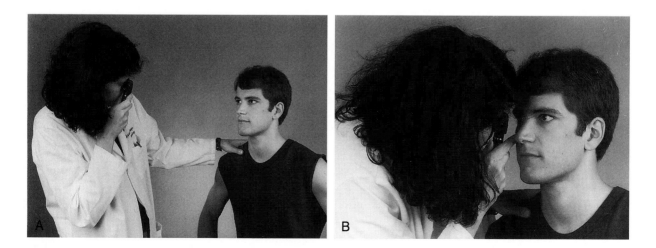

Figure 33–7. Ophthalmoscopic examination. The examiner locates the red reflex *(A)* and then focuses on the details of the optic nerve *(B)* through the pupil.

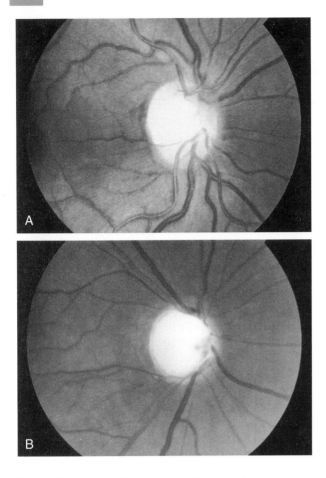

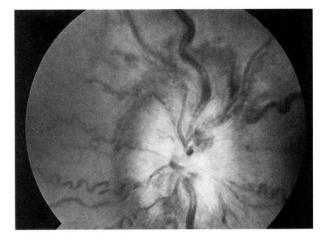

Figure 33–9. Papilledema with hemorrhages. Observe the tortuosity of vessels.

Figure 33–8. Normal optic disc *(A)* and an example of an abnormally enlarged central cup *(B)* in a patient with glaucoma.

slightly larger in diameter than arterioles. The presence of the *central cup,* or excavation, and its size should be documented (see Fig. 33–8*B*).

Papilledema, or swelling of the optic disc, is usually due to increased intracranial pressure, regardless of the cause of the pressure increase. Early signs of papilledema include disappearance of the normal cup, blurring of the disc margins, and arching and elevation of the vessels as they pass over the margin of the disc. As papilledema progresses, exudates and hemorrhages appear, as well as tortuosity of the vessels (Fig. 33–9).

Optic atrophy may be primary or secondary. Primary optic atrophy results from different processes involving the optic nerve such as retrobulbar optic nerve injury, compression by a tumor, or demyelination. Secondary optic atrophy is a consequence of chronic increased intracranial pressure, infarctions, or diseases such as syphilis (Fig. 33–10).

Cranial Nerves III, IV, and VI. The *oculomotor, trochlear, and abducens nerves (cranial nerves III, IV, and VI)* are usually examined as a group because they act together in controlling ocular muscles to ensure that the eyes remain parallel throughout their range of motion. A lesion affecting one or more of these nerves

results in weakness of the corresponding muscles, manifested by *diplopia* or *double vision.* Ocular motility is tested by having the patient follow the examiner's finger in upgaze and downgaze and from side to side (Fig. 33–11*A* and *C*).

The oculomotor nerve innervates the superior, medial, and inferior rectus muscles; the inferior oblique; and the constrictor of the pupil and the ciliary body as well as the levator of the eyelid (see Fig. 28–1). A complete lesion of the oculomotor nerve results in paralysis of the ipsilateral muscles innervated by the nerve and ptosis, pupillary dilation, and inability to look upward, downward, or inward. Aneurysms of the internal carotid artery or posterior communicating artery and pressure from herniation of the uncinate gyrus (uncus) in expanding lesions of the cerebral hemisphere are common causes of a peripheral complete third nerve palsy with pupillary involvement.

The trochlear nerve innervates the superior oblique muscle (see Fig. 28–1). When the fourth nerve is dam-

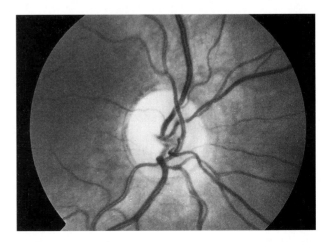

Figure 33–10. Optic atrophy. Note the pale appearance of the optic disc.

depress normally relative to the other eye. Thus, it is "higher" than the normal eye.

The abducens nerve controls the ipsilateral lateral rectus muscle, which makes the eye look outward (laterally). Defects in abduction (from a lesion of the sixth cranial nerve, for example) give the patient a "cross-eyed" appearance, because the normal eye is oriented straight ahead and the affected eye is rotated slightly inward (medially) owing to the un-opposed action of the medial rectus muscle on that side.

Cranial Nerve V. The *trigeminal nerve (cranial nerve V)* is both motor and sensory. In sensory testing, its innervation includes the face up to the vertex of the scalp but spares the angle of the mandible (which is innervated by C3). The sensation from the oral and nasal cavities is transmitted through the trigeminal nerve, although these areas are not usually included in the routine neurologic examination (see Chapter 18).

Pain and temperature should be tested in the three divisions of the fifth cranial nerve: the ophthalmic, the maxillary, and the mandibular (Fig. 33–12A and B; see also Figs. 18–4 and 18–13). The ophthalmic division innervates the scalp as far back as the vertex of the skull, forehead, cornea, conjunctiva, and skin of the side and tip of the nose. Corneal sensation is tested by gently touching the cornea with a cotton tip or tissue paper while the patient looks in the other direction (see Fig. 33–12C). This maneuver constitutes the afferent limb of the corneal reflex. The normal response is a rapid, partial or complete blinking movement of the eyelid elicited by the efferent limb of the corneal reflex via the facial nerve. The second trigeminal division, the maxillary nerve, conducts stimuli from the skin of the cheek, far lateral aspect of the nose, upper teeth, and jaw. The third division, the mandibular nerve, carries sensory and motor impulses. The sensory distribution is skin of the lower jaw, pinna of the ear, and lower teeth and gums as well as the side of the tongue.

The motor fibers supply the muscles of mastication: the temporal, masseter, and pterygoid muscles. The temporal and masseter muscles are examined by having the patient close the jaws together while the examiner palpates these muscles (Fig. 33–13A). The pterygoid muscles are responsible for side-to-side movements of the jaw, as well as aiding closure of the jaw (see Fig. 33–13B). The *jaw jerk reflex* is elicited by a gentle tap on the chin, with resultant closure of the jaw by the masticatory muscles. The afferent limb of this reflex is via receptors in the muscles of mastication that enter the brainstem on fibers of the mesencephalic tract, and the efferent limb is in response to collaterals of these fibers that bilaterally innervate the motor trigeminal muscles (see Fig. 14–12).

Cranial Nerve VII. The *facial nerve (cranial nerve VII)* is a complex nerve with motor, sensory, and parasympathetic (visceromotor) fibers. The motor portion of the nerve innervates the muscles of facial expression and is tested by instructing the patient to wrinkle the

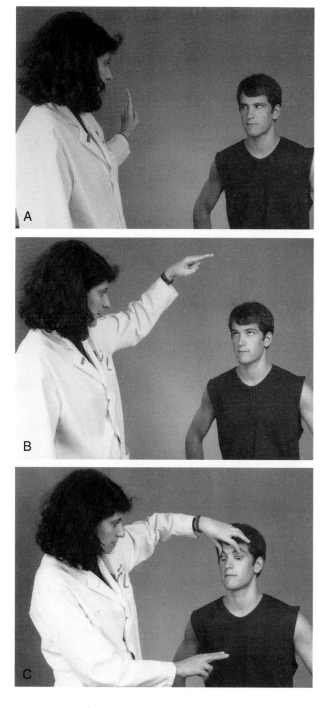

Figure 33–11. Test of ocular motility to evaluate the function of the extraocular muscles. The patient holds the head still and follows the examiner's fingers with the eyes. Examples here show the patient looking to his right *(A)*, upward *(B)*, and downward *(C)*.

aged, the affected ipsilateral eye is higher than the normal opposite eye, and it cannot be turned downward when the eye is rotated inward (adducted). The position of the globe (eyeball) is higher relative to the position of the other globe because the superior oblique muscle normally depresses the eyeball. When this muscle is weak or paralyzed, the eyeball will not

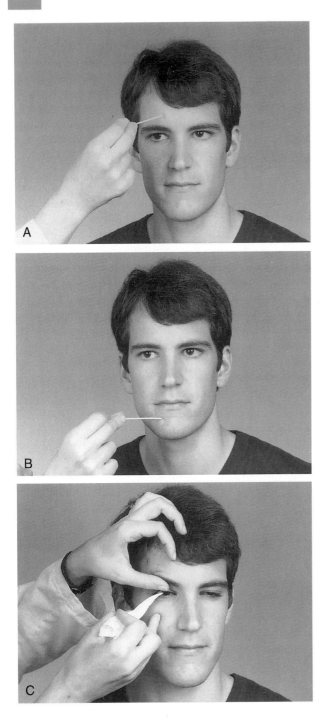

Figure 33–12. Testing sensory portions of the trigeminal nerve. Examples show a probe touching the ophthalmic *(A)* and mandibular *(B)* territories of the trigeminal nerve; the maxillary division is tested by touching the check below the eye. A wisp of tissue touched to the cornea *(C)* activates the afferent limb of the corneal reflex and results in closing of the eyes; the efferent limb is mediated by the facial nerve.

or corticonuclear (corticobulbar) pathways, and the other with involvement of the lower motor neuron, or "peripheral" seventh nerve palsy. The "central" or upper motor neuron facial palsy is characterized by inability to retract the corner of the mouth, while forehead function and eyelid closure remain for the most part unaffected. Lesions in the facial nucleus or the nerve proper will cause paralysis of half of the entire face, with inability to wrinkle the forehead or to close the eyelids and lips on the affected side (Fig. 33–15*A* and *B*; see also Fig. 25–11).

The sensory portions of the seventh nerve originate from the taste buds in the anterior two thirds of the tongue and from the posterior wall of the external ear canal. Taste is examined using sugar, salt, or quinine solutions. The patient is instructed to protrude the tongue; then the test substance is applied with a cotton-tipped applicator on one side of the tongue. The patient must identify the test substance before drawing the tongue back into the mouth, where function of the posterior portion of the tongue or the contralateral side masks the result of the test. The

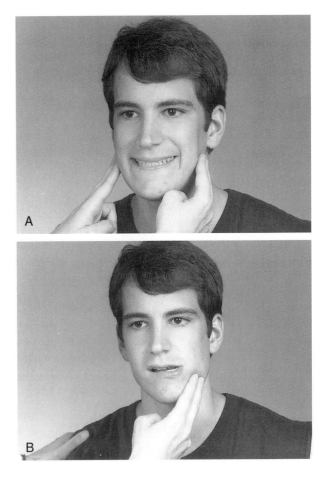

Figure 33–13. Testing the muscles of mastication. The patient clenches the masticatory muscles while they are palpated by the examiner *(A)* and deviates the jaw against resistance *(B)*; this maneuver tests the pterygoid muscles.

forehead, to close the eyelids tightly, to smile or grimace showing the teeth, and to whistle (Fig. 33–14*A–D*). There are two types of facial motor weakness, one with involvement of the upper motor neuron

Figure 33–14. Testing the muscles of facial expression. The patient is asked to tightly close the eyes *(A)*, smile *(B)*, purse the lips *(C)*, and wrinkle the forehead *(D)*. In each case, the examiner carefully assesses the symmetry of the face.

facial nerve also carries parasympathetic fibers to the maxillary and lacrimal glands (see Chapter 14).

Cranial Nerve VIII. The *eighth cranial nerve* is made up of two divisions: cochlear, subserving the sense of hearing, and vestibular, subserving the sense of balance—hence its common name, the *vestibulocochlear nerve.*

The cochlear division is usually tested using a tuning fork with a frequency of 256 vibrations per second to compare bone and air conduction (Fig. 33–16A–D). This examination is known as the *Rinne test.* In the normal ear, air conduction is greater than bone conduction. The vibrating tuning fork is placed against the mastoid bone (see Fig. 33–16C), and the patient is

instructed to indicate when he or she no longer senses the vibration. Then the tuning fork is placed near the external auditory canal (see Fig. 33–16D), and the time for air conduction is estimated. With a normal test result, the time for air conduction is about twice that for bone conduction. Bone conduction will be louder in conductive hearing loss; both air and bone conduction are abnormal in neurogenic hearing loss. The *Weber test* is performed by placing the tuning fork on the vertex of the skull (see Fig. 33–16B); normally, the vibration is perceived equally in both ears. If there is disease of the middle ear or the external ear is blocked (conductive hearing loss), the vibration is lateralized to the affected side. If the cochlear nerve is involved on

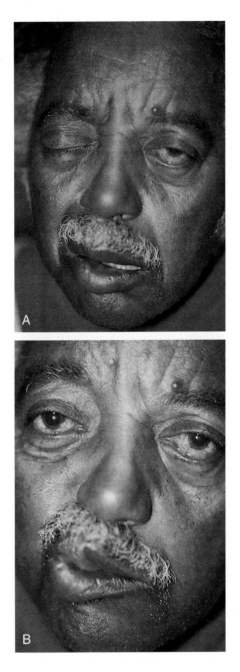

Figure 33–15. A patient with a lesion of the facial nerve. The patient has difficulty in closing his left eye and the left corner of his mouth droops (*A;* compare with Fig. 33–13*A* and *B*). The latter defect is especially evident when the patient attempts to purse his lips (*B;* compare with Fig. 33–13*C*).

one side (neurogenic hearing loss), the sound is heard better on the opposite or the normal side.

The vestibular division of the acoustic nerve is assessed using rotational and caloric stimuli to produce changes in the endolymph current in the semicircular canals (see Chapter 22). Typically, patients with vestibular dysfunction complain of vertigo, nausea and vomiting, and difficulty with balance, especially with move-

ment of the head. The *water caloric test*, or *Bárány test*, is done by irrigating the external auditory canal with 10 mL of cold water while holding the patient's head at 30 degrees above the horizontal. The patient is then examined for horizontal nystagmus with the slow component toward the side of the stimulus past the midline and the fast corrective phase of the nystagmus to the opposite side. The mnemonic COWS (*c*old—*o*pposite/ *w*arm—*s*ame) refers to the direction of the fast phase of the nystagmus.

Cranial Nerves IX and X. The *glossopharyngeal and vagus nerves (cranial nerves IX and X)* are usually examined at the same time. Touching the posterior wall of the pharynx with a tongue depressor tests the general sensory fibers of the ninth nerve. The normal response is the prompt contraction of the pharyngeal muscles, including the stylopharyngeus muscle. Afferent information conducted on the ninth nerve and the resultant contraction of the stylopharyngeus muscle constitute the circuit of the gag reflex.

Vagus nerve dysfunction will result in ipsilateral paralysis of the palatal, pharyngeal, and laryngeal muscles. In such cases, the voice is hoarse as a result of weakness of the vocal cord (and vocalis muscle), and the speech has a nasal sound. The soft palate should be observed while the patient says "Ah." Normally, the uvula remains in the midline, but in case of weakness of the palate on one side, it is pulled toward the contralateral side because of the unopposed action of muscles on the healthy side (see Fig. 25–12).

Cranial Nerve XI. The *spinal accessory nerve (cranial nerve XI)* innervates the ipsilateral sternocleidomastoid and trapezius muscles. It is examined having the patient turn the head forcibly against the examiner's hand away from the muscle being tested while the muscle is palpated (Fig. 33–17). Damage to this nerve causes inability to shrug the shoulder (weakness of the trapezius muscle) and winging of the scapula on the side of the lesion.

Cranial Nerve XII. The *hypoglossal nerve (cranial nerve XII)* supplies the extrinsic and intrinsic muscles of the tongue. To test the integrity of the hypoglossal nerve, the patient is asked to protrude the tongue in the midline and move it from side to side (Fig. 33–18*A–C*). In the presence of a lesion of the hypoglossal nerve, the tongue is seen to deviate toward the side of the lesion, toward the weak half, on attempted protrusion (see also Figs. 25–12 and 25–13). This deficit is due to a paralysis of the genioglossus muscle.

The Motor Examination

The motor examination includes a consideration of muscle tone and strength. Tone examination requires that the patient be as relaxed as possible. The patient may respond to suggestions such as "Go loose" or "Let

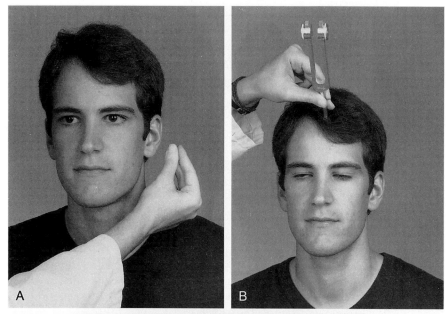

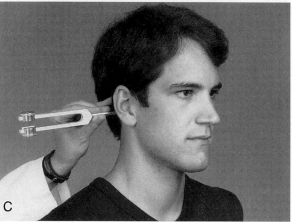

Figure 33–16. Test of auditory function (hearing). If the patient is unable to identify a sound made with the fingers (A), the examination then proceeds to a test of bone conduction for both ears together (B) and for each ear separately (C), and of air conduction for each ear (D).

Figure 33–17. Testing the strength of the sternocleidomastoid muscle by rotating the head against resistance. A test of the integrity of the accessory nerve also includes asking the patient to shrug the shoulders (trapezius muscle).

severe as to prevent passive movement. This increased tone is called *lead pipe rigidity* and is a feature of Parkinson disease (see Chapter 26).

Spasticity is a phasic change in muscle tone brought out by a rapid snap of the limb in extension or flexion. The spastic "catch" is an abrupt increase in the tone followed by a slow release, much as in the operation of the hydraulic hinge on the rear door of a hatchback automobile. Spasticity is seen with corticospinal tract lesions. *Hypotonia* is characterized by increased ease of passive movements, as exemplified by the pendular swing of a leg extended and released in the sitting position.

Muscle strength testing (Fig. 33–19*A–C*) requires the patient's cooperation. Results are usually graded as follows:

  0 = paralysis
+1 = minimal muscle contraction
+2 = muscle contracts but the patient is unable to lift the limb
+3 = able to hold the limb against gravity
+4 = able to hold against resistance but the examiner is able to overcome
+5 = not able to overcome resistance

The different muscle groups are examined in an organized fashion, proximal to distal in the upper and lower extremities, documenting the degree and pattern of strength or weakness observed (see Fig. 33–19*A–C*). A lesion in the cerebral hemisphere produces hemiparesis with weakness involving the face and upper and lower extremities on the contralateral side (see

me do all the work and go floppy," thereby permitting the examiner to move the patient's limbs freely. Normally, a mild resistance to movement is noted during the whole range of motion. In *hypertonicity*, increased resistance is present in extensor and flexor muscles. Flexor resistance can range between very mild to so

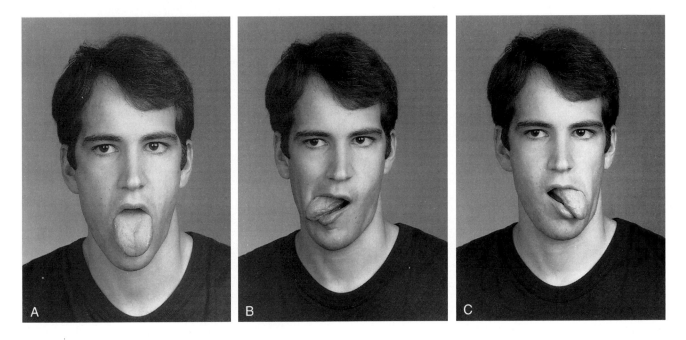

Figure 33–18. Testing the hypoglossal nerve. The patient is asked to protrude the tongue straight out *(A)*, to the right *(B)*, and to the left *(C)*. The examiner looks for asymmetry in these movements or for an inability to perform these movements.

Figure 33–19. Test of muscle strength. Many muscles can be used. The examples shown here are the biceps *(A)*, deltoid *(B)*, and quadriceps femoris *(C)* muscles.

Chapter 25). A mid-thoracic (or slightly lower) lesion in the spinal cord may produce weakness in both lower extremities *(paraplegia)*, with an associated *sensory deficit* and abnormal sphincter control. Weakness involving only one limb is called *monoparesis;* it is commonly, but not invariably, localized to a plexus or a peripheral nerve. A mid-cervical lesion of the spinal cord may result in *quadriplegia* (bilateral paralysis of both upper

and lower extremities) with a corresponding sensory loss; if the lesion is at the C1 or C2 level, the patient may also experience difficulty breathing without assistance. A lesion of one side of the spinal cord at mid-cervical levels may result in paralysis of the upper and lower extremities on that side; this deficit is a *hemiplegia* and is usually accompanied by characteristic sensory deficits (Brown-Séquard syndrome) (see Figs. 25–8 and 25–9).

Muscle Stretch Reflexes. The muscle stretch reflexes (also called "deep tendon" reflexes—a misnomer) are obtained by percussing the tendons of major muscles (Fig. 33–20A–D). The muscles are innervated by nerves from specific spinal cord levels. The afferent impulses are conducted to the spinal cord, or the brainstem, by the sensory fibers in the peripheral nerve and the corresponding posterior root or cranial nerve. The impulse then acts on the anterior horn cells of the cord (or motor cells of cranial nerves), and the action potential travels through the motor roots and peripheral nerve back to the muscle (see also Chapter 9). Normal reflexes indicate that the sensory-motor loop to and from the spinal cord (or brainstem) is intact.

Reflexes are modulated by down coming inhibitory and excitatory influences from the cortical, vestibular, and reticular regions of the cerebral hemispheres and brainstem (see Chapter 24). When the inhibitory influences are damaged, the resulting reflex elicited by tapping a tendon may be very brisk or hyperactive, called *hyperreflexia.* If the nerve leading to or from the muscle is injured, reflexes may be hypoactive *(hyporeflexia)* or absent *(areflexia).*

In the upper extremity, four reflexes are usually tested: the biceps reflex (see Fig. 33–20A), mediated by C5-C6 through the musculocutaneous nerve; the triceps reflex (see Fig. 33–20B), mediated by C7-C8 through the radial nerve; the brachioradialis reflex, mediated by C5-C6 radial nerve; and the finger flexor reflex, mediated by C7-C8 ulnar and median nerves. In the lower extremity, two reflexes are commonly tested: the quadriceps reflex, elicited by tapping the patellar tendon (see Fig. 33–20C) and mediated by L2-L4 through the femoral nerve, and the "Achilles reflex" or ankle jerk reflex, mediated by S1 through the sciatic (tibial) nerve and elicited by tapping the tendon of the gastrocnemius muscle (see Fig. 33–20D).

An example of a pathologic reflex is the *Babinski sign,* seen on stroking the lateral border of the sole of the foot from the heel to the base of the great toe (Fig. 33–21). This reflex consists of dorsiflexion of the great toe, sometimes with fanning of the other toes (see Fig. 33–21B). The normal response is flexion of all of the toes (see Fig. 33–21A). In an adult, the Babinski sign indicates some type of abnormal process, whereas this sign may be present in a normal infant. The incomplete myelination seen in newborns or infants is the likely explanation of this latter observation.

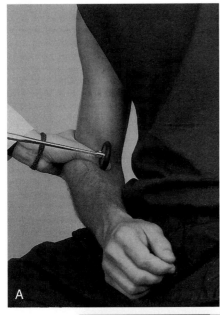
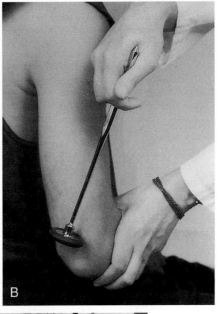
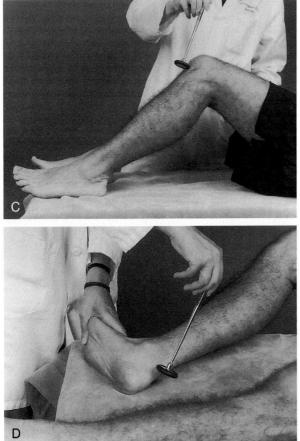

Figure 33–20. Examination for muscle stretch reflexes (tendon reflexes) of the biceps *(A)*, triceps *(B)*, the quadriceps femoris *(C, patellar tendon)*, and gastrocnemius *(D, Achilles tendon)* muscles.

Cerebellar Testing. Cerebellar testing can be thought of as a mix of motor and sensory testing that assesses the accuracy of movement. In addition to normal cerebellar function, the patient must have normal strength, tone, and sensory input in order to carry out coordinated movements. It is important to compare coordination of one side of the body with the other. The cerebellum is usually tested by having the patient perform a finger-to-nose-to-finger maneuver (in which the patient touches alternately the examiner's finger and

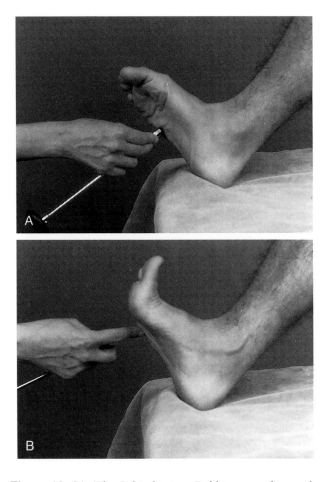

Figure 33–21. The Babinski sign. Rubbing a probe on the plantar aspect of the foot in a normal person results in a plantar flexion of the toes *(A)*. Dorsiflexion of the toes *(B)*—the Babinski sign—elicited by briskly rubbing the plantar surface of the foot is indicative of a lesion involving descending fibers from the cortex and brainstem that influence spinal motor neurons.

then his or her own nose rapidly) (Fig. 33–22*A* and *B*); the heel-knee-shin maneuver (in which the patient puts the heel on the opposite knee and runs it down the shin) is performed to test the accuracy of appendicular movement (see Fig. 33–22*C* and *D*). The inability to perform this maneuver is also called *limb ataxia*. These types of dysfunctions, largely relegated to the more distal parts of the body—the extremities—are

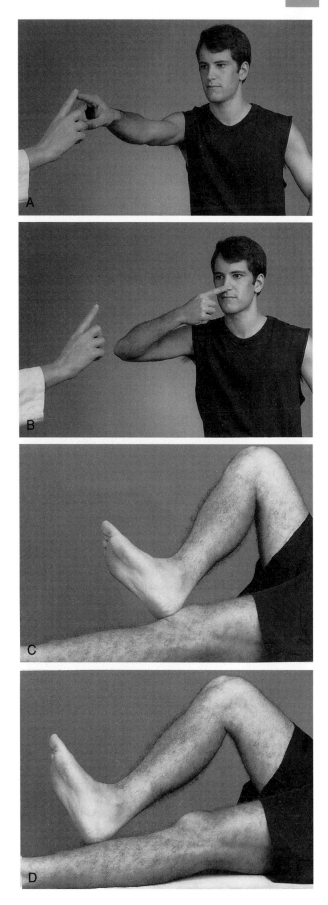

Figure 33–22. Testing of cerebellar function. The normal patient can touch the physician's finger and then his own nose and repeat the movement rapidly and without difficulty *(A, B)*. For the heel-to-shin test, the patient slides the heel of one foot down the shin of the other leg *(C, D)*.

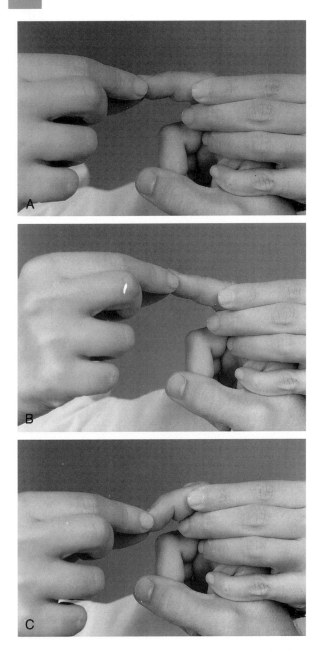

Figure 33–23. Test of proprioception/position sense. The physician holds the patient's finger *(A)* and, with the patient's eyes shut, asks the patient if the finger is being moved up *(B)* or down *(C)*. The same test can be conducted using, for example, the toe, hand, or foot.

indicative of damage to more lateral portions (the hemispheres) of the cerebellum (see also Figs. 27–18 and 27–19).

Truncal ataxia (titubation—from the Greek word meaning to stagger or lurch) is present when the patient exhibits unsteadiness while sitting, standing, or walking in tandem. This finding, in which primarily axial parts of the body are affected, is evidence of midline cerebellar dysfunction.

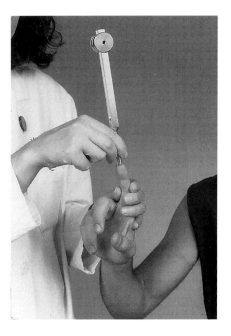

Figure 33–24. Test of vibratory sense. The tuning fork may be placed on the finger tip, tips of the toes, or bony prominences.

The Sensory Examination

Sensory testing is purely subjective; results obtained depend heavily on the patient's accuracy and cooperation. The sensory examination is most conveniently divided into *anterolateral system testing of pain and temperature sense* and *posterior column testing of vibration and position sense*.

The standard method of evaluating pain perception is to stimulate the skin with a pin and ask the patient

Figure 33–25. Test for shape and texture of an object (stereognosis). This portion of the neurologic examination evaluates cortical function.

if the stimulus is perceived as sharp. Because the entire body surface cannot be evaluated, the examination must be guided by the nature and location of signs and symptoms, such as numbness or tingling in a specific distribution. Temperature sensation may also be tested using a cold metallic object or a small tube of warm water.

To test *position sense*, the patient is instructed to relax and, with the eyes closed, to indicate whether he or she feels the finger, (or toe) moving up or down (Fig. 33–23*A–C*). The *Romberg test* evaluates the sense of position of the legs and trunk. While the patient stands with feet together and the eyes closed, the examiner looks for the presence of any sway or imbalance. The patient's ability to participate may be affected by vestibular, cerebellar, or peripheral nerve disease. Vibration sense is tested with a 128-Hz tuning fork, usually applied to the bony prominences of the terminal phalanges of the thumbs and great toes (Fig. 33–24). The patient is instructed to close the eyes and indicate whether a "buzzing" sensation is experienced.

Vibration changes may be seen in peripheral nerve disease and spinal cord problems, especially those affecting the posterior columns.

Cortical sensory function is evaluated only if there is no loss of primary sensation. Testing is done using familiar objects such as a quarter, a wooden cube, or a plastic pen placed in the patient's hand while the patient's eyes are closed (Fig. 33–25). The patient is then asked to identify the object(s). *Stereognosis* is the perception of the form and nature of an object. Two-point discrimination testing is also valuable. Using two pointed objects, the stimuli are applied at the same time, and the patient is asked whether one or two points are detected. *Agnosia* is a "percept stripped of its meaning." *Stereoagnosia* is an inability to identify objects by touch (*tactile agnosia*) or by sight (*visual agnosia*), sounds or words (*auditory agnosia*), colors (*color agnosia*), or the location or position of an extremity (*position agnosia*). This type of deficit results from lesions in the cerebral hemisphere.

Sources and Additional Reading

Adams AC: Neurology in Primary Care. FA Davis, Philadelphia, 2000.

Aminoff MJ, Greenberg DA, Simon RP: Clinical Neurology, 3rd ed. Appleton & Lange, Stamford, Conn, 1996.

Donaghy M: Neurology. Oxford University Press, New York, 1997.

Duus P: Topical Diagnosis in Neurology: Anatomy, Physiology, Signs, Symptoms, 2nd rev. ed. Thieme, Stuttgart, 1989.

Haerer AF: DeJong's The Neurologic Examination, 5th ed. JB Lippincott, Philadelphia, 1992.

Martin TJ, Corbett JJ: Neuro-Ophthalmology: The Requisites in Ophthalmology. Mosby, St. Louis, 2000.

Rowland LP: Merrits Neurology, 10th ed. Lippincott Williams & Wilkins, Philadelphia, 2000.

Victor M, Ropper AH: Adams and Victor's Principles of Neurology, 7th ed. McGraw-Hill, New York, 2001.

Index

Note: Page numbers followed by the letter f refer to figures; those followed by the letter t refer to tables.

Aα fiber(s), 260f
Aβ fiber(s), 143–144, 260f
Aδ fiber(s), 143, 260f, 277
  in anterolateral system receptors, 275
  in direct spinothalamic pathway, 279–280
  in reflexes, 146, 147
  in spinal trigeminal pathway, 285
  nociceptive sensations in, 277
Abducens motor nucleus, 154
Abducens nerve, 132, 132f, 152, 152f, 161, 203f, 212–213, 212f
  corticospinal fiber damage to, 391–392, 394f
  functional components of, 156, 156f, 175
  in eye movements, 447
    paralysis of, 452, 452f
  lesions of, 213
  motor nuclei of, 175
    origin of, 174–175, 174f
  of pons, 175–176, 175f, 176f
  testing of, 527
Abducens nucleus, 84, 178, 180f, 200, 201f, 212–213, 212f, 353, 353f, 450–452, 451f
  lesions of, 452, 452f
Accessory cuneate nucleus, 164, 165f
Accessory nerve(s), 152, 152f, 161
  anatomy and distribution of, 203, 203f, 204f, 205–206, 205f
  lesions of, 206
  relationship to vagus nerve, 206
Accessory nucleus
  corticonuclear fibers in, 396, 397f
  in medulla, 163, 164f
Accessory optic nucleus, 313, 459
Accessory optic system, 459
Accommodation
  axonal, 46
  lens, 448f, 449, 458
Acetylcholine, 68, 346
  as neurotransmitter, 27–28, 58, 59t
  in chemical synapses, 27
  in neuromuscular junctions, 375
  in parasympathetic system, 473
  in spinal motor neurons, 145
  synthesis of, 61
Acetylcholinesterase, 375
  degradation of acetylcholine by, 68
Achilles reflex, 533, 534f
Achromatopsia, 321
Acoustic canal, internal, 344f
Acoustic chiasm, functional, 329
Acoustic meatus, internal, 209, 209f, 210, 210f
Acoustic neuroma, 161, 162, 349, 349f
Acoustic startle reflex, 338f, 339
Acoustic stria, 329, 330f, 333f
  posterior (dorsal), 332
Acquired immunodeficiency syndrome (AIDS), microglial cells targeted in, 32
Acromegaly, 489, 489f
Action potential(s), 5, 256
  cable properties of membranes and, 40, 41f, 42, 45
  characteristics of, 25
  compound, 48–49, 49f
    in mixed and cutaneous nerves, 259, 260f
    peripheral nerve classification and, 259, 259t
  conduction of, 44–46, 44f, 45f, 46f

Action potential(s) (Continued)
  cutaneous mechanoreceptors and, 258f
  frequency of, 47, 48, 48f
  generation of, 42–46, 43f, 45f, 46f
    after hyperpolarization in, 43–44, 67
    all-or-none nature of, 42, 44, 44f, 45f
    depolarization cycle in, 42–43, 44f, 67
    excitability cycle in, 44, 44f, 45f
    ion channels in, 42–44, 43f, 44f
    receptor potential amplitude and, 46–48, 48f, 49f
    refractory periods in, 44, 44f, 45f
    repolarization cycle in, 43–44, 44f
    trigger zone of, 42, 47, 52, 53f, 54f, 67, 256
  graded, interaction of, 50
  in information transmission, 26, 27f
  in retinal ganglion cells, 309
  in taste transduction, 368–369, 369f
  postsynaptic mechanisms in, 51–52, 51f–53f
  presynaptic mechanisms in, 50–51, 51f
  propagation of, 44–46, 44f, 45f, 46f
    accommodation in, 46
    changes in, 42
    conduction block in, 46
    safety factor in, 45, 46
  receptor potential magnitude and, 47
  threshold in, 46, 256
Active zones, of presynaptic nerve terminal, 26, 26f, 60, 62, 62f
Acupuncture-like stimulation, for pain control, 292
Addiction, 504
Adenohypophysis, of pituitary gland, 220f, 221, 488
Adenomas, of pituitary, 488–491
  classification of, 489
  Cushing disease and, 490–491, 490f
  gonadotropic tumors as, 491
  incidence of, 489
  prolactin-secreting tumors as, 491, 491f
  secreting tumors as, 489–490, 489f
Adenosine, 58, 59t
Adenosine triphosphate, 58, 59t
Adenylyl cyclase, in G protein–coupled receptor action, 65–66, 65f, 67
Adie syndrome, 460
Affect, disorders of, 518
Ageusia, 371, 520
Aggressive behavior, 483
Agnosia, 236, 264, 537
  apperceptive, 321
  associative, 321
  auditory, 501, 537
  color, 537
  object, 321
  position, 537
  tactile, 501, 537
  visual, 501, 537
Agraphia, 517
Akinesia, 415, 420f
Alar plate. See Plate(s), alar.
Alcoholism, chronic, 483, 500
Alexia, 517
  without agraphia, 322
Allocortex, 494
Allodynia, 276
  in central pain syndrome, 291
Alpha cell(s), 312

Alpha rigidity, 385
Alveus, 497, 498f
Alzheimer disease, 28, 409, 500
Amblyopia, 321, 446
Amenorrhea, 491, 491f
L-Amino acid decarboxylase, 80, 80f
Amino acids, 27–28, 58, 59t. *See also* Neurotransmitter(s).
  in taste transduction, 368–369, 369f
Aminoaciduria, 421
Ammon's horn, 250, 496, 498f
Amnesia, 501
  anterograde, 483
  in Klüver-Bucy syndrome, 501
  in Korsakoff syndrome, 500
AMPA receptors, 67
Ampulla
  of hair cells, 345f, 346–347, 347f
  of semicircular canals, 344
Amputation
  of limb or digit, somatosensory cortex reorganization and, 268–269, 269f
  thalamic ventrocaudal nuclei changes and, 290
Amygdala, 236, 237, 250–251
  relationships of, 248f
  vasculature of, 251
Amygdaloid complex, 96f, 97, 98f, 500–502
  afferent and efferent connections of, 487, 501, 502f
  damage to, Klüver-Bucy syndrome and, 496, 501
  hypothalamic connections of, 485, 485f, 486f, 487
  nuclei of, 494, 495f
  structure of, 495f, 500–501
Amygdaloid nuclear complex (amygdala), 248f, 250
Amygdaloid nucleus, anterior cortical, 364f, 365
Analgesia, 191
  stimulation-produced, for chronic pain, 292
Anencephaly, 73, 74f
  with rachischisis, 138, 139f
Anesthesia, 274
  epidural, 292
  spinal, 113
  with injury of peripheral nerves, 138
Anesthetics, local, fibers affected by, 277
Aneurysm(s)
  at ophthalmic-carotid junction, 123
  intracranial, 122, 122f
    in vertebrobasilar system, 127
  of anterior communicating artery, 123
  of arteries affecting oculomotor nerve, 526
  of cavernous part of carotid artery, 132, 132f
  of ophthalmic artery, 315f
  ruptured, in subarachnoid hemorrhage, 119–120, 119f
Angina, 70
  pain patterns and pathways of, 297f, 298, 299f
Angular gyrus, 238f, 239
Anhidrosis, 149, 453, 470, 472f
Anisocoria, 452f, 460
Annulus of Zinn, 447f
Anodal block, 46
Anosmia, 362, 522
Anoxia, 277
Ansa lenticularis, 249, 249f, 409, 411f
Anterior chamber, 304, 304f
Anterior perforated substance, 239, 241f
Anterograde (wallerian) degeneration, of axons, 22
Anterolateral cordotomy, 149, 280, 283
Anterolateral system, 141, 148, 148f, 166, 166f, 180, 180f, 181, 181f, 197, 295
  blood supply of, 280, 281f
  central pathways in, 277–283, 279f–283f
  central sensitization in, 276
  function of, 192f, 193
  hyperalgesia in
    primary, 275–276
    secondary, 276
  in medulla, 162, 163, 163f, 164, 164f, 165f, 280, 281f, 282
  in midbrain, 191, 191f

Anterolateral system *(Continued)*
  in pons, 187, 187f
  in spinal cord, 141, 142f, 148, 148f
  in trauma or disease of spinal cord, 280, 282, 282f
  overview of, 274–275, 274f
  pain receptors in
    in muscles and joints, 276–277, 276t, 277f–279f
    in viscera, 276–277, 276t, 277f–279f, 296, 297f
  peripheral sensitization in, 275–276
  primary neurons in, 275
  receptors in, 275, 275t
  somatotopic fiber arrangement in, 280, 281f
  testing of, 277, 536–537, 536f
Antidiuretic hormone, 483
  effects of, 488
  in water balance reflex, 492
  synthesis and transport of, 487
Antigravity muscle(s), 380
  spastic, 389
Anxiety, hypothalamic lesions and, 491
Aortic arch, specialized receptors in, 299
Aortic bodies, chemoreceptors in, 294–295
Aphagia, 207
Aphasia, 236, 501
  auditory, 337
  Broca, 337, 516–517, 517f
  conduction, 517
  definition of, 516
  expressive, 516, 517f, 522
  fluent, 517, 522
  global, 517
  nonfluent, 516, 517f, 520
  receptive, 517, 522
  Wernicke, 337, 514, 516, 517, 517f
Aphonia, 207
Apnea, central, medullary compression and, 169
Apoptosis, 268
  of neurons, in cellular brain development, 87
Apraxia, 236, 390
  constructional, 518
  dressing, 518
Aqueduct, cerebral. *See* Cerebral aqueduct.
Aqueductal stenosis, hydrocephalus with, 104–105
Arachidonic acid, in second messenger G protein system, 52, 53f
Arachnoid barrier cell layer, 110f, 113
Arachnoid mater, 109f, 113–114, 114f
  early development of, 108f, 109
  of spinal cord, 140, 140f, 141f
  organization and structure of, 109f, 110, 110f
Arachnoid trabeculae, 103, 110, 110f, 113–114
Arachnoid villi
  absence of, hydrocephalus with, 105
  cerebrospinal fluid movement in, 103–104, 104f, 114, 114f
Archambault's loop, 318
Archicortex, 494, 507
Area(s). *See also* Brodmann area(s); Layer(s); Zone(s).
  association, 236
  basilar, of midbrain, 153, 153f
  Broca, 337
  parietal association, input to and lesions of, 264, 264f
  prerubral, 224f, 230
  septal, 250
  subcallosal, 242, 494
  supraoculomotor, 458
  tegmental, of midbrain, 153, 153f
  V1, 319, 322f
  ventral tegmental, 408f, 411, 412, 494, 495f
  vestibular, 154, 154f, 177
  visual association, 313
  Wernicke, 337
Area 2v, 355–356, 356f
Area postrema, 166, 166f
Area X, 141, 141f
Areflexia, 6, 146, 388, 533
Arnold-Chiari malformation, 73–74, 75f
Arterial vasocorona, 134f, 135, 142, 142f, 280, 281f, 393f, 394

Arteriovenous malformations, 122f, 123, 123f
  spinal, 135
Artery(ies)
  accessory meningeal, 111
  anterior cerebral, 122, 123, 124f, 125, 128, 128f, 232f, 233, 233f,
      242, 243f, 282, 316, 390, 496, 515
    A_1 segment of, 123, 243
    A_2 segment of, 123, 243
    damage to, 288
  anterior choroidal, 101, 123, 124f, 189, 197, 197f, 198, 245, 250,
      251, 316, 316f, 407, 407f, 496
  anterior communicating, 123, 124f, 242
    anteromedial branches of, 233, 316
  anterior inferior cerebellar, 102, 125, 126f, 127, 177f, 178, 184,
      184f, 185, 329, 427, 428f
    damage to, 329
  anterior (ventral) radicular, 134f, 135
  anterior spinal, 125, 126f, 134–135, 134f, 142, 142f, 162, 170
    branches of, 203
      lesions of, 169
    central branches of, 142, 142f
    penetrating branches of, 392, 393f
    sulcal branches of, 280, 281f, 393f, 394
    trauma to, 142
  ascending pharyngeal, meningeal branch of, 208
  basilar, 122, 124f, 125, 127, 177–178, 177f, 189, 329
    circumferential branches of, 184, 184f
      of midbrain, 197, 197f
      of pons, 184, 184f
    paramedian branches of, 184, 184f
      damage to, 391–392, 398–399
    short circumferential branches of, 329
  calcarine, 124f, 127f, 128, 242
  callosomarginal, 390
  central retinal, 123, 124f, 306, 314, 316
  cerebral
    border zones of, 515
    watershed infarcts and, 128, 128f, 515
  choroidal, 101
  ciliary, 306
  circle of Willis. *See* Circle of Willis.
  collicular (quadrigeminal), 197f, 198
  ethmoidal, 111
  hypophysial, 484, 484f
  internal auditory (labyrinthine), 329
  internal carotid, 111, 122, 123, 124f
    aneurysm of, 526
    cavernous part of, 132, 132f
      aneurysm of, 132, 132f
    cerebral part of, 123, 124f
  labyrinthine, 124f, 127, 210, 329, 344
  lacrimal, 111
  lateral posterior choroidal, 101, 126f, 128
  lenticulostriate, 125, 129, 239, 242, 243f, 245, 246f, 250
    lesions of, corticonuclear and corticospinal system damage with,
        398
  medial posterior choroidal, 101, 126f, 128, 189, 197–198, 232f,
      233f, 234
  medial striate, 245, 250, 407, 407f
    lenticulostriate branches of, 246f, 250
  middle cerebral, 122, 125, 125f, 128, 128f, 129, 129f, 242, 243f,
      251, 264, 282, 318, 390, 515
    branches of, 125, 126f
    damage to, 288, 517
    lenticulostriate branches of, 407, 407f
      lesions of, 391, 391f
    M_1 segment of, 125, 125f, 242, 243f
    M_2 segment of, 125, 125f, 242, 243f
    M_3 segment of, 125, 125f, 242, 243f
    M_4 segment of, 125, 125f, 242, 243f
    segments of, 125, 125f, 331
  middle meningeal, 111
  occipital, meningeal branch of, 208
  of base of brain, 124f
  of cerebral hemisphere
    coronal section of, 125f
    medial surface of, 124f

Artery(ies) *(Continued)*
  of dura, 111
  of spinal cord, 134–135, 134f
  of temporal lobe, 124f
  ophthalmic, 123, 124f, 306, 314, 315f, 316
  parieto-occipital, 124f, 128, 242
  pericallosal, 496
  pontine, 124f, 127
  posterior cerebral, 123, 124f, 127, 127f, 128, 128f, 129, 129f, 232f,
      233, 233f, 242, 251, 318, 496, 515
    branches of, 127f
    P_1 segment of, 127–128, 242
    P_2 segment of, 127–128, 242
    P_3 segment of, 127–128, 242
    P_4 segment of, 127–128, 242
    segments of, 126f, 127–128, 127f
    temporal branches of, 496
  posterior choroidal, 130, 233–234
  posterior communicating, 123, 124f, 129, 129f, 189, 198, 242
    aneurysm of, 526
    paramedian branches of, 391, 393f
    posteromedial branches of, 233
  posterior inferior cerebellar, 99f, 100f, 101, 125, 126f, 162, 169–
      170, 170, 171f, 176f, 177–178, 177f, 185, 427, 428f
    area served by, 170
    damage to, 286–287, 287f
    lesions of, 170, 171f
  posterior (dorsal) radicular, 134f, 135
  posterior spinal, 125, 126f, 134–135, 134f, 142, 142f, 162, 169,
      170, 170f
    in posterior column–medial lemniscal system, 260, 261f
    lesions of, 169, 260
  quadrigeminal, 126f, 127, 127f, 189–190, 197, 197f, 198, 329,
      456
  radicular, 142
  segmental
    central or sulcal branches of, 134–135, 134f
      spinal branches of, 134–135, 134f
  spinal medullary, 134f, 135, 142, 142f
  stylomastoid, 344
  superior cerebellar, 124f, 126f, 127, 177–178, 177f, 184, 184f,
      185, 189, 190, 197, 197f, 198, 329, 427, 428f
  temporal, 126f, 128
  thalamogeniculate, 126f, 128, 130, 232f, 233f, 234, 316, 331
  thalamoperforating, 126f, 127, 127f, 232f, 233, 484, 484f, 496
  to caudal thalamus, 126f
  to choroid plexuses of ventricles, 126f
  uncal, 496
  vertebral, 122, 125, 126f, 162, 169, 170, 170f
    branches of, 125, 126f
Artery of Adamkiewicz, 135, 142
Ascending reticular activating system, 197
Asimultagnosia, 322
Asomatognosia, 518
Aspartate
  as neurotransmitter, 58, 59t
  in cerebellar fibers, 427
  in medial superior olivary nucleus, 334
  in mossy fibers, 432
  in olfactory tract cells, 364
Asterixis, 421
Asterognosis, 264
Astrocytes, 28–30, 29f, 31f, 32f, 33f
  at blood-brain barrier, 30, 135, 135f
  end-feet of, 28, 29f, 30, 32f, 135, 135f
  fibrous, 28, 29f, 30
  of gray matter, 28, 29f
  of white matter, 28
  progenitors of, 81
  protoplasmic, 28, 29f, 30
  subependymal, 100f
Astrocytic scar, 29, 36
Ataxia, 441, 500
  cerebellar, 85
  Friedreich, 269
  limb, 535
  truncal, 435, 536

Ataxia (Continued)
  vestibulocochlear nerve injury and, 210
Athetosis, 417, 417f
Atrium, of lateral ventricle, 97, 98f
Atrophy, muscle, 388
Atropine, effects of, 475
Attention, 335
  auditory pathways serving, 339
  deficits in, 518, 519f
Auditory artery
  internal (labyrinthine), 329
Auditory brainstem response recording, 330f, 331
Auditory input, in thalamic nuclei, 228
Auditory meatus, internal, 325, 325f
Auditory nucleus
  blood supply of, 329–330
  brainstem, 331–336, 331f–336f
  hierarchy of, 329, 330f, 333f
Auditory system, 323–340
  acoustic startle reflex in, 338f, 339
  auditory and association cortices in, 336–337, 336f, 337f
  brainstem auditory nuclei and pathways in, 331–336, 331f–334f, 336f
  central auditory pathways in, 329–331, 330f
  descending auditory pathways in, 337–338
  ear anatomy and, 325–329, 325f, 326f, 328f
  information processing in, 329–333
  middle ear reflex in, 338, 338f
  orientation and attention in, 338f, 339
  overview of, 324
  sound waves and hearing in, 324–325, 324f. See also Sound.
  testing of, 529–530, 531f
Autologous transplants, for Parkinson disease, 421
Autonomic dysreflexia, 476
Autonomic nervous system, 4–6, 4f. See also Visceral motor system.
Aversion centers, 503
Axial muscle, control of, 437–439
Axon(s), 5, 5f, 16, 17f, 18f
  accommodation in, 46
  afferent
    rapidly adapting (phasic), 47
    slowly adapting (tonic), 47
  anterograde (wallerian) degeneration of, 22
  cable properties of, 40, 41f, 42, 45
  cerebellar efferent, 439
  cervicothalamic, 283
  classification of, into groups and types, 50, 50t
  collaterals of, in pyramidal cells, 509, 509f, 510
  conduction velocities of, 45–46, 48–49, 49f
    axon diameter and, 48–49, 49f
    functional grouping of, 49–50, 50t
  corticocortical, 510
  corticospinal, 389
  cutaneous sensory, receptive field of, 47
  development of, 87–88
    mental retardation and, 89
  dynamic range of, 48
  functional grouping of, 49–50, 50t
  length constants of, 42
  myelinated
    in central nervous system, 30–31, 35
    in peripheral nervous system, 33, 34f
  of Purkinje cells, 184
  of spinocerebellar tract, 269
  place codes and, 47
  postganglionic, 200
  preganglionic, in cranial motor nerve nuclei, 200
  recruitment of, 48, 49f
  regeneration of, 35–36
  spinomesencephalic, 282
  spinoreticular, 282
  spinothalamic, 282
  stimulus-response curves for, 48, 48f, 49f
  structure of, 17f, 20f, 21, 22f
  terminal arbor of, 16, 17f, 18f
  thalamocortical, 506

Axon(s) (Continued)
    from ventral posterolateral nucleus, 283
Axonal bouton, of chemical synapse, 26, 26f
Axonal sprouting, 36
Axonal transport, 21–22, 21t, 22f, 28
  anterograde, 21–22, 21t, 22f
  as research tool, 22
  for large dense-cored vesicles, 61, 61f
  retrograde, 21–22, 21t, 22f
Axonotmesis, 36

Babinski sign, 388, 394, 395f, 531, 533f
Balance
  control of, 380
  vestibulocerebellar influence on, 435, 436
Balint syndrome, 321–322
Ballismus, 417
Bands of Baillerger, outer and inner, 506f, 507
Bands of Büngner, 36
Bárány test, 530
Baroreceptor reflex, 300f, 301, 476, 476f, 492
Baroreceptors, 294–295, 294t, 295f
  of carotid sinus, visceral parasympathetic afferent pathways and, 299
Basal cell(s), of olfactory epithelium, 360, 361, 361f
Basal ganglia, 6f, 7. See also Nucleus(i), basal.
Basal lamina, 375
  of arachnoid barrier cell layer, 113
  of blood-brain barrier, 135, 135f
  of choroid plexus villi, 101, 102f, 103f
  of Schwann cells, 34f, 35, 35f
    in axonal regeneration, 36
Basal nuclei. See Nucleus(i), basal.
Basal plate(s). See Plate(s), basal.
Basal vein, of Rosenthal, 130, 131f
Basement membrane, pial, 116
Basilar area, of midbrain, 153, 153f
Basilar artery(ies), 122, 124f, 125, 127, 177–178, 177f, 189, 329
  circumferential branches of, 184, 184f
  of midbrain, 197, 197f
  of pons, 184, 184f
  paramedian branches of, 184, 184f
    damage to, 391–392, 398–399
  short circumferential branches of, 329
Basilar membrane(s), 325, 325f
  in cochlear tuning, 328
  in mechanoelectric transduction of sound, 328
Basis pedunculi, of midbrain, 190, 190f, 192f, 193, 194f, 197
Basket cell(s), 184, 184f, 429, 430f, 431, 434–435, 434f, 509
  of cerebral cortex, 508f
  of molecular layer of cerebellum, 84f, 85
Basophilic staining, of neurons, 17
Behavior
  aggressive, 491
  basal nuclei circuits and, 415–419
  caudolateral hypothalamic lesions and, 491
  emotional, 483
    cingulate gyrus and, 500
    limbic system and, 494, 496, 504
  passive, 491
  rostromedial hypothalamic lesions and, 491–492
Bell (facial) palsy, 396, 398f, 460, 528, 530f
Benign positional vertigo, 356–357
Bergmann glial cells, 81
Beta cell(s), 312
Beta-adrenergic receptors. See Receptor(s), β-adrenergic.
Betz cell(s), 389, 389f
  of cerebral cortex, 508
Biofeedback, hypothalamic responses and, 492
Biogenic amines
  as neurotransmitters, 27–28, 58, 59t
  in G protein receptor action, 65f, 66
Bipolar cell(s), 343, 344
  in spiral ganglion, 328–329, 328f

Bipolar cell(s) (Continued)
  of olfactory epithelium, 18, 18f, 20
  retinal, 306f, 310–311, 311f, 312t
    as edge-detectors, 310–311
Bipolar disorder, 28
Bladder
  autonomic control of, 477, 477f
  distention of, viscerosensory afferent input and, 298, 301–302
Bleeding. See Hemorrhage.
Blepharospasm, 460
Blind spot, 313f, 314, 318, 525
Blink reflex, 212, 215, 285, 446, 460, 462, 462f
  testing of, 527, 528f
Blood flow
  autonomic control of, 476–477, 476f
  sympathetic regulation of, 472
Blood osmolarity, antidiuretic hormone release and, 487–488
Blood pressure
  antidiuretic release and, 488
  autonomic control of, 476
  baroreceptor reflex and, 301, 492
Blood vessels. See also Artery(ies); Vein(s).
  glial cells and, 28, 29f
  perivascular feet covering, 32f
Blood-brain barrier, 135, 135f
  astrocytes at, 30, 32f
  carbidopa and, 28
Blood-cerebrospinal fluid barrier, 101, 103f
Blood-nerve barrier, 35
Body temperature, regulation of, 483, 491, 492
  fibers for, 163
Boutons en passant, 21
Bowel distention, viscerosensory afferent information on, 298, 301–302
Bowman's glands, 362
Brachium, 329, 330f
  of inferior colliculus, 189, 190f, 191, 195f, 197
  of superior colliculus, 189, 190f, 195f, 196f, 197
Brachium conjunctivum (superior cerebellar peduncle), 177, 181f, 182, 191, 192f, 193, 193f, 424, 424f, 427
Brachium pontis (middle cerebellar peduncle), 175, 175f, 177, 180, 181f, 182, 424, 424f
Bradycardia, 477
Bradykinesia, 415, 420f
Brain, 4, 4f. See also specific structures, e.g., Brainstem.
  blood supply to, 122, 130, 131f, 132
  development of, 72–79, 72f, 74f, 76f, 78f
    and ventricular system development, 94–97, 94f
    cellular events in, 87–89
    cerebral hemispheres in, 77
    critical periods in, 88
    diencephalon in, 77
    induction in, 72–73
    plasticity in, 268
    primary neurulation in, 73–75
      congenital defects with, 73–75
    secondary brain vesicles in, 77
    secondary neurulation in, 75, 77
  infectious diseases of, congenital defects and, 77–79
  plasticity of, 88, 268
  relation to spinal cord and meninges, 109–110, 109f
Brain-derived neurotrophic factor, 87
Brainstem, 151–157. See also Medulla oblongata; Mesencephalon; Pons.
  amygdaloid connections with, 501
  arteries on dorsal aspect of, 126f
  auditory nuclei and pathways in, 331–336, 331f–334f, 336f
  autonomic nuclei of, hypothalamic indirect links with, 487
  cortical input in, in pain transmission, 292
  cranial nerve nuclei of
    functional components of, 154–157, 155f–156f
    motor, 200–201, 201f
    sensory, 201–202, 202f
  cranial nerves of, 199–218
  descending projections from, 374, 383–384
  development of, 82–84

Brainstem (Continued)
  directions in, 9–10, 9f
  division(s) of, 152–153, 152f
    medulla oblongata in, 152, 152f
    midbrain in, 152–153, 152f
    pons in, 152, 152f
    tegmental and basilar areas in, 153, 153f
  importance of, 200
  in motor deficits, functional role of, 382f–384f, 383–385
  lesions of, cranial nerves and, 200
  primary sensory fibers in, diameter of, 259–262
  tectobulbospinal system of, 195
  veins of, 130f, 133f, 134
  ventricular spaces of, 153–154
Broca aphasia, 337, 514, 516, 517, 517f
Broca's area, 337
Brodmann, allocortex of, 496
Brodmann area(s), 264, 264f, 511, 512f
  3a, 356, 356f
  3b, 371
  17, 319, 322f
  18–21, 321, 322f
  22 (Wernicke's), 337
  41 (primary auditory cortex), 336
  42 (secondary auditory cortex), 336–337
  44 and 45, 337
  corticospinal neurons in, 390
  in Broca aphasia, 516, 517f
  in Wernicke aphasia, 516, 517f
  relation of, to thalamic nuclei, 513–514, 514f
Brown-Séquard syndrome, 150, 260, 394, 396f, 533
  anterolateral system and, 280, 282f
Bulb, olfactory, 238, 241f, 363–364, 363f, 364f
Bundle(s)
  association, 236
  commissural, 236
  medial forebrain, 224f, 230, 296–297, 480, 482f, 503, 503f
    connections of, to hypothalamus, 485, 486f
  olivocochlear, 337–338
  postcommissural, 484, 485f, 486f
  precommissural, 484, 485f, 486f
  ventral amygdaloid, 224f, 230
Butyrophenones, tardive dyskinesia and, 422

C fiber(s), 143, 260f, 277
  in anterolateral system receptors, 275
  in indirect spinothalamic pathway, 280
  in reflexes, 146, 147
  in spinal trigeminal pathway, 285
  nociceptive sensations in, 277
Calcarine artery(ies), 124f, 127f, 128, 242
Calcitonin gene–related peptide, 144, 277, 346
  in pars caudalis, 285
  in posterior horn, pain transmission and, 291
Calcium channels
  in Huntington disease, 420
  in information transmission, 26, 27f
  in membrane depolarization, 51, 51f
  proteins of, 60
  voltage-gated, 375
    in hair cell transduction, 328
    in taste receptor cells, 368, 369f
    in vestibular system, 346, 347
  voltage-sensitive, synaptic vesicle release and, 62, 62f
Calcium, in G protein receptor action, 65f, 66
Calcium spikes, 309
Calcium waves, astrocyte function and, 30
Callosomarginal artery(ies), 390
Caloric test, 354
Canal(s)
  Dorella, 451
  hypoglossal, 202
  neural, 73, 81, 94
  semicircular. See Semicircular canal(s).

Canals of Schlemm, 304, 304f
Cancer. *See also* Tumor(s).
  cerebrospinal fluid in, 103
Capacitor, 38–39
Capillaries, of choroid plexus villi, 101, 103f
Capsule
  external, 243, 245f
  extreme, 243, 245f
  internal, 7, 229, 229f, 236, 236f, 237, 244, 245f, 246f, 480, 481f
    anterior limb of, 244, 246f
    component parts of, 229, 229f, 244, 246f
    corticospinal fibers in, 391, 391f, 392f
    genu of, 244, 246f, 396
    lesions of, corticonuclear and corticospinal system damage with, 398
    posterior limb of, 244, 246f
    relationships of, 247f
    retrolenticular limb of, 244, 245f, 246f
    sublenticular limb of, 244, 245f, 246f, 329, 330f
    vascular lesions of thalamus and, 234
    vasculature of, 245, 246f
Carbidopa, for Parkinson disease, 28
Cardiac muscle, autonomic control of, 466
Cardiac nerve(s)
  in angina, 298
  sympathetic innervation of, 470
  viscerosensory afferent fibers in, 295, 296f
Cardiac pain, patterns and pathways of, 297f, 298, 299f
Cardiac thoracic nerve, in angina, 298
Cardiomegaly, 490
Cardiovascular disease, reserpine for, 69–70, 69f
Cardiovascular function, 230
  autonomic control of, 476–477, 476f
Carotid artery(ies)
  internal, 111, 122, 123, 124f
    aneurysm of, 526
    cavernous part of, 132, 132f
      aneurysm of, 132, 132f
    cerebral part of, 123, 124f
Carotid bodies
  chemoreceptors in, 294–295
  glossopharyngeal nerve and, 207–208
  visceral parasympathetic afferent pathways and, 299
Carotid sinus, baroreceptors of, visceral parasympathetic afferent pathways and, 299
Carotid-cavernous fistula, trauma and, 132, 132f
Cataracts, 305
  congenital, 321
Catechol-*O*-methyltransferase (COMT), for degradation of norepinephrine, 68, 69f
Cats, toxoplasmosis and, 77
Cauda equina, 117
  formation of, 82, 83f
Caudal central subdivision nucleus, 450, 451f
Caudal eminence, 75, 138
Caudal (visceral or cardiorespiratory) nucleus, 202, 202f, 370–371, 370f
Caudal pontine reticular nucleus, 453
Caudate nucleus, 197, 236f, 237, 245, 246, 247f, 248f, 407, 408f
  in Huntington disease, 419, 419f
  of lateral ventricle, 96f, 97, 98f
  parts of, 246, 248f
  relationships of, 248f
Causalgia, 473
Cavernous sinus(es), 113, 130, 131f, 132, 132f
Cell(s)
  alpha, 312
  amacrine retinal, 306f, 310, 311f, 312
  arachnoid barrier, 110f, 113
  arachnoid cap, 114, 114f
  basal, 360, 361, 361f
  basket, 184, 184f, 429, 430f, 431, 434–435, 434f, 509
    of cerebral cortex, 508f
    of molecular layer of cerebellum, 84f, 85
  beta, 312
  Betz, 389, 389f

Cell(s) *(Continued)*
  of cerebral cortex, 508
  bipolar, 343, 344
    in spiral ganglion, 328–329, 328f
    of olfactory epithelium, 18, 18f, 20
    retinal, 306f, 310–311, 311f, 312t
      as edge-detectors, 310–311
  body of, 5, 5f
  bushy, spherical and globular-shaped, 333
  chandelier, of cerebral cortex, 508f, 509
  cholinergic, in Alzheimer disease, 28
  choroid, 101, 102f, 103f
  choroid epithelial, 101, 102f, 103f
  complex, of visual cortex, 320, 320f
  delta, 312
  double pyramidal, 497, 498f
  endothelial, of blood-brain barrier, 135, 135f
  ependymal, 81, 94, 99, 100, 100f
    of cerebral cortex, 86, 87f
    of neural tube, 138
    of tela choroidea, 95, 95f, 96f
    of ventricular system, 77
  epsilon, 312
  gamma, 312
  ganglion
    cochlear, 328–329, 328f
    retinal, 306f, 310, 311, 311f, 312
      projections of, 312–316
      subdivisions of, 316, 317f, 322f
  glandular secretory, autonomic control of, 466
  glial, 4, 28–33
    Bergmann, 81
    differentiation of, 81
    radial
      in cerebellar development, 84f, 85
      in thalamic development, 85, 86f
    transporter proteins of, 68
    types and functions of, 16, 28, 29f, 29t
  Golgi, 84f, 85, 184, 184f, 270, 428, 430f, 431f, 434
  granule, 84f, 85, 184, 184f, 428–429, 429f, 430f, 431f, 434
    of olfactory bulb, 364
  hair, 325f, 326, 326f
    as transducers of sound, 327–328, 327f
    depolarization of, 327f, 328
    inner, 326, 327f
    of organ of Corti, 326, 327–328, 327f, 328f
    of semicircular canals, 345–347, 345f–347f, 347–349, 348f
    of vestibular sensory system, 349, 350f
      morphologic polarization of, 347, 347f
      morphology of, 345–346, 345f
      transduction by, 346–347, 346f, 347f
    outer, 326, 327f
      efferent feedback to, 338
    stereocilia in, 325f
  horizontal, retinal, 306f, 310
  in ventral posterolateral nucleus, classes of, 283
  juxtaglomerular, 363, 363f
  lateral olivocochlear efferent, 337–338
  low-threshold non-nociceptive, 283
  M, 312
  medial olivocochlear efferent, 337f, 338
  Merkel, 256, 257, 257f, 264
  microglial, 29f, 32
  microvillar, 361f, 362
  mitral, of olfactory bulb, 363, 363f, 364
  Müller, 81
  multinodal, 283
  multipolar, 333–334
  neural crest, 468
    of neural tube, 138
    structures derived from, 79–80, 79t, 80f
  neuroepithelial, 94
  nociceptive-specific, 283
  O2A progenitor, 81
  octopus, 334
  of dentate gyrus, 497, 498f

Cell(s) (*Continued*)
    of hippocampus, 497, 498f
    of olfactory tract, 364
    of subiculum, 497, 498f
    of visual cortex, 320, 320f
    organization of, in brain development, 87–89
    periglomerular, 363, 363f
    photoreceptor, 305, 306, 307f–309f, 308–309
        in visual processing, 309–312, 311f
    pillar, 326
    precursor
        glioblastic, 81
        neuroblastic, 81
    pseudounipolar, of dorsal root ganglia, somite development and, 80, 80f
    Purkinje, 84f, 85, 184, 184f–185f, 384, 385, 428, 429f, 430f, 431f, 433–435, 434f
        in decerebellate experiments, 384
        inhibitory influence of, in cerebellar cortex, 433–435, 433f, 434f
    pyramidal, 331, 333f, 389, 389f
        of cerebral cortex, 508–509, 508f, 509f
    Renshaw, 388f, 389
    satellite, 33
    Schwann, 33, 34f, 36, 81
    secretory, in membranous labyrinth, 344
    simple, of visual cortex, 320, 320f
    spinal border, 148, 271
    stellate, 184, 184f, 268, 429, 430f, 431, 434–435, 434f
        of cerebral cortex, 508f
        of molecular layer, 84f, 85
        spiny and aspiny, 509
    sustentacular, 360, 361f, 362
    tanycytes as, 81, 100, 100f
    taste receptor, 366–367, 366f
    trigeminal ganglion, 267
    tufted, of olfactory bulb, 363, 363f, 364
    W, 312
    wide dynamic range, 283
    X, 312
Cell adhesion molecules, in notochord, 73
Cell column(s)
    cortical, 267–268
        visual, 319–320
    general visceral efferent, 465. *See also under* Functional component(s).
    interomediolateral, 141, 142, 143f, 144f, 145, 305, 468
        neurons of, 232–233
    ocular dominance, 320, 320f, 513, 513f
    of alar and basal plates, development of, 155f, 156
    of brainstem, 82f, 83
        motor, 200–201, 201f
        sensory, 201–202, 202f
    of spinal cord, 81–82, 82f
Cell membrane. *See* Membrane(s).
Central cord syndrome, 142, 394
Central cup, on ophthalmoscopic examination, 526, 526f
Central facial paralysis, 396, 398f
Central gray, 99, 165, 165f
    in medulla, 163, 164f
    of midbrain, 190
        function of, 191, 192f, 193f
Central nervous system
    degeneration and regeneration of neurons in, 35–36
    development of, 81–87
        basic features of, 81
        brainstem in, 82–84
        cerebellum in, 84–85, 84f, 174–175, 174f
        cerebral cortex in, 85–86, 87f, 88f
        relationship of spinal cord to vertebral column in, 82, 83f
        spinal cord in, 81–82, 82f, 138–139, 139f
        thalamus in, 85, 86f, 87f
    directions in, 9–10, 9f
    effects on visceral motor system of, 475–477
    energy used in, 16
    general outline of, 4–6, 4f
    malformations of, 72

Central nervous system (*Continued*)
    number of cells in, 16
    regions of, 6–8, 6f
    tumors of, astrocytes and, 30, 32t, 33f
Central nucleus, 501
    of inferior colliculus, 191, 192f, 335
    of medulla, 168, 169f
Central pain syndrome, 291
Central seven, 396, 398f
Central superior nucleus, 183, 183f
Cerebellar artery(ies)
    anterior inferior, 102, 125, 126f, 127, 177f, 178, 184, 184f, 185, 329, 427, 428f
        damage to, 329
    posterior inferior, 99f, 100f, 101, 125, 126f, 162, 169–170, 170, 171f, 176f, 177–178, 177f, 185, 427, 428f
        area served by, 170
        damage to, 286–287, 287f
        lesions of, 170, 171f
    superior, 124f, 126f, 127, 177–178, 177f, 184, 184f, 185, 189, 190, 197, 197f, 198, 329, 427, 428f
Cerebellar ataxia, 269
Cerebellar nucleus, 84f, 85, 177, 177f, 184, 425–427, 427f
    emboliform, 426, 427f
    fastigial, 425–426, 427f
    globose, 426, 427f
    in vestibulospinal tract, 380
    medial, 425–426, 427f
Cerebellar vein(s)
    inferior, 134
    superior, 130f, 133f, 134
Cerebellopontine angle, 99, 161
Cerebellum, 6f, 7, 423–444
    blood supply of, 177–178, 177f, 185, 427–428, 428f
    cortex of, 184, 184f–185f, 428–435, 429f–431f, 432t, 433f–434f
    development of, 84–85, 84f, 174–175, 174f
    external features of, 176f, 177, 177f
    function of, testing of, 534–536, 535f, 536f
    functional modules of, 435–442, 436f–438f, 440f–442f
    interaction of, with motor areas of cortex, 401–402
    interior anatomy of, 184–185, 184f–185f
    interposed nuclei of, projections to red nucleus, 399
    lobes of, 177, 177f. *See also* Lobe(s), cerebellar.
    motor learning and, 443–444
    pontocerebellar module of, 439–441, 440f
    role of, in control of movement, 401–402
    spinocerebellar module of, 436–439
    structure of, 424–428, 424f–428f
    veins of, 130f, 133f, 134
    vestibulocerebellar module of, 435–436, 436f
    visceromotor functions and, 401–402, 442–443
Cerebral aqueduct, 77, 78f, 94f, 95, 96f, 97, 98–99, 98f, 99f, 153, 153f
    congenital atresia or stenosis of, 78f, 79
    forking of, 95
    in midbrain development, 188, 188f
    obstruction of, 79, 79f, 98–99
        during development, 94f, 95
Cerebral artery(ies)
    anterior, 122, 123, 124f, 125, 128, 128f, 232f, 233, 233f, 242, 243f, 282, 316, 390, 496, 515
        A₁ segment of, 123, 243
        A₂ segment of, 123, 243
        damage to, 288
    border zones of, 515
    middle, 122, 125, 125f, 128, 128f, 129, 129f, 242, 243f, 251, 264, 282, 318, 390, 515
        branches of, 125, 126f
        damage to, 288, 517
        lenticulostriate branches of, 407, 407f
        lesions of, 391, 391f
        M₁ segment of, 125, 125f, 242, 243f
        M₂ segment of, 125, 125f, 242, 243f
        M₃ segment of, 125, 125f, 242, 243f
        M₄ segment of, 125, 125f, 242, 243f
        segments of, 125, 125f, 331

Cerebral artery(ies) (Continued)
  posterior, 123, 124f, 127, 127f, 128, 128f, 129, 129f, 232f, 233,
      233f, 242, 251, 318, 496, 515
    branches of, 127f
    P₁ segment of, 127–128, 242
    P₂ segment of, 127–128, 242
    P₃ segment of, 127–128, 242
    P₄ segment of, 127–128, 242
    segments of, 126f, 127–128, 127f
    temporal branches of, 496
    watershed infarcts and, 128, 128f, 515
Cerebral cortex. See Cortex, cerebral.
Cerebral hemisphere(s)
  arteries on, 124f
  components of, 6f, 7
  development of, 77
  directions in, 9–10, 9f
  veins of, 130, 130f, 131f, 132
  white matter of, 242–245, 244f–246f
Cerebral infarction, 122, 122f
  watershed, 128, 128f, 515
Cerebral ischemia, 122, 122f
  transient, 500
Cerebral vein(s), 130, 130f, 131f
  anterior, 130, 131f
  deep middle, 130, 131f
  internal, 130, 131f, 132–134, 133f
  pathologic flow patterns of, 133
  superficial middle, 130, 130f, 131f
  temporal, 130, 130f
Cerebrospinal fluid, 94, 102–103, 103f
  abnormalities of, 102–103
  collection of, 102
  composition of, 102
  exchange with extracellular fluid of brain parenchyma, 101, 103f
  flow of, 103–104, 104f, 109f, 113, 114, 114f
  in early ventricular system, 78
  in subarachnoid space, 113–114
  obstruction of, 94, 95, 99, 104–106, 105f. See also Hydrocephalus.
    prenatal, 79, 79f
Cerebrovascular system, 121–136
Chemical coding
  in parasympathetic system, 475
  in postganglionic sympathetic neurons, 473, 473f
Chemical messengers. See also Neurotransmitter(s).
  neuronal communication by, 58
  synthesis, storage and release of, 60–63
Chemical neurotransmission
  basic data on, 58–60
  pharmacologic modification of synaptic transmission in, 68–70
  regulation of neuronal excitability in, 67
  signal transduction in, 63–67
  synthesis, storage and release of chemical messengers in, 60–63
  transduction mechanisms in, fast and slow, 58–59
Chemical synapse(s), 50–52, 51f, 52f, 53f
  delay in, 50, 58
  function of, 26–27, 27f
  Gray's type I and II, 26–27, 26f, 27t
  information flow across, 59–60, 60f
  morphologic characteristics of, 26–27, 27t
  regulatory mechanisms in, 59–60
  structure of, 25–27, 26f
Chemonociceptors, 274, 275, 275t
Chemoreceptor(s), 25, 46, 294–295, 294t, 295f. See also Recep-
    tor(s).
  of carotid body, visceral parasympathetic afferent pathways and,
    299
Chemoreceptor reflex, 477
Chewing, muscles for, 267, 271f, 272
  innervation of, 200, 214, 214f, 215
  testing of, 527, 528f
Childbirth, oxytocin effects in, 487
Chloride, relative concentrations of, in neuronal cytoplasm and extra-
    cellular fluid, 23, 24f
Cholecystokinin, 301
  in pain transmission, 292

Choline acetyltransferase
  in Alzheimer disease, 28
  in synthesis of acetylcholine, 61
Cholinergic cells, in Alzheimer disease, 28
Chorda tympani, 210f, 211
  taste neurons in, 369
Chorea, Sydenham, 421
Choreiform movements, 417
Choreoathetosis, 417
Choriocapillaris, 306
Choroid fissure, 95, 96f
Choroid, of eye, 305
Choroid plexus, 94, 100–101, 102f, 103f, 220f, 221
  arteries to, 126f
  blood supply of, 101–102
  cerebrospinal fluid produced by, 113
  formation of, 95, 96f, 97
  of third ventricle, 99f
  of ventricular system, 77
  papilloma of, hydrocephalus with, 106
Choroidal artery(ies), 101
  anterior, 101, 123, 124f, 189, 197, 197f, 198, 245, 250, 251, 316,
      316f, 407, 407f, 496
  lateral posterior, 101, 126f, 128
  medial posterior, 101, 126f, 128, 189, 197–198, 232f, 233f, 234
  posterior, 130, 233–234
Chromatolysis, 22
  of nerve cell bodies, 35–36
Cilia, olfactory, 360
Ciliary artery(ies), 306
Ciliary body, 304, 304f, 447, 448f, 449
Ciliary muscle(s), 216, 447, 448f
Cilium, of cone cells, 308
Cingulate gyrus, 239f, 242, 494, 495f
  blood supply of, 496
  in limbal system, 407, 407f
  isthmus of, 239f, 242, 494, 495f
  lesions of, emotional response and, 500
Cingulum, 243, 244f
Circadian rhythm, 232, 483
Circle of Willis (cerebral arterial circle), 124f, 128f, 129–130, 129f,
    484, 484f
  anteromedial branches of, 233
  central (performing or ganglionic) branches of, 232f, 233
  in diencephalic region, 232f, 233
  limbic system and, 496
  perforating branches of, 128f, 129–130, 129f, 484, 484f
  posteromedial branches of, 198, 233
Circumventricular organs, 221
Cistern(s), 117–119, 117f, 118t
  ambient, 117f, 118f, 127
  callosal, 123
  cerebellopontine, 127
  dorsal cerebellomedullary, 117, 117f
  dorsal cerebromedullary, 153, 153f
  interpeduncular, 117, 117f, 118f, 189
  lateral cerebellomedullary, 125
  lumbar, 109f, 117, 140, 140f
  of lamina terminalis, 117f, 118f, 123
  prepontine, 117f, 118f, 125
  quadrigeminal, 189
  structures associated with, 118, 118t
  subarachnoid, 113, 117–119, 117f, 118f
  sylvian, 125
Cisterna magna, 117–118, 117f, 118f, 125, 153, 153f
Cisternal puncture, 118
Clasp-knife effect, 389
Claude syndrome, 450, 450t
Claustrum, 243, 245f
Clinoid processes, 112f
Clonus, 388, 389
Clostridium tetani, toxin of, retrograde axonal transport and, 22
Cocaine, action of, 69f, 70
Coccygeal ligament(s), 109f, 113
Cochlea
  blood supply of, 329–330

Cochlea *(Continued)*
  membranous, 326, 326f
  position of, sound frequency and, 328–329, 328f
  primary afferent innervation in, 328–329, 328f
  tuning of, 328–329, 328f, 338
Cochlear duct, 326, 326f
Cochlear implants, 328
Cochlear microphonic, 328
Cochlear nerve
  damage to, 329
  functional components of, 156f, 157
Cochlear nucleus, 83, 85, 160f, 161, 208–209, 209f, 329, 330f, 333f
  anterior (ventral), 167, 167f, 168f, 202, 208, 331, 331f, 332–334
  cell types in, 332–334
  damage to, 209, 329
  in brainstem, 331–334, 331f–333f
  posterior, 167, 167f, 168f, 202, 208
  posterior (dorsal), 331, 331f, 332
Cochleotopic order, 328
Cognitive function, and limbic system, 504
Coincidence detection, 334
Collicular (quadrigeminal) artery, 197f, 198
Colliculus(i)
  facial, 154, 154f, 176f, 177, 178, 180f
  inferior, 153, 188, 188f, 189, 190f, 329, 330f, 333f
    ascending auditory pathways in, 333f, 335
    brachium of, 189, 190f, 191, 195f, 197, 329, 330f, 333f, 335
    central nucleus of, 335
    of midbrain, 191, 192f
    paracentral nuclei of, 333f, 335
  superior, 153, 188, 188f, 189, 190f, 216, 217, 313, 382f, 385, 455, 456f, 457f
    brachium of, 189, 190f, 195f, 196f, 197
    deep layers of, 338f, 339
    midbrain at, 194f
Color blindness, 309
Color perception, 321, 322f
Column(s). *See also* Cell column(s).
  cortical, 319
  definition of, 22, 23t
  ocular dominance, 320, 320f, 511, 511f
  orientation of, 320, 320f
  posterior (dorsal), 141, 163, 164, 165f. *See also* Fasciculus(i), cuneate; Fasciculus(i), gracile.
Commissural fibers, 237f, 239f, 240f, 243–245
Commissure
  anterior, 97, 99f, 236f, 237, 239f, 243–244
    anterior limb of, 363
  anterior white, 141
    lesions of, 141f, 144f, 150
  habenular, 233, 236f, 239f, 244
  hippocampal, 236f, 237, 239f, 243–244
  posterior, 98, 99f, 189, 190f, 236f, 239f, 244
    nucleus of, 197
  posterior tegmental, 335
Communication artery(ies)
  anterior, 123, 124f, 242
    anteromedial branches of, 233, 316
  posterior, 123, 124f, 129, 129f, 189, 198, 242
    aneurysm of, 526
    paramedian branches of, 391, 393f
    posteromedial branches of, 233
Compartments, supratentorial and infratentorial, 111
Complex(es)
  centromedian-parafascicular, in thalamic lesioning, 290
  inferior olivary, 161, 164, 166–167
  Merkel cell neurite, 256, 257f
  nigral, 411–412
  parabrachial nuclear, 287
  superior olivary, 331f
    in auditory information processing, 334–335, 334f
  ventrobasal, 262. *See also* Nucleus(i), ventral posterolateral.
Complex spike, 434, 434f
Compression injury, axonal sprouting in, 36
Computed tomography, 10–11, 10f, 11f, 12t
  appearance of tissues on, 11t

Concussion, 114
Conductance, 38, 45
  Goldman-Hodgkin-Katz equation as, 40
Conduction
  antidromic, 46
  electrotonic, 42
  orthodromic, 46
  saltatory, 31, 46
Conduction block, 46
Conduction velocity
  axon diameter and, 45–46, 48–49, 49f
    functional groupings of, 49–50, 50t
  groups of, 48–50, 49f, 50t
  of peripheral nerves, 259, 259t
Conductors, 38
Cones, retinal, 306, 307f–309f, 308–309
  types of, 308–309, 308f
Confabulation, 500
Congenital defects
  cataracts and, 305, 321
  defective neurulation and, 73–75
  infectious diseases causing, 77, 79, 79f, 80–81
Congenital dermal sinuses, development of, 108f, 109
Connective tissue, of choroid plexus villi, 101, 103f
Connexons, 50
Consciousness, regulation of, 197
Contralateral neglect, 518, 519f
Contrecoup injury, 114
Contusion, 114
Conus medullaris, 140, 140f
Copper metabolism, in Wilson disease, 421
Cornea
  anatomy of, 304, 304f
  nociceptors in, 285
  sensation in, testing of, 527, 528f
Corneal reflex, 212
  testing of, 527, 528f
  trigeminal nerve innervation of, 215
Corona radiata, 245
Corpora quadrigemina, 189
Corpus callosum, 96f, 97, 98f, 236, 236f, 237
  agenesis of, 237, 237f
  and hippocampal formation, 497, 497f
  components of, 237f, 239f, 240f, 243–244
Corpus cerebelli, 84f, 85, 175
Corpus striatum, 246, 246f
  development of, 236f, 237
Corpuscle(s)
  Meissner, 256, 257f
  pacinian, 251, 256, 257, 257f
Cortex
  anterior insular, 371
  auditory
    association, 337
    information processing in, 329, 330f, 333f
      binaural, 329, 332, 333, 333f
      monaural, 329, 332, 333f
    primary, 239, 242, 336–337, 336f, 337f
      tonotopic organization of, 336–337
    secondary, 336–337, 336f, 337f
  cerebellar, 428–435, 429f–431f, 432t, 433f–434f
    afferent and efferent fibers in, 432t
    layers of, 428–429, 429f, 431f
    neurons of, 18, 18f
    somatotopic localization in, 433, 433f
    synaptic interactions in, 433–435, 433f
    vestibular labyrinth connections to, 352, 352f
  cerebral, 7–8, 236, 505–520
    abducens nerve function and, 213
    agranular
      heterotypical, 511
      ventral insular, 365, 365f
    amygdaloid connections with, 501
    basal nuclear function and, 413, 417–419, 418f
    blood supply of, 241f, 242, 243f, 390, 514–515
    cell bodies in, 506–507, 506f, 507f, 509f

Cortex (*Continued*)
    columnar organization of, 511–513, 513f
    cytoarchitecture of, 511, 511f, 512f
    descending projections from, 374, 383–384
    development of, 85–86, 87f
      abnormalities of, 86, 88f
    diffuse inputs into, 510
    dominant hemisphere in, and language, 516–518
    functional categories of, 515, 516f
    granular, heterotypical, 511
    higher functions in, 515–520
    histology of, 506–507, 506f, 507f
    intrinsic circuitry of, 510–511, 510f
    laminar organization of, 510–511
    lobes of, 238–242, 238f–243f
    motor-related areas of, 389, 389f, 390
    neurons of, 18, 18f, 508–510, 508f, 509f
    organization and functions of, 506
  cingulate, 288, 401, 407, 407f
  entorhinal, 494
    in Alzheimer disease, 500
    lateral, 364f, 365
  frontal, in pain transmission, 292
  granular, 336
  limbic
    cytoarchitectural definitions of, 494
    in pain transmission, 292
  motor, 242. *See also* Motor system.
    cingulate, 288, 402, 407, 407f
    face, 395
    hand area of, 8, 8f
    in voluntary control of movement, 388, 389, 389f, 390. *See also*
      Motor deficit(s).
    premotor, 388, 390, 401
    primary, 239, 388, 390, 400, 400f
      blood supply of, 390
      cerebellar and globus pallidus influence on, 401
      control of movement and, 399–400
      corticofugal systems and, 387–404
      corticospinal neurons in, 400
      somatotopic organization of, 390, 390f
      vertical columns in, 399–400, 400f
    supplementary, 388, 390
      functional role of, 400–401, 401f
  multimodal association, 515, 516f
  oculomotor nerve function and, 217
  olfactory, 364–365, 364f, 365f
    projections of, 365–366, 365f
  orbitofrontal, 365, 365f
    lateral posterior, 371
  parietal association, 242
    eye movements and, 456
    intellectual functions mediated by, 518, 519f
  parietal, posterior, 401
  periamygdaloid, 364f, 365
  piriform, 364f, 365, 494
  prefrontal, 494, 495f
    intellectual functions mediated by, 518–520
  somatomotor, primary, thalamocortical relations with, 513–514, 514f
  somatosensory, 242, 415
    columnar organization of, 512–513, 513f
    imaging of, 268, 288
    nondiscriminative touch, thermal and pain signals to, 274, 274f
    pain perception and, 288, 289f
    pain transmission and, 292
    primary, 239, 263–264, 263f, 264f, 515, 516f
      blood supply of, 263f, 264
      contralateral, discriminative touch and position sense pathways
        in, 267f
      damage to, 264
      divisions of, 264, 264f
      homuncular representation of, 263f, 264
      plasticity and reorganization in, 268–269, 269f
      thalamocortical relations with, 514, 514f
    secondary, 263f, 264

Cortex (*Continued*)
    somatotopic organization of, 283, 283f
  striate, 319
  temporal, inferior, 321
  thalamocortical relations in, 513–514, 514f
  trochlear nucleus and, 216
  unimodal association, 515, 516f
    auditory, 515, 516f
    somatosensory, 515, 516f
    visual, 515, 516f
  vermal, vestibular connections of, 436, 437f
  vestibular, 355–356, 356f
  visual
    columnar organization of, 512–513, 513f
    connections in, critical period in, 88
    layers of, 317f
    other visual areas in, 321–322
    primary, 241, 242, 318, 319–320, 319f, 320f
      abnormal development of, 320–321
    secondary, 321–322, 322f
Cortical columns, 267–268
  visual, 319–320
Corticofugal systems, control of movement and, 387–404
Corticonuclear (corticobulbar) system
  course of, 395–396, 397f
  definition of, 395
  motor deficits in, 394–396, 397f, 398–399, 398f
  origin of, 394–395
  termination of, 396, 398, 398f, 399
Corticopontine system, 399
Corticoreticular system, 399
Corticorubral system, 399
Corticospinal system, 389–394, 389f–396f
Corticotropin, 488
  overproduction of, 490–491, 490f
Crania bifidum, 73
Cranial nerve(s). *See also specific nerves, e.g.,* Vagus nerve.
  development of, 79–80
  in spinal trigeminal pathway, 284f, 285
  mixed, 4
  nuclei of
    functional components of, 154–157, 155f–156f, 175. *See also*
      under Functional component(s).
      in brainstem, 156, 156f
    motor, 174–175, 174f
    of brainstem
      and functional components of, 156, 156f
      motor, 82f, 83–84, 174, 174f, 200–201, 201f
      sensory, 201–202, 202f
    of medulla, 7, 160f, 161
    of midbrain, 7
    of pons, 7
    of brainstem, 199–218
      motor, 82f, 83, 200–201, 201f
      sensory, 82f, 83, 201–202, 202f
    of medulla, 152, 152f, 202–208, 203f–205f, 207f
    of midbrain, 215–217
    of pons, 213–215, 214f
    of pons-medulla junction, 203f, 204f, 208–213, 209f, 210f, 212f
  parasympathetic afferent fibers in, 298–299, 300f, 301
  parasympathetic outflow pathways of, 474–475, 474t
  testing of, 522–530
Craniofacial structures, somatosensory information from, trigeminal system and, 264–265
Crista galli, 112f
Crista of hair cells, 345f, 346–347
Critical period, 268
  in brain development, 88
  in synaptic connection, 88–89
  in visual cortex development, 321
Crossed tectoreticulospinal system, 456, 457f
Crus cerebri, 153, 188, 188f
  corticospinal fibers in, 391, 392f, 393f
  of basilar pons, 175, 175f, 176f
  of midbrain, 189, 189f, 190f, 191, 192f, 193, 194f, 197

Cuneate nucleus, 160f, 161, 162, 162f, 163, 164, 164f, 165f, 271
    in posterior column–medial lemniscal system, 260, 261f, 262, 262f
    lateral, 271
Cuneiform nucleus, 196f, 197
Cuneus, 319, 322f
Cupola, of hair cells, 345f, 346
Cushing disease, 490–491, 490f
Cutaneous fibers, of facial nerve, from ear and auditory canal, 211
Cutaneous nerves, compound action potential in, 260f
Cuticular plate, 345, 345f
Cyclic adenosine monophosphate
    in olfactory transduction, 362
    in second messenger G protein system, 52, 53f, 65–66, 65f
    in slow chemical neurotransmission, 59
    in taste transduction, 368, 369f
Cyclic guanosine monophosphate
    in G protein–coupled receptor responses, 66
    in retinal transduction, 308
Cyclops, 77
Cytoarchitecture, of cerebral cortex, 506–507, 506f, 507f
Cytokines, secreted by astrocytes, 29
Cytomegalovirus, 77
Cytoplasm
    of oligodendrocyte, in myelin sheath formation, 33f
    of Schwann cell, in myelin sheath formation, 33, 35

Dandy-Walker malformation, 79, 79f
Deafferentation
    central pain syndrome and, 291
    ventrocaudal nuclei changes and, 289–290
Deafness. See also Hearing loss.
    central, 324
    conductive, 324
    monaural, 329
    sensorineural nerve, 209, 324, 328
Decerebellate rigidity, 382f
Decerebrate rigidity, 382f, 383, 383f
Decibels, 324
Decorticate posturing, 382f
Decorticate rigidity, 138, 384f, 385
Decussation
    hippocampal, 499
    motor (pyramidal), 161, 163, 164f, 392
        in caudal medulla, 163, 163f, 164f
    of superior cerebellar peduncle, 191, 192f, 193, 193f
    tegmental, ventral, 196, 381
Deep brain stimulation, for chronic pain treatment, 290
Dejerine-Roussy syndrome, 234
Delta cell(s), 312
Dementia, 501
Demyelinating diseases, 32
Dendrite(s), 5, 16, 17f, 18f, 19f
    apical, of pyramidal cells, 508–509, 508f, 509f
    basal, 508–509, 508f, 509f
    cable properties of, 40, 41f, 42, 45
    development of, mental retardation and, 89
    length constants of, 42
    of olfactory bulb, 363, 363f, 364
    of Purkinje cells, 184
    primary, 16
Dendritic digits, 428, 431f
Dendritic fields, in medium spiny neurons, 409, 411f
Dendritic spines, 16, 17f, 19f
    formation of, 89
Dense bars, 375
Dentate cerebellar nucleus, 177, 177f
Dentate gyrus, 250, 496–497, 498f
Dentate nucleus, 177f, 184
    relation to pontocerebellum, 439–441, 440f
Denticulate ligament(s), 110, 116–117, 140, 141f
Depolarization
    in action potential, 42–43, 43f, 44f
    of cell membrane, 24, 25, 42–43, 43f, 44f, 51, 51f
    of sensory neurons, 25

Depolarization (Continued)
    presynaptic, 51, 51f
    suprathreshold, 43, 44f
    threshold, 42–43, 44f
L-Deprenyl, for Parkinson disease, 421
Dermal sinus, congenital, development of, 108f, 109
Dermatome(s), 80f, 82, 259, 277, 277f, 278f
    facial, 265, 285, 286f
    innervation of, development of, 80, 80f, 82
Desensitization
    heterologous, 67
    homologous, 67
    of postsynaptic receptor action, 66–67
Desmosomes, 99, 100
Detrusor muscle, 477
Deuteranopia, 309
Diabetes insipidus, 488
Diabetes mellitus, oculomotor nerve lesions and, 217
Diacylglycerol (DAG), in G protein–coupled receptor responses, 52, 53f, 65f, 66
Diagonal band, 364
Diaphragma sella, 112f, 113
Diencephalon, 76f, 77, 78f, 94f, 95, 219–234. See also Epithalamus; Hypothalamus; Thalamus.
    basic organization of, 221–223, 222f–224f
    components of, 220
    development of, 85, 86f, 220f, 221, 221f
    junction of midbrain with, 189, 190f, 195f, 197, 221, 222f
    vasculature of, 232f, 233–234, 233f
Diffusion potential, 39
Digastric muscle, trigeminal nerve innervation of, 214, 214f
Digestive tract, visceromotor neurons of, 80
Dilator muscle, 304f, 305
Dilator pupillae muscle(s), 216, 447, 448f, 449
    sympathetic innervation of, 452, 452f
Diplopia, 217, 491, 526
Discriminative touch, 256–272
    impairment of, 260
    in trigeminal system, 264–265, 266f, 267
    pathways for, 267f
        posterior column–medial lemniscal, 256–264
        spinal trigeminal, 284f
        trigeminothalamic tract, 265–267
    receptive field properties and, 267–268
    testing of, 277, 537, 537f
    texture, 264, 536f, 537
    two-point, 265, 266f, 267, 392, 537, 537f
Disinhibition
    in basal nuclei, 406
    subthalamic, 413
    thalamic, 413
Dizziness, 356–357
    vestibulocochlear nerve injury and, 210
Docking complex, 62, 62f
L-Dopa, for Parkinson disease, 28, 420
Dopamine
    accumulation of, prevention of, 69, 69f
    as neurotransmitter, 58, 59t
    glial cell metabolism of, 68
    in connections between pars reticulata and pars compacta, 411
    in direct and indirect basal nuclear pathways, 417, 418f, 419
    in olfactory tract cells, 364
    loss of, in Parkinson disease, 28, 420
    receptors for, in schizophrenia, 412
Dopamine β-hydroxylase, 70
    in synthesis of norepinephrine, 61
Dopaminergic loop, 417, 418f, 419
Dorello canal, 451
Dorsal nucleus of Clarke, 141, 142, 143f, 270–271, 270f
Dorsolateral prefrontal loop, in basal nuclei parallel circuits, 413
Down syndrome, 89
Drugs, synaptic transmission modified by, 68–70, 69f
Duct
    cochlear, 326, 326f
    endolymphatic, 344
Ductus reuniens, 344

Dura mater, 110–113, 112f, 140, 140f, 141f
    blood supply of, 111
    border cell layer of, 110–111, 110f
    cranial vs. spinal, 113
    early development of, 108f, 109
    infoldings and sinuses of, 111, 112f, 113
    layers of, 110–111, 110f
    nerves of, 111
    organization and structure of, 109, 109f, 110f
    periosteal and meningeal, 110–111, 110f
Dural arteries, 111
Dural sac, spinal, 113
Dynein, in axonal transport, 21
Dysarthria, 206, 421, 441, 522
Dysdiadochokinesia, 441, 442f
Dysesthesia, in central pain syndrome, 291
Dysgeusia, 371, 522
Dysmetria, 441
Dysphagia, 206
Dyspnea, 207
Dysraphic defects, 73
Dystonia, athetotic, 417

Ear. See also Auditory system; Sound.
    anatomy of, 325–329, 325f–328f
    external, sound transmission in, 325, 325f
    inner, sound transmission in, 325f–327f, 326–329
    middle, sound transmission in, 325, 325f
    receptive fields in, 47
Eating, excessive, 501
Ectoderm, 72
Ectomeninx, 108, 108f, 109
Edinger-Westphal nucleus, 188, 188f, 195, 195f, 197, 200, 201f,
        212f, 216, 305, 450, 451f
    functional component of, 156, 156f
    in parasympathetic innervation, 471f, 474
    in pupillary light reflex, 460
    of midbrain, 84
Effector proteins, in G protein receptor action, 65f, 66
Eicosanoids, in G protein receptor action, 65f, 66
Electrical circuits, of neurons, 38–39, 38f
Electrical stimulation, deep brain, 290–291, 291
        transcutaneous, 291, 292
Electroencephalography, for imaging of pain, 288
Embolism, cerebral, 122, 122f
Eminence
    caudal, 75, 138
    inferior olivary, 161, 161f
Emotion(s)
    cingulate gyrus and, 500
    limbic system and, 494, 496, 504
Emotional behavior, 483
Encephalocele, 73, 74f
End-feet, astrocyte, 28, 29f, 32f, 135, 135f
    at blood-brain barrier, 30
Endocrine system
    hypothalamus in, 229, 480, 480f
    visceral motor system and, 466
Endocytosis, receptor-mediated, in axonal transport, 21
Endoderm, 72
Endolymph, 327, 343, 530
Endolymphatic duct, 344
Endolymphatic hydrops, 344–345
Endomeninx, 108, 108f
Endoneurium, of peripheral nerves, 35, 35f
Endoplasmic reticulum
    in biosynthesis of large dense-cored vesicles, 60–61, 61f
    of dendrites, 16, 17f, 19f
    rough, 17
    smooth, 17f
Endothelial cells, of blood-brain barrier, 135, 135f
Endothelium, of superior sagittal sinus, on arachnoid villi, 114, 114f
Energy
    in central nervous system, 16
    neuronal need for, 16

Enkephalin
    in medium spiny neurons, 409
    in stimulation-produced analgesia, for chronic pain, 292
    in striatopallidal neurons, 413
Enophthalmos, 149
Enteric nervous system, 472f, 475
Ependymal cells, 81, 94, 99–100, 100f
    of cerebral cortex, 86, 87f
    of neural tube, 138
    of tela choroidea, 95, 95f, 96f
    of ventricular system, 77
Ependymomas, 100–101, 101f
Epidural hematoma, 115, 116f
Epidural space, 113
    development of, 109, 109f
Epilepsy, Sommer's sector damage and, 500
Epinephrine, sympathetic receptors and, 473
Epineurium, of peripheral nerves, 35, 35f
Epiphysis, 220f, 221
Epipial layer, 116
Epithalamus, 223, 232–233, 232f. See also Diencephalon.
    anatomy of, 220
    development of, 85, 86f, 220f, 221
Epithelium
    anterior iridial, 448f
    olfactory, 360, 360f, 361f
        bipolar cells of, 18, 18f, 20
Epsilon cell(s), 312
Equilibrium potential, 39, 40
Ethanol, antidiuretic hormone release and, 488
Ethmoidal artery, 111
Ethmoidal nerve(s), 111
Excitability cycle, in action potential generation, 44, 44f, 45f
Exocytosis, of synaptic vesicles and large dense-cored vesicles, 62–
        63, 62f
Extraocular muscle(s), 446–449, 447f–449f. See also Eye movement(s).
    in rotational vestibulo-ocular reflex, 353–354
    lesions affecting, 212f, 215–217
    motor neurons of, 453, 453f
    testing of, 526–527, 527f
Eye. See also specific part, e.g., Retina.
    anatomy of, 304–305, 304f, 447, 448f, 449
    lesions of, with hemiparesis, 391
Eye blink, 212, 215, 285, 446, 460, 462, 462f
    testing of, 527, 528f
Eye fields
    cortical, 455–456, 455f
    frontal, 213, 216, 217, 395, 455–456, 455f
    hemifields as, 318
    parietal, 395, 455f, 456
    supplemental, 455f, 456
Eye movement(s)
    abducens nerve and, 213
    abduction, 447, 527
        paralysis of, 452, 452f
    adduction, 447, 527
    central structures involved in, 449–453, 449f, 450t, 451f–452f
    characteristics of, 446, 446t
    compensatory, 353–354, 353f, 446, 455. See also Head movement(s).
    conjugate, 446, 456–458
        deficits in, 452, 452f
    counter-roll, 350f, 354
    depression, 447
    disconjugate, 446
        definition of, 446
    elevation, 447
    extorsion, 447
    horizontal and vertical, 453–455
    in rotational vestibulo-ocular reflex, 353–354, 353f
    intorsion, 447
    nuclei involved in, 195, 196
    nystagmus as, 354, 354f. See also Nystagmus.
    oculomotor nerve and, 447
        injury of, 217, 450, 450t
    optokinetic, 458–459, 459f
    peripheral structures involved in, 446–449, 447f–449f

Eye movement(s) (Continued)
  reflex, 446, 458–462, 459f–462f
    testing of, 527, 528f
  saccades as, 446, 453–455, 453f
  smooth pursuit, 446, 456–458
  targeting, 453–458, 453f–458f
  types of, 446, 446t
  trochlear nerve and, 447
    injury of, 216
  vergence, 446, 453f, 458, 458f
  voluntary control of, 395
Eyelid
  anatomy of, 449, 449f
  movements of, 460, 462, 462f

Face
  flushing of, 470, 472f
  lower, corticonuclear fiber lesions affecting, 396, 398f
  sympathetic innervation of, 470
Facial defects
  forebrain development and, 77
  in holoprosencephaly, 77
Facial expression, 179
  muscles of
    innervation of, 200, 210f, 211
    testing of, 527–528, 529f
Facial motor cortex, 395
Facial nerve(s), 152, 152f, 161, 265
  anatomy and distribution of, 210–212, 210f
  development of, 79
  in muscles of facial expression, 200, 210f, 211
  in spinal trigeminal pathway, 284f, 285
  intermediate root of, 161
  internal genu of, 178, 180f, 210, 210f
  lesions of, 211–212
  motor nuclei of
    functional components of, 175
    origin of, 174–175, 174f
  of pons, 175–176, 175f, 176f
  parasympathetic outflow pathways of, 474, 474t
  relationship to abducens nerve, 180f
  relationship to trigeminal nerve, 206
  roots of, 176
  taste neurons in, 369
  testing of, 527–529, 528f–529f
Facial pain
  in tic douloureux, 215, 285
  onion skin pattern of, 285–286, 286f
  pars caudalis in, 285, 286
Facial palsy, 528, 530f
Falx cerebelli, 112f, 113
Falx cerebri, 109, 109f, 111, 112f
Familial dysautonomia, aberrant neural crest development and, 81
Fasciculus(i)
  arcuate, 243, 244f, 337
  cuneate, 140, 141, 141f, 148, 148f, 162, 162f, 163, 164, 164f, 165f, 259, 260
    in medulla, 162
  definition of, 22, 23t
  dorsal longitudinal, 296–297, 487
  gracile, 140, 141, 141f, 148, 148f, 162, 162f, 163, 164, 164f, 165f, 259, 260
    in medulla, 162
  inferior fronto-occipital, 243, 244f
  inferior longitudinal, 243, 244f
  lenticular, 224f, 231, 232, 249, 249f, 409, 411f
  mammillary, 485f, 486f, 487
  medial longitudinal, 149, 164, 165, 165f, 166f, 167, 167f, 178, 180f, 181, 181f, 192f, 193, 193f, 212, 212f, 213, 353, 449, 455
    in medulla, 163
    of midbrain, 191
    trochlear nerve and, 212f, 215, 216, 217
  posterior (dorsal) longitudinal, 296–297, 487
  posterolateral, 260f, 277, 284f, 285

Fasciculus(i) (Continued)
  subthalamic, 249, 249f
  superior longitudinal, 243, 244f
  thalamic, 223, 224f, 232, 409, 411f
  uncinate, 435
Feeding behavior, 483
Fiber(s). See also Tract(s).
  Aα, 260f
  Aβ, 143–144, 260f
  Aδ, 143, 260f, 277
    in anterolateral system receptors, 275
    in direct spinothalamic pathway, 279–280
    in reflexes, 146, 147
    in spinal trigeminal pathway, 285
    nociceptive sensations in, 277
  amygdalohypothalamic, 485, 485f, 486f
  anterior spinocerebellar, 182, 182f, 271
  anterior trigeminothalamic, 181, 194f, 197
    in pons, 187, 187f
  association, 243, 244f, 508, 509
    long, 243, 244f
    short, 243, 244f
  C, 143, 260f, 277
    in anterolateral system receptors, 275
    in indirect spinothalamic pathway, 280
    in reflexes, 146, 147
    in spinal trigeminal pathway, 285
    nociceptive sensations in, 277
  cable properties of, 40, 41f, 42, 45
  callosal, 508, 509
  centrifugal, 364
  cerebellar afferent, 431–432, 432t
  cerebellar corticonuclear, 439
  cerebellar corticovestibular, 352, 435
  cerebellar efferent, 426–427, 431–432, 432t, 439, 440f
  cerebellohypothalamic, 437f, 440f, 442
  cerebello-olivary, 439
  cerebelloreticular, 439
  cerebellorubral, 196, 382, 439
  cerebellothalamic, 196, 439
  cerebellovestibular, 352
  climbing, 184, 184f, 429, 430f, 434–435, 434f
    in cerebellar cortex, 432, 432t, 433–435
  cochlear afferent, tonotopic arrangement of, 332, 332f
  cochlear efferent, 327
  commissural, 237f, 239f, 240f, 243–245
  corticobulbar, 244
    of crus cerebri, 195
    of midbrain, 191, 191f
  corticofugal, 236, 244, 245, 245f
  corticohypothalamic, 486
  corticonigral, 410f, 411
  corticonuclear, 244, 245f
    of crus cerebri, 195
    of midbrain, 191, 191f
  corticopetal, 236, 244, 245
  corticopontine, 180, 180f, 244
    of crus cerebri, 195
    of midbrain, 191, 191f
  corticoreticular, 380
  corticorubral, 382
  corticospinal, 7, 8, 8f, 148, 178, 180, 180f, 244, 245f, 391–392, 391f–394f, 439
    in spinal cord and medulla, 163
    of crus cerebri, 195
    of midbrain, 191, 191f
  corticostriatal, 408, 410f, 415
  corticosubthalamic, 411
  corticothalamic, 244
  corticovestibular, 352, 435
  cranial parasympathetic, 298–299, 300f, 301
  cuneocerebellar, 164, 165f, 166, 166f, 167f, 271, 436
  cutaneous, of facial nerve, from ear and auditory canal, 211
  dentato-olivary, 439
  dentatorubral, 439
  dentatothalamic, 439

Fiber(s) (*Continued*)
  exteroceptive, 143–144, 144f, 269, 271
  extrafusal muscle, 374, 376, 376f
  fastigiospinal, 439
  fastigiovestibular, 352
  frontopontine
    of crus cerebri, 195
    of internal capsule, 244, 245f
  geniculocalcarine, 319f
  hypothalamic afferent, 230, 484–486, 485f, 486f
  hypothalamic efferent, 230
  hypothalamoamygdaloid, 487
  hypothalamocerebellar, 437f, 440f, 442
  hypothalamomedullary, 487
    injury to, 487
  hypothalamospinal, 487
    in pain transmission, 292
    injury to, 487
  hypothalamothalamic, 487
  in anterolateral system, somatotopic arrangement of, 280, 281f
  in organ of Corti, 327f, 328–329
  internal arcuate, 164, 165f, 262
  interoceptive, 144
  interstitiospinal, 149, 455
  intrafusal muscle, 145, 374, 376–378, 376f
  large-diameter, 259, 260f, 261f
    in spinal trigeminal pathway, 284f
  medial vestibulospinal, 149
  mossy, 184, 184f, 269
    in cerebellar cortex, 429, 431–432, 431f, 432t
  motor
    of cranial nerve nuclei, functional components of, 156–157, 156f, 473, 474t. *See also under* Functional component(s).
    of facial nerve, 179, 180f, 211, 212
    of spinal nerves, 145, 145f
    of vagus nerve, 206
  multilayered, 430f, 432, 432t
  muscle
    alpha-gamma coactivation of, 377–378, 377f, 378f
    dynamic bag, 376f, 377
    extrafusal, 374, 376, 376f
    gamma loop and, 377–378, 377f, 378f, 383
    in reflex circuits, 378–379
    intrafusal, 145, 374, 376–378, 376f
    nuclear bag, 376, 376f
    nuclear chain, 376, 376f
    sensory (afferent), 377
      annulospiral endings of, 377
      flower-spray endings of, 377
      secondary endings of, 377
      types Ia, Ib, II, 377, 378
    static bag, 376f, 377
  myelinated
    of central nervous system, 35
    of peripheral nervous system, 33, 34f
    of subcortical white matter, 236
  nigrostriatal, 408, 410f, 411
  nigrosubthalamic, 411
  nigrotectal, 411
  nigrothalamic, 411, 413
  nucleocortical, 427, 431
  occipitopontine, of crus cerebri, 195
  of accessory nerve, vagus fibers and, 206
  of anterolateral system, 148–149, 148f
  of central tegmental tract, 192f, 193, 193f
  olivocerebellar, 166, 166f, 167, 167f, 435, 439
  pallidonigral, 410
  pallidopallidal, 409, 410f
  pallidosubthalamic, 410, 411, 413
    in Parkinson disease, 419
  pallidothalamic, 409, 413
  parallel, in granule cell layer of cerebellum, 428, 430f, 431f, 434, 434f
  parietopontine, of crus cerebri, 195
  pedunculopontine, 410f, 411
  pedunculopontostriatal, 408–409, 410f

Fiber(s) (*Continued*)
  pontocerebellar, 180, 180f, 435, 439
  posterior spinocerebellar, 166, 166f, 167f
  posterior trigeminothalamic, 194f, 197
  postganglionic
    parasympathetic, in cranial motor nerve nuclei innervating visceral structures, 200
    sympathetic, 468–470, 469f
      with oculomotor nerve, 216
  postganglionic sympathetic, in visceral motor pathway, 466, 469f
  postsynaptic posterior column, 259
  preganglionic parasympathetic
    oculomotor nerve and, 217
    of Edinger-Westphal nucleus, 195
  preganglionic, sympathetic, 468–469, 469f
  preganglionic sympathetic, in visceral motor pathway, 466, 469f
  primary afferent, 259–262
    in spiral ganglion, 328–329, 328f
    nociceptive, 277, 278, 279f
    proprioceptive, in spinocerebellar tracts, 269–271
    pseudounipolar cell body of, 259
  projection, 508, 509
    internal capsule as, 244, 245f, 246f
  propriospinal, 148, 148f
  raphespinal, 169
    in pain transmission, 291f, 292
  reticulocerebellar, 166, 166f, 167f, 439
  reticulohypothalamic, 296–297
  reticulospinal, 149, 455
    in pain transmission, 291f, 292
    medullary, 380
  reticulothalamic, 282
  retinogeniculate, 318
  retinohypothalamic, 486
  retrolenticular, 244, 245f, 246f
  rubro-olivary, 179, 196, 382, 439
  rubrospinal, 181–182, 181f, 192f, 193, 195, 382, 439
  sacral parasympathetic, 298, 300f
  sensory (afferent)
    large-diameter, in spinal cord, 259, 260f, 261f, 284f
    of cranial nerve nuclei, functional components of, 154–157, 155f, 156f
    of facial nerve, 211
    of spinal nerves, 143–144, 144f
    of vagus nerve, 206
  small-diameter, 260f
    in spinal trigeminal pathway, 284f
  somatosensory, 138
  spinal
    afferent
      general somatic, 143–144, 144f. *See also under* Functional component(s).
      general visceral, 143–144, 144f. *See also under* Functional component(s).
      neurotransmitters of, 144–145
    descending, 138
    efferent
      general somatic, 145. *See also under* Functional component(s).
      general visceral, 145. *See also under* Functional component(s).
      types of, 145, 145f
    exteroceptive, 143–144, 144f, 269, 271
    interoceptive, 144, 144f
    motor, types of, 145, 145f
    proprioceptive, 144, 144f
    sensory, 143–144, 144f
    visceromotor (autonomic), 145, 145f
  spinobulbar, in anterolateral system, 274
  spinohypothalamic, 282
    in anterolateral system, 274
  spinomesencephalic, 148, 148f, 162, 193
    in anterolateral system, 274
  spino-olivary, 149, 162
    in anterolateral system, 274
  spinoreticular, 148, 148f, 149, 162, 178, 180, 180f, 193, 295, 380
    in anterolateral system, 274

Fiber(s) *(Continued)*
  in indirect spinothalamic pathway, 280
  spinotectal, 193, 194f, 197, 282
    in anterolateral system, 274
  spinothalamic, 148, 148f, 162
    in anterolateral system, 274
  spinovestibular, 149, 162, 352
  striatonigral, 413
  striatopallidal, 409, 413
  sublenticular, 244, 245f, 246f
  subthalamonigral, 410f, 411
  subthalamopallidal, 410, 411, 413
    in Parkinson disease, 419
  taste
    glossopharyngeal nerve and, 208
    trigeminal and facial nerves and, 206
  tectobulbospinal, 193, 338f, 339
  tecto-olivary, 195
  tectoreticular, 195
  tectospinal, 149, 195
  temporopontine, of crus cerebri, 195
  thalamocortical, 7, 244, 263, 497, 497f, 507
    from ventral posterolateral nucleus, 283
  thalamohypothalamic, 486
  thalamolenticular, 244, 245f, 246f
  thalamostriatal, 408, 410f
  trigeminal afferent, 214, 214f
  trigeminothalamic, 179, 181, 284f
    posterior (dorsal), 181
  unmyelinated, in peripheral nervous system, 34f
  vestibular afferent, 349–350, 350f, 352, 352f
  vestibular efferent, 346, 353
  vestibulocerebellar, 352, 435, 436f, 439
  vestibulospinal, 149
  viscerosensory, 138, 295
    parasympathetic, 298–301
    sympathetic, 295–298, 296f, 297f
      referred pain in, 297–298, 297f, 298f
      reticular formation projections of, 296–297, 297f
      thalamic projections of, 296, 297f
  W, of retinal ganglion cells, 316
  X(P), of retinal ganglion cells, 316, 317f, 322f
  Y(M), of retinal ganglion cells, 316, 317f, 322f
Fibroblast(s)
  of arachnoid barrier cell layer, 113
  of arachnoid trabeculae, 110f, 113
  of meninges and meningeal dura, 109, 110
  of periosteal dura, 110
Fibroblast growth factor, 87
Fibronectin, in somite development, 80, 80f
Field H, of Forel (prerubral area), 224f, 230–232, 409
Filum terminale externum, 109f, 113, 117, 140, 140f, 141f
Filum terminale internum, 110, 117, 140, 140f, 141f
Fimbria, 248f, 250
  of hippocampus, 497, 498f
Final common path, 374, 388
Fingertips, two-point tactile discrimination in, 265
Fissure(s). *See also* Sulcus(i).
  anterior (ventral) median, 161
    of spinal cord, 140, 141f
  cerebellar, 84f, 85
  choroid, 95, 96f
  interhemispheric, 86
  longitudinal cerebral, 236
  posterolateral, 84f, 85, 174f, 175
    of cerebellum, 176f, 177, 177f, 424, 425f–427f
  primary, 84f, 85, 174f, 175
    of cerebellum, 176f, 177, 177f, 424, 425f–427f
  superior orbital, 213
  sylvian, 86
  transverse cerebral, 86
Flocculus, 457
Fluorescent substance, axonal transport and, 22
Flutter-vibration sense, 256, 266f, 267f, 284f, 392
Folic acid, neural tube defects and, 73
Follicle-stimulating hormone, 480, 491

Folstein Mini-Mental Status test, 522, 523f
Food intake (satiety) center, of hypothalamus, 230, 483
Foramen (foramina)
  infraorbital, 213
  interventricular of Monro, 77, 78f, 94f, 95, 96f, 97, 98f, 220f, 221, 236f, 237
  intervertebral, development of, 80f, 82
  jugular, 205f, 208
    in viscerosensory parasympathetic pathways, 300f, 301
  of fourth ventricle, 95, 95f
  of Luschka, 77, 78f, 95, 95f, 99, 100f, 153, 204f
  of Magendie, 77, 78f, 95, 95f, 99, 100f, 153, 153f
  optic, 123, 124f
  stylomastoid, 210f, 211
Foramen magnum, 73, 139
Foramen rotundum, 213
Forceps
  major (occipital), of corpus callosum, 243
  minor (frontal), of corpus callosum, 243
Forebrain
  development of, 77
  directions in, 9–10, 9f
Forel, field H of (prerubral area), 224f, 230–232, 409
Form perception, 321, 322f
Fornix, 96f, 97, 98f, 224f, 230, 237, 248f, 250, 497, 498f
  column of, 248f, 250
  hypothalamic connections of, 482f, 484, 485, 485f, 486f
  of limbic system, 494
  postcommissural, 499
  precommissural, 499
  relationships of, 248f
Fossa
  interpeduncular, 153, 189, 189f
  pterygopalatine, 213
  rhomboid, 153–154, 154f
    of pons, 176–177, 176f
Fovea
  inferior, 154, 154f
  superior, 154, 154f
Fovea centralis, 309, 309f
Foveation, 458
Frequency coding, 48, 48f, 256, 271
  in cochlear nerve, 329
Friedreich ataxia, 269
Frontal gyrus
  inferior, 238, 238f
    pars opercularis of, 238, 238f, 240f, 337
    pars triangularis of, 238, 238f, 240f, 337
  superior, 238, 238f
Frontal pole, 238, 238f
Functional acoustic chiasm, 329
Functional component(s)
  development of, 156
  general somatic afferent, 81, 82, 82f, 143–144, 144f, 156f, 157, 175, 256
    cranial nerves carrying, 202
    of glossopharyngeal nerve, 207
    of vagus nerve, 206
  general somatic efferent, 145, 156, 156f, 175, 188
    of cranial motor nerve nuclei, 200, 201f
    of medulla, 160
    of oculomotor nucleus, 195, 195f
  general visceral afferent, 82, 82f, 143–144, 144f, 156, 156f, 175
    of glossopharyngeal nerve, 207
    of medulla, 161
    of vagus nerve, 206
  general visceral efferent, 81, 82f, 145, 156, 156f, 175, 188
    and viscerosensory parasympathetic fibers, 300f
    of cranial nerves III, VII, IX, X, 473, 474t
    of Edinger-Westphal nucleus, 195, 195f
    of medulla, 160
    of vagus nerve, 206, 473, 474t
    preganglionic parasympathetic, 200, 201f, 210, 211
  of brainstem, 82f, 83, 156, 156f
  of cranial nerve motor and sensory nuclei, 175, 200, 201f
  of cranial nerve nuclei of brainstem, 156, 156f

Functional component(s) (Continued)
   of facial nerve, 210, 211, 473, 474t
   of glossopharyngeal nerve, 207, 473, 474t
   of lower motor neurons in brainstem, 200, 201f
   of spinal nerves, 143–146
   of trochlear nucleus, 193, 193f
   of vagus nerve, 206, 473, 474t
   special somatic afferent, 156f, 157, 175
      of brainstem, 82f, 83
      of medulla, 161
   special visceral afferent, 156, 156f, 175
      of brainstem, 82f, 83
      of glossopharyngeal nerve, 207
      of medulla, 161
      of vagus nerve, 206
   special visceral efferent, 156, 156f
      for muscles arising in pharyngeal arches, 200–201, 201f
      of brainstem, 82f, 83
      of medulla, 160
      of vagus nerve, 206
Functional systems, related to regions, 8–9, 8f
Funiculus(i)
   anterior, 141f
      fibers in, 148f, 149
   ipsilateral lateral, 283
   lateral, 141, 141f, 271
      descending tracts in, 148f, 149
   posterior (dorsal), 140–141, 141f
Fusiform gyrus, lesions of, 321
Fusion pore, formation of, 62–63, 62f

G protein
   in postsynaptic receptors in slow chemical neurotransmission, 59
   in second-messenger synapse, 51–52, 53f
   olfactory-specific, 362
   subunits of, 64, 66
G protein–coupled receptors, 362. See also Receptor(s),
      β-adrenergic.
   for binding of chemical messengers, 63–66, 65f, 67
   structure of, 65–66, 65f
Gag reflex, 299
Gait
   in Parkinson disease, 420
   lurching, 435, 436
   unsteady, 441
      vestibulocochlear nerve injury and, 210
Galactorrhea, 491, 491f
Gamma cell(s), 312
Gamma loop, 377–378, 377f, 378f, 383
Gamma rigidity, 384
Gamma-aminobutyric acid (GABA), 335
   as neurotransmitter, 27–28, 58, 59t, 67
   in chemical synapses, 27
   in intrinsic neurons, 509
   in medium spiny neurons, 409
   in striatal and striatopallidal neurons, 413
Ganglion (ganglia)
   aorticorenal, 470
   autonomic, 468
   basal, 245. See Nucleus(i), basal.
   celiac, 470
   cervical, 305, 470, 470f, 471f
      neurons of, sympathetic input to, 468–469, 470f
   ciliary, 216, 305, 450
   coccygeal, 469
   definitions of, 23, 23t
   gasserian, 265, 266f, 267f
   geniculate, 210f, 211, 369
      in spinal trigeminal pathway, 284f, 285
      origin of, 79
   inferior
      in viscerosensory parasympathetic pathways, 300f, 301
      of vagus nerve, 206
   inferior mesenteric, 470–471

Ganglion (ganglia) (Continued)
   inferior nodose, 370
   inferior petrosal, 370
   jugular, origin of, 79
   nodose, 79, 370
   of cranial nerves, development of, 79–80
   of glossopharyngeal nerve, 79, 207, 207f
   otic, 474
   posterior (dorsal) root, 80, 80f, 138, 139f
      in meningeal development, 108f
      neurons of, 18, 18f
      section of, decerebrate rigidity and, 383
      spinal fibers of
         lateral, 143, 144f
            pain fibers in, 260f, 277
         medial, 144, 144f
   prevertebral, 80, 469–471, 471f
   pterygopalatine, 210f, 211, 474
   Scarpa, 343–344, 344f
   semilunar, 79, 265, 266f, 267f
   sensory, 79–80
   spiral, 208, 209f, 325f, 326–327, 326f
      primary afferent neurons in, 328–329, 328f
   submandibular, 210f, 211, 472
   superior
      in viscerosensory parasympathetic pathways, 300f, 301
      of cranial nerves IX and X, in spinal trigeminal pathway, 284f,
         285
      of vagus nerve, 206
   superior cervical, 305, 470, 470f, 471f
      orbital innervation and, 452, 452f
   superior mesenteric, 470
   sympathetic
      functional and chemical coding in, 472–473
      internal organization of, 471–473
      prevertebral, 469–471, 471f
   sympathetic chain (paravertebral), 80, 469–471, 471f
   terminal (intramural), of vagus nerve, 206
   trigeminal, 265, 266f, 267f
      in spinal trigeminal pathway, 284f, 285
      origin of, 79
   vestibular, 209, 209f, 343–344, 344f
Ganglion cell(s)
   cochlear, 328–329, 328f
   retinal, 305, 306f, 310, 311, 311f, 312
      projections of, 312–316
      subdivisions of, 316, 317f, 322f
Ganglion impar, 469
Gap junctions, 467
Gaze centers
   horizontal, 213, 453–455, 454f
   vertical, 216, 217, 454f, 455
Gaze palsies, 500
   horizontal, 453–454
   vertical, 455
Gaze systems, 446
   abducens nerve lesions and, 213
   oculomotor nerve injury and, 217
   trochlear nerve injury and, 216
GDP (guanosine diphosphate), in signal transduction, 65–66,
      65f
Gemmules, 428
Gender differences, in pain, 288
Generator potential, 25, 42, 362
Genes, homeobox (Hox), 84
Geniculate body(ies)
   lateral, of diencephalon, 189, 190f, 223, 224f, 226f, 316, 317f
   medial, of diencephalon, 189, 190f, 223, 224f, 226f, 316, 317f
Genioglossus muscle(s)
   corticonuclear fiber lesions affecting, 396, 398f
   hypoglossal nerve function and, 202, 203
Gennari, stria of, 319, 319f, 506f, 507
Gigantism, 489, 489f
Glaucoma, 304
Glia
   radial, 81, 84f, 85

Glia (Continued)
    in thalamic development, 85, 86f
    progenitor of, 81
Glia limitans, 28, 29f
Glial cell(s), 4, 28–33
    Bergmann, 81
    differentiation of, 81
    radial
        in cerebellar development, 84f, 85
        in thalamic development, 85, 86f
    transporter proteins of, 68
    types and functions of, 16, 28, 29f, 29t
Glial fibrillary acidic protein, 28, 29f
Glial scarring, 95
Glioblasts, 138
Gliosis, 95
Globus pallidus, 236f, 237, 246, 246f, 247f, 248f, 407, 408f, 409–410, 415
    interaction with motor areas of cortex, 401–402
    role in control of movement, 401–402
    surgery on, for Parkinson disease, 421
Glomerulus(i)
    cerebellar, 429, 430f
    olfactory, 363–364, 363f, 364f
Glomus
    calcified, 97, 97f
    in atrium of lateral ventricle, 96f, 97
Glomus tumor, 349, 349f
Glossopharyngeal nerve, 152, 152f, 161, 265
    anatomy and distribution of, 203f, 204f, 205f, 207–208, 207f
    development of, 79
    functional components of, 207, 207f
    in spinal trigeminal pathway, 284f, 285
    innervating stylopharyngeus muscle, 200
    lesions of, 208
    lingual-tonsillar branch of, 369
    parasympathetic outflow pathways of, 474, 474t
    taste neurons in, 369
    testing of, 530
    visceral parasympathetic afferent fibers in, 299, 300f, 301
Glossopharyngeal neuralgia, 208
Glutamate, 144
    as neurotransmitter, 58, 59t
    astrocyte function and, 30, 31f
    excitotoxicity of, in Huntington disease, 420
    in cerebellar fibers, 427
    in long-term potentiation, 500
    in medial superior olivary nucleus, 334
    in mossy fibers, 431
    in olfactory tract cells, 364
    in pain transmission, 291, 292
    in photoreceptor cells, 308, 310–312
    in subthalamic neurons, 411
Glutamine synthetase, 30
Glycine
    as neurotransmitter, 58, 59t, 67
    in superior olivary complex, 334
Glycoprotein, transmembrane, in receptors and channel subunits, 64–65, 64f
Goldman-Hodgkin-Katz equation, 40
Golgi cell(s), 84f, 85, 184, 184f, 270, 428, 430f, 431f, 434
Golgi complex
    in biosynthesis of large dense-cored vesicles, 60–61, 61f
    of neurons, 17, 17f
Golgi method, 507
Golgi tendon organ, 270
    in autonomic inhibition, 146, 146f
    in muscle innervation, 377f, 378
    proprioceptive receptors on, 258
Gonadotropic tumors, 491
Gonadotropin(s), 488
Gonadotropin-releasing hormone, 480
Granulations, arachnoid, 114
Gratification centers, 503
Gravitoinertial acceleration, 347

Gray matter, 5
    around cerebral aqueduct, 99
    astrocytes of, 30
    central, 99, 165, 165f
        of medulla, 163, 164f
        of midbrain, 190, 191, 192f, 193f
        of spinal cord, 81, 139, 141–142, 141f, 143f
    oligodendrocytes in, 29f, 30
    periaqueductal, 99, 153, 480, 481f, 484, 494, 495f
        hypothalamic efferent projections to, 487
        in pain transmission, 292
        of midbrain, 190, 191, 192f, 193f
Great cerebral vein of Galen, 130, 131f, 133, 133f
    malformations of, 133, 133f
Growth cone, 81
    in cellular brain development, 87–88
Growth factor(s)
    apoptosis and, 87
    astrocyte, 29
    fibroblast, 87
    in glial cell differentiation, 81
Growth hormone, 488
    excess production of, 489, 489f
Guanethidine, action of, 69f, 70
Guanosine diphosphate (GDP), in G protein receptor action, 65–66, 65f
Guanosine nucleotide–binding protein (G protein). See G protein.
Guanosine triphosphate (GTP), in G protein receptor action, 65–66, 65f
Guanylate cyclase, in G protein–coupled receptor responses, 66
Gustation, 360. See also Taste.
Gyrus(i), 7–8
    angular, 238f, 239, 337
    anterior paracentral, 239, 239f
    cingulate, 239f, 242, 494, 495f
        blood supply of, 496
        in limbal system, 407, 407f
        isthmus of, 239f, 242, 494, 495f
        lesions of, emotional response and, 500
    dentate, 250, 496–497, 498f
    fusiform, lesions of, 321
    inferior frontal, 238, 238f
        pars opercularis of, 238, 238f, 240f, 337
        pars triangularis of, 238, 238f, 240f, 337
    inferior temporal, 239, 240f
    lingual, 239f, 241, 319, 322f
    middle temporal, 239, 240f
    occipital, 239f, 241
    occipitotemporal, 239, 240f
        lesions of, 321
    of cerebral cortex, 236
    orbital, 238, 241f
    parahippocampal, 239f, 242, 494, 495f, 497
        blood supply of, 496
    paraterminal, 250
    postcentral, 238f, 239, 240f, 242, 283
        visceral nociceptive input to, 296, 297f
    posterior paracentral, 239, 239f
    precentral, 238f, 239, 240f, 242, 389f, 390
    superior frontal, 238, 238f
    superior temporal, 239, 240f
    supramarginal, 238f, 239, 337
    transverse temporal (of Heschl), 239, 241, 241f, 242, 336
Gyrus breves, 239, 241f
Gyrus longi, 239, 241f
Gyrus rectus, 238, 241f

Habenula, 220f, 221
Hair bundle, 326, 327, 327f
Hair cell(s), 325f, 326, 326f
    as transducers of sound, 327–328, 327f
    depolarization of, 327f, 328
    inner, 326, 327f
    of organ of Corti, 326, 327–328, 327f, 328f

Hair cell(s) (Continued)
  of semicircular canals, 345–347, 345f–347f, 347–349, 348f
  of vestibular sensory system, 349, 350f
    morphologic polarization of, 347, 347f
    morphology of, 345–346, 345f
    transduction by, 346–347, 346f, 347f
  outer, 326, 327f
    efferent feedback to, 338
  stereocilia in, 325f
Head
  orientation of, 343, 380
  sympathetic innervation of, 470
Head movement(s)
  in postural adjustment, 343, 344f, 354
  linear, otolith organs and, 347
  rotational, 347–348, 348f
    compensatory eye movements with, 353–354, 353f
    nystagmus in, 354, 354f
  semicircular canals and, 347–349
  tilting, 436
    ocular counter-roll in, 350f, 354
    otolith organ response to, 350f
  vestibular function and, 343, 344f
Head trauma
  meningeal hemorrhage and, 115
  vestibular afferent response to, 349, 349f
Hearing, testing of, 529–530, 531f
Hearing loss. See also Auditory system; Ear; Sound.
  conductive, 209, 324, 325, 529
  neurogenic, 529
  sensorineural, 209, 324
    cochlear implants and, 328
  testing for, 529–530, 531f
Heat dissipation/conservation, hypothalamic control of, 492
Helicotrema, 325, 325f
Hematoma
  "dural border," 116, 116f
  epidural, 115, 116f
Hemianalgesia, in Wallenberg syndrome, 286
Hemianesthesia, 234
  alternating, 138
Hemianopia, 525
  bitemporal, 315, 315f
  contralateral left or right, 319, 525
  homonymous, 316, 316f, 517, 525
    with motor deficits, 391
  nasal, 316
Hemiballismus, 231, 417
Hemifields, of eye, 318
Hemiparesis, 150, 391
  contralateral, 234
  of extremities, contralateral, 392
Hemiplegia, 150, 391, 533
  alternating, 391, 392, 394f, 451
  contralateral, 451
  inferior alternating, 392
  lateral corticospinal tract lesions and, 149
  middle alternating, 392, 394f
  spastic, 394, 395f
  superior alternating, 391
Hemisphere(s)
  cerebellar, 177, 177f, 424–425, 425f–427f
  dominant and nondominant, 516
Hemithermoanesthesia, in Wallenberg syndrome, 286
Hemorrhage
  extradural, 115–116, 116f
  intracranial, causes of, 122–123, 122f
  intraventricular, 79, 122
    cerebral aqueduct obstruction and, 79, 98–99
    hydrocephalus with, 104
  meningeal, 122
  parenchymatous, 122
  subarachnoid, 113, 119–120, 119f, 122
    cerebrospinal fluid composition in, 102–103
    signs of, 119–120

Hemorrhage (Continued)
  subdural, 115–116, 116f
  triventricular, hydrocephalus with, 104, 105f
Hepatolenticular degeneration, 421
Herniation
  tonsillar, 111
  uncal, 111, 460
Herpes zoster, 277, 279f, 285
Herring bodies, 487
Hippocampal complex, 237
Hippocampal decussation, 499
Hippocampal formation, 494, 495f, 496–497, 497f–499f
  and corpus callosum, 497, 497f
  bilateral damage to, 500
  function of, 499–500
  of limbic system, 496–497, 497f–499f
Hippocampus, 223, 236f, 237, 250–251, 484, 485f, 496–497, 498f
  afferents and efferents of, 497, 499
  development of, 247f, 250
  olfactory connections to, 366
  regions of, 497
  relationships of, 248f
  vasculature of, 251
Hirschsprung disease, 80–81, 466
Histamine, as neurotransmitter, 58, 59t
Hodgkin cycle, 43, 44
Holoprosencephaly, 77, 78f
  alobar, 77, 78f
  semilobar, 77, 78f
Homunculus, motor, 390, 390f
Hormone(s)
  follicle-stimulating, 480, 491
  gonadotropin-releasing, 480
  gonadotropins as, 488
  growth, 488, 489, 489f
  luteinizing, 480, 491
  releasing, 483, 484
    in tuberoinfundibular tract, 488
  thyrotropin as, 500
  thyrotropin-releasing, 233
Horn
  Ammon's, 250, 496, 498f
  anterior, 138–139, 139f
    brainstem and spinal influence on, 379–386
    laminae of, 141f, 142
    motor neurons of, 145, 145f. See also Neuron(s), motor, lower.
    peripheral sensory input to, 376–379, 376f–378f
  anterior (ventral), of spinal cord, 80f, 81, 141, 142
  lateral, 81, 142
  medullary posterior, 285
  of lateral ventricle
    anterior, 96f, 97, 98f
    inferior, 96f, 97, 98f
    posterior, 96f, 97, 98f
  posterior (dorsal), 138–139, 139f
    in pain transmission, 291f, 292
    in viscerosensory afferent pathways, 296f, 298, 300f
    laminae of, 141, 141f, 280f
    major sensory inputs and outputs of, 280f
    neurons of, 276, 278–279
    of spinal cord, 80f, 81, 141, 142
    referred pain and, 297, 298
Horner syndrome, 149, 170, 452–453, 452f, 470, 472f, 487
Horseradish peroxidase, axonal transport and, 22
Human embryonic transplants, for therapy for Parkinson disease, 421
Human immunodeficiency virus (HIV), microglial cells targeted by, 32
Humoral effects, on visceral motor system, 466
Hunger, 491
Huntington disease, 409, 417
  basal nuclear dysfunction and, 419–420, 419f
Hydrocephalus, 94, 104–106, 105f
  communicating, 104–106
  congenital, 79, 79f
  normal pressure, 106
  obstructive, 104

Hydrocephalus *(Continued)*
  triventricular, 99
  with meningomyelocele, 75
Hydrocephalus ex vacuo, 104, 106
Hygroma, 116
Hyperacusis, 212, 369
Hyperalgesia, 274
  in central pain syndrome, 291
  primary, 275–276
  secondary, 276
Hypermetamorphosis, 501
Hypermetria, 441
Hyperopia, 321
Hyperorality, 501
Hyperphagia, 230, 501
Hyperpolarization
  at inhibitory synapses, 52, 54f
  light-induced, 308, 311–312, 311f
  of cell membrane, 25
  of sensory neurons, 25
Hyperpolarization block, 46
Hyperprolactinemia, 491
Hyperreflexia, 6, 146, 388, 391, 533
Hypersexuality, 501
Hypersomnia, 197
Hypertension
  baroreceptor reflex and, 301, 492
  idiopathic intracranial, hydrocephalus with, 106
  reserpine for, 69–70, 69f
Hypertonia, 388, 391, 532
Hypesthesia, 274, 277
Hypogeusia, 371
Hypoglossal canal, 202
Hypoglossal nerve, 152, 152f, 161, 164
  anatomy and distribution of, 202–203, 203f–204f
  functional components of, 156, 156f
  impaired function of, 203
  testing of, 530, 532f
Hypoglossus muscle, hypoglossal nerve function and, 202
Hypometria, 441
Hypophysial artery, 484, 484f
Hypophysial portal system, 482f
Hyporeflexia, 6, 146, 388, 533
Hyposmia, 362
Hypotelorism, 77
Hypotension
  baroreceptor reflex and, 301
  orthostatic, 70, 476
  watershed infarcts and, in regions of cerebral arteries, 128, 128f
Hypothalamus, 223, 229–230, 230f, 231f, 479–492. *See also* Diencephalon.
  afferent connections of, 484–486, 485f, 486f
    viscosensory, 298, 300f
  anatomy of, 220
  blood supply of, 485, 485f
  boundaries of, 480, 481f
  caudolateral, 486, 491
    lesions of, 491
  cortical input into, in pain transmission, 292
  development of, 85, 86f
  divisions of, 480, 482f, 483–484
  efferent connections of, 486–487
  function of, 6f, 7, 229, 480
    regional, 491–492
    visceromotor, 442–443
  intrinsic connections of, 487–488
  lateral zone of, 229–230, 231f, 480, 482f, 483
    olfactory connections to, 365–366
  medial zone of, 230, 231f, 482f, 483–484
    chiasmatic region of, 230, 231f
    mammillary region of, 230, 231f, 484
    supraoptic region of, 484
    tuberal region of, 230, 231f, 484
  periventricular zone of, 484
  pituitary links to, 488
  pituitary tumors and, 488–491, 489f, 490f

Hypothalamus *(Continued)*
  preoptic area of, 480, 482f
  reflexes of, 492
  releasing hormones of, 488
  role in autonomic function, 475, 480, 480f
  role in endocrine and limbic systems, 480, 480f
  stria terminalis connection with, 501
Hypotonia, 388, 441, 532
  cerebellar, 85
Hypovolemia, 488
Hypoxia, perinatal, axonal and synaptic development and, 89

Imaging
  of brain and skull, 10–12, 10f
  of pain in somatosensory pathways, 268, 288
Impaired check, 441
Impedance matching, 325
Incus, 325, 325f
Induction
  central, 77
  of notochord, 73
Indusium griseum, 497
Infarcts
  cerebral, 122, 122f
  watershed, 515
    in regions of cerebral arteries, 128, 128f
Infections
  congenital nervous system defects and, 77
  mycotic, subacute meningitis with, 120
Inferior oblique muscle(s), 447, 447f
Inferior rectus muscle(s), 447, 447f
Inflammation
  central nervous system, microglial cells in, 32
  joint pain with, 276
Information processing
  electrochemical basis of, 38
  in central auditory pathways, 329–331, 330f
  in cerebellar cortex, 433–435, 434f
  information input and transmission in, by sensory neurons, 25–27, 26f, 27f, 27t
  information representation in, 47–48, 48f, 49f
    by frequency coding, 48, 48f, 256, 271, 329
    by place codes, 47, 329
    by recruitment, 48, 49f
  information storage in, 38
  localization of function in, 38
  organization of, 38
  parallel, 38
  redundancy in, 38
  synaptic, 5, 38, 52–53, 53f–55f
Infundibulum, 97, 99f, 221
Inhibiting factors, 230
Inhibition
  autonomic (inverse myotatic reflex), 146, 146f
  lateral surround, 256
  reciprocal, 146, 146f, 378
  recurrent circuit for, 388f
Inhibitory postsynaptic potentials, 67
Injury. *See* Trauma.
Inositol 1,4,5-triphosphate
  in G protein receptor action, 52, 53f, 65f, 66
  in olfactory transduction, 362, 362f
  in taste transduction, 368, 369f
Inspiratory stridor, 207
Insula, 238, 241f
  anterior, 288
  central sulcus of, 238, 241f
Insulators, 38
Integrins, in somite development, 80, 80f
Intellectual function, defects of
  in lesions of parietal association cortex, 518, 519f
  in lesions of prefrontal cortex, 518–520
Intensity coding, in cochlear nerve, 329
Intercavernous sinus(es), 113, 132

Interleukin-1, secreted by microglial cells, 32
Intermediate zone, 81
  of spinal cord gray matter, 81, 139, 141
Internal capsule. *See* Capsule, internal.
Internal carotid system, 123, 124f, 125, 125f, 126f
Interneuron(s)
  anterior horn, 374
  corticospinal fibers synapsing with, 394
  definition of, 23
  excitatory glutaminergic, 146, 146f, 147, 147f
  Ia, 378
  inhibitory, 388f, 389
    GABAergic, 263
    glycinergic, 146, 146f, 147, 147f
  local circuit, of ventral posterior nucleus, 263
  of cerebral cortex, 509–510, 510f
  of olfactory bulb, 363
  of spinal cord, 138, 277
Internodes, 32
Internuclear ophthalmoplegia, 213
Interpeduncular fossa, 153, 189, 189f
Interthalamic adhesion, 85, 221, 222f, 228
Intracranial pressure. *See also* Hydrocephalus.
  abnormalities of, 104–106, 105f
  supratentorial and infratentorial compartments and, 111, 113
Intraocular muscle(s), 447, 448f, 449
Ion channel(s)
  binding sites in, 65
  calcium, in Huntington disease, 420
  G protein–coupled, 66
  in action potential generation, 42–44, 43f, 44f
  in chemical neurotransmission, 59
  in hair cell mechanoelectric transduction, 327f, 328
  in nerve cell membrane, 23–25, 24f
  in receptor potentials, 46–47
  in taste receptors, 368, 369f
  ligand-gated, 24, 24f, 375
    gene superfamilies of, 65
    structure of, 64–65, 64f
  membrane potential and, 24–25, 39–40, 39f, 41f
  of astrocytes, 29–30
  permeability of, 42, 47, 51, 51f, 52f
  potassium. *See* Potassium channel.
  sodium. *See* Sodium channel.
  subunits of, 64–65, 64f
  voltage-gated, 24, 24f
    action potential generation and, 42–44, 43f, 44f
    blockage of, 46
    structure of, 64
Iris, 304–305, 304f
Ischemia, cerebral, 122, 122f
  transient, 500

Jaw jerk reflex, 267
  trigeminal nerve innervation of, 215
Jaw muscle(s), 267
  innervation of, 267, 271f, 272
  pars caudalis and, 286
  testing of, 527, 528f
  trigeminocerebellar connections and, 215, 271f, 272
Joint(s)
  pain receptors in, 276–277, 276t, 277f–279f
  proprioceptors in, 258, 258f
  temporomandibular, 272
    disorders of, pars caudalis and, 286
    innervation of, 285
Jugular vein, internal, 130, 132, 208
Juxtarestiform body, 162, 166, 166f, 167f, 178, 179f, 180f, 424, 424f, 427
  vestibulocerebellar fibers in, 352

Kainate receptors, 67
Kayser-Fleischer ring, 421

Kidneys, antidiuretic hormone and, 488
Kinesin, in axonal transport, 21
Kinesthesia, 256, 258, 264
Kinocilium, 345, 345f–347f
Klüver-Bucy syndrome, 496
  characteristics of, 500
Knee jerk reflex, 6, 146, 146f
Korsakoff syndrome, 483–484, 499–500

Labyrinth
  blood supply of, 344
  bony, 342, 342f
  membranous, 325, 325f, 343, 343f, 344–345
  vestibular
    cerebellar afferent connections with, 352, 352f
    peripheral, 342–345, 342f–345f
Labyrinthine artery(ies), 124f, 127, 210, 329, 344
Lacrimal artery(ies), 111
Lacunae, lateral (venous), 114
Lambert-Eaton syndrome, 24
Lamina(ae)
  arcuate, 263
  definition of, 22, 23t
  external medullary, 223, 224f
  fibrodendritic, 335
  internal medullary, 223, 226f
  isofrequency, 333f, 335
  medullary, 282, 409
  of spinal cord, 141, 141f
  posterior horn, major sensory inputs and outputs of, 280f
  Rexed's, 141, 141f, 278, 468
  spinal I-IX
    collaterals from, in anterolateral system in medulla, 282
    pain fiber input into, 278–279
Lamina propria, 360, 361f, 362
Lamina terminalis, 73, 97, 99f, 236f, 237, 480, 481f
  organum vasculosum of, 221, 221f
Laminin, in somite development, 80, 80f
L-amino acid decarboxylase, 69
Language
  assessment of, 522
  impairment of, 516–518. *See also* Aphasia.
  nonverbal, aphasias of, 518
  Wernicke's area and, 337
Laryngeal muscle(s), intrinsic, innervation of, 200
Laryngeal nerve(s), superior, 369
Lateral geniculate body, 189, 190f, 223, 224f, 226f, 316, 317f
Lateral medullary syndrome (Wallenberg syndrome), 170, 171f, 215, 286–287, 287f
Lateral orbitofrontal loop, in basal nuclei parallel circuits, 413
Lateral rectus muscle(s), 353, 447, 447f
  abducens nerve function and, 213
Laterality, concept of, 389
Law of specific nerve energy, 47
Layer(s). *See also* Lamina(ae); Zone(s).
  arachnoid barrier cell, 110f, 113
  definition of, 22, 23t
  dural border cell, meningeal hemorrhage and, 115–116, 116f
  epipial, 116
  ganglion cell, 306, 306f, 309
  germinal. *See* Layer(s), granule cell.
  glomerular, 363–364, 363f
  Golgi cell, 84f, 85
    of cerebellum, 428, 430f, 431f, 434
  granule cell
    of cerebellum, 184, 184f, 428–429, 429f, 430f, 431f
      external, 84f, 85
      internal, 84f, 85
    of cerebral cortex
      external, 506f, 507, 508f
      internal, 506f, 507, 508f
    of hippocampus, 497, 498f
    of olfactory bulb, 363, 363f, 364
  magnicellular, of retina, 316, 317f, 322f

Layer(s) *(Continued)*
 mantle, 81, 138, 139, 188
  in meningeal development, 108f
 marginal, 188
 mitral cell, 363, 363f, 364
 molecular
  of cerebellum, 84f, 85, 184, 184f, 429, 429f, 430f, 431, 431f
  of cerebral cortex, 506f, 507, 508f
  of hippocampus, 497, 498f
 multiform, of cerebral cortex, 506f, 507, 508f
 nerve fiber, of retina, 306, 306f, 313–314
 nuclear
  inner, 306, 306f
  outer, 306, 306f
 of cerebral cortex, 86, 87f, 506f, 507, 508f
 of hippocampus, 497, 498f
 of lateral geniculate nucleus, 316, 317f
 of olfactory bulb, 363, 363f
 parvocellular, of retina, 316, 317f, 322f
 photoreceptor cell, outer and inner segments, 306, 306f, 307f, 309f
 plexiform
  external, 363, 363f, 364
  inner, 306, 306f, 310
  outer, 306, 306f, 310
 polymorphic, of hippocampus, 497, 498f
 Purkinje cell
  of cerebellar cortex, 184, 184f
  of cerebellum, 84f, 85, 428, 429f, 431f
 pyramidal cell
  of cerebral cortex, 506f, 507, 508f
  of hippocampus, 497, 498f
 retinal, 306, 306f, 307f, 309f, 310, 313–314, 316, 317f, 322f
 ventricular, of neural tube, 138, 139f
L-cones, 308, 308f
L-deprenyl, for Parkinson disease, 421
L-dopa, for Parkinson disease, 28, 420
Lead pipe rigidity, 532
Lemniscus(i)
 definition of, 22, 23t
 lateral, 180, 181f, 182, 329, 330f, 333f
  nuclei of, 333f, 335
  of inferior colliculus, 191, 192f
  of midbrain, 191
 medial, 162, 163f, 164, 165, 165f, 166f, 167, 167f, 168, 168f, 180, 180f, 181, 181f, 182f, 192f, 193, 194f, 197
  in pons, 187, 187f
  of midbrain, 191, 191f
  orientation of, 181f, 193, 193f
  somatotopic organization of, 262, 262f
 trigeminal, 265, 266f, 267f
Length constant, 42
Lens, 304f, 305, 448f, 449
 accommodation of, 448f, 449, 458
Lenticulostriate artery(ies), 125, 129, 239, 242, 243f, 245, 246f, 250
 lesions of, corticonuclear and corticospinal system damage with, 398
Leptomeninges, 110, 116
 early development of, 108f, 109
Leptomeningitis, 120
Leukodystrophy, myelination and, 89
Levator palpebrae superioris muscle(s), 449, 449f
Ligament(s)
 coccygeal, 109f, 113
 denticulate, 110, 116–117, 140, 141f
 periodontal
  afferent fibers of, 267, 272
  receptors in, 267
 suspensory, 448f, 449
Ligand-dependent regulators of nuclear transcription, for binding of chemical messengers, 63
Ligand-gated ion channels, 24, 24f, 63, 64–65, 64f, 375. *See also* Ion channel(s), ligand-gated.
Light stimulus
 in G protein receptor action, 65f, 66
 retinal response to, 308, 311–312, 311f, 313
 ultraviolet, 305

Limb ataxia, 535
Limb extensor muscle(s), control of, 380, 437–439
Limb flexor muscle(s)
 control of, 437–439
 proximal, 382
 rubrospinal tract and, 385
Limbic loop, in basal nuclei parallel circuits, 413
Limbic system, 241, 282, 493–504
 amygdaloid complex and, 500–501, 502f. *See also* Amygdaloid complex.
 blood supply to, 496
 cognitive function and, 504
 cortical areas relating to, 514
 disorders of, 421
 early functional concepts of, 494, 496
 emotions and, 494, 496, 504
 hippocampal formation in, 496, 497, 497f, 499f
 hypothalamus in, 480, 480f
 long-term potentiation and memory and, 500
 nucleus accumbens and, 503f, 504
 septal region of, 503–504, 503f
 structure and organization of, 494
 subcortical nuclei of, 494, 495f
Limen insula, 125, 241, 242
Lingual gyrus, 239f, 241, 319, 322f
Lipid bilayer, 39f, 60
 in release of synaptic vesicles, 62–63
Lissencephaly, 86, 88f
Lithium carbonate, for bipolar disorder, 28
Lobe(s)
 cerebellar, 424–425, 425f–427f
  anterior, 85, 174f, 175, 176f, 177, 177f, 269, 270f
   in decerebrate experiments, 384–385
  flocculonodular, 84f, 85, 175, 177, 177f, 435
  posterior, 85, 174f, 175, 176f, 177, 177f
  paravermis of, 269, 270f
 frontal, 238–239, 238f, 240f, 241f
  thalamocortical projections to, 513–514, 514f
 insular, 239, 240f, 241, 241f
 limbic, 238, 239f, 241–242, 494, 495f, 496f
  function of, 242
  thalamocortical projections to, 514, 514f
 medial temporal (hippocampus), 223
 occipital, 238, 238f, 239f, 241
  thalamocortical projections to, 514, 514f
 of cerebral cortex, 238–242, 238f–243f
 of pituitary gland, 220f, 221
 parietal, 238, 238f, 239, 239f
  testing of, 522, 524f
  thalamocortical projections to, 514, 514f
 temporal, 238, 238f, 239, 239f, 240f, 241f
  arteries of, 124f
  Klüver-Bucy syndrome and, 496, 501
  lesions of, 250
  thalamocortical projections to, 514, 514f
Lobotomy, prefrontal, 520
Lobule(s)
 cerebellar, 424–425, 425f–427f
 inferior parietal, 238f, 239, 337
 posterior paracentral, 283
 superior parietal, 238f, 239
Local circuit cells, definition of, 23
Localizing signs, of neurologic deficits, 8–9, 8f
Locus ceruleus, 181, 181f, 182, 364, 510
Lumbar cistern, 109f, 117
Lumbar puncture, 102, 117, 140, 140f
Luteinizing hormone, 480, 491

M cell(s), 312
M stream, 321, 322f
Macula
 ocular, 318f
  sparing of, 319
 of otolith organs, 346, 347f, 350f

Macula lutea, 309
Magnetic resonance imaging, 10f, 11–12, 11t, 12t
    functional, 268, 288
Magnetoencephalography, 268
    for imaging of pain, 288
Malleus, 325, 325f
Malnutrition, perinatal, axonal and synaptic development and, 89
Mammillary body
    in Korsakoff syndrome, 500
    nuclei of, 494
Mandibular nerve(s), 265
    branches of, 111
MAO inhibitors, Parkinson disease progression and, 420–421
Marginal layer, 138, 139
Massa intermedia, 85, 99f, 221, 222f, 228
Mastication, muscles of, 267, 271f, 272
    innervation of, 200, 214, 214f, 215
    testing of, 527, 528f
Matrix, of striatal complex, 408
Maxillary nerve(s), 265
    branches of, 111
M-cones, 308–309, 308f
Mechanonociceptors, 274
Mechanoreceptor(s), 25. See also Receptor(s).
    cutaneous tactile, 269, 275, 275t
    deep tactile, 257, 257t, 258f
    high-threshold, 274, 275, 275t, 276f
    modulation of channel permeability in, 46
    peripheral, 256–259
    visceral
        rapidly adapting, 294, 294t
        slowly adapting, 294, 294t
Medial aperture (foramen of Magendie), 77, 78f, 95, 95f, 99, 100f, 153, 153f
Medial geniculate body, 189, 190f, 223, 224f, 226f, 336, 337f
Medial rectus muscle(s), 353, 447, 447f
    innervation of, 212f, 213
Medulla oblongata, 6f, 7, 82–83, 82f, 152, 152f, 159–172
    alar and basal plate structures in, 155f
    anterolateral system fibers in, 280, 281f, 282
    ascending and descending pathways of, 162–163, 163f
    autonomic nuclei of, hypothalamic link to, 487
    caudal, sensory and motor decussations at, 163–166, 164f–166f
    contiguous areas of, 153, 153f
    corticospinal fibers in, damage to, 392
    cranial nerves of, 203–208, 203f–205f, 207f
    development of, 160–161, 160f
    dorsal, 162, 162f
    external features of, 161–162, 161f, 162f
    internal anatomy of, 162–170, 163f–171f
    junction of, with pons, 167–169, 167f–169f
    lateral, 161–162, 162f
    mid level of, 166–167, 166f, 167f
    posterior column–medial lemniscus system in, 262
    pyrimidal decussation of, 152, 152f
    rostral, 167–169, 167f–169f, 476
    spinal cord transition to, 163, 164f
    vasculature of, 162, 169–170, 170f, 171f
    ventral, 161, 161f
    ventrolateral, 476
        role in autonomic function, 476
Megacolon, congenital, 80–81, 468
Meissner corpuscles, 256, 257f, 264, 265
Melatonin, 232
Membrane(s)
    basilar, 325, 325f
        in cochlear tuning, 328
        in mechanoelectric transduction of sound, 328
    cable properties of, 40, 41f, 42, 45
    depolarization of, 24
    electrical properties of, 38–40, 39f
    hair cell, in sound transduction, 327–328
    hyperpolarization of, 24–25
    ion channels in, 23–25, 24f
    limiting, of retina, 306, 306f

Membrane(s) (Continued)
    lipid bilayer, of vesicles, 39f, 60, 62–63
    of vesicles, composition of, 60
    postsynaptic, 375
    tectorial, 326, 328
    tympanic, 325, 325f
    vestibular (Reissner), 325, 325f
Membrane capacitance, 45
Membrane channels, 39, 39f. See also Ion channel(s).
    in olfactory transduction, 362, 362f
    permeability of, 42
        in sensory receptors, 47
        postsynaptic, 51, 51f, 52f
    voltage-gated, conduction block in, 46
Membrane potential
    action, 25. See Action potential(s).
    depolarization of, 24, 25, 42–43, 43f, 44f
        presynaptic, 51, 51f
        suprathreshold, 43, 44f
        threshold, 42–43, 44f
    excitatory postsynaptic, 52, 53f, 54f, 67
    graded, 25, 42, 256
        in retinal cells, 309
    hyperpolarization of, 25, 46
    inhibitory postsynaptic, 67
    ion channel-mediated changes in, 24–25, 39–40, 39f, 41f
    receptor potential and, 46–47, 48, 48f, 49f
    resting, 23, 25, 39–40, 39f
    types of changes in, 25
Membrane pump, 39
    of vesicles, 60
Memory
    associative, disorders of, 419
    long-term and short-term, 483
        hippocampal lesions and, 499–500
        long-term potentiation and, 500
    storage sites of, cerebellar role in, 443
    testing of, 520
Ménière disease, 344–345
    vestibulocochlear nerve injury and, 210
Meningeal artery
    accessory, 111
Meningeal artery(ies)
    middle, 111
Meninges, 107–120. See also Arachnoid mater; Dura mater; Pia mater.
    definition of, 108
    development of, 108–109, 108f–109f
    function of, 108
    innervation of, 285
    nociceptors in, 285
    relation of, to brain and spinal cord, 109–110, 109f
    spinal, 140, 140f, 141f
    structure and layers of, 110f
Meningioma(s), 115, 115f
    hydrocephalus with, 104
    meningeal hemorrhages and, 115–116, 115f, 116f
    olfactory groove, 115, 115f
Meningitis, 120
    bacterial, 120
        cerebrospinal fluid in, 102
        inflammation in, 32
    signs of, 120
    viral, 120
Meningocele, 73, 74–75, 74f, 75f, 138, 139f
Meningoencephalocele, 73, 74f
Meningohydroencephalocele, 73, 74f
Meningomyelocele, 74–75, 75f, 138, 139f
Mental status examination, 522
Merkel cell(s), 256, 257, 257f, 264
Merkel cell neurite complex, 256, 257, 257f
Merkel cells, 264
Mesencephalon, 76f, 77, 78f, 82f, 83, 94, 94f, 95, 152–153, 187–198. See also Midbrain.
    development of, 188, 188f
Mesenchyme, meningeal development from, 108

Mesoderm, 72
Metathalamus, 223
Metencephalon, 76f, 77, 94f, 95, 174. *See also* Cerebellum; Pons.
    development of, 174–175, 174f
α-Methyltyrosine, action of, 68–69, 69f
Metyrosine, action of, 69, 69f
Meyer's loop, 318, 319f
Microglial cell(s), 29f, 32
Microgyria, 86, 88f
Microtubules
    axonal, 17f, 21, 22f
    dendritic, 16, 17f, 19f
Micturition, autonomic control of, 477, 477f
Midbrain, 6f, 7, 82f, 83, 187–198
    alar and basal plate origin of, 155f
    auditory information processing in, 329 330f, 333f
    cranial nerves of, 215–217
    development of, 188, 188f
    diencephalon junction with, 189, 190f, 195f, 196f, 197, 221, 222f
    dorsal, 189, 190f
    external features of, 189–190, 189f–190f
    in brainstem, 152–153, 152f
    internal anatomy of, 190–198
        ascending pathways of, 191, 191f
        basis pedunculi of, 190–191, 190f
        caudal levels of, 191, 192f–194f, 193, 195
        descending pathways of, 191, 191f
        diencephalon junction with, 190f, 195f, 196f, 197, 221, 222f
        regions of, 190–191, 190f
        reticular and raphe nuclei of, 196f, 197
        rostral levels of, 194f–196f, 195, 196, 197
        tectum of, 190–191, 190f
        tegmentum of, 190–191, 190f, 480, 481f
        vasculature of, 189–190, 190f, 197–198, 197f
    pons junction with, 189, 190f
    supraoculomotor area in, 458
    tumor in, hydrocephalus with, 104
    ventral, 189, 189f
Middle ear muscle(s), 325
Midline shift, 111
Migraine headache, 285
Miosis, 149, 305, 449, 452f, 453, 470, 472f
Mitochondria
    of chemical synapse, 26, 26f
    of dendrites, 16, 17f, 19f
Modules, of cerebellum, 435–439
Monoamine oxidase
    degradation of norepinephrine by, 68, 69f
    inhibitors of, Parkinson disease and, 420–421
Monoparesis, 533
Morphine, for chronic pain, 292
Mossy fiber(s), 184, 184f, 269, 429, 431–432, 431f, 432t
Mossy fiber rosette, 430f, 431, 431f
Motor cortex. *See* Cortex, motor.
Motor deficit(s). *See also specific deficits, e.g.,* Apraxia.
    brainstem role in, 382f–384f, 383–385
    general features of, 388–389
    in corticonuclear system, 394–396, 397f, 398–399, 398f
    in corticopontine system, 399
    in corticoreticular system, 399
    in corticorubral system, 399
    in corticospinal system, 389–394, 389f–396f
    internal capsular infarcts and, 391, 392f
    lower motor neuron signs in, 388
    pontocerebellar dysfunction and, 441–442
    spasticity in, 388–389
    spinocerebellar structural deficits and, 439
    upper motor neuron signs in, 388
    vestibulocerebellar dysfunction and, 435–436
Motor end plate, 5, 374, 375, 375f
Motor fiber(s)
    of cranial nerve nuclei, functional components of, 156–157, 156f,
        471, 472t. *See also* Functional component(s).
    of facial nerve, 179, 180f, 211, 212
    of spinal nerves, 145, 145f
    of vagus nerve, 206

Motor homunculus, 390, 390f
Motor learning, cerebellar influence on, 443–444
Motor loop, in basal nuclear parallel circuits, 413, 415, 415f
Motor outflow, somatic, vs. parasympathetic and sympathetic, 464,
    465f, 465t
Motor system
    anterior horn neurons in
        brainstem and spinal influence on, 379–386
        peripheral sensory influence on, 373–379
    brainstem and spinal cord role in, 382f–384f, 383–385. *See also*
        Motor deficit(s).
    corticofugal systems and, 387–404
    hierarchical organization vs. parallel processing in, 402, 402f
    pyramidal vs. extrapyramidal, 406
    testing of, 530, 532–536, 532f–535f
    visceral. *See* Visceral motor system.
    visual, 445–463. *See also* Eye movement(s).
Motor units, 375–376, 375f
    size principle of, 376
    types of, 376
Mouse, mutant, weaver, 85
Movement(s). *See also* Eye movement(s); Head movement(s).
    athetoid, 417, 417f
    choreiform, 417
    coordinated, testing of, 534–536, 535f, 536f
    decomposition of (uncoordinated), 441
    dentate nucleus and, 440, 440f
    direction of, coding for, by primary motor cortex, 400
    of limbs
        planning and control of, by pontocerebellum, 439
        spinocerebellar processing of, 437–439
    perception of, 355–356
    pitch and, 347, 349
    regulation of, disturbances in, 269
    roll and, 347, 349
    voluntary
        brainstem and spinal influence on, 379–382
        corticofugal systems and, 387–404
        preparation for, 401, 415, 439
        reticulospinal tracts and, 380, 381f
        rubrospinal tracts and, 380–382, 381f
        spinocerebellar processing of, 437–439
        vestibulospinal tracts and, 379–380
MPTP, 68
    Parkinson disease and, 420–421
Mucosa, olfactory, 360
Müller cell(s), 81
Multiple sclerosis, 32
    cerebrospinal fluid in, 102
Multisensory integration, 335
Muscle(s)
    antigravity, 380
        spastic, 389
    axial, control of, 437–439
    cardiac, autonomic control of, 466
    ciliary, 216, 447, 448f
    constrictors of pharynx, innervation of, 200
    contraction, 375, 375f, 376–378
        input and output circuits in, 377–378, 377f, 378f
    detrusor, 477
    digastric, trigeminal nerve innervation of, 214, 214f
    dilator, 304f, 305
    dilator pupillae, 216, 447, 448f, 449
        sympathetic innervation of, 452, 452f
    extraocular, 446–449, 447f–449f. *See also* Eye movement(s).
        in rotational vestibulo-ocular reflex, 353–354
        lesions affecting, 212f, 215–217
        motor neurons of, 453, 453f
        testing of, 526–527, 527f
    fibrillations of, 388
    genioglossus
        corticonuclear fiber lesions affecting, 396, 398f
        hypoglossal nerve function and, 202, 203
    hyperreflexia of, 6, 146, 388, 391
    hypertonia of, 388, 391
    hypoglossus, hypoglossal nerve function and, 202

Muscle(s) (Continued)
  hyporeflexia of, 6, 146, 388
  hypotonia of, 388, 441
  inferior oblique, 447, 447f
  inferior rectus, 447, 447f
  intraocular, 447, 448f, 449
  intrinsic laryngeal, innervation of, 200
  jaw, 267
    pars caudalis and, 286
    testing of, 527, 528f
    trigeminocerebellar connections and, 215, 271f, 272
  lateral rectus, 353, 447, 447f
    abducens nerve function and, 213
  levator palpebrae superioris, 449, 449f
  limb extensor, control of, 380, 437–439
  limb flexor
    control of, 437–439
    proximal, 382
    rubrospinal tract and, 385
  lower motor neuron deficits in, signs of, 388
  medial rectus, 353, 447, 447f
    innervation of, 212f, 213
  middle ear, 325
  neck, 380
  oculomotor nerve innervation of, 212f, 216–217
  of facial expression
    innervation of, 200, 210f, 211
    testing of, 527–528, 529f
  of lower face, corticonuclear fiber lesions affecting, 396, 398f
  of mastication, 267, 271f, 272
    innervation of, 200, 214, 214f, 215
    testing of, 527, 528f
  of soft palate, corticonuclear fiber lesions affecting, 396, 398f
  of uvula, corticonuclear fiber lesions affecting, 396, 398f
  orbicularis oculi, 449, 449f
  pain receptors in, 276–277, 276t, 277f–279f
  palatine, innervation of, 200
  palatoglossus, hypoglossal nerve function and, 202
  paravertebral, control of, 380
  proprioreceptors on, 258, 258f
  pterygoid, testing of, 525, 526f
  short ciliary, 216
  slow-twitch and fast-twitch, 376
  smooth, autonomic control of, 464
  spasticity of, 388
  sphincter, 304f, 305
  sphincter pupillae, 447, 448f, 449
  stapedius, 325, 325f, 338, 338f
  sternocleidomastoid, accessory nerve lesions and, 206
  stylopharyngeus
    glossopharyngeal nerve and, 208
    innervation of, 200
  superior oblique, 447, 447f
  superior rectus, 447, 447f
  superior tarsal, 216
  superior tarsal (of Müller), 449, 449f, 468
  tarsal, eyelid movements and, 460, 462
  temporal, in jaw jerk reflex, 267
  tensor tympani, 325, 325f, 338
  tensor veli palatini, 214, 214f
  trapezius, accessory nerve lesions and, 206
  trigeminal nerve innervation of, 213–215, 214f
  trochlear nerve innervation of, 212f, 215–216
  urethral external, 475
  vocal, innervation of, 200
  weakness of, 388
Muscle fiber(s)
  alpha-gamma coactivation of, 377–378, 377f, 378f
  dynamic bag, 376f, 377
  extrafusal, 374, 376, 376f
  gamma loop and, 377–378, 377f, 378f, 383
  in reflex circuits, 378–379
  intrafusal, 145, 374, 376–378, 376f
  nuclear bag, 376, 376f
  nuclear chain, 376, 376f
  sensory (afferent), 377
    annulospiral endings of, 377

Muscle fiber(s) (Continued)
  flower-spray endings of, 377
  secondary endings of, 377
  types Ia, Ib, II, 377, 378
  static bag, 376f, 377
Muscle spindles, 145, 374, 376–377, 376f
  afferent fibers in, 257f, 258
    in reflex circuits, 378–379
  cells of, in spinocerebellar tracts, 270–271
  in jaw muscles, 267
  in muscle movement and gamma loop, 377–378, 377f, 378f
  proprioceptive receptors on, 257f, 258
Muscle strength, testing of, 532–533, 533f
Muscle stretch reflexes, testing of, 533, 534f
Muscle tone, 388, 441
  examination of, 530, 532
Mutism, 516
  akinetic, 500
Myasthenia gravis, 24
  characteristics of, 146
  neurotransmitter blockage in, 145–146
  somatic motor neurons in, 138
Mycotic infections, subacute meningitis with, 120
Mydriasis, 217, 305, 449, 450
Myelencephalon, 76f, 77, 78f, 82–83, 94f, 95, 152, 152f, 160. See also
    Medulla oblongata.
Myelin sheath, 5
  conduction velocity and, 46
  formation of, 30, 33f
  in central nervous system, 35
  in internodal segments, 32
  in nervous system development, 89
  in peripheral nervous system, 33, 34f, 35
  loss of, 32
  nerve conduction and, 30–31
  production of, 30
Myelinated fiber(s)
  of central nervous system, 35
  of peripheral nervous system, 33, 34f
  of subcortical white matter, 236
Myelodysplasia, 75
Myeloschisis, 74, 75f
Mylohyoid, trigeminal nerve innervation of, 214, 214f
Myopia, 321, 449
Myotome, formation of, 80f, 82

Nausea, vestibulocochlear nerve injury and, 210
Near triad, 458
Neck
  lesions of, results of, 394, 395f
  muscles of, 380
  sympathetic innervation of, 470
Neocortex, 494, 507
Neostriatum, 246, 246f, 407
  cell loss in, in Huntington disease, 419, 419f
  connections of, 410f
  lesions of, hypokinetic disorders as, 415, 416f
Nernst equation, 40
Nerve(s). See also Fiber(s); Nucleus(i).
  abducens, 132, 132f, 152, 152f, 161, 203f, 212–213, 212f
    corticospinal fiber damage to, 391–392, 394f
    functional components of, 156, 156f, 175
    in eye movements, 447
      paralysis of, 452, 452f
    lesions of, 213
    motor nuclei of, 175
      origin of, 174–175, 174f
    of pons, 175–176, 175f, 176f
    testing of, 527
  accessory, 152, 152f, 161
    anatomy and distribution of, 203, 203f, 204f, 205–206, 205f
    lesions of, 206
    relationship of, to vagus nerve, 206
  cardiac
    in angina, 298

Nerve(s) (Continued)
 sympathetic innervation of, 470
 viscerosensory afferent fibers in, 295, 296f
 cardiac thoracic, in angina, 298
 cochlear
  damage to, 329
  functional components of, 156f, 157
 cranial. See also specific nerves, e.g., Vagus nerve.
  development of, 79–80
  in spinal trigeminal pathway, 284f, 285
  mixed, 4
  nuclei of
   functional components of, 154–157, 155f–156f
    in brainstem, 156, 156f
    motor, 174–175, 174f
   of brainstem, 199–218
    motor, 82f, 83, 200–201, 201f
    sensory, 82f, 83, 201–202, 202f
   of medulla, 152, 152f, 202–208, 203f–205f, 207f
   of midbrain, 215–217
   of pons, 213–215, 214f
   of pons-medulla junction, 203f, 204f, 208–213, 209f,
    210f, 212f
  parasympathetic afferent fibers in, 298–299, 300f, 301
  parasympathetic outflow pathways of, 474–475, 474t
  testing of, 522–530
 cutaneous, compound action potential in, 260f
 deep petrosal, 210f, 211
 ethmoidal, 111
 facial, 152, 152f, 161, 265
  anatomy and distribution of, 210–212, 210f
  development of, 79
  in muscles of facial expression, 200, 210f, 211
  in spinal trigeminal pathway, 284f, 285
  intermediate root of, 161
  internal genu of, 178, 180f, 210, 210f
  lesions of, 211–212
  motor nuclei of
   functional components of, 175
   origin of, 174–175, 174f
  of pons, 175–176, 175f, 176f
  parasympathetic outflow pathways of, 474, 474t
  relationship of, to abducens nerve, 180f
  relationship of, to trigeminal nerve, 206
  roots of, 176
  taste neurons in, 369
  testing of, 527–529, 528f–529f
 glossopharyngeal, 152, 152f, 161, 265
  anatomy and distribution of, 203f, 204f, 205f, 207–208,
   207f
  development of, 79
  functional components of, 207, 207f
  in spinal trigeminal pathway, 284f, 285
  innervating stylopharyngeus muscle, 200
  lesions of, 208
  lingual-tonsillar branch of, 369
  parasympathetic outflow pathways of, 474, 474t
  taste neurons in, 369
  testing of, 530
  visceral parasympathetic afferent fibers in, 299, 300f, 301
 greater petrosal, 211
 greater superficial petrosal, 369
 hypoglossal, 152, 152f, 161, 164
  anatomy and distribution of, 202–203, 203f–204f
  functional components of, 156, 156f
  impaired function of, 203
  testing of, 530, 532f
 intermediate, 176, 180f, 369
  in parasympathetic innervation, 471f, 474
 long ciliary, 452
 mandibular, 265
  branches of, 111
 maxillary, 265
  branches of, 111
 mixed, 4, 4f, 260f
 oculomotor, 132, 132f, 152f, 153
  anatomy and distribution of, 212f, 216–217

Nerve(s) (Continued)
  aneurysm and, 526
  basilar artery branches and, 127
  functional components of, 156, 156f
  in eye movements, 447
  lesions affecting, 217, 450, 450t
  of midbrain, 189, 189f
  parasympathetic outflow pathways of, 474, 474t
  posterior cerebral arteries and, 124f, 127
  testing of, 526, 527f
 of pterygoid canal, 211
 olfactory, 360–361, 361f
  testing of, 522, 524, 524f
 ophthalmic, 265
 optic, 306, 312, 313–316, 313f
  damage to, 314, 314f, 315f
  testing of, 524–526, 524f–526f
 parasympathetic, viscerosensory afferent fibers in, 295, 298–301
 pelvic, 475
  viscerosensory fibers from, 298
 peripheral
  causalgia and, 473
  groups of, diameters and conduction velocities of, 259, 259t
  injury to, motor or sensory deficits with, 138
 pulmonary, sympathetic innervation of, 470
 salivatory, superior, motor nuclei of
  functional components of, 175
  origin of, 174–175, 174f
 short ciliary, 450
 spinal, 4, 4f, 143–146, 144f
  anterior root of, C pain fibers in, 277
  connections of, to sympathetic chain ganglia, 469–470, 469f
  dermatomes of, 277, 277f, 278f
  development of, 80f, 82
  fibers of, 143
  levels of, 143
  motor components of, 145, 145f
  posterior root ganglion of, pain fibers in, 260f, 277
  sensory component of, 143–144
  somite development and, 80, 80f
 spinal accessory, testing of, 530, 532f
 splanchnic, 144, 468, 469f
  viscerosensory afferent fibers in, 295, 296f
 superior laryngeal, 369
 sympathetic
  nociceptors in, 295
  viscerosensory afferent fibers in, 295–298, 296f, 297f
 tentorial, 111
 to dura, 111
 trigeminal, 111, 152, 152f
  anatomy and organization of, 213–215, 214f
  dermatomal maps of, 278f
  development of, 79
  distribution of, 285
  divisions of, 265, 266f, 267f
  functional components of, 175
  in blink reflex, 460
  in muscles of mastication, 200, 213, 214f, 215, 267, 271f,
   272
  in spinal trigeminal pathway, 284f, 285
  injury of, 215, 285
  mandibular division of, 213, 214f
   testing of, 527, 528f
  maxillary division of, 132, 132f, 213, 214f
   testing of, 527, 528f
  mesencephalic nucleus of, 79
  motor nuclei of, 174, 174f, 214, 214f
  of pons, 175–176, 175f, 176f
  ophthalmic division of, 132, 132f, 213, 214f
   testing of, 527, 528f
  peripheral distribution of, 285, 286f
  roots of (portio major and portio minor), 213, 265, 266f
  sensory nuclei of, 213–214, 214f
  spinal (descending) tract of, 265, 266f, 285
  testing of, 527, 528f
 trochlear, 132, 132f, 152f, 153, 176f, 189, 190f, 212f, 215–216
  functional components of, 156, 156f

Nerve(s) (Continued)
    in eye movements, 447
    injury to, 216
    testing of, 527
    vagus, 152, 152f, 161, 265
        anatomy and distribution of, 203f–205f, 206–207
        development of, 79
        dorsal motor nucleus of, 206
        function of, 206
        in spinal trigeminal pathway, 284f, 285
        lesions of, 206–207
        meningeal branch of, 208
        parasympathetic outflow pathways of, 472–473, 472t
        sensory fibers of, 206
        taste neurons in, 369
        testing of, 530
        visceral parasympathetic afferent fibers in, 299, 300f, 301
    vestibular, functional components of, 156f, 157
    vestibulocochlear, 152, 152f, 161, 202
        cochlear portion of, 327
        function of, 343–344, 344f
        in pons-medulla junction, 203f, 204f, 208–210, 209f
        injury to, 209–210
        of pons, 175–176, 175f, 176f
        origin of, 79–80
        sensory nuclei of, functional components of, 175
        testing of, 529–530, 531f
Nerve blocks, pain fibers affected by, 277
Nerve calyx, of hair cells, 345f, 346
Nerve endings, free (naked), 257f, 267, 275
Nerve growth factor, 87, 468, 468f
Nervi erigentes, 475
Nervous system
    autonomic. See Visceral motor system.
    central. See Central nervous system.
    components of, 4
    development of, 71–89, 72f
    localization of function in, 38
    organization of information processing in, 38
    peripheral. See Peripheral nervous system.
    redundancy in, 38
    vegetative, 4–6, 4f
    visceromotor. See Visceral motor system.
Neural canal, 73, 81, 94
Neural crest, 138, 139f
    cells of, 468
    meningeal development from, 108, 108f
    structures derived from, 79–80, 79t, 80f
Neural crest cell(s), 468
    of neural tube, 138
    structures derived from, 79–80, 79t, 80f
Neural folds, 73
    development of, 138, 139f
Neural plate
    development of, 138, 139f
    formation of, 72f, 73
Neural tube, 94, 95
    defects of, folic acid and, 73
    development of, 81, 138–139, 139f
    formation of, 72f, 73
    in meningeal development, 108f
    in neurulation, 73, 75
    marginal zone of, 81
Neuralgia
    glossopharyngeal, 208
    postherpetic, 277
    trigeminal (tic douloureux), 215, 285
Neurapraxia, nerve regeneration and, 35
Neuroblast(s)
    cortical, 85–86, 87f
    in neurulation, 73
    migration of, in cerebellar development, 84f, 85
    of spinal cord, 138, 139
Neuroectoderm, 72, 138, 139f
Neurofibrillary tangles, in Alzheimer disease, 28
Neurofibroma, 35

Neurofilaments
    axonal, 17f, 21, 22f
    dendritic, 16, 17f, 19f
Neurohypophysis, of pituitary gland, 220f, 221
Neuroimaging techniques, 268, 288
Neuroleptic drugs, tardive dyskinesia and, 422
Neurologic deficits, localizing signs and, 8–9, 8f
Neurologic examination, 521–537
    instruments needed for, 522, 522f
Neuroma, 36
    acoustic, 161–162, 349, 349f
Neuromediator, 5
Neuromodulator, 5
Neuromuscular junction(s), 5, 374, 375f
    in myasthenia gravis, 24
Neuron(s), 4. See also Fiber(s); Nucleus(i).
    abducens internuclear, 213, 452, 454f
    acetylcholine-containing, 409, 417
    afferent, 23
        primary nociceptive, central target of, 278, 279f
    as information receivers, 25
    as information transmitters, 25–27, 26f, 27f, 27t
    basic structure of, 4–6, 5f
    biology of, 16–28
    bipolar, 18f, 20, 20t
    classification of, 22–23, 23t
    cortical
        functional properties of, 268
        intrinsic, 509
        receptive field properties of, 267–268
    corticospinal, 389f, 390
    deep, 283
    degeneration and regeneration of, 35–36
    dopaminergic, 23
    efferent, 23
    electrical properties of, 23–25, 24f, 24t, 38–39, 38f
    electrochemical integration in, 37–55
    energy needs of, 16
    enteric, 475
    excitability of, regulation of, 67
    excitatory burst, 453, 454
    fastigial, Purkinje cell inhibition of, 384
    first-order, in posterior column–medial lemniscal system, 260
    functions of, 16
    fusiform, 506f, 507
    general somatic afferent/efferent. See under Functional component(s).
    general visceral afferent/efferent. See under Functional component(s).
    glial cells and, 28, 29f
    glutaminergic, 23
    groups of, definitions of, 22–23, 23t
    inhibitory burst, 453, 454
    local circuit, 508, 510
    medium spiny, 409, 411f, 417
        in Huntington disease, 419
    motor
        alpha, 374, 377, 384, 385, 467, 467f
        anterior (ventral) horn, 81, 145, 145f, 374–376, 374f, 375f
            brainstem and spinal systems influencing, 379–382, 379f–381f
            motor units of, 375–376, 375f
            neuromuscular junctions and, 375, 375f
            of spinal cord, 138, 145, 145f
            peripheral (sensory) input to, 376–379, 376f, 377f, 378f
            types and distribution of, 373–374, 375f
        extraocular muscle, burst-tonic firing pattern of, 453, 453f, 454
        facial, 338
        gamma, 374, 377–378, 377f, 378f, 383, 384, 389
        lower
            in brainstem, functional components of, 200, 201f
            motor units of, 375–376, 375f
            neuromuscular junctions of, 375, 375f
            peripheral sensory and supraspinal influences on, 373–386
            sensory feedback to, 374
            signs of motor deficits in, 388

Neuron(s) *(Continued)*
    topographic arrangement of, 374
    types and distribution of, 373–374, 375f
    of brainstem, 83
    of lateral horn, 81
    trigeminal, 338
    upper, 374
        signs of damage to, 388, 391–392, 394, 395f
    motor visceral
        of spinal cord, 138
        postganglionic
            development of, 468
            sympathetic, 468–470, 469f
        preganglionic
            development of, 467–468
            sympathetic, 468–469, 469f
        vs. somatic motor neurons, 467, 467f
    multipolar, 16, 17f, 18, 18f, 19f, 20t
    of cerebral cortex, sensory submodalities of, 512–513
    of cervical ganglia, sympathetic input to, 468–469, 470f
    of interomediolateral cell column, 232–233
    of posterior column–medial lemniscal system, 260, 262, 262f, 263
    of spinal trigeminal pathway, 284f, 285
    of supraoptic and paraventricular nuclei, 483
    of trigeminal nerve, 265
    of ventral posterior nucleus, 263
    olfactory receptor, 360, 361, 361f, 362, 362f
    organization of, in brain development, 87–89
    pallidosubthalamic, in Parkinson disease, 419
    parasympathetic
        postganglionic, 473–474, 474t
        preganglionic, 473–474, 474t
            functional and chemical coding of, 475
    peptidergic, formation and transport of large dense-cored vesicles
        in, 61, 61f
    posterior horn
        central processing of sensory information in, nociceptive input
            and, 276
        deep, 278
        functional properties of, 278–279
        low-threshold, 278
        nociceptive-specific, 278
        referred pain and, 297, 298
        wide dynamic range, 278
    postsynaptic
        dynamic range of, 52–53, 55f
        fast and slow neurotransmission in, 58–59
        regulation of chemical messenger in, 59–60
    premotor, in smooth pursuit movements, 457
    presynaptic, dynamic range of, 52–53, 55f
    projection, definition of, 23
    pseudounipolar, 18, 18f, 20, 20t
    pyramidal, 388, 506f, 507, 508f
        of cerebral cortex, 508–509, 508f, 509f
    sensory
        neurotransmitters of, 144–145
        of brainstem, 82f, 83
        primary, 25
    spinal
        general somatic afferent/efferent. *See under* Functional compo-
            nent(s).
        general visceral afferent/efferent. *See under* Functional compo-
            nent(s).
    stellate
        smooth (aspiny), 506f, 507, 508f
        spiny, 506f, 507, 508f
    striatal, 413
    striatopallidal, 413
    structure of, 16–20, 17f–20f, 22f
        cell body of, 16–17, 17f, 18f, 19f
    subthalamopallidal, in Parkinson disease, 419
    sympathetic
        postganglionic
            chemical coding in, 473, 473f
            in visceral motor pathway, 466
            of spinal cord, 144f, 145

Neuron(s) *(Continued)*
    preganglionic
        functional and chemical coding of, 472–473, 473f
        in visceral motor pathway, 466
        of spinal cord, 144f, 145
    thalamocortical, 263, 265
    trigeminothalamic second-order, in spinal trigeminal nucleus, 287
    types of, 18, 18f, 20t
        in cerebral cortex, 508–510, 508f
    vasopressor, 476
Neuronal communication, chemical basis of, 57–70
Neuronal integration, electrochemical basis of, 37–55
Neuropeptide Y, 473
Neuropeptides
    as neurotransmitters, 58, 59t. *See also* Neurotransmitter(s).
    biosynthesis of, 60, 61f
Neuropore(s)
    anterior, 73, 94, 138, 139f
        defects in, 73–74, 74f–75f
    posterior, 73, 94, 138, 139f
        defects in, 74, 75f
Neurotmesis, nerve regeneration and, 36
Neurotransmitter(s)
    composition of, 27–28
    criteria defining, 58, 58t
    excitatory, 67
    in cerebellar cortex, 428–429
    in chemical synapses, 50–51, 51f
    in hair cell mechanoelectric transduction, 328
    in information transmission, 5, 26–27, 27f
    in myasthenia gravis, 145–146
    in pain transmission, 291
    in parasympathetic system, 475
    in pars caudalis, 285
    in primary sensory neurons, 144–145
    in regulation of neuronal excitability, 67
    in spinal motor neurons, 145–146
    in sympathetic system, 473
    in synaptic cleft
        enzymatic degradation of, 68
        removal of, 51, 51f
    in visceral motor neurons, 467, 467f
    inhibitory, 67
    ligand-gated ion channels and, 24
    metabolism of
        astrocyte function and, 30, 31f
        disorders of, 28
    neuronal communication by, 58
    presynaptic receptor autoregulation of, 59
    quantum of, 26, 51, 51f
    small-molecule, 60–62
    substances acting as, 58, 59t
Neurotrophins, 468, 468f
Neurulation
    primary, 73–75
        congenital defects associated with, 73
    secondary, 75
        congenital defects associated with, 75
Nigral complex, 411–412
Nissl substance, 17, 506
Nitric oxide
    as neurotransmitter, 58, 59t
    in long-term potentiation, 500
    pain transmission and, 291
NMDA receptors, 67, 420, 500
Nociceptive input
    in insular cortex, 241
    in spinal trigeminal pathway, 284f, 285
    primary afferent fiber types carrying, 277
    role of pars caudalis in, 285–286
    viscerosensory
        ascending parasympathetic pathways of, 298–301
        ascending sympathetic pathways of, 295–298, 296f, 297f
        referred, patterns and circuits of, 297–298, 297f–299f
Nociceptor(s), 25. *See also* Thermonociceptors.
    chemonociceptors as, 275, 275t

Nociceptor(s) (*Continued*)
  cutaneous, 275, 275t. *See also* Mechanoreceptor(s).
    Aδ mechanical, 275, 275t
    C-polymodal, 275, 275t
    for cold and heat, 275, 275t
    high-threshold, 275, 275t, 276f
    thermonociceptors as, 275, 275t, 276f
  deep, 276–277, 276t
  for visceral organs, 276–277, 276t, 294, 294t
  in sympathetic nerves, 295
  joint, 276–277, 276t
  mechanonociceptors as, 274
  muscle, 276–277, 276t
  peripheral, sensitization of, 275–276
  visceral, 276–277, 276t
Nodes of Ranvier, 30–31
  action potential propagation and, 46
Nondiscriminative touch, 256, 274
  in postsynaptic posterior column pathway, 262, 262f
  in spinal trigeminal pathway, 284f
  pars caudalis in, 285–286
  receptors for, 275, 275t, 276f
Noradrenergic synapse, 68–70, 69f
Norepinephrine, 233, 473
  as neurotransmitter, 58, 59t
  in central nervous system and peripheral nervous system, 68
  receptors for, 67
  reuptake of, 68, 69f
  synthesis of, 61
    drug action and, 68–70, 69f
Notch
  preoccipital, 238, 238f
  supraorbital, 213
  tentorial, 111, 112f
Notochord, 72, 72f
Nucleolus, of neuron, 17
Nucleotides, as neurotransmitters, 58, 59t
Nucleus(i), 5. *See also* Fiber(s); Nerve(s).
  abducens, 84, 178, 180f, 200, 201f, 212–213, 212f, 353, 353f,
      450–452, 451f
    lesions of, 452, 452f
  abducens motor, 154
  accessory
    corticonuclear fibers in, 396, 397f
    in medulla, 163, 164f
  accessory cuneate, 164, 165f
  accessory optic, 313, 459
  amygdaloid, anterior cortical, 364f, 365
  anterior, 482f, 483
    of medial hypothalamus, 230, 231f
  arcuate, 483
    of medial hypothalamus, 230, 231f
  auditory
    blood supply of, 329–330
    brainstem, 331–336, 331f–336f
    hierarchy of, 329, 330f, 333f
  basal, 236, 245–250, 246f–249f, 405–422
    blood supply of, 246f
    components of, 245–246, 246f, 406f–411f, 407–412
    concepts of, 406–407
    connections to, 249–250, 249f
    direct and indirect pathways of, 412–413, 412f, 414f
    dorsal and ventral divisions of, 245, 407
    eye movements and, 456
    integrated function of, 246, 417–419, 418f
    lesions of
      aphasia with, 518
      behavioral consequences of, 415–419, 417f, 418f
      etiology of, 419–422, 419f–421f
      hyperkinetic disturbances with, 416f, 417
      hypokinetic disturbances with, 415–417, 416f, 420f
      motor loop and, 415, 416f, 417
    of Meynert (substantia innominata), 245–246, 246f, 249, 249f,
        407, 409, 510
    parallel circuits in, 413, 415, 415f
      closed component of, 413

Nucleus(i) (*Continued*)
    motor loop in, 413, 415, 415f
    open component of, 413
  relationships of, 247f
  striatal complex in, 407, 408f, 409
  vasculature of, 243f, 246f, 250
  ventral, 412
  basilar pontine, 174f, 175, 439
  cardiorespiratory, 202, 202f, 370–371, 370f
  caudal (visceral or cardiorespiratory), 202, 202f, 370–371, 370f
  caudal central subdivision, 450, 451f
  caudal pontine reticular, 453
  caudate, 197, 236f, 237, 245, 246, 247f, 248f, 407, 408f
    in Huntington disease, 419, 419f
    of lateral ventricle, 96f, 97, 98f
    parts of, 246, 248f
    relationships of, 248f
  central, 501
    of inferior colliculus, 191, 192f, 335
    of medulla, 168, 169f
  central superior, 183, 183f
  cerebellar, 84f, 85, 177, 177f, 184, 425–427, 427f
    dentate, 177, 177f, 184, 426, 427f
    emboliform, 177, 177f, 426, 427f
    fastigial, 425–426, 427f
    globose, 426, 427f
    in vestibulospinal tract, 380
    medial, 425–426, 427f
  cochlear, 83, 85, 160f, 161, 208–209, 209f, 329, 330f, 333f
    anterior (ventral), 167, 167f, 168f, 202, 208, 331, 331f, 332–334
    cell types in, 332–334
    damage to, 209, 329
    in brainstem, 331–334, 331f–333f
    posterior, 167, 167f, 168f, 202, 208
    posterior (dorsal), 331, 331f, 332
  cranial nerve
    functional components of, 154–157, 155f–156f, 175. *See also*
        under Functional component(s).
    of brainstem
      and functional components of, 156, 156f
      motor, 82f, 83–84, 174, 174f, 200–201, 201f
      sensory, 201–202, 202f
    of medulla, 7, 160f, 161
    of midbrain, 7
    of pons, 7
  cuneate, 160f, 161, 162, 162f, 163, 164, 164f, 165f, 271
    in posterior column–medial lemniscal system, 260, 261f, 262,
        262f
    lateral, 271
  cuneiform, 196f, 197
  definition of, 22, 23t
  dentate, 177f, 184
    relation of, to pontocerebellum, 439–441, 440f
  dorsal, of Clarke, 141, 142, 143f, 270–271, 270f
  dorsal proper sensory, 141
  dorsal tegmental, hypothalamic efferent projections to, 487
  dorsal vagal
    in baroreceptor reflex, 301
    in viscerosensory parasympathetic pathways, 300f, 301
  dorsomedial, 481
    of dorsal thalamus, 223, 225f, 226f, 227, 227f
    of medial hypothalamus, 230, 231f
    of thalamus, 365, 365f, 494, 495f
  Edinger-Westphal, 188, 188f, 195, 195f, 197, 200, 201f, 212f, 216,
      305, 450, 451f
    functional component of, 156, 156f
    in parasympathetic innervation, 471f, 474
    in pupillary light reflex, 460
    of midbrain, 84
  emboliform, 177f, 184, 437, 438f, 439
  facial
    functional component of, 156, 156f
    of pons, 84
  facial motor, 167, 168f, 179, 179f, 180f, 200, 201f, 210, 210f
    corticonuclear fibers in, 396, 397f
  fastigial, 177f, 184, 435, 436, 437, 437f, 439

Nucleus(i) *(Continued)*
    cerebellar, 177, 177f
        in vestibulospinal tract, 380
        projections of, 439
    globose, 177f, 184, 437, 438f, 439
    globose cerebellar, 177, 177f
    gracile, 160f, 161, 162, 162f, 163, 164, 164f, 165f
        in posterior column–medial lemniscal system, 260, 261f, 262, 262f
    gustatory, 179, 202, 202f, 370–371, 370f
    habenular, 232, 232f, 233, 494, 495f, 501, 503f
    hypoglossal, 83–84, 160, 165, 165f, 166, 166f, 167f, 200, 201f, 202–203, 203f–204f
        corticonuclear fibers in, 396, 397f
    hypothalamic
        in visceromotor functions, 442
        lateral, 483
        posterior, 481f, 482f, 483, 484
    in autonomic function, 475–476
    intermediate, 210, 210f
    intermediolateral, 141, 468
    interpeduncular, 192f, 193
    interstitial, of Cajal, 194f, 195, 454f, 455
    intralaminar, 280
        in anterolateral system, 282
        of thalamus, 274, 274f, 287
    lateral cervical, 283
    lateral geniculate, 85, 195f, 197, 224f–226f, 228, 316, 317f, 318, 322f
        layers of, 316, 317f, 318
        retinal projections to, 313, 313f
    lenticular, 236f, 237, 245, 246, 247f, 248f
        parts of, 246, 248f
    mammillary, 223, 224f, 483, 494
        intermediate, 481f, 482f, 483
        lateral, 481, 482f, 483
        medial, 482f, 483, 499, 499f
        of medial hypothalamus, 230, 231f
    medial dorsal, 290
    medial geniculate, 195f, 197, 224f–226f, 228, 329, 330f, 335–336
        dorsal division of, 336
        medial (magnocellular) division of, 336
        ventral division of, 336
    mesencephalic, 181, 181f, 182, 182f, 191, 192f, 202, 213, 214, 214f
        functional components of, 156f, 157
        of trigeminal nerve, 266f, 267, 271f
    motor, of cranial nerves, 82f, 83
        functional components of, 175
        in brainstem, 82f, 83–84
        in pons, 174, 174f
    motor vagal, dorsal, 154, 160, 165, 165f, 166, 166f, 167f, 200, 201f, 471f, 474
        functional components of, 156, 156f
        in visceromotor functions, 442
        of medulla, 84
    oculomotor, 84, 194f, 195, 195f, 200, 201f, 212f, 216–217, 353, 353f, 449–450, 449f, 451f
        lesions affecting, 217
    of amygdaloid complex, 494, 495f, 501
    of Darkschewitsch, 195, 196f
    of inferior colliculus, 329, 330f, 333f
        central, 191, 192f, 335
        external (lateral), 191, 192f, 333f, 335
        pericentral (dorsal), 191, 192f, 333f, 335
    of lateral lemniscus, 329, 330f, 333f, 335
    of neuron, 17, 17f
    of optic tract, 459
    of posterior commissure, 196, 196f, 197
    of posterior thalamic group, in anterolateral system, 282
    of substantia innominata, 494, 495f. *See also* Nucleus(i), basal, of Meynert (substantia innominata).
    of superior colliculus, 194f, 195
    olfactory, anterior, 363, 364, 364f, 365
    olivary
        inferior, 85, 161, 162f, 164
        lateral superior, 335

Nucleus(i) *(Continued)*
    medial accessory, 166, 166f, 167f, 439
    posterior (dorsal) accessory, 166, 166f, 167f
    pretectal, 313, 460
    principal, 166, 166f, 167f
    principal inferior, 439
    superior, 168, 168f, 179, 179f, 180f, 182, 182f, 329, 330f
        lateral, 334
    parabrachial
        lateral, 182, 183f
        medial, 182, 183f
    paraventricular, 230, 231f, 482f, 483
        in water balance reflex, 492
        lesions of, 488
        oxytocin synthesis in, 487
    parvocellular, 169, 169f
    periolivary, 333f, 334
    periventricular, of hypothalamus, in pain transmission, 292
    pontine, 85, 174f, 175
        basilar, 180
    posterior, 279
        of medial hypothalamus, 230, 231f
        of thalamus, 228, 274, 274f, 287
    posterior column, 163, 164f, 260, 261f, 262, 262f
        organization of, 262, 262f
        submodality segregation of inputs in, 260, 262, 262f
    posterolateral, in anterolateral system, somatotopic organization of, 282–283, 283f
    posteromarginal, 141
        pain fiber input into, 278
    preoptic, 230, 231f, 480
        lateral, 480
        medial, 480
    prepositus hypoglossal, 167, 168f, 454f, 455
    pretectal, 196f, 197
    pulvinar, 224f, 225f, 226f, 227, 227f, 228
    raphe, 168, 364
        in pain transmission, 292
        of pons, 182, 183f
    red, 188, 188f, 194f, 195, 195f
        cortical projections to, 399
        function and efferents of, 194f, 195, 195f, 196, 196f
        magnicellular division of, 381–382, 439
        parvicellular division of, 381–382, 439
        parvicellular part of, 439
    reticular, 196f, 197
        caudal pontine, 182, 183f
        gigantocellular, 168, 169f, 182, 183f, 380
        lateral, 164, 165f, 166, 166f, 167f, 169, 169f
        of brainstem, 168
        of pons, 182, 183f
        oral, 380
        oral pontine, 182, 183f, 453
        pontine, 380
        thalamic, 223, 224f, 317f
    reticularis tegmenti pontis, 459
    reticulotegmental, 181f, 182, 182f
    rostral (gustatory), 179, 202, 202f, 370–371, 370f
    rostral interstitial, 455
        of medial longitudinal fasciculus, 216, 217
    sacral parasympathetic, 475
    sacral visceromotor, 82, 145
    salivatory
        inferior, 160, 167, 168f, 200, 201f
            functional component of, 156, 156f
            in parasympathetic innervation, 471f, 474
            of medulla, 84
            on glossopharyngeal nerve, 207
        superior, 178, 180f
            functional component of, 156, 156f
            of pons, 84
            parasympathetic, 471f, 474
    sensory
        of cranial nerves, 82f, 83
        in pons, 174f, 175
        posterior proper, non-noxious fiber input into, 278

Nucleus(i) *(Continued)*
 principal, 202, 213, 214f, 265, 266f, 267, 267f, 284f, 285
  divisions of, 265, 266f, 267, 267f
  role in trigeminal system, 267
 septal, 249f, 250, 494, 495f, 503, 503f
  afferent and efferent pathways to, 503, 503f
  stria terminalis connection with, 501
 sexually dimorphic, 480
 solitary, 83, 160f, 161, 165, 165f, 166, 166f, 167f, 174f, 175, 178, 179, 179f, 180f, 201, 201f, 202f
  blood pressure regulation and, 492
  functional component of, 156–157, 156f. *See also under Functional component(s).*
  in baroreceptor reflex, 301, 476, 476f
  in taste pathways, 369–371, 370f
  in visceromotor functions, 442
  in viscerosensory parasympathetic pathways, 300f, 301
 spinal trigeminal, 83, 160f, 161, 179, 179f, 180f, 202, 213, 214f, 265, 266f, 271f, 272, 284f, 285
  divisions of, 157
  functional components of, 156f, 157. *See also under Functional component(s).*
  pars caudalis of, 154f, 157, 163, 164, 164f, 165f, 202, 213, 214f, 272, 285, 286f
  pars interpolaris of, 154f, 157, 166, 166f, 167f, 202, 213, 272, 285, 286, 287
  pars oralis of, 157, 167, 168f, 202, 213, 285, 287
  second-order trigeminothalamic neurons in, 287
  somatotopic arrangement of hemiface in, 285, 286f
 spinal vestibular, 202
 subcortical, of limbic system, 494, 495f
 subcuneiform, 196f, 197
 substantia nigra and, 406. *See also* Nucleus(i), basal.
 subthalamic, 224f, 230, 231, 236, 246, 406, 407, 410–411. *See also* Nucleus(i), basal.
  basal nuclei and, 249, 249f
  lesions of, 231
   hyperkinetic disorders as, 416f, 417
 suprachiasmatic, 312, 482f, 483
  of hypothalamus, 232
  of medial hypothalamus, 230, 231f
 suprageniculate, of thalamus, 228
 supraoptic, 230, 231f, 482f, 483
  in water balance reflex, 492
  lesions of, 488
  oxytocin synthesis in, 487
 tegmental
  anterior (of Tsai), 197
  pedunculopontine, 406, 407, 412. *See also* Nucleus(i), basal.
 thalamic, 220
  anterior, 223, 224f, 225f, 226f, 494, 495f, 499, 499f
  association, 228
  centromedian, 226f, 228
  classification of, 228–229
  groups of, 223, 226f
  intralaminar, 228
  lateral, 224f, 225f, 226f, 227, 227f, 228
   dorsal and posterior, 224f–227f, 227
  medial, 223, 225f, 226f, 227
  medial geniculate, 224f–226f, 228
  midline, 223, 226f, 228
  nonspecific, 510
  parafascicular, 226f, 228
  paratenial, 228
  posterior complex of, 228
  relation of, to cerebral cortex, 227, 227f, 513–514, 514f
  relay, 85, 228
  reticular, 224f, 228
  specific or nonspecific, 228–229
 thoracic, posterior (dorsal nucleus of Clarke), 141, 142, 143f, 270–271, 270f
 trigeminal
  mesencephalic, 265, 266f, 267
  motor, 181, 181f, 182f, 200, 201f, 266f, 267, 271f
   corticonuclear fibers in, 396, 397f
   functional component of, 156, 156f

Nucleus(i) *(Continued)*
 of pons, 84
 principal, 287
 principal sensory, 83, 181, 181f, 182f
 sensory
  components of, 202
  in brainstem, 265
 trochlear, 84, 200, 201f, 212f, 215, 450, 451f
  general somatic efferent bodies of, 193, 193f
  of midbrain, 191, 192f
 tuberal, 481
 ventral anterior, 224f–227f, 227
 ventral basal, 412
 ventral caudal, in pain perception, 288–290
 ventral intermediate, 228, 421
 ventral posterior, 227, 227f, 262–263, 263f
  blood supply of, 263f
  somatotopic organization of, 263, 263f
 ventral posterior inferior, 264
 ventral posteroinferior, in vestibulo-thalamo-cortical network, 355
 ventral posterolateral, 227, 227f, 263, 267f, 274, 274f, 279
  blood supply of, 263
  in anterolateral system, somatotopic organization of, 282–283, 283f
  in pain perception, 288–290
  in vestibulo-thalamo-cortical network, 355
  visceral nociceptive input to, 296, 297f
 ventral posteromedial, 227, 227f, 228, 263, 265, 266f, 267f, 287
  blood supply of, 263
  in pain perception, 288–290
  in spinal trigeminal pathway, 284f
  parvicellular division of, 370f, 371
 ventral tegmental, hypothalamic efferent projections to, 487
 ventromedial, 483
  of medial hypothalamus, 230, 231f
 vestibular, 83, 154, 160f, 161, 166, 166f, 167f, 178, 179f, 349–353, 351f, 352f
  afferents to, 349–350, 352, 352f, 353
  central, 342
  cerebellar connections to, 352, 352f
  commissural connections to, 352
  efferent connections to, 353
  functions of, 208
  in smooth pursuit movements, 457
  inferior, 166, 166f, 167, 167f, 168f, 349, 351f
  lateral, 202, 208, 209f, 349, 351f
  medial, 166, 166f, 167, 167f, 168f, 202, 208, 209f, 349, 351f
  Purkinje cell inhibition of, 384
  superior, 202, 349, 351f
Nucleus accumbens, 245, 246, 247f–249f, 249, 407, 408f, 494, 495f, 503f, 504
 stria terminalis connection with, 501
Nucleus ambiguus, 160, 164, 165f, 166, 166f, 167f, 200, 201f, 206, 207, 471f, 474
 functional component of, 156, 156f
 in baroreceptor reflex, 301
 in viscerosensory parasympathetic pathways, 300f, 301
 motor, corticonuclear fibers in, 396, 397f
 of medulla, 84
Nucleus limitans, 228
Nucleus pigmentosus pontis, 181
Nucleus proprius, 141, 278
Nucleus pulposus, 72f, 73
Nucleus raphe dorsalis, 183, 183f, 191, 192f, 196f, 197
Nucleus raphe magnus, 169, 169f, 182, 183f, 292
Nucleus raphe obscurus, 169, 169f
Nucleus raphe pallidus, 169, 169f
Nucleus raphe pontis, 183, 183f
Nucleus z, of posterior column, 262, 262f
Nursing, oxytocin effects in, 487
Nystagmus, 354, 354f, 436, 442
 horizontal, testing for, 530
 optokinetic, 459, 459f
 pendular, 442
 vestibulocochlear nerve injury and, 210

O2A progenitor cells, 81
Oat cell carcinoma, 24
Obex, 160, 161
  medullary structures at, 166, 166f
Occipital artery(ies)
  meningeal branch of, 208
Occipital gyrus, 239f, 241
Occipital pole, 239f, 241
Occipitotemporal gyrus, 239, 240f
  lesions of, 321
Ocular counter-roll, 350f, 354
Ocular dominance columns, 320, 320f
Oculomotor complex, 195, 195f
Oculomotor loop, in basal nuclei, 413, 415
Oculomotor nerve, 132, 132f, 152f, 153
  anatomy and distribution of, 212f, 216–217
  aneurysm and, 526
  basilar artery branches and, 127
  functional components of, 156, 156f
  in eye movements, 447
  lesions affecting, 217, 450, 450t
  of midbrain, 189, 189f
  parasympathetic outflow pathways of, 474, 474t
  posterior cerebral arteries and, 124f, 127
  testing of, 526, 527f
Oculomotor system, 446
  central structures involved in, 449–453, 449f, 450t, 451f–452f
  peripheral structures involved in, 446–449, 447f–449f
  reflex movements in, 458–462, 459f–462f
  targeting movements in, 453–458
Odor perception, 360
Ohm's law, 38
Olfaction
  definition of, 360
  disorders of, 362
Olfactory bulb, 238, 241f, 363–364, 363f, 364f
  efferent projections of, 364–365, 364f
Olfactory epithelium, 360, 360f, 361f
  bipolar cells of, 18, 18f, 20
Olfactory fila, 360
Olfactory nerve, 360–361, 361f
  testing of, 522, 524, 524f
Olfactory system
  declines in, 366
  disorders in, 366
Olfactory tract, 238, 241f, 363
Olfactory trigone, 238, 241f
Olfactory tubercle, 245, 364f, 365, 406f, 407, 408f
Oligodendrocytes, 29f, 30–33, 33f
  function of, 30, 33f
Olivary, medial superior, 331f, 333f, 334–335, 334f
Olivary nucleus. See Nucleus(i), olivary.
Olive, 161, 161f
  inferior, 161, 161f, 459
Olivocochlear bundle, 337–338
Operculum(a)
  frontal, 241, 241f, 371
  parietal, 241, 241f
  temporal, 241, 241f
Ophthalmic artery(ies), 123, 124f, 306, 314, 315f, 316
Ophthalmic nerve, 265
Ophthalmoplegia, internuclear, 452, 454f
Ophthalmoscope, 314
Ophthalmoscopic examination, 525–526, 525f, 526f
Opsin, cone, 308
Optic atrophy, 526, 526f
Optic chiasm, 97, 99f, 313–316, 313f
Optic disc, 306, 312, 313
  ophthalmoscopic examination of, 525, 525f, 526f
Optic foramen, 123, 124f
Optic nerve, 306, 312, 313–316, 313f
  damage to, 314, 314f, 315f
  testing of, 524–526, 524f–526f
Optic radiations, 245, 245f, 313f, 316, 318, 318f, 319f
  blood supply of, 318

Optic radiations (Continued)
  bundles of, 318
  lesion of, 319f
Optic system, accessory, 352
Optic tectum, 456, 457f
Optokinetic reflex, 446
Orbicularis oculi muscle(s), 449, 449f
Orbit, sympathetic innervation of, 452–453, 452f
Orbital gyrus, 238, 241f
Organ of Corti, 326, 327–328, 327f, 328f
Organum vasculosum, of lamina terminalis, 221, 221f
Orientation
  auditory pathways serving, 338f, 339
  of head
    eye movements and. See Eye movement(s); Head movement(s).
    vestibular function and, 343, 344f, 380
  spatial, 355–356
  testing of, 522
Orientation columns, 320, 320f
Orthostatic hypotension, 70, 476
Osmoreceptors, 294–295, 294t
Ossicles, 325, 325f
Otic placode, 83
  vestibulocochlear nerve origin from, 79–80
Otitis media, 325
Otoconia, 346, 349
  of otolith organs, 346
Otolith membrane, 346, 347
Otolith organs, 342, 342f, 347, 349, 350f
  function of, 349, 350f
  hair cells of, 346–347, 346f–347f, 349, 350f
Otosclerosis, 325
Oval window, 325, 325f
Oxytocin, 483
  release and effects of, 487
  synthesis and transport of, 487

P stream, 321, 322f
Pacchionian bodies, 114
Pachygyria, 86, 88f
Pachymeninx, 109, 110. See also Dura mater.
Pacinian corpuscles, 47, 257f, 258
Paget disease, 106
Pain, 274
  anterolateral system and, 163
  cardiac, patterns and pathways of, 297f, 298, 299f
  central or thalamic, 291
  chronic
    thalamic ventrocaudal nuclei changes and, 288–290
    therapy for, 290–291
  control of, 191, 291–292
  dental, 285
  facial
    atypical, 286
    in tic douloureux, 215, 285
    onion-skin pattern of, 285–286, 286f
    pars caudalis in, 285–286
  gender differences in, 288
  imaging studies of, 268, 288
  in trigeminal nerve, testing of, 527
  muscle and joint, 276–277, 276t
  perception of, 277, 288, 289f
    and suffering, 288, 289f
    in somatosensory thalamus, 288–291
  phantom limb, 290
  receptors for, 275, 275t, 276, 276f, 277. See also Nociceptor(s).
  referred, 297–298, 297f–299f
  sensation of, testing of, 537
  spinal trigeminal nucleus and, 202
  spinal trigeminal tract and, 163
  sympathetic innervation and, 473
  thalamic, 234
  transmission of, 290f, 291–292, 291f
    neural modulation of, 290f, 291–292, 291f

Pain (*Continued*)
  visceral, 276t, 277, 294, 295
    ascending parasympathetic pathways of, 298–301
    ascending sympathetic pathways of, 295–298, 296f, 297f
    poor localization of, 296
    referred, patterns and circuits of, 297–298, 297f–299f
Palatine muscles, innervation of, 200
Palatoglossus muscle, hypoglossal nerve function and, 202
Paleocortex, 365, 494, 507
Paleostriatum, 246, 246f, 407
Pallidal complex, 408f, 409–410
Pallidum, ventral, 245–246, 246f, 408f, 409–410
Papez circuit, 487, 496, 499, 499f
Papillae
  circumvallate, 367f, 368
  filiform, 367f, 368
  foliate, 367f, 368
  fungiform, 367, 367f
  on tongue, 367–368, 367f
Papilledema, 314, 314f, 526, 526f
Papilloma, of choroid plexus, hydrocephalus with, 106
Paracentral gyrus
  anterior, 239, 239f
  posterior, 239, 239f
Paraflocculus, 457
Parageusia, 371
Parahippocampal gyrus, 239f, 242, 494, 495f, 497
  blood supply of, 496
Paralysis
  flaccid, 138, 388
    of tongue, 392, 399f
  of upper and lower extremities, 149
Paraphasia, 517
Paraplegia, 533
Paraterminal gyrus, 250
Paravermis areas, of cerebellum, 177, 177f
Paravertebral muscle, control of, 380
Paresthesia, 274
  from peripheral nerve injury, 138
  from trigeminal nerve injury, 285
  in central pain syndrome, 291
Pargyline, action of, 69f, 70
Parieto-occipital artery(ies), 124f, 128, 242
Parkinson disease, 28, 417, 418f, 419, 532
  akinesia and bradykinesia in, 415, 420f
  basal nuclear dysfunction in, 420–421, 420f
  eye movements and, 456
  surgical treatments for, 421
Parosmia, 366, 522
Pars caudalis
  magnocellular region of, 285, 286f
  of spinal trigeminal nucleus, 154f, 157, 163, 164f, 165f, 202, 213,
    214f, 272, 285, 286f
Pars compacta, 249
  of substantia nigra, 411–412
    lesions of, 418f
Pars intermedia, of pituitary gland, 221
Pars interpolaris, of spinal trigeminal nucleus, 154f, 157, 166, 166f,
  167f, 202, 213, 272, 285, 286, 287
Pars lateralis, of substantia nigra, 411
Pars opercularis, of inferior frontal gyrus, 238, 238f, 240f, 337
Pars oralis, of spinal trigeminal nucleus, 154f, 157, 167, 168f, 202,
  213, 285, 287
Pars orbitalis, of inferior frontal gyrus, 238, 238f, 240f
Pars reticulata, of substantia nigra, 249, 411–412
Pars triangularis, of inferior frontal gyrus, 238, 238f, 240f, 337
Patches, 407, 409f
Path, final common, 374, 388
Pathway(s), 5f, 6. *See also* System(s); Tract(s).
  ascending, 7
    of medulla, 162, 163f
    of midbrain, 191, 191f
    of pons, 178, 178f
    of spinal cord, 148–149, 148f
    viscerosensory
      parasympathetic, 298–301

Pathway(s) (*Continued*)
    sympathetic, 295–298, 296f, 297f
  auditory
    ascending
      central, 329–331, 330f
      in brainstem, 331–336, 333f
      binaural, 333f, 334–335, 334f
      descending, 337–338
      monaural, 333f, 334, 335
  autonomic, neurons in, 144f, 145
  basal nuclear
    direct, 409
    indirect, 409
  descending, 7
    auditory, 337–338
    of medulla, 162–163, 163f
    of midbrain, 191, 191f
    of pons, 178, 178f
    of spinal cord, 148–149, 148f
  enkephalinergic, in pain transmission, 292
  for crossed extension reflex, 147–148, 147f
  for discriminative touch, 267f
  for flexor reflex, 146–147, 147f
  for tendon reflex, 146, 146f
  geniculostriate, 318f
  neospinothalamic, 275
    pain localization and, 288, 289f
  of anterolateral system, 277–283, 279f–283f
  olfactory, central, 363–366, 364f–365f
  paleospinothalamic, 275
    suffering and, 288, 289f
  perforant, 497, 499f
  posterior column–medial lemniscal, 256–264
  retinogeniculate, 318f
  retinogeniculostriate, 457f
  spinal trigeminal, 284f, 285–287, 286f–287f, 287f
    central pathway in, 284f, 285–287, 286f, 287f
    primary neurons in, 284f, 285, 286f
  spinocerebellar, nonconscious proprioception in, 269–272
  spinocervicothalamic, 283, 283f
  spinothalamic
    direct (neospinothalamic), 279–280
    indirect (paleospinothalamic), 280
  taste
    central, 370–371, 370f
    peripheral, 369–370, 370f
  trigeminal, divisions of, 265
  trigeminocerebellar, 271f, 272
  trigeminothalamic, anterior, 274, 274f
  ventral amygdalofugal, 248f, 250, 485, 486f, 494, 501, 502f
  viscerosensory, 293–302
    ascending
      for parasympathetic afferents, 298–301
      for sympathetic afferents, 295–298, 296f, 297f
    reticular activating system in, 301–302
  visual, 313, 313f
Pedicle, of cone cells, 308, 310
Peduncle(s)
  cerebellar, 152, 152f, 176f, 177, 424, 424f
    inferior, 162, 166, 166f, 167f, 179, 180f
    middle, 175, 175f, 177, 180, 181f, 182, 424, 424f. *See also*
      Brachium pontis (middle cerebellar peduncle).
      trigeminal nerve and, 213
    superior, 181f, 182, 271f, 272, 424, 424f, 427. *See also* Brachium
      conjunctivum (superior cerebellar peduncle).
      decussation of, 177, 191, 192f, 193, 193f
  cerebral, 191
  mammillary, 296–297, 483, 485–486
Peptidase, degradation of neuropeptides by, 68
Peptides
  in posterior horn, pain transmission and, 291
  neuroactive, 233
  sympathetic, 473
Periaqueductal gray matter, 99, 153, 480, 481f, 484, 494, 495f. *See
  also* Gray matter.
  hypothalamic efferent projections to, 487

Periaqueductal gray matter *(Continued)*
  in pain transmission, 292
  of midbrain, 190, 191, 192f, 193f
Pericallosal artery(ies), 496
Periglomerular cell(s), 363, 363f
Perikaryon, 5, 16
Perilymph, 327, 328, 343
Perineurium, of peripheral nerves, 35, 35f
Periodontal ligament(s)
  afferent fibers of, 267, 272
  receptors in, 267
Periosteum, innervation of, 285
Peripheral nervous system
  development of, 79–81
  general outline of, 4–6, 4f
  glial cells in, 28, 29f, 29t
  supporting cells of, 33, 34f, 35
  tumors of, 35
Perivascular space, 116
Periventricular gray, in hypothalamus, in pain transmission, 292
Periventricular zone, of hypothalamus, 488
Petrosal nerve(s)
  deep, 210f, 211
  greater, 211
  greater superficial, 369
Petrosal sinus(es)
  inferior, 130f, 131f, 132, 208
  superior, 111, 112f, 130f, 131f, 132
pH, of extracellular space, astrocytic function and, 30
Phantom pain, 36
Pharyngeal arch
  cranial motor nerves innervating, 200
  muscles originating from, trigeminal nerve innervation of, 214, 214f, 215
Pharyngeal artery
  ascending, meningneal branch of, 208
Phenothiazines, tardive dyskinesia and, 422
Phosphatidylinositol, 28, 66
Phospholipases, in G protein–coupled receptor responses, 52, 53f, 66
Phosphorylation, in postsynaptic receptor action, 66–67
Photons, 25
Photoreceptor cell(s), 305, 306, 307f–309f, 308–309
  in visual processing, 309–312, 311f
Photoreceptor cells, 25, 305, 306, 307f–309f, 308–309
  in visual processing, 309–312, 311f
Pia intima, 116
Pia mater, 110f, 116–117
  early development of, 108f, 109
  glial cells and, 28, 29f
  of choroid plexus, 96f
  of spinal cord, 140, 140f, 141f
  of ventricular system, 77
  organization and structure of, 109f, 110, 110f
Pigment epithelium, 305–306, 306f
  retinal, 305–306, 306f
Pillar cell(s), 326
Pineal gland, 189, 190f, 221, 232, 232f
  tumors of, 99, 104, 189, 233
Pinealocytes, functions of, 232–233
Pinealoma, 233
  blocking cerebral aqueduct, 99
  hydrocephalus with, 104
Pinna (of ear), 325
Pinocytic transport, at blood-brain barrier, 30
Pinocytosis, in axonal transport, 21
Pitocin, 487
Pituitary gland, 480
  anterior, regulation of, hypothalamic, 229, 230
  development of, 220f, 221
  hormones released by, 487–488
  hypothalamic links to, 229, 230, 482f, 487
  tumors of, 488–491, 489f, 490f
    hormonally active, 489–490, 489f
    secreting, 489–490, 489f
    visual field deficits and, 315f

Place codes, 47
  in information representation, 47
  of cochlea, 329
Placidity, 501
Placode(s)
  cranial nerve development and, 79, 80
  otic, 83
  vestibulocochlear nerve origin from, 79–80
Planum temporale, 336
Plasmalemma, of Schwann cells, in myelin sheath formation, 34f
Plasticity, of early brain, 88, 268
Plate(s)
  alar, 138, 139f
    cranial sensory nerve nuclei and, 155f, 156, 201–202, 202f
    development of, 80f, 81, 82f
    in brainstem development, 83
    in midbrain development, 188, 188f
    in pons and cerebellum development, 174–175, 174f
    of medulla, 160f, 161
    ventricular spaces and, 155f
  basal, 138, 139f
    development of, 80f, 81, 82f
      and cranial nerve nuclei, 155f, 156
    in brainstem development, 83
    in midbrain development, 188, 188f
    in pons and cerebellum development, 174–175, 174f
    motor cell columns formed by, 200
    of medulla, 160–161, 160f
    ventricular spaces and, 155f
  cerebellar, 85, 174f, 175
  cortical, formation of, 86, 87f
  cuticular, 345, 345f
  quadrigeminal, 188
Plexus(es)
  Auerbach's, 80
  basilar, 132
  Meissner's, 80
  myenteric, 475
  of fenestrated capillaries, 488
  submucosal, 475
Poliomyelitis, somatic motor neurons in, 138
Polydipsia, 488
Polyribosomes, of dendrites, 16
Polyuria, 488
Pons, 6f, 7, 82–83, 82f
  alar and basal plate structures in, 155f, 174, 174f
  basilar, 152, 174f, 175–176, 175f–176f, 189, 189f
    corticospinal fibers in, 391–392
    trigeminal nerve and, 213
  cranial nerves of, 213–215, 214f
  development of, 174–175, 174f
  dorsolateral, 457
  external features of, 175–177, 175f–176f
  in brainstem, 152, 152f
  internal anatomy of, 178–184, 178f–184f
    at caudal level, 178–180, 179f–180f
    at mid level, 181–182, 181f, 182f
    at rostral level, 181f, 182
  posterior column–medial lemniscus system in, 262
  rhomboid fossa of, 176–177, 176f
  vasculature of, 183–184, 184f
Pons-medulla junction, cranial nerves of, 203f, 204f, 208–213, 209f, 210f, 212f
Pons-midbrain junction, 189, 190f
Pontine artery(ies), 124f, 127
Pontine flexure, 76f, 77
Pontine tegmentum, 152, 174f, 175
Pontocerebellum, 439–441, 440f
  dysfunction of, 441–442
  functions of, 439f
Population coding, 256
Position sense
  impairment of, 260
  in trigeminal system, 266f, 267f
  nerve pathways for, 267f
  testing of, 536, 536f, 537

Positron emission tomography, 268
  for imaging of pain, 288
Postcentral gyrus, 238f, 239, 240f, 242, 283
  visceral nociceptive input to, 296, 297f
Posterior chamber, 304, 304f
Posterior column(s), 148, 148f, 259
  chronic stimulation of, for pain control, 292
  injury of, 148, 260
  nuclei of, 260, 261f, 262, 262f
  organization of fibers in, 259–262
  postsynaptic fibers of, 149
Posterior column–medial lemniscal system, 256–264
  characteristics of, 256
  discriminative touch pathways in, 267f
  peripheral mechanoreceptors in, 256–259
  primary afferent fibers in, 259
  primary somatosensory cortex in, 263–264, 263f, 264f
  resolution in, 256
  spinal cord and brainstem in, 259–262
  testing of, 277, 536–537, 536f, 537f
  ventral posterolateral nucleus in, 262–263, 263f
Posterior inferior cerebellar artery syndrome, 170, 215, 286–287, 287f
Posterior perforated substance, 189
Posterior rhizotomy, 277
Postsynaptic element, 5, 5f
  of chemical synapse, 26, 26f
Postsynaptic potentials
  excitatory, 52, 53f, 54f, 67
  inhibitory, 67
Posture
  control of, 380
  vestibulocerebellar influence on, 435
  vestibulospinal network and, 354–355
Posturing, in Parkinson disease, 420, 420f
Potassium
  in endolymph, 327, 327f, 328, 343, 346
  in taste transduction, 368, 369f
  relative concentrations of, in neuronal cytoplasm and extracellular fluid, 23, 24f
Potassium channel
  in resting membrane potential, 39–40, 39f
  voltage-modulated, 42, 43f
Potassium ions, astrocyte function and, 30
Precentral gyrus, 238f, 239, 240f, 242, 389f, 390
Precuneus, 238f, 239, 242
Precursor cell(s)
  glioblastic, 81
  neuroblastic, 81
Presbyopia, 305, 449
Presynaptic element, 5, 5f, 375, 375f
  of chemical synapse, 26, 26f
Presynaptic terminal, 88
Proboscis, 77
Probst tract, 267
Prolactin, 488
  tumors secreting, 491, 491f
Propranolol, action of, 69f, 70
Proprioception, 256
  nonconscious, 269–272
  testing of, 536, 536f
Proprioceptive input
  from trigeminal mesencephalic nucleus, 267
  in Brodmann areas, 264
  in jaw muscle use, 267, 271f, 272
  in spinal trigeminal pathway, 284f
Propriospinal connections, 279
Prosencephalization, 77
Prosencephalon, 76f, 77, 87f, 94, 94f
Prosis, 149
Prosopagnosia, 321
Protanopia, 309
Protein(s)
  cerebrospinal fluid, 103, 105
  effector, in G protein receptor action, 65f, 66
  in transport and docking of synaptic vesicles, 62–63, 62f

Protein(s) (Continued)
  odorant-binding, 362
  of vesicle membranes, 60
  transport, for uptake of neurotransmitter from synaptic cleft, 68
Protein G
  in postsynaptic receptors in slow chemical neurotransmission, 59
  in second-messenger synapse, 51–52, 53f
  olfactory-specific, 362
  subunits of, 64, 66
Protein kinase(s)
  cAMP-dependent, 67
  in G protein–coupled receptor responses, 66
Pseudotumor cerebri, hydrocephalus with, 106
Pseudounipolar, of dorsal root ganglia, somite development and cell(s), 80, 80f
Psychiatric disorders, 419
Pterygoid canal
  nerves of, 211
Pterygopalatine fossa, 213
Ptosis, 217, 452f, 453, 470, 472f
Puberty, precocious, 233
Pulvinar, 195f, 197, 224f–227f, 227, 313
  of diencephalon, 189, 190f
Pupil, 304, 304f
  Argyll Robertson, 460
  changes in, as sign of brain damage, 460, 461f
  relative afferent defect in, 461f
Pupillary light reflex, 197, 305, 446, 459–450, 460f
  pathways for, 460, 461f
Purkinje cell(s), 84f, 85, 184, 184f–185f, 384, 385, 428, 429f, 430f, 431f, 433–435, 434f
  in decerebellate experiments, 384
  inhibitory influence of, in cerebellar cortex, 433–435, 433f, 434f
Putamen, 197, 236f, 237, 246, 246f, 247f, 248f, 407, 408f, 415
Pyramid(s)
  corticospinal fibers in, 392
  motor decussation of, 163, 164f
  of medulla, 160f, 161, 164, 175, 175f
Pyramidal cell(s), 331, 333f, 389, 389f
  of cerebral cortex, 508–509, 508f, 509f

Quadrantanopia
  contralateral inferior, 318
  contralateral superior, 318
  left superior homonymous, 319f
Quadreplegia, 533
Quadriceps stretch reflex, 146, 146f
  testing of, 533, 534f
Quadrigeminal artery(ies), 126f, 127, 127f, 189–190, 197, 197f, 198, 329, 456
Quantum, of neurotransmitter, 26, 51, 51f

Rabies virus, retrograde axonal transport and, 21–22
Rachischisis, 138, 139f
Radiations
  geniculocalcarine, 244, 245f
  geniculotemporal (auditory), 244, 245f
  optic, 245, 245f, 313f, 316, 318, 318f, 319f
  thalamic
    anterior, 244, 245f
    central, 244, 245f
Radicular artery, 142
  anterior (ventral), 134f, 135
  posterior (dorsal), 134f, 135
Ramus(i)
  gray communicating, 144f, 145, 469
  white communicating, 144, 144f, 145, 468, 469
Ranvier, nodes of, 30–31
Rathke pouch, 221
Rebound phenomenon, 441
Receptive fields, 258–259, 258f
  cutaneous, 47

Receptive fields (Continued)
  in ear, 47
  of retinal ganglion cells, 316
  properties of, in cortical neurons, 267–268
  visual, 309, 310f, 311, 311f, 319–320, 320f
Receptor(s). See also Mechanoreceptor(s).
  acetylcholine (cholinergic), 63
  adaptation of, 47
  adrenergic
    β-adrenergic
      action in, 65–66, 65f, 67
      antagonists of, 69f, 70
      desensitization of, 66–67
      structure and function of, 66
      structure of, 64, 64f
      subtypes of, 63, 473
    agonists and antagonists of, 63
  as transducers, 46–47, 63–64
  categories of, 63
  cough, 294
  cutaneous, 256–257, 257f, 257t, 258f, 258t
    of face and oral cavity, 285
    rapidly adapting, 257, 257f, 258f
    slowly adapting, 257, 257f, 258f
  dopaminergic
    in schizophrenia, 412
    on postsynaptic striatal neurons, 417, 418f, 419
  downregulation of, 67
  epinephrine (adrenergic), 63
  for nondiscriminative touch/thermal sense and pain, 275, 275t
  for testosterone, 63
  function of, 63
  G protein–coupled, 362. See also Receptor(s), β-adrenergic.
    for binding of chemical messengers, 63–66, 65f, 67
    structure of, 65–66, 65f
  GABA_A, 64
  glutamate, ionotropic excitatory, 67
  graded potentials of, 46–47
  hair follicle, 256, 257, 257f
  in information transmission, 26, 27f
  in lips and perioral skin, 265
  in parasympathetic targets, 475
  in periodontal ligament, 265
  in peripheral vestibular labyrinth, 342–345, 342f–345f
  in sympathetic target cells, 473
  internal thermal, 294–295, 294t
  intrinsic, in hypothalamus, 492
  ionic mechanisms in, 47
  kainate, 67
  modalities and submodalities of, 47
  muscarinic, 63
  nicotinic cholinergic, 63, 375
    in spinal motor nerves, 145
    structure of, 64–65, 64f
  NMDA, 67
    in Huntington disease, 420
  NMDA-type glutamate, in long-term potentiation, 500
  norepinephrine, 67
  olfactory, 360–362, 360f, 361f
  pain. See Nociceptor(s).
  peripheral
    of sensations in teeth, 165
    of tactile sensations in head, 265, 266f, 267
  postsynaptic
    α-adrenergic or β-adrenergic, drugs affecting, 68–70, 69f
    in chemical neurotransmission, 59
    neurotransmitter action and, 67
    regulation of, 66–67
  presynaptic, α_2-adrenergic, drugs affecting, 69f, 70
  proprioceptive, 257f, 258, 258t
  sensory
    as transducers, 46–47
    for semicircular canals and utricle, 344
    ionic mechanisms of, 47
    place codes of, 47
    submodalities of, 47

Receptor(s) (Continued)
  structure and function of, 64, 64f
  subtypes of, 63
  tactile, 47
    peripheral, 165, 265, 266f, 267
  taste, 366–367, 366f
    distribution of, 367–368, 367f
    transduction in, 368–369, 369f
  transmembrane, 64, 64f
  vestibular sensory, 345–347, 345f–347f
    peripheral apparatus of, 342
  viscerosensory, 294–295
    physiologic, 294, 294t, 295
      in parasympathetic nerves, 295
    specialized, 294–295, 294t
  visual, 47
Receptor organs, vestibular, 343–344, 344f
Receptor potential(s), 46–47
  action potential generation and, 47
  amplitude of, 48, 48f, 49f
Recess(es)
  infundibular, 97, 99f
  lateral, 99, 153
  of third ventricle, 97–98, 99f, 222, 222f
  pineal, 97, 99f
  supraoptic, 97, 99f
  suprapineal, 97, 99f
Reciprocal inhibition, 146, 146f, 378
Rectus muscle(s)
  inferior, 447, 447f
  lateral, 353, 447, 447f
    abducens nerve function and, 213
  medial, 353, 447, 447f
    innervation of, 212f, 213
  superior, 447, 447f
Recurrent inhibition circuit, 388f
Reflex(es)
  acoustic startle, 338f, 339
  ankle jerk, testing of, 533, 534f
  baroreceptor, 300f, 301, 476, 476f
    hypothalamic influence on, 492
  biceps, testing of, 533, 534f
  blink, 212, 215, 285, 446, 460, 462, 462f
    testing of, 527, 528f
  brachioradialis, testing of, 533, 534f
  cerebellar influence on, 443–444
  chemoreceptor, 477
  compensatory, definition of, 446
  corneal blink, 212, 285, 446, 460, 462, 462f
    testing of, 527, 528f
    trigeminal nerve innervation of, 215
  crossed extensor, 147–148, 147f, 378
  flexor, 146–147, 147f
    testing of, 533, 534f
  flexor withdrawal, 277
  gag, 299
  hypothalamic, 492
  inverse myotatic, 146, 146f
  inverted plantar, 388
  jaw jerk, 267
    testing of, 527, 528f
    trigeminal nerve innervation of, 215
  knee jerk, 6, 146, 146f
  light, 391
  linear, 354
  long-latency, 400, 400f
  middle ear, 338, 338f
  motor, auditory, 335
  myotatic, 146, 146f
  nociceptive, 146–147, 147f
  opticokinetic, 446
  patellar, 6, 146, 146f
  pupillary light, 197, 305, 459–450, 460f, 461f
    definition of, 446
  quadriceps stretch, 146, 146f
    testing of, 533, 534f

Reflex(es) (Continued)
  simple, 5f, 6
  spinal, 138, 143, 146–148, 146f, 147f, 378–379
  spinal autonomic, in viscerosensory afferent pathways, 298
  stretch
    monosynaptic, 146, 146f
    muscle, 377–378
      testing of, 533, 534f
    quadriceps, 146, 146f
    tendon, 378
  supraspinal autonomic, in viscerosensory afferent pathways, 298
  temperature regulation, hypothalamic influence on, 492
  tendon, 146, 146f, 378
  triceps, testing of, 533, 534f
  vestibulocolic, 355
  vestibulo-ocular, 353, 353f, 446, 458
    rotational, 353–354, 353f
  visceral, 294
  water balance, hypothalamic influence on, 492
  withdrawal, 146–147, 147f
Reflex arc
  basic, 6
  monosynaptic, 6
Region(s), relation of, to functional systems, 8–9, 8f
Regional cerebral blood flow studies, 268
Regional neurobiology, 8
Releasing factors, 230
Releasing hormones, 483, 484
Renshaw cell(s), 388f, 389
Repolarization, in action potential, 43–44, 44f
Reserpine, action of, 69, 69f
Resistance, 38, 45
Restiform body, 161, 162, 162f, 164, 165f, 166, 166f, 167, 167f,
    168f, 179, 179f, 180f, 424, 424f
  in pons, 178
  in spinocerebellar tracts, 270f–271f, 271
Resting potential, sodium-potassium pump and, 25
Reticular activating system, ascending, visceral input to, 301–302
Reticular area
  lateral medullary, 168, 169f
  medial medullary, 168, 169f
  ventrolateral, 169, 169f
Reticular formation, 168, 287
  in anterolateral system, 282
  in autonomic function, 476
  lateral, 182, 182f
  midbrain, 196f, 197, 395
  parabrachial pontine, 407, 412
  paramedian pontine, 395, 453–455, 454f
    abducens nerve function and, 213
  visceral nociceptive input to, 296–297
  viscerosensory parasympathetic afferent pathways and, 301, 302
Reticular nucleus. See Nucleus(i), reticular.
Retina, 304f, 305
  bipolar cells of, 18, 18f, 20
  blind spot in, 313f, 314, 318, 523
  central artery of, 123, 124f
  detached, 305, 314f
  disparity in, 458
  layers of, 306, 306f
  nasal, 313f, 314, 316
  neural, 305–306, 306f
    blood supply of, 306
  processing of visual input in, 306f, 309–312, 311f
  projections of, 312–316
  receptive fields in, 47
  refraction of light by, 313
  temporal, 313f, 314, 316
  visual acuity in, 314, 314f
Retinal artery, central, 123, 124f, 306, 314, 316
Retinal pigment epithelium, 305–306, 306f
Retinal slip, 458
Retinogeniculate projections, 313, 313f
Retinotopic map, 85, 313, 320
Rhinencephalon, 494
Rhinitis, 362

Rhodopsin, 308
Rhombencephalon, 76f, 77, 78f, 94, 94f, 152, 153
Rhombic lip, 84–85, 175
Rhomboid fossa, 99, 153–154, 154f
  of pons, 176–177, 176f
Rhombomere, 84
Ribosomes
  of dendrites, 16
  of neurons, 17
Rigidity, 421
  lead pipe, in Parkinson disease, 532
Riley-Day syndrome, 466
Rinne test, 529
Rods, 306, 307f–309f, 308–309
Rolandic vein, of central sulcus, 130, 130f
Romberg test, 537
Root(s)
  anterior, in meningeal development, 108f
  dorsal (posterior)
    ganglia of, neurons of, 18, 18f
    of peripheral nervous system, 4, 4f
  trigeminal sensory, 284f, 285
  ventral (anterior)
    of peripheral nervous system, 4, 4f
    of spinal cord, 468
Round window, 325, 325f
Rubella virus, 77, 305
Ruffini endings, 257, 257f, 258, 267

Sac, endolymphatic, 344
Saccades, 453–455, 453f, 454f
  definition of, 446
  orienting head movements with, 455
  supranuclear control of, 455–456, 455f, 456f
Saccule, 342f, 343
Sagittal sinus(es)
  inferior, 111, 112f, 130, 130f, 131f
  superior, 111, 112f, 130, 130f, 131f
Salivary glands, von Ebner's, 368
Salivatory nucleus. See Nucleus(i), salivatory.
Saltatory conduction, 31, 46
Satellite cells, 33
Satiety, 491
Satiety center, of hypothalamus, 230, 483
Scala media, 325, 325f
Scala tympani, 325, 325f
Scala vestibuli, 325, 325f
Scalp, sympathetic innervation of, 470
Scarpa ganglion, 343–344, 344f
Schizencephaly, 86, 88f
Schizophrenia, dopamine receptors in, 412
Schmidt-Lanterman clefts, 34f, 35
Schwann cell(s), 33, 34f, 36, 81
  in axonal regeneration, 36
Schwannoma, 35
Sclera, of eye, 305
Sclerotome, formation of, 80f, 82
S-cones, 308f, 309
Scotoma, 525
  central, 314f
Second messengers, 65f, 66
Segmental artery(ies)
  central or sulcal branches of, 134–135, 134f
  spinal branches of, 134–135, 134f
Sella turcica, 112f
Semicircular canal(s), 209, 209f, 342, 342f, 344f, 347–349, 348f, 350f
  dehiscence of, 345, 345f
  ducts of, 344
  function of, 347–349, 348f
  hair cells of, 345–347, 345f–347f, 347–349, 348f
Senile plaques, in Alzheimer's disease, 28
Sensitization
  central, 276
  peripheral, 275–276

Sensory (afferent) fiber(s)
  large-diameter, in spinal cord, 259, 260f, 261f, 284f
  of cranial nerve nuclei, functional components of, 154–157, 155f, 156f
  of facial nerve, 211
  of spinal nerves, 143–144, 144f
  of vagus nerve, 206
Sensory input
  neurons receiving, 25
  peripheral, to anterior horn, 376–379, 376f–378f
  testing of, 536–537, 536f–537f
  thalamus and, 7
  vestibular, in vestibulospinal tract, 380
Sensory receptors
  adaptation of, 47
  as transducers, 46–47
  ionic mechanisms of, 47
  modality of, 47
  place codes of, 47
  potentials of, 46
  submodalities of, 47
  tactile, 47
  visual, 47
Septum
  posterior intermediate, of spinal cord, 140, 141f
  posterior median, of spinal cord, 140, 141f
Septum pellucidum, 96f, 97, 98f, 250
Serotonin, 58, 59t, 232, 292
Sham rage, 483
Shape discrimination
  loss of, 264
  testing of, 536f, 537
Sheath(s), connective tissue, of peripheral nerves, 35, 35f
Shingles, 277, 279f, 285
Sigmoid sinus(es), 130, 130f, 131f, 132, 208
Simple spike, 434, 434f
Sinus(es)
  cavernous, 113, 130, 131f, 132, 132f
  confluence of, 112f, 130, 131f
  congenital dermal, development of, 108f, 109
  grooves for, 112f
  intercavernous, 113, 132
  of dura mater, 111, 112f, 113
  petrosal
    inferior, 130f, 131f, 132, 208
    superior, 111, 112f, 130f, 131f, 132
  sagittal
    inferior, 111, 112f, 130, 130f, 131f
    superior, 111, 112f, 130, 130f, 131f
  sigmoid, 130, 130f, 131f, 132, 208
  straight, 111, 112f, 130, 130f, 131f, 133, 133f
  tentorium cerebelli and, 112f
  transverse, 111, 112f, 130, 130f, 131f
  venous, 109, 109f, 113, 130–134, 130f–133f
    arachnoid villi and, 114f
    cerebrospinal fluid movement into, 103–104, 104f
Sinusitis, 362
Skin
  glabrous, 256
  hairy, 256
  perioral, sensation in, peripheral receptors for, 265
  receptive fields in, for location of axonal stimulus, 47
Slow-twitch and fast-twitch muscle(s), 376
Small cell carcinoma, 24
Smell, disorders of, 362
Smooth muscles, autonomic control of, 464
Smooth pursuit eye movements, 446, 456–458
Snellen chart, 524, 524f
Sodium
  in neuronal cytoplasm and extracellular fluid, 23, 24f
  in perilymph, 343
  in taste transduction, 368, 369f
Sodium channel
  in rods, 308
  membrane potential and, 40, 41f
  voltage-gated, 46

Sodium channel (Continued)
  voltage-modulated, 42, 43f
Sodium-potassium pump, 25
Soft palate, corticonuclear fiber lesions affecting, 396, 398f
Soft palate muscles, corticonuclear fiber lesions affecting, 396, 398f
Soma (cell body), 5
  of neuron, 16, 17f, 18f, 19f
Somatosensory system. See also Cortex, somatosensory.
  position sense in, 255–272
  tactile discrimination in, 255–272
  touch/thermal sense, and pain in, 273–292
Somatosensory thalamus, pain perception in, 288–291
Somatostatin, in pain transmission, 292
Somatotopy
  fractured, 433, 433f
  in cerebellar cortex, 433, 433f
  of primary motor cortex, 390, 390f
Somite(s), innervation of, development of, 80, 80f
Sommer's sector, 500
Sound
  brain localization of, 324–325, 324f, 334–335, 334f
    binaural, 333f, 334, 334f, 335
    monaural, 325, 333f, 334, 335
  frequency of, 324, 324f, 329
    cells responding to, 335
    cochlear tuning of, 328, 329
  intensity of, 324, 324f, 329
    cells responding to, 335
  measurement of, 324
  pitch of, 324
  processing of, 325–329, 325f, 326f, 328f
    in central auditory pathways, 329–331, 330f, 333f
  reflexive and learned responses to, 338f, 339
  tonotopic representation of, 328
  transmission of
    in external ear, 325, 325f
    in inner ear, 325f–327f, 326–329
    in middle ear, 325, 325f
Space constant (lambda), 42
Spasticity, 388–389, 391, 530
Spasticity, of muscles, 388
Speech
  disorders of, testing of, 522
  fluent paraphasic, 517
  in Broca aphasia, 516–517, 517f
  nasal, 530
  prosody of, 518
  scanning, 441
  telegraphic, 516
  Wernicke's area and, 337
Spherule, of rods, 308, 310
Sphincter, of iris, 304f, 305
Sphincter muscle(s), 304f, 305
Sphincter pupillae, 216
Sphincter pupillae muscle(s), 447, 448f, 449
Spina bifida, 74, 75f
Spina bifida aperta, 74
Spina bifida cystica, 74, 75f
Spina bifida occulta, 74, 138
Spinal accessory nerve(s), testing of, 530, 532f
Spinal artery(ies)
  anterior, 125, 126f, 134–135, 134f, 142, 142f, 162, 170
    branches of, 203
      lesions of, 169
    central branches of, 142, 142f
    penetrating branches of, 392, 393f
    sulcal branches of, 280, 281f, 393f, 394
    trauma to, 142
  posterior, 125, 126f, 134–135, 134f, 142, 142f, 162, 169, 170, 170f
    in posterior column–medial lemniscal system, 260, 261f
    lesions of, 169, 260
Spinal cord, 4, 4f, 137–150
  anterior horn of, 80f, 81, 138–139, 139f, 141, 141f, 142
  arteries of, 134–135, 134f
  autonomic nuclei of, hypothalamic link to, 487

Spinal cord (Continued)
  blood supply of, 142, 142f
  central canal of, 94, 94f
  cervical, enlargement of, motor signals with, 394, 395f
  cervical levels of, 142, 143f
  development of, 80f, 81–82, 82f, 138–139, 139f
  directions in, 9–10, 9f
  function of, 6–7, 6f, 138
  functional components of, 81–82, 82f
  hemisection of
    anterolateral system and, 280, 282f
    functional, 150, 394, 396f
  intermediate zone of, 81
  lesions of
    anterolateral system and, 280, 282f
    autonomic function and, 476
    cervical, 394, 395f
    general features of, 150
    lumbosacral, 394, 395f
    weakness of extremities with, 533
  lumbar levels of, 142–143, 143f
  medullary transition with, 163, 164f
  nerves of, 143–146, 144f–145f
  pathways and tracts of, 148–149
  posterior horn of, 80f, 81, 141, 142
  posterior (dorsal) root entry zone of, 140, 141f
  primary sensory fibers in, diameter of, 259–262, 260f, 261f
  reflexes of, 146–148
  regional characteristics of, 142–143, 143f
  relation to brain and meninges, 109–110, 109f
  relation to vertebral column, 82, 83f
  roots of, related to vertebrae, 139–140, 140f
  sacral levels of, 143, 143f
  structure of, 139–142, 140f–143f
  surface features of, 139–140, 140f, 141f
  thoracic levels of, 142, 143f
  veins of, 134f, 135
  ventral root of, 468
  viscerosensory afferent fibers entering, 296
Spinal cord artery(ies), 134–135, 134f
Spinal fiber(s)
  afferent
    general somatic, 143–144, 144f. See also under Functional component(s).
    general visceral, 143–144, 144f. See also under Functional component(s).
    neurotransmitters of, 144–145
  descending, 138
  efferent
    general somatic, 145. See also under Functional component(s).
    general visceral, 145. See also under Functional component(s).
    types of, 145, 145f
  exteroceptive, 143–144, 144f, 269, 271
  interoceptive, 144, 144f
  motor, types of, 145, 145f
  proprioceptive, 144, 144f
  sensory, 143–144, 144f
  visceromotor (autonomic), 145, 145f
Spinal medullary artery(ies), 134f, 135, 142, 142f
Spinal nerve(s), 4, 4f, 143–146, 144f
  anterior root of, C pain fibers in, 277
  connections to sympathetic chain ganglia, 469–470, 469f
  dermatomes of, 277, 277f, 278f
  development of, 80f, 82
  fibers of, 143
  levels of, 143
  motor components of, 145, 145f
  posterior root ganglion of, pain fibers in, 260f, 277
  sensory component of, 143–144
  somite development and, 80, 80f
Spinal systems, in motor deficits, 382f–384f, 383–385
Spinal tap (lumbar puncture), 117, 140, 140f
Spinal vein(s)
  anterior, 134f, 135
  posterior, 134f, 135
Spinocerebellar module, 436–439, 437f, 438f

Spinocerebellum, 269
  damage to, 439
Splanchnic nerve, 144, 468, 469f
  viscerosensory afferent fibers in, 295, 296f
Stain(s)
  basophilic, 17
  of cells of cerebral cortex, 506–507, 506f, 507f
Stapedius muscle(s), 325, 325f, 338, 338f
Stapes, 325, 325f
Stereocilia
  in sound transmission, 325f
  of hair cells, 326, 327f, 328
  of vestibular sensory system, 345–347, 345f–347f
Stereognosis, 256
  oral, 267
  testing of, 536f, 537
Stereotaxic surgery, for chronic pain, 288
Sternocleidomastoid muscle, accessory nerve lesions and, 206
Stimulus
  dynamic range of, 48
  location of, place codes for, 47
  tactile, accuracy of detection of, 258–259
Stimulus amplitude
  action potential discharge frequency and, 48, 48f, 49f
  successive transformations of, 55f
Strabismus, 321, 446
Straight sinus(es), 111, 112f, 130, 130f, 131f, 133, 133f
Stratum, definition of, 22, 23t
Stress response, neuropeptides in, 63
Stretch marks, 491
Stria(ae)
  acoustic, 329, 330f, 332, 333f
  lateral longitudinal, 497
  lateral olfactory, 238, 241f, 365
  medial longitudinal, 497
  medial olfactory, 238, 241f
  of Gennari, 319, 319f, 506f, 507
Stria medullaris, 154, 154f
Stria medullaris thalami, 222f, 224f, 228, 232, 233, 501
Stria terminalis, 224f, 230, 237, 248f, 250, 485, 486f, 501, 502f
  of limbic system, 494
  relationships of, 248f
Stria vascularis, 326, 326f, 327, 328
Striatal complex, 407, 408f, 409
  in Huntington disease, 419, 419f
Striate artery(ies)
  medial, 245, 250, 407, 407f
    lenticulostriate branches of, 246f, 250
Striola, 347, 347f, 350f
Striosomes, 407–408, 409, 409f, 412
Stroma, of iris, 304–305
Stylomastoid artery(ies), 344
Stylopharyngeus muscle(s)
  glossopharyngeal nerve and, 208
  innervation of, 200
Subarachnoid hemorrhage, 113, 119–120, 119f, 122
  cerebrospinal fluid composition in, 102–103
  signs of, 119–120
Subarachnoid space, 108, 109f, 110f, 113–114, 117
  cerebrospinal fluid movement in, 103, 104f
  of interpeduncular fossa, 189
  of spinal cord, 140, 140f, 141f
  relationships of, 110
Subcommissural organ, 221, 221f
Subforniceal organ, 221
Subiculum, 250, 484, 485f, 496–497, 498f
  in Alzheimer disease, 500
Subjunctional folds, 375
Subpial space, 116
Subplate, of cerebral cortex, 86, 87f
Substance P, 144, 277, 301
  in medium spiny neurons, 409
  in olfactory tract cells, 364
  in pars caudalis, 285
  in posterior horn, pain transmission and, 291, 292
  in striatal neurons, 413

Substantia gelatinosa, 141, 141f, 278, 285
Substantia innominata, 245–246, 246f, 249, 249f, 250, 407, 409
    nuclei of, 494, 495f
Substantia nigra, 28, 188, 188f, 236, 246, 407, 408f, 410, 411
    basal nuclei and, 249, 249f
    of midbrain, 190f, 191, 192f, 193, 194f, 197
    pars reticulata, 415
Subthalamus, 230–232. See also Thalamus, ventral.
    connections of, 410f
Suffering, pathways serving, 288, 289f
Sulcus(i), 7
    anterolateral, 140, 141f
    calcarine, 239f, 241
    callosal, 239f, 242
    central, 238, 238f, 239, 239f
        of insula, 238, 239, 241f
    cingulate, 238, 239f, 242
        marginal ramus of, 239, 239f
    collateral, 238, 239, 239f, 240f, 241f, 242
    epithalamic, 85
    hypothalamic, 85, 220f, 221, 480
    inferior frontal, 238, 238f
    inferior temporal, 239, 241f
    intraparietal, 238f, 239
    lateral, 238, 238f, 239, 240f, 241f
    marginal, 239, 239f
    median, 154
    of cerebral hemispheres, 238, 238f
    olfactory, 238, 241f
    paracentral, 239, 239f
    parieto-occipital, 238, 238f, 239f
    postcentral, 238f, 239
    posterior intermediate, 140, 141f
    posterior median, of spinal cord, 140, 141f
    posterolateral, of spinal cord, 140, 141f
    postolivary, 161
    preolivary, 161
    rhinal, 241f, 242
    superior, 238, 238f
    superior temporal, 239, 240f, 241f
    sylvian, 238, 238f
Sulcus limitans, 138, 139f, 154, 154f, 155f, 166, 166f, 167f, 174–175, 174f
    development of, 80, 82f
    in brainstem development, 83, 155f, 156
    superior fovea of, 176f, 177
Superior canal dehiscence syndrome, 345, 345f
Superior oblique muscle(s), 447, 447f
Superior rectus muscle(s), 447, 447f
Superior tarsal (of Müller) muscle(s), 449, 449f, 470
Superior tarsal muscle(s), 216
Supramarginal gyrus, 238f, 239, 337
Suspensory ligament(s), 448f, 449
SV2, 60
Sweat glands, sympathetic innervation of, 473
Sweating, 491–492
Sydenham chorea, 421
Sympathetic trunk, 468
Synapse(s), 5, 21, 58, 374, 375f
    basic structure of, 5, 5f
    chemical, 50–52, 51f, 52f, 53f
        delay in, 50, 58
        function of, 26–27, 27f
        Gray's type I and II, 26–27, 26f, 27t
        information flow across, 59–60, 60f
        morphologic characteristics of, 26–27, 27t
        regulatory mechanisms in, 59–60
        structure of, 25–27, 26f
    currents and potentials of, 52, 53f, 54f
    definition of, 25
    electrical, 5, 5f, 50, 51f
    embryonic development of, 88
    excitatory, 52, 53f, 54f, 67
    glutaminergic, 363, 364, 364f
    in cerebellar cortex, 433–435, 434f
    in myasthenia gravis, 24

Synapse(s) (Continued)
    in olfactory bulb, 363–364, 364f
    information processing at, 52–53, 53f–55f
        action potential generation and, 52–53, 55f
    inhibitory (GABAergic), 52, 54f, 310, 363, 364, 364f
        hyperpolarization in, 52, 54f, 67
        shunt path in, 52, 54f
    noradrenergic, 68–70, 69f
    of sensory neurons, 25
    retinal, 310
    second-messenger, 51–52, 53f, 59
    signal transmission and, 16
    single-messenger, 51–52, 52f
Synapsin, 62
Synaptic cleft, 5, 5f, 58, 375
    in chemical synapses, 26, 26f, 50, 51f
    maintenance of environment of, 68
    neurotransmitters in
        enzymatic degradation of, 68
        transporter-mediated uptake of, 68
    removal of neurotransmitter from, 51, 51f
Synaptic connections
    critical periods in, 88–89
    of dendrites, 16, 17f, 19f
Synaptic potentials, 52–53, 53f–55f, 67
Synaptic ribbons, 307f, 308
Synaptic stabilization, 88
Synaptic transmission, fast and slow, 58–59
Synaptic vesicles, 5, 26, 26f, 27, 50–51, 51f
    biosynthesis of, 60, 61f
    formation, transport and recycling of, 61, 62f
    localization of, 62, 62f
    release of, 62–63, 62f
Synaptobrevin, 60
Synaptogenesis, 88
Synaptophysin, 60
Synaptotagmin, 60
Syphilis, 77
    meningeal, cerebrospinal fluid in, 102
Syringobulbia, 74
Syringomyelia, 74, 150, 280
System(s). See also Pathway(s); Tract(s).
    accessory optic, 459
    anterolateral. See Anterolateral system.
    auditory, 323–340
    cerebrovascular, 121–136
    corticonuclear (corticobulbar), 394–399
    corticopontine, 399
    corticoreticular, 399
    corticorubral, 399
    corticospinal, 389–390, 389f, 390f
    internal carotid, 123, 124f, 125, 125f, 126f
    limbic. See Limbic system.
    motor. See Motor system.
    nervous. See Nervous system.
    oculomotor. See Oculomotor system.
    olivocochlear cochlear efferent, 337–338
    posterior column–medial lemniscal, 256–264
        characteristics of, 256
        peripheral mechanoreceptors in, 256–259
        primary afferent fibers in, 259
        primary somatosensory cortex in, 263–264, 263f, 264f
        resolution in, 256
        spinal cord and brainstem in, 259–262
        ventral posterolateral nucleus in, 262–263, 263f
    reticular activating, ascending, 301–302
    somatosensory. See also Cortex, somatosensory.
        position sense in, 255–272
        tactile discrimination in, 255–272
        touch, thermal sense, and pain in, 273–292
    supraspinal, 383–384
        role of, in motor deficits, 382f–384f, 383–385
    tectobulbospinal, 164, 165, 165f, 166f, 167, 167f, 178, 180f, 181, 181f, 195
    tectoreticulospinal, crossed, 456, 457f
    trigeminal, 264–265, 266f, 267f

System(s) (Continued)
  ventricular
    cerebrospinal fluid movement in, 103–104, 104f
    development of, 77–79, 78f, 79f, 94–99, 94f–100f
  vertebobasilar, 125, 126f, 127–128, 127f
  vestibular, 341–358
    components of, 342
  visceral motor. See Visceral motor system.
  viscerosensory, 293–302
  visual, 302–322
  visual motor, 445–462

Tabes dorsalis, 460
Tachycardia, 477
Tactile sensation. See also Mechanoreceptor(s); Touch.
  cutaneous, 269, 275, 275t
  deep, 257, 257t, 258–259, 258f
  impairment of, 260
  in head, peripheral receptors of, 265, 266f, 267
Tanycytes, 81, 100, 100f
Tapetum, 243
Tardive dyskinesia, 421–422
Tarsal muscles, eyelid movements and, 460, 462
Taste, 165
  components of, 360
  definition of, 360
  disorders of, 371
  fibers conveying, 201–202, 202f
  loss of, 369–370
  orbitofrontal cortex connections and, 365
  testing of, 526–527
  transduction of, 368, 369f
Taste buds, 366–367, 366f, 367f
  extralingual, 368
  lingual, 367–368, 367f
Taste fibers
  glossopharyngeal nerve and, 208
  of facial nerve, 206, 211, 212
  of trigeminal nerve, 206
  of vagus nerve, 206
Tectobulbospinal system, 164, 165, 165f, 166f, 167, 167f, 178, 180f,
    181, 181f
  of brainstem, 195
  of medulla, 163, 164f
Tectoreticulospinal system, crossed, 456, 457f
Tectorial membrane, 326, 328
Tectum, 153, 153f
  of midbrain, 190, 190f
Teeth, sensation in
  nociceptors in, 385
  peripheral receptors for, 165
Tegmentum
  of midbrain, 153, 153f, 190, 190f
  pontine, 174f, 175, 176f, 177
Tela choroidea, 95, 95f, 96f, 98, 99, 99f, 100f, 153, 153f
Telangiectasia, in myelodysplasia, 75
Telencephalic flexure, 76f, 77, 78f
Telencephalic (cerebral) vesicles, 76f, 77, 78f
Telencephalon, 76f, 77, 78f, 94f, 95, 235–251
  amygdala of, 250–251
  basal nuclei of, 245–250, 246f–249f
  cerebral cortex lobes in, 238–242, 238f–243f
  components of, 236–237, 236f. See also Capsule, internal; Cerebral
      hemisphere(s); Cortex, cerebral.
  development of, 236f, 237, 237f
  hippocampus of, 250–251
  white matter of cerebral hemisphere in, 242–245, 244f–246f
Temperature, body
  fibers for
    in anterolateral system, 163
    in spinal trigeminal tract, 163
  regulation of, 483, 491, 492
Temporal artery(ies), 126f, 128
Temporal bone, 344f, 345f

Temporal gyrus
  inferior, 239, 240f
  middle, 239, 240f
  superior, 239, 240f
  transverse (of Heschl), 239, 241, 241f, 242, 336
Temporal lobe, 238, 239, 239f, 240f, 241f
  arteries of, 124f
  lesions of, 250
    Klüver-Bucy syndrome and, 496, 501
    thalamocortical projections to, 514, 514f
Temporal muscle, in jaw jerk reflex, 267
Temporal pole, 239, 239f
Temporomandibular joint, 272, 285
  disorders of, pars caudalis and, 286
Tensor tympani, 325, 325f, 338
  trigeminal nerve innervation of, 214, 214f
Tensor tympani muscle(s), 325, 325f, 338
Tensor veli palatini muscle(s), 214, 214f
  trigeminal nerve innervation of, 214, 214f
Tentorial notch, 111, 112f
Tentorium cerebelli, 109, 109f, 111, 112f
Terminal arbor, of axons, 16, 17f, 18f, 19f, 20f, 21
Terminal boutons, of axons, 19f, 20f, 21
Testosterone, intracellular receptors for, 63
Tetanus toxin, retrograde axonal transport and, 22
Tethered cord syndrome, 75
Tetrodotoxin, conduction block with, 46
Thalamic nucleus. See Nucleus(i), thalamic.
Thalamogeniculate artery(ies), 126f, 128, 130, 232f, 233f, 234, 316, 331
Thalamoperforating artery(ies), 126f, 127, 127f, 232f, 233, 484, 484f,
    496
Thalamus
  anatomy of, 220
  basal nuclear direct and indirect pathways to, 412–413, 412f, 414f
  blood supply of, 126f, 246f
  development of, 85, 86f, 87f
  dorsal, 220, 222–223, 224f–227f, 225f, 226f, 227–229. See also
      Diencephalon.
    basic anatomy of, 223, 224f, 225f, 226f
    organization of, 227f, 228–229
    thalamic nuclei in, 223, 226f
  function of, 6f, 7, 223
  hemorrhage into or tumors of, Wernicke aphasia and, 517
  hypothalamic efferent projections to, 487
  in pain perception, 288–291
  lateral ventricle and, 97, 98f
  lesioning of, for treatment of chronic pain, 290
  motor areas of, projections of, to motor cortex, 401–402
  radiations of, 244, 245f
  relationships of, 247f
  vascular lesions of, 234
  ventral, 223, 224f, 230–232. See also Diencephalon.
    anatomy of, 220
  vestibular, 355, 356f
  visceral nociceptive input to, 296, 297f
Thermal sensation
  in spinal trigeminal pathway, 202, 284f
  in trigeminal nerve, testing of, 527
  innocuous, 274
  painful, 274
  role of pars caudalis in, 285–286
  testing of, 537
Thermonociceptors, 274, 275, 275t, 276f
Thermoreceptors, 25, 256, 274
  cutaneous, 275, 275t
  non-nociceptive, 275, 275t, 276f
Thiamine deficiency, 483
  in Korsakoff syndrome, 500
Thrombus, 122
Thyrotropin hormone, overproduction of, 490
Thyrotropin-releasing hormone, 233
Tic douloureux (trigeminal neuralgia), 215, 285
Tight junctions
  of arachnoid barrier cell layer, 113
  of blood-brain barrier, 135, 135f
  of choroid epithelial cells, 101, 103f

Time constant, 40
Titubation, 436
Tongue, 165
    corticonuclear fiber lesions affecting, 396, 399f
    deviation of, testing of, 530, 532f
    flaccid paralysis of, 392
    hypoglossal nerve function and, 202, 203
Tonotopic map
    of auditory cortex, 328, 336–337
    of cochlear fibers, 332
Tonsillar herniation, 169
Touch
    discriminative, 256–272
        impairment of, 260
        in trigeminal system, 264–265, 266f, 267
        pathway(s) for, 267f
            posterior column–medial lemniscal, 256–264
            spinal trigeminal, 284f
            trigeminothalamic tract as, 265–267
        receptive field properties and, 267–268
        testing of, 277, 537, 537f
        texture, 264, 536f, 537
        two-point, 265, 266f, 267, 392, 537, 537f
    flutter-vibration sense in, 256, 266f, 267f, 284f, 392
    nociception in, 256. See also Nociceptor(s).
    nondiscriminative, 256, 274
        in postsynaptic posterior column pathway, 262, 262f
        in spinal trigeminal pathway, 284f
        pars caudalis in, 285–286
        receptors for, 275, 275t, 276f
    pathways for, anterolateral system vs. posterior column–medial
        lemniscal system, 274
    submodalities of, 256
    thermal sensation in, 256. See also Thermal sensation.
Toxin(s)
    perinatal environmental, axonal and synaptic development and, 89
    tetanus, 22
Toxoplasmosis, congenital defects and, 77
Tract(s), 7. See also Fasciculus(i); Fiber(s); Pathway(s); System(s).
    ascending, of spinal cord, 148–149, 148f
    central tegmental, 168, 168f, 178, 179, 180f, 192f, 193, 193f, 196,
        370f, 371
    corticonuclear, 390
    corticospinal, 162, 163f
        anterior, 149, 392
        course of, 391–392, 391f–394f
        in decerebrate rigidity, 382f, 383, 383f
        in upper cervical cord, damage to, 394, 395f
        lateral, 141, 149, 163, 163f, 164f, 392
            blood supply to, 393f, 394
            lesions of, 149
        origin of, 390
        termination of, 392, 394
    cuneocerebellar, 270f, 271
    definition of, 22, 23t
    descending, of spinal cord, 148–149, 148f
    fastigiospinal, 149
    habenulointerpeduncular, 232f, 233, 501, 503f
    hypothalamospinal, 149
    Lissauer's, 141, 143, 144f, 260f, 277, 284f, 285
    mammillotegmental, 485f, 487
    mammillothalamic, 223, 224f, 482f, 483, 485f, 487, 494, 499, 499f
    mesencephalic, 181, 181f, 182, 182f, 191, 192f, 213, 214f
        functional components of, 156f, 157
        of trigeminal nerve, 266f, 267
    of caudal midbrain tegmentum, 194f, 196
    olfactory, 238, 241f, 363
        lateral, 363, 364
    olivocochlear, 209
    optic, 313–316, 313f, 316f
        lesions of, 316, 316f, 318
    posterolateral, 143, 144f
    posterolateral (dorsolateral), 141
    Probst, 267
    raphespinal, 149
    reticulospinal, 149, 162, 163f, 380, 381f

Tract(s) (Continued)
    in decerebrate rigidity, 384
    lateral, 380, 381f
    rubrospinal, 149, 162, 163, 163f, 164, 164f, 165f, 166, 166f, 167f,
        178, 180, 180f, 196, 380–382, 381f
    solitary, 165, 165f, 166, 166f, 167f, 201, 201f, 202f
        functional component of, 156–157, 156f
        in taste pathways, 369
        vestibular efferent connections to, 353
        viscerosensory parasympathetic pathways and, 300f, 301
    spinal
        ascending, 148–149, 148f
        descending, 148–149, 148f
        of trigeminal nerve, 265, 266f
    spinal trigeminal, 163, 164, 164f, 165f, 166, 166f, 167f, 179, 180f,
        214, 214f, 284f, 285
        functional components of, 156f, 157
    spinocerebellar, 148
        anterior, 163, 164f, 166, 166f, 167f, 270f, 271
            in medulla, 162, 163f
            in pons, 187, 187f
        degeneration of, 269
        dorsal, 436
        organization of, 269, 270f
        posterior, 163, 164f, 269–271, 270f
            in medulla, 162, 163f
        primary afferent proprioceptive fibers in, 269–271
        rostral, 270f, 271–272
    spinocervicothalamic, 149
    spinomesencephalic, pain localization and, 288, 289f
    spinoreticular, pain localization and, 288, 289f
    spinothalamic, pain localization and, 288, 289f
    supraopticohypophysial, 482f, 483, 487–488
        lesions of, 488
    tectospinal, 162, 163f
    trigeminothalamic
        anterior (ventral), 163, 164f, 192f, 193, 214, 265, 266f, 267,
            267f, 284f, 285–287, 286f–287f
            of midbrain, 191, 191f
        posterior (dorsal), 192f, 193, 214, 265, 266f, 267, 267f
            of midbrain, 191, 191f
    tuberoinfundibular, 480, 482f, 484, 488
    ventral spinocerebellar, 436
    vestibulospinal, 162, 163f, 379–380, 379f
        in decerebellate rigidity, 384–385
        lateral, 354–355, 355f, 379f, 380
        medial, 355, 355f, 379–380, 379f
Tract of Lissauer, 141
Tractotomy, 285
Transcutaneous electrical nerve stimulation, 291, 292
Transduction, 46–47
    definition of, 25
    in formation of action potential, 256
    of hair cells of vestibular system, 346–347, 346f, 347f
    of head movements, 343
    of sound, mechanoelectrical, 327–328, 327f
    olfactory, 362–363, 362f
    signal, 58–59, 63–67
    taste, 368–369, 369f
    visual, 308–312
Transient ischemic attack, 123
Transplants, autologous, for Parkinson disease, 421
Transporter(s)
    for degradation of norepinephrine, 68, 69f
    for small chemical messengers, 61–62
    for uptake of neurotransmitters from synaptic cleft, 68
    of vesicles, 60
    proton-coupled, for chemical messengers, 60
Transverse sinus(es), 111, 112f, 130, 130f, 131f
Trapezius muscle, accessory nerve lesions and, 206
Trapezoid body, 168, 168f, 180, 180f, 182f, 329, 330f, 331f, 332,
    333f, 334
Trauma
    carotid-cavernous fistula with, 132, 132f
    central nervous system, astrocyte function and, 28–29
    head, 115, 349, 349f

Trauma (*Continued*)
  meningeal hemorrhage and, 115, 116
  nerve degeneration and regeneration and, 35–36
  subarachnoid hemorrhage and, 119–120
  to anterior spinal artery, 142
  to anterolateral system fibers, 149
  vestibular afferent response to, 349, 349f
Tremor, 421, 436, 441, 441f
  cerebellar, 85
  kinetic (intention), 441, 441f
  resting, 441
    in Parkinson disease, 420, 420f
  static, 441
Trigeminal nerve, 111, 152, 152f
  anatomy and organization of, 213–215, 214f
  dermatomal maps of, 278f
  development of, 79
  distribution of, 285
  divisions of, 265, 266f, 267f
  functional components of, 175
  in blink reflex, 460
  in muscles of mastication, 200, 213, 214f, 215, 267, 271f, 272
  in spinal trigeminal pathway, 284f, 285
  injury of, 215, 285
  mandibular division of, 213, 214f
    testing of, 527, 528f
  maxillary division of, 132, 132f, 213, 214f
    testing of, 527, 528f
  mesencephalic nucleus of, 79
  motor nuclei of, 174, 174f, 214, 214f
  of pons, 175–176, 175f, 176f
  ophthalmic division of, 132, 132f, 213, 214f
    testing of, 527, 528f
  peripheral distribution of, 285, 286f
  roots of (portio major and minor), 213, 265, 266f
  sensory nuclei of, 213–214, 214f
  spinal (descending) tract of, 265, 266f, 285
  testing of, 527, 528f
Trigeminal neuralgia (tic douloureux), 215, 285
Trigeminal nucleus. See Nucleus(i), trigeminal.
Trigeminal system, 264–265, 266f, 267f
Trigeminal tubercle, 162, 163, 164f, 284f, 285
Trigeminocerebellar connections, 271f, 272
Trigger zone(s)
  for trigeminal neuralgia, 285
  of action potential generation, 42, 47, 52, 53f, 54f, 67, 256
Trigone(s)
  hypoglossal, 154, 154f
  olfactory, 238, 241f
  vagal, 154, 154f
Trochlear nerve, 132, 132f, 152f, 153, 176f, 189, 190f, 212f, 215–216
  functional components of, 156, 156f
  in eye movements, 447
  injury to, 216
  testing of, 527
Trochlear nucleus. See Nucleus(i), trochlear.
Trophic factors, in cellular brain development, 88
Tropic factors, in cellular brain development, 88
Tubercle(s)
  anterior thalamic, 223
  cuneate, 162, 162f
  gracile, 162, 162f
  olfactory, 245, 364f, 365
  trigeminal, 162, 163, 164f, 284f, 285
Tuberculosis, subacute meningitis with, 120
Tuberculum cinereum, 162, 162f, 163, 164f, 284f, 285
Tumor(s). See also specific tumor, e.g., Meningioma(s).
  astrocyte, 30, 32t, 33f
  blocking cerebral aqueduct, 99
  brain, blood-brain barrier and, 135
  brainstem, anterolateral system effects of, 280, 282
  central nervous system, originating from glial cells, 32t
  glomus, 349, 349f
  gonadotropic, 491
  midbrain, hydrocephalus with, 104
  oat cell carcinoma, 24

Tumor(s) (*Continued*)
  of peripheral nerve origin, 35
  pineal gland, 189
  pituitary, 488–491, 489f, 490f
    visual field deficits and, 315f
  Rathke's pouch, 221
  small cell carcinoma, 24
Tumor necrosis factor-α, secreted by microglial cells, 32
Tuning fork, in test of eighth cranial nerve, 529
Tunnel of Corti, 326, 327f
Tympanic cavity, 325, 325f
Tympanic membrane, 325, 325f
Tyramine, action of, 69f, 70
Tyrosine, in synthesis of norepinephrine, 68–69, 69f
Tyrosine hydroxylase, in synthesis of norepinephrine, 68–69, 69f

Uncal artery(ies), 496
Uncus, 239f, 242, 494, 495f
Urinary bladder activity, autonomic control of, 477, 477f
Urination, autonomic control of, 477, 477f
Urine, antidiuretic hormone and, 488
Uterus, postpartum, oxytocin effects in, 487
Utricle, 342f, 343, 344f
Uvea, 305
Uveitis, 305
Uvula, 530
  corticonuclear fiber lesions affecting, 396, 398f
Uvular muscles, corticonuclear fiber lesions affecting, 396, 398f

V1 area, 319, 322f. See also Cortex, visual, primary.
Vagus nerve, 152, 152f, 161, 265
  anatomy and distribution of, 203f–205f, 206–207
  development of, 79
  dorsal motor nucleus of, 206
  function of, 206
  in spinal trigeminal pathway, 284f, 285
  lesions of, 206–207
  meningeal branch of, 208
  parasympathetic outflow pathways of, 474–475, 474t
  sensory fibers of, 206
  taste neurons in, 369
  testing of, 530
  visceral parasympathetic afferent fibers in, 299, 300f, 301
Varicosities, of axons, 21
Vascular lesions
  brainstem, anterolateral system effects of, 280, 282
  causes of, 122–123
Vasodepressor response, viscerosensory parasympathetic pathways
    and, 301
Vasopressin, 483. See also Antidiuretic hormone.
Vasopressor response, viscerosensory parasympathetic pathways and,
    301
Vein(s)
  anastomotic
    inferior, of Labbé, 130, 130f
    superior, of Trolard, 130, 130f
  basal, of Rosenthal, 130, 131f
  cerebellar
    inferior, 134
    superior, 130f, 133f, 134
  cerebral, 130, 130f, 131f
    anterior, 130, 131f
    deep middle, 130, 131f
    internal, 130, 131f, 132–134, 133f
    pathologic flow patterns of, 133
    superficial middle, 130, 130f, 131f
    temporal, 130, 130f
  great cerebral, of Galen, 130, 131f, 133, 133f
    malformations of, 133, 133f
  internal, of cerebral hemisphere, 132–134
  jugular, internal, 130

Vein(s) (Continued)
  of brain, 122, 130–134, 130f–133f
  of brainstem, 134
  of cerebellum, 130f, 133f, 134
  of cerebral hemispheres, 130, 130f, 131f
  of spinal cord, 134f, 135
  spinal
    anterior, 134f, 135
    posterior, 134f, 135
  terminal, 132–133, 133f
  thalamostriate, 132–133, 133f
  transverse caudate, 133, 133f
Velum, anterior medullary, 153, 153f
Venous sinus(es), 109, 109f, 113, 130–134, 130f–133f
  arachnoid villi and, 114f
  cerebrospinal fluid movement into, 103–104, 104f
Ventral striatum, 245, 246f
Ventricle(s)
  early development of, 94–95, 94f
  fourth, 96f, 98f, 99, 99f, 100f, 153, 153f
    at medulla, 166, 166f–167f
    enlarged, 79, 79f
    foramen of, 95, 95f
    formation of, 77, 78f
  lateral, 96f, 97, 97f, 98f
    early development of, 94f, 95
    enlarged, 79, 79f
    formation of, 77, 78f
    relationships of, 248f
  of brainstem, 153–154, 153f
  third, 96f, 97–98, 98f, 99f, 220f, 221, 480, 481f
    as cavity of diencephalon, 221–222, 223f, 224f
    boundaries of, 97–98, 99f
    early development of, 94f, 95
    formation of, 77, 78f
Ventricular system
  cerebrospinal fluid movement in, 103–104, 104f
  development of, 77–79, 78f, 79f, 94–99, 94f–100f
Vergence movements, 446, 453f, 458f, 458f
Vermis
  cerebellar, 177, 177f, 424–425, 425f–427f
    aplasia of, 79, 79f
  vestibular connections of, 436
Vertebra(ae)
  development of, 82
  relation of, to spinal cord, 82, 83f
Vertebral artery(ies), 122, 125, 126f, 162, 169, 170, 170f
  branches of, 125, 126f
Vertebrobasilar insufficiency, 125
Vertebrobasilar system, 125, 126f, 127–128, 127f
Vertical gaze center, 216, 217, 454f, 455
Vertigo, 209–210, 356–357
Vesicle(s)
  biosynthesis of, 60, 61f
  brain
    development of, 75, 76f, 77, 94–95, 94f
    primary, 75, 76f, 77
    secondary, 76f, 77
  large dense-cored, 60
    biosynthesis of, 60–61, 61f
    exocytosis of, 62–63, 62f
      vs. exocytosis in synaptic vesicles, 61f, 62f, 63
    formation and transport of, 61, 61f
  lateral, 236f, 237
  membranes of, composition of, 60
  neurosecretory, 60
  olfactory, 360
  secretory, neurotransmitters stored in, 58
  small, biogenic amines in, 60
  synaptic. See Synaptic vesicle(s).
  telencephalic (cerebral), 76f, 77, 78f, 236f, 237
  third, 236f, 237
Vestibular compensation, 352
Vestibular function, testing of, 530
Vestibular labyrinth
  cerebellar afferent connections with, 352, 352f

Vestibular labyrinth (Continued)
  peripheral, 342–345, 342f–345f
Vestibular nerve, functional components of, 156f, 157
Vestibular nucleus. See Nucleus(i), vestibular.
Vestibular sensory input, posture and balance control and, 380
Vestibular system, 341–358
Vestibulocerebellum
  components of, 435, 436f
  damage to, 435–436
  function of, 435
Vestibulocochlear nerve, 152, 152f, 161, 202
  cochlear portion of, 327
  function of, 343–344, 344f
  in pons-medulla junction, 203f, 204f, 208–210, 209f
  injury to, 209–210
  of pons, 175–176, 175f, 176f
  origin of, 79–80
  sensory nuclei of, functional components of, 175
  testing of, 529–530, 531f
Vestibulo-ocular network, 353–354, 353f, 354f
Vestibulo-ocular reflex, 446
Vestibulospinal network, 354–355, 355f
Vestibulo-thalamo-cortical network, 355–356, 356f
Vibratory sense, 256
  damage to, 392
  in spinal trigeminal pathway, 284f
  in trigeminal system, 266f, 267f
  testing of, 536f, 537
Villus(i)
  arachnoid, 113, 114, 114f
    cerebrospinal fluid movement in, 103–104, 104f
  of choroid plexus, 95, 101, 102f, 103f
Virchow-Robin space, 116
Viscera
  motor and sensory fibers in, on vagus nerve, 206
  pain receptors in, 276–277, 276t, 277f–279f
Visceral motor system (autonomic), 4–6, 4f, 465–478
  brainstem reticular formation in, 476
  cardiovascular system control and, 476–477, 476f
  central autonomic network and, 475
  cerebellar influence on, 442–443
  development of, 80–81, 467–468, 468f
  enteric nervous system and, 475
  functions of, 220
  hypothalamic role in, 220, 229, 475, 480, 480f
  organization of, 466–467, 466t, 467f, 467t
  outflow regulation in, 475–477, 476f, 477f
  outflow targets in, 469f–471f, 470–471
    parasympathetic, 466–467, 466t, 473, 474t
  parasympathetic division of, 473–475, 474t, 491
  peripheral, 466–467
  rostral ventrolateral medulla in, 476
  sensory input in, 475–477, 476f, 477f
  solitary nucleus in, 475–476
  supraspinal control of autonomic function in, 476
  sympathetic division of, 468–473, 469f, 473f, 491
    organization of, 469–471, 469f, 470f, 471f
    outflow targets in, 466–467, 466t, 473
Visceral organs and tissues, 4
Visceroceptor(s), classification of, 294, 294t
Viscerosensory fiber(s), 138, 295
  parasympathetic, 298–301
  sympathetic, 295–298, 296f, 297f
    referred pain in, 297–298, 297f, 298f
    reticular formation projections of, 296–297, 297f
    thalamic projections of, 296, 297f
Viscerosensory pathways, 241, 293–302
Viscerosensory transducer, hypothalamus as, 229
Vision, stereoscopic, 447
Visual acuity, 314, 314f
  testing of, 524, 524f
Visual deficits, with pituitary tumors, 488–489
Visual fields, 313, 313f, 314
  bitemporal loss of, 524
  deficits of, 313, 315, 315f, 316, 316f, 319f
  input from, 228, 318f

Visual fields *(Continued)*
    processing of, 318–322, 322f
    testing of, 524–525, 525f
Visual motor systems, 445–462
Visual processing, 321–322, 322f
    eye anatomy and, 304–306, 304f, 306f
    lesions of, 321–322
    receptive fields in, 309, 310f, 311, 311f
    retinal, 309–312, 311f
Visual relay center, development of, 85
Visual system, 302–322. *See also under* Retina.
Vitreous body, 304, 304f
Vocal muscles, innervation of, 200
Vomiting, vestibulocochlear nerve injury and, 210

W cell(s), 312
W fibers, of retinal ganglion cells, 316
Wallenberg syndrome (lateral medullary syndrome), 170, 171f, 215,
    286–287, 287f, 291
Water balance, regulation of, hypothalamic reflex controlling, 490
Water caloric test, 530
Weber syndrome, 198, 450, 450t
Weber test, 529, 531f
Wernicke aphasia, 514, 516, 517, 517f
Wernicke area, 337
Wernicke-Korsakoff syndrome, 500
Wheat germ agglutinin–horseradish peroxidase, anterograde trans-
    port and, 22
White matter, 5
    astrocytes of, 28, 30
    of cerebral cortex, 86, 87f
    of cerebral hemisphere, 242–245, 244f–246f

White matter *(Continued)*
    of spinal cord, 139, 140–141, 141f
    oligodendrocytes in, 29f, 30
    subcortical, 7, 236
        of cerebellar cortex, 184, 184f
Wilson disease, 421, 421f

X cell(s), 312
X(P) fibers, of retinal ganglion cells, 316, 317f, 322f

Y(M) fibers, of retinal ganglion cells, 316, 317f, 322f
Yohimbine, action of, 69f, 70

Zinn
    annulus of, 447f
    zonule of, 448f, 449
Zona incerta, 223, 224f, 230, 409, 411f
Zone(s)
    cerebellar, 424–425, 425f–427f
    intermediate, of spinal cord, 81, 139, 141
    of cerebellar primordium, 84f, 85
    of cerebral cortex, 86, 87f
    periventricular, 488
    ventricular, of neural tube, 138, 139f
Zonulae adherentes, 99
Zonulae occludentes, of choroid epithelial cells, 101, 103f
Zonule of Zinn, 448f, 449